D1134686

# MICROBIAL BIOTECHNOLOGY

Knowledge in microbiology is growing exponentially through the determination of genomic sequences of hundreds of microorganisms and the invention of new technologies, such as genomics, transcriptomics, and proteomics, to deal with this avalanche of information.

These genomic data are now exploited in thousands of applications, ranging from medicine, agriculture, organic chemistry, public health, and biomass conversion, to biomining. *Microbial Biotechnology* focuses on uses of major societal importance, enabling an in-depth analysis of these critically important applications. Some, such as wastewater treatment, have changed only modestly over time; others, such as directed molecular evolution, or "green" chemistry, are as current as today's headlines.

This fully revised second edition provides an exciting interdisciplinary journey through the rapidly changing landscape of discovery in microbial biotechnology. An ideal text for courses in applied microbiology and biotechnology, this book will also serve as an invaluable overview of recent advances in this field for professional life scientists and for the diverse community of other professionals with interests in biotechnology.

Alexander N. Glazer is a biochemist and molecular biologist and has been on the faculty of the University of California since 1964. He is a Professor of the Graduate School in the Department of Molecular and Cell Biology at the University of California, Berkeley. Dr. Glazer is a member of the National Academy of Sciences and a Fellow of the American Academy of Arts and Sciences, the American Academy of Microbiology, the American Association for the Advancement of Science, and the California Academy of Sciences. He was twice the recipient of a Guggenheim Fellowship. He was the recipient of the Botanical Society of America Darbaker Prize, 1980 and the National Academy of Sciences Scientific Reviewing Prize, 1991, a lecturer of the Foundation for Microbiology, 1996–98; and a National Guest Lecturer, New Zealand Institute of Chemistry, 1999. Dr. Glazer has authored over 250 research papers and reviews. He is a co-inventor on more than 40 U.S. patents. Since 1996, he has served as a member of the Editorial Affairs Committee of Annual Reviews, Inc.

Hiroshi Nikaido is a biochemist and microbiologist. He received his M.D. from Keio University in Japan in 1955 and became a faculty member at Harvard Medical School in 1963, before moving to University of California in 1969. He is a Professor of Biochemistry and Molecular Biology in the Department of Molecular and Cell Biology at the University of California, Berkeley. Dr. Nikaido is a Fellow of the American Academy of Arts and Sciences and the American Academy of Microbiology. He was the recipient of a Guggenheim Fellowship, NIH Senior International Fellowship, Paul Ehrlich prize (1969), Hoechst-Roussel Award of American Society for Microbiology (1984), and Freedom-to-Discover Award for Distinguished Research in Infectious Diseases from Bristol-Myers Squibb (2004). He was an Editor of *Journal of Bacteriology* from 1998 to 2002. Dr. Nikaido has authored nearly 300 research papers and reviews.

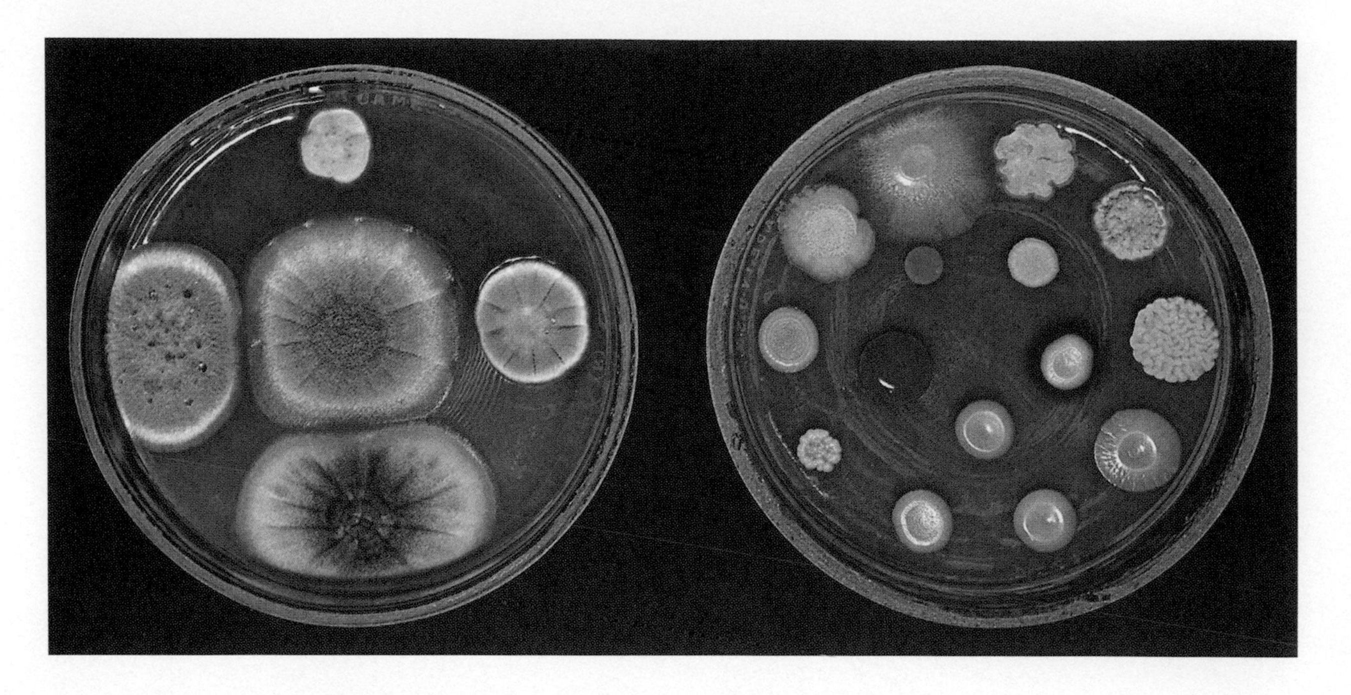

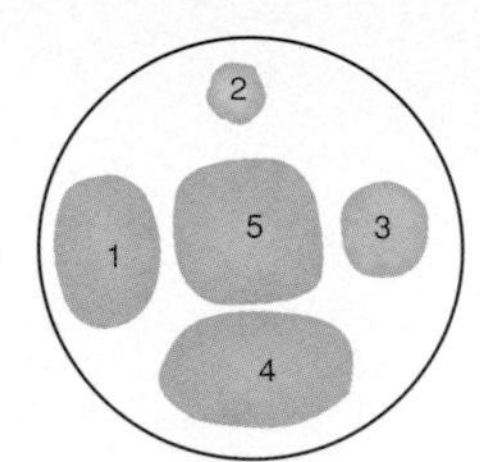


MOLDS
1 *Penicillium chrysogenum*
2 *Monascus purpurea*
3 *Penicillium notatum*
4 *Aspergillus niger*
5 *Aspergillus oryzae*

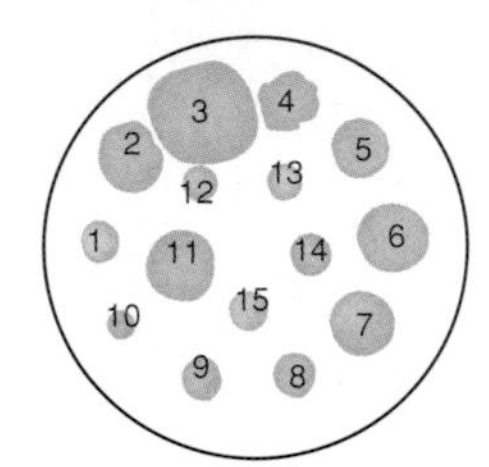


YEASTS
1 *Saccharomyces cerevisiae*
2 *Candida utilis*
3 *Aureobasidium pullulans*
4 *Trichosporon cutaneum*
5 *Saccharomycopsis capsularis*
6 *Saccharomycopsis lipolytica*
7 *Hanseniaspora guilliermondii*
8 *Hansenula capsulata*
9 *Saccharomyces carlsbergensis*
10 *Saccharomyces rouxii*
11 *Rhodotorula rubra*
12 *Phaffia rhodozyma*
13 *Cryptococcus laurentii*
14 *Metschnikowia pulcherrima*
15 *Rhodotorula pallida*

Cultures of molds and yeasts on nutrient agar in glass Petri dishes. From H. Phaff, Industrial microorganisms, Scientific American, September 1981. Copyright © 1981 by Scientific American, Inc. All rights reserved.

# MICROBIAL BIOTECHNOLOGY

## Fundamentals of Applied Microbiology, Second Edition

**Alexander N. Glazer**

University of California, Berkeley

**Hiroshi Nikaido**

University of California, Berkeley

CAMBRIDGE UNIVERSITY PRESS
Cambridge, New York, Melbourne, Madrid, Cape Town, Singapore, São Paulo

Cambridge University Press
32 Avenue of the Americas, New York, NY 10013-2473, USA

www.cambridge.org
Information on this title: www.cambridge.org/9780521842105



First published 2007
First edition © W.H. Freeman and Company 1995
Second edition © Alexander N. Glazer and Hiroshi Nikaido 2007
First edition published by W.H. Freeman and Company 1995
Second edition published by Cambridge University Press 2007

Printed in the United States of America

*A catalog record for this publication is available from the British Library.*

*Library of Congress Cataloging in Publication Data*

Glazer, Alexander N.
Microbial biotechnology : fundamentals of applied microbiology /
Alexander N. Glazer and Hiroshi Nikaido. — 2nd ed.
p. ; cm.
Includes bibliographical references.
ISBN-13: 978-0-521-84210-5 (hardcover)
1. Microbial biotechnology. I. Nikaido, Hiroshi. II. Title.
[DNLM: 1. Biotechnology. 2. Microbiology. TP 248.27.M53 G553m 2007]
TP248.27.M53G57 2007
660.6′2—dc22 2007016151

ISBN 978-0-521-84210-5 hardback

*We dedicate this book to Eva and Kishiko,*

*for the gift of years of support, tolerance, and patience.*

# Contents in Brief



Advances of particular relevance and importance will be posted periodically on the website www.cambridge.org/glazer.

# Contents

# Preamble

Il n'y a pas des sciences appliquées . . . mais il y'a des applications de la science. (*There are no applied sciences . . . but there are the applications of science.)*

– Louis Pasteur

Microorganisms are the most versatile and adaptable forms of life on Earth, and they have existed here for some 3.5 billion years. Indeed, for the first 2 billion years of their existence, prokaryotes alone ruled the biosphere, colonizing every accessible ecological niche, from glacial ice to the hydrothermal vents of the deep-sea bottoms. As these early prokaryotes evolved, they developed the major metabolic pathways characteristic of all living organisms today, as well as various other metabolic processes, such as nitrogen fixation, still restricted to prokaryotes alone. Over their long period of global dominance, prokaryotes also changed the earth, transforming its anaerobic atmosphere to one rich in oxygen and generating massive amounts of organic compounds. Eventually, they created an environment suited to the maintenance of more complex forms of life.

Today, the biochemistry and physiology of bacteria and other microorganisms provide a living record of several billion years' worth of genetic responses to an ever-changing world. At the same time, their physiologic and metabolic versatility and their ability to survive in small niches cause them to be much less affected by the changes in the biosphere than are larger, more complex forms of life. Thus, it is likely that representatives of most of the microbial species that existed before humans are still here to be explored.

Such an exploration is by no means a purely academic pursuit. The many thousands of microorganisms already available in pure culture and the thousands of others yet to be cultured or discovered represent a large fraction of the total gene pool of the living world, and this tremendous genetic diversity is the raw material of genetic engineering, the direct manipulation of the heritable characteristics of living organisms. Biologists are now able to greatly accelerate the acquisition of desired traits in an organism by directly modifying its genetic makeup through the manipulation of its DNA, rather than through the traditional methods of breeding and selection at the level of the whole organism. The various techniques of manipulation summarized under the rubric of "recombinant DNA technology" can take the form of removing genes, adding genes from a different organism, modifying genetic control mechanisms, and introducing synthetic DNA, sometimes enabling a cell to perform functions that are totally new to the living world. In these ways, new stable heritable traits have by now been introduced into all forms of life. One result has been a significant enhancement of the already considerable practical value of applied microbiology. Applied microbiology covers a broad spectrum of activities, contributing to medicine, agriculture, "green" chemistry, exploitation of sources of renewable energy, wastewater treatment, and bioremediation, to name but a few. The ability to manipulate the genetic makeup of organisms has led to explosive progress in all areas of this field.

The purpose of this book is to provide a rigorous, unified treatment of all facets of microbial biotechnology, freely crossing the boundaries of formal disciplines in order to do so: microbiology supplies the raw materials; genomics, transcriptomics, and proteomics provide the blueprints; biochemistry, chemistry, and process engineering provide the tools; and many other scientific fields serve as important reservoirs of information. Moreover, unlike a textbook of biochemistry, microbiology, molecular biology, organic chemistry, or some other vast basic field, which must concentrate solely on teaching general principles and patterns in order to provide an overview, this one will continually emphasize the importance of diversity and uniqueness. In applied microbiology, one is frequently likely to seek the unusual: a producer of a novel antibiotic, a parasitic organism that specifically infects a particularly widespread and noxious pest, a hyperthermophilic bacterium that might serve as a source of enzymes active above 100°C. In sum, this book examines the fundamental principles and facts that underlie current practical applications of bacteria, fungi, and other microorganisms; describes those applications; and examines future prospects for related technologies.

The stage on which microbial biotechnology performs today is vastly different from that portrayed in the first edition of this book, published 12 years ago. The second edition has been extensively rewritten to incorporate the avalanche of new knowledge. What are some of the most influential of these recent advances?

- Hundreds of prokaryotic and fungal genomes have been fully sequenced, and partial genomic information is available for many more organisms available in pure culture.
- The understanding of the phylogenetic and evolutionary relationships among microorganisms now rests on the objective foundation provided by this large body of sequence data. These data have also revealed the mosaic and dynamic aspects of microbial genomes.
- Environmental DNA libraries offer a glimpse of the immensity and functional diversity of the microbial world and provide rapid access to genes from tens of thousands of yet-uncultured microorganisms.

- Extensive databases of annotated sequences along with sophisticated computational tools allow rapid access to the burgeoning body of information and reveal potential functions of new sequences.
- The polymerase chain reaction coupled with versatile techniques for the generation of recombinant organisms allows exploitation of sequence information to create new molecules or organisms with desired properties.
- Genomics, transcriptomics, and metabolomics use powerful new techniques to map how complex cell functions arise from coordinated regulation of multiple genes to give rise to the interdependent pathways of metabolism and to the integration of the sensory inputs that ensure proper functioning of cells in responding to environmental change.
- In the past 10 years, these developments have also changed the processes used in all of the "classical" areas of biotechnology – for instance, in the production of amino acids, antibiotics, polymers, and vaccines.
- The growing human population of the earth, equipped with the ability to effect massive environmental change by applying ever-increasing technological sophistication, is placing huge and unsustainable demands on natural resources. Microbial biotechnology is of increasing importance in contributing to the generation of crops with resistance to particular insect pests, tolerance to herbicides, and improved ability to survive drought and high levels of salt. The urgent need to minimize the discharge of organic chemical pollutants into the environment along with the need to conserve declining reserves of petrochemicals has led to the advent of "green" chemistry with attendant rapid growth in the use of biocatalysts. The future of the use of biomass as a renewable source of energy is critically dependent on progress in efficient direct microbial conversion of complex mixtures of polysaccharides to ethanol. The treatment of wastewater, a critical contribution of microorganisms to maintaining the life-support systems of the planet, is an important area for future innovation.

The application of biotechnology to medicine, agriculture, the chemical industry, and the environment is changing all aspects of everyday life, and the pace of that change is increasing. Thus, basic understanding of the many facets of microbial biotechnology is important to scientists and nonscientists alike. We hope that both will find this book a useful source of information. Although a strong technical background may be necessary to assimilate the fine points described herein, we have tried to make the fundamental concepts and issues accessible to readers whose background in the life sciences is quite modest. The attempt is vital, for only an *informed* public can distinguish desirable biotechnological options from the undesirable, those likely to succeed from those likely to result in costly failure.

# Acknowledgments

We are grateful to our colleagues who read various chapters, to Moira Lerner for her helpful developmental editing of three of the chapters, and to the many scientists and publishers who allowed us to reproduce illustrations and other material and generously provided their original images and electronic files for this purpose.

We are indebted to Kirk Jensen for his interest in our plans for this book and for introducing us to Cambridge University Press. Working with the Cambridge staff has been a pleasure. Dr. Katrina Halliday provided encouragement and steady editorial guidance from the early stages of this project through the completion of the manuscript. We are particularly grateful to Clare Georgy and Alison Evans for their careful review of the manuscript and for undertaking the arduous task of securing permissions to reproduce many illustrations and other material. We thank Marielle Poss for her oversight of the production process, and are grateful to Alan Gold for designing the creative and elegant layout for the book. We thank Ken Karpinski at Aptara for his oversight and meticulous attention to detail in the production of this book and his unfailing gracious help when there were snags in the process. Finally, we thank Georgette Koslovsky for her precise and thoughtful copy editing.

The combined efforts of all of these individuals have contributed a great deal to the accuracy and aesthetic quality of this book. The authors are responsible for any imperfections that remain.

ONE

# Microbial Diversity

Molecular phylogenies divide all living organisms into three domains – Bacteria ("true bacteria"), Archaea, and Eukarya (eukaryotes: protists, fungi, plants, animals). The place of viruses (Box 1.1) in the phylogenetic tree of life is uncertain. In this book, we focus on the contributions of Bacteria, Archaea, and Fungi to microbial biotechnology. In so doing, we include organisms from all three domains. We also devote some attention to the uses of viruses as well as to the problems they pose in certain technological contexts.

The domains of Bacteria and Archaea encompass a huge diversity of organisms that differ in their sources of energy, their sources of cell carbon or nitrogen, their metabolic pathways, the end products of their metabolism, and their ability to attack various naturally occurring organic compounds. Different bacteria and archaea have adapted to every available climate and microenvironment on Earth. Halophilic microorganisms grow in brine ponds encrusted with salt, thermophilic microorganisms grow on smoldering coal piles or in volcanic hot springs, and barophilic microorganisms live under enormous pressure in the depths of the seas. Some bacteria are symbionts of plants; other bacteria live as intracellular parasites inside mammalian cells or form stable consortia with other microorganisms. The seemingly limitless diversity of the microorganisms provides an immense pool of raw material for applied microbiology.

The morphological variety of organisms classified as fungi rivals that of the bacteria and archaea. Fungi are particularly effective in colonizing dry wood and are responsible for most of the decomposition of plant materials by secreting powerful extracellular enzymes to degrade biopolymers (proteins, polysaccharides, and lignin). They produce a huge number of small organic molecules of unusual structure, including many important antibiotics. On the other hand, fungi as a group lack some of the metabolic capabilities of the bacteria. In particular, fungi do not carry out photosynthesis or nitrogen fixation and are unable to exploit the oxidation of inorganic compounds as a source of energy. Fungi are unable to use inorganic compounds other than oxygen as terminal electron acceptors in respiration. Fungi as a group are also less versatile than bacteria in the range of organic compounds they can

Viruses differ from all other organisms in three major respects: they contain only one kind of nucleic acid, either deoxyribonucleic acid (DNA) or ribonucleic acid (RNA); only the nucleic acid is necessary for their reproduction; and they are unable to reproduce outside of a host's living cell. Viruses are not described further in this chapter but will be encountered later in the discussion of vaccines (Chapter 5)

**BOX 1.1**

use as sole sources of cell carbon. Frequently, fungi and bacteria complement each other's abilities in degrading complex organic materials.

A consortium is a system of several organisms (frequently two) in which each organism contributes something needed by the others. Many fundamental processes in nature are the outcome of such interactions among microorganisms influencing the biosphere on a worldwide scale. For example, consortia of bacteria and fungi play an indispensable role in the cycling of organic matter. By decomposing the organic by-products and the remains of plants and animals, they release nutrients that sustain the growth of all living things. The top six inches of fertile soil may contain over two tons of fungi and bacteria per acre. In fact, the respiration of bacteria and fungi has been estimated to account for over 90% of the carbon dioxide production in the biosphere. Technology, too, takes advantage of the special abilities of mixed cultures of microorganisms, employing them in beverage, food, and dairy fermentations, for example, and in biotreatment processes for wastewater.

Lately, the challenges posed by the need to clean up massive oil spills and to decontaminate toxic waste sites with minimum permanent damage to the environment have directed attention to the powerful degradative capabilities of consortia of microorganisms. Experience suggests that encouraging the growth of natural mixed microbial populations at the site of contamination can contribute more successfully to the degradation of undesirable organic compounds in diverse ecological settings than can the introduction of a single ingeniously engineered recombinant microorganism with new metabolic capabilities. We are still far from an adequate understanding of microbial interactions in natural environments.

This chapter has a dual purpose: to provide a guide to the relative placements of important microorganisms on the taxonomic map of the microbial world and to explore the importance of the diversity of microorganisms to biotechnology.

## PROKARYOTES AND EUKARYOTES

Cellular organisms fall into two classes that differ from each other in the fundamental internal organization of their cells. The cells of eukaryotes contain a true membrane-bounded nucleus (karyon), which in turn contains a set of chromosomes that serve as the major repositories of genetic information in the cell. Eukaryotic cells also contain other membrane-bounded organelles that possess genetic information, namely mitochondria and chloroplasts. In the prokaryotes, the chromosome (nucleoid) is a closed circular DNA molecule, which lies in the cytoplasm, is not surrounded by a nuclear membrane, and contains all of the information necessary for the reproduction of the cell. Prokaryotes also have no other membrane-bounded organelles whatsoever. Bacteria and archaea are prokaryotes, whereas fungi are eukaryotes. The choice of a fungus (such as the yeast *Saccharomyces cerevisiae*) or a bacterium (such as *Escherichia coli*) for a particular application is often

**TABLE 1.1 A comparison of Bacterial, Archaeal, and Eukaryal cells**

| | Bacteria | Archaea | Eukarya |
|---|---|---|---|
| STRUCTURAL FEATURES | | | |
| Chromosome number | One | One | More than one |
| Nuclear membrane | Absent | Absent | Present |
| Nucleolus | Absent | Absent | Present |
| Mitotic apparatus | Absent | Absent | Present |
| Microtubules | Absent | Absent | Present |
| Membrane lipids | Glycerol diesters | Glycerol diethers or glycerol tetraethers | Glycerol diesters |
| Membrane sterols | Rare | Rare | Nearly universal |
| Peptidoglycan | Present | Absent | Absent |
| GENE STRUCTURE, TRANSCRIPTION, AND TRANSLATION | | | |
| Introns in genes | Rare | Rare | Common |
| Transcription coupled with translation | Yes | May occur | No |
| Polygenic mRNA | Yes | Yes | No |
| Terminal polyadenylation of mRNA | Absent | Present | Present |
| Ribosome subunit sizes (sedimentation coefficient) | 30S, 50S | 30S, 50S | 40S, 60S (cytoplasmic) |
| Amino acid carried by initiator tRNA | Formylmethionine | Methionine | Methionine |
| METABOLIC PROCESSES | | | |
| Oxidative phosphorylation | Membrane dependent | Membrane dependent | In mitochondria |
| Photosynthesis | Membrane dependent | Membrane dependent | In chloroplasts |
| Reduced inorganic compounds as energy source | May be used | May be used | Not used |
| Nonglycolytic pathways for anaerobic energy generation | May occur | May occur | Do not occur |
| Poly-$\beta$-hydroxybutyrate as organic reserve material | Occurs | Occurs | Does not occur |
| Nitrogen fixation | Occurs | Occurs | Does not occur |
| OTHER PROCESSES | | | |
| Exo- and endocytosis | Does not occur | Does not occur | May occur |
| Amoeboid movement | Does not occur | Does not occur | May occur |

mRNA, messenger RNA; tRNA, transfer RNA.

dictated by the basic genetic, biochemical, and physiological differences between prokaryotes and eukaryotes.

## THE TWO GROUPS OF PROKARYOTES

Among prokaryotes, a general distinction is made between the bacteria and the archaea. The evolutionary distance between the bacteria, the archaea, and the eukaryotes, estimated from the divergence in their ribosomal RNA (rRNA) sequences, is so great that it is believed that these three groups may have diverged from an ancient progenitor rather than evolving from one another. With respect to many molecular features, the archaea are almost as different from the bacteria as the latter are from eukaryotes (Table 1.1). For

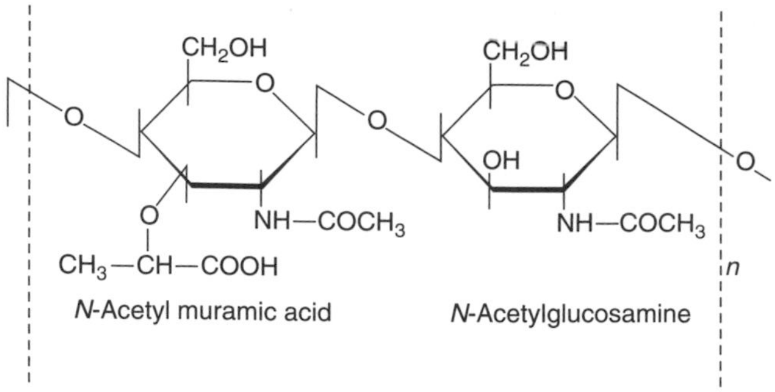

FIGURE 1.1

Repeating unit of the polysaccharide backbone of the peptidoglycan layer in the cell wall of bacteria.

$CH_2OH$ O $CH_2OH$ O O OH O O $NH—COCH_3$ $NH—COCH_3$ $CH_3—CH—COOH$ n
*N*-Acetyl muramic acid *N*-Acetylglucosamine

example, the cell wall structure of bacteria is based on a cross-linked polymer called peptidoglycan with an *N*-acetylglucosamine–*N*-acetylmuramic acid repeating unit (Figure 1.1). Because of the virtually universal presence of peptidoglycan in bacteria and its absence in eukaryotes, the presence of muramic acid is considered a bacterial "signature." The different archaea have a variety of cell wall polymers, but none of them incorporates muramic acid. The most dramatic difference between these organisms is in the nature of the glycerol lipids that make up the cytoplasmic membrane. The hydrophobic moieties in the archaea are ether-linked and branched aliphatic chains, whereas those of bacteria and eukaryotes are ester-linked straight aliphatic chains (Figure 1.2).

Initially, the archaea were believed to be typical of extreme environments tolerated by few bacteria and fewer eukaryotes. The archaea include three distinct kinds of microorganisms, all found in extreme environments: the methanogens, the extreme halophiles, and the thermoacidophiles. The methanogens live only in oxygen-free environments and generate methane by the reduction of carbon dioxide. The halophiles require very high concentrations of salt to survive and are found in natural habitats such as the Great Salt Lake and the Dead Sea as well as in man-made salt evaporation ponds. The thermoacidophiles are found in hot sulfur springs at temperatures above 80°C in strongly acidic environments (pH < 2). However, analyses of 16S rDNA analyzed in environmental samples show archaea to be present in marine sediments, in coastal and open ocean waters, and in freshwater sediments and soils. Planktonic members of the Crenarchaeota phylum are reported to represent about 20% of all of the bacterial and archaeal cells found in the oceans. An archaeal symbiont, *Crenarchaeum symbiosum*, lives in the tissues of the marine sponge *Axinella mexicana* in coastal waters of about 10°C. It now appears that bacteria and archaea have many types of habitats in common.

## GRAM STAIN METHOD

The Gram stain procedure was described by the Danish physician Hans Christian Gram in 1884 and has survived in virtually unmodified form. Gram worked at the morgue of the City Hospital of Berlin, where he developed a method to detect bacteria in tissues by differential staining. In a widely used

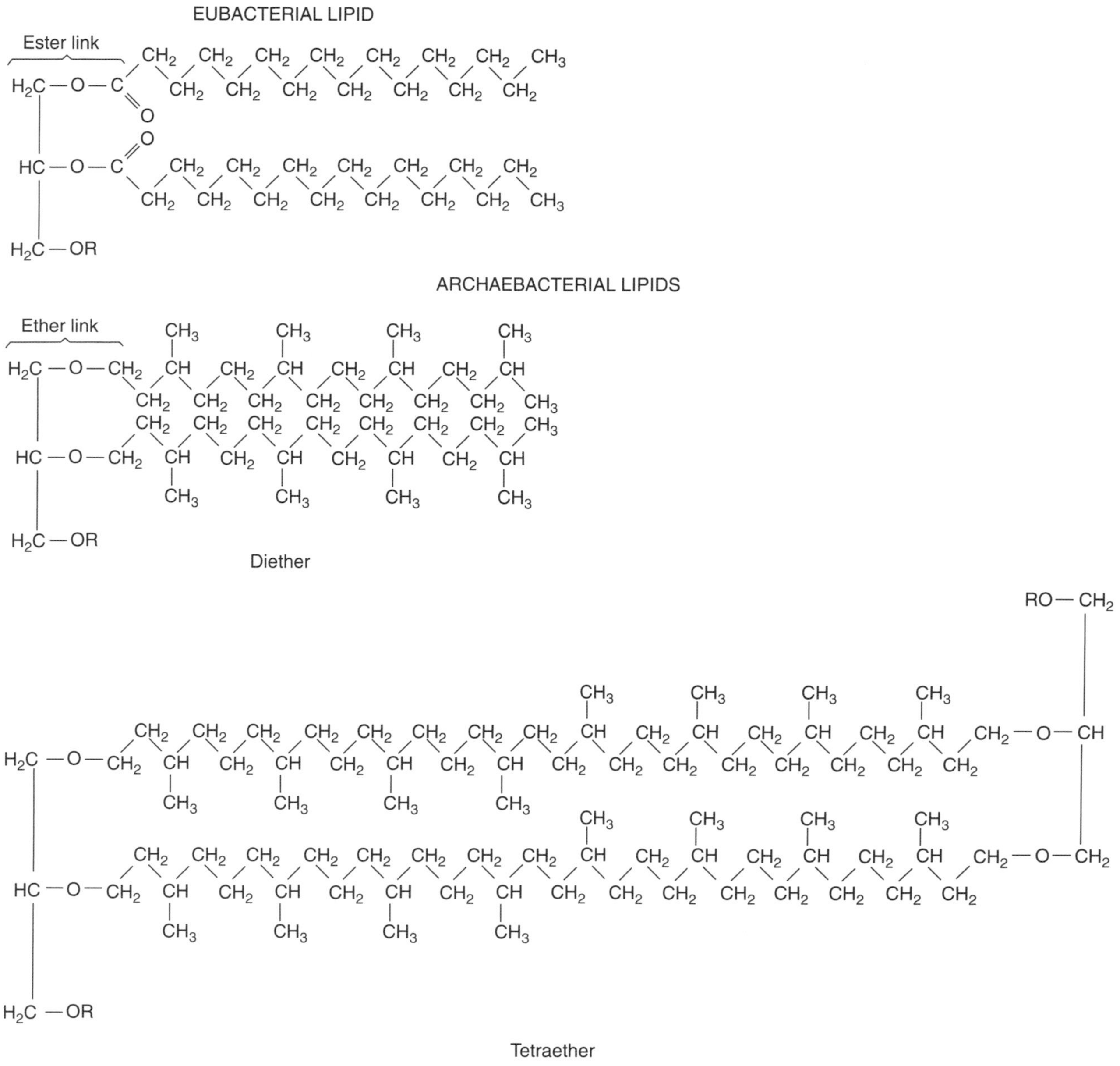

EUBACTERIAL LIPID

Ester link

ARCHAEBACTERIAL LIPIDS

Ether link

Diether

Tetraether

version of his empirical procedure, a heat-fixed tissue sample or smear of bacteria on a glass slide is stained first with a solution of the dye crystal violet and then with a dilute solution of iodine to form an insoluble crystal violet-iodine complex. The preparation is then washed with either alcohol or acetone. Bacteria that are rapidly decolorized by this means are said to be Gram-negative; those that remain violet are said to be Gram-positive. The ease of dye elution, and consequently the Gram staining behavior of bacteria, correlates with the structure of the cell walls. Gram-positive bacteria have a thick cell wall of highly cross-linked peptidoglycan, whereas Gram-negative bacteria usually have a thin peptidoglycan layer covered by an outer

**FIGURE 1.2**

Membrane lipids of bacteria and eukaryotes are glycerol esters of straight-chain fatty acids such as palmitate. Archaeal membrane lipids are diethers or tetraethers in which the glycerol unit is linked by an ether link to phytanols, branched-chain hydrocarbons. Moreover, the configuration about the central carbon of the glycerol unit is D in the ester-linked lipids but L in the ether-linked lipids. R is phosphate or phosphate esters in phospholipids and sugars in glycolipids.

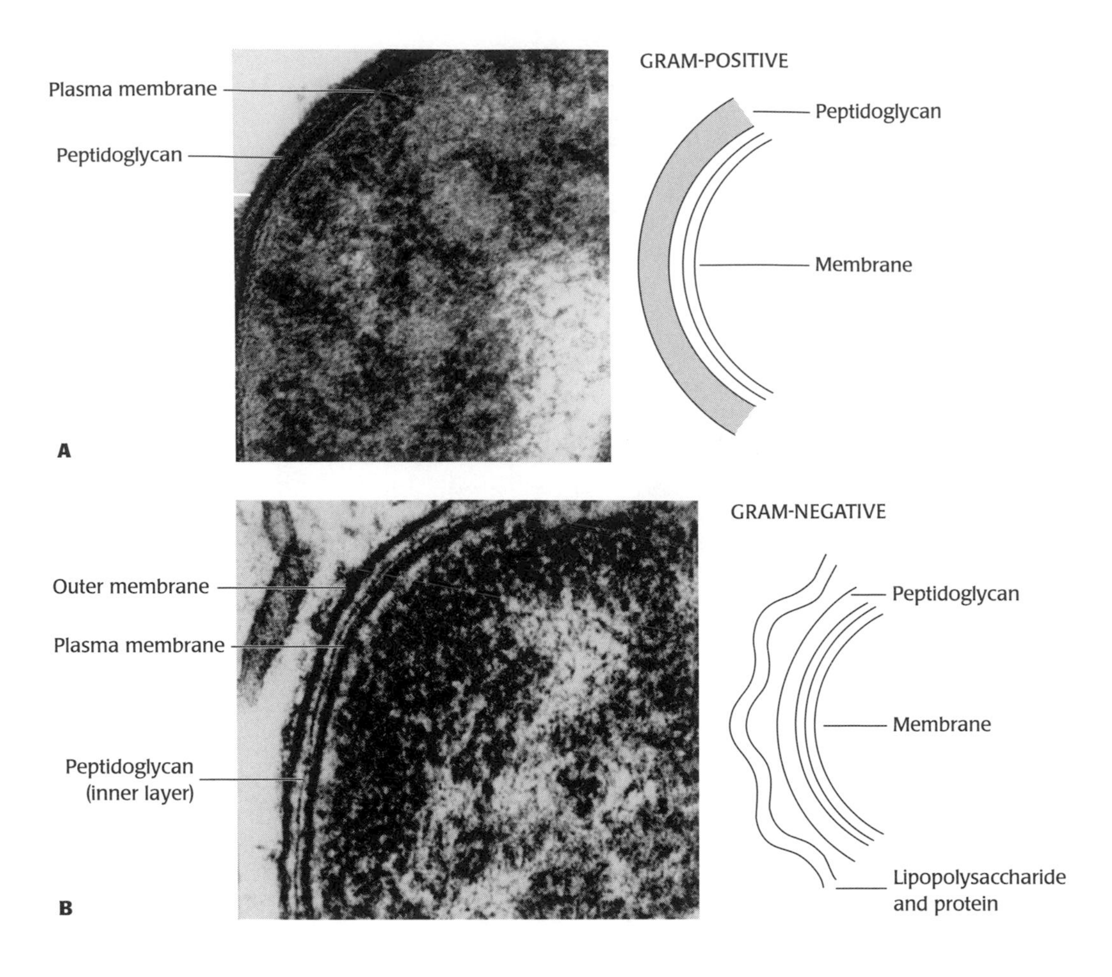


**FIGURE 1.3**

Electron micrographs of bacterial cell walls. (**A**) Gram-positive, *Arthrobacter crystallopoietes*. Magnification, 126,000×. (**B**) Gram-negative, *Leucothrix mucor*. Magnification, 165,000×. [Reproduced with permission from Brock, T. D., and Madigan, M. T. (1988). *Biology of Microorganisms*, 5th Edition, Englewood Cliffs, NJ: Prentice Hall, Figure 3.22.]

membrane. The outer membrane is an asymmetric lipid bilayer membrane: a lipopolysaccharide forms the exterior layer and phospholipid forms the inner layer (Figure 1.3).

The presence of the outer membrane on Gram-negative bacteria confers a higher resistance to antibiotics, such as penicillin, and to degradative enzymes, such as lysozyme. Eubacteria are almost equally divided between Gram-positive and Gram-negative types, and the result of the Gram stain remains a valuable character in bacterial classification.

## PRINCIPAL MODES OF METABOLISM

Organisms that use organic compounds as their major source of cell carbon are called *heterotrophs*; those that use carbon dioxide as the major source are called *autotrophs*. Organisms that use chemical bond energy for the generation of adenosine triphosphate (ATP) are called *chemotrophs*, whereas those that use light energy for this purpose are called *phototrophs*. These descriptions lead to the division of microorganisms into the four types listed in Table 1.2. Those chemoautotrophs that obtain energy from the oxidation of inorganic compounds are also called *chemolithotrophs*.

All organisms need energy and reducing power in order to conduct the biosynthetic reactions required for growth. In all cases, the energy-generating processes produce ATP (a molecule with high phosphate group donor potential); reducing power is stored in nicotinamide adenine dinucleotides (NADH and NADPH; molecules with high electron donor potential). Prokaryotes exhibit a wider range of energy-generating schemes than do eukaryotes. The three types of processes that lead to the formation of ATP in prokaryotes are reviewed very briefly below and summarized in Table 1.3.

**TABLE 1.2 Principal modes of metabolism**

| Type | Prokaryotes | Eukaryotes |
|---|---|---|
| Chemoautotrophs | + | none |
| Chemoheterotrophs | + | + ("animals," fungi) |
| Photoautotrophs | + | + ("plants") |
| Photoheterotrophs | + | none |

## Abstraction of Chemical Bond Energy from Preformed Organic Compounds (Chemoheterotrophy)

*Catabolic pathways* are sequences of chemical reactions in which carbon compounds are degraded. The molecules are altered or broken into small fragments, usually by reactions involving the removal of electrons (that is, by oxidations). The enzymes that catalyze catabolic reactions are usually located in the cytoplasm. There are two classes of energy-producing catabolic pathways: fermentations and respirations.

*Fermentations* are catabolic pathways that operate when no exogenous electron acceptor is present and in which the structures of carbon compounds are rearranged, thereby releasing free energy, which is used to make ATP. It is essential to distinguish between the biological meaning of fermentation as presented here and its meaning in the common parlance of applied microbiology. To the biotechnologist, a fermentation is any process mediated by microorganisms that involves a transformation of organic substances. The rigorous, chemical definition of a fermentation is that it is a process in which no net oxidation–reduction occurs; the electrons of the substrate are distributed among the products. For example, in a lactic acid fermentation, one mole of glucose is converted to two moles of lactic acid (Figure 1.4). The process whereby some of the released free energy is conserved in activated compounds formed in the course of catabolism and then used to generate ATP is called *substrate-level phosphorylation.*

*Respirations* are catabolic pathways by which organic compounds can be completely oxidized to carbon dioxide (mainly via the tricarboxylic acid cycle) because an exogenous terminal electron acceptor is present. Released free energy is conserved in the form of a protonic potential, or a proton motive force, generated by the vectorial (unidirectional) translocation of protons across a membrane within which components of an electron transport chain are contained. The vectorial translocation of protons is driven by the passage of electrons along the electron transport chain to the molecule that serves as the terminal electron acceptor. ATP is generated at the expense of the proton gradient upon return of the protons through a transmembrane enzyme complex, an $F_oF_1$-type adenosine triphosphatase (ATPase). This process is called *oxidative phosphorylation.*

**TABLE 1.3 Summary of the principal modes of microbial metabolism**

| Source of energy utilized | Major source of carbon assimilated | Generation of ATP and NADH (NADPH): Process | Electron donor ↓ $-e^-$ oxidized donor | Electron acceptor ↓ $+e^-$ reduced acceptor | Physiological group of microorganisms |
|---|---|---|---|---|---|
| Chemical bond energy ("chemotrophs") | Organic compounds ("chemoorganotrophs") | Fermentation | Organic compund ↓ Oxidized organic compound (and, in some cases, $CO_2$) | Organic compund ↓ Reduced organic compound (and, in some cases, $H_2$) | Many obligately anaerobic and many facultative chemoorganotrophic bacteria; some fungi, such as yeasts |
| | | Respiration | Organic compound ↓ $CO_2$ | $O_2$ ↓ $H_2O$ | Many obligately aerobic and many facultative chemoorganotrophic bacteria; many fungi and protozoa |
| | | Anaerobic respiration | | $NO_3^-$ ↓ $NO_2^-$ | Nitrate reducers* |
| | | | | $NO_2^-$ ↓ $N_2$ | Denitrifiers* |
| | | | | $SO_4^{2-}$ ↓ $H_2S$ | Sulfate reducers |
| | $CO_2$ ("chemolithotrophs") | Respiration | $H_2$ ↓ $H_2O$ | $O_2$ ↓ $H_2O$ | Hydrogen bacteria |
| | | | $NH_3$ ↓ $NO_2^-$ | | Ammonia oxidizers (e.g., *Nitrosomonas*) |
| | | | $NO_2^-$ ↓ $NO_3^-$ | | Nitrite oxidizers (e.g., *Nitrobacter*) |
| | | | $H_2S$ ↓ S or S ↓ $SO_4^{2-}$ | | Sulfur oxidizers (e.g., *Thiobacillus*) |
| | | Anaerobic respiration | $H_2$ ↓ $H_2O$ | $CO_2$ ↓ $CH_4$ | Methanogenic bacteria |
| Radiant light energy ("phototrolphs") | Organic compound ("photoorganotrophs") | Phototransduction | Organic compound ↓ Oxidized organic compound | Bacteriorhodopsin | *Halobacterium** |
| | | Photosynthesis | | | Purple nonsulfur* and gliding green* bacteria |
| | $CO_2$ ("photolithotrophs") | | $H_2S$ ↓ S or S ↓ $SO_4^{2-}$ | NADP ↓ NADPH | Green sulfur and purple sulfur bacteria |
| | | | $H_2O$ ↓ $O_2$ | NADP ↓ NADPH | Cyanobacteria (blue-green algae, eukaryotic algae, some protozoa) |

* These bacteria utilize the alternative pathways of metabolism indicated in the table when they are in the absence of oxygen ($O_2$).

In *aerobic* respiration, molecular oxygen ($O_2$) is utilized as the terminal electron acceptor. In *anaerobic* respiration, other oxidized substances are used as terminal electron acceptors for electron transport chains. Such molecules include nitrate ($NO_3^-$), sulfur (S), sulfate ($SO_4^{2-}$), carbonate ($CO_3^{2-}$), ferric ion ($Fe^{3+}$), and even organic compounds such as fumarate ion, and trimethylamine *N*-oxide.

$$C_6H_{12}O_6 \longrightarrow 2\ CH_3CHOHCOOH$$

Glucose Lactic acid

FIGURE 1.4

Overall equation for the fermentation reaction sequence, in which glucose is converted to lactic acid (homolactic fermentation).

### Abstraction of Chemical Bond Energy from Inorganic Compounds (Chemolithotrophy)

Certain prokaryotes use reduced inorganic compounds such as hydrogen ($H_2$), $Fe^{2+}$, ammonia ($NH_3$), nitrite ($NO_2^-$), sulfur, or hydrogen sulfide ($H_2S$) as electron donors to specific electron transfer chains, commonly with $O_2$ as terminal electron acceptor but in some instances with $CO_2$ or sulfate, to generate ATP by oxidative phosphorylation.

### Conversion of Light Energy to Chemical Energy (Phototrophy)

*Photosynthesis* is performed within membrane-bound macromolecular complexes containing pigments (bacteriochlorophylls, chlorophylls, carotenoids, bilins) that absorb light energy. The absorbed energy is conveyed to reaction centers, where it produces a charge separation in a special pair of chlorophyll (or bacteriochlorophyll) molecules. Reaction centers are specialized electron transport chains. The charge separation initiates electron flow within reaction centers, and the light-energy driven electron flow generates a vectorial proton gradient in a manner analogous to that described above for respiratory electron flow.

Some bacteria perform photosynthesis only under anaerobic conditions. This is termed *anoxygenic photosynthesis*. In other bacteria, photosynthesis is accompanied by the light-driven evolution of oxygen (similar to the photosynthesis in chloroplasts). Such photosynthesis is termed *oxygenic photosynthesis*.

Halobacteria perform a unique type of photosynthesis when the oxygen partial pressure is low. In the late 1960s, the cytoplasmic membrane of these organisms was found to contain an intrinsic membrane protein, *bacteriorhodopsin*, with a covalently attached carotenoid, retinal, as a chromophore. Absorption of light drives the isomerization of the retinal, after which the retinal rapidly returns to its original conformation. The retinal photocycle results in a vectorial pumping of protons by bacteriorhodopsin to the exterior of the cell with the generation of a proton motive force. ATP is generated at the expense of the proton gradient. Extensive screening of environmental samples shows that photosynthesis based on bacteriorhodopsin homologs appears to be widespread in many genera of marine planktonic bacteria and most likely in bacteria in other environments as well.

Different prokaryotes use one or another of the above processes as a preferred mode of energy generation. However, almost all prokaryotes are able to switch from one form of energy production to another, depending on the nature of the available substrates and on the environmental conditions. For example, purple nonsulfur bacteria grow on a variety of organic acids as substrates and obtain energy from respiration when oxygen is present. However, under anaerobic conditions and in the presence of light, these organisms synthesize intracellular membranes that possess the complexes needed for photosynthesis, and then they use light energy to generate ATP. Under aerobic conditions, the enteric bacterium *E. coli* oxidizes substrates such as succinate and lactate and utilizes an electron transport system with ubiquinone, cytochrome *b*, and cytochrome *o* as components and $O_2$ as a terminal electron acceptor. Under anaerobic conditions, with formate as a substrate, *E. coli* utilizes an electron transport system with ubiquinone and cytochrome *b* as components and nitrate as a terminal electron acceptor. When *E. coli* is growing on oxaloacetate as a substrate under anaerobic conditions, the sequence of carriers is NADH, flavoprotein, menaquinone, and cytochrome *b*, and fumarate is the terminal electron acceptor. There are hundreds of other well-defined examples of such metabolic versatility among prokaryotes. This flexibility in modes of energy generation is limited to the prokaryotes and gives these organisms a virtual monopoly on the colonization of certain ecological niches.

## THE IMPORTANCE OF THE IDENTIFICATION AND CLASSIFICATION OF MICROORGANISMS

In the search for organisms to assist in a technical process or to produce unusual metabolites, each time a new organism can be placed within a well-studied genus, strong and readily testable predictions can be made concerning many of its genetic, biochemical, and physiological characteristics (Box 1.2).

### CLASSIFICATION AND PHYLOGENY

Taxonomic systems for biological organisms are hierarchical. The most inclusive unit of classification is a kingdom (or domain), followed by phylum (or division), class, order, family, genus, species, and subspecies. By convention, the scientific names of genera and species of organisms are italicized or are underlined (Table 1.4). An additional rank below the subspecies level – pathovar, serovar, or biotype – is added when it is desired to distinguish a strain by a special character that it possesses. For example, the rank of a pathovar (or pathotype) is applied to an organism with pathogenic properties for a certain host or hosts, as exemplified by *Xanthomonas campestris* pv *vesicatoria*, the causal agent of bacterial spot of pepper and tomato. Serovar (or serotype) refers to distinctive antigenic properties, and biovar (or biotype) is applied to strains with special biochemical or physiological properties.

"Taxonomy (the science of classification) is often undervalued as a glorified form of filing – with each species in its folder, like a stamp in its prescribed place in an album; but taxonomy is a fundamental and dynamic science, dedicated to exploring the causes of relationships and similarities among organisms. Classifications are theories about the basis of natural order, not dull catalogues compiled only to avoid chaos."

*Source*: Gould, S. J. (1989). *Wonderful Life. The Burgess Shale and the Nature of History*, New York: W. W. Norton & Co.

**BOX 1.2**

In principle, any group of organisms can be classified according to any set of criteria, as long as the scheme results in reproducible identification of new

**TABLE 1.4 Ranking of taxonomic categories**

| Category | | Examples | |
|---|---|---|---|
| Domain | Archaea | Bacteria | Fungi |
| Phylum | Crenarchaeota | Proteobacteria | Ascomycota |
| Class | Thermoprotei | $\alpha$-Proteobacteria | Saccharomycetes |
| Order | Sulfolobales | Legionellales | Saccharomycetales |
| Family | Sulfolobaceae | Legionellaceae | Saccharomycetaceae |
| Genus | *Sulfolobus* | *Legionella* | *Saccharomyces* |
| Species | *Sulfolobus acidocaldarius* | *Legionella pneumophila* | *Saccharomyces cerevisiae* |

strains. However, a classification scheme based on totally arbitrary criteria is likely to be of very limited practical use. Thus taxonomists may group together apparently similar, presumably related species into a genus and presumably related genera into a family in the hope that this classification accurately reflects the evolutionary or phylogenetic relationships among various organisms. A hierarchical classification of this type was still being used by the recognized authority in prokaryote taxonomy, *Bergey's Manual of Determinative Bacteriology* (ninth edition), in 1994.

But how does one build such a taxonomic scheme? To classify a microorganism in this manner, one must first obtain a large uniform population of individuals, a *pure culture*. In the traditional methods of taxonomy, one then examines the organism's phenotypic characters – that is, the properties that result from the expression of its genotype, which is defined as the complete set of genes that it possesses. Phenotype includes morphological characteristics such as the size and shape of individual cells and their arrangement in multicellular clusters, the occurrence and arrangement of flagella, and the nature of membrane and cell wall layers; behavioral characteristics such as motility and chemotactic or phototactic responses; and cultural characteristics such as colony shape and size, optimal growth temperature and pH range, tolerance of the presence of oxygen and of high concentrations of salts, and the ability to resist adverse conditions by the formation of spores. The range of compounds that support the growth of a given organism, the way these compounds are degraded, and the nature of the end products (including the involvement of oxygen in the process) represent an important set of phenotypic characters.

It is customary to examine dozens of characters; in the computer-based method of numerical taxonomy, hundreds of characters may be examined. For identification of bacteria, armed with such information, one could then consult the ninth edition of *Bergey's Manual of Systematic Bacteriology*. The identification of a bacterium is thus a relatively straightforward matter. However, some difficulty is encountered when one wants to deduce phylogenetic relationships between organisms on the basis of the classification scheme presented in that edition of *Bergey's Manual*. A series of comments parallel to those made concerning prokaryotes can be made about the classification of fungi.

In basing a classification scheme on phenotypic characters a taxonomist must decide which characters are more fundamental and thus useful for dividing organisms into major groups, such as families, and which characters are more variable and thus suitable for dividing the major groups into smaller ones, such as species. In traditional taxonomy, the shape of the bacterial cell, for example, has been used for dividing bacteria into large groups. Thus of the lactic acid bacteria (which, as we will see later, characteristically obtain energy by fermenting hexoses into lactic acid plus sometimes ethanol and carbon dioxide), those with round cells and those with rod-shaped cells were placed in two completely different groups in the ninth edition of *Bergey's Manual.*

More recent quantitative information on the phylogenetic relationships among organisms has become available through comparison of their DNA sequences. Because the prokaryote world is so diverse, however, this method is only useful for comparing species of bacteria that are very closely related. Otherwise, the DNA sequences will be so dissimilar that no data of significance will be obtained. Thus it was the use of rRNA sequences for comparison, pioneered by Carl Woese in the early 1970s, that revolutionized the field. rRNA is present and performs an identical function in every cellular organism, and more importantly, its sequence has changed extremely slowly during the course of evolution. It is therefore an ideal marker for comparing distantly related organisms. Characteristic sequences of nucleotides, or "signature" sequences, may be conserved for a long time in a given branch of the phylogenetic tree and enable scientists to assign organisms on different branches with great confidence.

Returning to the classification of lactic acid bacteria, although the round-shaped lactic acid bacteria were placed far away from the rod-shaped ones in the 1994 *Bergey's Manual*, their rRNA sequences show that many of the former are actually very closely related to the latter.

We have now entered the era of phylogenetic systems of classification. The 2001 edition of *Bergey's Manual of Determinative Bacteriology* (second edition) "follows a phylogenetic framework, based on analysis of the nucleotide sequence of the ribosomal small subunit RNA, rather than a phenotypic structure." We must always keep in mind the vast time scale we are dealing with when we consider the evolution of bacteria. Even bacteria that are thought to be closely related phylogenetically can be quite distant on the evolutionary time scale, relative to the changes that have taken place among higher organisms. Thus, if we are looking at characteristics that change rapidly during the course of evolution, then the phylogenetic relationship may not offer much help. However, it will certainly help us in the study of slowly changing characters. An example is the organization and regulation of biosynthetic pathways. Because the prokaryotic world is so diverse, different pathways are seen in the biosynthesis of even such common compounds as amino acids. The distribution and the mechanism of control of these pathways, which we need to know in order to use bacteria to produce amino acids (see Chapter 9), clearly follow the 16S rRNA phylogenetic lines.

## INFORMATION CONTENT OF 16S rRNA

The 16S rRNA is a component of the small ribosomal subunit (30S ribosomal subunit) and is sometimes referred to as SSU rRNA. The predicted secondary structure of 16S RNA is shown in Figure 1.5. This structure was based on the analysis of approximately 7000 16S RNA sequences and was in about 98% accord with the crystal structure of the 16S RNA as seen in the high-resolution crystal structure of the 30S ribosomal subunit. Thus a common core of secondary or higher order structures is preserved throughout evolution, with some 67% of the bases involved in helix formation by intramolecular base pairing. Functional roles of the 16S RNA, conserved throughout evolution, doubtless dictate this high level of structure conservation.

Several websites provide databases of aligned 16S ribosomal DNA (rDNA) sequences (see references at the end of this chapter). Phylogenetic relationships are inferred from the number and character of positional differences between the aligned sequences (see Box 1.3). These primary data are then subjected to analysis by one of several tree-building algorithms. A tree is constructed from the results of such an analysis in which the terminal nodes (the 16S rDNA sequences) represent a particular organism and the internal nodes (the inferred common ancestor 16S rDNA sequences) are connected by branches. The branching pattern indicates the path of evolution, and the combined lengths of the peripheral and internal branches connecting two terminal nodes are a measure of the phylogenetic distance between two 16S rDNA sequences that serve as the surrogates for the source organisms.

On the basis of analyses of the relationships between 16S RNA gene sequences, two phyla are recognized within *Archaea* and 23 phyla within *Bacteria*. The evolutionary relationships between these phyla are illustrated in Figure 1.6. The archaea cluster into two phyla, Crenarchaeota and Euryarcheota. The bacterial phyla cluster into three broad groups: deep-rooted bacterial groups, particularly thermophiles; the Gram-negative bacteria; and the Gram-positive bacteria. Figure 1.7 shows the relationship between these phyla and the major phenotypic groups of prokaryotes selected as the basis of the classification in the earlier version of *Bergey's Manual of Systematic Bacteriology* (ninth edition). The comparison illustrates vividly how a classification based on phenotypic criteria can split into multiple groups species that belong within a single phylogenetic group.

## LIMITATIONS OF 16S rRNA PHYLOGENY

All biological classifications are human-imposed subdivisions upon the reality of the paucity of sharp discontinuities among the species in nature (Box 1.4). Moreover, a classification, based on a single character even one as rich in information as the 16S rRNA sequence, is bound to suffer from other shortcomings as well. This is evident from the following observations.

- The divergence of present-day rRNA sequences allows us to establish the succession of common ancestral sequences. However, it does not allow a direct correlation to a time scale.

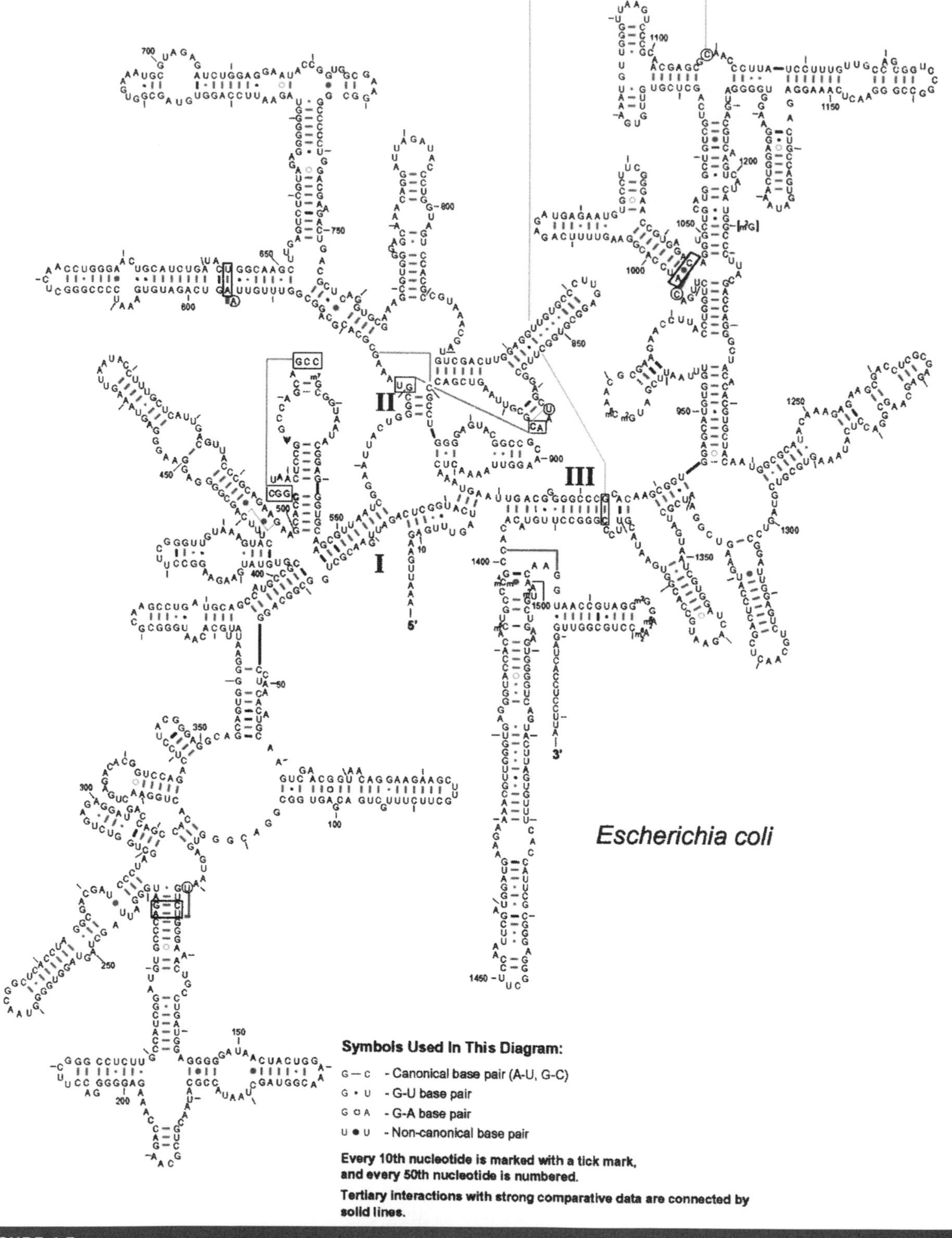


**FIGURE 1.5**

Predicted secondary structure of 16S rRNA. [Data from http://www.rna.icmb.utexas.edu/ and Cannone, J. J., Subramanian, S., Schnare, M. N., Collett, J. R., D'Souza, L. M., Du, Y., Feng, B., Lin, N., Madabusi, L. V., Muller, K. M., Pande, N., Shang, Z., Yu, N., and Gutell, R. R. (2002). The comparative RNA web (CRW) site: an online database of comparative sequence and structure information for ribosomal, intron, and other RNAs. *BMC Bioinformatics*, 3, 2; correction: *BMC Bioinformatics*, 3, 15.]

- A similarity in 16S rRNA gene sequence between strains that exceeds 97% is used to assign them to the same genus. However, the genomes of some organisms contain multiple copies of rRNA sequences. In certain of these organisms, a significant degree of sequence divergence exists between the multiple homologous genes. For example, the actinomycete *Thermospora bispora bispora* contains two copies of the 16S rRNA gene on the same chromosome within the same cell that differ from each other at the sequence level by 6.4%. The archaeon *Haloarcula marismortui* contains two rRNA operons, which show a sequence divergence of 5%. Such a situation poses problems for assignment of 16S rRNA gene-based relationships for these organisms.

**Information Content of 16S rDNA**

There are 974 (63.2%) variable (informative) positions in the 16S rDNA of Bacteria and 971 (63%) in that of Archaea. Four nucleotides may occupy a given position, and the maximum information content per position is defined by the number of possible character states (potential deletion or insertion is not considered).

Hence, the possible number of information bits is $\log_2 n \times p$, where $n$ is the number of character states and $p$ is the number of informative positions. This yields 1948 bits of information for Bacteria and 1942 for Archaea. However, empirically it is found that the number of allowed character states varies from position to position as follows:

| Number of nucleotides per position | Bacteria | Archaea |
|---|---|---|
| Four | 407 (26.4%) | 301 (19.5%) |
| Three | 209 (13.6%) | 233 (15.2%) |
| Two | 358 (23.2%) | 437 (28.3%) |

Taking the above data into account, the information content is reduced to 1506 bits for Bacteria and 1385 bits for Archaea.

*Source*: Ludwig, W., and Klenk, H-P. Overview: a phylogenetic backbone and taxonomic framework for prokaryotic systematics. (2001). In *Bergey's Manual of Systematic Bacteriology*, 2nd Edition, Volume 1, G. M. Garrity (ed.), pp. 49–65, New York: Springer-Verlag.

BOX 1.3

- Some organisms have identical 16S rRNA sequences but differ more at the whole genome level than do other organisms whose rRNAs differ at several variable positions.
- Sequencing of complete genomes shows that lateral gene transfer (discussed later in this chapter) and recombination have played a significant role in the evolution of prokaryote genomes. There is clear evidence in bacteria classified within the genera *Bradyrhizobium*, *Mesorhizobium*, and *Sinorhizobium* that distinct segments along the 16S rRNA gene sequences were introduced by lateral gene transfer followed by recombination (Figure 1.8). This resulted in incorrect tree topology and genus assignments and raises the strong possibility that other phylogenetic placements based solely on 16S rRNA gene sequence divergence may need to be reassessed in the future as more genomic information becomes available.
- It is widely agreed that 16S rRNA phylogenetic relationships are of limited value in predicting adequately the phenotypic capabilities of microorganisms.

## DNA–DNA HYBRIDIZATION

It is now evident that there is insufficient difference between 16S rRNA sequences to distinguish between closely related species and that interstrain DNA–DNA hybridization is the method of choice for assigning strains to a species. This method measures levels of homology between complete genomes. The phylogenetic definition of a species by this technique is as

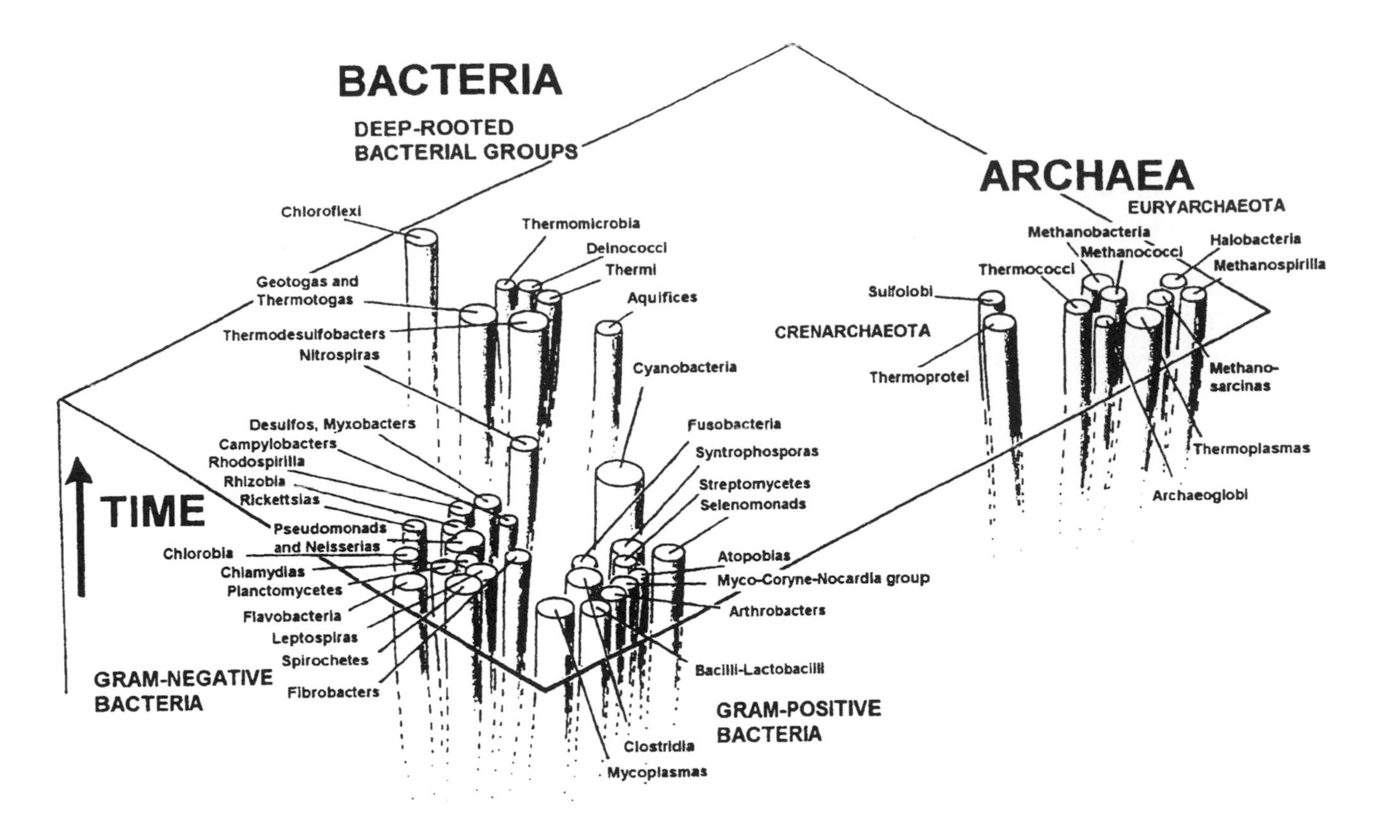


FIGURE 1.6

A two-dimensional projection of the phylogenetic tree of the major prokaryotic groups. Groups that lie close to together are more likely to have a recent common ancestry than are those that are well separated. The *dashed lines* in the time dimension below the plane indicate the still uncertain evolutionary origins of these groups. The computational procedure used to generate such two-dimensional projections of the genomic sequence data is outlined by G. M. Garrity and J. G. Holt (2001) in *Bergey's Manual of Systematic Bacteriology*, 2nd Edition, Volume 1, Garrity, G. M. (ed.), pp. 119–123, New York: Springer-Verlag. (Courtesy of Peter H. A. Sneath.)

"strains with approximately 70% or greater DNA–DNA relatedness and with 5°C or less $\Delta T_m$. Both values must be considered." (Source: Wayne, L. G., et al. (1987). Report of the ad hoc committee on the reconciliation of approaches to bacterial systematics. International Journal of Systematic Bacteriology, 37, 463–46). $T_m$ is the melting temperature of the hybrid DNA duplexes as measured by stepwise denaturation by heating (see Figure 1.9). $\Delta T_m$ is the difference in °C between homologous and heterologous hybrid duplexes formed under standard conditions.

## PLASMIDS AND THE CLASSIFICATION OF BACTERIA

The genetic information of a bacterial cell is contained not only in the main chromosome but also in extrachromosomal DNA elements called plasmids. Plasmids are self-replicating within a cell, and many plasmids have a block of genes that enable them to move from one bacterial cell to another. Loss of its plasmids has no effect on the essential functions of a bacterial cell. Consequently, the cell is seen to act as *host* to the plasmids. Similar to bacterial chromosomes, but much smaller, plasmids are circular double-stranded DNA molecules. Plasmid DNA often replicates at a different rate and sometimes on a different schedule from those of chromosomal DNA, and cells may contain multiple copies of specific plasmids. Some plasmids encode resistance to certain antibiotics or heavy metal ions or to ultraviolet

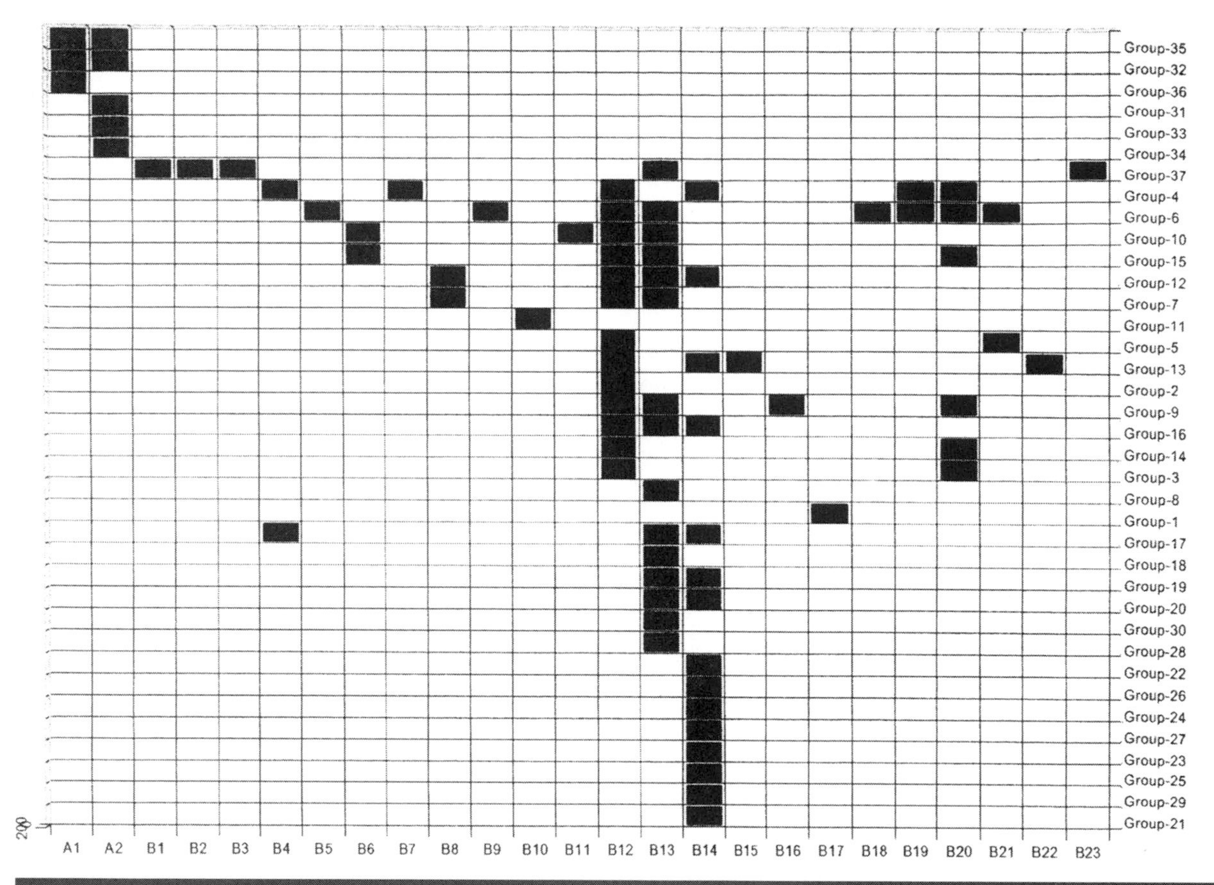


**FIGURE 1.7**

Occurrence of major phenotypic groups within the 25 prokaryotic phyla. This figure illustrates the relationship between these phyla and the major phenotypic groups of prokaryotes selected as the basis of the classification in the earlier version of *Bergey's Manual of Systematic Bacteriology* (ninth edition). [Reproduced with permission from Garrity, G.M., and Holt, J.G. (2001), The road map to the manual. In *Bergey's Manual of Systematic Bacteriology*, 2nd Edition, Volume 1, Garrity, G.M. (ed.) p. 124, New York: Springer-Verlag.]

**Prokaryotic phyla**[1]

A1 *Crenarcheota*
A2 *Euryarcheota*
B1 *Aquificae*
B2 *Thermotogae*
B3 *Thermodesulfobacteria*
B4 *"Deinococcus-Thermus"*
B5 *Chrysiogenetes*
B6 *Chloroflexi*
B7 *Thermomicrobia*
B8 *Nitrospirae*
B9 *Deferrobacteres*
B10 *Cyanobacteria*
B11 *Chlorobi*
B12 *Proteobacteria*
B13 *Firmicutes*
B14 *Actinobacteria*
B15 *Planctomycetes*
B16 *Chlamydiae*
B17 *Spirochaetes*
B18 *Fibrobacteres*
B19 *Acidobacteria*
B20 *Bacteroidetes*
B21 *Fusobacteria*
B22 *Verrucomicrobia*
B23 *Dictyoglomi*

**Major phenotypic groups of prokaryotes**[2]

Group 1 Spirochetes
Group 2 Aerobic/microaerophilic, motile, helical/vibrioid, Gram negative bacteria
Group 3 Nonmotile or rarely motile, curved Gram-negative bacteria
Group 4 Gram-negative aerobic/microaerophilic rods and cocci
Group 5 Facultatively anaerobic Gram-negative rods
Group 6 Anaerobic, straight, curved, and helical Gram-negative rods
Group 7 Dissimilatory sulfate- or sulfite-reducing bacteria
Group 8 Anaerobic Gram-negative cocci
Group 9 Symbiotic and parasitic bacteria of vertebrate and invertebrate species
Group 10 Anoxygenic phototrophic bacteria
Group 11 Oxygenic phototrophic bacteria
Group 12 Aerobic chemolithotropic bacteria and associated genera
Group 13 Budding and/or appendaged bacteria
Group 14 Sheathed bacteria
Group 15 Nonphotosynthetic, nonfruiting, gliding bacteria
Group 16 Fruiting gliding bacteria: the myxobacteria
Group 17 Gram positive cocci
Group 18 Endospore-forming Gram-positive rods and cocci
Group 19 Regular, nonsporulating, Gram-positive rods
Group 20 Irregular, nonsporulating, Gram-positive rods
Group 21 Mycobacteria
Group 22 Nocardioform actinomycetes
Group 23 Actinomycetes with multilocular sporangia
Group 24 Actinoplanetes
Group 25 *Streptomycetes* and related genera
Group 26 Maduromycetes
Group 27 *Thermomonospora* and related genera
Group 28 *Thermoactinomycetes*
Group 29 Other actinomycete genera
Group 30 Mycoplasmas
Group 31 The methanogens
Group 32 *Archaeal* sulfate reducers
Group 33 Extremely halophilic *Archaea*
Group 34 *Archaea* lacking a cell wall
Group 35 Extremely thermophilic and hyperthermophilic S-metabolizing *Archaea*
Group 36 Hyperthermophilic non–S-metabolizing *Archaea*
Group 37 Thermophilic and hyperthermophilic bacteria

1 Two phyla (A1 and A2) occur within the *Archaea* and B1-B23 within the *Bacteria*. These two prokaryotic domains were subdivided into these phyla on the basis of DNA sequence data, principally 16S and 23S rDNA. Earlier treatment of prokaryote taxonomy subdivided some 590 genera into major phenotypic groups (represented above as Groups 1-37). Assignment to these phenotypic groups was based on readily recognizable phenotypic or metabolic characters that could be used for the presumptive identification of species [see Holt, J.G. et al. (eds.) (1994). *Bergey's Manual of Determinative Bacteriology*, 9th Edition, Baltimore: Williams & Wilkins].

2 The group number refers to the phenotypic group used in the *Bergey's Manual of Determinative Bacteriology*, 9th Edition.

"... [I]t is becoming clear that the biodiversity is much greater than expected. When numerous strains are analyzed and grouped by various methods, as in *Xanthomonas*, it appears that this genus constitutes a continuum of geno- and phenotypes with cloudy condensed nodes representing ecologically more successful types. Thus, any attempt to divide biological populations into discrete taxa, as is done in the current classification systems, will always be more or less artificial because of its inconsistency with the real continuous nature of biodiversity. Obviously, this situation will be more pronounced in one genus than another."

*Source*: Vauterin, L., and Swings, J. (1997). Are classification and phytopathological diversity compatible in *Xanthomonas*? *Journal of Industrial Microbiology & Biotechnology*, 19, 77–82.

**BOX 1.4**

radiation. Others, surprisingly, carry genes coding for functions that have been thought to be a distinguishing characteristic of the host species. For example, the most characteristic trait of the fluorescent *Pseudomonas* (see below) is thought to be its ability to degrade a wide range of organic compounds; however, many of the genes that make these degradations possible are located on plasmids. The same is true of the genes for nitrogen fixation in the species that carries out much of the biological nitrogen fixation on Earth – *Rhizobium* – and of the genes for disease-causing factors (toxins, proteases, or hemolysins; i.e., the proteins that lyse red blood cells and other animal cells) in many pathogenic bacteria. Because plasmids sometimes confer highly noticeable phenotypic traits on their hosts, they may influence the classification of the host organism. For example, certain strains of *Streptococcus lactis*, classified as *S. lactis* subsp. *diacetylactis*, carry a plasmid that allows them to utilize citrate. These are the strains responsible for the characteristic aroma of cultured butter, which results from the diacetyl they produce when fermenting citrate in milk.

Some plasmids have the ability to transfer themselves from one bacterial host cell into another. Sometimes the host is of a different species or genus. On the other hand, the plasmid genes can become integrated into the host's chromosome and become a part of the permanently inherited genetic makeup of the cell. This "lateral" transfer of genetic information into different groups of bacteria, if it were to occur frequently, would make every bacterium into an extremely complex hodgepodge of genes coming from many different sources. Experimental studies, however, have shown that lateral exchange certainly has not occurred to the extent of obliterating the phylogenetic lines of descent of various organisms.

The ability of plasmids to replicate themselves has been utilized in the construction of cloning vectors, many of which contain a replication function derived from plasmids and can therefore be maintained indefinitely

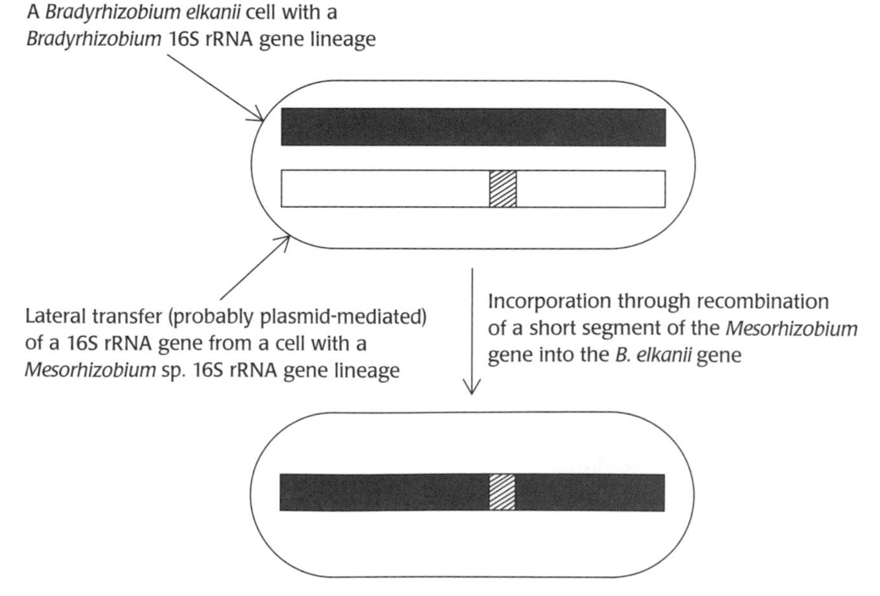


**FIGURE 1.8**

Diagrammatic representation of lateral gene transfer and recombination events leading to the incorporation of a short segment of the 16S rRNA gene of *Mesorhizobium mediterraneum* (Upm-Ca 36) into the 16S rRNA gene of *Bradyrhizobium elkanii* to produce the present day *B. elkanii* (USDA 76) 16S rRNA gene. [Based on data from van Berkum, P., et al. (2003). Discordant phylogenies within the *rrn* loci of Rhizobia. *Journal of Bacteriology*, 185, 2988–2998.]

in the cytoplasm of the host bacteria. However, in many cases, the replication of plasmids requires the participation of host functions too. This is one of the reasons why plasmids can survive in only a limited range of hosts. One way to construct a vector that can replicate in a wide range of hosts is to use the replication genes from a plasmid that has a broad host range. Here again, a knowledge of phylogenetic relationships will help us in predicting the range of host bacteria that would support the replication of such vectors. For example, many broad–host range plasmids isolated from the Gram-negative bacteria of the "purple bacteria" group (see below) are likely to replicate in most of the members of this group, or at least in the members of the same subgroup.

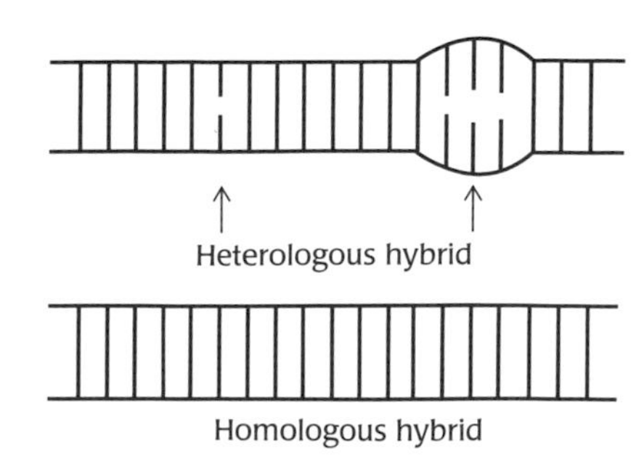


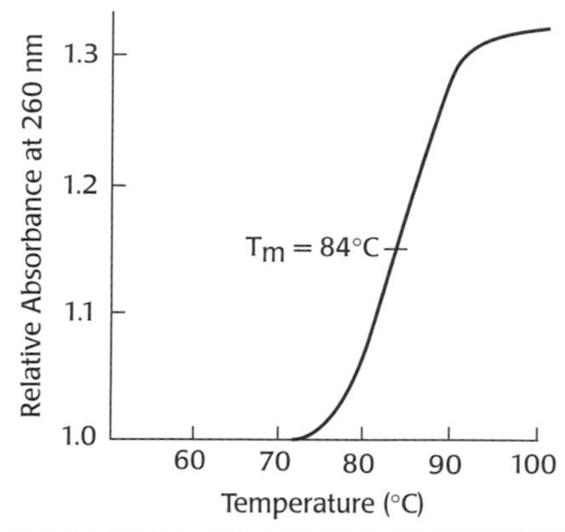


**FIGURE 1.9**

Temperature dependence of the absorbance of a solution of a perfectly complementary DNA hybrid duplex at 260 nm ($A_{260}$). The separation of the two strands (also termed the "melting" of the DNA) is accompanied by an increase in the absorbance at 260 nm. The temperature at which the change in $A_{260}$ is 50% complete is designated as the melting temperature ($T_m$). The $T_m$ is sensitive to the pH and ionic strength of the buffer. In the representation of a heterologous hybrid (*upper diagram*), the *arrows* point to noncomplementary positions in the two DNA sequences. Such a hybrid would have a much lower $T_m$ than the perfectly complementary hybrid duplex whose melting curve is shown here.

## ANALYSIS OF MICROBIAL POPULATIONS IN NATURAL ENVIRONMENTS

We have seen above how the sequence of 16S rRNA is utilized to classify prokaryotes and to assign phylogenetic relationships. The universal acceptance of this molecular marker has resulted in the determination of a very large number of 16S rDNA sequences. As of March 1, 2007, the Ribosomal Database Project provides over 335,800 small ribosomal subunit rRNA sequences from a wide variety of prokaryotic taxa. By amplifying and sequencing the 16S rDNA of an unknown organism, it is now possible to determine its phylogenetic relationship to the 16S rDNA sequences characteristic of the known genera of prokaryotes.

As of March 20, 2007, the curated Swiss-Prot database contains over 1,500,000 distinct protein sequences from all kingdoms of organisms. The complete sequences of over 480 prokaryote genomes have been determined as of January, 2007, and contribute substantially to the total content of this database. Comparative sequence information on protein coding genes allows the use of molecular markers other than 16S rDNA to explore in complementary ways the taxonomy, phylogeny, and functional diversity of prokaryotes and has opened the way to powerful *in situ* analyses of microbial populations in natural environments.

### NUCLEIC ACID SEQUENCE–BASED METHODS IN ENVIRONMENTAL MICROBIOLOGY

We examine three powerful methods selected from among many sequence-based approaches to the study of natural microbial populations. *Labeled nucleic acid probes* allow sensitive detection and enumeration of cells in a mixed population that contain a particular nucleic acid sequence. Where the sequences of flanking regions of a DNA sequence of interest are known, the *polymerase chain reaction* (PCR) allows completely selective amplification of that sequence from a complex mixture of nucleic acids. Finally, *whole-genome shotgun sequencing* of the DNA of microbial populations provides unique insights into both the complexity of natural microbial populations

**1** Perfect match between probe and target

Probe 3'-5' GAGAACGGUAGCCUACAC•
Target 5'-3' ---CGGGCCUCUUGCCAUCGGAUGUGCCCAGAU---

**2** One mismatch between probe and target near 5' end of probe

Probe 3'-5' GAGAACGGUAGCCUACAC•
Target 5'-3' ---CGGGCCUCUUGCCAUCGGAU**C**UGCCCAGAU---

**3** Internal mismatch between probe and target

Probe 3'-5' GAGAACGGUAGCCUACAC•
Target 5'-3' ---CGGGCCUCUUGC**G**AUCGGAUGUGCCCAGAU---

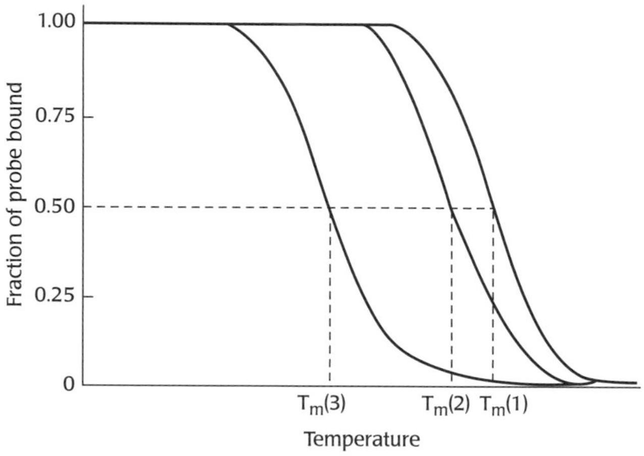

Fraction of probe bound
1.00
0.75
0.50
0.25
0
$T_m(3)$ $T_m(2)$ $T_m(1)$
Temperature

**FIGURE 1.10**

Hybridization of oligonucleotide probes to a target DNA sequence. The sequence of probe 1 is perfectly complementary to that of the target DNA, whereas there is one mismatch (position indicated in boldface in the sequence of the target DNA) in each of the oligonucleotide probes 2 and 3. The *black dot* at the 5′ end of each probe indicates a covalently attached fluorescent label. The *lower panel* illustrates the dissociation profile of each of the hybrids. Note that the higher the $T_m$ the higher the stringency of hybridization.

and the functions of some of the organisms present. Together, these methods provide information on the microbial diversity, gene content, and relative abundance of organisms in environmental samples.

## NUCLEIC ACID PROBES

The interactions between three related oligonucleotide probes and a target DNA sequence are illustrated in Figure 1.10. There is a perfect complementarity between the probe and the target in case 1 and a single mismatch, in different positions, between the probe and the target in cases 2 and 3. As illustrated, under appropriate conditions, each of these probes can base-pair (*hybridize*) with the denatured target DNA. However, the hybrids with mismatches will be less stable than the one with the perfectly complementary probe and as illustrated in the lower panel of Figure 1.10, will dissociate at a lower temperature ($T_m$) than the perfectly matched probe. The stability of the hybrid is affected by both the location and the character of the mismatch. Appropriate selection of the composition of the hybridization buffer and the temperature is critical to optimize the *stringency* of the hybridization, that is, the selection of the conditions under which the perfectly matched probe will remain bound to immobilized target nucleic acid while hybrids with mismatches either will not form or will dissociate upon washing with probe-free hybridization buffer. The optimum temperature for hybridization is thus slightly below the $T_m$. Note that the probes illustrated in Figure 1.10 bear fluorescent labels that allow optical detection with very high sensitivity.

The huge database of 16S rDNA sequences guides the design of PCR oligonucleotide primers that are complementary to different sets of target sequences. *Universal primers* can be selected that are complementary to a region of 16S rDNA sequence that is perfectly conserved in all 16S rDNA genes. Using the same approach, *primers unique to archaeal and eukaryal sequences* or those complementary only to *bacterial sequences* can be chosen and enable the selective amplification of these sets of sequences. Any one of such PCR primers can be labeled with a fluorophore (or other label) and utilized as a probe for the target sequence. At the highest level of selectivity, if a 16S rDNA sequence is fully known, an oligonucleotide label can be synthesized that will bind with high stringency exclusively to this unique target sequence.

## CATALOGS OF 16S rDNA SEQUENCES: MEASURES OF MICROBIAL DIVERSITY IN NATURAL ENVIRONMENTS

Total DNA can be readily obtained from diverse natural environments. PCR primers, chosen as described above, are then utilized for the amplification of the 16S rDNA sequences present in the DNA sample. The amplified sequences are then cloned and sequenced, and the resulting catalog of 16S rDNA sequences is a set of molecular signatures for the organisms present in the environment under study and a measure of the diversity of the population. The individual 16S rDNA sequences also provide information on the taxonomic affinities and phylogenetic relationships between these organisms and well-studied microorganisms. A minute fraction of microorganisms present in any natural environment is available in pure culture. The catalog of 16S rDNA sequences obtained from that environment provides a revealing glimpse of the organismal complexity of the actual microbial population. Such analyses on a wide range of natural environments have shown that microorganisms available in pure culture represent a minute percentage of those present in nature. Moreover, in many natural environments, the most abundant microorganisms have yet to be cultured.

## ISOLATION OF A HYPERTHERMOPHILIC ARCHAEAL ORGANISM PREDICTED BY *IN SITU* RNA ANALYSIS: A CASE HISTORY

This case history traces the path from a particular 16S rDNA sequence derived from an analysis of a hot pool sample to a pure culture of the source organism (the literature citation is provided in the reference list at the end of the chapter).

The finding of a novel 16S rDNA sequence indicates the presence of a new organism. The rDNA sequence by itself gives information on the affinities of the source organism to known organisms but no information on its morphology, physiology, or biochemistry. However, as we shall see, knowledge of the unique 16S rDNA sequence may suffice to isolate the source organism in pure culture.

Let us consider the application of a labeled probe complementary to a 16S rDNA sequence to the detection of cells containing the target sequence. Prokaryotic cells contain many ribosomes and correspondingly many copies of the small ribosomal subunit 16S rRNA sequence. Consequently, when a permeabilized cell is exposed to the labeled probe, the probe will hybridize with the many target molecules within the cell and the multiply labeled cell may be readily detected by fluorescence microscopy, in flow cytometry, or by other methods.

This case history begins with *in situ* analyses of the 16S rDNA sequences in samples from the Obsidian hot pool in Yellowstone National Park in the United States. The water in this pool is at pH 6.7, and the temperature at different sites ranges from 73°C to 93°C. The analyses revealed the presence of numerous archaeal sequences different from those of previously isolated species. The following steps were taken to isolate in pure culture the source organism of one these new archaeal sequences.

- Aerobic and anaerobic samples of water and sediment were taken from different sites in the pool.
- Aerobic and anaerobic enrichment cultures obtained in the laboratory contained mixtures of coccoid, rod-, and plate-shaped cells differing in length and diameter.
- Whole-cell hybridization of these cultures with fluorescently labeled probes targeted to an *archaea-specific region* within the 16S rRNA showed that archaeal cells were present in over half of the enrichment cultures.
- A fluorescently labeled probe was designed with specificity toward one of the new archaeal 16S rDNA sequences (designated pSL91) identified in the initial *in situ* analysis. Whole-cell hybridization with the probe gave a positive signal in one of the enrichment cultures. This culture had been grown anaerobically at 83°C with cell extracts as a heterogeneous energy source. (These cell extracts were prepared from the mixed cell population in the initial enrichment cultures to maximize the probability that needed nutrients, normally present in the source natural environment, would be present.) Fluorescence microscopy showed that the labeling was confined to rare grapelike aggregates of coccoid cells not previously seen in terrestrial hot springs. Within this enrichment culture, the dominant cells were filamentous cells resembling those of *Thermophilum* and *Thermoproteus* species. These were more abundant than the labeled coccoid cells by over four orders of magnitude. Conventional isolation of the cells of interest by plating was not feasible.
- With the morphology of the cells of interest as a guide, they were isolated from an unlabeled enrichment culture by an "optical tweezers" technique. The microorganism was cloned by micromanipulation in a computer-controlled inverse microscope equipped with a strongly focused infrared laser. Cells were separated from the enrichment culture in a 10-cm long

square capillary connected to a sterile syringe. The single cells were injected into sterile anaerobic culture medium.

- Upon incubation under anaerobic conditions at 83°C, approximately 30% of the cloned cells gave rise to pure cultures of grapelike aggregates.
- The sequence of the 16S rDNA of the isolate was indistinguishable from that of pSL91. The researchers noted, "In this way, we have shown for the first time the identity of a sequence determined without cultivation from the native habitat with that of a pure culture."

## 'ENVIRONMENTAL GENOME' SHOTGUN SEQUENCING

In the "whole-genome shotgun sequencing" method, the whole genome of the organism under study is shredded, the fragments cloned in a sequencing vector, the resulting plasmid clones sequenced from both ends, and the sequences assembled. The method debuted in July 1995 with the report of the determination of the complete nucleotide sequence (1,830,137 base pairs [bp]) of the genome of the bacterium *Haemophilus influenzae* Rd. At that time, this was the first complete genome sequence from a free-living organism. The method was applied successfully to many other prokaryotes and several eukaryotes, and its culminating achievement, reported in February 2001, was the determination of the 2.91 billion–bp consensus sequence of the euchromatin portion of the human genome.

This powerful method has now been applied to exploring the genomic complexity of microbial populations in natural environments. The samples for the first such study were collected from the Sargasso Sea near Bermuda. This is an oligotrophic (nutrient-poor) body of water. Microbial populations were collected on filters of 0.1 to 3.0 $\mu$m from surface seawater samples of 170 to 200 L. Because this initial study concentrated on prokaryotes, the analysis of genomic DNA focused on the material collected from the size fraction of 0.2 to 0.8 $\mu$m. Analysis of the DNA extracted from about 1500 L of seawater yielded 1.045 billion bp of nonredundant sequence. The analysis of these sequence data provides a sobering glimpse of the complexity of the microbial population that inhabits the waters of the study area. Comparison with known sequences indicated that the DNA originated from at least 1800 distinct genomes. Of these, 148 had no relatives among known prokaryotes. More than 1.2 million genes were identified in the samples. The total number of entries in the Swiss-Prot database at the time of this study was around 140,000. The data set had valuable new information on the diversity and taxonomic distribution of particular proteins. For example, 782 new proteorhodopsin-like proteins were identified. Proteorhodopsin-like proteins are of particular interest because some families of these proteins are involved in phototrophy, light harvesting–dependent energy generation, as described earlier in this chapter. Because the sequencing approach used here frequently links a gene encoding a protein with a particular biological function to a phylogenetically informative marker, it provides information on the phylogenetic diversity of the organisms that contain the gene with

**TABLE 1.5 rRNA operon copy number in various bacteria**

| Bacterium | Major phylogenetic division | Operon copy number |
|---|---|---|
| *Vibrio cholerae* | $\gamma$-Proteobacteria | 9 |
| *Escherichia coli* | $\gamma$-Proteobacteria | 7 |
| *Haemophilus influenzae* | $\gamma$-Proteobacteria | 6 |
| *Pseudomonas putida* | $\gamma$-Proteobacteria | 6 |
| *Pseudomonas stutzeri* | $\gamma$-Proteobacteria | 4 |
| *Variovorax* sp. | $\beta$-Proteobacteria | 1 |
| *Agrobacterium tumefaciens* | $\alpha$-Proteobacteria | 4 |
| *Bradyrhizobium japonicum* | $\alpha$-Proteobacteria | 1 |
| *Bacteroides uniformis* | *Cytophaga/Flexibacter/Bacteroides* | 4 |
| *Synechococcus* PCC6301 | Cyanobacteria | 2 |
| *Borrelia burgdoferi* | Spirochaetes | 1 |
| *Streptomyces venezuelae* | Gram-positive bacteria (high G+C) | 7 |
| *Mycobacterium leprae* | Gram-positive bacteria (high G+C) | 1 |
| *Clostridium beijerinckii* | Gram-positive bacteria (low G+C) | 13 |
| *Mycoplasma pneumoniae* | Gram-positive bacteria (low G+C) | 1 |
| *Thermus thermophilus* | Thermophiles | 2 |
| *Aquifex pyrophilus* | Thermophiles | 6 |

*Source*: Klappenbach, J. A., Dunbar, J. M., and Schmidt, T. M. (2003). rRNA operon copy number reflects ecological strategies of bacteria. *Applied Environmental Microbiology*, 66, 1328–1333.

that biological function. In the case of the proteorhodopsin-like genes, their distribution was found to be much broader than indicated by earlier surveys.

The wealth of data produced by the shotgun sequencing approach offers other advantages. It is not possible to rely exclusively on 16S rDNA as a measure of species diversity and of their relative abundance in environmental samples because the number of copies of rRNA genes varies by more than an order of magnitude among prokaryotes (Table 1.5). In this study, six proteins (AtpD, GyrB, Hsp70, RecA, RpoB, TufA) that are encoded by one gene only in virtually all known bacteria were used as phylogenetic markers to determine species abundance.

Depending on assumptions about the fraction of organisms present at very low individual abundance in the population analyzed in this study, the total prokaryote diversity in the Sargasso Sea sample may be as high as around 47,700 species.

## CULTIVATION OF PROKARYOTES

Researchers have applied oligonucleotide probes based on the sequence of 16S rDNA to samples from a wide variety of natural environments ranging from sites in the open ocean and hydrothermal vents to the guts of termites, the rumen of cattle, and rice paddies. In all these natural environments and many others, the probes reveal the presence of a multitude of different prokaryotes widely separated from one another on the 16S rDNA–based phylogenetic tree. Organisms encoding a particular 16S rDNA sequence are referred to as "ribotypes." It is estimated that more than 99% of the ribotypes detected by the oligonucleotide probes represent organisms different from those available in pure culture.

Sometimes this is the case for the predominant ribotype within the population in a particular environment. A much-cited example is a marine $\alpha$-proteobacterial ribotype designated SAR11. Organisms encoding this ribotype are estimated to represent around 50% of the microbial cells in many open-ocean systems. SAR11 is present at 500,000 cells/ml in some samples taken from the surface of the Sargasso Sea. About ten years after the initial description of the SAR11 ribotype, the organism was brought into pure culture.

The above observations led some to conclude that a great majority of free-living microorganisms were "uncultivable." This view is not supported by the facts. With sufficient effort, the source organisms of various ribotypes have been successfully cultured. Leadbetter (see reference at the end of this chapter) in a recent review stated the reasons for failure in cultivation succinctly: "First, many microbes will not grow in the laboratory, primarily because we have an insufficient knowledge or imagination of the chemistry of their native, extracellular milieu, and so we are unable to recreate viable laboratory conditions for them. . . . Second, the impatient laboratory scientist might have overlooked the fact that an organism has actually grown under his or her very watch, because obvious turbidity or colonies had not developed." With SAR11, cultivation was successfully accomplished in sterilized seawater media supplemented with low amounts of ammonium and phosphate, with the recognition that this organism grew very slowly with doubling times on the order of one to two days and that the cultures grew to densities only of about $10^4$/ml. The factor(s) that limit the growth of the cultures to such low densities are yet to be determined.

A general method described for the isolation and cultivation in pure culture of a wide diversity of prokaryotes employs a massively parallel approach. In this procedure, microbial cells are first separated from environmental samples (seawater, soil) by density gradient centrifugation. They are then encapsulated in agarose gel microdroplets at one cell per droplet. The gel microdroplets are packed in a growth column in which the upward flow of low-nutrient medium washes out free bacterial cells and promotes growth of cells within the gel microdroplets (Figure 1.11). The gel microdroplets containing colonies are detected and separated by flow cytometry into 96-cell microtiter plates containing a rich organic medium (Figure 1.12). The clonal cultures develop within the microtiter plate wells.

## TAXONOMIC DIVERSITY OF BACTERIA WITH USES IN BIOTECHNOLOGY

Within each of the formal subdivisions in the taxonomic outlines of the major prokaryotic groups shown in Figures 1.6 and 1.7 there is a world of

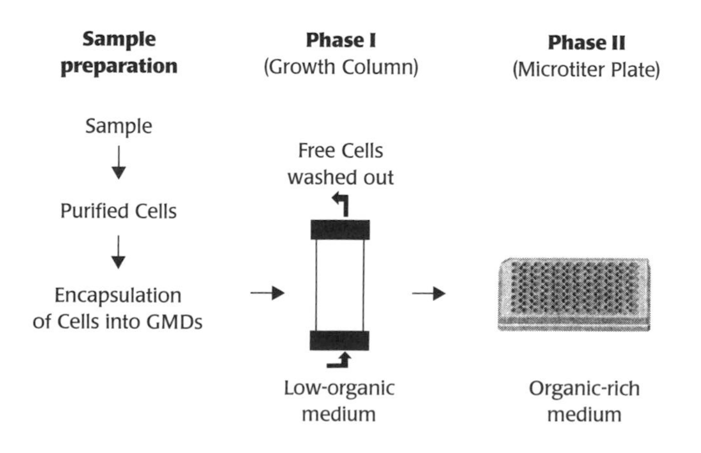


**FIGURE 1.11**

Cells captured from environmental samples were encapsulated into gel microdroplets (GMDs) and incubated in growth columns (phase I). GMDs containing colonies were detected and separated by flow cytometry into 96-well microtiter plates containing a rich organic medium (phase II). [Reproduced with permission from Zengler, K., Toledo, G., Rappe, M., Elkins, J., Mathur, E. J., Short, J. M., and Keller, M. (2002). Cultivating the uncultivated. *Proceedings of the National Academy of Sciences U.S.A.*, 99, 15681–15686.]

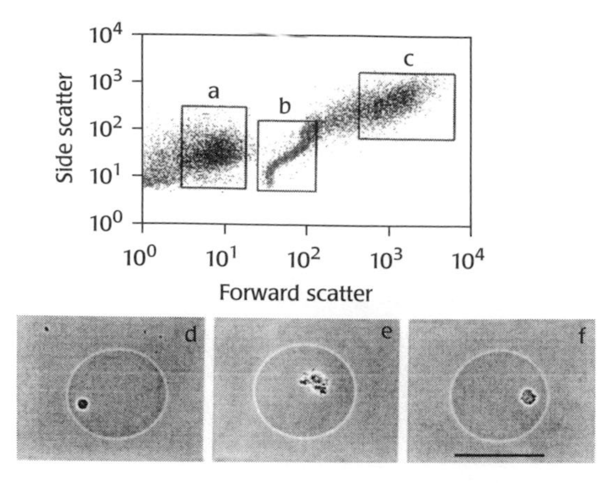


FIGURE 1.12

Discrimination among **(A)** free-living cells, **(B)** singly occupied or empty GMDs, and **(C)** GMDs containing microcolonies was accomplished by flow cytometry in forward and side light-scatter mode. **(D, E, and F)** Phase contrast photomicrographs of separated GMDs containing microcolonies. Bar = 50 $\mu$m. [Reproduced with permission from Zengler, K., Toledo, G., Rappe, M., Elkins, J., Mathur, E. J., Short, J. M., and Keller, M. (2002). Cultivating the uncultivated. *Proceedings of the National Academy of Sciences U.S.A.*, 99, 15681–15686.]

microorganisms whose immense diversity and complexity have only recently been revealed. Below we introduce some of the groups of prokaryotes that are particularly important in biotechnology.

## TAXONOMIC DIVERSITY OF USEFUL BACTERIA

As mentioned at the beginning of this chapter, Bacteria as a domain show tremendous metabolic versatility. Archaea, although frequently adapted to extreme environments and often obtaining energy in rather unexpected ways, are less versatile in the diversity of their metabolism. Consequently, most of the prokaryotes so far utilized for biotechnological applications belong to Bacteria. Below we list some of the groups of bacteria that are particularly important in biotechnology to familiarize readers with their names and properties and to demonstrate the importance of bacterial diversity to biotechnology. For each of the organisms mentioned, we indicate the phylum to which it has been assigned using the designations provided in Figure 1.7.

### Deinococcus-Thermus (Phylum B4)

The phylum Deinococcus-Thermus is subdivided into two orders, each of which contains a single family. *Deinococcus*, a representative of the first family, is an unusual organism with an extremely high resistance to ionizing radiation. A Gram-positive chemoheterotroph, *Deinococcus* has been isolated from soil, ground meat, and dust. The cells are bright red or pink because of their high carotenoid content and are surrounded by an outer membrane layer, normally absent from Gram-positive bacteria. However, this outer membrane is chemically distinctive in that it does not contain the lipopolysaccharide characteristic of the outer membranes of Gram-negative bacteria. *Thermus*, the sole genus of the second family, consists of Gram-negative straight rods or filaments. These organisms are thermophilic, aerobic heterotrophs or chemoheterotrophs with a strictly respiratory metabolism.

**The Genus *Thermus*.** Cells of *Thermus* strains are nonmotile. The organism was first found in hot springs in 1969 and was one of the most thermophilic bacteria then known. Most species have an optimum temperature for growth of 70°C to 72°C and can grow at significantly higher temperatures. Their habitat is not limited to hot springs, however; one investigator found that the best source for isolation is the hot water tanks in homes and institutions.

*Thermus aquaticus* is currently used as the source of a thermostable DNA polymerase (Taq polymerase), valuable in the amplification of genes by PCR. This enzyme has been cloned and expressed in *E. coli* and is manufactured on a large scale.

### Proteobacteria (Phylum B12)

**Gamma ($\gamma$) Division of the Proteobacteria.** This division contains both the family of enteric bacteria (containing the well-known *E. coli*) and some of the best-known *Pseudomonas* species. It also contains the family Acidithiobacillaceae, made up of widely distributed chemoautotrophic bacteria.

*E. coli*

This inhabitant of the intestinal tract of higher animals is the most extensively studied living organism. It is one of a family of closely related organisms called enterics or Enterobacteriaceae, most of which are rod-shaped Gram-negative bacteria with peritrichous flagella (*peritrichous* indicates that the flagella are more or less uniformly distributed over the surface of the cell). The hallmark of *E. coli* metabolism is that they can generate ATP either by oxidative degradation of organic compounds in the presence of air or by fermentation of simple sugars under anaerobic conditions. The range of compounds they can metabolize is limited, as expected for intestinal bacteria that encounter only food of the type ingested by the host animal. However, their metabolism is extremely well regulated. This too is a necessity for intestinal bacteria, as they live a "feast-or-famine" existence and have to compete with many other organisms in a confined environment. These are important points: Much of our knowledge of metabolic regulation was derived from *E. coli*, and we tend to think, often incorrectly, that the *E. coli* model applies to other organisms living in totally different ways (see discussion of amino acid production in Chapter 9).

Because it grows rapidly on well-defined simple media and because of the wealth of information on its genetics, biochemistry, and physiology, *E. coli* has been a favorite choice for the production of foreign proteins by means of recombinant DNA technology. Human insulin and growth hormone are prime examples (see Chapter 2).

*Fluorescent pseudomonads*

Molecular phylogenetic studies of the very disparate species of organisms formerly classified in the genus *Pseudomonas* now divide them among the $\alpha$, $\beta$, and $\gamma$ subgroups of the proteobacteria. The Gram-negative bacteria included in the genus *Pseudomonas* in the $\gamma$ division of the proteobacteria (such as the species *Pseudomonas aeruginosa*, *Pseudomonas putida*, *Pseudomonas fluorescens*, and *Pseudomonas syringae*) differ from *E. coli* in many ways. Although they are also rod-shaped like *E. coli*, their flagella are polarly located. The fluorescent pseudomonads excrete yellowish green fluorescent compounds into the culture medium. They cannot make ATP by fermentation; they are obligate aerobes. In marked contrast to *E. coli*, some of the species are famous for using a wide range of organic compounds as energy sources – in some cases, more than a hundred. These properties are ideally suited to the pseudomonads' existence as "generalists" in soil and water.

Many members of the fluorescent group degrade compounds such as camphor, toluene, and octane, as well as certain man-made substances, such as halogenated aromatic compounds. Thus naturally occurring and laboratory-engineered fluorescent strains are the subjects of active study as possible candidates for the reclamation of sites that have been contaminated with high levels of toxic organic compounds. Interestingly, the genes for the enzymes that degrade camphor or octane are usually found on plasmids, although the enzymes that degrade aromatic compounds through the "classical" or "ortho cleavage" pathway, which involves opening a catechol ring between the two hydroxyl groups of catechol, are apparently coded for by chromosomal genes.

The fluorescent group includes one plant pathogen, *P. syringae*. The outer membrane of this pseudomonad contains a protein complex that nucleates the formation of ice crystals. These organisms cause frost to form on leaves at temperatures only slightly below the freezing point, $-3°C$ rather than $-10°C$, damaging the outer tissue so that the bacteria can invade the plant. The resulting damage to crops is estimated to exceed $1 billion annually in the United States alone. To decrease this damage, "ice minus" *P. syringae* strains were made by using recombinant DNA methods to inactivate one of the genes necessary for the formation of the ice nucleation complex. When the mutant organisms are sprayed on plants, they compete with the wild type and decrease the chances of ice formation. On the other hand, the wild-type *P. syringae* strains have been put to practical use manufacturing snow at ski resorts, with appreciable savings in the cost of cooling. Genetic engineering has produced strains able to form ice crystals at even higher temperatures; thus science has found ways both to increase this microbe's powers and to disarm it.

*The genus Xanthomonas*

This plant pathogen is related to the fluorescent pseudomonads and produces characteristic yellow pigments, which gave the genus its name (Greek *xanthos*, meaning "yellow"). Like many other animal and plant pathogens, this organism secretes polysaccharide into the medium. The *Xanthomonas campestris* polysaccharide, xanthan, has been put to many uses in the food industry as well as in enhanced oil recovery (see Chapter 8).

*The genus Acidithiobacillus*

The species within this genus are small aerobic, Gram-negative rods, obligately acidophilic (optimum $pH < 4.0$), that oxidize reduced sulfur compounds or Fe[II] for energy generation. *Acidithiobacillus thiooxidans* (formerly *Thiobacillus thiooxidans*) utilizes reduced sulfur compounds for growth but cannot oxidize $Fe^{2+}$ whereas *Acidithiobacillus ferrooxidans* (formerly *Thiobacillus ferrooxidans*) grows on $Fe^{2+}$ as the sole energy substrate. In the absence of oxygen, this organism is able to oxidize reduced forms of sulfur using ferric iron as an alternative electron acceptor. *Acidithiobacillus* species find wide use in the heap leaching of metal ores.

**Alpha ($\alpha$) Division of the Proteobacteria.** The $\alpha$ division contains organisms with some unusual properties. One of its subgroups consists of three members – *Agrobacterium*, *Rhizobium*, and *Rickettsia* – that all interact closely with eukaryotic hosts, the first two with plants and the last one with animal cells. Interestingly, rRNA sequence data show that the ancestor of the mitochondria in animal cells was an organism belonging to the $\alpha$ division of the purple bacteria branch.

*The genus Rhizobium*

These bacteria are flagellated Gram-negative rods. *Rhizobium* strains are aerobic chemoheterotrophs that live in the soil and invade the root hairs of leguminous plants, where they form root nodules within which they fix nitrogen largely for the plant's benefit. The recognition between a *Rhizobium* species and its plant host is very specific and is discussed in Chapter 6. The practical importance of this genus is evident from the fact that although a total of about 100 million metric tons of synthetic nitrogen fertilizers are produced per year, nitrogen-fixing microorganisms yearly convert about 200 million tons of nitrogen to ammonia, and the major portion of this biological nitrogen fixation is carried out by the symbiotic nitrogen fixers, such as *Rhizobium*.

*The genus Agrobacterium*

These flagellated Gram-negative rods are also aerobic chemoheterotrophs abundant in soil. They carry a large plasmid, the Ti plasmid, which encodes various functions for the transfer of a small portion of the plasmid DNA, called T-DNA, into plant cells. The T-DNA becomes integrated into the plant chromosomal DNA and stimulates the synthesis of a plant growth hormone, thereby causing the growth of galls or tumors in the host plant.

The ability of these strains to transfer genes into plant cells is, to date, the only known example of natural gene transfer between a prokaryote and a eukaryote. It is a phenomenon of immense potential importance in biotechnology because it opens the door to the stable transfer of foreign genes into crop plants. One can imagine the Ti plasmid being used to introduce genes for engineered storage proteins, enriched in certain essential amino acids into cereal grains; to transfer genes for fixing nitrogen; or to introduce resistance genes for specific diseases or herbicides into plants. The Ti plasmid system is described in detail in Chapter 6.

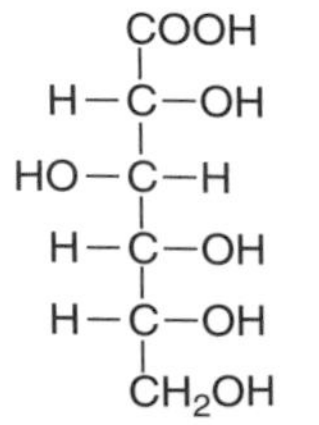

COOH
H–C–OH
HO–C–H
H–C–OH
H–C–OH
$CH_2OH$

FIGURE 1.13
Gluconic acid.

*The genus Zymomonas*

These Gram-negative rods with polar flagella are found in sugar-rich fermenting plant extracts such as palm wine (which is made from palm sap), sugar cane extract, and apple ciders. They can grow either by fermentation or by respiration. However, sugars are fermented not by the Embden–Meyerhof pathway used by the enteric species, but by the Entner–Doudoroff pathway, with ethanol as virtually the only end product rather than a mixture of lactate, ethanol, formate, acetate, and other end products. In certain respects, discussed in Chapter 13, *Zymomonas* offers advantages over yeasts in large-scale ethanol production.

*The genus Gluconobacter*

The cells of this genus are ellipsoidal to rod-shaped, and many strains are motile, with polar flagella. They are obligately aerobic chemoheterotrophs that characteristically obtain energy by oxidizing ethanol to acetic acid – but the acetic acid is not oxidized further. This is remarkable, for in almost every oxidative degradation pathway in other organisms, the substrate is always oxidized completely to $CO_2$. Thanks to this property, *Gluconobacter* is very useful in the manufacture of vinegar. It can also oxidize glucose to gluconic acid (Figure 1.13), a product of considerable commercial importance.

### Firmicutes (Phylum B13)

This phylum contains the Gram-positive bacteria with a low DNA mol% G + C content. Many branches within this phylum contain endospore-forming, obligate anaerobes traditionally classified in the Gram-positive genus *Clostridium*. Endospores are thick-walled spores formed within the bacterial cell (Greek *endon*, meaning "within"). The production of the endospore is an extremely complex process and appears to have been "invented" only once during biological evolution because it is found only in the Gram-positive branch. It is thus reasonable to hypothesize that the ancestor of this branch was an obligately anaerobic chemoheterotroph capable of endospore production and that some members later became adapted for an aerobic mode of life and some lost the capacity for sporulation. Another observation of interest is that when defined on the basis of rRNA sequence, this branch turns out to contain two sub-branches with Gram-negative–type cell walls. As Gram-negative cell walls are also present in all other branches of bacteria, again it may be assumed that the ancestral bacteria had a cell wall of Gram-negative type and that the Gram-positive cell wall arose by the loss of the outer membrane structure.

Phylum B13 has been subdivided into three classes: the Clostridia, the Mollicutes, and the Bacilli. The genus *Clostridium* is extremely heterogeneous. The evolutionary distance between one *Clostridium* species and another may be as great as the distance between animals and plants. The other line of low-GC organisms described below contains mostly facultatively aerobic organisms such as *Bacillus*, lactic acid bacteria, and *Staphylococcus*.

***Clostridium.*** As mentioned, *Clostridium* species are rod-shaped, usually flagellated, Gram-positive bacteria that are strictly anaerobic and form endospores under unfavorable conditions. Phylogenetic data suggest that the ancestral organisms of this genus arose long ago and that some of them preserve the fermentation pathways that were prevalent when the earth's atmosphere was largely devoid of oxygen and that have since disappeared in other branches of life. These pathways are obviously of interest to comparative biochemists. Some are useful in biotechnological applications because they terminate in useful products such as ethanol, acetylmethylcarbinol, butanol, and acetone.

Circumstances brought about by the outbreak of World War I led to a practical interest in clostridial fermentations. At that time, acetone was an important ingredient for the manufacture of smokeless powder (cordite), and before 1914, acetone was prepared starting from wood. Dry distillation (pyrolysis) of wood yielded a liquid distillate containing 10% acetic acid as well as other volatile products. Acetic acid was separated by distillation into a calcium hydroxide solution to form calcium acetate. Dried calcium acetate was then decomposed by heating to produce acetone and calcium carbonate. The wartime demand for acetone far outstripped the supply available from this process. Chaim Weizmann, a chemist with the firm Strange and Graham Ltd. in Manchester, England, happened to be working on the microbial production of acetone and butanol by bacterial fermentation of starch to obtain butanol for the manufacture of rubber. Among the organisms he screened, Weizmann discovered a bacterium, later named *Clostridium acetobutylicum*, that produced 12 tons of acetone from 100 tons of molasses. *C. acetobutylicum* fermentation became a major source of acetone by 1916. Later, the production of organic solvents by the petroleum industry slowly eroded the market for the fermentation product, and in 1982 the last operating clostridial fermentation plant, in South Africa, was closed down. However, the advent of genetic engineering has raised the possibility that clostridial fermentation will be reinstated as an important source of acetone.

**The *Lactobacillus-Staphylococcus-Bacillus* Cluster.** Unlike clostridia, most of the organisms in this cluster can be classified as facultative anaerobes, growing in the absence as well as presence of oxygen. However, their relationship with oxygen varies from that of the lactic acid bacteria, which tolerate its presence but carry out the same, fermentative, metabolism of sugar regardless of the presence or absence of air, to that of *Staphylococcus* and *Bacillus*, which switch from fermentative to respiratory metabolism in response to oxygen level. Among them, only *Bacillus* still retains the presumed ancestral capability of forming endospores. Four major groups are found, all of interest to biotechnologists.

*The genera Lactobacillus, Pediococcus, and Leuconostoc*

These genera form part of the group commonly called "lactic acid bacteria." They obtain energy by fermentation of simple sugars such as glucose, producing lactic acid in some *Lactobacillus* species and *Pediococcus* and lactic acid, ethanol, and carbon dioxide in *Leuconostoc* and the so-called heterofermentative species of *Lactobacillus*. All lack flagella. *Lactobacillus* cells are rod-shaped, whereas *Leuconostoc* and *Pediococcus* cells are round. Although cell shape is one of the basic categorizing criteria in traditional taxonomy schemes, 16S RNA studies reveal that the organisms in these genera are very closely related. They grow best at low oxygen tension in habitats rich in soluble sugars, peptides, purines, pyrimidines, and vitamins. These bacteria tolerate acidic conditions well and are not inhibited by the drop in pH that accompanies the conversion of glucose to lactic acid. The growth of many other bacteria slows down when pH is low, thus they present minimal competition to the lactobacilli under acidic conditions.

Various strains of these genera are used in starter cultures, together with appropriate strains of *Streptococcus* (see below), to produce cheeses and fermented milk products such as butter, buttermilk, and yogurt.

*The genus Streptococcus*

The cells of streptococci are spherical, and they generate ATP by converting glucose into two molecules of lactic acid. Unlike *Leuconostoc* and *Pediococcus*, this genus is distantly related to the genus *Lactobacillus*. Some streptococci are associated with higher animals, and some are pathogens. Others occur in association with plants. *Streptococcus cremoris* is the main organism used for the manufacture of hard-pressed cheeses such as Gouda and cheddar as well as soft-ripened cheeses such as Camembert. Streptococci are also important in the production of other fermented milk products. Together with *Lactobacillus* and its relatives, the streptococci account for a world output of dairy products in excess of 20 million metric tons per year and a value of about $50 billion. The strains of streptococci and lactobacilli are supplied to dairy and meat industry by commercial enterprises that specialize in the production of starter cultures.

*The genus Bacillus*

These are rod-shaped, motile organisms that form endospores when conditions are unfavorable for growth. It was the latter property that first called attention to this genus: Robert Koch showed, in classic studies culminating in 1876, that *Bacillus anthracis* was the causative organism of anthrax, the killer of cattle and sheep, and that the long persistence of anthrax infections in certain pastures was due to the resistance of the spores of *B. anthracis* to drying and to prolonged residence in soil. The majority of *Bacillus* strains are harmless saprophytes (organisms that feed on decaying organic matter).

*Bacillus* strains are all chemoheterotrophic, and they can grow in the presence of air, in contrast to the other group of endospore-forming bacteria, *Clostridium*. Many *Bacillus* strains can switch between fermentative and respiratory modes of metabolism, whereas others employ strictly respiratory metabolism. Many are inhabitants of soil, have rather simple nutritional requirements, and grow rapidly in synthetic media. Some strains are thermophilic and grow well at 65°C to 75°C. A number of *Bacillus* species produce extracellular hydrolytic enzymes that break down proteins, nucleic acids, polysaccharides, and lipids. Some of these enzymes are produced commercially in large amounts: the proteolytic enzymes are used in laundry detergents, and the polysaccharide-hydrolyzing enzymes are used in the degradation of starch (see Chapter 13). Some species are insect pathogens, and one of these, *Bacillus thuringiensis*, is the only bacterium exploited on a large scale as a biological insecticide (see Chapter 7). Antibiotics synthesized by some *Bacillus* strains are produced on a commercial scale, for example, bacitracin from *Bacillus subtilis* and polymyxin from *Bacillus polymyxa*.

*The genus Staphylococcus*

These bacteria are spherical, nonmotile cells that grow in irregular clusters. They are related to *Bacillus* but do not form endospores. They can switch between fermentative and respiratory modes of metabolism, and they use sugars as the chief source of energy. The main habitat of *Staphylococcus* is the skin of humans and animals; their remarkable tolerance of high salt concentration makes this possible (the drying of sweat is likely to concentrate salt on the skin). Because they are relatively resistant to drying, they are also found in secondary locations such as meat, poultry, animal feeds, and dust and air inside homes. *Staphylococcus aureus* is the species that causes skin infections as well as other, more serious diseases, including endocarditis and osteomyelitis.

One of the virulence factors produced by *S. aureus* is called protein A. As is well known, much of an animal's defense against bacterial infection depends on its production of antibodies, proteins with an antigen-binding domain that recognizes and binds to specific structures found on the surface of invading bacteria. Many of the beneficial protective consequences of the binding of antibody to antigen, however, are evoked through conformational changes at the other, nonspecific, end of the antibody molecule, called the Fc region. Protein A prevents one class of antibody, the immunoglobulin G class, from causing these effects, by binding tightly to its Fc domain. Protein A is used extensively for protein purification and analytical procedures, because if an antibody is available to a molecule one wishes to isolate, either protein A or whole-cell preparations of *S. aureus* can be used to selectively bind to the complex formed by the antibody and the target molecule.

## Actinobacteria (Phylum B14)

The Actinobacteria are a phylum of Gram-positive bacteria whose DNA has a high-GC content. Most are essentially aerobic soil bacteria with respiratory metabolisms; most lack flagella and are rod-shaped, are often slender and long, and have a tendency to divide irregularly and form branched filaments. That these organisms are quite closely related to one another was apparent even to the practitioners of the traditional taxonomic methods.

Nevertheless, there are significant differences among the organisms belonging to the different genera that are contained within this phylum. One cluster includes the genera *Arthrobacter* and *Cellulomonas*, organisms with slender rod-shaped cells, as well as a genus with spherical cells, *Micrococcus*. A second cluster encompasses the *Corynebacterium–Mycobacterium–Nocardia* group, which are basically aerobic soil organisms with a very characteristic cell wall: its polysaccharide, arabinogalactan, is substituted with fatty acids of exceptionally long chain lengths called mycolic acids. Together with *Pseudomonas* species, these organisms are suspected to be very important in the degradation of unusual organic compounds in the soil. Unfortunately, our knowledge of this group is quite limited, except for those atypical species that cause human diseases. Another group includes *Streptomyces* and its relatives, organisms that grow as clusters of highly branched filaments. Important members from each of the groups are briefly described below.

**The Genus *Cellulomonas*.** Like some other members of their subgroup, they are irregular rods with a respiratory metabolism. The main biochemical distinguishing feature of *Cellulomonas* strains is their ability to decompose cellulose. The cellulose-degrading enzymes of *Cellulomonas* have been closely studied in recent years because of interest in the use of cellulose-rich plant matter as a source of feedstock for the production of alcohol and proteins (Chapter 13).

**The Genus *Corynebacterium*.** One species, *Corynebacterium glutamicum*, became famous when it was discovered to have the ability to convert a very large fraction of its feedstock into glutamic acid and excrete it into the medium. This process, which involves some remarkable features in its regulation of amino acid biosynthetic pathways combined with an accidental undersupply of a vitamin (biotin) and of oxygen, is described in detail in Chapter 9. This organism and its relatives appear to have a somewhat simpler regulatory mechanism for amino acid biosynthesis under ordinary conditions, and this has been exploited for the production of other amino acids.

**The Genus *Streptomyces*.** *Streptomyces* strains grow as branching filaments called hyphae, which form convoluted networks called mycelia. As the mycelium ages, filaments called sporophores, or aerial hyphae, form that project above the surface of the colony. The aerial hyphae divide by forming internal cross-walls, and the individual cells mature into spores (conidia). These spores are quite different from the endospores formed within the cells of clostridia and bacilli. Although streptomycetes look very much like fungi both macroscopically and microscopically, they are totally different organisms, *Streptomyces* being prokaryotic.

*Streptomyces*, like most members of the actinomycete line, are inhabitants of soil. Several traits have made them successful in this habitat. They degrade polymeric substrates such as polysaccharides (starch, pectin, and chitin) as well as proteins. They have simple growth requirements, and their alternating spore–mycelium–spore life cycle allows them to survive the rapid changes in moisture, temperature, and aeration and allows them to become dispersed to optimal sites by wind.

In 1943, Selman Waksman and his collaborators discovered a potent antibacterial substance, streptomycin, released into the growth medium by *Streptomyces griseus*. This was the second antibiotic of very high utility to be characterized, soon after the characterization of penicillin. Since then, many other antibiotics have been isolated from streptomycetes, including tetracycline, erythromycin, neomycin, and gentamicin. The subject is treated further in Chapter 10.

## CHARACTERISTICS OF THE FUNGI

The kingdom Fungi encompasses an extraordinary diversity of organisms: bread molds, yeasts, powdery mildews, cup and sponge fungi, smuts, rusts, puffballs, and mushrooms. Some are invisible to the naked eye. Others grow to over two feet in diameter. Whatever their differences, however, all fungi have certain important properties in common (see glossary in Box 1.5):

- They are eukaryotic.
- They produce spores by means of sexual and asexual reproduction.
- They grow as hyphae or as yeasts, the hyphae exhibiting apical growth.
- They are heterotrophic and do not perform photosynthesis. Most fungi are saprophytes or symbionts, but some are parasites of man, other animals, or plants.
- They absorb nutrients through their cell membranes. Phagotrophy (ingestion of solid food particles) is a very rare property among the fungi. To utilize particulate or high molecular weight substrates, the fungi secrete various degradative, extracellular enzymes.
- They generally have rigid, polysaccharide-rich cell walls.

## CLASSIFICATION OF THE FUNGI

The classification of the kingdom Fungi recognizes five divisions: Chytridiomycota, Glomeromycota, Zygomycota, Ascomycota, and Basidiomycota. More than 70,000 species of fungi have been described, but it has been suggested that there may be as many as 1,500,000. Phylogenetic analyses based upon 18S rDNA and protein sequences indicate that the fungi are more closely related to animals than to plants or to the algae. The identity of the common ancestor of animals and fungi is still a matter of speculation.

Fungi are assigned to one of the five divisions primarily on the basis of molecular analyses. Additional important considerations are the differences in the morphology of their reproductive structures, the nature of their reproductive stages (Figure 1.14), and the composition of their cell walls, which typically contain 80% to 90% polysaccharide polymers, with most of the

**Fungi: A Glossary of Pertinent Terms**

**ascocarp** (a type of **sporocarp**, see below) – ascus-bearing structure (see **ascus** below) or "fruiting body."

**ascus** – in ascomycetes, the ascospores are produced within the cell wall and membrane of a diploid cell that has undergone meiosis and sporulation; the resulting saclike structure that surrounds the ascospores is called the ascus.

**asexual reproduction** – production of progeny identical to the parent by mitotic cell division.

**basidium** – a fungal cell that bears spores terminally and singly in extensions of its wall after karyogamy (see Figure 1.14) and meiosis.

**conidium** – a type of asexual spore that represents a separable portion of a hypha (see **thallus**).

**diploid** – having two sets of chromosomes ($2\times$), as opposed to one set (**haploid**) or more (polyploid).

**gametangium** (plural, **gametangia**) – a cell producing **gametes** (a gamete is a sex cell capable of fusing with another gamete, generally of opposite mating type, to form a zygote).

**mycotoxins** – in general, low molecular weight fungal metabolites capable of eliciting a toxic response in humans and animals.

**phagotroph** – an organism that ingests solid food particles.

**saprophyte** – an organism that lives on decaying organic matter.

**septa** – transverse walls dividing hyphae into compartments.

**sporocarps** – certain fungi reproduce sexually by a process of conjugation resulting in the formation of **zygospores**. Structures in which the zygospores occur in clusters surrounded by sterile hyphae are called sporocarps.

**thallus** – the part of a fungus that grows and absorbs nutrients and eventually produces the reproductive part, the "fruiting body." The thallus is composed of microscopic tubular vegetative filaments that branch and rebranch. These vegetative filaments are called **hyphae**, and a thallus made up of hyphae is termed the **mycelium**.

**yeast** – a fungus that is mainly unicellular.

**zygospore** – thick-walled resting spore resulting from conjugation of two cells of opposite sex (mating type).

**BOX 1.5**

remainder consisting of protein and lipid. The features typical of the five subdivisions of the Fungi are described briefly below.

## CHYTRIDIOMYCOTA

Chytrids are a morphologically diverse group with approximately 1000 known species. Their cell walls contain chitin (a $\beta$-(1-4)-linked polymer of *N*-acetylglucosamine), the "signature" polysaccharide of fungi. Their reproductive cells (gametes) have a flagellum that enables them to swim. No other fungi have flagella. Chytrids are predominantly aquatic and are found in both

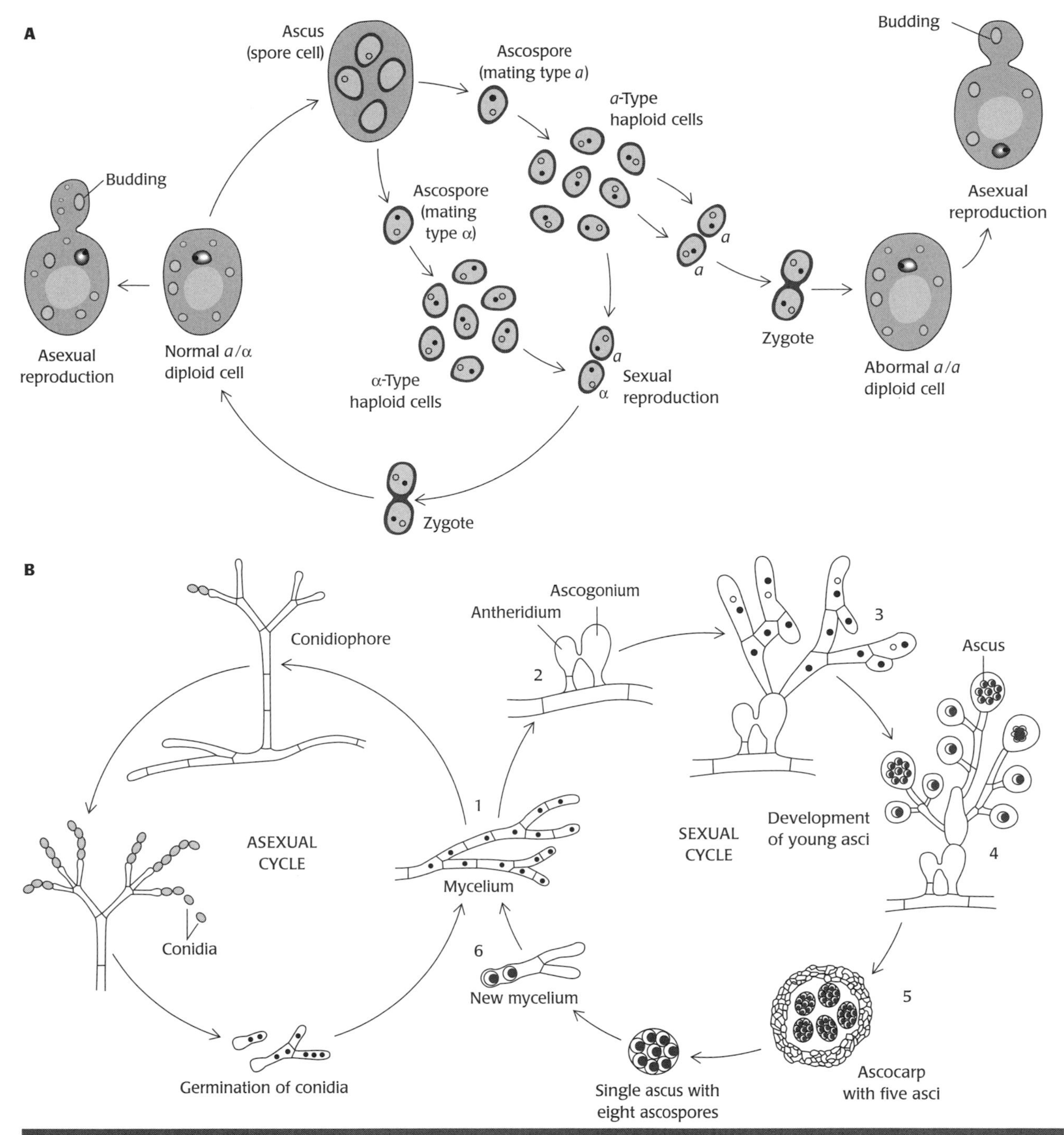


**FIGURE 1.14**

(**A**) The reproduction of yeast is normally asexual, proceeding by the formation of buds on the cell surface, but sexual reproduction can be induced under special conditions. In the sexual cycle, a normal diploid cell by meiosis and sporulation gives rise to asci, or spore cells, that contain four haploid ascospores. The ascospores are of two mating types: *a* and $\alpha$. Each type can develop by budding into other haploid cells. The mating of an *a* haploid cell and an $\alpha$ haploid cell yields a normal $a/\alpha$ diploid cell. Haploid cells of the same sex can also unite occasionally to form abnormal diploid cells ($a/a$ or $\alpha/\alpha$) that can reproduce only asexually, by budding in the usual way. The majority of industrial yeasts reproduce by budding. (**B**) Reproduction of a multicellular fungus, such as one of the higher ascomycetes, can be asexual or sexual. The details vary with genus and species. The branched vegetative structure common to both reproductive cycles is the mycelium, composed of hyphae (**1**). In the asexual cycle, the mycelium gives rise to conidiophores that bear the spores called conidia, which are dispersed by the wind. In the sexual cycle, the mycelium develops gametangial structures (**2**), each consisting of an antheridium (containing "+" nuclei) and an ascogonium (containing "−" nuclei). The nuclei pair in the ascogonium but do not fuse. Ascogenous binucleate hyphae develop from the fertilized ascogonium (**3**), and the pairs of nuclei undergo mitosis, which replicates the newly paired chromosomes. Finally, some pairs of nuclei fuse, a process called karyogamy (**4**), at the tips of the ascogenous hyphae. This is the only diploid stage in the life cycle. Soon afterward, the diploid nuclei (*large dots*) undergo meiosis, or reduction division. The result is eight haploid nuclei (*small dots*), each of which develops into an ascospore. At the same time, the developing asci are enclosed by mycelial hyphae in an ascocarp (**5**). In the example shown here, the ascocarp is a cleistothecium, a closed structure. Ascospores germinate to yield binucleate or multinucleate mycelium (**6**). [After Phaff, H. (1981). Industrial microorganisms. *Scientific American*, 245, 76–89.]

freshwater and marine environments. Many are saprophytic. For example, *Rhizophlyctis rosea* is a commonly encountered decomposer of cellulose in soils. Others are parasites on plants, insects, and certain amphibians.

## GLOMEROMYCOTA

The arbuscular mycorrhizal (AM) fungi are placed in this division. AM fungi are obligately symbiotic, asexual organisms. The term *mycorrhiza* (plural, *mycorrhizas*) refers to a close physical association of the fungal mycelium with the roots of a plant. Plants with mycorrhizas are particularly successful in infertile soils. A mycelial network may extend to a length 20,000 km (~12,400 miles) in one cubic meter of soil. The extensive mycelium absorbs inorganic nutrients, most importantly phosphate, from a large volume of soil or other substratum and provides them to the plant, whereas the latter provides carbohydrates to the fungus. AM fungi are of great ecological and economic importance. Over 80% of vascular land plant families participate in mycorrhizas. In the case of some plants, the association with their AM fungal partner is essential to normal development. AM fungi enhance plant biodiversity and help control pests such as nematodes and fungal pathogens.

## ZYGOMYCOTA

Members of Zygomycota produce nonmotile asexual spores (zygospores) formed in a sporocarp. The thallus is usually mycelial and typically aseptate (lacking cross-walls). The cell wall is composed of chitosan (a poorly acetylated or nonacetylated polymer of glucosamine) and chitin. Representative organisms are the soil saprophytes *Mucor* and *Rhizopus*. *Rhizopus nigricans* has long been used in the production of citric acid. *Entomophthora* is an important common parasite of insects such as house flies and aphids.

## ASCOMYCOTA

Ascomycota is the largest subdivision of fungi, containing some 15,000 species. The vegetative structure consists of either single cells (as in yeasts) or septate (segmented) filaments, most segments containing several nuclei. The cell walls are composed of chitin and glucans (and mannan, in many yeasts). Sexual reproduction leads to the formation of spores in an ascus. Two organisms from this subdivision, *Neurospora* (bread mold) and *Saccharomyces* (baker's and brewer's yeast), are especially familiar to geneticists (others include *Schizosaccharomyces*, *Ceratocystis ulmi* (the cause of Dutch elm disease), and *Erysiphe graminis*, a powdery mildew fungus that parasitizes cereals.

More than 40 species of *Saccharomyces* are recognized. *Saccharomyces cerevisiae* strains grow on the surface of grapes and other sugar-rich plants. These unicellular organisms multiply by budding. *S. cerevisiae* stocks are used extensively in the fermentation of certain beers and wines, in the production of baker's yeast, and in many biotechnological applications. Pombe, an African beer made from millet, and arrack, an Asian beer made from molasses, rice, or cocoa palm sap, are both products of fermentation by *Schizosaccharomyces pombe. Schizosaccharomyces* yeasts divide by binary fission and are termed *fission yeasts.*

## BASIDIOMYCOTA

Basidiomycetes form sexual spores on a special cell known as the basidium. As in the Ascomycota, the vegetative structure is either unicellular (yeasts) or a septate mycelium. The cell walls are composed of glucans and chitin. Representatives of this subdivision include *Serpula lacrymans,* a dry-rot fungus causing wood decay; rust and smut fungi such as *Puccinia graminis,* the cause of black stem rust in grasses and cereals; and *Ustilago maydis,* which afflicts corn plants. The *Agaricus* species, the mushrooms most commonly cultivated for human consumption in the Western world, are included among the Basidiomycota. Mushrooms are grown commercially on organic composts.

## DEUTEROMYCETES

This artificial group was created to accommodate those fungi that were known only in their asexual stage. Because their sexual state is absent, unknown, or lost and only vegetative reproduction carried out by asexual reproductive structures known as conidia is present, these fungi are also known as "fungi imperfecti." Their vegetative structure is either unicellular (yeasts) or a septate mycelium similar to those of ascomycetes and basidiomycetes. The polysaccharides of the cell wall are glucans and chitin. With molecular sequencing data, the closest sexual relatives of many of these asexual forms have been found. Members of genera of considerable economic importance, such as *Aspergillus* and *Penicillium,* are included in the deuteromycetes.

*Aspergillus niger* is used in the production of citric and gluconic acids. *Aspergillus oryzae* is used in the food industry in fermentations of rice and soya products and in the industrial production of proteolytic and amylolytic enzymes. However, some strains of *Aspergillus* are pathogens of plants – for example, crown rot of groundnuts and boll rot of cotton. Infestation of dried fruits, groundnut meal, or peanuts by *Aspergillus flavus* may result in the production of aflatoxin $B_1$ (Figure 1.15), a mycotoxin known to induce liver cancer in humans and poultry. *Penicillium* species grow on all kinds of decaying materials and are cosmopolitan in their distribution. Their spores are almost universally present in the air and frequently contaminate cultures of other microorganisms. In 1928, Alexander Fleming found that a Petri dish in which he was culturing staphylococci had become contaminated by a growth of *Penicillium.* He noticed that the growth of the staphylococci was inhibited in the region of the plate close to the fungal colony. Studies stimulated by this phenomenon led to isolation and purification of penicillin and

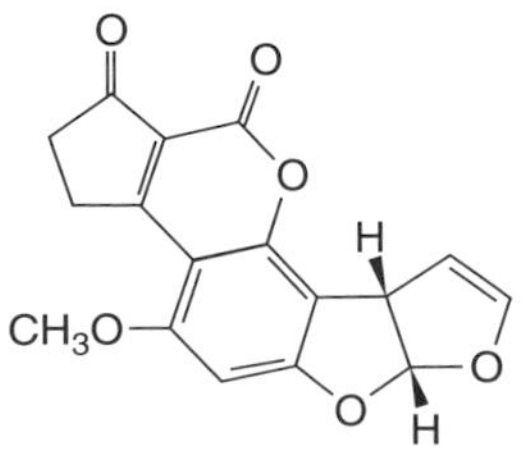

**FIGURE 1.15**
Aflatoxin $B_1$.

laid the foundation of antibiotic therapy (see Chapter 10; Figure 10.1.). The *Penicillium notatum* strain purified by Fleming was the first used for penicillin production, although as a result of intensive screening, it has since been replaced by other strains. *Penicillium griseofulvum* is a source of griseofulvin, which is given orally to treat fungal infections of the skin or nails. (In fungi sensitive to griseofulvin, the antibiotic binds to proteins involved in the assembly of tubulin into microtubules. This prevents the separation of the chromosomes in mitosis, and hyphal growth ceases.)

Several species of *Penicillium* are important in the food industry; for example, *Penicillium camemberti* and *Penicillium roqueforti* are used in the manufacture of the cheeses that bear their names. Not all *Penicillium* species are sources of benefit, however: *Penicillium italicum* and *Penicillium digitatum* cause rotting of citrus fruit, and *Penicillium expansum* causes a brown-rot of apples. *Penicillium* species are also major producers of mycotoxins.

## YEASTS, THE MOST EXPLOITED OF FUNGI

The term *yeast* is not specific to a formal taxonomic group but rather encompasses organisms with a growth form shown by a range of unrelated fungi. Many hundreds of thousands of tons of yeast are grown yearly. Many of these unicellular fungi are put to practical use in wine making, brewing, and baking and as sources of enzymes. Yeasts recovered as by-products of alcohol fermentations are sold for animal feed. *Torulopsis* and *Candida* strains are grown specifically for feed on molasses or on the spent sulfite liquor that is a by-product of paper pulp manufacture. Yeasts that utilize hydrocarbons and methanol are grown for the production of protein. Baker's yeast, *Saccharomyces cerevisiae*, is produced in large amounts.

Yeasts are classified on the basis of (1) the sequences of their 18S rDNA and other molecular markers, (2) the microscopic appearance of the cells, (3) the mode of sexual reproduction, (4) certain physiological features (especially metabolic capabilities and nutritional requirements), and (5) biochemical features (cell wall chemistry, and type of ubiquinone present in the mitochondrial respiratory electron transport chain).

The physiological features that distinguish different yeasts include the range of carbohydrates (mono-, di-, tri-, and polysaccharides) that a given organism can use as a source of carbon and energy under semianaerobic and aerobic conditions, the relative ability to grow in the presence of 50% to 60% (weight-to-volume [w/v]) D-glucose or 10% (w/v) sodium chloride plus 5% (w/v) glucose (a measure of osmotolerance), and the relative ability to hydrolyze and utilize lipids. These properties help investigators to determine which yeast strains merit investigation for a particular application. Thus, as with the prokaryotes, detailed taxonomic studies of yeasts and other fungi are of considerable importance.

Yeasts grow well at lower pH values than those optimal for most bacteria and are insensitive to antibiotics that inhibit bacterial growth. Consequently, large-scale cultures of yeasts can be kept free from contamination by fast-growing bacteria. Because of their larger size, yeasts are more easily and cheaply harvested than bacteria. Industrial yeasts in current use do not present public health problems. With these advantages and with the advent of genetic engineering, the range of applications of yeasts is expanding rapidly.

## CULTURE COLLECTIONS AND THE PRESERVATION OF MICROORGANISMS

Users of microorganisms require reliable sources of pure, authenticated cultures. Worldwide, there are over 500 culture collections that make strains of bacteria and fungi available, generally for a modest fee (Box 1.6). These collections obtain most of their strains from microbiologists working in universities or research institutes; other strains come from industries that no longer have a use for them. Moreover, the law now requires that if a process that uses a microorganism is to be patented, a culture of the microorganism must be deposited with a recognized culture collection. In the United Kingdom alone, the national culture collections hold more than 27,000 strains of bacteria and fungi. The American Type Culture Collections include more than 35,000 strains of bacteria, fungi, yeasts, viruses, and plasmids.

No single preservation procedure is appropriate for all organisms. Instead, there are four basic methods that differ in cost and convenience. Microbial cells can be maintained for short periods on slants and stabs of appropriate nutrient-containing agar or stored for longer periods in freeze-dried or other frozen form. For organisms that produce spores, the latter can be preserved in dry form on solid supports.

- The simplest procedure is to transfer cultures periodically to fresh solid slants of agar in the appropriate medium and to incubate them at a suitable growth temperature. Once the slant cultures are well established, they are kept in a refrigerator at 5°C enclosed in a container to avoid desiccation. This is the least expensive procedure and keeps cells viable for many months, but there is a danger that mutants or contaminants may accumulate in such cultures.

- Lyophilization (freeze-drying) is a particularly convenient preservation method. Microbial cells are mixed with a medium containing skim milk powder (at 20% w/v) or sucrose (at 12% w/v) and frozen, after which the water is removed from them by sublimation under partial vacuum. Lyophilized samples remain viable for many years and can be shipped without refrigeration.

- Cells can be stored for prolonged periods of time at liquid nitrogen temperature. In this procedure, the cells are placed in ampules with media containing (by volume) either 10% glycerol or 5% dimethylsulfoxide and are slowly frozen; their temperature is decreased by 1°C to 2°C per minute until it reaches about −50°C. The ampules are then stored at −156°C to −196°C in a liquid nitrogen refrigerator. Additives such as skim milk, sucrose, glycerol, and dimethylsulfoxide minimize damage to the cells by preventing ice crystals from forming during the freezing process.

According to the Bacteriological Code of 1990, when a new taxon is proposed, the author must designate a specific strain to be the nomenclatural type in the publication describing the new genus or species. Thus the type strain is composed of viable cells that are descended from the nomenclatural type. The Judicial Commission of August 1999 stated that a viable culture of a type strain must be deposited in at least two permanently established culture collections.

The Bacteriological Code of 1990 requires that a publication that proposes a new taxon (a family, genus, or species) must designate a specific strain to be the nomenclatural type for that taxon. A viable culture of the type strain must be deposited in at least two permanently established culture collections. The American Type Culture Collection (ATCC) holds more than 3600 type cultures of validly described species.

All substantial culture collections belong to the World Federation of Culture Collections. The home pages of these collections are listed at http://wdcm.nig.ac.jp/hpcc.html.

**BOX 1.6**

■ Many spore-forming bacteria and fungi can be preserved by slowly air-drying the spores at ambient temperature on the surface of sterilized soil, silica gel, or glass beads.

## SUMMARY

The terms *prokaryotes* and *fungi* describe a huge number of organisms that differ in their sources of energy, cell carbon and nitrogen, metabolic pathways, end products of metabolism, and ability to attack various naturally occurring compounds. Cellular organisms fall into two classes that differ from each other in the fundamental internal organization of their cells. Prokaryotes have no membrane-bounded organelles, whereas eukaryotes contain membrane-bounded nuclei as well as organelles (mitochondria, chloroplasts) that also possess genetic information. Organisms in the kingdoms Bacteria and Archaea are prokaryotes, whereas Fungi are eukaryotes. With respect to many molecular features, the archaea are almost as different from the bacteria as the latter are from eukaryotes. The archaea include three distinct kinds of bacteria found in extreme environments – the methanogens, the extreme halophiles, and the thermoacidophiles – as well as widely distributed organisms that are not extremophiles but flourish under many different conditions. Living organisms can be subdivided into four classes on the basis of their principal modes of metabolism. Those that use organic compounds as their major source of cell carbon are called *heterotrophs*; those that use carbon dioxide as the major source are called *autotrophs*. Organisms that use chemical bond energy for the generation of ATP are called *chemotrophs*, whereas those that use light energy for this purpose are called *phototrophs*. Various bacteria perform one or more of these four modes of metabolism. In contrast, plants utilize light energy for the generation of ATP and use carbon dioxide as the major source of cell carbon, and are exclusively *photoautotrophs*. Fungi and animals use organic compounds as the major source of cell carbon and the chemical bond energy of such compounds for the generation of ATP, and are exclusively *chemoheterotrophs*. Correct identification and classification of prokaryotes and fungi is important because the unintentional rediscovery and renaming of previously described organisms and the redetermination of their properties represent an unnecessary duplication of effort. On the other hand, each time a new organism can be placed within a well-studied genus, strong and readily testable predictions can be made concerning many of its genetic, biochemical, and physiological characteristics. Taxonomy of microorganisms now relies largely on genomic DNA sequence comparisons. Sequences of slowly evolving macromolecules (rRNAs) allow classification of distantly related microorganisms. Molecular analyses of samples from diverse environments show that only a very small fraction of the prokaryotes and fungi present have been cultured. New general methods have been developed for isolation of previously uncultured microorganisms from such samples. Prokaryotes useful in biotechnology come from many different branches of the phylogenetic tree based on 16S RNA sequences. Plasmids, self-replicating extrachromosomal

DNA elements within bacterial cells, sometimes confer highly noticeable phenotypic traits on their hosts and may complicate the classification of the host organism. The use of molecular analyses has greatly advanced the classification of fungi. Five divisions make up the kingdom Fungi. Yeasts are the most exploited of the fungi and are grown in the hundreds of tons in wine making, brewing, and baking, and as sources of enzymes. Only a small fraction of known prokaryotes and fungi have been studied extensively and a still smaller fraction put to practical uses. Culture collections make available pure cultures of tens of thousands of different bacterial and fungal strains. Methods have been developed for the long-term preservation of bacteria and of fungal spores.

## SELECTED REFERENCES AND ONLINE RESOURCES


### General Background: Prokaryotes

The Authors. (2007) Crystal Ball – 2007. *Environmental Microbiology*, 9, 1–11.

Ingraham, J. L., and Ingraham, C. A. (2004). *Introduction to Microbiology: A Case History Approach*, 3rd Edition, Pacific Grove, CA: Brooks/Cole.

Gest, H. (2003). *Microbes: An Invisible Universe*, Revised Edition, Washington, DC: ASM Press.

Dyer, B. D. (2003). *A Field Guide to Bacteria*, Ithaca: Cornell University Press.

Madigan, M. T., Martinko, J. M., and Parker, J. (2003). *Brock Biology of Microorganisms*, 10th Edition, Upper Saddle River, NJ: Prentice Hall.

Lengeler, J. W., Drews, G., and Schlegel, H. G. (eds.) (1999). *Biology of the Prokaryotes*, Malden, MA: Thieme.

Microbiology Online Resources http://www.nature.com/nrmicro/info/links.html.

### Classification and Phylogeny

Garrity, G. M. (ed.) (2001). *Bergey's Manual of Systematic Bacteriology*, 2nd Edition, New York: Springer-Verlag.

Koonin, E. V., Makarova, K. S., and Aravind, L. (2001). Horizontal gene transfer in prokaryotes: quantification and classification. *Annual Review of Microbiology*, 55, 709–742.

Achenbach, L. A., and Coates, J. D. (2000). Disparity between bacterial phylogeny and physiology. *ASM News*, 66, 714–715.

### Nucleic Acid Probes in Environmental Microbiology

Amann, R., and Schleifer, K.-H. (2001). Nucleic acid probes and their application in environmental microbiology. In *Bergey's Manual of Systematic Bacteriology*, 2nd Edition, Volume 1, *The Archaea and the Deeply Branching and Phototrophic Bacteria*, G. M. Garrity (ed.), pp. 67–82, New York: Springer-Verlag.

Zhang, Z., Willson, R. C., and Fox, G. E. (2002). Identification of characteristic oligonucleotides in the bacterial 16S ribosomal RNA sequence dataset. *Bioinformatics*, 18, 244–250.

Loy, A., et al. (2002). Oligonucleotide microarray for 16S rRNA gene-based detection of all recognized lineages of sulfate-reducing prokaryotes in the environment. *Applied Environmental Microbiology*, 68, 5064–5081.

Sebat, J. L., Colwell, F. S., and Crawford, R. L. (2003). Metagenomic profiling: microarray analysis of an environmental genomic library. *Applied Environmental Microbiology*, 69, 4927–4934.

Burggraf S., Mayer, T., Barns, S. M., Rossnagel, P., and Stetter, K. O. (1995). Isolation of a hyperthermophilic archaeum predicted by in situ RNA analysis. *Nature*, 376, 57–58.

**Ribosomal RNA Databases: Sequences and Structural Information**

European Ribosomal RNA Database http://oberon.fvms.ugent.be:8080/rRNA/ or http://www.psb.ugent.be/rRNA/.

Ribosomal Database Project II http://rdp.cme.msu.edu/.

Comparative RNA website http://rna.icmb.utexas.edu/.

**'Environmental Genome' Shotgun Sequencing**

Venter, J. C., et al. (2004). Environmental genome shotgun sequencing of the Sargasso Sea. *Science*, 304, 66–74.

**Isolating Microorganisms in Pure Culture**

Huber, R., Burggraf, S., Mayer, T., Barns, S. M., Rossnagel, P., and Stetter, K. O. (1995). Isolation of a hyperthermophilic archaeum predicted by *in situ* RNA analysis. *Nature*, 376, 57–58.

Leadbetter, J. R. (2003). Cultivation of recalcitrant microbes: cells are alive, well and revealing their secrets in the 21st century laboratory. *Current Opinions in Microbiology*, 6, 274–281.

Rappe, M. S., Connon, S. A., Vergin, K. L., and Giovannoni, S. J. (2002). Cultivation of the ubiquitous SAR11 marine bacterioplankton clade. *Nature*, 418, 630–633.

Sait, M., Hugenholtz, P., and Janssen, P. H. (2002). Cultivation of globally distributed soil bacteria from phylogenetic lineages previously only detected in cultivation-independent surveys. *Environmental Microbiology*, 4, 654–666.

Zengler, K., Toledo, G., Rappe, M., Elkins, J., Mathur, E. J., Short, J. M., and Keller, M. (2002). Cultivating the uncultivated. *Proceedings of the National Academy of Sciences U.S.A.*, 99, 15681–15686.

Keller, M., and Zengler, K. (2004). Tapping into microbial diversity *Nature Reviews. Microbiology*, 2, 141–150.

**General Background: Fungi**

Esser, K. (ed.) (2004). *The Mycota: A Comprehensive Treatise on Fungi as Experimental Systems for Basic and Applied Research*, 2nd Edition, Berlin; New York: Springer-Verlag.

Watling, R. (2003). *Fungi*, London: Natural History Museum.

Dighton, J. (2003). *Fungi in Ecosystem Processes*, New York: M. Dekker.

Alexopoulos, C. J., Mims, C. W., and Blackwell, M. (1996). *Introductory Mycology*, 4th Edition, New York: John Wiley.

Schüssler, A., Schwarzott, D., and Walker, C. (2001). A new fungal phylum, the *Glomeromycota*: phylogeny and evolution. *Mycological Research*, 105, 1413–1421.

Guarro, J., Gené, J., and Stchigel, A. M. (1999). Developments in fungal taxonomy. *Clinical Microbiology Reviews*, 12, 454–500.

Berbee, M. L., and Taylor, J. W. (1993). Dating the evolutionary radiations of the true fungi. *Canadian Journal of Botany*, 71, 1114–1127.

Pennisi, E. (2004). The secret life of fungi. *Science*, 304, 1620–1622.

Wardle, D. A., Bardgett, R. D., Klironomos, J. N., Setäla, H. , van der Putten, W. H., and Wall, D. H. (2004). Ecological linkages between aboveground and belowground biota. *Science*, 304, 1629–1633.

Tree of Life Web Project http://tolweb.org/tree?group=Fungi&contgroup=Eukaryotes#TOC6.

**Culture Collections and the Preservation of Microorganisms**

Smith, D., and Onions, A. H. S. (1994). *The Preservation and Maintenance of Living Fungi*, Wallingford, U.K.: CAB International.

TWO

# Microbial Biotechnology: Scope, Techniques, Examples

One can be a good biologist without necessarily knowing much about microorganisms, but one cannot be a good microbiologist without a fair basic knowledge of biology!

– Stanier, R. Y., Doudoroff, M., and Adelberg, E. A. (1957). *The Microbial World*. p. vii, Englewood Cliffs, NJ: Prentice-Hall, Inc.

Microorganisms, whether cultured or represented only in environmental DNA samples, constitute the natural resource base of microbial biotechnology. Numerous prokaryotic and fungal genomes have been completely sequenced and the functions of many genes established. For a newly sequenced prokaryotic genome, functions for over 60% of the open reading frames can be provisionally assigned by sequence homology with genes of known function. Knowledge of the ecology, genetics, physiology, and metabolism of thousands of prokaryotes and fungi provides an indispensable complement to the sequence database.

This is an era of explosive growth of analysis and manipulation of microbial genomes as well as of invention of many new, creative ways in which both microorganisms and their genetic endowment are utilized. Microbial biotechnology is riding the crest of the wave of genomics.

The umbrella of microbial biotechnology covers many scientific activities, ranging from production of recombinant human hormones to that of microbial insecticides, from mineral leaching to bioremediation of toxic wastes. In this chapter, we sketch the complex terrain of microbial biotechnology. The purpose of this chapter is to convey the impact, the extraordinary breadth of applications, and the multidisciplinary nature of this technology. The common denominator to the subjects discussed is that in all instances, prokaryotes or fungi provide the indispensable component. Topics addressed in later chapters of this book are treated briefly. Those not described elsewhere are discussed here in some detail.

## HUMAN THERAPEUTICS

### PRODUCTION OF HETEROLOGOUS PROTEINS

One of the most dramatic and immediate impacts of genetic engineering was the production in bacteria of large amounts of proteins encoded by human genes. In 1982, insulin, expressed from human insulin genes on plasmids inserted into *Escherichia coli*, was the first genetically engineered therapeutic agent to be approved for clinical use in humans. Bacterially produced insulin, used widely in the treatment of diabetes, is indistinguishable in its structure and clinical effects from natural insulin. Human growth hormone (hGH), a protein made naturally by the pituitary gland, was the second such product. Inadequate secretion of hGH in children results in dwarfism. Before the advent of recombinant DNA technology, hGH was prepared from pituitaries removed from human cadavers. The supply of such preparations was limited and the cost prohibitive. Furthermore, there were dangers in their administration that led to withdrawal from the market. Some patients treated with injections of pituitary hGH developed a disease caused by a contaminating slow virus, Jakob–Creutzfeldt syndrome, which leads to dementia and death. hGH can be produced in genetically engineered *E. coli* in large amounts, at relatively little cost, and free from such contaminants.

Human tissue plasminogen activator (tPA), a proteolytic enzyme (a "serine" protease) with an affinity for fibrin clots, is another therapeutic agent made available in large amounts as a consequence of recombinant DNA technology. At the surface of fibrin clots, tPA cleaves a single peptide bond in plasminogen to form another serine protease, plasmin, which then degrades the clots. This clot-degrading property of tPA makes it a life-saving drug in the treatment of patients with acute myocardial infarction (damage to heart muscle due to arterial blockage).

Recombinant human insulin and hGH offered impressive proof of the clinical efficacy and safety of human proteins made by engineered microorganisms. As exemplified by the list in Table 2.1, the list of recombinant human gene products expressed in bacteria or fungi continues to grow rapidly. We devote Chapters 3 and 5 to discussion of the production of heterologous proteins and vaccines in these organisms.

### DNA VACCINES

In the early 1990s, attention focused on the potential wide-ranging opportunities offered by DNA vaccines. DNA vaccines consist of appropriately engineered plasmid DNA prepared on a large scale in *E. coli*. The obvious advantages of DNA plasmid vaccines are that they are not infectious, do not replicate, and encode only the protein(s) of interest. Unlike other types of vaccines, there is no protein component, and hence induction of an immune response against subsequent immunizations is minimized.

A vaccine plasmid includes the following major components: a strong promoter system for expression in eukaryotic cells of an antigenic protein

**TABLE 2.1 Examples of human proteins cloned in *E. coli:* their biological functions and current or envisaged therapeutic use**

| Protein | Function(s) | Therapeutic use(s) |
|---|---|---|
| $\alpha_1$-Antitrypsin | Protease inhibitor | Treatment of emphysema |
| Calcitonin | Influences $Ca^{2+}$ and phosphate metabolism | Treatment of osteomalacia |
| Colony stimulating factors | Stimulate hematopoiesis | Antitumor |
| Epidermal growth factor | Epithelial cell growth, tooth eruption | Wound healing |
| Erythropoietin | Stimulates hematopoiesis | Treatment of anemia |
| Factor VIII | Blood clotting factor | Prevention of bleeding in hemophiliacs |
| Factor IX | Blood clotting factor | Prevention of bleeding in hemophiliacs |
| Growth hormone releasing factor | Stimulates growth hormone secretion | Growth promotion |
| Interferons ($\alpha$, $\beta$,$\gamma$) | A family of 20 to 25 low molecular weight proteins that cause cells to become resistant to the growth of a wide variety of viruses | Antiviral, antitumor, anti-inflammatory |
| Interleukins 1, 2, and 3 | Stimulators of cells in the immune system | Antitumor; treatment of immune disorders |
| Lymphotoxin | A bone-resorbing factor produced by leukocytes | Antitumor |
| Somatomedin C (IGF-I) | Sulfate uptake by cartilage | Growth promotion |
| Serum albumin | Major protein constituent of plasma | Plasma supplement |
| Superoxide dismutase | Decomposes superoxide free radicals in the blood | Prevention of damage when $O_2$-rich blood enters $O_2$-deprived tissues; has applications in cardiac treatment and organ transplantation |
| Tumor necrosis factor | A product of mononuclear phagocytes cytotoxic to certain tumor cell lines | Antitumor |
| Urogastrone | Control of gastrointestinal secretion | Antiulcerative |
| Urokinase | Plasminogen activator | Anticoagulant (dissolution of blood clots) |

(e.g., a viral coat protein), the immediate early promoter of cytomegalovirus is frequently used; a cloning site for the insertion of the gene encoding the antigenic protein; and an appropriately located polyadenylation termination sequence. Most eukaryotic mRNAs contain a polyadenylate (polyA) tail at the 3′ end that appears to be important to the translation efficiency and the stability of the mRNA. The plasmid also includes a prokaryotic origin of replication for its production in *E. coli* and a selectable marker, such as the ampicillin resistance gene, to allow selection of bacterial cells that contain the plasmid.

DNA vaccines are generally introduced by intramuscular injection. It is still not known how cells internalize the DNA after the injection. The encoded antigen is then expressed *in situ* in the cells of the vaccine recipient and elicits an immune response.

Such vaccines have attractive features. The immunizing antigens may be derived from viruses, bacteria, parasites, or tumors. Antigens can be expressed singly or in multiple combinations. In one case, the DNA vaccine contained multiple variants of a highly mutable gene, for example, the gene encoding gp120, a glycoprotein located on the external surface of HIV.

Like all T cells, Th cells arise in the thymus. Th1 cells belong to the $CD4^+$ subset of lymphocytes that participate in cell-mediated immunity. They are essential for combating intracellular pathogens such as viruses and certain bacteria – for example, *Listeria monocytogenes* (causative organism of listeriosis) and *Mycobacterium tuberculosis* (the organism that causes tuberculosis).

BOX 2.1

In other vaccines, the entire genome of the infectious microorganism was introduced into a common plasmid backbone by "shotgun cloning."

DNA vaccines induce both *humoral* responses (the appearance of serum antibodies against the antigen) and *cellular* responses (activation of various T cells). These responses have been documented in animal models of disease in which protection is mediated by such responses.

Important issues remain to be resolved before DNA vaccines can take a regular place alongside other types of vaccines. In clinical trials, vaccines for malaria, hepatitis B, HIV, and influenza elicited only moderate response in human volunteers. An assessment of DNA vaccines encoding certain highly conserved influenza virus proteins concluded that there is a need for considerable enhancement of the immune response to DNA immunization before such vaccines become a promising approach for humans. Moreover, the plasmid DNA itself stimulates T helper 1 (Th1) cells and thereby might contribute to the development or worsening of Th1-mediated organ-specific autoimmunity disorders (see Box 2.1). Other potential concerns have also been identified.

## SECONDARY METABOLITES AS A SOURCE OF DRUGS

Microorganisms produce a huge number of small molecular weight compounds that are broadly described as *secondary metabolites*. A traditional approach to the discovery of new, naturally occurring bioactive molecules utilizes "screens." A screen is an assay procedure that allows testing of numerous compounds for a particular activity. Tens of thousands of secondary metabolites and other compounds have been examined for biological activity in various organisms and many have proved invaluable as *antibacterial* or *antifungal agents, anticancer drugs, immunosuppressants, herbicides, tools for research*, and the like (Table 2.2).

Genetically modified microorganisms have been engineered to produce such compounds in large amounts. Among these, antibiotics are the secondary metabolites considered among the most important to human therapeutics, and the most extensive use of screens is in the search for compounds with selective toxicity for bacteria, fungi, or protozoa. It is estimated that natural microbial antibiotics provide the starting point for over 75% of marketed antimicrobial agents. Chapter 10 is devoted to an extensive discussion of antibiotics. The three examples that follow illustrate the exceptional importance of natural products in other important therapeutic applications.

## AVERMECTINS

Many microorganisms indigenous to the soil, especially actinomycete bacteria and many fungi, produce biologically active secondary metabolites. Intensive screening of culture supernatants (usually called "fermentation broths"), rich in secondary metabolites, has led to the discovery of numerous clinically valuable antibiotics, with penicillin as the most famous example, but of many other types of valuable compounds as well. The structures of newly characterized compounds with herbicidal, insecticidal, and

**TABLE 2.2 Bacterial and fungal secondary metabolites**

| Compound | Source organism[a] | Comments |
|---|---|---|
| Actinomycin | *Streptomyces chrysomallus* | Actinomycin, bleomycin, and griseofulvin, inhibit DNA replication |
| Bleomycin | *Streptomyces verticillus* | |
| Griseofulvin | ***Penicillium griseofulvum*** | |
| Rifamycin | *Amycolatopsis mediterranei* | Inhibits transcription by inhibiting DNA-dependent RNA polymerase. Valuable in the treatment of tuberculosis |
| Chloramphenicol | *Streptomyces venezuelae* | Chloramphenicol, tetracycline, lincomycin, and erythromycin inhibit translation by 70S ribosomes |
| Tetracycline | *Streptomyces aureofaciens* | |
| Lincomycin | *Streptomyces lincolnensis* | |
| Erythromycin | *Streptomyces erythreus* | |
| Cycloheximide | *Streptomyces griseus* | Inhibits translation by 80S ribosomes |
| Puromycin | *Streptomyces alboniger* | Puromycin and fusidic acid inhibit translation by 70S and 80S ribosomes |
| Fusidic acid | ***Acremonium fusidioides*** | |
| Cycloserine | *Streptomyces* sp. | Cycloserine, bacitracin, penicillin, cephalosporin, vancomycin, and teicoplanin inhibit peptidoglycan synthesis |
| Bacitracin | *Bacillus licheniformis* | |
| Penicillin | ***Penicillium chrysogenum*** | |
| Cephalosporin | ***Cephalosporium acremonium*** | |
| Vancomycin | *Amycolatopsis orientalis* | |
| Teicoplanin | *Actinoplanes teichomyceticus* | |
| Polymyxin | *Paenibacillus polymyxa* | Polymyxin and amphotericin are polyether surfactants that perturb cell membranes |
| Amphotericin | *Streptomyces nodosus* | |
| Gramicidin | *Bacillus brevis* | Channel-forming ionophore |
| Monensin | *Streptomyces cinnamonensis* | Mobile carrier ionophore; coccidiotic agent |
| Avermectins | *Streptomyces avermitilis* | Avermectins have high activity against helminths and arthropods |
| Clavulanic acid | *Streptomyces clavuligerus* | A penicillinase inhibitor that protects penicillin from inactivation by resistant pathogens; used in conjunction with penicillin |
| Kasugamycin | *Streptomyces kasugaensis* | Kasugamycin and polyoxins are fungicides |
| Polyoxins | *Streptomyces cacaoi* | |
| Nikkomycin | *Streptomyces tendae* | Nikkomycin and spinocins are insecticides |
| Spinosins | *Saccharopolyspora spinosa* | |
| Bialaphos | *Streptomyces hygroscopicus* | Herbicide |
| Cyclosporin A | ***Tolypocladium inflatum*** | Cyclosporin A, FK-506, and rapamycin are immunosuppressants for organ transplant recipients |
| FK-506 (tacrolimus) | *Saccharopolyspora erythrea* | |
| Rapamycin | *Streptomyces hygroscopicus* | |
| Doxorubicin | *Streptomyces peucetius* | An anticancer drug used in treating late-stage tumors |
| Ergot alkaloids | ***Claviceps purpurea*** | Uterocontractants |
| Lovastatin (mevinolin) | ***Aspergillus terreus*** | Cholesterol-lowering agent in humans and animals |
| Acarbose | *Actinoplanes* sp. | Inhibits human intestinal glucosidase |
| Gibberellins | ***Gibberella fujikuroi*** | Plant growth regulators |
| Zearalenone | ***Gibberella zeae*** | Anabolic agent used in farm animals |

[a] Fungi are shown in boldface.

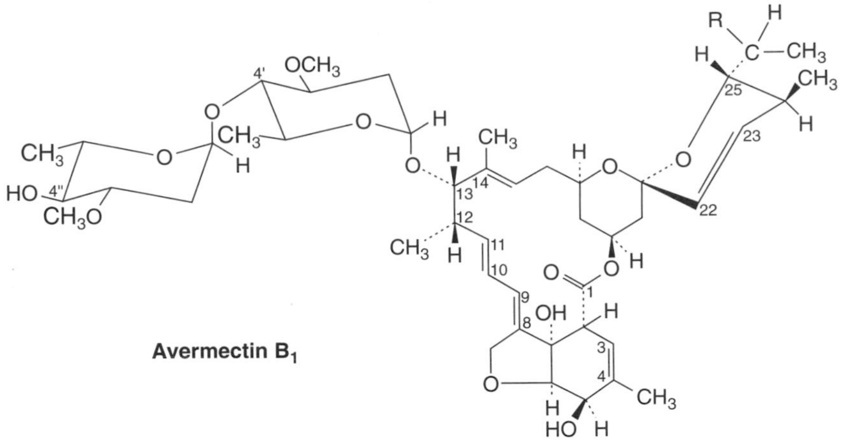

**Avermectin $B_1$**

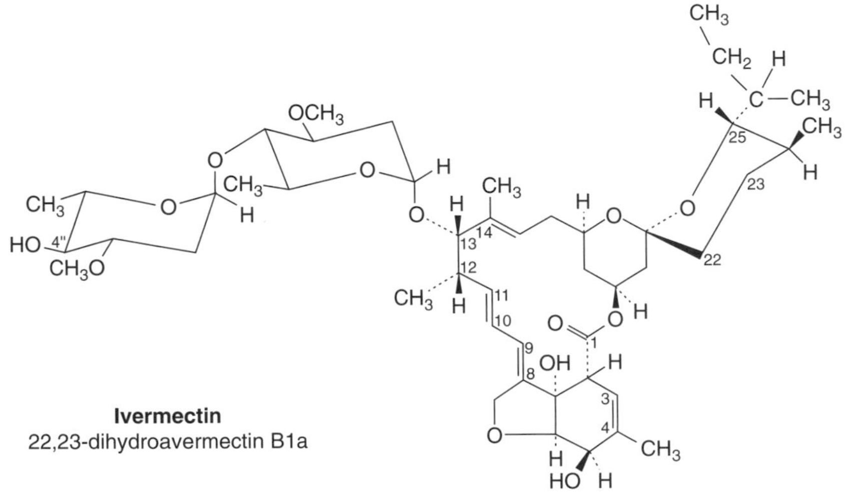

**Ivermectin**
22,23-dihydroavermectin B1a

**FIGURE 2.1**

Avermectin $B_1$. This compound is the major macrocyclic lactone produced by *Streptomyces avermitilis*. Ivermectin is a synthetic derivative of avermectin $B_1$.

nematocidal activities from soil microorganisms are described in the scientific literature at a rate of several hundred each year.

The avermectins were discovered in the early 1980s as a result of a deliberate search for antihelminthic compounds produced by soil microorganisms. Helminths are parasitic worms that infect the intestines of any animal unfortunate enough to ingest their eggs. There were two particularly notable features of the screening program. First, the microbial fermentation broths were tested by being administered in the diet to mice infested with the nematode *Nematospiroides dubius*. Nematodes are a subclass of helminths that includes roundworms or threadworms. Although such an *in vivo* assay was expensive, it simultaneously tested for efficacy of the preparation against the nematode and toxicity to the host. Second, to increase the chance of discovering new types of compounds, the selection of microorganisms for testing was biased toward those with unusual morphological traits and nutritional requirements. The morphological characteristics of *Streptomyces avermitilis*, the producer of avermectins, were unlike those of other known *Streptomyces* species. *S. avermitilis* produces a family of closely related macrocyclic lactones (Figure 2.1), compounds that are active against certain nematodes and arthropods at extremely low doses, but have relatively low toxicity to mammals. These avermectins and their derivatives, as the compounds came to be called, are highly effective in veterinary use and in treating infestations in humans.

Avermectins act on invertebrates by activating glutamate-gated chloride channels in their nerves and muscles, disrupting pharyngeal function and locomotion. The paralyzed parasite most likely starves to death. Their selective toxicity – they do not harm vertebrates – has led to the conclusion that avermectins affect a specific cellular target either absent or inaccessible in the resistant organisms. The avermectins do not migrate in soils from the site of application and are subject to both rapid photodegradation and microbial decomposition. Consequently, avermectins are not expected to persist for a long time in the feces of treated animals. The biological activity and selective toxicity of the avermectins could not have been anticipated even if the structures of these compounds had been known.

The structure of a naturally occurring small molecule with desirable biological activity is generally used as the starting point for the design and preparation of semisynthetic derivatives with improved activity, selectivity, and stability characteristics. This has proved to be the case for avermectins. Ivermectin (IVM; 22,23-dihydroavermectin $B_{1a}$, Figure 2.1), a semisynthetic derivative of avermectin $B_{1a}$, is an indispensable drug in mass treatment programs to eradicate two widespread serious diseases that affect millions of people and that are caused by nematodes: river blindness (onchocerciasis) and lymphatic filariasis (Box 2.2).

**River Blindness (Onchocerciasis) and Lymphatic Filariasis**

Onchocerciasis, first described in 1875, is caused by a filarial nematode (*Onchocerca volvulus*), a parasite transmitted by the bite of infected blackflies of the genus *Simulium*. Onchocerciasis is a leading cause of eye disease in Africa, the Eastern Mediterranean area, and Latin America. In 2002, it was estimated that 17.7 million people were infected; of these, about 250,000 went blind and another 250,000 suffered significant visual impairment. Ivermectin kills the infectious larvae of *O. volvulus* but not the adult worms. The disease is controlled by an annual dose of IVM of 150 $\mu$g/kg.

Lymphatic filariasis is caused by the nematodes *Wuchereria bancrofti*, *Brugia malayi*, and *Brugia timori*. The disease is endemic in most of the warm, humid regions of the world, including South America, Africa, Asia, and the Pacific Islands. The principal vectors are mosquitoes. Infections may lead to a wide variety of symptoms, including acute recurrent fever, lymphadenitis, and blood disorders. IVM controls lymphatic filariasis in a manner similar to that described for onchocerciasis.

Unexpectedly, endosymbiotic bacteria make the decisive contribution to the onset of river blindness. Bacteria of the genus *Wolbachia* are essential endosymbionts in all the pathogenic nematodes mentioned above. In humans infected with *O. volvulus*, adult worms survive for up to 14 years in subcutaneous nodules and release millions of microfilariae over this time. The microfilariae migrate through the skin and enter the eye. When some of these filariae die, the host response may result in eye inflammation that causes progressive loss of vision and ultimately leads to blindness. The host immune response plays a critical role in the inflammatory response associated with the pathogenesis of river blindness. This response is initiated by the release from the dead and degenerating worms of endotoxin-like molecules originating in the *Wolbachia* endosymbionts. Consequently, elimination of *Wolbachia* by antibiotic treatment may prevent onchocerciasis.

*Sources:* Benenson, A. S. (ed.) (1990). *Control of Communicable Diseases in Man*, 15th Edition, Washington, D.C.: American Public Health Association; Cooper, P. J., and Nutman, T. B. (2002). Onchocerciasis. *Current Treatment Options in Infectious Diseases*, 4, 327–335; Brown, R. K., Ricci, F. M., and Ottesen, E. A. (2000). Ivermectin: effectiveness in lymphatic filariasis. *Parasitology*, 121, S133–S146; Saint André, A., et al. (2002). The role of endosymbiotic *Wolbachia* bacteria in the pathogenesis of river blindness. *Science*, 295, 1892–1895.

BOX 2.2

## ZARAGOZIC ACIDS (SQUALESTATINS)

Over 93% of the cholesterol in the human body is located in cells, where it performs indispensable structural and metabolic roles. The remaining 7% circulates in the plasma, where it contributes to atherosclerosis (formation of

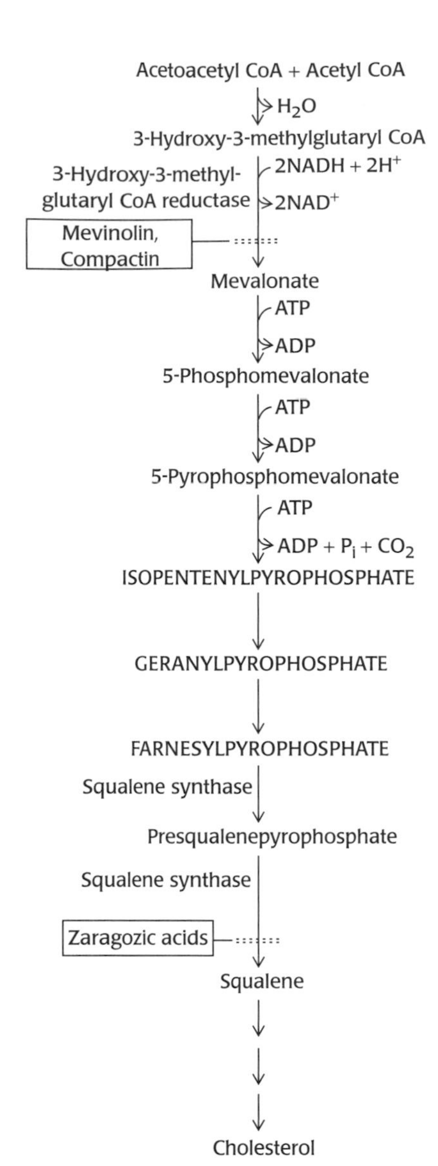


**FIGURE 2.2**

Biosynthetic pathway leading to cholesterol in humans. Isopentenyl, geranyl, and farnesyl pyrophosphate are precursors not only of sterols but also of several important isoprenoid derivatives. The fungal fermentation products mevinolin (from *Aspergillus terreus*) and compactin (from *Penicillium* spp.) are highly effective drugs used to reduce serum cholesterol in humans. These compounds are potent inhibitors of 3-hydroxy-3-methylglutaryl-CoA reductase and block formation of all products of the mammalian polyisoprenoid pathway. In contrast, the zaragozic acids inhibit squalene synthase, which catalyzes the first committed step in sterol synthesis, and do not affect the formation of other isoprenoids.

plaques on the walls of the arteries supplying the heart, the brain, and other vital organs). For delivery to tissues, plasma cholesterol is packaged in lipoprotein particles; two thirds is associated with low-density lipoprotein (LDL) and the balance with high-density lipoprotein.

The disorder *familial hypercholesterolemia* occurs in one in 500 of the population and results in elevated plasma levels of cholesterol-bearing LDL. Male heterozygotes with dominant familial hypercholesterolemia have an 85% chance of occurrence of heart attacks (myocardial infarction) before the age of 60. (Homozygotes of either sex die of heart disease at an early age). A much larger number of people, who do not have familial hypercholesterolemia, have plasma levels of LDL at the upper limit of the normal range and are also at high risk for atherosclerosis. The goal of therapy in these subjects is to reduce the level of LDL without impairing cholesterol delivery to cells. This is achieved by partial inhibition of cholesterol biosynthesis.

Cholesterol is a product of the isoprenoid pathway in mammals. In addition to cholesterol and other steroids, this pathway produces several key metabolic intermediates essential to cells – dolichol, ubiquinone, the farnesyl and geranylgeranyl moieties of prenylated proteins, and the isopentenyl side chain of isopentenyl adenine. The pathways for the synthesis of these compounds diverge from the synthesis of cholesterol either at or before the farnesyl diphosphate branch point (Figure 2.2). The first committed step in cholesterol biosynthesis is the squalene synthase–catalyzed conversion of two moles of farnesyl pyrophosphate to one mole of squalene. Therefore, squalene synthase is an attractive target for selective inhibition of cholesterol biosynthesis.

Screening of fungal cultures led to the discovery of three structurally related and very potent inhibitors of squalene synthase. Zaragozic acid A (squalestatin S1; Figure 2.3) was obtained from an unidentified fungus found in a water sample taken from the Jalon River in Zaragoza, Spain, hence the name. Soon after, zaragozic acids B and C were obtained from fungi isolated elsewhere: *Sporomiella intermedia*, a coprophilous fungus isolated from cottontail rabbit dung in Tucson, Arizona, and *Leptodontium elatius*, isolated from wood in a forest in North Carolina, respectively.

Squalene synthase catalyzes a two-step reaction. Farnesyl pyrophosphate is converted to presqualene diphosphate and then to squalene. The zaragozic acids are potent inhibitors of squalene synthase competitive with farnesyl pyrophosphate. Their inhibition constants ($K_i$s) are extraordinarily low, about $10^{-11}$ M, and they are at least $10^3$ times more potent inhibitors of the catalytic activity of squalene synthase than any previously described compound. Structural comparisons suggest that the zaragozic acids bind to squalene synthase in a manner similar to that of presqualene pyrophosphate (Figure 2.2). Experiments in laboratory animals indicate that zaragozic acids are promising therapeutic agents for hypercholesterolemia. They have also proved valuable as specific inhibitors of squalene synthase in studies of the regulation of hydroxymethylglutaryl–coenzyme A (CoA) reductase and of other aspects of lipoprotein metabolism.

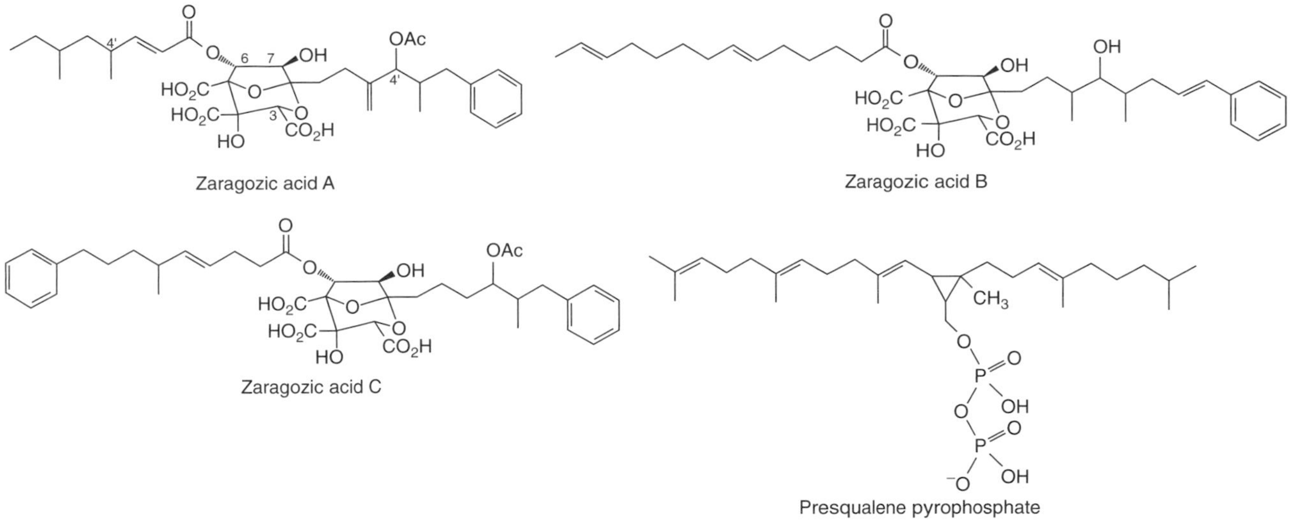

Zaragozic acid A

Zaragozic acid B

Zaragozic acid C

Presqualene pyrophosphate

FIGURE 2.3

Structure of zaragozic acids and of presqualene pyrophosphate. [From Wilson, K. E., Burk, B. M., Biftu, T., Ball, R. G., and Hoogsteen, K. (1992). Zaragozic acid A, a potent inhibitor of squalene synthase: initial chemistry and absolute stereochemistry. *Journal of Organic Chemistry*, 57, 7151–7158.]

A recent study reveals that zaragozic acids have unexpected promise in other therapeutic applications. Squalestatin was shown to cure prion-infected neurons and to protect against prion neurotoxicity. Prion diseases (or transmissible spongiform encephalopathies) are fatal neurodegenerative disorders that include kuru and Creutzfeldt–Jakob disease in humans. In prion diseases, the normal cellular prion, $PrP^{c}$, is converted into a $\beta$-sheet–rich conformer, $PrP^{Sc}$, whose aggregation is believed to lead to neurodegeneration. Low concentrations of squalestatin reduced the cholesterol content of the neurons and prevented the formation of $PrP^{Sc}$. These observations suggest that squalestatin is a potential drug for the treatment of prion diseases.

## TAXOL

Microbial endophytes (bacteria and fungi) are an enormous, highly diverse component of the microbial world. Plant endophytes live in plant tissues between living plant cells but generally can be isolated and cultured independent of the host. For some endophytes, there is evidence that genetic exchange takes place in both directions between the plant and the endophyte. Such exchange raises the possibility that higher plant pathways for the synthesis of complex organic molecules that have desirable biological activities might be transferred to their endophytes.

The story of the highly effective anticancer drug taxol provides proof of the validity of this notion. Taxol, a highly substituted diterpenoid with multiple asymmetric centers (Figure 2.4) was isolated in 1965 from the Pacific yew (*Taxus brevifolia*). In human cells, taxol prevents the depolymerization of microtubules during cell division. It has the same effect in fungi. Consequently, in nature, taxol is a fungicide.

FIGURE 2.4

Taxol.

Taxol proved to be an exceptionally effective anticancer drug, and demand far exceeded the amount that could be produced from the Pacific yew

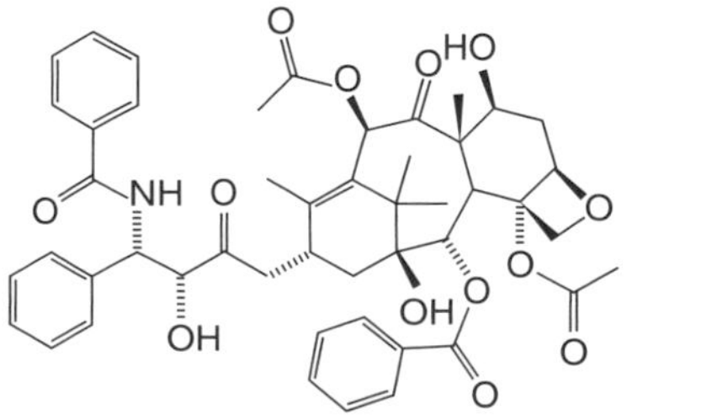

"When the NCI-USDA (National Cancer Institute – U.S. Department of Agriculture) screening program finally was shut down in 1981, taxol was about all the government had to show for more than 20 arduous years of sifting through natural products. From 1960 to 1981, the program had screened 114,045 plant extracts and more than 16,000 extracts from animals. Yet of all these exquisite molecules made by nature, in the rarefied air of advanced testing, taxol stood alone."

*Source:* Stephenson, F. (2002). *A Tale of Taxol*. Florida State University Office of Research http://www.research.fsu.edu/researchr/fall2002/taxol.html.

**BOX 2.3**

(Box 2.3). Moreover, the level at which these slow-growing trees were being utilized for taxol production threatened them with extinction. The development in 1989 of a commercially viable organic synthesis of taxol resolved the problem. In the early 2000s, a plant cell fermentation process for taxol production displaced the chemical synthesis. Here, calluses of a specific *Taxus* cell line are propagated on a simple defined medium to produce taxol.

Even so, it would be advantageous if taxol could be produced by an inexpensive microbial fermentation. The Pacific yew is not the only tree that produces taxol. This compound is in fact found in each of the world's *Taxus* species. The possibility was then explored that a taxol-producing endophyte might be discovered in a *Taxus* species. In 1993, a taxol-producing endophytic fungus, *Taxomyces andreanae*, was discovered in *T. brevifolia*. Subsequently, fungal endophytes in a wide variety of higher plants were found to make taxol. In culture, these endophytes make taxol in submicrogram per liter amounts. A great deal of work remains to be done to achieve high levels of microbial taxol production.

## AGRICULTURE

Methods dependent on microbial biotechnology greatly increase the diversity of genes that can be incorporated into crop plants and dramatically shorten the time required for the production of new varieties of plants. It is now possible to transfer foreign genes into plant cells. Transgenic plants that are viable and fertile can be regenerated from these transformed cells, and the genes that have been introduced into these transgenic plants are as stable as other genes in the plant nuclei and show a normal pattern of inheritance. Transgenic plants are most commonly generated by exploiting a plasmid vector carried by *Agrobacterium tumefaciens*, a bacterium that we discuss in detail in Chapter 6. Foreign DNA carrying from one to 50 genes can be introduced into plants in this manner, with the donor DNA originating from different plant species, animal cells, or microorganisms.

Higher plants have genes whose expression shows precise temporal and spatial regulation in various parts of plants – for example, leaves, floral organs, and seeds that appear at specific times during plant development and/or at specific locations, or whose expression is regulated by light. Other plant genes respond to different stimuli, such as plant hormones, nutrients, lack of oxygen (anaerobiosis), heat shock, and wounding. It is therefore possible to insert the control sequence(s) from such genes into transgenic plants to confine the expression of foreign genes to specific organelles or tissues and to determine the initiation and duration of such expression. Microorganisms that live on or within plants can be manipulated to control insect pests and fungal disease or to establish new symbioses, such as those between nitrogen-fixing bacteria and plants.

In bacteria and yeast, trehalose-6-phosphate is synthesized from UDP-glucose and glucose-6-phosphate in a reaction catalyzed by trehalose-6-phosphate synthase (OtsA). Trehalose-6-phosphate phosphatase (OtsB) then converts trehalose-6-phosphate to trehalose.

UDPGlc + α-Glc-6-phosphate →(OtsA, −UDP) trehalose-6-phosphate →(OtsB, $-PO_4^{2-}$) trehalose

Even though they do not accumulate trehalose in significant amounts, higher plants contain genes homologous to OtsA and OtsB.

**BOX 2.4**

What are some of the objectives and concerns of plant microbial biotechnology, and how are they being addressed? We provide an overview here and follow with a detailed discussion in Chapter 6.

## ABILITY TO GROW IN HARSH ENVIRONMENTS

Extending the habitat range for plants may be achieved by imparting traits such as cold, heat, and drought tolerance; ability to withstand high moisture or high salt concentrations; and resistance to iron deficiency in very alkaline soils. Tolerances toward environmental stresses are likely to be polygenic traits and as a consequence may be difficult to transfer from one kind of organism to another. However, there are some successes, as illustrated by the following example.

Trehalose, a disaccharide of glucose, acts as a compatible solute that stabilizes and protects proteins and biological membranes in bacteria, fungi, and invertebrates from damage during desiccation. Except for highly desiccation-tolerant "resurrection plants," most plants do not accumulate detectable amounts of trehalose. *E. coli* genes *otsA* and *otsB* for trehalose biosynthesis (Box 2.4) were introduced into *indica* rice. An *otsA–otsB* fusion gene was generated so that only a single transformation event would be necessary and to achieve a higher catalytic efficiency of trehalose formation. To obtain either tissue-specific or stress-inducible expression, two different constructs were made. In one, the fusion gene, equipped with a transit peptide, was placed under the control of the promoter of *rbcS*, the gene encoding the small subunit of ribulose bisphosphate carboxylase, to direct the gene product to the chloroplast. In the second, the gene was placed under the control of an abscisic acid–inducible promoter. Here, the OtsA–OtsB enzyme fusion remains in the cytosol. The constructs were introduced into rice using *Agrobacterium*-mediated gene transfer.

**TABLE 2.3 Global area of transgenic crops in 2003 by trait**

| Crop and trait(s)[a] | Area (millions of hectares) |
|---|---|
| Herbicide-tolerant soybean | 41.4 |
| *Bt* maize | 9.1 |
| Herbicide-tolerant canola | 3.6 |
| *Bt*/herbicide-tolerant maize | 3.2 |
| Herbicide-tolerant maize | 3.2 |
| *Bt* cotton | 3.1 |
| *Bt*/herbicide-tolerant cotton | 2.6 |
| Herbicide-tolerant cotton | 1.5 |

[a] *Bt* designates a transgenic crop that expresses a *Bacillus thuringiensis* insecticidal protein. The herbicide Roundup [glyphosate; *N*-(phosphonomethyl)glycine] inhibits 5-enolpyruvylshikimate-3-phosphate synthase (EPSPS), an enzyme in the pathway for the biosynthesis of phenylalanine, tyrosine, and tryptophan. EPSPS is present in plants, fungi, and bacteria but is not found in animals. Transgenic plants exhibiting herbicide tolerance to Roundup contain a form of EPSPS that has a low affinity for binding glyphosate. Roundup-tolerant transgenic canola plants also contain the enzyme glyphosate oxidoreductase, which rapidly inactivates Roundup by converting it to glyoxylate and aminomethylphosphonic acid. The glyphosate oxidoreductase originates from a soil proteobacterium, *Achromobacter* species strain LBAA (*Ochrobactrum anthropi*).

*Source*: The source of the data on global area of transgenic crops in 2003 by trait is the International Service for the Acquisition of Agri-biotech Applications, http://www.isaaa.org/.

Compared with nontransgenic rice, several independent transgenic lines showed sustained plant growth under drought, salt, or low temperature stress conditions. The transgenic rice contained three- to ninefold greater levels of trehalose than the nontransgenic rice. However, the striking finding was that the trehalose level did not exceed 1 mg/g wet weight of tissue under any conditions. Consequently, in rice, trehalose must exert its protective effect indirectly rather than primarily through affecting the bulk properties of water within the plant cells. A detailed analysis of the transgenic rice with each of the constructs showed less photooxidative damage to photosystem II (allowing maintenance of higher capacity for photosynthesis), higher levels of soluble carbohydrate, and greater ability to control $K^+/Na^+$ balance in the roots under the stress conditions, than seen in nontransgenic rice controls. These results indicate that in rice, trehalose acts as a regulatory molecule that affects the expression of genes associated with carbon metabolism and those involved with ion uptake and possibly other processes as well. This example offers a valuable lesson. The presence of homologous genes in widely diverged organisms that catalyze the synthesis of the same product offers no guarantee of a universal identical role for the product.

Initial field trials on the transgenic rice are promising and offer the prospect of growing rice in saline soils, or in areas where availability of water would depend on intermittent rainfall.

## HERBICIDE TOLERANCE

Many otherwise effective broad-spectrum herbicides do not distinguish between weeds and crops. Crop plants can be modified to become resistant to particular herbicides. When applied to a weed-infested field of such genetically modified plants, these herbicides act as selective weed killers.

## RESISTANCE TO INSECT PESTS

Certain strains of the bacterium *Bacillus thuringiensis* produce protein endotoxins that permeabilize the epithelial cells in the gut of the larvae of lepidopteran insects, moths, and butterflies (Chapter 7). Genes encoding particular *B. thuringiensis* endotoxins have been transferred into and expressed in tobacco, cotton, and tomato. In field tests, the transgenic tomato and tobacco plants were only slightly damaged by caterpillar larvae under conditions that led to total defoliation of control plants. Transgenic maize and cotton, containing *B. thuringiensis cry* genes that encode insecticidal proteins, accounted for over 26% of the global area of transgenic crops in 2003 (Table 2.3). A different approach to achieve the same end was to transfer a *B. thuringiensis* endotoxin gene into bacteria such as *Clavibacter xyli* subsp. *cynodontis*, which colonizes the interior of plants. This organism is generally found inside Bermuda grass plants but can reach population sizes in excess of $10^8$/gram of stem tissue if purposefully inoculated into other monocotyledonous species, such as corn. Recombinant *C. xyli* strains expressing the endotoxin show promise in controlling leaf- and stem-feeding lepidopteran larvae.

## CONTROL OF PATHOGENIC BACTERIA, FUNGI, AND PARASITIC NEMATODES

The cell walls of many plant pests, such as insects and fungi, contain chitin (poly-*N*-acetylglucosamine) as a major structural component. Many bacteria (e.g., species of *Serratia*, *Streptomyces*, and *Vibrio*) produce chitin-degrading enzymes (chitinases). The control of some fungal diseases by such bacteria has been correlated with the production of chitinases. Genes encoding chitinases from several different soil bacteria have been cloned into *Pseudomonas fluorescens*, an efficient colonizer of plant roots. The effectiveness of these recombinant strains in controlling fungal disease is not yet known.

## *BACILLUS SUBTILIS* STRAINS AS BROAD-SPECTRUM MICROBIAL PESTICIDES

Selected *B. subtilis* strains are widely accepted as broad-spectrum microbial pesticides. Strains of the common soil bacterium *B. subtilis* secrete a formidable array of compounds, which together display antifungal, antibacterial, and even insecticidal activities. These include two different classes of lipopeptides designated iturins and plipastatins; a surfactant called surfactin; 2,3-dihydroxybenzoylglycine, an iron-chelating agent; and potent proteases with broad specificity. Iturin lipopeptides consist of one $\beta$-amino fatty acid and seven $\alpha$-amino acids, whereas surfactins and plipastatins consist of one $\beta$-hydroxy fatty acid and seven and 10 $\alpha$-amino acids, respectively (Figure 2.5).

Large amounts of a *B. subtilis* strain capable of producing this potent mixture of products can be obtained by solid-state fermentation with soybean curd residue as substrate. Under these cultivation conditions, the cells produce greatly elevated levels of the lipopeptides. When the mixture of cells and metabolites obtained by the solid-state fermentation is directly applied to soil, it suppresses the growth of various plant pathogens.

A patented strain, *B. subtilis* QST-713, isolated from soil taken from a California orchard, produces more than 30 iturin- and plipastatin-type lipopeptides, including two agrastatins (previously undescribed members of the plipastatin family; Figure 2.5). *B. subtilis* QST-713 is grown to high density, and the aqueous fermentation broth containing the bacterial cells, spores, and lipopeptides is concentrated and spray dried. The resulting powder is sold as a biofungicide either in dry form or as an aqueous suspension. When the biofungicide is applied to plants, it coats leaf surfaces, preventing the attachment of pathogens. The three types of lipopeptides act in an interdependent manner through mixed micelle formation at very low concentrations (~25 ppm) to destroy fungal cells and spores by permeabilizing their membranes.

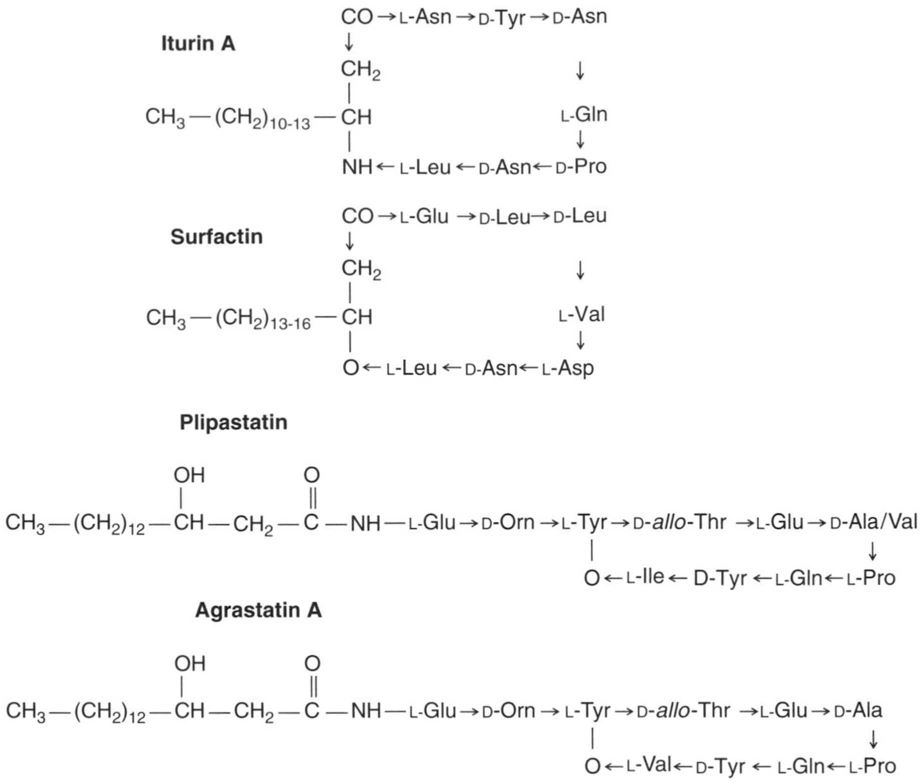

**Iturin A**

CO → L-Asn → D-Tyr → D-Asn ↓ L-Gln ↓ D-Pro → D-Asn → L-Leu → NH

$CH_3—(CH_2)_{10\text{-}13}—CH(—CH_2—CO)(—NH)$

**Surfactin**

CO → L-Glu → D-Leu → D-Leu ↓ L-Val ↓ L-Asp → D-Asn → L-Leu → O

$CH_3—(CH_2)_{13\text{-}16}—CH(—CH_2—CO)(—O)$

**Plipastatin**

$CH_3—(CH_2)_{12}—CH(OH)—CH_2—C(=O)—NH$—L-Glu → D-Orn → L-Tyr → D-*allo*-Thr → L-Glu → D-Ala/Val ↓ L-Pro → L-Gln → D-Tyr → L-Ile → O (to L-Tyr)

**Agrastatin A**

$CH_3—(CH_2)_{12}—CH(OH)—CH_2—C(=O)—NH$—L-Glu → D-Orn → L-Tyr → D-*allo*-Thr → L-Glu → D-Ala ↓ L-Pro → L-Gln → D-Tyr → L-Val → O (to L-Tyr)

FIGURE 2.5

Lipopeptides produced by *Bacillus subtilis* QST-713.

The *B. subtilis* QST-713 fungicide is widely used on commercial fruit, nut, and vegetable crops such as tomatoes, lettuce, and wine grapes, and in private gardens.

## RESISTANCE TO VIRAL DISEASES

Plant virus diseases are difficult to control. Research in the mid-1980s showed that transgenic tobacco expressing the coat protein (capsid) gene of tobacco mosaic virus (TMV) is resistant to TMV, and it was speculated that the resistance is the result of the interference with virus uncoating by the expressed coat protein. Similar coat protein transgene-mediated protection was reported for a number of other related plant RNA viruses, TMV, cucumber mosaic virus, alfalfa mosaic virus, and several potato viruses. The protection is now known to be the result of RNA silencing, a cell-based sequence-specific posttranscriptional RNA degradation system that is programmed by the transgene-encoded RNA sequence (described in Chapter 6). In Hawaii, papaya ranks as the second most important fruit crop. This crop was subject to severe damage caused by papaya ringspot virus (PRSV). The introduction in 1998 of transgenic papaya cultivars with a transgene that expressed a PRSV coat protein saved the Hawaiian papaya industry.

In recent years, transgenic plants have been engineered with a variety of other sequences, encoding either viral proteins or RNAs that confer virus resistance.

### NITROGEN FIXATION

**TABLE 2.4 Examples of fermented foods and of fermenting microorganisms**

| Product | Starting material | Fermenting organisms |
|---|---|---|
| Beer | Barley and hops | *Saccharomyces carlsbergensis* |
| Cheeses | Milk | Various |
| Cider | Apples | *Saccharomyces* spp. |
| Kimchi | Cabbage | Lactic acid bacteria |
| Olives | Green olives | *Lactobacillus plantarum*<br>*Pediococcus dextrinicus* |
| Pickles | Cucumbers | *Lactobacillus plantarum*<br>*Pediococcus dextrinicus* |
| Vinegar | Cider or wine | *Acetobacter* spp. |
| Whisky | Corn, rye | *Saccharomyces cerevisiae* |
| Wine | Grapes | *Saccharomyces* spp. |
| Yogurt | Milk | *Lactobacillus bulgaricus*<br>*Streptococcus thermophilus* |

Leguminous plants, including important crops such as soybeans, form symbiotic associations with species of *Rhizobium*, *Bradyrhizobium*, and *Frankia* that fix atmospheric molecular nitrogen. Free-living rhizobia are found in the soil. Natural infection of host plants by the bacteria leads to formation of root nodules within which the rhizobia proliferate. It has been a practice for almost a hundred years to add commercially produced rhizobia to soil as legume inoculants to reduce the need for nitrogenous fertilizer. No adverse effects of such applications have been observed. Consequently, no adverse consequences should attend large-scale applications of genetically engineered strains of rhizobia.

Strains of *Bradyrhizobium japonicum* and *Rhizobium meliloti*, engineered to increase the expression of certain genes important to nitrogen fixation, were shown to give greater biomass increases of their respective host plants under greenhouse conditions compared with the wild-type bacterial strains. Because of the very high population of free-living rhizobia in the soil, newly introduced strains have to be introduced at very high concentrations to overcome competition from the resident bacteria. This leads to high inoculant cost. Studies of the mechanism of infection and of the biochemical determinants of *Rhizobium* competitiveness may reveal ways of resolving this difficulty.

Transfer of the genes for nodule formation to *Agrobacterium* enables the recombinant organism to initiate nodulation on non-legumes, which suggests that it may be possible to extend nitrogen fixation to non-leguminous plants. This goal will require manipulation of the host plant as well as of the bacterial genes.

Exploitation of microbial biotechnology in agriculture is driven by the realization that agricultural practices that rely heavily on expensive nitrogenous fertilizers and widespread use of pesticides are no longer sustainable.

## FOOD TECHNOLOGY

### PREPARATION OF FERMENTED FOODS

The use of microorganisms to produce fermented foods has a very long history. Microbial fermentation is essential to production of wine, beer, bologna, buttermilk, cheeses, kefir, olives, salami, sauerkraut, and many more (Table 2.4; Box 2.5). The metabolic end products produced by the microorganisms flavor fermented foods. For example, mold-ripened cheeses owe their distinctive flavors to the mixture of aldehydes, ketones, and short-chain fatty acids produced by the fungi.

"Throughout history and around the world, human societies at every level of complexity discovered how to make fermented beverages from sugar sources available in their local habitats.[1] This nearly universal phenomenon of fermented beverage production is explained by ethanol's combined analgesic, disinfectant, and profound mind-altering effects. . . . By using a combined chemical, archaeobotanical, and archaeological approach, we present evidence here that ancient Chinese fermented beverage production does indeed extend back nearly nine millennia. Moreover, our analyses of unique liquid samples from tightly lidded bronze vessels, dated to the Shang/Western Zhou Dynasties (*ca.* 1250–1000 B.C.), reveal that refinements in beverage production took place over the ensuing 5,000 years, including the development of a special saccharification (amylolysis) fermentation system in which fungi break down the polysaccharides in rice and millet."

[1] McGovern, P. E. (2003). *Ancient Wine: The Search for the Origins of Viniculture*, Princeton: Princeton University Press.

*Source:* Quoted from McGovern, P. E., et al. (2004). Fermented beverages of pre- and protohistoric China. *Proceedings of the National Academy of Sciences* USA, 101, 17593–17598.

**BOX 2.5**

Lactic acid bacteria are widely used to produce fermented foods. These organisms are also of particular importance in the food fermentation industry because they produce peptides and proteins (bacteriocins) that inhibit the growth of undesirable organisms that cause food spoilage and the multiplication of foodborne pathogens. The latter include *Clostridium botulinum* (the cause of botulism) and *Listeria monocytogenes* (which produces meningoencephalitis, meningitis, perinatal septicemia, and other disorders in humans).

## NISIN

Nisin, an antimicrobial peptide produced by strains of *Lactococcus lactis*, is widely used as a preservative at low concentrations (up to 250 ppm in the finished product) primarily in heat-processed and low pH foods. Nisin inhibits the growth of a wide range of Gram-positive bacteria, including *Listeria*, *Clostridium*, *Bacillus*, and enterococci, but is not effective against Gram-negative bacteria, yeasts, and molds. The antibacterial activity of nisin is the combined outcome of its high-affinity interaction with lipid II at the outer leaflet of the bacterial cytoplasmic membrane and permeabilization of the membrane through pore formation (see Box 2.6).

Nisin is designated as a Generally Regarded as Safe (GRAS) food preservative in the United States and in many other countries around the world. It is used in many food products, including pasteurized cheese spreads with fruits, vegetables, or meats; liquid egg products; dressings and sauces; fresh and recombined milk; some beers; canned foods; and frozen dessert.

## *LACTOBACILLUS SAKEI*: A PROMISING BIOPRESERVATIVE

*L. sakei*, a psychrophilic lactic acid bacterium, was first isolated from *sake*, a Japanese rice beer that is produced partly by lactic acid fermentation. Subsequently, *L. sakei* strains were found to dominate the spontaneous fermentation of meat in the manufacture of salami and other dry fermented sausages. Such strains are also major components of the microbial flora of processed food products stored at cold temperature. *L. sakei* starter cultures have come to be widely used in the manufacture of fermented meats, and this organism has been shown to prevent the growth of spoilage organisms and pathogens. *L. sakei* is also a transient inhabitant of the human gut.

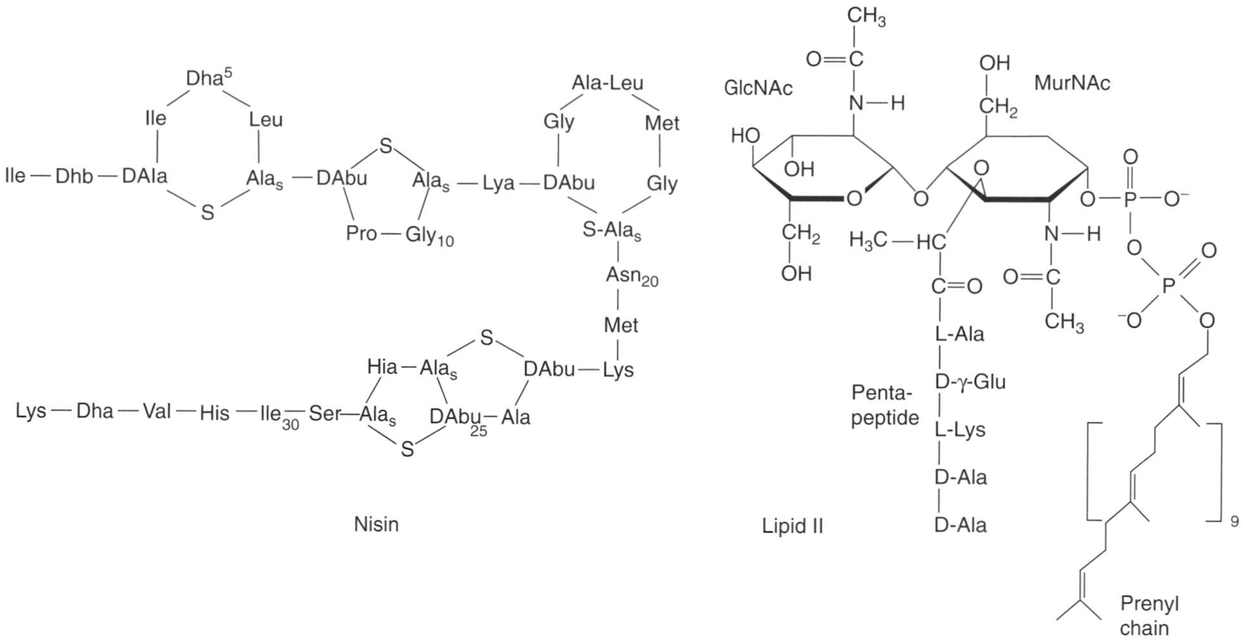

Nisin is a 34-residue cationic peptide produced by *Lactobacillus lactis*. The precursor peptide is gene encoded and synthesized on ribosomes. Specific serine and threonine residues are dehydrated through posttranslational modification to dehydroalanine (Dha) and dehydrobutyrine (Dhb). *Meso*-lanthionine and 3-methyl-lanthionine residues are indicated by DAla-S-$Ala_S$ and DAbu-S-$Ala_S$, respectively (in which the amino-terminal moieties have the D configuration). Nisin is a member of a class of antibiotics called *lantibiotics* because they contain lanthionine.

The structure of lipid II is made up of a membrane-incorporated undecaprenol (a $C_{55}$ polyisoprenol) to which the amino sugar *N*-acetylmuramic acid (MurNAc) carrying a pentapeptide is attached through a pyrophophosphate. The composition of the pentapeptide differs among bacterial genera. The final building block of lipid II is *N*-acetylglucosamine (GlcNAc).

In a key first step, nisin binds to lipid II, an essential intermediate in peptidoglycan synthesis, at the outer surface of the bacterial cytoplasmic membrane and then forms a pore. Nisin was the first antibiotic shown to kill Gram-positive bacteria by a binding site–specific pore formation. It is effective in the nanomolar concentration range.

**BOX 2.6**

A number of other lactic acid bacteria are either transient or permanent members of the human gastrointestinal flora, including *Lactobacillus acidophilus*. In that setting, these organisms – called *probiotic* species – stimulate the immune response and suppress the growth of potentially pathogenic bacteria. Recently, the genome of *L. sakei* 23K, isolated from a French sausage, was completely sequenced and was 43% identical to *L. acidophilus*. There is much interest in using safe bacteria as *biopreservatives*, and for the various reasons outlined above, *L. sakei* is an excellent candidate.

The availability of the complete genome of *L. sakei* 23K allows one to formulate testable hypotheses as to the attributes of this organism that enable it to flourish on fresh meat and to survive stressful conditions it encounters during meat fermentation and storage. Such challenges include high levels of oxidative stress, high salt, and low temperatures.

The *L. sakei* genome codes for four proteins predicted to be involved in cell–cell interaction and in binding to collagen exposed on the surface of meat. Such proteins are absent from other lactobacilli. Two other gene clusters are predicted to function in the production of surface polysaccharides that may contribute to the attachment of the bacterium to the meat surface. These protein and polysaccharide surface components might mediate the aggregation of *L. sakei* and formation of a biofilm on the meat surface that would exclude other microorganisms.

Meat undergoes autoproteolysis on aging with release of amino acids. *L. sakei* is auxotrophic for all amino acids (except glutamic and aspartic), and thus the meat surface is an excellent ecological niche.

Meat storage frequently requires refrigeration and salts (up to 9% NaCl). *L. sakei* is well adapted to both low temperature and the osmotic stresses encountered at high salt concentrations. It has a larger number of putative cold stress proteins than other lactobacilli. It also has uptake systems for the efficient accumulation of osmo- and cryoprotective solutes such as betaine and carnitine. *L. sakei* is also well equipped with enzymes that detoxify reactive oxygen species such as superoxide or organic hydroperoxides generated during meat processing.

Finally, *L. sakei* requires and takes up both heme and iron from the meat. The competition for iron may represent yet another important factor in the ability of *L. sakei* to exclude other organisms from the meat surface.

## MONENSIN

Monensin is the most widely used compound fed to cattle to increase feed efficiency. In feedlot cattle, a dosage of 350 mg/day led to an improvement in feed efficiency of approximately 6%. In grazing cattle, the average daily gain increased by 15%. Monensin produces these outcomes by changing the makeup of the bacterial population in the rumen, thereby influencing the balance of the end products of ruminal fermentation metabolism.

Monensin is produced by the bacterium *Streptomyces cinnamonensis*. It is a member of a large and important class of polyketides, the polyether ionophores (Table 2.5). The compound is toxic to many bacteria, fungi, protozoa, and higher organisms. The $pK_a$ of the carboxyl group in monensin is 7.95, so at the acidic pH of the rumen, the uncharged lipophilic molecule accumulates in cell membranes of bacteria sensitive to this ionophore. Monensin forms cyclic complexes with alkali metal cations ($Na^+$, $K^+$, $Rb^+$) with a preference for $Na^+$, with six oxygen atoms serving as ligands to the cation (Figure 2.6). The ratio of $Na^+/K^+$ concentrations in the rumen ranges from 2 to 10. The direction of metal ion and proton movement across a cell membrane is directed by the magnitude of the existing ion concentration gradient. Monensin acts as an "antiporter" that releases a proton at the inner face of the cytoplasmic membrane as it picks up $K^+$. At the outer face of the cytoplasmic membrane, it releases the $K^+$ and picks up either $H^+$ or $Na^+$. The cell responds to these ion fluxes by utilizing its Na/K and $H^+$ ATPases to maintain ion balance and intracellular pH. Depending on the extent of exhaustion of ATP, and of the resulting membrane depolarization, the cells cease to grow and reproduce, and may die.

In the anaerobic environment of the rumen, ruminal microorganisms generate the energy and nutrients for their growth by fermenting carbohydrates (primarily cellulose) and proteins. The major resulting products, volatile fatty acids (acetic, propionic, and butyric) and microbial protein, serve as the sources of energy and nutrients for the cow. The fatty acids pass through the rumen wall into the bloodstream. The cow derives most of its energy from the oxidation of these compounds. Degradation of the microbial cells in the gastrointestinal tract provides amino acids. However, other bacterial fermentation end products, particularly methane and ammonia that are released to the environment, represent loss to the cow of a sizeable fraction of the potential energy and protein sources from the feed.

The major end products of the fermentative metabolism of the Gram-positive bacteria in the rumen are acetate, butyrate, formate, lactate, hydrogen, and ammonia. The methanogenic bacteria in the rumen are not able to

**TABLE 2.5 Ionophores used as anticoccidial and growth-promoting feed additives**

| Ionophore[a] | Species | Indication |
|---|---|---|
| Monensin | Beef cattle | Improved feed efficiency and increased rate of weight gain |
| Salinomycin | Nonlactating dairy cattle | |
| | Broiler chickens; turkeys | Prevention of coccidiosis[b] |
| Lasalocid | Beef cattle | Improved feed efficiency and increased rate of weight gain |
| | Nonlactating dairy cattle | |
| | Sheep | Prevention of coccidiosis |
| Laidlomycin | Beef cattle | Improved feed efficiency and increased rate of weight gain |

[a] Ionophores are highly lipophilic polyethers produced by *Streptomyces* species that form neutral complexes with $Na^+$ and $K^+$ (see Figure 2.6).

[b] Coccidiosis is caused by species of intracellular protozoan parasites of the genus *Eimeria*. It is economically one of the most costly diseases of poultry. The worldwide annual cost of the "failure to thrive" of infected poultry and of the prophylactic administration of antibiotics is several hundred million dollars. For extensive information on *Eimeria*, see http://www.iah.bbsrc.ac.uk/eimeria/biology.htm.

**FIGURE 2.6**

Monensin A, salinomycin, and lasalocid are polyether ionophores produced by *Streptomyces* species. The structure of the neutral monensin A complex with $Na^+$ is also shown.

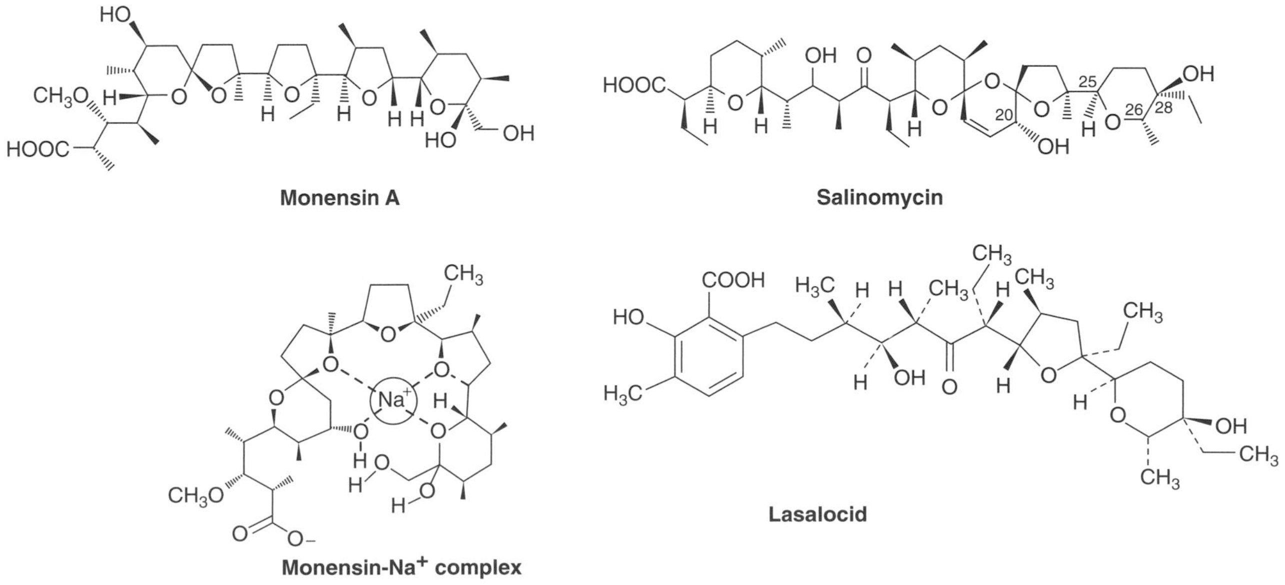

Monensin A

Salinomycin

Monensin-$Na^+$ complex

Lasalocid

use complex organic compounds. They obtain energy by utilizing formate, acetate, carbon dioxide, and hydrogen to generate methane as follows:

$$4HCOOH \rightarrow CH_4 + 3CO_2 + 2H_2O$$
$$CH_3COOH \rightarrow CH_4 + CO_2$$
$$CO_2 + 4H_2 \rightarrow CH_4 + 2H_2O$$

The effects of monensin on ruminal fermentation are as follows. Much less methane is produced. The ratio of propionate to acetate is higher. Less ammonia is produced, and the amount of protein N available to the cow is greater. How does monensin modulate the fermentative metabolism in the rumen?

The recommended daily dosage of monensin is 350 mg, the mass of the monensin–$Na^+$ complex is 693, and the rumen volume of cattle is approximately 70 L. Thus, the initial ruminal concentration of unbound monensin–$Na^+$ is 7 $\mu$M. At such a low concentration, monensin–$Na^+$ rapidly partitions into the membranes of the most sensitive bacteria. However, studies with radiolabeled monensin show that binding also takes place to feed particles, protozoa, and ionophore-resistant bacteria. The potential binding sites are far from saturated at this monensin concentration. Gram-positive ruminal bacteria are more sensitive to monensin than are Gram-negative ones. In general, bacteria with outer membranes and/or associated extracellular polysaccharide are more resistant, presumably because of the hindrance of access of monensin to the cytoplasmic membrane.

Under these conditions, monensin does not inhibit methanogenic bacteria but does inhibit the Gram-positive $H_2$-producing bacteria that supply the methanogens with $H_2$ and that also produce acetate, butyrate, and formate. The result is a decrease in methane production. The fermentative pathways of ruminal Gram-negative bacteria lead to propionate and succinate. These organisms are not inhibited by monensin. The overall result is an increase in the propionate-to-acetate ratio, in essence an increase in the energy source for the cow. The ruminal obligate amino acid–fermenting bacteria are monensin sensitive. The inhibition of these bacteria produces the large observed decrease in ammonia production. The consequence is that more protein N is available to the cow.

In summary, monensin modulates ruminal fermentative metabolism by selective inhibition of the metabolic activities of particular groups of bacteria.

## SINGLE-CELL PROTEIN

The term *single-cell protein*, or SCP, describes the protein-rich cell mass derived from microorganisms grown on a large scale for either animal or human consumption. SCP has a high content of protein containing all the essential amino acids. Microorganisms are an excellent source of SCP because of their rapid growth rate, their ability to use very inexpensive raw materials as carbon sources, and the uniquely high efficiency, expressed as grams of protein produced per kilogram of raw material, with which they transform these carbon sources to protein.

In spite of these advantages, only one SCP product approved for human consumption has reached the market. This product is "mycoprotein," the processed cell mass preparation from the filamentous fungus *Fusarium venenatum*. We consider here the positive nutritional properties of this product and examine the many concerns that needed to be examined and addressed before this product gained regulatory approvals.

The source organism, *F. venenatum* strain PTA-2684, was cultured from a soil sample obtained from Buckinghamshire, United Kingdom. Marlow Foods Ltd. chose this strain of *F. venenatum* from more than 3000 organisms obtained from around the world. The manufacturing process for mycoprotein is designed to ensure the absence of undesirable constituents of fungal cells from the final product.

Mycotoxins are synthesized by *Fusarium* species as well as by members of other genera of filamentous fungi, such as *Aspergillus* and *Penicillium*. Mycotoxins are products of fungal *secondary metabolism*. Thus, they are not essential to the energy-producing or biosynthetic metabolism of the fungus, or to fungal reproduction. Rather, under growth-limiting or stress conditions, they appear to give the fungus an advantage over other fungi and bacteria with which it may be competing. Mycotoxins are nearly all cytotoxic. They disrupt cell membranes and interfere with protein, RNA, and DNA synthesis. Their toxicity extends beyond microorganisms to the cells of higher plants and animals, including humans.

*Fusarium* species produce different classes of mycotoxins, trichothecenes and fusarins. Deoxynivalenol, also known as vomitoxin, is one of about 150 related trichothecene compounds that are formed by a number of species of *Fusarium* and some other fungi. Deoxynivalenol is nearly always formed before harvest when crops are invaded by certain species of *Fusarium* closely related to *Fusarium venenatum*. These *Fusarium* species are important plant pathogens that cause heat blight in wheat. Deoxynivalenol is heat stable and persists in stored grain.

The general structure of trichothecenes is shown below. In deoxynivalenol, $R^1$ is OH, $R^2$ is H, $R^3$ is OH, $R^4$ is OH, and $R^5$ is O.

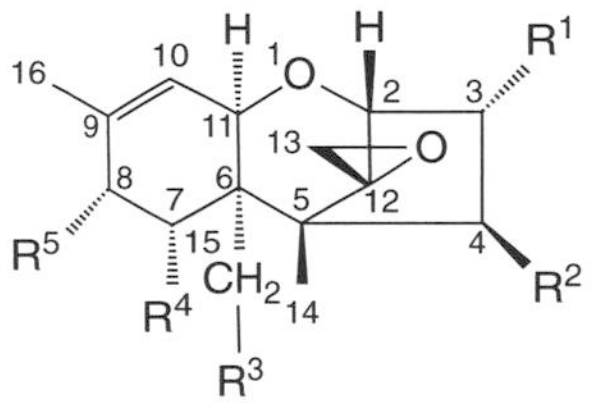

**BOX 2.7**

*F. venenatum* is grown with aeration under steady-state conditions maintained by continuous feed of nutrient medium and concomitant removal of the culture. These fermentation conditions were chosen to prevent the production of the highly toxic mycotoxins (Box 2.7). *Fusarium* species produce trichothecene and fusarin mycotoxins when growth is limited by nutrient limitation, a high ratio of carbon to nitrogen nutrients, low oxygen tension, or the lack of a micronutrient.

To prevent mycotoxin synthesis, the production strain is grown at a high rate without any nutritional limitations. The culture is supplied with a nutritionally balanced, chemically defined fermentation medium, with glucose as the sole carbon source. The medium is provided at a rate that allows the cells to grow at a specific rate of at least 0.17 per hour. To monitor the levels of mycotoxins, the final product is analyzed for these compounds by high-performance liquid chromatography with mass spectrometric detection. The detection limits per kilogram wet weight of product are 2 $\mu$g for individual trichothecenes and 5 $\mu$g for fusarin mycotoxins. With these sensitivity levels, no mycotoxins are detected in the final product.

Rapidly growing bacterial and fungal cells are rich in RNA. RNA in the diet is broken down into purines and pyrimidines. Purines are converted

**TABLE 2.6 Nutritional analysis of *Fusarium venenatum* single-cell protein (Quorn mycoprotein[a])**

| Nutrient | g/100 g dry weight |
|---|---|
| Protein (amino acid N × 6.22) | 48 |
| Fat | 12 |
| Fatty acids: | |
| Palmitic ($C_{16}$) | 1.6 |
| Stearic ($C_{18}$) | 0.3 |
| Oleic ($C_{18:1}$) | 1.4 |
| Linoleic ($C_{18:2}$) | 4.3 |
| $\alpha$-Linolenic ($C_{18:3}$) | 1.0 |
| Dietary fiber[b] | 25 |
| Carbohydrate | 12 |
| Water | 0 |

[a] The term *single-cell protein* is commonly used to describe the protein-rich cell mass derived from microorganisms. The SCP characterized above is the product of Marlow Foods Ltd., marketed as Quorn mycoprotein.

[b] This largely insoluble fraction is derived from the cell wall of *F. venenatum*. The cell wall is composed of chitin (poly-*N*-acetylglucosamine) and $\beta$-glucans (with $\beta$-1:3 and $\beta$-1:6 glucosidic linkages).

*Source*: Data from Miller, S. A., and Dwyer, J. T. (2001). Evaluating the safety and nutritional value of mycoprotein. *Food Technology*, 55, 42–47.

**TABLE 2.7 Essential amino acid content of *Fusarium venenatum* single-cell protein (Quorn mycoprotein) compared with that of other protein-containing foods**

| | Amino acid content (g/100 g of edible portion) | | | | | |
|---|---|---|---|---|---|---|
| Essential amino acid | Mycoprotein | Cow's milk[a] | Egg[b] | Beef[c] | Soybeans[d] (dry) | Wheat[e] |
| Histidine | 0.39 | 0.09 | 0.3 | 0.66 | 0.98 | 0.32 |
| Isoleucine | 0.57 | 0.20 | 0.68 | 0.87 | 1.77 | 0.53 |
| Leucine | 0.95 | 0.32 | 1.1 | 1.53 | 2.97 | 0.93 |
| Lysine | 0.91 | 0.26 | 0.90 | 1.6 | 2.4 | 0.30 |
| Methionine | 0.23 | 0.08 | 0.39 | 0.5 | 0.49 | 0.22 |
| Phenylalanine | 0.54 | 0.16 | 0.66 | 0.76 | 1.91 | 0.68 |
| Tryptophan | 0.18 | 0.05 | 0.16 | 0.22 | 0.53 | 0.18 |
| Threonine | 0.61 | 0.15 | 0.6 | 0.84 | 1.59 | 0.37 |
| Valine | 0.6 | 0.22 | 0.76 | 0.94 | 1.82 | 0.59 |

[a] Whole fluid milk (3.3% fat)
[b] Raw fresh egg
[c] Ground beef (regular, baked-medium)
[d] Raw peanuts (all types)
[e] Durum wheat

*Source*: Data from Miller, S. A., and Dwyer, J. T. (2001). Evaluating the safety and nutritional value of mycoprotein. *Food Technology*, 55, 42–47.

to uric acid and add to the serum uric acid derived from the metabolism of endogenous purines. Elevated uric acid increases the risk of developing gout and kidney stones in susceptible individuals. To address this problem, a United Nations Protein Advisory Group recommended in 1972 that SCPs intended for human consumption provide no more than 2 g of RNA per day.

The fermentation broth containing the fungal biomass removed from the fermenter is rapidly heated by injection of steam. The rapid heating process kills the cells, with concomitant degradation of RNA. The fermentation broth is subsequently separated from the cell mass by centrifugation, and the RNA degradation products are discarded with the supernatant. These steps reduce the content of RNA in the cell mass from about 10% in viable cells to about 0.5% to a maximum of 2% in mycoprotein on a dry weight basis. With estimated limits of dietary intake of mycoprotein of 17 to 33 g/person/day on a dry weight basis, the intake of RNA from consumption of mycoprotein would range from 0.35 to 0.7 g/person/day, well below the level recommended by the United Nations Protein Advisory Group.

Table 2.6 summarizes the composition of mycoprotein. The mat-like filamentous fungal mass is described as giving mycoprotein a meatlike texture. Close to 50% of the dry weight is protein containing all the essential amino acids (Table 2.7). Just like milk casein or egg white proteins, this protein is fully digestible. About 25% of the dry weight consists of cell wall components, chitin, and $\beta$-glucans. This fraction is insoluble and indigestible, properties characteristic of dietary fiber. Fat represents about 12% of the dry weight. With its low ratio of saturated to unsaturated fatty acids (see

Table 2.6), it is more like vegetable than animal fat. Mycoprotein contains significant amounts of ergosterol but no cholesterol.

Animal studies have shown that mycoprotein does not cause chronic toxicity, is not a reproductive toxicant, is not a teratogen, and is not carcinogenic. It does not interfere with the absorption of calcium, iron, or other essential inorganic nutrients. Marlow Foods Ltd. reported that mycoprotein is much less allergenic in humans than are many commonly consumed foods, such as those containing shellfish or peanuts. Anecdotal reports hint at higher numbers of adverse reactions.

Mycoprotein has been commercially available in the United Kingdom since 1985, in other countries in Europe since 1991, and in the United States since 2002. Products marketed in Europe include meat-free burgers and fillets and prepared meals, such as stir-fries, curries, and pasta dishes, in which mycoprotein is the central component. In the United Kingdom and Europe, the acceptance of mycoprotein as a meat substitute in a wide variety of foods has been significant, with a reported 15 million customers. The story of mycoprotein illustrates the long road of regulatory approvals and customer acceptance that a new SCP product must travel.

**TABLE 2.8 Components of the global water supply**

| Component | Estimated % |
|---|---|
| Saltwater | 97.5 |
| Freshwater | 2.5 |
| Distribution of freshwater | |
| Glaciers and permanent snow cover | 68.9 |
| Fresh groundwater | 29.9 |
| Freshwater lakes and rivers | 0.3 |
| Soil moisture, marshlands, permafrost, etc. | 0.9 |

*Source: World Water Resources at the Beginning of the 21st Century*, http://webworld.unesco.org/water/ihp/db/shiklomanov/summary/html/summary.html.

## ENVIRONMENTAL APPLICATIONS OF MICROORGANISMS

Microorganisms mitigate a multitude of impacts that result from human use of the natural resources of the planet. First and foremost, the essential role of microorganisms in the treatment of wastewater is critical to the well-being of life on Earth. Bioremediation, biomining, and microbial desulfurization of coal are other large-scale processes in which important positive environmental outcomes are achieved by directly exploiting the combined metabolic capabilities of naturally occurring communities of microorganisms. In such applications, the functioning of a particular microbial community can be influenced through the manipulation of conditions (e.g., nutrients, oxygen tension, temperature, agitation).

### WASTEWATER TREATMENT

Living organisms consist of about 70% water. A human being, for instance, has to consume an average of 1.5 L/day to survive. Freshwater represents only about 2.5% of the water on the planet (Table 2.8) and is now a scarce resource in many parts of the world. The volume of water being contaminated and the need to reclaim wastewater are increasing with the growth in population and industrial use. Wastewater originates from four primary sources: sewage, industrial effluents, agricultural runoff, and storm water and urban runoff. Treatment of wastewater is essential to prevent contamination

**TABLE 2.9 Inputs of metals and organic contaminants to urban wastewater**

| |
|---|
| Heavy metals (Cd, Hg, Cu, Ni, Pd, Pb, Zn, Ag, As, Se) |
| Polycyclic aromatic hydrocarbons |
| Chlorinated biphenyls |
| 2-(2-Ethylhexyl)phthalate |
| Nonylphenol |
| Nitrosamines |
| Anionic and nonionic surfactants |
| Aliphatic hydrocarbons |
| Monocyclic aromatic hydrocarbons |
| Polyaromatic hydrocarbons |
| Chlorophenols and chlorobenzenes |
| Polychlorinated dibenzo-*p*-dioxins and dibenzofurans |
| Solvents (both chlorinated and nonchlorinated) |

*Source*: *Pollutants in Urban Wastewater and Sewage Sludge*. (2001). Luxembourg: Office for Official Publications of the European Communities. Input of contaminants to the urban wastewater system occurs from three generic sources: domestic and commercial waste and urban runoff. The above publication provides detailed information on the major sources and on the levels of the pollutants listed here.

of drinking water and the entry of pathogens and contaminants into the food chain. Given the number and variety of the substantial contaminants, such as those listed in Table 2.9, the success of some of the current treatments for the reclamation of pure water is little short of amazing.

Primary treatment of sewage consists of removal of suspended solids. The secondary treatment of sewage reduces the biochemical oxygen demand (Box 2.8). This is accomplished by lowering the organic compound content of the effluent from the primary treatment through microbial oxidation by an incompletely characterized community of microorganisms in "activated sludge." Bacteria of *Zoogloea* species play an important role in the aerobic secondary stage of sewage treatment. These organisms produce abundant extracellular polysaccharide and, as a result, form aggregates called flocs. Such aggregates efficiently adsorb organic matter, part of which is then metabolized by the bacteria. The flocs settle out and are transferred to an anaerobic digester, where other bacteria complete the degradation of the adsorbed organic matter.

The microbial communities in a water treatment plant convert organic carbon to carbon dioxide, water, and sludge; convert some 80% of the ammonia and nitrate to molecular nitrogen; remove some soluble phosphate through incorporation into the sludge, either as polyphosphate granules within bacterial cells or as struvite (crystalline $MgNH_4PO_4$); and remove pathogenic bacteria.

However, serious challenges in wastewater treatment have yet to be fully addressed. The level of residual fixed nitrogen compounds and of phosphate in the effluent is still high enough to pose risks of eutrophication in the receiving bodies of water. Residues of many widely used pharmaceuticals present in municipal wastewater are incompletely removed and emerge in the effluent. Some of these compounds are biologically active at nanograms per liter and have demonstrable undesirable environmental effects. The chemical industry uses thousands of synthetic organic compounds in huge amounts, and many of these (or their degradation products) pose a similar concern to the pharmaceutical ones. Finally, wastewater treatment consumes energy but converts much of the ammonia and nitrate to nitrogen gas, and a significant amount of the phosphate remains in the effluent. Alternative processes that would recover the fixed nitrogen compounds and phosphate would offset the energy and economic costs of manufacturing the corresponding amount of chemical fertilizer and lessen the acute environmental problems associated with elevated nutrient levels in aquatic ecosystems. Chapter 14 explores wastewater treatment and the above issues in some detail.

**Biological Oxygen Demand**

Maintenance of high oxygen concentration in aquatic ecosystems is essential for the survival of fish and other aquatic organisms. Decomposition of organic matter may rapidly deplete the oxygen. When organic matter such as untreated sewage is added to an aquatic ecosystem, it is degraded by bacteria that consume oxygen in the process. The biological oxygen demand (BOD) is related to the amount of organic matter in the water. Usually, the oxygen consumption is measured over a period of five days and is abbreviated $BOD_5$. $BOD_5$ for municipal wastewater generally ranges from 80 to 250 mg $O_2$ per liter. Appropriate secondary treatment decreases the $BOD_5$ to less than 20 mg $O_2$ per liter.

BOX 2.8

## BIOREMEDIATION

Bioremediation depends on the activities of living organisms to clean up pollutants dispersed in the environment. Physical or chemical treatments, such as vaporization, extraction, or adsorption, relocate rather than remove pollutants. In contrast, there are many instances in which biodegradation converts organic pollutants to harmless inorganic products, including carbon dioxide, water, and halide ions. Other advantages are that bioremediation is generally inexpensive and causes little disturbance to the environment. Naturally occurring consortia, frequently dominated by bacteria, have the capacity to degrade a wide spectrum of environmental pollutants.

Notably, such consortia are responsible for the cleanup of massive oil spills. There is a long list of oil spills with serious environmental impact. Following are three of many examples of this type of widely dispersed pollution. In March 1989, some 41 million liters (>10.5 million gallons) of crude oil escaped from the tanker *Exxon Valdez* and contaminated more than 2000 km (~1250 miles) of rocky intertidal coastline in Alaska. In 1991, during the Gulf War, huge amounts of oil were released into the marine environment, with devastating impact on marine life. In 1997, more than 5000 tons of heavy oil leaked from the Russian tanker *Nakhodka*, which ran aground and sank in the Sea of Japan. The oil contaminated more than 500 km (~310 miles) of the coastline. Over time, in all of these cases, the endogenous microbial community largely degraded the oil. In the case of the *Exxon Valdez*, the activity of the naturally occurring hydrocarbon-degrading bacteria at the spill site was enhanced by the addition of fertilizer containing organic nitrogen compounds and inorganic phosphorus compounds (Chapter 14, page 507).

Many thousands of organic and inorganic compounds are used daily around the world in hundreds of thousands of products. These compounds are introduced either accidentally or on purpose into the soil and groundwater. The seriousness of the problem posed by the introduction of human-made contaminants into the environment is highlighted by the following pronouncement by the Danish government in 2003:

> The government's most important goal with regard to chemicals is that by 2020 there should no longer be any products or goods on the market

containing chemicals with particularly problematic health or environmental impacts. (For reference, see page 88).

Among such pollutants, highly chlorinated compounds have received particular attention because of their known and potential adverse environmental and health impacts. One class of such compounds includes highly chlorinated aliphatics such as tetrachloroethene, trichloroethene, 1,1,1-trichloroethane, and carbon tetrachloride, which are used as dry cleaning fluids and degreasing solvents. Another class is represented by highly chlorinated aromatics such as pentachlorophenol (wood preservative), polychlorobiphenyls (insulators, heat exchangers), and dioxins (combustion by-products). These compounds are either fully or partially degraded by the combined activities of various endogenous microorganisms under aerobic or anaerobic conditions. By and large, the natural attenuation of chlorinated organic compounds at many different sites by the action of endogenous microbial populations, whether under aerobic or anaerobic conditions, is slow, is incomplete, and, in some cases, has resulted in the formation of toxic products. The complex subject of the biodegradation of these and other organic compounds is explored in detail in Chapter 14.

Cleanup of sites contaminated by radionuclides poses an exceptionally challenging problem of great importance. A U.S. Department of Energy (DOE) report summarizes the situation succinctly.

> With the end of the Cold War threat in the early '90s and the subsequent shutdown of all nuclear weapons production reactors in the United States, DOE has shifted its emphasis to remediation, decommissioning, and decontamination of the immense volumes of contaminated water and soils, and the over 7,000 structures spread over 120 sites (7,280 square kilometers) in 36 states and territories. DOE's environmental legacy includes 1.7 trillion gallons of contaminated ground water in 5,700 distinct plumes, 40 million cubic meters of contaminated soil and debris, and 3 million cubic meters of waste buried in landfills, trenches, and spill areas." Source: U.S. Department of Energy. (2003). *Bioremediation of Metals and Radionuclides. What Is It and How It Works*, LBNL-42595, 2nd Edition, p. 5, Washington, D.C.: Office of Biological and Environmental Research, Office of Science, U.S. Department of Energy.

Subsurface bioremediation of such sites has attracted much attention. A key objective is to stabilize the buried wastes in place to prevent leaching and widespread contamination of groundwater. The most common radioactive components in these wastes are uranium (U), strontium (Sr), plutonium (Pu), cesium (Cs), and technetium (Tc). Some important physical and chemical properties of these radionuclides are summarized in Table 2.10. Uranium can exist in the oxidation states +3, +4, +5, and +6. U(IV) is generally water insoluble and precipitates as uraninite, $(UO_2)^0$, or coffinite (a silicate mineral). U(VI) is water soluble and readily forms the soluble uranyl ion $(UO)_2{}^{2+}$. In contrast, U(IV) phosphates are quite insoluble. These properties of uranium form the basis for the two distinct microorganism-mediated *in situ* immobilization approaches explored below.

**TABLE 2.10 Long-lived and mobile radionuclides present in ground water, sediments, and soils at contaminated sites at a number of U.S. Department of Energy facilities**

| Element | Radioactive isotopes, emissions, and oxidation states | Comments |
|---|---|---|
| Uranium (U) | 17 isotopes; U-226 to U-242 with half-lives ranging from $10^5$ to $10^9$ years; $\alpha$, $\beta$, and $\gamma$ emissions Oxid. states: +3, +4, +5, +6 | U-235 is used for nuclear weapon production and as a source of energy in some nuclear reactors. Has a half-life of $7.13 \times 10^6$ years |
| Plutonium (Pu) | 15 isotopes; Pu-232 to Pu-246 with half-lives of $10^2$ to $10^3$ years; $\alpha$ and $\gamma$ emissions Oxid. states: +3, +4, +5, +6, +7 | Pu-239, with a half-life of 24,100 years is used in the production of nuclear fuel and nuclear weapons. Extremely radiotoxic if inhaled or injected. |
| Technetium (Tc) | 25 isotopes; Tc-90 to Tc-108; Tc-99 has a half-life of 212,000 years; $\alpha$ and $\beta$ emissions Oxid. states: +7 to 0 | Derived from U and Pu fission. Produced in kilogram amounts as a fission product in nuclear reactors. The Tc(VII) pertechnetate ion ($TcO_4^-$) is very stable in water under oxic conditions |
| Strontium (Sr) | The artificial isotope Sr-90 has a half-life of 28 years; $\beta$ emission Oxid. state: +2 | Because of its chemical similarity to calcium, Sr-90 can enter the food chain and concentrate in bones and teeth |
| Cesium (Cs) | 20 Cs isotopes; Cs-137 has a half-life of 30 years; $\beta$ emission Oxid. state: +1 | Because of its chemical similarity to potassium, it is taken up by organisms in the same manner |

*Source*: U.S. Department of Energy. (2003). *Bioremediation of Metals and Radionuclides. What Is It and How It Works*, 2nd Edition, LBNL-42595 (2003), A NABIR Primer, Washington, D.C.: Office of Biological and Environmental Research, Office of Science, U.S. Department of Energy.

### Uranyl Ion Immobilization Through Subsurface Reduction of U(VI) to U(IV) by *Geobacter* Species

*Geobacter* species ($\delta$-Proteobacteria) are important members of the subsurface biota. The complete genome sequence of *Geobacter sulfurreducens* is known. This organism generates ATP by oxidizing acetate to $CO_2$ using an electron transfer pathway with Fe(III) (present in abundance in the subsurface environment as Fe(III) oxides) as the terminal electron acceptor. Acetate is generated by other members of the subsurface biota (Figure 2.7). In the same manner, *G. sulfurreducens* can use U(VI) as an electron acceptor and generate the U(IV) that then forms insoluble compounds (see above).

### *In Situ* Immobilization of Uranyl Ion Mediated by *Pseudomonas aeruginosa* Polyphosphate Metabolism

The phosphate polymer, polyphosphate, with chain lengths made up of up to a few hundred phosphate monomers, is involved in heavy metal tolerance and removal in many microorganisms. As shown in Figure 2.8, polyphosphate is reversibly and processively synthesized by polyphosphate kinase (PPK), generally with ATP as the phosphoryl donor, and irreversibly and processively hydrolyzed by exopolyphosphatase (PPX).

*P. aeruginosa* strain HPN854 was engineered to express the *ppk* gene under the control of an inducible promoter. Upon overexpression of *ppk*,

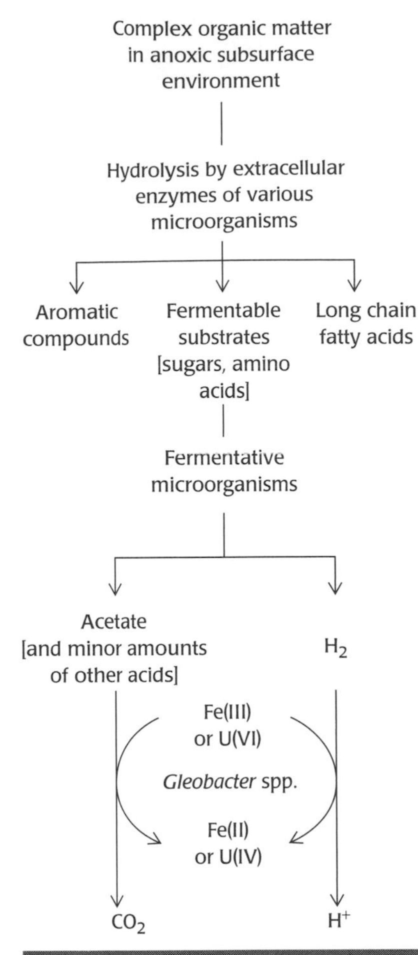


FIGURE 2.7

Subsurface microbial metabolism of complex organic matter under anoxic conditions generating acetate and hydrogen that are used as substrates by *Geobacter* species with Fe(III) or other metals, such as U(VI), as terminal electron acceptors.

this strain accumulated 100-fold higher levels of polyphosphate than did wild-type cells. When such polyphosphate-filled cells were suspended in a medium free of carbon substrates, phosphate was efficiently released. The degradation of the polymer was not mediated by either *ppk* or *ppx*, but rather was linked to the degradation of glycogen.

When 1 mM uranyl nitrate was added to a dense culture under the above conditions, a substantial amount of uranyl ion was immediately bound to hydroxyl groups on the surface of the bacterial cells. With gradual release of phosphate from the cells, the surface-bound uranyl was transformed to uranyl phosphate. The cells accumulated more than 40% of their dry weight as uranyl ion, with the balance of the uranyl ion forming small crystals of uranyl phosphate. Interestingly, the rate of release of phosphate was unaltered when the cells were exposed to a lethal dose of $^{60}Co$ radiation.

## BIOMINING: HEAVY METAL EXTRACTION USING MICROORGANISMS

Biomining utilizes naturally occurring prokaryotic communities. Here, microorganisms are used to leach metals, principally copper but also nickel and zinc, from low-grade sulfide- and/or iron-containing ores. The process exploits the energy metabolism of various acidophilic chemolithoautotrophs that utilize inorganic compounds as energy sources and $CO_2$ as the source of carbon. These organisms use either ferrous iron or sulfide as an electron donor and oxygen as an electron acceptor with the formation of ferric iron or sulfuric acid. In the first case, the subsequent reaction of $Fe^{3+}$ with insoluble metal sulfides yields soluble metal sulfates; in the second, metal sulfides are oxidized directly to metal sulfates. The metals are readily recovered from the leachate by electrolytic procedures, and the residual solution is recycled.

Gold is inert to microbial action. However, bioleaching of sulfidic gold-containing ores under acidic conditions opens up the interior of the ore particles to solvent. After bioleaching, the ore is rinsed with water and the gold is solubilized with a cyanide solution.

Current research on biomining is directed to improving understanding of the microbiology of the leaching process and to exploring the use of microbes that grow at high temperatures. Biomining is discussed in detail in Chapter 14.

## MICROBIAL DESULFURIZATION OF COAL

Coal contains substantial amounts of sulfur, both in pyrite ($FeS_2$) and in organic sulfur compounds (predominantly thiophene derivatives). The composition of coal varies considerably depending on the source. For example, Texas lignite coal contains 0.4% pyrite S and 0.8% organic S, whereas Illinois coal contains 1.2% pyrite S and 3.2% organic S, by weight. When coal is burned, most of this sulfur is converted to $SO_2$. The $SO_2$ combines with moisture in the atmosphere to form sulfurous acid ($H_2SO_3$), a major component of acid smog and acid rain.

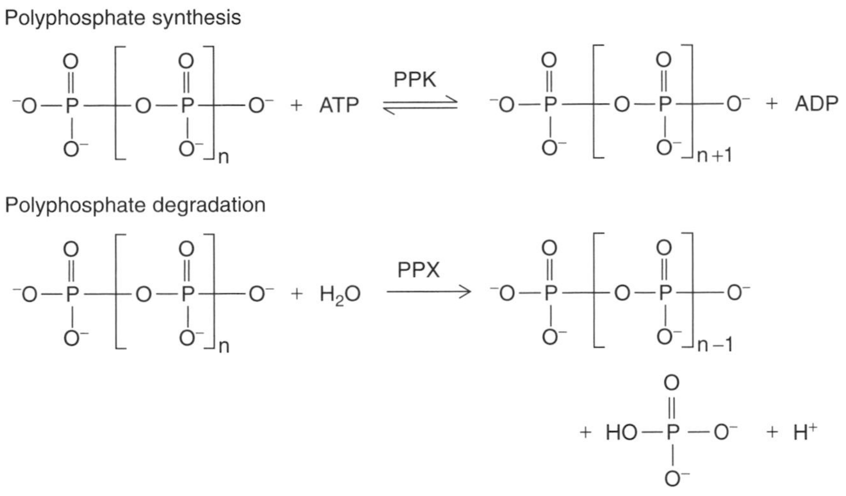

Polyphosphate synthesis

PPK, + ATP, + ADP

Polyphosphate degradation

PPX, + $H_2O$, + $H^+$

**FIGURE 2.8**

Microbial biosynthesis of polyphosphate catalyzed by polyphosphate kinase (PPK) and its degradation catalyzed by polyphosphate phosphatase (PPX). [Based on Lovley, D. R. (2003). In *The Prokaryotes*, Release 3.4. *The Prokaryotes* website at http://141.150.157.117.8080/prokPUB/chaprender/jsp/showchap.jsp?chapnum = 279.]

Microbial desulfurization of coal, by converting the pyrite to ferric sulfate and leaching it out of the coal (see "Biomining" earlier), provides one way of ameliorating this problem. As much as one or two weeks are required to complete the desulfurization, and large areas of land are required for the leach heaps and the storage of coal.

## FUNGAL REMOVAL OF PITCH IN PAPER PULP MANUFACTURING

In the paper manufacturing industry, treatment of wood with certain white rot fungi to degrade certain wood extractives before pulping substantially decreases the toxicity of pulp mill effluent toward aquatic organisms. Compounds that are extractable from wood with organic solvents make up between 1.5% and 5.5% of the dry weight of softwoods (angiosperms) and hardwoods (gymnosperms). These compounds, called *wood extractives*, consist mainly of triglycerides, fatty acids, diterpenoid resin acids (Figure 2.9), sterols, waxes, and sterol esters. Resin acids are present in most softwoods but are generally absent or are minor components in hardwood species (Table 2.11). During wood pulping and refining of paper pulp, the wood extractives are released, forming colloidal particles commonly referred to as *pitch* or *resin*. These colloidal particles form deposits in the pulp and in the machinery. These deposits can cause mill shutdowns and various quality defects in the finished paper products. Moreover, the resin constituents in pulp mill effluent show acute toxicity toward fish and aquatic organisms.

Pretreatment of the wood with fungi to degrade some of the wood extractives before pulping has met with considerable success. Basidiomycete fungi and *Ophiostoma* species colonize living and recently dead wood. Many of the species in this genus are referred to as sap-staining or blue-staining fungi because they stain colonized wood. To avoid this problem, a commercial fungal product, Cartapip, utilizes an "albino" strain of *Ophiostoma piliferum*. When applied to wood chip piles, this fungus has been particularly effective in degrading triglycerides and fatty acids in both softwoods and hardwoods,

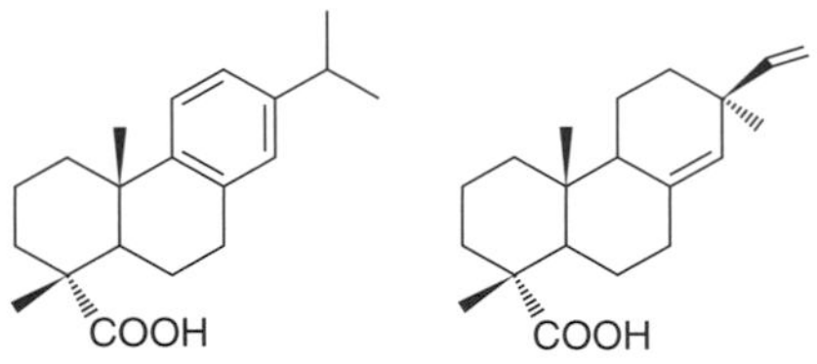

COOH COOH

**FIGURE 2.9**

Resin acids, found in wood extractives, are classified as abietanes and pimeranes. Abietanes have an isopropyl side chain at the C-13 carbon atom, whereas pimeranes have vinyl and methyl substituents at these positions. Members of these two classes are exemplified above by dehydroabietic acid (on the left) and pimaric acid (on the right). The majority of the acute toxicity of wastewater produced in the course of pulping of wood to extract fibers for paper manufacture is attributable to resin acids. [Source: Martindagger, V. J. J., Yu, Z., and Mohn, W. W. (1999). Recent advances in understanding resin acid biodegradation: microbial diversity and metabolism. *Archives of Microbiology*, 172, 131–138.]

**TABLE 2.11 Major components of wood extractives in softwoods and hardwoods**

| | Softwoods | | | Hardwoods | |
|---|---|---|---|---|---|
| | ***Pinus sylvestris*** **Pine** | ***Picea abies*** **Spruce** | **mg/g** | ***Populus tremula*** **Poplar** | ***Eucalyptus globulus*** **Eucalyptus** |
| Free fatty acids | 1.73 | 0.78 | | 1.06 | 0.28 |
| Resin acids | 6.65 | 2.85 | | 0.17 | 0 |
| Hydrocarbons | 0.74 | 0.19 | | 1.14 | 0.17 |
| Waxes or sterol esters | 0.83 | 0.87 | | 3.07 | 0.57 |
| Monoglycerides | 0.18 | 0.55 | | 1.18 | 0.02 |
| Diglycerides | 0.32 | 0.55 | | 0.58 | 0.02 |
| Triglycerides | 8.74 | 1.94 | | 10.37 | 0.13 |
| Higher alcohols or sterols | 1.39 | 1.00 | | 2.40 | 0.68 |

*Source*: Data from Gutiérrez, A., del Rio, J. C., Martinez, M. J., and Martinez, A. T. (2001). The biotechnological control of pitch in paper pulp manufacturing. *Trends in Biotechnology*, 19, 340–348.

but only partially effective in the removal of other pitch-forming compounds (sterols, sterol esters, and waxes) or the biotoxic resin acids. After four weeks of treatment at a moisture level of 70% on a wet wood weight basis at 27°C, *O. piliferum* produced up to a 50% reduction in the pitch content of softwoods, with less than a 5% loss of woody mass (Table 2.12). Moreover, the effluent biotoxicity was reduced 11- to 14-fold compared with untreated controls.

A number of white rot basidiomycete fungi are able to degrade the sterol esters and waxes. Several different bacteria, isolated by enrichment of pulp mill effluent, are able to degrade resin acids. There is now a substantial amount of work that demonstrates that fungi and bacteria, as well as enzymes derived from these organisms, are capable of minimizing pitch deposition during the pulping process and substantially decreasing the toxicity of the effluents.

## MICROBIAL WHOLE-CELL BIOREPORTERS

Over a quarter-century ago, luminescent bacteria were introduced as biosensors for the rapid assessment of toxic compounds in aquatic environments. The use of these organisms has now become "institutionalized" for a wide range of toxicological assays. These assays are versatile because the change in signal (bioluminescence) is linked directly to change in the global metabolism of the cell independent of the cause.

The advent of genetic manipulation by recombinant DNA technology has created a broad range of specific microbial biosensors. The great majority of these are genetically engineered bacteria within which a promoter–operator (the sensing element) responds to the stress condition (toxic organic or inorganic compound, DNA damage, etc.) and changes the level of expression of a reporter gene that codes for a protein (the signal). The protein

**TABLE 2.12 Total extractives (g/100 g) after two-week seasoning or Cartapip treatment of different pulpwoods (number in parentheses is the percentage decrease)**

| | Control | Seasoning[a] | Cartapip |
|---|---|---|---|
| *Pinus contorta* (Lodgepole pine) | 2.3 | 2.3 (0%) | 1.9 (17%) |
| *Populus tremuloides* (Quaking aspen) | 3.1 | 2.9 (6%) | 2.2 (29%) |
| *Pinus taeda* (Loblolly pine) | 2.6 | 2.0 (23%) | 1.4 (46%) |
| *Eucalyptus globulus* (Blue gum eucalyptus) | 1.5 | 1.2 (20%) | 0.8 (50%) |

[a] *Seasoning* refers to the storage of wood chips in a wood yard. During storage, wood extractives are lost through hydrolysis by plant enzymes, oxidative processes, and alteration by wood-colonizing organisms.

*Source*: Data from Gutiérrez, A., del Rio, J. C., Martinez, M. J., and Martinez, A. T. (2001). The biotechnological control of pitch in paper pulp manufacturing. *Trends in Biotechnology*, 19, 340–348.

may be detected either directly (e.g., green fluorescent protein) or through its catalytic activity (e.g., formation of a fluorescent or chemiluminescent product).

## *VIBRIO FISCHERI* CYTOTOXICITY TEST

Toxicological assays that depend on the bioluminescence of *V. fischeri* NRRLB-11177 are used widely in detecting contaminants in aquatic environments, monitoring wastewater treatment, and generally in assessing the relative cytotoxicity of a wide range of compounds that are released into the environment as a direct or indirect consequence of human activity.

The *V. fischeri* cytotoxicity assay is described in Box 2.9 and illustrated in Figure 2.10. Because the intensity of bioluminescence is dependent on the intracellular levels of ATP and NADPH, the assay effectively monitors the metabolic status of the cell. Consequently, damage to the cytoplasmic membrane, interference with transport processes that bring metabolites into the cell, interference with electron transport systems, and other perturbation of the ion gradients across the cytoplasmic membrane all result in a decrease in bioluminescence. This strong point of the assay is also a weakness. In itself, the assay provides no information on the nature of the toxic effect or the molecular target affected by the analyte. However, the assay has proved to serve as a useful indicator of toxicity of a wide variety of compounds to aquatic organisms.

## REPORTER GENE BIOASSAYS

The above limitations are addressed by assays designed to detect specific molecules. Following are three examples from among dozens of such assays.

Certain *Staphylococcus aureus* strains carry plasmid pI258, which contains the operon *cadAcadC*. This operon confers resistance to $Cd^{2+}$ and $Zn^{2+}$, and $Cd^{2+}$ also acts as an inducer. The bioluminescence of *S. aureus*,

*Vibrio fischeri* NRRLB-11177 is a naturally bioluminescent marine bacterium. The bioluminescence results from an oxidoreductase (reaction 1) and a luciferase (reaction 2) catalyzed reaction sequence:

1. $NADPH + FMN \Leftrightarrow NADP^+ + FMNH_2$
2. $FMNH_2 + RCHO + O_2 \Rightarrow FMN + RCO_2H + H_2O +$ **light**,

and RCHO is palmitaldehyde.

Palmitaldehyde is regenerated by the following reaction:

$$RCO_2^- + NADPH + 2H^+ + ATP \Rightarrow RCHO + NADP^+ + H_2O + AMP + PPi.$$

The **cytotoxicity assay** is performed as follows:

1. Freeze-dried cells are reconstituted in buffer and incubated at the desired assay temperature.
2. Equal volumes of solution of analyte at different concentrations are added to equal aliquots of bacterial suspension.
3. Luminescence of these solutions and of a control solution (lacking analyte) is measured as shown in Figure 2.10.
4. The percent inhibition (I) is given by

$$I = [(I_c - I_a)/I_a] \times 100,$$

where $I_c$ is the bioluminescence of the control solution and $I_a$ that of a solution containing analyte. The analyte concentration that gives 50% inhibition of bioluminescence (designated $EC_{50}$) provides a quantitative measure of toxicity under the conditions of this assay.

BOX 2.9

engineered to carry a construct in which the luciferase genes, *luxAB* of *Vibrio harveyi*, are placed under the control of the *cadA* promoter, allows detection of $Cd^{2+}$ over a concentration range of 1 to 100 $\mu$M.

The *Pseudomonas oleovorans* pathway for octane sensing consists of a transcriptional activator, encoded by *alkS*, which activates the *alkB* promoter in the presence of linear alkanes with chain lengths ranging from $C_6$ to $C_{12}$. This activator/promoter system can be utilized to express green fluorescent protein (GFP) in *E. coli*. Wild-type GFP is quickly degraded, so a particularly stable mutant of GFP was used in the engineered octane-sensing *E. coli* strain. The bioluminescence of the octane-sensing *E. coli* showed a dose-dependent response range from 0.01 to 0.1 $\mu$M octane and allowed monitoring of mass transfer of octane through the gas phase or by diffusion from microdroplets through water.

*Erwinia herbicola* 299R is a colonizer of the plant leaf surface. This epiphytic bacterium was converted to a whole-cell sensor for local sugar availability. The bacterium was transformed with a plasmid, $pP_{fruB}$-gfp[AAV], in which the promoter region of the operon responsible for fructose utilization in *E. coli* was fused to a variant of GFP that folds faster than wild-type GFP, gives a brighter fluorescence, and has a significantly reduced stability. These properties make the fluorescence of the engineered *E. herbicola* strain track closely the rate and level of the GFP gene expression. The

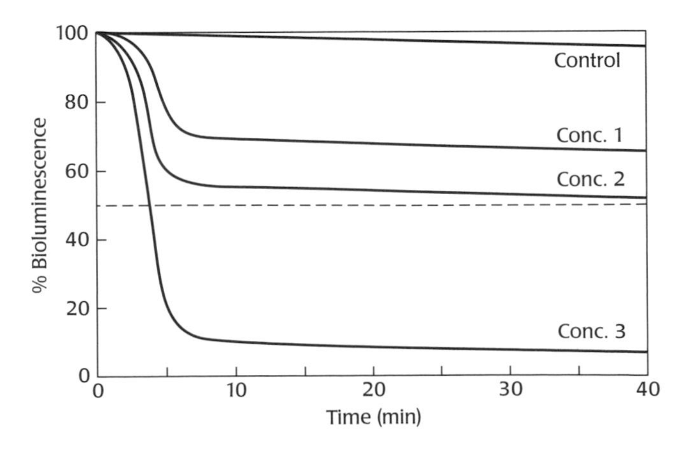


FIGURE 2.10

Time-course and extent of bioluminescence reduction of cells of *Vibrio fischeri* NRRLB-11177 under the conditions of the *V. fischeri* cytotoxicity test (see Box 2.9) in the presence of increasing concentration (from Conc. 1 to Conc. 3) of the analyte being assayed.

engineered strain was sprayed on the surface of bean plant leaves and collected by rinsing sample leaves at intervals after one to 24 hours. The intensity of fluorescence emission from individual cells, measured by epifluorescence microscopy, provided information on the level of sugar on the leaves at the various times. The results showed that the sugar level was relatively high in the initial phases of the experiment and then declined as the bacteria multiplied. Such single-cell sensors have enormous potential in the study of interactions between microorganisms and their hosts, as well as those among microorganisms (Box 2.10).

"From the perspective of a bacterium 2 $\mu$m in length, the surface of a matchbox represents an area roughly the size of Rhode island, or the Grand Duchy of Luxembourg. To the same bacterium, an Italian espresso compares in volume to Lake Baikal, the largest freshwater lake on Earth, while a hot air balloon takes on the proportions of the Earth itself."

*Source:* Leveau, J. H. J., and Lindow, S. E. (2002). Bioreporters in microbial ecology. *Current Opinion in Microbiology*, 5, 259–265.

**BOX 2.10**

## ORGANIC CHEMISTRY

The capabilities of microorganisms to catalyze chemical reactions are immense. The sum total of microbial biosynthetic pathways generates an extraordinary diversity and number of organic compounds, simple and complex, low molecular weight and polymeric. Moreover, microorganisms are able to degrade all these "natural products" to compounds that support the growth of living organisms. The chemical and pharmaceutical industries face problems of persistent, widespread environmental pollution by synthetic organic compounds, the depletion of nonrenewable oil resources coupled with the rising cost of petroleum-based products, and the rapid increase in the concentration of greenhouse gases in the atmosphere. For solutions to these problems, chemists have greatly increased their use of the microbial toolbox of metabolites and enzymes. Recent developments show that organic chemistry will increasingly and broadly look to microbial biotechnology for catalysts and processes that enable original ways of solving acute environmental problems and that are more efficient and cheaper than those used by the industry throughout the twentieth century.

### FEEDSTOCK CHEMICALS

Feedstock chemicals are the basic building blocks that serve as the raw materials used to synthesize other chemicals, ranging from small molecules to plastics and rubber, or that are used as solvents in a variety of industrial processes. The primary products of petroleum refining, such as ethylene, propylene, benzene, toluene, and xylenes, are the dominant feedstocks for the chemical industry. These compounds and their derivatives account for over 97% of synthetic organic chemicals; their production in the United States exceeds 200 billion pounds. Approximately 7% of petroleum is used to make chemicals. Alternative renewable sources of feedstock chemicals are needed to conserve world oil reserves and, because of concern about global warming, to minimize the increase in atmospheric carbon dioxide. What are the prospects of our finding alternative abundant sources of feedstock chemicals?

The biomass produced by photosynthetic organisms is estimated at 200 billion tons annually; of this, humans use only 3% to 4% as food or to other

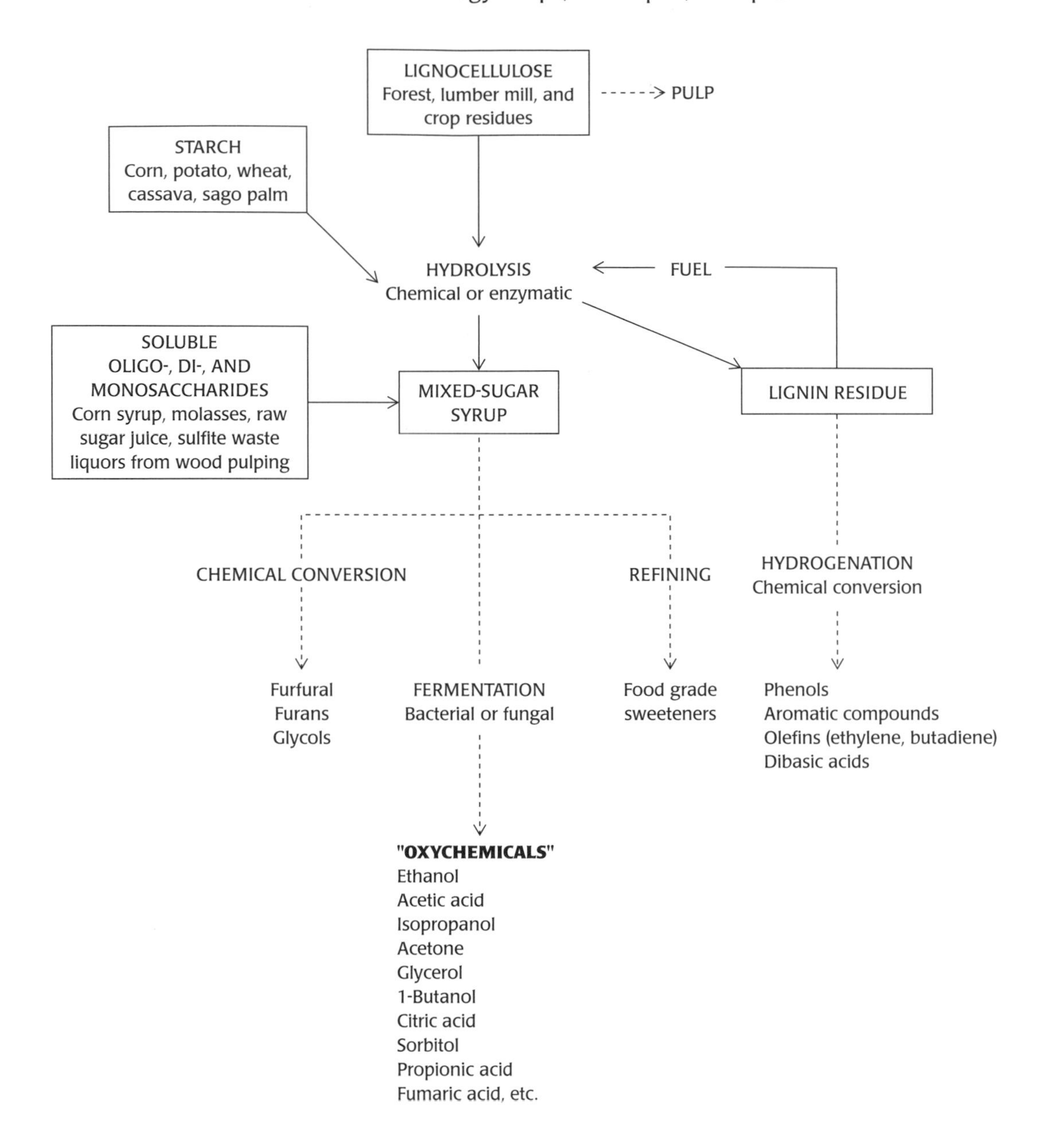


**FIGURE 2.11**

Renewable sources of key feedstock chemicals. [Based on Busche, R. M. (1985). The business of biomass. *Biotechnology Progress*, 1, 165–180.]

ends. Plants produce an immense supply of carbohydrate-rich materials, including lignocellulose, the main structural component of wood; the starch in corn, wheat, potatoes, cassava, and so on; and the sugars in corn syrup and molasses. In principle, plant matter represents an abundant, inexpensive source of organic matter that could be converted to primary feedstock chemicals by a combination of microbial fermentation and chemical processes (Figure 2.11). In several countries, under abnormal conditions of supply and demand during World Wars I and II, certain organic chemicals were produced on a large scale by microbial fermentation. In 1975, in Brazil, a combination of high oil prices, the need to conserve foreign currency, and a greatly depressed world market price for sugar triggered the creation of a large-scale government-sponsored industry to produce ethanol by fermentation. Worldwide, however, only a very small fraction of available biomass is actually utilized to such ends.

The major primary feedstocks (ethylene, propylene, benzene, toluene, xylenes, propane, ethane, ethylene, butadiene, *n*-butane cyclohexane, isobutene, and isoprene) are all hydrocarbons. The end products obtained by direct bacterial or fungal degradation of biomass are all "oxychemicals" (compounds containing oxygen as well as hydrogen and carbon; Figure 2.11). Therefore, a dehydration step is required in the conversion of a fermentation product (such as ethanol) to a hydrocarbon feedstock (such as ethylene). In order for such use of organic matter to compete with petrochemicals, the combined cost of the fermentation process, including recovery of the fermentation product, and of the subsequent dehydration process must not exceed that of using petrochemicals. At the prevailing oil and biomass feedstock prices, the petrochemical process is cheaper.

When the oxychemicals themselves are desirable as feedstocks, the economic picture is more favorable. Some useful or potentially useful ones are listed in Figure 2.11. Although all are obtainable by fermentation, a number of these compounds are produced from petrochemicals at lower cost. For example, fumaric acid, once manufactured by large-scale fermentation with a strain of the fungus *Rhizopus*, is produced more cheaply via the catalytic oxidation of benzene or butane. Microbiological processes are currently used in the large-scale industrial manufacture of some chemicals. These are exemplified by ethanol, monosodium glutamate, citric acid, lysine, acrylamide, fructose, malic acid, and aspartic acid. However, these chemicals are but a small fraction of the universe of products of the organic chemical industry.

The compounds listed in Figure 2.11 and their derivatives represent, by weight, close to one half the amount of the 100 industrial organic chemicals made in the largest quantity. Future shift toward greater production by a biomass-based chemical industry will depend strictly on economics rather than on feasibility. The example that follows suggests that the present situation may largely persist for some 15 to 20 years into the future.

## INDUSTRIAL MANUFACTURE OF ACETIC ACID

Acetic acid endows vinegar with its characteristic odor and sour taste. Vinegar contains about 4% to 8% acetic acid and is prepared from wine or other dilute solutions of alcohol. Acetic acid bacteria belonging to the genera *Acetobacter* and *Gluconobacter* are unique organisms that tolerate high acetic acid and ethanol concentrations and, in the presence of oxygen, generate vinegar by oxidizing ethanol to acetic acid:

$$CH_3CH_2OH + O_2 \rightarrow CH_3COOH + H_2O.$$

Because of its high freezing point of 16.7°C, pure acetic acid, in either its solid or liquid state, is referred to as glacial (icelike). Glacial acetic acid has a boiling point of 118°C and can be isolated from vinegar in pure form by distillation. However, for industrial use, glacial acetic acid is still synthesized more cheaply chemically from carbon monoxide and methanol. The global production of acetic acid is about 8 million tons per year. Acetic acid is used primarily in the manufacture of a variety of acetate esters. Among many others, these include cellulose acetate (used to make films and textiles), vinyl acetate (the building block of polyvinyl acetate), and aspirin (the acetate ester of salicylic acid). Acetic anhydride ($CH_3C(O)$-O-$(O)CCH_3$), synthesized from acetic acid, is the reagent used in the synthesis of the majority of acetate esters.

In the most widely used industrial chemical process, acetic acid is synthesized from methanol and carbon monoxide:

$$CH_3OH + CO \xrightarrow[\text{150°C to 250°C, 1.0 to 1.5 atm}]{\text{Rhodium iodide catalyst}} CH_3COOH.$$

A new plant built by the German chemical company Celanese in Nanjing in eastern China will produce 600,000 tons per year of acetic acid from methane by a high throughput process. In the first step, oxidation of methane with oxygen yields methanol and carbon monoxide:

$$2CH_4 + 2O_2 \xrightarrow{\text{350°C to 500°C, 60 to 100 atm}} CH_3OH + CO + 2H_2O.$$

In the second step, the methanol and carbon dioxide, with rhodium iodide as the catalyst, react to form acetic acid, as shown earlier. A 15-year agreement is in place for a supply of methane to the plant. Obviously, in the near term, it is not expected that a microbial process for acetic acid production will replace the chemical route.

## GREEN CHEMISTRY: A PARADIGM SHIFT

> Second Law of Thermodynamics: "Heat cannot of itself, without the intervention of any external agency, pass from a colder to a hotter body." – Clausius, R. (1854).
>
> "In a world rapidly running out of fossil fuel, the second law of thermodynamics may well turn out to be the central scientific truth of the 21st century." – Gutstein, D. (1994). Chance and necessity. *Nature*. 368, 598.

As the twentieth century entered its final decade, a number of nagging concerns coalesced to give birth to the subdiscipline of "green chemistry." Before exploring the foundations of green chemistry, let us examine the drivers behind its creation.

The twentieth century saw rapid population growth paralleled by an unprecedented rate of technological innovation. The population growth made necessary, and new technologies made possible, an explosive growth in the chemical industry. By the mid-twentieth century, petroleum became the source of the great majority of organic compounds. Today, the global chemical and pharmaceutical industries employ more than 10 million people and contribute about 9% of world trade.

In the latter part of the twentieth century, several pressing problems demanded urgent attention. Foremost was the growing rate of energy consumption in the face of increasingly rapid depletion of nonrenewable resources. Then there was the realization that global climate change, driven in significant part by carbon dioxide emissions from the combustion of fossil fuels, posed a major threat. Another problem was the omnipresence of human-generated toxic materials in the environment. At the same time, increasing regulatory requirements imposed increasing demands on industry. In the oil and chemical industries, waste treatment and disposal, site remediation, environmental health and safety, and legal costs have risen to represent 15% to 30% of capital expenditures.

In 1990, in the United States, the Pollution Prevention Act established a national policy to prevent or reduce pollution at its source when feasible. In 1991, the U.S. Environmental Agency Office of Pollution Prevention and Toxics launched a research grant program called "Alternative Synthetic Pathways for Pollution Prevention." This program supported research projects that included pollution prevention as an objective in the design and synthesis of chemicals. This program essentially defined green chemistry as "the design of chemical products and processes that reduce or eliminate the use and generation of hazardous substances." In an influential book published in 1998, *Green Chemistry: Theory and Practice*, Paul Anastas and John Warner formulated a set of 12 principles of green chemistry (Table 2.13). These principles gained rapid and wide acceptance. The concept of "atom economy" deserves special comment. It emphasizes the importance of designing reactions to maximize the amounts of all starting materials that finish up in the product. In the example shown in Figure 2.12, conversion of cyclohexanone to methylenecycloxane by the Wittig reaction gives a product yield of 86%. However, the outcome of this reaction from the perspective of atom economy is very poor;

**TABLE 2.13 The 12 principles of green chemistry**

**1.** It is better to prevent waste than to treat or clean up waste after it is formed.

**2.** Synthetic methods should be designed to maximize the incorporation of all materials used in the process into the final product.

**3.** Wherever practicable, synthetic methods should be designed to use and generate substances that possess little or no toxicity to human health and the environment.

**4.** Chemical products should be designed to preserve efficacy of function while reducing toxicity.

**5.** The use of auxiliary substances (e.g., solvents, separation agents) should be made unnecessary wherever possible, and innocuous when used.

**6.** Energy requirements should be recognized for their environmental and economic impacts and should be minimized. Synthetic methods should be conducted at ambient temperature and pressure.

**7.** A raw material or feedstock should be renewable rather than depleting whenever technically and economically practicable.

**8.** Unnecessary derivatization (blocking groups, protection/deprotection, temporary modification of physical/chemical processes) should be avoided whenever possible.

**9.** Catalytic reagents (as selective as possible) are superior to stoichiometric reagents.

**10.** Chemical products should be designed so that at the end of their function they do not persist in the environment and break down into innocuous degradation products.

**11.** Analytical methodologies need to be further developed to allow for real-time, in-process monitoring and control prior to the formation of hazardous substances.

**12.** Substances and the form of a substance used in a chemical process should be chosen to minimize the potential for chemical accidents, including releases, explosions, and fires.

*Source*: Anastas, P. T., Warner, J. C. (1998). *Green Chemistry: Theory and Practice*, p. 30, Fig. 4.1, New York: Oxford University Press; by permission of Oxford University Press.

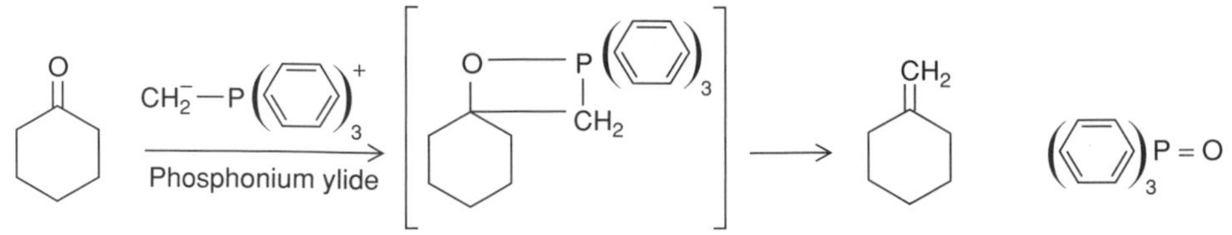

| | Cyclohexanone | Phosphonium ylide | Methylene-cyclohexane | Phosphine oxide by-product |
|---|---|---|---|---|
| Formula | $C_6H_{10}O$ | $C_{19}H_{17}P$ | $C_7H_{12}$ | $C_{18}H_{15}PO$ |
| Mass/g mol$^{-1}$ | 98 | 276 | 96 | 278 |

$$\% \text{ atom economy} = \frac{\text{formula weight of all atoms used}}{\text{formula weight of all reactants used}} \times \frac{100}{1}$$
$$= \frac{C_7H_{12}}{C_6H_{10}O + C_{19}H_{17}P} \times \frac{100}{1}$$
$$= \frac{96}{98 \quad 276} \times \frac{100}{1}$$
$$= 26\%$$

**FIGURE 2.12**

Atom economy for the conversion of cyclohexanone to methylene-cyclohexane by the Wittig reaction. [From Grant, S., Freer, A. A., Winfield, J. M., Gray, C., and Lennon, D. (2005). Introducing undergraduates to green chemistry: an interactive teaching exercise. *Green Chemistry*, 7, 121–128.]

the stoichiometric formation of the by-product phosphine oxide results in an atom economy for this reaction of only 26%.

Broad adherence to the principles of green chemistry strongly favors the use of the tools of microbial biotechnology. Microbial processes and reactions that exploit biocatalysis conform to these principles. For example, these processes and reactions are generally run in water, mostly at or near room temperature, and at atmospheric pressure. Relatively low amounts of energy are needed. Toxic metal ions are not employed, and the by-products are readily biodegradable. Generally, no protective groups need be used.

Indeed, there is now a clear trend to design processes and synthetic schemes that take advantage of the above attributes of microbial biotechnology, and, in particular, of the high specificity exhibited in enzyme-catalyzed reactions. Below, we provide two examples of the influence of green chemistry on industrial processes.

## POLYLACTIDE PRODUCTION FROM AGRICULTURAL FEEDSTOCKS

We noted above that some 7% of petroleum is used to make organic chemicals. A large fraction of these chemicals is used to provide the monomeric building blocks for a wide variety of polymers. Products made from polymers or that contain polymers are omnipresent in the modern world. More than 150 million tons of plastics are synthesized annually. Following are a few examples of the many widely encountered synthetic polymers.

*Styrene-butadiene* rubber is one of the most widely used polymers. A partial list of its uses includes the production of tires, conveyor belts, brake and clutch pads, extruded gaskets, hard rubber battery box cases, shoe soles and heels, molded rubber goods, cable insulation, and food packaging. *Polystyrene* is a hard, rigid solid. Foamed polystyrene is used extensively

| Polymer | Formula | Monomer(s) | |
|---|---|---|---|
| Styrene-butadiene rubber | 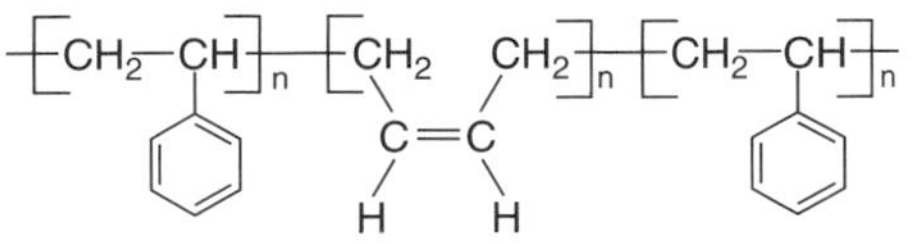 $-[CH_2-CH(C_6H_5)]_n-[CH_2-CH=CH-CH_2]_n-[CH_2-CH(C_6H_5)]_n-$ | styrene | $H_2C=CHC_6H_5$ |
| | | butadiene | $H_2C=CH-CH=CH_2$ |
| Polystyrene | 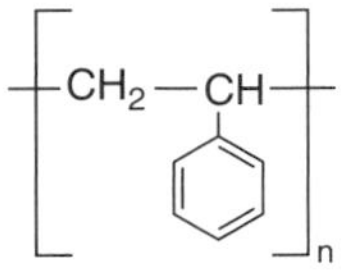 $-[CH_2-CH(C_6H_5)]_n-$ | styrene | $H_2C=CHC_6H_5$ |
| Polyvinylchloride | 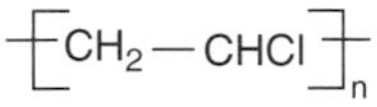 $-[CH_2-CHCl]_n-$ | vinylchloride | $H_2C=CHCl$ |
| Polytetrafluoroethylene | 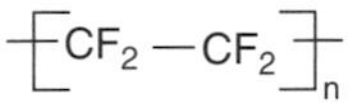 $-[CF_2-CF_2]_n-$ | tetrafluoroethylene | $CF_2=CF_2$ |
| Polyethylene terephthalate | 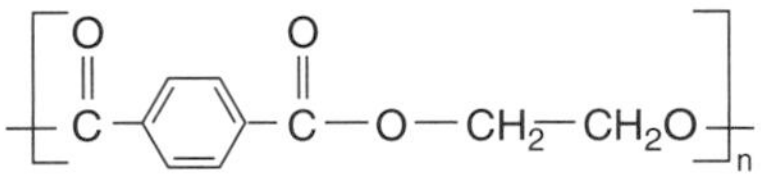 $-[C(=O)-C_6H_4-C(=O)-O-CH_2-CH_2O]_n-$ | ethylene glycol | $HOH_2C-CH_2OH$ |
| | | terephthalic acid | $HOOC-C_6H_4-COOH$ |

FIGURE 2.13

Formulae and monomer building blocks of some widely used chemically synthesized polymers.

in the production of packaging. This material persists in the environment and is a highly visible component of the waste that washes up on beaches. Polystyrene is also used in injection molding, insulation, and lamination. *Polyvinyl chloride*, a widely used plastic, is notable for the fact that the approximately 25 million tons manufactured annually account for some 40% of global chlorine consumption. Polyvinyl chloride is produced as fiber, foam, or film. It finds many outdoor applications because it is water repellent and resists weathering. *Polytetrafluoroethylene*, better known as Teflon, is used to make no-stick surfaces and electrical insulation. *Polyethylene terephthalate* (PET) is one of the most widely used polyester polymers. Lightweight, recyclable water and soft drink bottles are made from PET.

Note that the first four of the above polymers do not contain oxygen (Figure 2.13). As discussed earlier, the oxychemicals produced by microorganisms cannot compete with petrochemicals as feedstocks for the manufacture of such polymers. However, it should be possible to arrive at oxychemical precursors for novel polymers that can be used for some of the applications that currently employ polymers such as polyethylene terephthalate and that, in addition, may have properties that allow them to partially replace the other types of polymers as well. We examine polylactic acid (PLA) from that perspective.

The starting material for the Cargill Dow PLA manufacturing process is cornstarch. The starch is broken down to glucose with microbial enzymes.

FIGURE 2.14

Preparation of polylactide polymer (L-PLA) from L-lactic acid. [From Vink, E. T. H., Rábago, K. R, Glassner, D. A., and Gruber, P. R. (2003). Applications of life cycle assessment to NatureWorks polylactide (PLA) production. *Polymer Degradation and Stability*, 80, 403–419.]

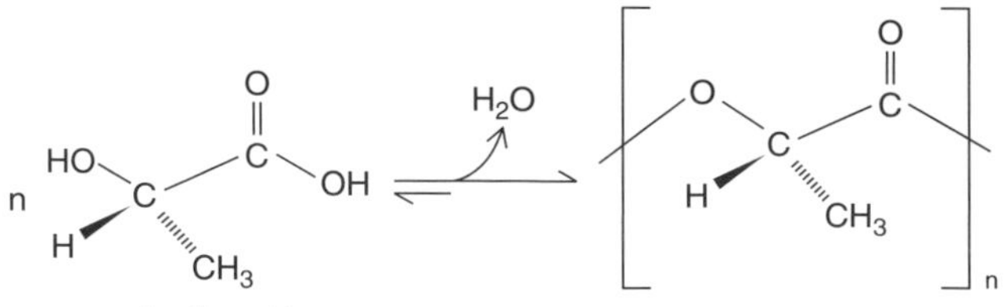

The glucose is then fermented to lactic acid by incubation with an acid-tolerant homolactic bacterial strain. This bacterium was isolated from the corn steep water at a commercial corn milling facility. In a nutrient medium containing initially 50 g/L of carbohydrate, it is able to generate about 40 g/L of lactic acid at final incubation pH of about 4.0. The lactic acid produced by this fermentation consists of 99.5% of the L-isomer and 0.5% of the D-isomer, whereas chemical synthesis of lactic acid yields the racemic mixture of 50% L and 50% D.

The polymer, L-PLA, is made by polymerization of L-lactic acid (Figure 2.14). L-PLA fibers and films are fire retardant and stain resistant and have many uses, such as packaging, paper coating, apparel, furnishings, fiberfill, and carpets. Bottles and other containers are produced from molded L-PLA. Discarded products can be readily recycled. Moreover, when composted at 60°C, the degradation of L-PLA is complete in 40 days, as determined from quantitative $CO_2$ recovery. Thus, this polymer is appropriate for many applications, such as agricultural mulch films and bags, in which recovery of the product is not practical.

L-PLA exemplifies a product that meets many of the requirements of green chemistry. It is synthesized entirely from renewable materials (sugars from corn, sugar beets, etc.). In the L-lactic acid polymerization, water is produced as the only by-product (Figure 2.14). The polymer is readily recycled back to the monomer and can be quantitatively biodegraded to $CO_2$ and water by bacteria and fungi in the natural environment.

## COMPARISON OF CHEMICAL AND ENZYMATIC SYNTHESES OF 6-AMINOPENICILLANIC ACID

Enzyme catalysis has gained a wide acceptance and an ever-expanding role in organic chemistry. From the perspective of green chemistry, enzyme-catalyzed reactions offer many highly desirable features. Enzymes as catalysts are efficient and generally show both stereo- and regioselectivity. Moreover, they can be produced in large quantity by recombinant DNA technology and are biodegradable. Enzyme-catalyzed reactions mostly proceed at moderate temperatures and near-neutral pH. Chemical functional group activation is generally not needed, and the protection and deprotection steps characteristic of chemical organic synthesis are avoided. Consequently, enzyme-catalyzed reactions show astounding reagent economy. Some of these features are illustrated by a simple example, a comparison of the chemical versus the enzymatic cleavage of the amide linkage in penicillin G

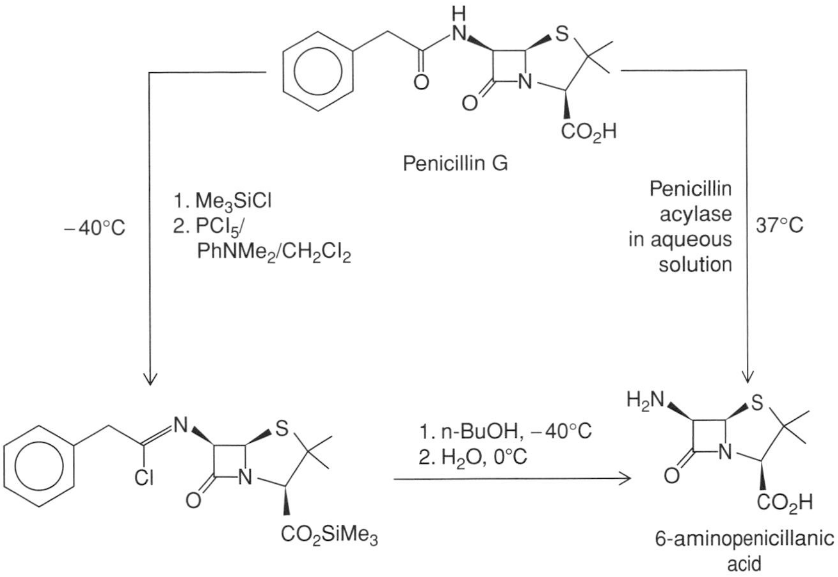

**FIGURE 2.15**

Comparison of chemical and enzymatic processes for the conversion of penicillin G to 6-aminopenicillanic acid. [From Sheldon, R. A., and van Rantwijk, F. (2004). Biocatalysis for sustainable organic synthesis. *Australian Journal of Chemistry*, 57, 281–289.]

(produced by fermentation with *Penicillium chrysogenum*) to yield 6-aminopenicillanic acid (6-APA), a key intermediate in the synthesis of semisynthetic penicillins.

The multistep chemical conversion is shown in Figure 2.15 along with the one-step conversion catalyzed by the enzyme penicillin acylase. The enzyme-catalyzed reaction superseded the chemical synthesis in the 1980s, when highly stable penicillin acylases became available, were produced in large amounts, and were immobilized on solid supports to allow reuse of the enzyme. The enzyme-catalyzed synthesis was much cheaper than the chemical one and, as detailed below, resulted in an impressive decrease in chemical waste.

The chemical synthesis of 1 kg of 6-APA requires 0.6 kg of trimethylsilylchloride ($Me_3SiCl$), 1.2 kg of phosphorus pentachloride ($PCl_5$), 1.6 kg of N,N′-dimethylaniline ($PhNMe_2$), 0.2 kg of ammonia ($NH_3$), 8.4 L of *n*-butanol (*n*-BuOH), and 8.4 L of dichloromethylene ($CH_2Cl_2$). The reactions are carried out at −40°C.

The enzyme-catalyzed synthesis of 1 kg of 6-APA is performed in 2 L of water and 0.09 kg of ammonia. In contrast to the numerous waste products from the chemical synthesis that need to be either recovered or discarded, the waste components of the enzyme-catalyzed reaction, ammonia and phenylacetic acid, are both readily utilized by living organisms. Whereas the chemical reactions are carried out −40°C, the enzyme-catalyzed reaction proceeds at 37°C, thus imposing a very much smaller energy requirement.

## SUMMARY

Microbial biotechnology crossed a threshold into a world of new possibilities with the advent of whole-genome shotgun sequencing. The number of fully sequenced genomes of prokaryotes and fungi greatly exceeds the sum of all other known genomes. The same statement holds for the number of individual genes of known sequence. In parallel with the explosive growth in the database of genomic information is the advent of techniques that allow rapid analysis of gene expression patterns in organisms exposed to different challenges, whether environmental or chemical.

In sum, the wealth of the pathways and products of microbial metabolism is enormous. Microorganisms are efficient factories for the production of macromolecules and a multitude of unique small molecules. It should come as no surprise that prokaryotes and fungi are growing ever more central in contributing to human therapeutics, agriculture, food technology, environmental procedures (wastewater treatment, bioremediation, heavy metal extraction, etc.), and organic chemistry, and as uniquely versatile whole-cell bioreporters in toxicology and other contexts. For each of these areas, case histories are provided to illustrate the diverse key contributions of microbial biotechnology.

A range of challenges posed by the threats of rapid global warming, of growing and widespread environmental pollution by toxic synthetic organic compounds and heavy metals, and of the depletion of petroleum reserves have led to the acceptance of the inevitability of a transition to green chemistry. "Green chemistry is the use of chemistry for pollution prevention. More specifically, green chemistry is the design of chemical products and processes that reduce or eliminate the use and generation of hazardous substances" (http://www.epa.gov/greenchemistry/whats_gc.html). The gradual implementation of green chemistry has resulted in the rapid growth in the number of large-scale microbial fermentations and the rising dominance in the fraction of chemical syntheses that employ enzymes as catalysts.

## SELECTED REFERENCES AND ONLINE RESOURCES

### General

Lederberg, J. (ed.) (2000). *Encyclopedia of Microbiology*, San Diego: Academic Press.

Ratledge, C., and Kristiansen, B. (eds.) (2001). *Basic Biotechnology*, 2nd Edition, Cambridge: Cambridge University Press.

Laird, S. A., and ten Kate, K. (1999). *The Commercial Use of Biodiversity. Access to Genetic Resources and Benefit-Sharing*, London: Earthscan Publications Ltd.

Demain, A. L., and Davies, J. E. (eds.) (1999). *Manual of Industrial Microbiology and Biotechnology*, 2nd Edition, Washington, D.C.: ASM Press.

Demain, A. L. (1999). Pharmaceutically active secondary metabolites of microorganisms. *Applied Microbiology and Biotechnology*, 52, 455–463.

### Human Therapeutics

Swartz, J. R. (2001). Advances in *Escherichia coli* production of therapeutic proteins. *Current Opinion in Biotechnology*, 12, 195–201.

Prather, K. J., Sagar, S., Murphy, J., and Chartrain, M. (2003). Industrial scale production of plasmid DNA for vaccine and gene therapy: plasmid design, production and purification. *Enzyme and Microbial Technology*, 33, 865–883.

Cordell, G. A. (2000). Biodiversity and drug discovery – a symbiotic relationship. *Phytochemistry*, 55, 463–480.

Ikeda, H., Nonomiya, T., Usami, M., Ohta, T., and Omura, S. (1999). Organization of the biosynthetic gene cluster for the polyketide anthelminthic macrolide avermectin in *Streptomyces avermitilis*. *Proceedings of the National Academy of Sciences USA*, 96, 9509–9514.

Ikeda, H., et al. (2003). Complete genome sequence and comparative analysis of the industrial microorganism *Streptomyces avermitilis*. *Nature Biotechnology*, 21, 526–531.

Strobel, G. A. (2002). Rainforest endophytes and bioactive products. *Critical Reviews in Biotechnology*, 22, 315–333.

Orr, G. A., Verdier-Pinard, P., McDaid, H., and Horwitz, S. B. (2003). Mechanisms of taxol resistance related to microtubules. *Oncogene*, 22, 7280–7295.

### Agriculture

Slater, A., Scott N. W., and Fowler, M. R. (2003). *Plant Biotechnology: the Genetic Manipulation of Plants*, Oxford: Oxford University Press.

Garg, A. K., et al. (2002). Trehalose accumulation in rice plants confers high tolerance levels to different abiotic stresses. *PNAS*, 99, 15898–15903.

Gonsalves, D. (1988). Resistance to papaya ringspot virus. *Annual Review of Phytopathology*, 36, 415–437.

Gonsalves, D. (2002). Coat protein transgenic papaya: "acquired" immunity for controlling papaya ringspot virus. *Current Topics in Microbiology and Immunology*, 266, 73–83.

Lindbo, J. A., and Dougherty, W. G. (2005). Plant pathology and RNAi: a brief history. *Annual Review of Phytopathology*, 43, 191–204.

Russell, B. J., and Houlihan, A. J. (2003). Ionophore resistance of ruminal bacteria and its potential impact on human health. *FEMS Microbiology Reviews*, 27, 65–74.

### Food Technology

Twomey, D., Ross, R. P., Ryan, M., Meaney, B., and Hill, C. (2002). Lantibiotics produced by lactic acid bacteria: structure, function and applications. *Antonie van Leeuwenhoek*, 82, 165–185.

Hansen, J. N. (1994). Nisin as a model food preservative. *Critical Reviews of Food Science and Nutrition*, 34, 69–93.

Breukink, E., Wiedemann, I., van Kraaij, C., Kuipers, O. P., Sahl, H.-G., and de Kruiff, B. (1999). Use of the cell wall precursor lipid II by a pore-forming peptide antibiotic. *Science*, 286, 2361–2364.

Hsu, S-T., et al. (2004). The nisin-lipid II complex reveals a pyrophosphate cage that provides a blueprint for novel antibiotics. *Nature Structural and Molecular Biology*, 11, 963–967.

Eijsink, V. G. H. (2005). Bacterial lessons in sausage making. *Nature Biotechnology*, 23, 1494–1495.

Chillou, S., et al. (2005). The complete genome sequence of the meat-borne lactic acid bacterium *Lactobacillus sakei* 23K. *Nature Biotechnology*, 23, 1527–1533.

### Single-Cell Protein

Quorn™ International website http://www.quorn.com.

Miller, S. A. and Dwyer, J. T. (2001). Evaluating the safety and nutritional value of mycoprotein. *Food Technology*, 55, 42–47.

Hoff, M., Trüeb, R. M., Ballmer-Weber, B. K., Vieths, S., and Wuetrich, B. (2003). Immediate-type hypersensitivity reaction to ingestion of mycoprotein (Quorn) in a patient allergic to molds caused by acidic ribosomal protein P2. *Journal of Allergy and Clinical Immunology*, 111, 1106–1110.

**Environmental Applications of Microorganisms**

Wackett, L. P., and Hershberger, C. D. (2001). *Biocatalysis and Biodegradation: Microbial Transformation of Organic Compounds*, Washington, D.C.: ASM Press.

Hurst, C. J. (ed.). (2002). *Manual of Environmental Microbiology*, 2nd Edition, Washington, D.C.: ASM Press.

Atlas, R. M., and Philp, J. C. (eds.) (2005). *Bioremediation: Applied Microbiology Solutions for Real-World Environmental Cleanup*, Washington, D.C.: ASM Press.

The Danish Government. (2003). Making markets work for environmental policies – achieving cost-effective solutions, http://www.mst.dk.

Watanabe, K., and Baker, P. W. (2000). Environmentally relevant microorganisms. *Journal of Bioscience and Bioengineering*, 89, 1–11.

Wilsenach, J. A., Maurer, M., Larsen, T. A., and van Loosdrecht, M. C. M. (2003). From waste treatment to integrated resource management. *Water Science and Technology*, 48, 1–9.

Bosecker, K. (2001). Microbial leaching in environmental clean-up programmes. *Hydrometallurgy*, 59, 245–248.

Watanabe, K., and Baker, P. W. (2000). Environmentally relevant microorganisms. *Journal of Bioscience and Bioengineering*, 89, 1–11.

**Microbial Whole-Cell Bioreporters**

Farré, M., and Barceló, D. (2003). Toxicity testing of wastewater and sewage sludge by biosensors, bioassays and chemical analysis. *Trends in Analytical Chemistry*, 22, 299–310.

Köhler, S., Belkin, S., and Schmid, R. D. (2000). Reporter gene bioassays in environmental analysis. *Fresenius Journal of Analytical Chemistry*, 366, 769–779.

Belkin, S. (2003). Microbial whole-cell sensing systems of environmental pollutants. *Current Opinion in Microbiology*, 6, 206–212.

Leveau, J. H. J., and Lindow, S. E. (2002). Bioreporters in microbial ecology. *Current Opinion in Microbiology*, 5, 259–285.

Yoon, K. P., Misra, T. K., and Silver, S. (1991). Regulation of the *cadA* cadmium resistance determinant of *Staphylococcus aureus* plasmid pI258. *Journal of Bacteriology*, 173, 7643–7649.

Jaspers, M. C. M., Meier, C., Zehnder, A. J. B., Harms, H., and van der Meer, J. F. (2001). Measuring mass transfer processes of octane with the help of an *alkS-alkB::gfp*-tagged *Escherichia coli*. *Environmental Microbiology*, 3, 512–524.

Leveau, J. H. J., and Lindow, S. E. (2001). Appetite of an epiphyte: quantitative monitoring of bacterial sugar consumption in the phyllosphere. *PNAS*, 98, 3446–3453.

**Organic Chemistry**

Zeikus, J. G. (2000). Biobased industrial products: back to the future for agriculture. In *The Biobased Economy of the Twenty-First Century: Agriculture Expanding into Health, Energy, Chemicals, and Materials*, A. Eaglesham, W. F. Brown, and R. W. F. Hardy (eds.), Ithaca, NY: National Agricultural Biotechnology Council.

Warner, J. C., Cannon, A. S., and Dye, K. M. (2004). Green chemistry. *Environmental Impact Assessment Review*, 24, 775–799.

Trost, B. M. (1995). Atom economy – a challenge for organic synthesis: homogenous catalysis leads the way. *Angewandte Chemie International Edition English*, 34, 259–281.

Jenck, J. F., Agterberg, F., and Droescher, M. J. (2004). Products and processes for a sustainable chemical industry: a review of achievements and prospects. *Green Chemistry*, 6, 544–556.

Zaks, A. (2001). Industrial biocatalysis. *Current Opinion in Chemical Biology*, 5, 130–136.

Böschen, S., Lenoir, D., and Scheringer, M. (2003). Sustainable chemistry: starting points and prospects. *Naturwissenschaften*, 90, 93–102.

Prust, C., et al. (2005). Complete genome sequence of the acetic acid bacterium *Gluconobacter oxydans*. *Nature Biotechnology*, 23, 195–200.

Vink, E. T. H., Rábago, K. R., Glassner, D. A., and Gruber, P. R. (2003). Applications of life cycle assessment to Nature Works$^{TM}$ polylactide (PLA) production. *Polymer Degradation and Stability*, 80, 403–419.

Drumright, R. E., Gruber, P. R., and Henton, D. E. (2000). Polylactic acid technology. *Advanced Materials*, 12, 1841–1846.

THREE

# Production of Proteins in Bacteria and Yeast

The human body functions properly only when thousands of bioactive peptides and proteins – hormones, lymphokines, interferons, various enzymes – are produced in precisely regulated amounts, and serious diseases result whenever any of these macromolecules are in short supply. Until 1982, however, the only available pharmaceutical preparations of these peptides and proteins for the treatment of such diseases were obtained from animal sources, and they were sometimes prohibitively expensive. Bioactive proteins and peptides typically occur at low concentrations in animal tissues, so it was difficult to purify significant amounts for medical use. Some important proteins, such as pituitary growth hormone, differ in animals and humans to the extent that a preparation of animal origin is useless for treating humans. Finally, it was extremely difficult to isolate labile macromolecules from human and animal tissues without running some risk that the products might be contaminated by viral particles and viral nucleic acids.

The introduction of recombinant DNA techniques brought about a revolution in the production of these compounds (Chapter 2). It is now possible to clone a DNA segment coding for a protein and introduce the cloned fragment into a suitable microorganism, such as *Escherichia coli* or the yeast *Saccharomyces cerevisiae*. The "engineered" microorganism then works as a living factory, producing very large amounts of rare peptides and proteins from the inexpensive ingredients of the culture medium. And with such products obtained in this way from pure cultures of microorganisms, there is no chance of contamination by viruses harmful to humans.

## PRODUCTION OF PROTEINS IN BACTERIA

For several reasons, bacteria were the first microorganisms to be chosen for use as living factories. To begin with, a great deal was known about their genetics, physiology, and biochemistry. After *Homo sapiens*, the bacterium *E. coli* is the most thoroughly studied and best-understood organism in the living world. Furthermore, it is easy to culture bacteria in large amounts in inexpensive media, and bacteria can multiply very rapidly. For example,

*E. coli* doubles its mass every 20 minutes or so in a rich medium. Finally, bacteria are so small that up to a billion cells can fit on a single Petri dish only 10 cm in diameter. This permits us to test very large populations in order to find extremely rare mutants or recombinants – an enormous help at many stages of genetic and recombinant DNA manipulations.

## INTRODUCTION OF DNA INTO BACTERIA

The field of bacterial genetics grew explosively in the mid-twentieth century, laying much of the groundwork for the development of procedures that efficiently introduce foreign DNA into bacteria. The three basic approaches take advantage of the three modes by which bacteria are known to exchange genetic information. There are two aspects of a genetic exchange: DNA (1) leaves a donor cell and (2) enters a recipient cell. It is the latter process, the uptake of DNA by a cell, that is all-important to biotechnologists.

### Direct Introduction by Transformation

*Transformation* was the first process of genetic exchange to be discovered in bacteria. In 1928, Frederick Griffith injected living cells of *noncapsulated* pneumococcus (*Streptococcus pneumoniae*) together with heat-killed cells of a *capsulated* pneumococcus strain into mice and found that the non-capsulated strain then acquired, presumably from the capsulated strain, the ability to produce a capsule. These experiments thus showed that genetic information can be transferred into living bacterial cells from a preparation containing no living donor cells. In 1944, the substance that carried the genetic information in the transformation process was identified as DNA in the famous work of Oswald T. Avery, Colin M. MacLeod, and Maclyn McCarty. This discovery led to the development of modern molecular biology.

We now know of several species of bacteria that, like pneumococcus, have a natural ability to undergo transformation, such as *Bacillus subtilis*, *Neisseria gonorrhoeae*, and *Haemophilus influenzae*. In some of these organisms, DNA is known to be taken up via elaborate machinery produced by the recipient cell, suggesting that the uptake is an active process. The ability to take up DNA, which is called *competence*, is typically developed only under special conditions.

The genetics and physiology of naturally transformable species are not well known, however, with the exception perhaps of *Bacillus subtilis*. Thus it was fortunate for biotechnological applications that the best-studied bacterium, *E. coli*, was found to accept exogenous DNA in an artificial transformation process. In the classical process, *E. coli* cells are first converted into a competent state by resuspension in buffer solutions containing very high concentrations (typically 30 mM) of $CaCl_2$ at 0°C. The effect of $Ca^{2+}$ on a membrane bilayer with a high content of acidic lipids is to "freeze" the hydrocarbon interior, presumably by binding tightly to the negatively charged head groups of the lipids. Because the outer membrane of Gram-negative bacteria such as *E. coli* (see Figure 1.3B) contains a large number of acidic groups (in the form of lipopolysaccharide [LPS]) at a very high density, this membrane becomes frozen and brittle, with cracks through which macromolecules, including DNA, can pass. After DNA is added to the suspension, the cells are heated to 42°C and then chilled. Under these conditions, cells have been found to take up pieces of DNA through the cytoplasmic membrane, but the molecular mechanisms of the process still remain obscure.

Transformation can be achieved by similar means in certain other bacteria, but there are many species for which this method does not work. One method that works with many organisms (also including *E. coli*) is *electroporation*. In this process, we apply short electrical pulses of very high voltage, which is believed to reorient asymmetric membrane components that carry charged groups, thus creating transient holes in the membrane. DNA fragments can then enter through these openings, either by spontaneous diffusion or driven by the electric charge.

## Introduction by Conjugation

We have said that it is difficult to introduce DNA directly into certain species of bacteria. In such cases, taking an indirect route sometimes achieves the desired result. First, a piece of DNA is introduced into an organism (such as *E. coli*) that *can* receive DNA by transformation. This piece of DNA is then transferred from the *E. coli* into the species of interest by another form of genetic exchange in bacteria, conjugation.

The *conjugational transfer* of genes in bacteria was discovered by Joshua Lederberg and Edward L. Tatum in 1946. Subsequent work has shown it to be a unidirectional transfer from a cell containing a sex plasmid, or F-plasmid (for "fertility"), into a cell lacking that plasmid. The transfer of chromosomal genes by conjugation occurs only in rare donor cells, in which the sex plasmid has become integrated into the chromosome. A more frequent process, which occurs with nearly 100% efficiency, is the transfer of just the F-plasmid from a donor to a recipient (Figure 3.1). Conjugation requires that the donor and recipient cells join to form a stable pair connected, at least in the beginning, by a filamentous apparatus (sex pilus).

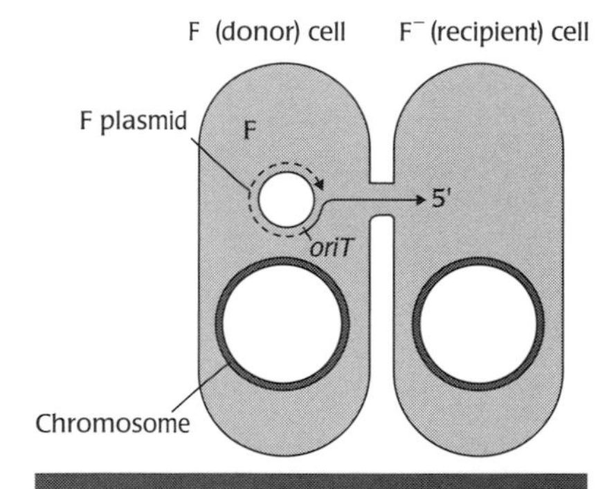


**FIGURE 3.1**

Conjugational transfer of the F-plasmid. One of the strands of the F-plasmid is cut at a specific position (oriT, for "origin of transfer"). This strand becomes elongated by rolling-circle replication (*broken line*), gradually displacing the old part of this strand, which enters into the $F^-$ cell 5′-end first. A complementary strand is synthesized in the cytoplasm of the recipient cell, and the plasmid is then circularized, converting the recipient cell from $F^-$ to $F^+$.

Redrawn based on artwork from the first edition (1995), published by W.H. Freeman.

As we shall see, the first step in the cloning of a fragment of DNA is to insert it into a suitable *vector DNA*, and plasmids are the most frequently used vectors. However, the unmodified sex plasmids are *not* used as vectors. If they were, the job of transferring the recombinant plasmids to other strains and species would be easy, because all the proteins needed for such a transfer are encoded on the plasmid itself. But the procedure could also be potentially dangerous, because if a plasmid-containing strain were to escape into the environment, the recombinant plasmid with the foreign DNA could conceivably start to spread into other, naturally occurring bacteria. The current practice, therefore, is to use as vectors only *nonconjugative* or *non–self-transferring* plasmids (plasmids that lack the information for the cell-to-cell transfer). For these plasmids to be transferred by conjugation, the missing information must be supplied from another plasmid. This procedure is called *plasmid mobilization*. It is useful when DNA must be transferred into strains that cannot be made to receive it at high efficiency by transformation.

### Injection of Bacteriophage DNA and Transduction

A problem with the transformation process is its low efficiency. With *E. coli* as the recipient, the usual frequency of transformation suggests that only one out of hundreds of thousands of the exogenous DNA molecules enters the cell. In contrast, when bacteriophage (bacterial virus) infects bacterial cells, *every* virus particle adsorbs to a susceptible host cell and injects it with the DNA contained in the virus head at very high efficiency, often close to 100%. (The general features of the bacteriophage replication cycle are described in Figure 3.2.) Scientists have been able to take advantage of this natural process to inject foreign DNA into bacterial cells, thanks to a third type of genetic exchange in bacteria, transduction.

In *generalized transduction*, a piece of bacterial chromosome is transferred into a recipient cell by means of a bacteriophage. The chromosomal

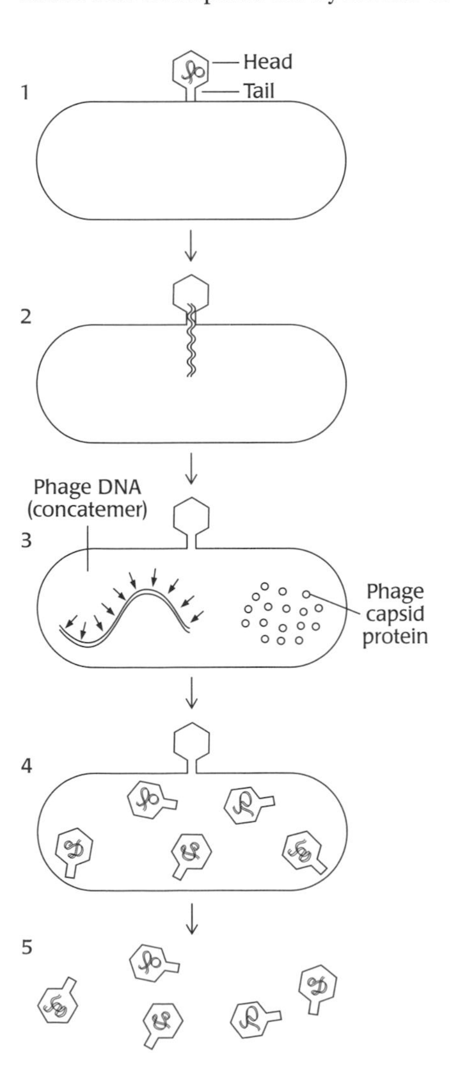


**FIGURE 3.2**

Multiplication of a virulent bacteriophage (bacterial virus) within a bacterial cell. The bacteriophage first adsorbs to a specific structure on the cell surface (step 1). The phage DNA is then injected into the cytoplasm, in some cases driven by contraction of the tail sheath (step 2). Within the cytoplasm, phage DNA and phage capsid (head as well as tail) proteins are synthesized separately (step 3). With most phages, DNA is synthesized as a concatemer containing many repeats of the genomic sequence. Finally, the DNA is cut to the length that corresponds to one phage genome (*arrows* in step 3) and becomes packaged into phage heads (step 4). The cell is then lysed (step 5). Thus, when a mixture of phages and a larger number of host bacterial cells is spread as a lawn on the surface of a solid medium, phages released by the lysis of one cell infect neighboring cells, causing cycles of lysis and infection and finally producing a small area of clearing (a plaque) where most of the host cells have lysed. This course of events occurs with *virulent* phages, which always cause lytic infections. With *temperate* phages, such as λ or P1, the infection may result in the *lysogenic response*, in which the phage DNA is replicated in step with the host genome without exhibiting the runaway replication of the lytic response. Temperate phages usually produce turbid plaques because some host cells within the plaques survive as lysogenic bacteria.

Redrawn based on artwork from the first edition (1995), published by W.H. Freeman.

DNA gets into the phage head by the mechanism illustrated in Figure 3.3. Once there, the fragment is injected into the cytoplasm of a new host cell in exactly the same way as the phage DNA. The phage head simply injects any DNA it happens to be carrying, regardless of the nature or the source of that DNA. Recombinant DNA technologies utilize this feature of the virus infection process by packaging recombinant DNA into phage heads *in vitro*. The specific vectors used for this type of delivery, phage λ and cosmids, are described in more detail below.

## USE OF VECTORS

Let us assume that we have isolated a fragment of DNA coding for a commercially valuable protein and we want to convert *E. coli* into a factory that produces large amounts of this protein. Our first inclination might be to inject this piece of foreign DNA directly into *E. coli* cells by using one of the methods just described. Unfortunately, that approach would not work. A random piece of DNA floating in the cytoplasm would not be replicated. Only DNA that contains a special *replication origin* sequence is recognized and replicated by *E. coli*, and there is almost no chance that a fragment of foreign DNA will contain such a sequence. It is true that the foreign DNA fragment would be replicated if it got inserted into the bacterial chromosome and became a part of it – that is, if it became successfully "integrated" into the chromosome. (We rely on a similar process of integration when we introduce fragments of foreign DNA into higher plants and higher animals to create *transgenic* plants and animals.) In bacteria, however, the chromosomal integration of unrelated pieces of DNA is a rare event. Even if our fragment did become integrated into some part of the bacterial chromosome, the genes in the fragment would exist in the cell as single copies only, so they would not be expressed very strongly. Furthermore, the large size of the chromosome would prevent us from manipulating the fragment further – for example, by cutting it out for *subcloning*.

For these reasons, it is usually necessary to insert a cloned foreign gene into a vector – typically a plasmid or phage DNA that is much smaller in size than the bacterial chromosomes – that replicates autonomously in host microorganisms and acts as a carrier of the inserted foreign DNA sequence. There are hundreds of cloning vectors now available, each with its advantages and disadvantages. However, before we discuss the properties of each type of cloning vector, we must start by drawing a general picture of the cloning process itself.

### Strategy for Shotgun Cloning

Say that we are going to clone, in *E. coli*, a gene *X* coding for a protein X from a "foreign" organism (i.e., an organism other than *E. coli*). The coding region of an average prokaryotic gene is only 1 or 2 kilobases (kb) long. In contrast, the genome of a bacterium has a length of thousands of kilobases, and that of a higher eukaryote a total length of millions of kilobases. Thus, gene *X* makes

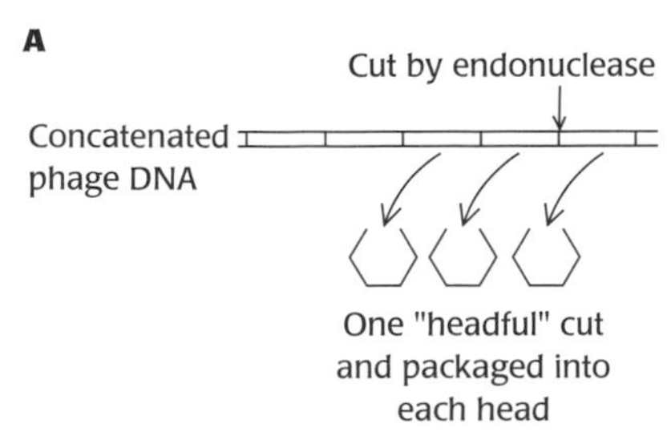


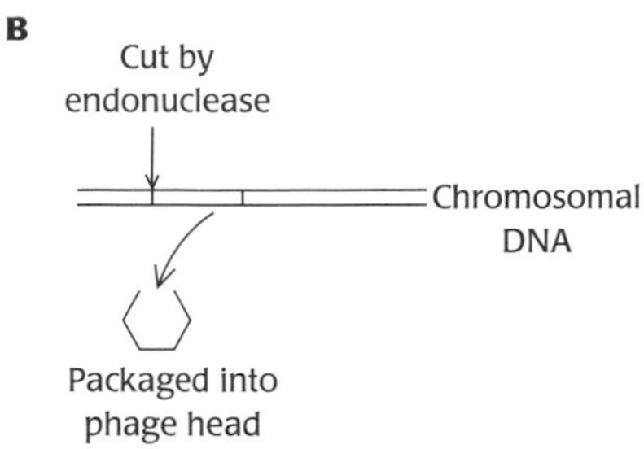


FIGURE 3.3

The generation of transducing phage particles. (**A**) In the normal infection cycle, the DNA of such phages as P22 is synthesized as a long concatemer, which is then cleaved by a phage-coded endonuclease at a specific site, the *pac* site. Then each newly assembled phage head becomes packed with a "headful" length of DNA. (**B**) If the chromosomal DNA is cut by an endonuclease – perhaps because it carries a sequence similar to that of the *pac* site, perhaps for other reasons – the headful packaging mechanism incorporates fragments of chromosomal DNA into newly assembled phage heads. In *generalized transduction*, these transducing particles subsequently inject such fragments of host DNA into other bacteria, and the DNA recombines with the chromosomes of the recipients to generate *transductants*.

Redrawn based on artwork from the first edition (1995), published by W.H. Freeman.

up only a small part (one in thousands to one in millions) of the genome. The usual first step in a cloning effort is therefore to clone random segments of the genome of the source organism (this is often called *shotgun cloning*) so that the subsequent isolation and identification of a clone containing the gene *X*, but not much else, will become possible (Figure 3.4). At this stage, it is advantageous to use vectors that can accommodate large DNA fragments because that dramatically decreases the number of recombinant DNA clones that must be examined in order to find the one containing the gene *X* (Box 3.1).

The large fragment cloned in this first step – the primary cloning – contains many genes in addition to gene *X*. Such complex pieces of DNA are not suitable for use in expression, sequencing, or site-directed mutagenesis. This is why it is necessary to pull out a small portion of the DNA, corresponding to only a little more than gene *X*. This essential step is called *subcloning*, and several different types of vectors are available for the purpose.

Genes from higher eukaryotes usually contain one or more intervening sequences, or *introns*, that do not code for the amino acid sequence of the protein product (Figure 3.5). As a rather extreme example, the gene for thyroglobulin has a size of 300 kb, but that includes 36 introns; the actual coding regions represent only 3% of the total gene length. When RNA transcripts are made from the DNA sequence, they still contain the sequences corresponding to introns. These sequences are then removed from the transcripts by *splicing*, and the mature mRNA molecules that leave the nucleus and enter the cytoplasm do not contain the intervening sequences. The mRNAs are also modified usually at the 3′ terminus by the addition of polyadenylate "tails" (see Figure 3.5).

To determine the nucleotide sequence of a particular gene (say for the purpose of identifying genetic defects in an inherited disease), it is necessary to clone the gene from the genomic DNA so that the intron sequences are included as well. This cloning of intron-containing genes

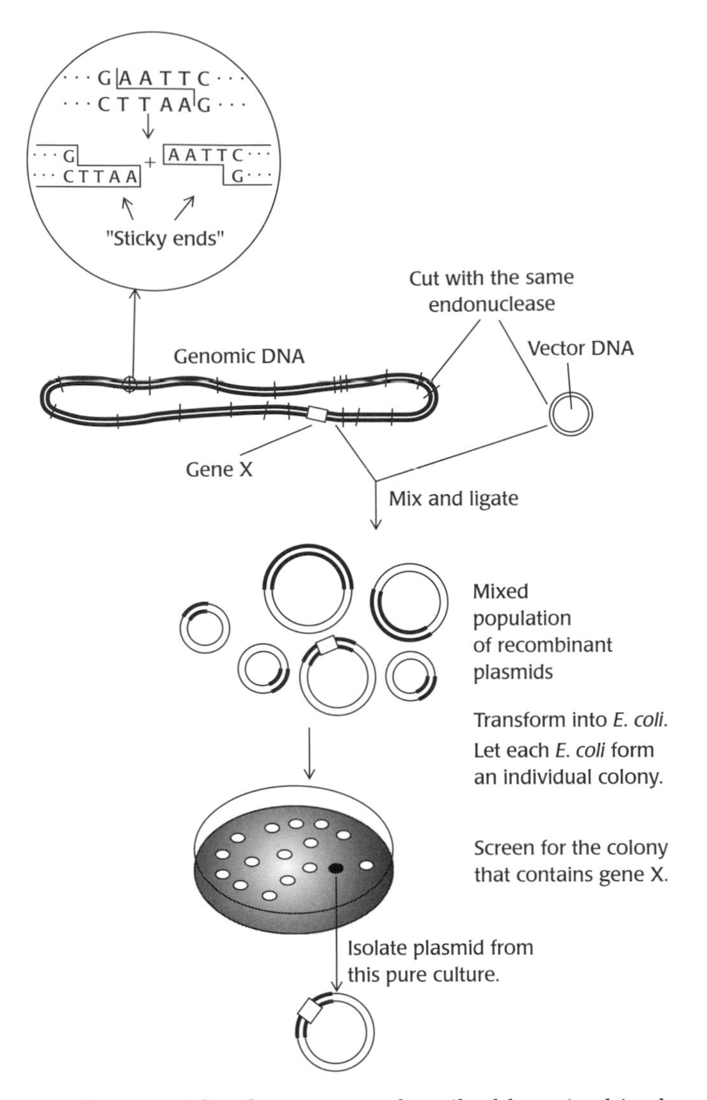


**FIGURE 3.4**

Shotgun cloning of genomic DNA in *E. coli.* In the first step, one restriction endonuclease is used both to cut and open the vector plasmid DNA and to create fragments of genomic DNA. With most endonucleases, this procedure creates complementary "sticky ends" (see enlargement, here illustrating the ends created by restriction endonuclease EcoRI), which facilitate the end-to-end attachment of fragments by the complementary annealing of hanging protrusions. In the second step, the opened vector DNA is mixed with the fragments of the donor DNA. Many of the ends of the donor fragments will then anneal to the open ends of the vector DNA because of the complementary overhanging sequences. Addition of DNA ligase results in the covalent connection between the ends of DNA strands, producing a library of recombinant DNA. In the next step, the recombinant DNA pieces are introduced into *E. coli*, and the bacteria are spread on an agar plate containing a suitable growth medium so that each bacterium will produce a colony – a pure clone – well separated from other colonies. When the vector contains an antibiotic resistance gene, the antibiotic is added to the medium so that only those *E. coli* cells that have received the recombinant plasmid (or the resealed vector plasmid) will grow to produce colonies. Because transformation is a rare event, each clone will contain only one plasmid species. The colony containing the desired gene can then be identified by one of the methods discussed in the text. A pure preparation of the recombinant plasmid, amplified to billions of copies, can now be isolated from this *E. coli* strain, and the fragment can be "subcloned" further in different vectors for the purpose of expression, sequencing, or mutagenesis.

Redrawn based on artwork from the first edition (1995), published by W.H. Freeman.

requires specialized vectors, as described later in this chapter; luckily, for most biotechnological applications, it is also undesirable. Bacterial DNAs do not contain introns, and bacteria cannot carry out the splicing reactions. We will see later that even a eukaryotic microorganism such as yeast cannot be relied on to recognize all the splicing signals to be found in the RNA transcripts of genes of higher animals and plants. These eukaryotic genes, therefore, may not be expressed properly in microorganisms. Consequently, in these cases a better template for cloning is usually the mature mRNA, which does not contain the intervening sequences. In such a procedure, the mRNA is first converted to a double-stranded DNA through use of the enzyme reverse transcriptase, which was originally found as a product of RNA viruses (Figure 3.6). Because each eukaryotic mRNA usually contains coding information for only one protein, each of these DNAs, called cDNAs (for "complementary DNA"), also codes for one protein. For this reason, cDNA molecules can then be inserted directly into specialized vectors, such as expression vectors, often circumventing the need for subcloning. Importantly,

**Fragment Size and the Probability of Finding a Desired Gene in a Set of DNA Fragments**

Let us assume that we use a vector that can accommodate up to 40 kb DNA to clone fragments of a 4000-kb bacterial genome. We use a restriction endonuclease with rare recognition sites so that the genomic DNA is cut into about 100 distinct fragments with an average size of 40 kb. Among these, only one fragment (say, fragment 29) contains the gene *X*. So when we randomly examine clones to find the one containing gene *X*, how certain can we be of success? If we had the 100 fragments from the single chromosome of one bacterium in a box, then that set of 100 would certainly contain fragment 29. In actual practice, however, we will be using fragments generated from a mixture of many DNA molecules obtained from billions (or even more) of bacteria. Thus, when we pick just 100 fragments of these molecules (or 100 clones) at random, we are likely to have gathered multiple copies of some fragments and no copies of others (possibly including fragment 29). Statistical calculation shows that in order to have a probability *P* of finding the fragment containing *X*, one has to examine N clones, which is expressed by

$$N = \ln(1 - P) / \ln(1 - R),$$

where *R* is the ratio of the fragment size (here 40 kb) to the genome size (4000 kb). If one wants a 99% probability ($P = 0.99$) of fragment 29 being included in the collection, one has to examine 465 clones. This equation shows that if the size of the fragments cloned into vectors is 10 times smaller (4 kb), then the number of clones that must be examined increases to 4500. It is thus advantageous in a *primary cloning* (i.e., in the production of a "genomic library") to use a vector with a large insert size.

BOX 3.1

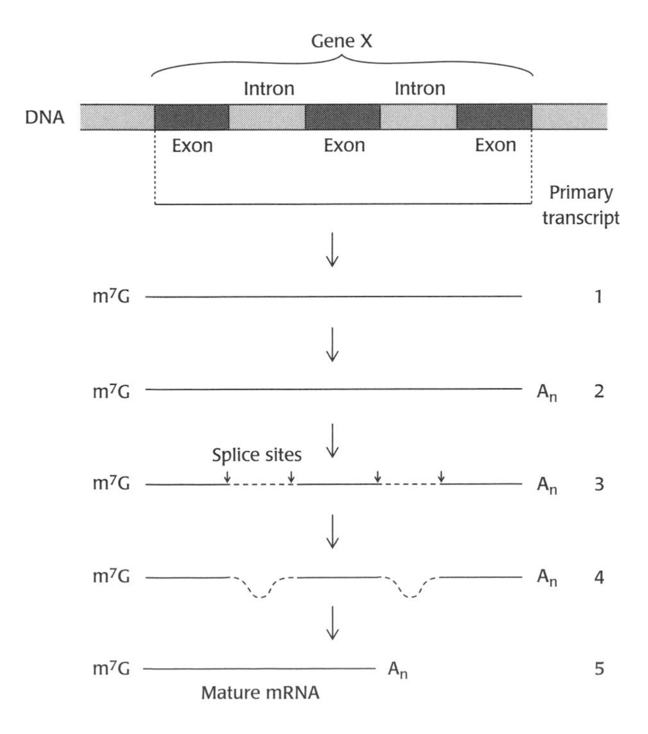


**FIGURE 3.5**

The processing of RNA transcripts in eukaryotes. A eukaryotic gene, especially one from a higher animal, is likely to contain many intervening sequences, or *introns*. Primary RNA transcripts of eukaryotic genes are processed first by "capping" – that is, by the addition of 7-methyl-guanosine monophosphate units at the 5′-end through 5′-5′ linkage – and by the shortening of the 3′-end (stage 1). A polyA tail is then added to the 3′-end (stage 2). Finally, the RNA sequences that correspond to the introns in the DNA are spliced out (stage 3), producing the mature mRNA.

Redrawn based on artwork from the first edition (1995), published by W.H. Freeman.

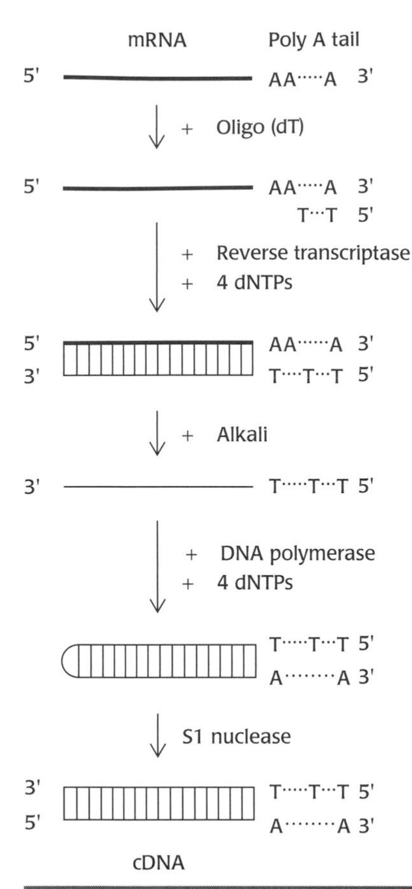


FIGURE 3.6

Production of cDNA from mRNA. With oligo(dT) as the primer, reverse transcriptase is used to synthesize a single strand of DNA. The template mRNA is then degraded with alkali, and DNA polymerase is used to synthesize a complementary DNA sequence on the first strand. Finally, treatment with S1 nuclease cuts the looped end of the DNA, generating a double-stranded cDNA.

Redrawn based on artwork from the first edition (1995), published by W.H. Freeman.

PCR-based amplification of individual genes, now made possible because of the presence of enormous amounts of information on gene sequences in thousands of organisms, also allows us to bypass the classical shotgun cloning method.

## Cloning Vectors

Some cloning vectors are used only for general-purpose cloning, such as the primary cloning and identification of the coding segments. Plasmids are used most commonly for such purposes, although phage λ–derived vectors and cosmids are advantageous in situations that require the cloning of large segments of DNA. Vectors derived from single-stranded DNA phages are used for some special purposes. We discuss below some features of these vectors. Expression vectors, which are used for the high-level expression of cloned genes, are addressed later in this chapter (pages 115).

**Plasmids.** One of the first generation of plasmid vectors is pBR322 (Figure 3.7). It is still very frequently used, and many other plasmid vectors have been derived from it by the introduction of additional desirable properties. In the following description, we shall use pBR322 as an example and shall examine the various features that make it a good, general-purpose cloning vector.

The first important feature of this and any cloning vector is the presence of an origin of replication (*ori* in Figure 3.7), obtained for pBR322 from a naturally occurring colicin plasmid (Box 3.2). This origin of replication is recognized by the *E. coli* DNA replication machinery, which then initiates replication of the vector (and its foreign DNA inserts). A second feature of pBR322, and indeed of practically all the plasmid vectors, is the presence of an antibiotic resistance gene. In fact, pBR322 contains two such genes: *bla*, coding for $\beta$-lactamase, which degrades penicillins (including ampicillin) and cephalosporins and thereby produces resistance against these compounds, and *tet*, which codes for a membrane protein that acts as an exit pump for tetracycline, thus producing resistance to tetracycline and its relatives. These resistance markers are needed because, when plasmid DNA is introduced into *E. coli* cells by transformation, only one out of tens of thousands of cells receives a plasmid. Isolating this extremely rare cell would be practically impossible if there were no genetic markers to facilitate its selection (Box 3.3) out of the large excess of cells that failed to acquire the plasmid. Antibiotic resistance is an ideal positive selection marker, because all one has to do after transformation is to spread a large population of cells onto plates containing adequate concentrations of the antibiotics (Figure 3.8). The only cells to survive will be those that have acquired the plasmid, with its resistance genes.

The antibiotic resistance genes also serve a second purpose in pBR322. During the attempt to insert a piece of foreign DNA into a vector DNA that has been opened up by a restriction enzyme (see Figure 3.8), the vector DNA very often recircularizes (closes up again) without incorporating the foreign DNA. This is because unimolecular reactions, which are required for recircularization, occur much more frequently than the bimolecular reactions that are needed for the insertion of another piece of DNA. Reclosure of the vector DNA can be minimized by treating the opened vector with phosphatase (Figure 3.9), but it is difficult to prevent recircularization entirely. Thus, it is important to have a quick way of telling, from the phenotype of the transformed strains (transformants), whether the plasmids contain any inserted foreign DNA. Again, the resistance markers in pBR322 provide the needed information. For example, if one opens up the vector DNA by using BamHI or SalI restriction endonuclease (the cleavage sites for which lie within the *tet* tetracycline resistance gene), then the successful insertion of the cloned DNA will interrupt that gene and create transformants that are susceptible to tetracycline (see Figure 3.8). Screening for such transformants can be achieved conveniently by replica plating (Box 3.4). (By selecting for ampicillin resistance, we can still select for transformants that have successfully acquired plasmids.)

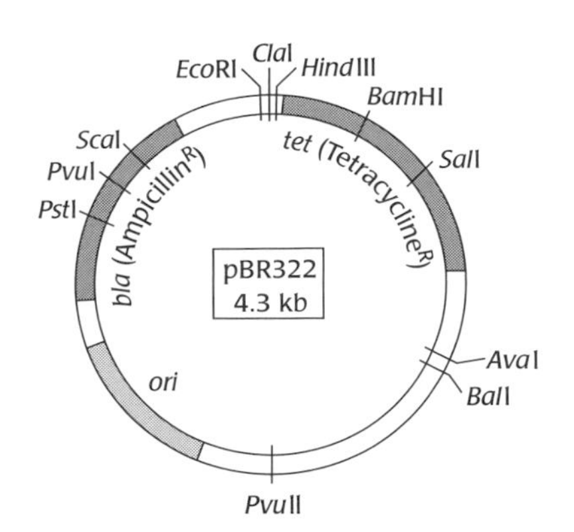


FIGURE 3.7

Structure of a plasmid vector, pBR322. Note that the vector has only a single susceptible site for each of the commonly used restriction endonucleases, such as EcoRI, BamHI, and so on.

Redrawn based on artwork from the first edition (1995), published by W.H. Freeman.

**Colicin Plasmids**

Many *E. coli* strains produce extracellular proteins, called colicins, that are able to kill a range of other bacteria. In most of these colicin-producing strains, the gene coding for the colicin protein is found on a plasmid (colicin plasmid), often along with genes that endow the colicin-producing strain with immunity against the colicin.

**BOX 3.2**

The third characteristic of pBR322 that makes it so useful as a cloning vector is that it contains only one site of cleavage for many commonly used restriction enzymes. (The precursor plasmid to pBR322 did contain multiple restriction sites for some of these enzymes, and the extra sites were eliminated.) This feature is found in most of the widely used cloning vectors and is very important. If the vector contained, say, three sites for the restriction enzyme EcoRI, religation of a mixture containing the three fragments produced from the vector and one fragment of foreign DNA will create many species of recombinant products (Figure 3.10). In contrast, with pBR322 containing a single EcoRI site, a large proportion of the product will be the desired recombinant plasmid, containing complete sequences of the vector and the foreign DNA (see Figure 3.10). Commonly, the foreign DNA is cut using the same restriction enzyme that is used in cutting the vector. Then all the ends of DNA will have the same hanging protrusions ("sticky ends"), which base-pair exactly with each other, increasing the chance of insertion of the foreign DNA (see Figure 3.9).

With plasmid vectors, specially constructed host strains of *E. coli* are often used. One feature of such strains is the defect in the restriction system (e.g., through mutations in the *hsdR* or *hsdS* gene), so that foreign DNA is not destroyed by the restriction enzyme of *E. coli*. Another feature is the defect in the homologous recombination system (e.g., through mutations in the *recA* gene), so as to prevent the alteration in the recombinant plasmids in the host strain. Examples of such strains are DH5$\alpha$, HB101, and JM109.

**$\lambda$ Phage Vectors.** As we have seen already, plasmids are convenient vectors. However, they are not ideal for every application. For example, when very large (>20 kb) pieces of DNA are inserted into the common plasmid vectors, it becomes difficult to introduce the large, recombinant plasmid into a host by transformation and to maintain such plasmids in successive generations of host cells. This is a problem when one wants to clone random fragments of genomic DNA in search of a particular gene, because the odds that the gene of interest will appear in any given fragment plummet when the average size of the cloned fragment decreases (see Box 3.1). The need to clone large fragments becomes especially acute when one is working with the genomic DNA of higher animals and plants, because such eukaryotic genes are interrupted frequently by introns, and so only very large pieces of DNA can contain a complete gene. Some of the λ-derived vectors are more useful than plasmids for this type of situation.

Phage λ is a well-known temperate bacteriophage (Box 3.5) containing linear, double-stranded DNA. It was originally discovered in some strains of *E. coli* K-12, the standard strain used in bacterial genetics. The entire λ phage

**Selection and Screening**

In bacterial genetics, these terms have distinctly different meanings. A procedure in which a number of colonies are examined one at a time for a specific property – say, for their ability to hydrolyze certain substrates – is considered a *screening* (or sometimes scoring) procedure. In contrast, a procedure in which, out of many millions of initial population, only cells with certain properties are allowed to grow and form colonies is called a *selection* procedure. An example is the spreading of large numbers of bacteria on plates containing a particular antibiotic to select for the rare, antibiotic-resistant cells. Selection procedures are much more efficient than screening procedures because they enable us to examine much larger numbers of cells in a single experiment.

BOX 3.3

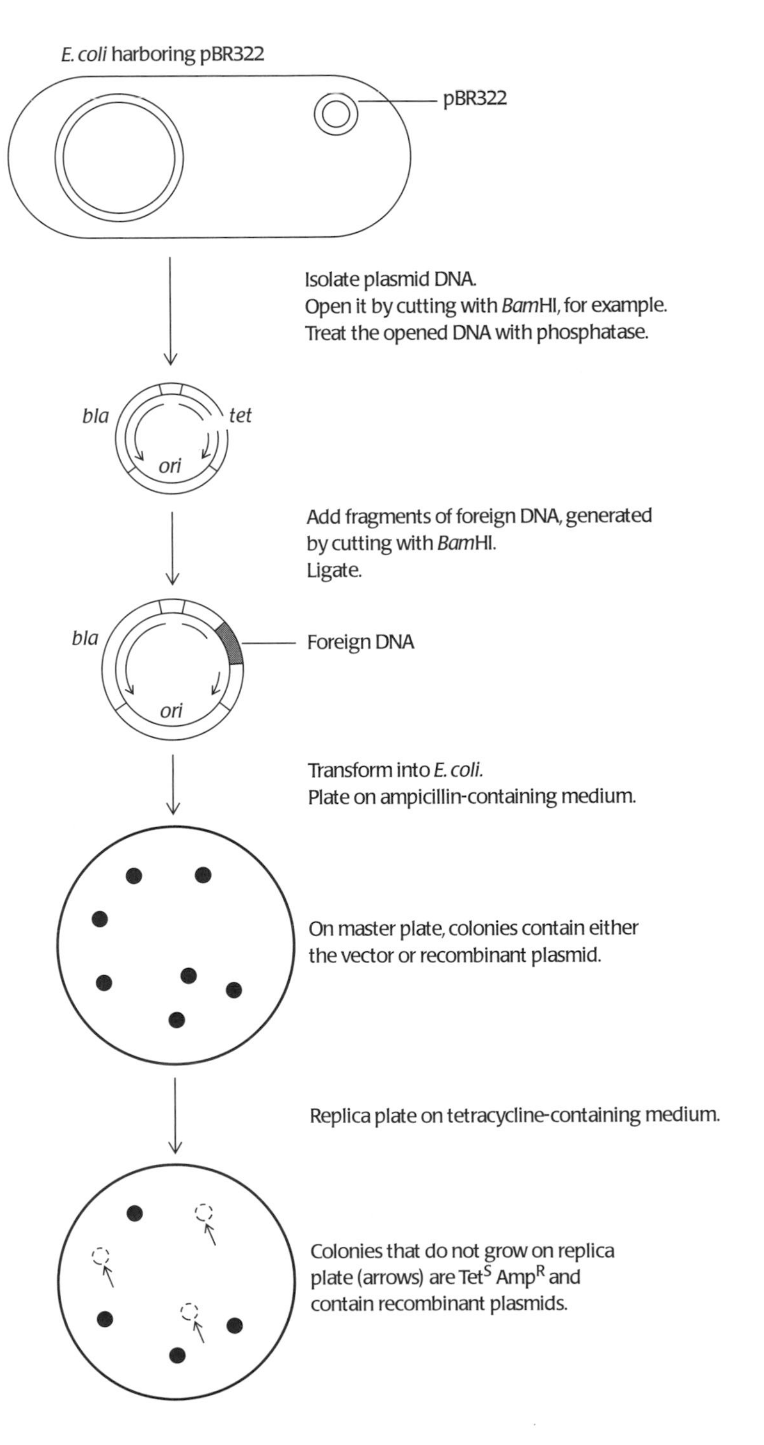


FIGURE 3.8

Cloning of foreign DNA segments in pBR322. The vector DNA is cut open by a restriction endonuclease and then treated with phosphatase (see Figure 3.10) in order to prevent its religation. The addition of foreign DNA cut with the same restriction endonuclease results in the annealing of the foreign DNA to the complementary ends of the cut vector. After ligation and transformation into *E. coli*, the cells are plated on a suitable selective medium. In the example shown, the insert was cloned into the BamHI site, thus destroying the *tet* gene. The plasmid-containing cells were therefore selected on ampicillin-containing plates (by using their ampicillin-resistant – $Amp^R$ – phenotype), and the presence of inserts in the plasmids was detected by the inability of certain colonies to grow on tetracycline-containing plates (by using their tetracycline-susceptible – $Tet^S$ – phenotype). This screening can be conveniently accomplished by the replica-plating technique (see Box 3.4). When sites within *bla* genes (such as PstI or PvuI) are used in cloning, the tetracycline-resistant ($Tet^R$) cells that contain the recombinant plasmids are selected on tetracycline-containing plates, and the presence of inserts is scored on ampicillin-containing plates.

Redrawn based on artwork from the first edition (1995), published by W.H. Freeman.

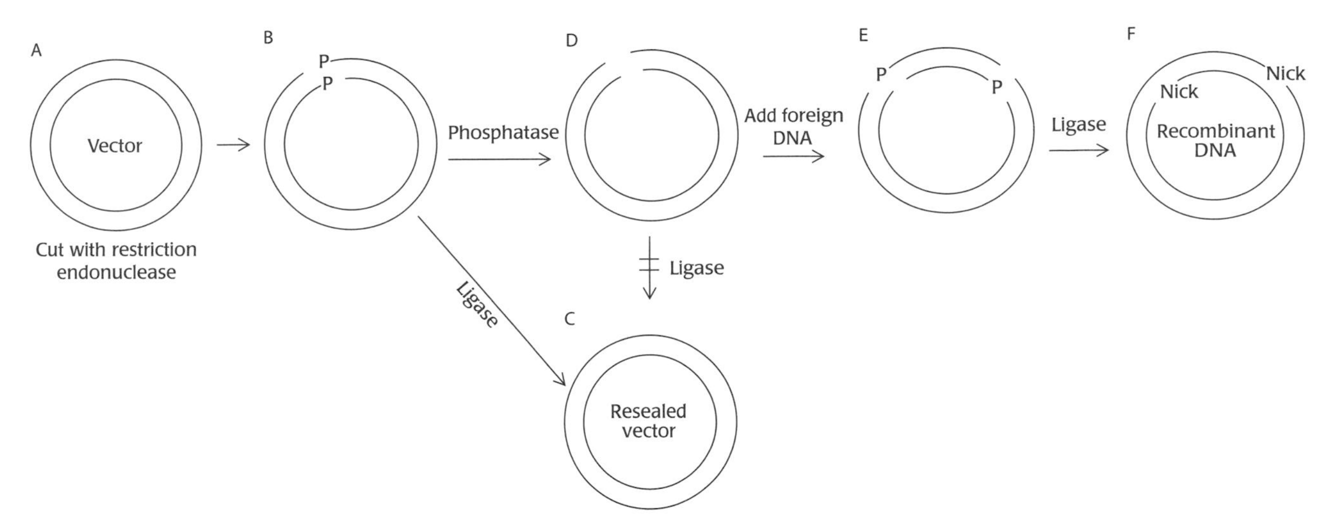


**FIGURE 3.9**

Preventing the religation of opened vector DNA. When the vector DNA (**A**) is opened by cutting with a restriction endonuclease, the open ends are usually staggered, with the 5′-phosphate groups still in place (**B**). These 5′-phosphate groups can react with 3′-OH ends of other DNA strands in the presence of DNA ligase, producing closed strands linked with phosphodiester bonds. Without further treatments, it is difficult to use this DNA in the construction of recombinant DNA, because ligation will cause much of the vector DNA simply to reseal (**C**). To prevent this, the opened vector DNA is treated with phosphatase. The treated vector DNA (**D**) cannot reseal on itself, because it lacks the 5′-phosphate groups needed for the formation of phosphodiester bonds. If foreign DNA cut with the same endonuclease is added, its staggered ends, with phosphate groups attached, become annealed with the staggered ends of the vector DNA (**E**). Finally, ligase connects the foreign DNA strands at the end containing the 5′-phosphate (**F**). Although the recombinant DNA created still contains nicks, these are readily repaired once it is transformed into the host cell.

Redrawn based on artwork from the first edition (1995), published by W.H. Freeman.

genome is 50 kb long, but an 8-kb region of this DNA (the *b*-region, Figure 3.11) has no known function. Another, adjoining region about 7 kb long and containing *att*, *int*, and *xis* (see Figure 3.11) is not needed for the lytic growth (see Box 3.5) of the phage. These two segments can be removed and replaced with a segment of foreign DNA without affecting the phage multiplication. In a λ-derived vector such as EMBL3 (see Figure 3.11), the insert can be significantly longer (up to 20 kb or even slightly more) than the length of these two deleted λ fragments (15 kb) for two reasons. (1) Two additional short segments (KH54 and *nin5*), totaling about 5 kb and representing regions not needed for lytic growth, were deleted to create EMBL3. (2) The head of the λ phage can package a piece of DNA that is slightly longer (by about 2 kb) than the length of the normal λ DNA.

As with most bacteriophages, λ phage particles are produced during the last stage of infection: The phage DNA is packaged into proteinaceous phage capsids that have been assembled in the cytoplasm of the infected cell (see Figure 3.2). We take advantage of this packaging reaction in using λ-derived cloning vectors. In practice, the fragment of foreign DNA is inserted into the vector DNA by cutting with restriction endonuclease, annealing the ends,

**Replica Plating**

This method, developed by Joshua and Esther Lederberg, permits the screening of many colonies in one operation. For example, if we want to screen a population of *E. coli* for their susceptibility to tetracycline, we first spread the population on an agar medium without tetracycline so that a few hundred colonies arise, after incubation, on a single plate (the master plate). The surface of the master plate is lightly "stamped" with a flat, sterile piece of velvet, and then the velvet is momentarily placed on the surface of a new plate that contains tetracycline (the replica plate). From each colony on the master plate, a few cells are transferred onto the replica plate by this operation. After incubation of the replica plate, colonies that exist on the master plate but do not develop at corresponding locations on the replica plate are noted: They correspond to tetracycline-susceptible clones.

**BOX 3.4**

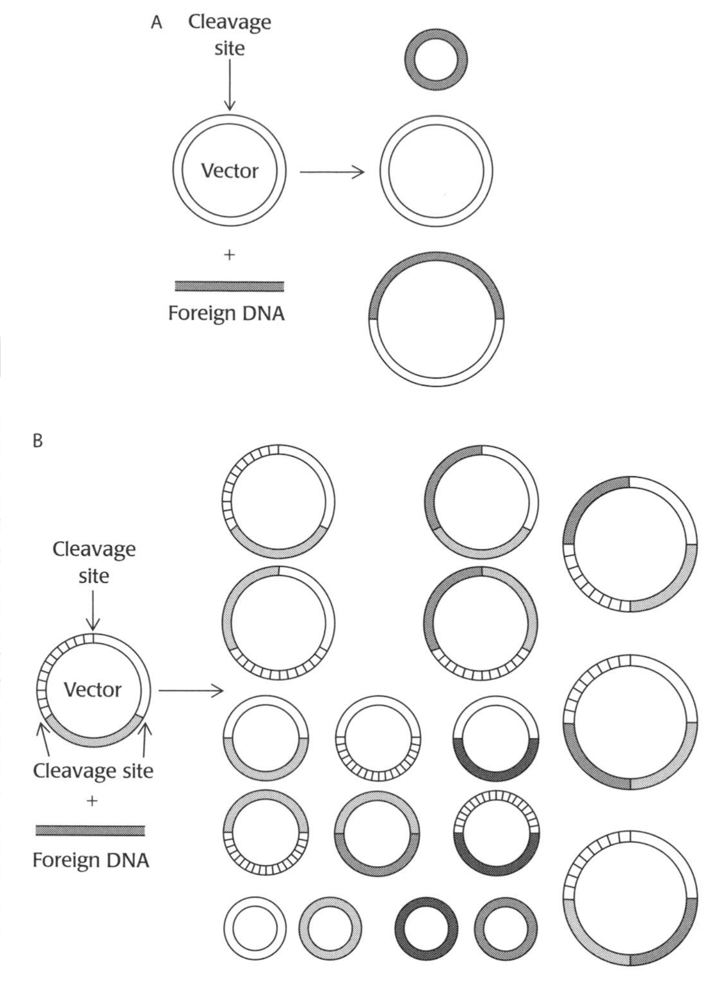


**FIGURE 3.10**

Vectors containing single or multiple cleavage sites for a restriction endonuclease. If a vector contains a single cleavage site for an endonuclease (**A**), then annealing and ligation with a segment of foreign DNA produce only three species of circular DNA, one of which is the desired recombinant containing the foreign DNA and the vector sequence. In contrast, if a vector is cut at three places by an endonuclease, annealing and ligation with foreign DNA produce many species of circular DNA (**B**), only a small fraction of which is the desired recombinant species. The situation is far worse in reality because for simplicity, the figure does not show the species in which multiple copies of one fragment are present within a single molecule. Clearly, it is a major disadvantage for a vector to have more than one cleavage site for each of the commonly used endonucleases.

Redrawn based on artwork from the first edition (1995), published by W.H. Freeman.

and then ligating with DNA ligase. When one mixes the recombinant DNA thus produced with a mixture of the proteins that form the phage capsids, the capsid is assembled and the DNA is packaged spontaneously into λ particles *in vitro*. After packaging, the new phages containing recombinant DNA are used to infect the host bacteria, a process in which DNA enters the bacteria with an efficiency of nearly 100% (rather than 0.001% or less, which is typical

**Lytic and Lysogenic Responses in Phage Infection**

Bacteriophages are classified as either virulent (such as T4 and T5) or temperate (such as P1, P22, and λ). When a bacterial host is infected by a *virulent* phage, a lytic response is inevitable: The phage multiplies extensively within the cell, which ultimately bursts (lyses) and dies. Infection by a *temperate* phage brings either a lytic or a lysogenic response. In the latter, replication of the phage genome is limited, and the phage genome continues to coexist within the host as a "prophage," either a separate, plasmidlike piece of DNA (as in the case of P1) or a part of the host chromosome (as in the case of phage λ). Many prophages can be "induced" to initiate a lytic cycle by inactivation of repressor proteins.

**BOX 3.5**

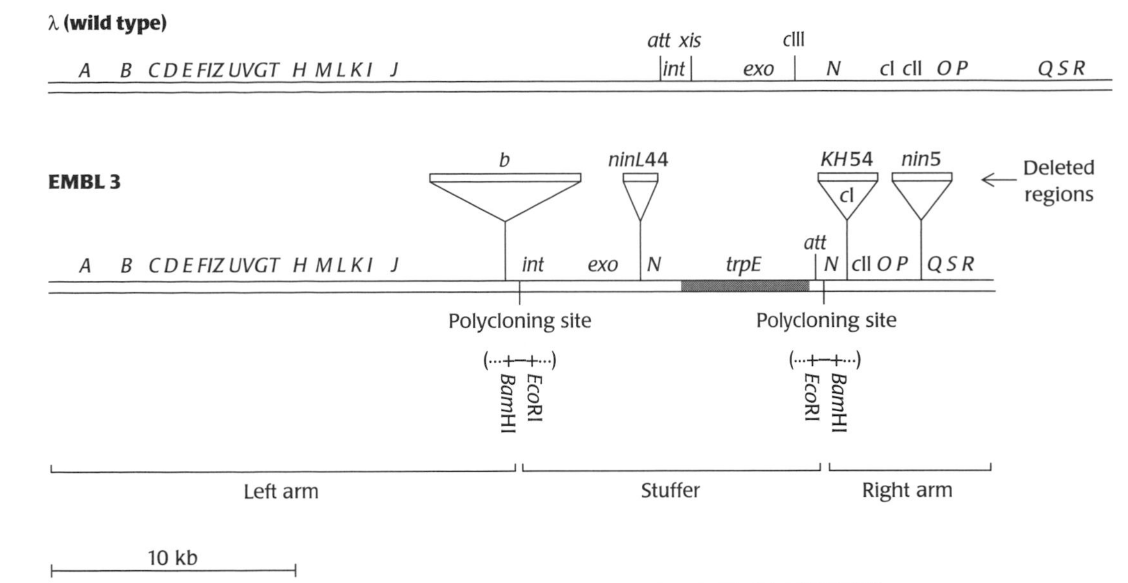


**FIGURE 3.11**

Phage λ and the EMBL3 vector. The λ-based vectors require that both the vector DNA itself and the recombinant DNA be packed efficiently into the λ phage heads. Packaging demands that the DNA have a length between 78% and 105% of the length of the normal λ phage DNA, so *replacement vectors* such as EMBL3 contain a *stuffer* segment, which is replaced by a foreign DNA segment of the same or somewhat larger size in the recombinant DNA constructs. More specifically, to create EMBL3, several deletions and one insertion (the *trpE* gene) were made in the λ genome. The stuffer sequence between the two polycloning sites increases the size of the vector DNA itself so that it is packaged efficiently into phage heads, allowing workers to prepare sufficient quantities of vector DNA by propagating the vector as a phage. The cloning is performed by cutting the vector DNA at the two polycloning sites, preferably with BamHI, removing the stuffer fragment, and then ligating the insert DNA (cut partially with Sau3A, which generates the same overhanging ends as BamHI) in between the two polycloning sites. The inserted fragment thus replaces the stuffer fragment in the vector, producing recombinant DNA large enough to be packaged into phage heads. Instead of physically removing the stuffer fragment by electrophoresis, one can cut the mixture of the three fragments of vector DNA further with EcoRI (there is no other EcoRI site in the vector), thereby preventing the stuffer, now with EcoRI ends, from becoming religated in the middle of the vector. In the recombinant DNA, the sequences necessary for lysogenic integration into chromosomal DNA (*att* and *int*) are deleted with the stuffer segment. Thus, the "phage" particle containing the recombinant DNA can cause only lytic infection of the host. [Modified from Sambrook, J., Fritsch, F. F., and Maniatis, T. (1989). *Molecular Cloning: A Laboratory Manual*, 2nd Edition, Cold Spring Harbor, NY: Cold Spring Harbor Laboratory Press.]

of the transformation process). With some vectors, the recombinant DNA may become integrated into the host chromosome as a prophage and can be stably maintained as such until the prophage is induced to initiate the lytic cycle. With others, however, the part of the phage genome that is required for integration has been deleted (for an example in EMBL3, see Figure 3.11), and all infection events result in extensive multiplication of the phage, followed by cell lysis.

Some products of foreign genes are very toxic to the host, and it is difficult to clone such genes by using a plasmid vector, even when the plasmid exists in small numbers of copies per cell ("has a low copy number"). This is because we can isolate and identify plasmid-containing bacterial strains

**Termination of mRNA Transcription in *E. coli***

The elongation of mRNA is terminated by two major mechanisms. In *rho-dependent termination*, a protein factor, rho, recognizes a complex nucleotide sequence and releases the RNA polymerase from the DNA helix. In *rho-independent termination*, a simpler nucleotide sequence, forming a short loop followed by a succession of U in the mRNA, is recognized by the RNA polymerase itself as the termination signal.

BOX 3.6

only when the plasmids coexist with the host bacteria for many generations. For the cloning of such deleterious genes, the λ phage vectors of the nonintegrating type are ideal; with such vectors the infected host cells are soon killed anyway, and the toxicity of the cloned protein does not make much difference. Lambda-based vectors are also very effective at expressing foreign genes, because some promoters in the lambda genome are quite powerful, and because lambda produces an antiterminator protein N, so that rho-dependent termination of transcription (Box 3.6) can be suppressed. Phage λgt11 is an example of a vector that is useful when the screening of the clones is dependent on the expression of foreign genes.

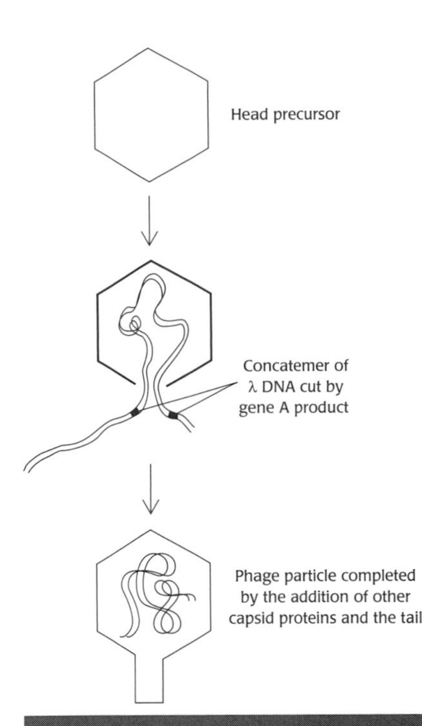


FIGURE 3.12

Packaging of DNA into phage head. Normally, λ DNA is produced as concatemers. An enzyme associated with the phage head (gene *A* product) cuts the DNA at each *cos* site, and the linear DNA is then packaged into the phage particle, together with the tail that has been assembled separately.

Redrawn based on artwork from the first edition (1995), published by W.H. Freeman.

**Cosmids.** λ DNA is synthesized in the cytoplasm of the infected cells as a polysequence, or concatemer, containing several repeats of the λ genome. A λ-coded enzyme recognizes the *cos* (or *co*hesive *s*ite) sequences that correspond to the proper ends of the genome and cuts the DNA at these points, preparing it to be packaged into the head (Figure 3.12). *Cosmid vectors* are vectors that contain λ *cos* sites but little other material derived from the λ genome. Foreign DNA inserts are cloned between the two *cos* sequences, which then initiate the *in vitro* packaging of the recombinant DNA, composed of the cosmid and its insert, into λ phage heads. Cosmids also contain a plasmid origin of replication, so that they can be replicated as plasmids, and an antibiotic resistance marker, so that cosmid-containing cells can be selected for (Figure 3.13). Because the cosmid vector is so small (typically only several kilobases), it is possible to clone up to 40 kb of foreign DNA into cosmids and deliver the recombinant DNA very efficiently via phagelike particles assembled *in vitro*. Because of their ability to incorporate larger pieces of foreign DNA, cosmids are significantly better than λ vectors for cloning genomic DNA of higher eukaryotes. However, because cosmids have to be propagated as plasmids, it is difficult to use them for cloning genes (or cDNAs) that code for proteins that are toxic for the *E. coli* hosts.

**Bacterial Artificial Chromosome.** Cosmids allow the cloning of DNA sequences up to about 40 kb. However, some genes of higher eukaryotes, containing many introns, are larger. Furthermore, in sequencing the genomes of higher animals and plants, it is necessary to begin with clones of very large segments of DNA, containing hundreds of kilobases. For such purposes, yeast artificial chromosomes, or YACs (described later in this chapter), were the standard vector. However, more recently, bacterial artificial chromosomes (BACs) are the vectors that are most often used. BACs are plasmid vectors, with the F-factor origin of replication and with genes that ensure the partition of the plasmid into both of the daughter cells. BACs are maintained at a very low copy number (1 or 2 per cell), just like the

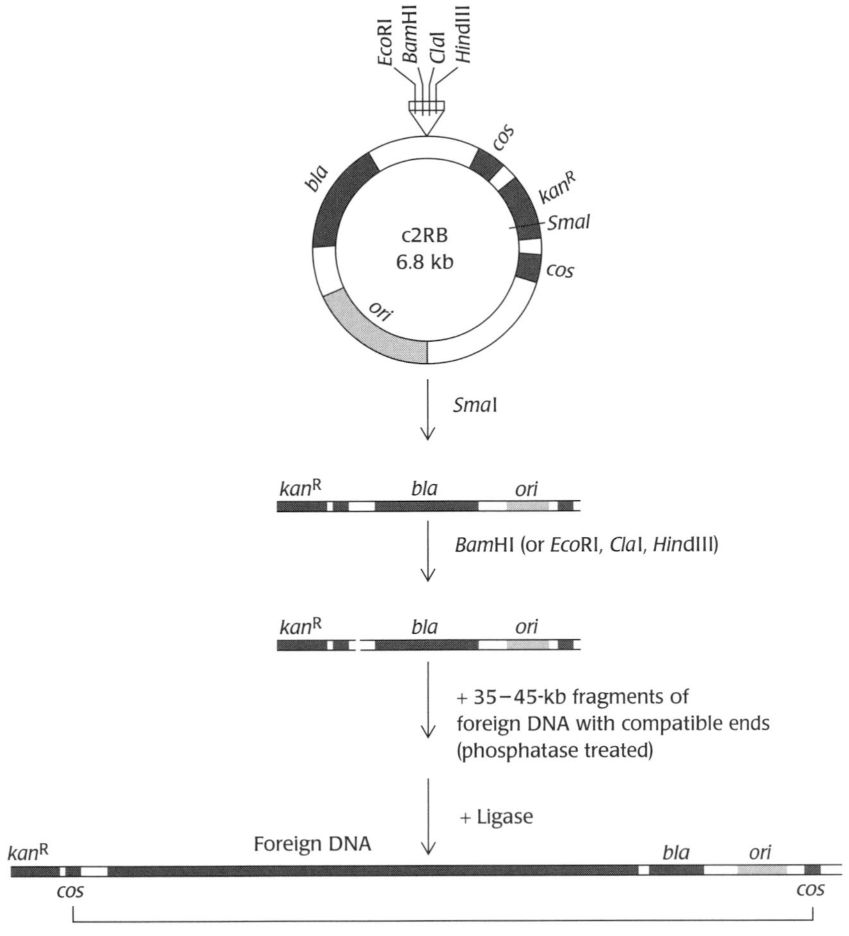


FIGURE 3.13

Cloning with a cosmid vector. The cosmid vector c2RB, shown as an example here, contains two *cos* sites, a plasmid origin of replication, a polycloning site (a short stretch of DNA containing cleavage sites for several restriction endonucleases), and two antibiotic resistance markers ($Amp^R$ and $Kan^R$). Cutting the cosmid with, say, SmaI and BamHI produces two cosmid halves. Ligation with 40-kb fragments of foreign DNA partially digested with MboI or Sau3A (which produce the ends complementary to those produced by BamHI) creates the construct shown. This is then packaged *in vitro* and introduced into *E. coli*. The strains containing recombinant DNA should be both ampicillin resistant (because of the *bla* gene) and kanamycin sensitive, because packaging eliminates the kanamycin resistance gene. The latter feature is useful for eliminating plasmids made of multiple copies of the fragments of the vector. [Modified from Sambrook, J., Fritsch, F. F., and Maniatis, T. (1989). *Molecular Cloning: A Laboratory Manual*, 2nd Edition, Cold Spring Harbor, NY: Cold Spring Harbor Laboratory Press.]

F-factor, and this also helps the stable maintenance of BAC-based constructs in *E. coli*. The large portion of F-factor, coding for the cell-to-cell transfer of this DNA through conjugation, has been removed so that it will not spread to other cells. YAC DNA is difficult to separate from yeast chromosomal DNA because it behaves almost exactly like other yeast chromosomes. In contrast, the BAC plasmid is easy to isolate away from the bacterial chromosome. Another major advantage of the BAC system is that it is rare for a BAC-based recombinant plasmid to contain more than one piece of cloned DNA, in contrast to YAC-based constructs that tend to contain chimeric pieces of foreign DNA at a very high frequency.

**Derivatives of Single-Stranded DNA Phages.** One closely related family of phages (fl, fd, M13) infects only those *E. coli* cells that contain the F sex factor. A remarkable feature of these phages is that they continuously produce progeny phages within the growing cells without causing the lysis and death of the host. These phages contain a circular, single-stranded DNA about 6.4 kb long that is replicated as a double-stranded, plasmid-like entity in the *E. coli* cell. The phage particle itself is filamentous, so insertion of foreign DNA

**Sequencing of DNA by the Dideoxy Chain Termination Method**

This ingenious method, developed by Frederick Sanger, makes possible the sequencing of fairly long stretches of DNA. The first step is to anneal an oligonucleotide primer to the single-stranded DNA one wishes to sequence. DNA polymerase then synthesizes the complementary strand as a 3′-extension of the primer. To each of four such reaction mixtures, one adds low concentrations of an unnatural nucleoside triphosphate containing 2,3-dideoxyribose rather than 2-deoxyribose. This causes chain elongation to stop on those occasions when the unnatural nucleotide is incorporated into the DNA strand. If the template strand contains, for example, C at positions 50, 55, and 60, then the newly made complementary strand becomes truncated when dideoxyguanosine phosphate is incorporated at the corresponding positions. Thus, DNA of 50, 55, and 60 nucleotides in length will be made only in the reaction mixture to which dideoxyguanosine triphosphate was added. Analysis of the products by gel electrophoresis, on the basis of their length, thus permits unequivocal sequencing of the DNA.

BOX 3.7

at an intergenic site within the phage DNA simply results in an elongation of the phage particle.

These vectors, when they were developed, were essential for DNA sequencing by the dideoxy chain termination method (Box 3.7). However, sequencing is now carried out nearly entirely by using double-stranded DNA. The single-stranded DNA phage vectors were also the vectors of choice for site-directed mutagenesis, but now this can be carried out also by using double-stranded DNA constructs (Figure 3.14). One area where such vectors are still useful is the "phage display" of mutated proteins (Figure 3.15). In this strategy, a DNA sequence coding for a foreign protein of interest is inserted into the 5′-terminal domain of the phage gene coding for protein III. This protein is located at the tip of the filamentous phage, and its N-terminal domain extends into the medium. When the foreign gene is mutated by a site-directed, or random, mutagenesis procedure, each phage particle will express one specific mutated version of the protein. These phages can then be selected out by their affinity to a target. Thus, if the foreign gene codes for an antibody, then the phage expressing a higher affinity antibody can be "fished out" of a mixture of millions of phages, and because the gene coding for this desired mutant is located within the phage, it can be easily recovered. This physical connection between the mutated protein and the coding gene makes this approach extremely useful, especially in an effort to "evolve" proteins of interest through a random mutagenesis approach. More recently, the *ribosome display* method, which exploits the physical connection between the translating ribosomes and the mRNA, has been introduced.

Two convenient features that were first introduced into M13 vectors (Figure 3.16) are now present in many vectors of other types. The first is a system for distinguishing between recombinant clones and the original vectors. It consists of a fragment of the *lacZ* gene that contains the portion coding for the N-terminal fifth of the LacZ protein. When this truncated LacZ fragment is expressed in a host cell that contains a *lacZ* gene lacking the 5′-terminal part of the gene, both fragments can assemble together spontaneously to produce a functioning enzyme (alpha-complementation). Thus, when a cell harboring this vector phage is placed on a plate containing 5-bromo-4-chloro-3-indolyl-$\beta$-D-galactopyranoside (X-gal), hydrolysis of X-gal by $\beta$-galactosidase (LacZ protein) produces indoxyl, which is oxidized to indigo that stains the colony blue. When a segment of foreign DNA becomes inserted into the cloning site, the coding sequence of the truncated *lacZ* gene is interrupted, the functional N-terminal LacZ fragment is

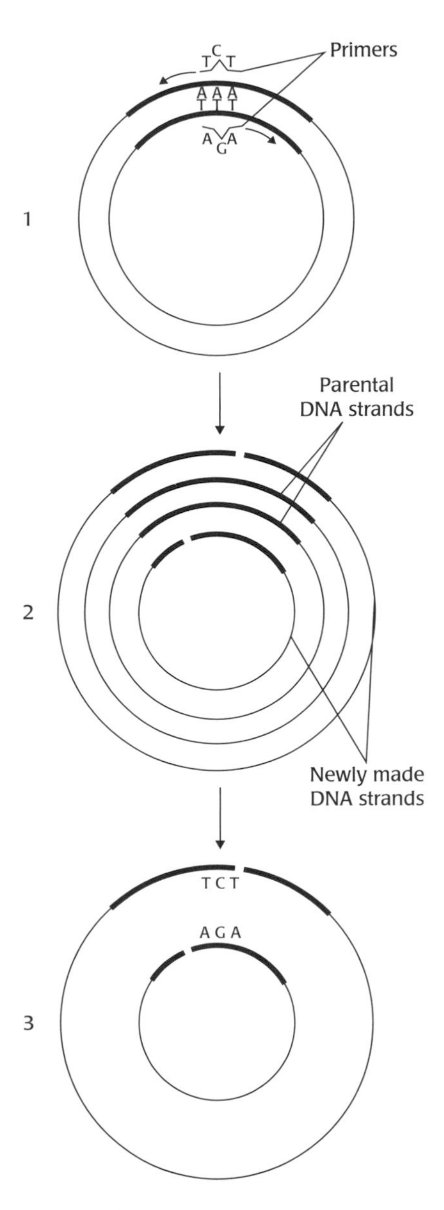


**FIGURE 3.14**

Site-directed mutagenesis. This figure illustrates the QuickChange method developed by Stratagene. Starting from a recombinant plasmid containing the target gene insert (*thicker line*), two primers covering the same overlapping area are used. If one wants to change a lysine residue in the protein (coded by the AAA codon) into arginine (coded by the AGA codon), the primers will contain a single mismatched nucleotide residue (AGA and TCT, respectively). These are indicated by protrusions in the figure, step 1. PCR is used to elongate the primer and cover the entire plasmid, as well as to amplify the DNA, resulting in the structure shown as step 2, which contains both the newly synthesized strands containing the desired mutation as well as the parental strands not containing the mutations. The latter were made in *E. coli* cells, and therefore some of the bases are methylated. Treatment with the restriction endonuclease DpnI, which specifically cleaves DNA containing 6-methylated guanine, cleaves the parental strands, leaving behind the *in vitro* synthesized, and therefore unmethylated, strands that contain the desired mutation (step 3).

not produced, and the colony stays white. (In principle, the same effect can be achieved by inserting the whole *lacZ* gene in the vector. However, *lacZ* is a large gene, and introducing large DNA fragments makes the recombinant M13 construction rather unstable.) The second feature is the insertion, close to the beginning of the *lacZ* gene, of a short sequence called *polylinker*, or *multiple cloning site*, designed to contain cleavage sites for many popular restriction enzymes. This sequence serves as a convenient site of insertion of foreign DNA. Its proximity to the efficient *lac* promoter ensures good expression of the cloned gene, as long as the gene is in the correct orientation. (The advantage of this construction for gene expression is further discussed in the section dealing with expression vectors).

*Phagemids* are a variant on these vectors. These chimeric vectors contain two origins of replication, one from a plasmid and the other from fl or some

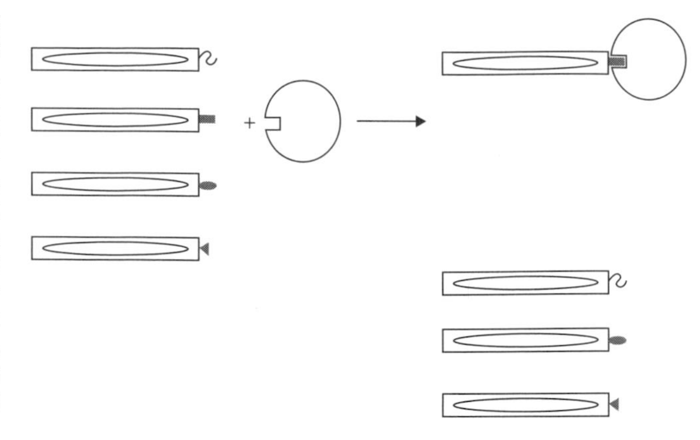

FIGURE 3.15

Phage display technology. In this approach, the DNA sequence coding for the protein to be mutated is inserted in the exposed domain of the protein III (PIII), which is located at the tip of the filamentous phage such as M13. This sequence is subjected to random changes (e.g., by using, in the chemical synthesis of DNA, a mixture of four deoxynucleoside triphosphates, rather than the single one, at given positions in the sequence), and the replicative form DNA is transformed into *E. coli*. An assembly of phages producing various mutated forms of the protein will emerge from the host cells. This mixture is then subjected to an affinity selection, for example, with a protein that may be expected to interact with the protein at the phage tip. Only the phage with the tip protein domain that "fits" with the protein used for selection will be retained. The precise mutation in the gene can be recovered from the genome of the phage that has been retained.

other phage. These vectors multiply as plasmids in the host cell because they lack genes needed for replication as phage DNA and for the assembly of phage particles. However, once the missing phage functions are supplied by superinfecting the host with helper phages, they are replicated as phage DNA, packaged, and released into the medium as phagelike particles.

## DETECTION OF THE CLONE CONTAINING THE DESIRED FRAGMENT

Cloning fragments of genomic DNA is not a difficult task. Many restriction endonucleases are commercially available, as are numerous vectors of sophisticated design, such as those we have described. Usually the most challenging step in the shotgun cloning is the detection, among many clones, of the ones that contain the fragment of interest. The magnitude of this task becomes clear when we realize that even with vectors that can accept a 20-kb piece of DNA, and even when the source is a bacterial genome (5000 kb), to have a 99% probability of finding one clone with the desired gene, we have to examine about 1000 clones (see Box 3.1). The thought of attempting the same task with the genome of a higher eukaryote, which could be almost three orders of magnitude larger than that of *E. coli*, is daunting to say the least. One would have to examine almost a million recombinant clones in order to be 99% certain of recovering the gene of interest. Thus, careful strategic planning becomes necessary for the identification of desired clones.

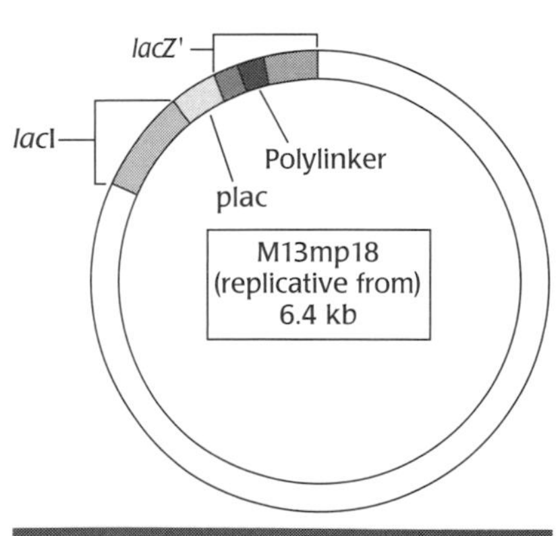


FIGURE 3.16

An example of M13-derived vectors. The M13mpl8 vector contains a polylinker sequence near the 5′-terminus of the sequence coding for a fragment of the *lacZ* gene called *lacZ′*. Precise sequences of the promoter region and one version of the polylinker of this type are given in Figure 3.21.

Redrawn based on artwork from the first edition (1995), published by W.H. Freeman.

### Importance of Using a Better Template

If one wants to express eukaryotic proteins in bacteria, which cannot carry out the splicing reaction, it is best to use mRNA as the template, because the sequences corresponding to the intron sequences are already spliced out, as described earlier. In such cases, the usual procedure is to obtain the specific types of cells in which the target gene is expressed strongly and then to use the mRNA from those cells as the source of genetic information. This approach exploits a source in which the sequence of interest has been very strongly amplified. In such cases, the recombinant DNA constructs will be highly enriched in the target sequence, so a minimal amount of screening will be needed to isolate the desired construct. Large amounts of stable RNA (ribosomal RNAs, transfer RNAs, and so on) are also present in any cell, but they can be removed easily by taking advantage of the fact that eukaryotic mRNA molecules have a polyA "tail" at the 3′-terminus. Once the mRNA fraction is isolated, it can be purified according to size in order to obtain a fraction enriched in the sequence of interest. The mRNA is then converted into double-stranded cDNA as described above (see Figure 3.6) for insertion into a cloning vector. Most of the sequences coding for the production of animal and human peptides and proteins have been cloned by using mRNA preparations.

All of the recombinant constructs should contain the desired piece of DNA, if it was created by the PCR amplification (see below); this allows us to totally circumvent the clone identification step as well as all the other steps of the shotgun cloning.

### Clone Identification Based on Protein Products

When a cloned gene is expected to be transcribed and translated in the host bacterium (see the discussion of expression vectors that follows), the task of identification, and perhaps even selection (see Box 3.3), of the cells containing the right clone is fairly straightforward. In the simplest case, one can test for the function of the protein coded by the cloned gene. Let us assume that we want to clone from some organism (call it Organism A) the gene for anthranilate synthase, an enzyme involved in the synthesis of tryptophan, for the purpose of improving the commercial production of this important amino acid (see Chapter 9). *E. coli*, like most bacteria, can synthesize all the usual amino acids from simple carbon sources and ammonia, and it contains a gene, *trpE*, that codes for anthranilate synthase. We first introduce a mutation into the *E. coli trpE* gene. The mutant strain cannot synthesize tryptophan and thus cannot grow unless we add tryptophan to the growth medium. We now introduce into this strain, by transformation, recombinant plasmids containing fragments of the DNA of Organism A and spread a large number of transformant cells on a solid medium that does not contain tryptophan. Most of the cells contain either no plasmid or plasmids with irrelevant pieces of DNA and are unable to grow. The only cells that grow and form visible colonies are those that contain the rare recombinant plasmid with the *trpE* homolog from Organism A. In this manner, we achieve an efficient selection of these rare plasmids.

In the foregoing example, the desired gene had a function required in many microorganisms. In some cases, however, the desired gene would have a significant function in the source organism, Organism A, but not in *E. coli*. An example is an attempt to clone a gene coding for one of the enzymes of the xylene degradation pathway from *Pseudomonas putida*. Many strains of this organism contain a series of enzymes that lead to the complete oxidation of an aromatic hydrocarbon, xylene, but one of these enzymes can perform no useful function in *E. coli*, which does not contain any other enzymes of this series. *Shuttle vectors*, which contain origins of replication of both *E. coli* and some other microorganism, are useful in such situations. We can then screen for the clone that expresses the desired function in a mutant of Organism A that lacks this function, because the recombinant plasmids will be replicated in this organism. At the same time, we can propagate the plasmids in *E. coli*, in which subcloning and other procedures can be carried out more easily.

In many cases, though, a *complementation assay* such as the one described would be difficult or even impossible to perform. For example, if we are trying to clone a eukaryotic gene coding for a hormone that has no homologs in unicellular bacteria, complementation cannot be used as a method of detection. A frequently used approach in these cases is to detect production of the desired protein by its reactivity with specific antibodies. Unfortunately, this usually involves screening, rather than selection, of the recombinant clones. However, if the screening can be carried out on plates with hundreds of colonies on each, it is not so difficult to test tens of thousands of recombinant clones in a single experiment. λ Phage vectors are especially convenient for this method, because within each plaque (see the legend of Figure 3.2) generated by lytic infection by a recombinant phage or by induced lysis of an *E. coli* strain lysogenic for a recombinant phage, the cells will have been lysed already, releasing into the medium the proteins expressed from the recombinant fragment. Furthermore, because a single lysing cell contains hundreds of copies of the phage genome, each including the cloned piece, the expression of the cloned genes is strongly enhanced. The λ gt11 expression vector was especially constructed for screening of this type.

### Clone Identification Based on DNA Sequence

The methods we have examined depend on successful expression of the cloned genes. But this is not always assured, especially when the cloned DNA comes from a source phylogenetically distant from the bacterial host. The RNA polymerase of the host bacteria does not recognize the promoter and other regulatory elements of eukaryotic genes, or even those of remotely related bacteria. Pieces of cDNA lack such regulatory "upstream" sequences altogether, and genomic DNA from eukaryotes will not result in the production of whole proteins because of the presence of introns.

Because of these problems, it often becomes necessary to identify the clone containing the desired fragment by its DNA sequence. Scoring for such clones can be done by hybridization with suitable DNA probes, labeled either with a radioactive isotope or with chemical substituents that can be detected by nonradioactive methods, such as fluorescence. The major hurdle in this procedure is finding the requisite DNA probe, especially when the exact sequence of the clone is not yet known. This is not an insurmountable problem, however. If the sought-after gene has homologs in related organisms, and if their sequences are known, it is possible to design probes that correspond to the most conserved regions of the aligned sequences and use conditions of low stringency for hybridization. In fact, this is probably the most frequently used method for cloning genes and cDNAs from eukaryotes, because the evolutionary divergence between higher eukaryotes tends to be quite small in comparison with that between prokaryotic groups. Alternatively, if at least a partial sequence of the protein is known, one can deduce the DNA sequences that would code for such an amino acid sequence and use a "degenerate" probe that contains a mixture of these possible DNA sequences. Using this approach is particularly advantageous when the peptide sequence does not contain amino acids that, like leucine or arginine, are coded by many codons.

In practice, the cells containing recombinant plasmids are spread on plates so that there will be a few hundred colonies per plate. These colonies are replica-plated onto a filter and placed on a fresh plate. After the cells have grown, the filter is lifted out and treated with an NaOH solution to lyse the cells and denature the DNA. The proteins are digested by a protease, and the DNA is fixed onto the filter by "baking" at 80°C. The filter is then incubated with the labeled probe DNA, and any probe that anneals to the DNA on the filter is detected after suitable washing (Figure 3.17). Although this is only a

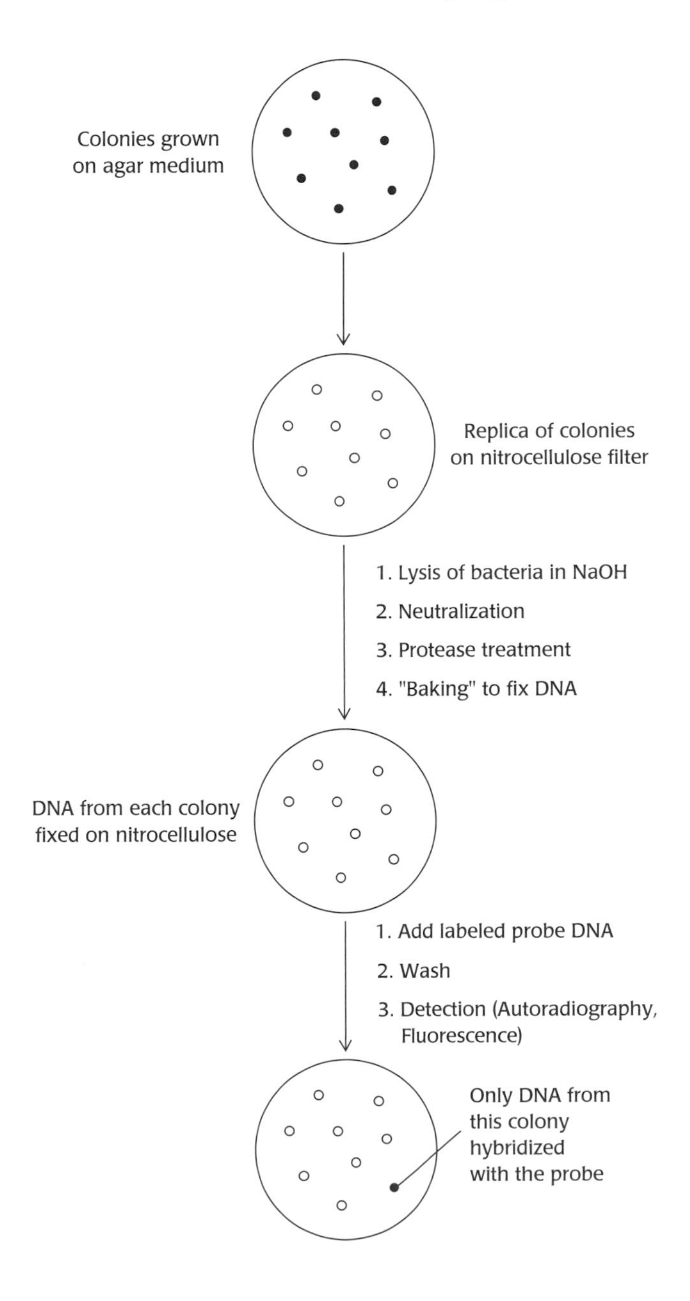


**FIGURE 3.17**

Use of a DNA probe to detect the desired recombinant clone.

Redrawn based on artwork from the first edition (1995), published by W.H. Freeman.

screening method, in this way one can test a fairly large number of colonies in a short time.

### Combined Detection of the DNA Sequence and the Protein Product

In some cases, it is possible to combine the two methods we have outlined. For example, to clone gene *X* from a bacterium very distantly related to *E. coli*, one might begin by making random transposon insertions (Box 3.8) in the chromosomes of the bacterium containing gene *X*. If a transposon inserts into gene *X*, it will disrupt this gene, generally resulting in a recognizable phenotype. Genomic DNA from this mutant organism is then cloned into a plasmid vector, and the recombinant plasmids are introduced into an *E. coli* host strain by transformation. Most transposons contain a resistance marker that codes for antibiotic resistance or resistance to other toxic compounds, such as mercury (Figure 3.18). Moreover, because a transposon is a piece of "selfish" DNA that propagates itself in diverse species of bacteria (see Box 3.8), its resistance genes are designed to be expressed efficiently in many bacterial species. Thus the resistance gene, located within the transposon in the recombinant plasmid, will be expressed in the *E. coli* host, and it should therefore be possible to select for this plasmid.

A clone of gene *X* still exists in two pieces within the plasmid, flanking the transposon. The cloned DNA can be cut out of the plasmid, and the fragments of gene *X* DNA can be used as probes in the next phase. One then clones random fragments from the wild-type genome (which does not contain the transposon) into a plasmid or other vector. Screening of these recombinant clones with the DNA probes of gene *X* described above will lead to identification of the clone that contains the wild-type version of the gene, uninterrupted by the transposon sequence. Alternatively, the sequence of the wild-type gene can be retrieved by the procedure called inverse PCR (see below).

**Transposons**

A transposon is a segment of DNA that has the ability to insert its copy at random sites in the genome. Thus, a transposon has a natural tendency to increase the number of its copies under favorable conditions and can be considered an example of a piece of "selfish DNA." It is bounded by inverted repeat sequences, and it usually contains a drug resistance gene. It also contains gene(s) for enzymes that catalyze the transpositional insertion. Because of these properties, transposons played a major part in the natural formation of "R plasmids," which code for resistance to many antibacterial agents. Transposons are also very useful as genetic tools. For example, introduction of a transposon into a new cell results in the insertion of copies of the transposon at various places on the chromosome. Insertion of such large fragments in the middle of a gene totally inactivates the gene, so *transposon mutagenesis* is a convenient method for generating null mutants (mutants in which the gene function is utterly obliterated). Furthermore, the mutations are easy to analyze genetically, and the alleles can be easily cloned because of the presence of antibiotic resistance markers within the transposon sequences.

**BOX 3.8**

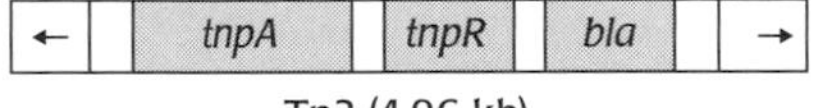

Tn3 (4.96 kb)

**FIGURE 3.18**

An example of a transposon, Tn3. The genes *tnpA* and *tnpR* code for transposase (cointegrase) and resolvase, two enzymes needed for the insertion of a copy of the transposon into a new site on the DNA duplex. [For the mechanism, see Grindley, N. D. F., and Reed, R. R. (1985). Transpositional recombination in prokaryotes. *Annual Review of Biochemistry*, 54, 863–896.] The gene *bla* codes for a $\beta$-lactamase, which produces resistance to $\beta$-lactam antibiotics such as ampicillin and cephalothin. The ends of the transposon contain inverted repeats (*arrows*).

Redrawn based on artwork from the first edition (1995), published by W.H. Freeman.

## POLYMERASE CHAIN REACTION AND THE UTILITY OF GENOMIC DATABASES

In cases in which we know at least short stretches of nucleotide sequence either within the gene of interest or in an area flanking that gene, it is possible to isolate the desired clone without going through the painstaking shotgun cloning procedures we have described. This is done with a technique known as polymerase chain reaction (PCR), and in 1993 its inventor, Kary Mullis, received a Nobel Prize in chemistry for devising it. In this method, we first synthesize oligonucleotide primers that are complementary to opposite strands of these short stretches of DNA (Figure 3.19). We then add a large excess of these primers to a denatured preparation of genomic DNA or cDNA and let the primers anneal to the complementary sequences. Adding a heat-resistant DNA polymerase and a mixture of deoxyribonucleoside triphosphates results in the elongation of primers into complementary strands of DNA, as shown in Figure 3.19, step 1. The mixture is next heated to denature the DNA again, and then it is rapidly cooled. Under these conditions, annealing occurs predominantly between primers and DNA strands (some of which are newly synthesized), rather than between long DNA strands, because the latter takes place more slowly. Because the polymerase is heat resistant, DNA synthesis begins again by utilizing the primers (Figure 3.19, step 2). In the first round of DNA synthesis, the newly made DNA strands have random ends. In the second round, the mixture becomes enriched for strands that begin and end at sequences corresponding to the two primers, because some primers have annealed to the strands synthesized in the first round (Figure 3.19, step 2). After the DNA synthesis and DNA denaturation/annealing steps are continued for many cycles, most of the newly made strands will have a finite length and will correspond only to the limited region of the DNA between the two primers – that is, to the gene of interest if the primers corresponding to the regions flanking the gene were used.

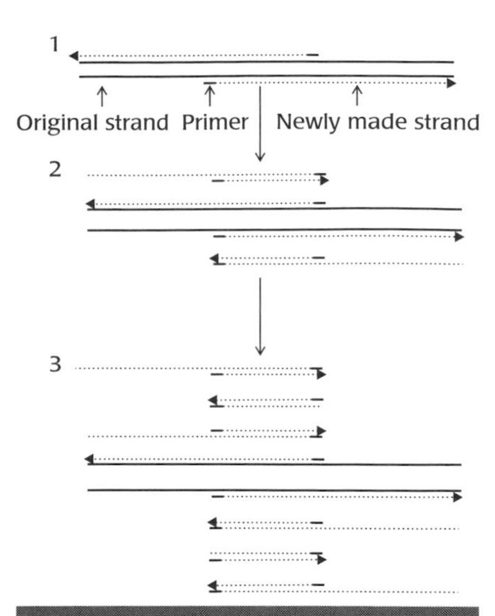


**FIGURE 3.19**

Amplification of a defined segment of genome by PCR. In step 1, primers (*short lines*) are annealed to complementary sequences of genomic DNA (*continuous lines*). Addition of DNA polymerase and deoxyribonucleoside triphosphates results in elongation of the primer (*dotted lines*). In step 2, the reaction mixture is heat denatured and then renatured, causing some of the primers to anneal to newly made strands (*dotted lines*). Elongation produces new strands, two of which now have a limited length, terminating at positions corresponding to the primer sequences. In the third cycle, after denaturation, renaturation, and elongation again, eight out of the total of 16 strands have this limited length. After further cycles, practically all of the newly made DNA will have the finite, short length.

Redrawn based on artwork from the first edition (1995), published by W.H. Freeman.

The crucial factor for the success of PCR was the discovery of a thermostable DNA polymerase that can withstand many cycles of heating and cooling. In theory, the usual heat-labile enzyme should suffice if it is added fresh at the beginning of every cycle; however, a large number of cycles are required to achieve a high degree of amplification, and impurities brought in each time the enzyme is added eventually inhibit the reaction. The heat-stable enzyme commonly used (Taq polymerase) is derived from a thermophilic Gram-negative eubacterium, *Thermus aquaticus*, which grows optimally at around 70°C to 80°C. Lately, even more thermostable DNA polymerases, isolated from archaebacteria living in marine thermal vents at temperatures of 98°C to 104°C, are being used in PCR procedures.

There are several ways to clone the PCR amplification product into vectors. The Taq polymerase creates products with a one-residue overhang of deoxyadenosine at the 3′-end. Thus, the product can be cloned into a cleaved site of a vector, which contains the complementary one-residue overhang of deoxythymidine at the 5′-end. Alternatively, the 3′-5′ exonuclease activity of the Klenow fragment of DNA polymerase can be used to remove the 3′-overhang of the PCR product, to create flush or "blunt" ends. The product can then be cloned into vectors, which were cleaved with endonucleases known to create blunt ends, by a process called "blunt end ligation." Perhaps the most efficient procedure is to use primers that contain 5′-extensions corresponding to the restriction sites of endonucleases to be used. The presence of such extra sequences does not inhibit the PCR process. The product is then cleaved with the restriction endonuclease(s), to generate sticky ends that will anneal with the complementary sticky ends of the vector, created by cleavage by the same enzymes.

One potential problem with the use of Taq polymerase is that it lacks the 3′-5′ exonuclease activity that is used in "proofreading" the newly made strand. Thus, errors occur during DNA synthesis, and if they occur in the early cycles, the amplified DNA may differ in sequence from that of the original template. However, the frequency of error is strongly reduced with the newer archaebacterial enzymes, some of which contain the 3′-5′ exonuclease activity (Chapter 11).

The PCR procedure offers important advantages. As we have seen, one can totally circumvent the complicated cloning steps as well as the steps involved in identifying the clone that contains the desired gene. Moreover, because the degree of amplification is very large, the process is extremely sensitive. In theory, this procedure could amplify a single copy of a gene, and in many experiments the results have approached this limit. This has led to the development of many diagnostic tools: whereas it took many days (or even weeks) to culture and identify pathogenic bacteria infecting patients, it now takes only a few hours to show the presence of such pathogens by amplifying specific DNA sequences of each pathogen. There are several commercial systems that were approved by the Food and Drug Administration (FDA) for detection of *Mycobacterium tuberculosis*, which causes tuberculosis and grows extremely slowly. Diagnostic PCR is not limited to detection of pathogens. Some types of cancer cells are marked by characteristic changes in the genome, and these can be detected by PCR in a sensitive way.

PCR depends on knowing the sequence either within or around the gene of interest. One way to satisfy this requirement is to isolate the protein product and to determine the amino acid sequences of the N-terminus and of internal peptides generated by the enzymatic or chemical cleavage of the protein. Primers are then made on the basis of these amino acid sequences. These primers are mixtures containing various degenerate codons.

In recent years, however, there has been an explosive growth in our knowledge of genome sequences. The Comprehensive Microbial Resource webpage at The Institute for Genomic Research (http://www.tigr.org) lists, at this writing, more than 400 complete or nearly complete genome sequences of microorganisms. The sequences of individual genes and fragments deposited in GenBank and other databases exceed, by far, the sequences in complete genomes, and the sum of both types of sequences reached 100 gigabases (Gb), or $10^{11}$ bases, in August 2005. This huge size of information on genes of diverse organisms, coding for almost any imaginable function, now allows us to construct primers on almost any gene. As a hypothetical example, let us suppose that you are interested in the biological oxidation of MTBE (methyl *tert*-butylether), a gasoline additive, which is not easily biodegraded and is polluting the environment. In your search for organisms and enzymes that degrade this compound, you find an article reporting that propane monooxygenase of *Mycobacterium vaccae* rapidly oxidizes MTBE to convert it into innocuous products. However, all *Mycobacterium* species are classified as potential human pathogens, and therefore there is no chance that you can use this bacterial species directly for environmental cleanup. You will thus have to clone the gene coding for this enzyme from *M. vaccae*. However, the sequence of this gene is not known. You know, however, that a propane monooxygenase gene has been sequenced from a related genus, *Gordonia*. In such a situation, you can take advantage of the enormous amount of our knowledge on DNA sequences by first pulling out the *protein* sequences that are most related to the *Gordonia* protein sequence. (Nucleotide sequences change too rapidly during evolution, and it is more useful to rely on protein sequences in order to find homologs). This can be done by using the program BLAST, at the website of National Center for Biotechnology Information (http://www.ncbi.nlm.nih.gov). You can then align the sequence of the *Gordonia* enzyme with those of several homologous enzymes, preferably coming from different genera. This will show two internal segments in which stretches of at least five amino acids are completely conserved. You can design primers from these sequences, taking the codon degeneracy into account, and amplify an internal fragment of the desired gene from *M. vaccae*.

In the example above, the PCR procedure amplified only a fragment of the desired gene. A procedure called inverse PCR is a convenient starting point for the cloning of the entire gene. As shown in Figure 3.20, one cuts the chromosome with a restriction enzyme, and then self-ligates the fragments to make them circular. Use of primers from the already cloned small fragment, going in divergent directions, will result in the amplification of the entire sequence of the larger chromosomal fragment. This can be sequenced to elucidate the exact sequences of the 5′- and 3′-termini of the complete gene, and these can then be used to design primers for the amplification of the complete gene.

Finally, error-free chemical synthesis of long (up to 40 kb) DNA is now possible. Thus the entire DNA sequence including promoter, operator, RBS, coding sequence (with optimal codon usage), and terminator at optimal locations and sometimes even including many genes, can be synthesized. Such an approach may soon replace most of the cloning and PCR methods described, if the cost of synthesis becomes competitive.

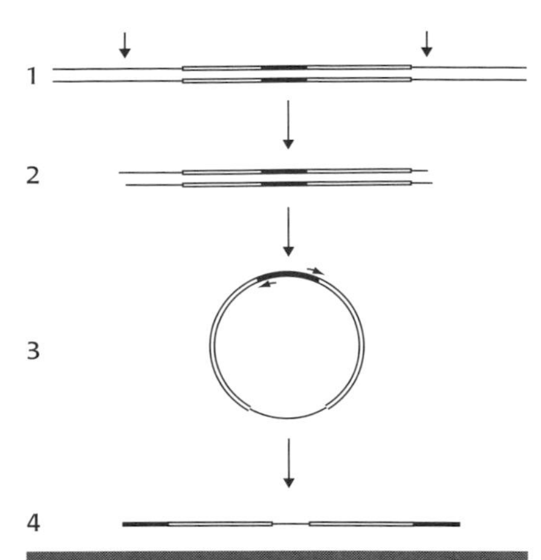


**FIGURE 3.20**

Inverse PCR. If we know the sequence of only a segment (represented by a *black section* in step 1) of a gene of interest (the rest of the gene is represented by an *empty box* in step 1), it is possible to recover the rest of the gene from the genomic DNA. (The known sequence could be that of a transposon that has inserted into the gene, see p. 112). We first cut the genomic DNA by using a restriction endonuclease (*vertical arrows* in step 1), generating fragments with complementary overhanging ends (step 2). Annealing of the ends and ligation produces circular DNA fragments (step 3). These DNA circles cannot replicate in intact cells as they lack the origin of replication. However, they can be replicated *in vitro* as linear pieces of DNA by using PCR. For this purpose, we use a set of primers that are directed outward from the known segment of the gene (see the *small arrows* in step 3). Because the normal PCR procedures use a forward primer and a reverse primer that face each other, this procedure is called *inverse* PCR, which results in the recovery of the flanking parts of the gene of interest (step 4). Many modifications have been devised for this approach.

## EXPRESSION OF CLONED GENES

The usual reason for cloning a gene is to obtain the protein product in substantial quantities. Even when that is not the case, identification of the correct clone often requires expression of the cloned gene. However, many of the general-purpose cloning vectors are not designed for strong expression of cloned genes. There are many reasons why genes in the fragments cloned into pBR322, for instance, are often expressed only at a low level. Frequently, the foreign promoter in a fragment from another organism is not efficiently recognized by *E. coli* RNA polymerase. In such a case, successful transcription must start from promoters recognized well by *E. coli* – those for the *tet* and *bla* genes – and continue onto the cloned segment of the recombinant plasmid. The problem is that there may be sequences in between that act as transcription terminators. Even if the mRNA is successfully produced, it may not contain the proper ribosome-binding sequence (see below) at a proper place. These difficulties indicate that a different arrangement is needed to ensure a high level of expression of foreign genes in a reproducible manner.

Vectors of a special class called *expression vectors* are designed for this purpose. Most expression vectors are plasmids because multiple copies of plasmids can exist stably in the cell. More plasmids carrying a given gene result in a higher production of specific mRNA because each copy of the gene is transcribed independently. This principle is sometimes called the *gene dosage effect*.

**Consensus Sequence**

Promoters of various genes share a common stretch of nucleotide sequence, with only minor variations. Such homologous sequences, and often their idealized versions, are called consensus sequences. Actual sequences may deviate significantly from idealized sequences: for example, the sequence 10 bases upstream from the transcription initiation site (−10 sequences) for *E. coli trp* operon (TTAACT) and *lac* operon (TATGTT) are similar but not completely identical to the consensus TATAAT.

BOX 3.9

Expression vectors must have a strong promoter. *E. coli* promoters contain two "consensus" sequences (Box 3.9): TTGACA, about 35 nucleotides upstream from the transcription start site, and TATAAT, about 10 nucleotides upstream. These two sequences are therefore separated by a 16- to 18-bp intervening region. When the vector's promoter has sequences that closely resemble the host bacteria's, the genes downstream of it tend to be expressed strongly. Strong promoters are also found in phage genomes, because phage life cycle depends on its proteins being produced in very large amounts during the short period of phage infection. The system using phage T7 promoter is described later in this chapter.

Strong promoters introduce a problem, however. When *E. coli* cells produce a very large amount of a protein that does not contribute to cell growth, such a situation tends to be deleterious. Thus, cells that have lost the plasmid, and cells whose plasmids have been altered and have ceased to produce the protein, have a competitive advantage and will eventually become the predominant members of the population. This instability can be a severe problem in industrial-scale production, because extensive scale-up means that *E. coli* must go through a proportionately larger number of generations, significantly increasing the likelihood that nonproducing cells will appear. For this reason, it is preferable, and in most cases necessary, to use promoters whose expression can be regulated so that the production of the foreign protein can be delayed until the culture has reached a high density. Common regulatable promoters used in *E. coli* include pLac (from the lactose operon), pTrp (from the tryptophan operon), pTac (a man-made hybrid between pLac and pTrp, used to produce a much higher level expression than that of its parents) and pAra (from the arabinose operon). The lactose promoter is easy to induce, but its uninduced (basal) level of transcription is often significant and may create problems when the foreign gene products being expressed are strongly toxic to *E. coli*. Accordingly, it is a common practice to use this promoter in the presence of the *lacI^q^* allele, which leads to the increased production of LacI repressor, thanks to a mutation in the *lacI* promoter, to suppress efficiently the uninduced level of transcription. Furthermore, the lactose promoter commonly used contains a mutation called UV5, which abrogates the catabolite repression so that the cloned gene can be expressed in a rich medium.

Good expression vectors need to have a Shine–Dalgarno sequence, or ribosome-binding sequence (RBS), typically AAGGA, a sequence complementary to a part of the 3′-terminal segment of 16S rRNA. This complementarity allows the mRNA to associate with the 30S ribosomal subunit of *E. coli*. The proper distance between the RBS and the first codon, ATG, of the gene is critical for an efficient initiation of translation: in one example, decreasing the distance from the optimal one (seven nucleotides in between these sequences) by only two nucleotides decreased the expression level by more than 90%. The Shine–Dalgarno sequence is absent in eukaryotic mRNA. If the cloned fragment comes from such an organism, it is necessary to insert the *E. coli* RBS into the vector and to place the 5′-terminus of the cloned gene close to this RBS to ensure the efficient translation of the mRNA.

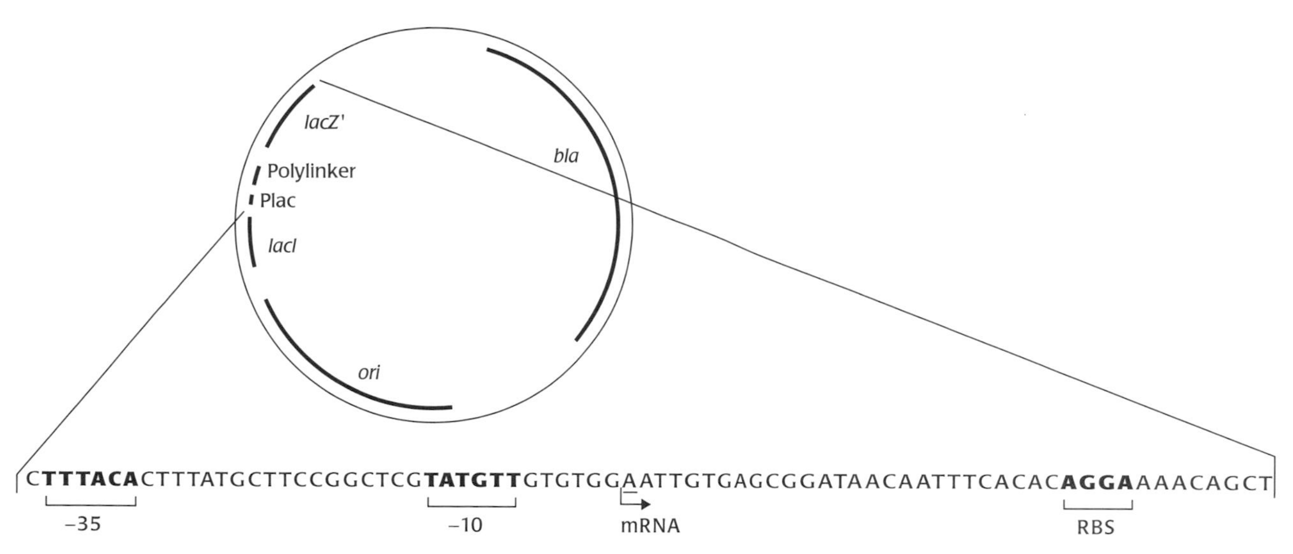


**FIGURE 3.21**

Structure of the *E. coli* expression vector pUC18. In addition to the origin of replication (*ori*) and an antibiotic resistance marker (*bla*), this vector contains a portion of the *E. coli lac* operon. The latter includes the repressor gene (*lacI*), the promoter region (Plac), and the 5′-terminal portion of the *lacZ* gene, coding for about 60 amino acid residues. As shown at the bottom, a polylinker region (containing restriction sites for more than 10 endonucleases) is inserted inside the *lacZ* gene. The amino acids present in the LacZ protein are shown in boldface type, those coded by the polylinker sequence in standard type. The polylinker does not contain any nonsense codons and is inserted in phase, so the vector codes for a complete N-terminal fragment of LacZ with an 18–amino acid insert. The -35 and -10 regions of the promoter, as well as the RBS, are indicated. Note that these sequences deviate somewhat from the consensus sequences. The catabolite-activator-protein (or cAMP-binding protein) (CAP)-binding sequence of the *lac* promoter is located upstream of the sequence shown here.

Some of these features are illustrated by the vectors of the pUC series (Figure 3.21). The segment that contains the regulatable promoter, pLac, and the 5′-terminal portion of the *lacZ* gene comes intact from the lactose operon of *E. coli*. Thus, the promoter is at a proper (natural) distance from the transcription initiation site. The Shine–Dalgarno sequence is located at its natural distance from the initiation codon of the *lacZ* gene. At the very beginning of the *lacZ* gene, the polylinker developed earlier for the M13 series of vectors (see page 107) provides many cloning sites within a very short stretch of DNA. Because of this arrangement, it is possible to express the cloned protein very efficiently as a fusion protein containing only a few of the N-terminal amino acids of LacZ. A final convenient feature is that the disruption by the cloned fragment of the 5′-terminal fragment of the *lacZ* gene makes it possible to distinguish cells containing recombinant clones from those containing resealed vectors only, as explained in connection with the M13 vectors.

Another example of widely used expression vectors is the pET series (Figure 3.22), developed by Novagen on the basis of studies on T7 phage biology by F. William Studier. The gene to be expressed is cloned within a polylinker behind a very strong T7 promoter. Because this promoter is recognized by the T7 RNA polymerase but not at all by the *E. coli* polymerase, the uninduced level of expression can be kept exceptionally low. When the culture reaches a high density and the cloned gene is ready to be expressed, the T7 RNA polymerase gene, cloned behind pLacUV5 promoter in the host strain, is induced by using IPTG. If the protein to be expressed is exceptionally toxic to the host cell, host strains expressing T7 lysozyme at a low level are used. Because lysozymes become necessary to lyse the host cells only at the last stage of phage infection, in which transcription of other phage genes is

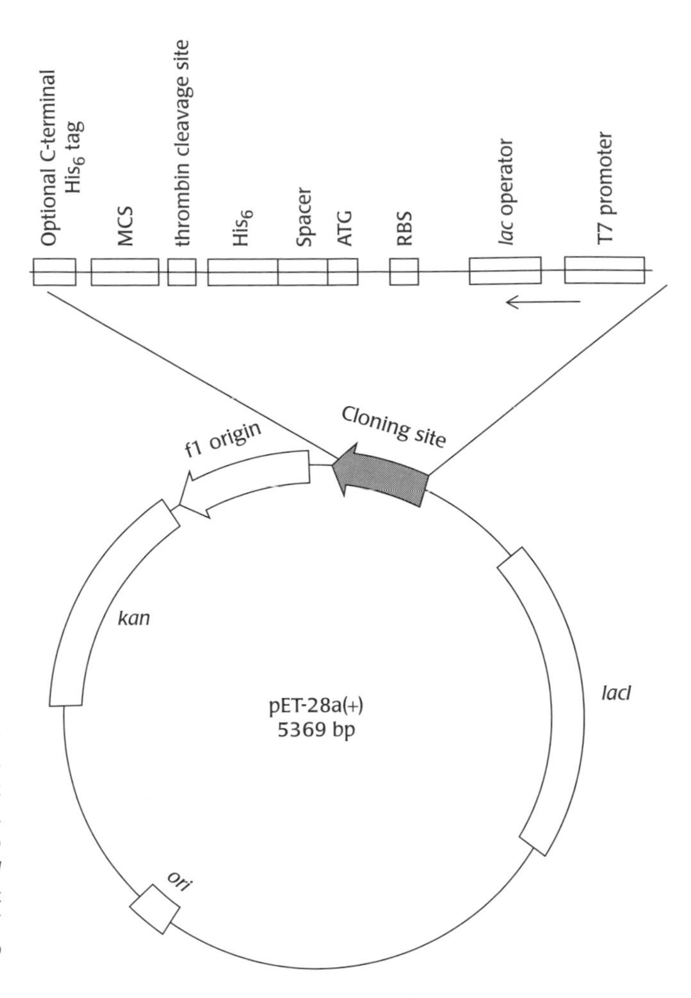


FIGURE 3.22

Structure of the *E. coli* expression vector pET-28a(+). The cloning site (shown by a *thick black arrow*) contains a T7 promoter, followed by the upstream region of the *lac* operon containing the RBS as well as the *lac* operator (similar to the pUC18, Figure 3.21), by the initiation codon ATG, a three-codon spacer, and a hexahistidine tag sequence, then by a cleavage site by thrombin, and finally by the multiple cloning site. Thus, the target protein will be produced with an N-terminal hexahistidine tag (see p. 119), which can be removed after affinity purification by treatment with thrombin. (The cloning site even contains an additional hexahistidine sequence, which can be used to attach a C-terminal tag to the protein). The plasmid, in addition to the antibiotic resistance marker *kan*, which produces kanamycin resistance, also contains the origin of replication for phage f1. Thus, this is a *phagemid* (see text), and the construct can be recovered as a single-stranded DNA by infection with a helper phage. Note also that this multicopy plasmid contains the *lacI* gene, which produces lactose repressor that prevents, in the absence of IPTG, the "leaky" expression of the target gene by binding to the *lac* operator site in the cloning site.

not needed, T7 lysozyme acts as a natural inhibitor of T7 RNA polymerase. Thus, the transcription of the cloned gene by the low levels of T7 polymerase, which could be produced by the baseline level transcription of its gene in the absence of IPTG, becomes nearly completely inhibited. With the pET series, an expression level of a cloned protein approaching 40% to 50% of the total cellular protein is sometimes reported.

Finally, when the cloned gene comes from an organism that is not closely related to *E. coli*, one should pay close attention to the codon usage. For example, arginine codons AGA and AGG are rarely found in *E. coli* genes but are frequent in eukaryotes. Expression of such genes in *E. coli* often leads to translational arrest, with subsequent degradation of mRNA. Such codons should be altered by the site-directed mutagenesis to enhance translation in *E. coli*.

## RECOVERY AND PURIFICATION OF EXPRESSED PROTEINS

Even when the cloned gene is successfully expressed in a bacterial host, product recovery is not always a simple matter. Potential problems and some approaches to solving them are discussed below.

**TABLE 3.1 Specific cleavage reactions**

| Cleavage effector | Cleavage site |
|---|---|
| Acidic pH | –Asp↓—Pro– |
| Hydroxylamine | –Asn↓—Gly– |
| CNBr | –Met↓—Xaa– |
| Trypsin | –Arg (or Lys)↓—Xaa– |
| Clostripain | –Arg↓—Xaa– |
| Collagenase | –Pro–Xaa↓—Gly–Pro–Yaa↓— |
| Factor Xa | Ile–Glu–Gly–Arg↓—Xaa– |
| Enterokinase | –Asp–Asp–Asp–Asp–Lys↓—Gly– |
| Tobacco etch virus (TEV) protease | –Glu–Xaa–Xaa–Tyr–Xaa–Gln↓—Ser(or Gly) |

Xaa and Yaa indicate any amino acid residue, and the vertical arrow indicates position of cleaved peptide bond.

### Expression of Fusion Proteins

When short peptides are expressed in *E. coli*, they are likely to be rapidly degraded by the various and plentiful peptidases in the bacterial cytoplasm. To protect these products, the DNA sequences coding for them are usually fused to genes that code for proteins endogenous to *E. coli*. On expression of the resulting fusion protein, the small foreign peptide is folded as a portion of the large endogenous protein and generally escapes proteolytic degradation.

Selective site-specific cleavage of the fusion protein is required to separate these peptides from the "carrier" proteins. Some of the conditions and reagents that cleave proteins at specific sites are listed in Table 3.1. When the peptides do not contain internal bonds that would be cleaved by trypsin, CNBr, or acid, it is safe to generate a cleavage site for one of these agents at the peptide–carrier junction by altering DNA sequence. Peptides that are fairly large are likely to contain sites susceptible to such simple agents; in these cases a protease, such as factor Xa, with its very stringent amino acid sequence specificity, is used to cleave at the desired site.

Another advantage of expressing peptides and proteins as fusion products is that it facilitates product purification. For example, if the foreign gene is fused with a sequence coding for an immunoglobulin G (IgG) antibody–binding domain of protein A from *Staphylococcus aureus*, the fusion protein can be recovered by simply passing the cellular extract through a column of immobilized IgG. Other schemes fuse products to glutathione S-transferase (GST), which allows purification of the product with an affinity column (Box 3.10) of immobilized glutathione, or fuse them to a stretch of histidine residues, and then purify them by exploiting the metal complexation of histidine. Creation of fusion proteins is also an important strategy for avoiding the aggregation of the expressed protein, discussed below.

**Affinity Columns**

Traditional methods of protein purification rely on gross, physicochemical properties of the proteins, such as electrical charge (in ion exchange chromatography), size (gel filtration chromatography), and hydrophobicity (hydrophobic chromatography). A single fraction obtained after such purification procedures tends to contain many proteins, if the starting material is a complex mixture. In contrast, affinity chromatography relies on *specific* interaction of the proteins with specific ligand molecules. For example, to purify an enzyme, either the substrate of the enzyme or a substrate analog is covalently linked to a granular matrix material, and the granules are packed into a column. When a crude mixture of hundreds of proteins is passed through this affinity column, only the enzyme is bound specifically to its substrate immobilized in the column; all the other proteins pass through the column unretarded. Elution of the column by procedures that decrease the affinity of the enzyme to the substrate, perhaps by altering the enzyme conformation, results in the one-step purification of the desired enzyme.

Recombinant DNA technology now allows us to put any of the many available "tags" to the protein of interest. For example, the hexahistidine tag allows purification by $Ni^{2+}$ columns, the maltose-binding protein tag by amylose columns, the glutathione *S*-transferase (GST) tag by immobilized glutathione columns, and so forth.

BOX 3.10

## Formation of Inclusion Bodies

When expressed at high levels in *E. coli* cytoplasm, many foreign proteins, especially those of eukaryotic origin, form insoluble aggregates called *inclusion bodies*. They are presumed to form where high concentrations of the overproduced, nascent proteins favor intermolecular interactions between the hydrophobic stretches of incompletely folded polypeptide chains, and they lead to aggregation and misfolding of these proteins (Figure 3.23).

These high, localized concentrations of nascent proteins are partly a consequence of the use of overexpression systems with their high gene dosage and powerful promoters. They are also partly the result of the prokaryotic structure of *E. coli*. Under the typical eukaryotic conditions of synthesis, many nascent human and animal proteins would be sequestered into compartments separated from the cytosol, such as the lumen of the endoplasmic reticulum. In *E. coli*, however, the newly synthesized proteins must remain at large in the undifferentiated bacterial cytoplasm. Furthermore, several factors tend to retard the folding of foreign proteins in *E. coli*, thus increasing the chances of intermolecular association and aggregation: (1) The conditions in the *E. coli* cytoplasm – for example, pH, ionic strength, and redox potential – are different from the normal environment in which these proteins are folded into their final conformations. Many secreted proteins of eukaryotic origin cannot fold in the highly reducing cytoplasm *of E. coli* because disulfide bonds, which are normally formed in the oxidizing environment of the endoplasmic reticulum and help the folding process, are not produced. (2) The correct folding of many polypeptides is facilitated by various helper proteins (Box 3.11). These include peptidyl-proline *cis/trans* isomerase, which facilitates the interconversion of two forms of proline groups, protein disulfide isomerase, which catalyzes the exchange of disulfide linkages in the substrate protein, thereby facilitating the production of the form with correct disulfide pairs, and a group of *molecular chaperones*, which also enhance the folding process or at least prevent the premature formation of aggregates of denatured proteins. The nature and the concentration of such helper proteins in *E. coli* obviously differ from those in the various compartments of the eukaryotic cells.

In some cases, the formation of inclusion bodies can be avoided, as described in the next section. Even when this is difficult, however, we may be able to use inclusion bodies to advantage in the purification of recombinant proteins. The cells are broken, the extracts centrifuged, and the inclusion bodies recovered as a sediment. Because the sediment also contains membrane fragments, it is customary to wash it by resuspension in detergent solutions (to dissolve and remove membrane components) and by recentrifugation. In this manner, the complex and tedious process of protein purification can be almost completely bypassed. Finally, the inclusion bodies are solubilized with protein denaturants, such as 6 M urea or 8 M guanidinium

### Foldases and Molecular Chaperones

Christian B. Anfinsen's group showed in 1957 that a completely denatured ribonuclease A can be spontaneously renatured *in vitro* into its native conformation with the concomitant formation of its four disulfide bonds with correctly paired cysteine residues. This famous discovery was interpreted by many to imply that all proteins become folded spontaneously, without assistance from any other cellular component. However, the fact that certain reactions occur spontaneously does not mean that cells do not use helper proteins to facilitate those processes. Indeed, recent years have witnessed the discovery of two classes of proteins that assist the folding of newly made proteins.

The members of one class of such proteins possess enzyme activities in the classical sense, and are sometimes called "foldases." These include peptidyl prolyl *cis-trans* isomerase and enzymes involved in the formation and isomerization of disulfide bonds. The former enzyme helps in the folding process by facilitating the interconversion between *cis-* and *trans-* configurations of the peptide bond linking the nitrogen atom of proline and the carboxyl group of the preceding amino acid residue. Practically all of the peptide bonds in proteins have the *trans* configuration, but bonds involving proline are the exception. Spontaneous *cis-trans* isomerization of such bonds occurs slowly. Enzymes that facilitate the formation of protein disulfide bonds, and their isomerization, are important in the folding of proteins that contain such bonds. If, in the course of folding, disulfide bonds are not formed, or formed between incorrect pairs of cysteine residues, the protein is likely to become misfolded.

Proteins of the second class that assist in the folding process are molecular chaperones, which appear to perform more subtle and complex functions. The structure of these chaperones has been conserved strongly during evolution, and most of them belong to the *heat-shock proteins* – either to the Hsp70 or the Hsp60 class. (They are called heat-shock proteins because in many organisms they are overproduced when the organism experiences high temperatures; these proteins are thought to facilitate the unfolding and proper refolding of heat-denatured proteins.) In *E. coli*, most of the nascent polypeptides fold with the help of a ribosome-associated protein called Trigger Factor, which is a chaperone that also has prolyl *cis-trans* isomerase activity. The slowly folding proteins, with their exposed hydrophobic patches, are then bound by DnaK, a representative of the Hsp70 class. DnaK shields the hydrophobic patches of the nascent proteins so that they do not interact with other nascent proteins and form insoluble aggregates, or inclusion bodies. DnaK works with two other proteins, DnaJ and GrpE, so that the release of the nascent protein is timed by ATP hydrolysis. If the protein is still incompletely folded, it will be bound to DnaK again, and the cycle will be repeated. The proteins that are most difficult to fold are handled by GroEL, a representative of the Hsp60 class, also called *chaperonin*. In *E. coli*, GroEL occurs as a 14-mer in a double-doughnut configuration with two large cavities. The incompletely folded protein enters one of the cavities, which is closed by a 7-mer of an associated protein GroES. This allows the slow folding of the protein without deleterious interaction with other nascent proteins. The process is again timed by the hydrolysis of many ATP molecules.

**BOX 3.11**

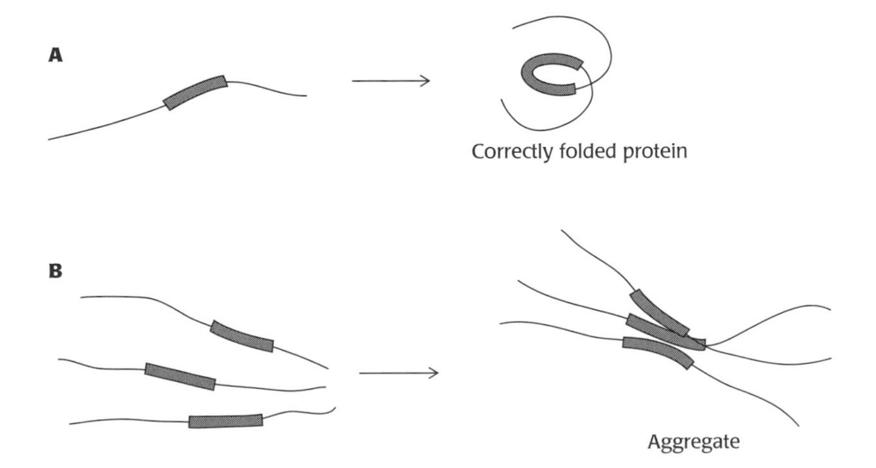


**FIGURE 3.23**

Presumed mechanism for the aggregation of overexpressed proteins. (**A**) Normally, nascent polypeptides fold into a globular conformation, with hydrophobic stretches (*thick line*) hidden in the interior. (**B**) However, when concentrations of nascent polypeptides are very high, there is increased likelihood that an exposed hydrophobic region on one molecule will interact with that on another molecule before the individual chains have a chance to fold properly. These intermolecular interactions between nascent chains result in aggregation and in an irreversible misfolding of the protein, producing inclusion bodies.

Redrawn based on artwork from the first edition (1995), published by W.H. Freeman.

hydrochloride, and the proteins are renatured by the gradual removal of denaturants (Box 3.12). Procedures of this type have been successfully used for the purification of many proteins. Although theoretical considerations indicate that *in vitro* renaturation should be done at low concentrations of the protein to minimize intermolecular interactions, in practice some systems tolerate fairly high concentrations, presumably because even these concentrations are quite low compared with those reached in overproducing cells.

**Renaturation of Proteins Containing Disulfide Bonds**

If the protein contains disulfide bonds, its complete unfolding requires their cleavage. This can be done by reduction (with either dithiothreitol or mercaptoethanol). In this case, the solubilized denatured protein must be purified, always in the presence of reducing agents. Alternatively, the disulfide bond can be cleaved by converting cysteine sulfur atoms into S-sulfonates with the addition of sodium sulfite. S-sulfonates are stable at neutral or acidic pH, and thus the solubilized proteins can be conveniently purified if alkalinization of the samples is avoided. After purification, S-sulfonates can be reconverted to sulfhydryl groups by the addition of mercaptoethanol. In both procedures, the renaturation of the protein is accomplished by removal of the reducing agent and of the denaturing agent, and the oxidation of the cysteine residues into disulfides is accomplished by exposure to air. More recently, successful attempts have been made to facilitate renaturation by adding foldases and even chaperones; however, the high cost of the additional proteins would constitute a major problem.

BOX 3.12

### Preventing the Formation of Inclusion Bodies

Although inclusion bodies are a convenient starting material for purification, the denaturation and the controlled renaturation steps are costly. It is especially problematic that the renaturation process usually works best at low protein concentrations, a requirement that increases cost and decreases yield. Careful cost analysis shows that the expense of the renaturation step is the main reason why the commercial production of large proteins such as tissue plasminogen activator and factor VIII by recombinant DNA technology is carried out in animal cell cultures rather than in *E. coli*. In animal cell cultures, the recombinant protein folds spontaneously into the native conformation, and inclusion bodies are not formed.

Production costs for these proteins would be further reduced if they could be produced in the native conformation in a microbial host. Much effort has therefore been devoted to finding conditions that would decrease the extent of inclusion body formation in *E. coli*. So far, a technique that universally and drastically decreases inclusion body formation has not been discovered. Among the approaches tried, lowering of the growth temperature was effective in many cases. Attempts have also been made, with success, to co-express chaperones and foldases to improve the correct protein folding in *E. coli*. In some cases, folding of foreign proteins was improved if *E. coli* was grown in the presence of low concentrations of ethanol (usually 2% to 3%). A plausible explanation of this result is that ethanol induces the "heat shock response" in *E. coli*, which leads to increased production of foldases and chaperones. Another approach is to fuse the coding sequences of foreign proteins to the 3′-terminus of genes coding for "solubilizer" proteins, such as *E. coli* thioredoxin or the mature form of maltose-binding protein. In many cases, the fused proteins were found to be produced in a totally soluble form (Box 3.13).

### Secretion Vectors

Whether the recombinant proteins form inclusion bodies in the cytoplasm of the host bacterium or remain in soluble form, their purification is always a challenge. One way to simplify the task of separating the recombinant protein from the myriad of host proteins, at least in principle, is to cause the recombinant proteins to be secreted into the culture medium. After that, purification would become quite straightforward, because bacteria are usually grown in simple, protein-free media. For this reason, much effort has been spent on the construction of *secretion vectors*.

In both prokaryotes and eukaryotes, proteins destined to be secreted from the cell are synthesized with an extra sequence, a *leader* (or *signal*) *sequence*, of about two dozen residues at the N-terminus. This sequence guides the nascent protein to the secretory apparatus in the cytoplasmic membrane and is split off by leader peptidase after the polypeptide is translocated across the membrane. The presence of the leader sequence is a necessary, but not always a sufficient, condition for secretion: Some artificial constructs composed of leader sequences fused to soluble, cytosolic proteins fail to be secreted, presumably because the mature part of the protein folds quickly to a stable, globular conformation and cannot be translocated in that condition. This suggests that the secretion vector strategy will work best when the products are unlikely to fold rapidly into a tight, stable conformation.

Indeed, this strategy proved useful in the production of insulin-like growth factor I (IGF-1), a peptide composed of about 70 amino acid residues, in the early days of biotechnology. The cDNA for IGF-1 was cloned behind the sequence coding for the leader sequence of protein A, a secreted, IgG-binding protein from *S. aureus*, and the plasmid was transformed into *E. coli* HB101. In addition, two copies of the sequence coding for the IgG-binding domain of protein A were inserted between the leader sequence and the cDNA for IGF-1 to facilitate the purification and to inhibit proteolytic degradation (see page 119). An "affinity handle" such as this is important because when proteins are secreted, they must be purified from the culture supernatant, with its very large volume. The affinity handle provides a way of rapidly and efficiently concentrating the desired product (see Box 3.10). In this case, the culture supernatant was passed through a column of IgG-Sepharose, which adsorbed all of the secreted proteins containing the IgG-binding protein A sequence. The fusion protein was then cleaved with hydroxylamine by taking advantage of the hydroxylamine-sensitive Asn-Gly sequence introduced just in front of the IGF-1 sequence. The IGF-1 peptide was then purified by conventional column chromatography methods.

Although secretion vectors have often proved useful, secretion is not yet a universally applicable approach. Some proteins fail to be secreted even when fused to a leader sequence, as mentioned earlier. Unfortunately, *E. coli* cells are surrounded by the outer membrane, and thus the export from the cytoplasm results in secretion into another cellular compartment, the periplasm between outer and inner membranes. In the case mentioned above, apparently a raised temperature (44°C) needed to induce the protein A promoter also permeabilized the outer membrane, causing a large fraction of the periplasmic protein to leak out into the medium. However, this is a rare phenomenon, and *E. coli* strains that are leaky and at the same time grow in a robust manner have not yet been developed. This fact led to attempts to use Gram-positive bacteria, such as *B. subtilis*, as the host for production of recombinant proteins; however, secretion of powerful proteases by such bacteria has so far hampered this effort. In any case, periplasm has several features that are attractive for the correct folding of foreign proteins. It is

**Thioredoxin and Maltose-Binding Protein Fusions**

*E. coli* thioredoxin is a small protein (molecular weight 11,675) with two cysteine residues in close proximity. It appears to fold efficiently, since its expression at a very high level (up to 40% of the total *E. coli* protein) still does not cause the formation of inclusion bodies. When foreign genes are fused to the 3′-terminus of the thioredoxin gene, the fusion protein indeed seemed to fold much better than the foreign protein expressed alone, presumably because the initial folding of the thioredoxin domain facilitates the subsequent folding of the following foreign protein. Similarly, fusion of foreign protein with the mature sequence of *E. coli* maltose-binding protein has been used with many examples of success. In this case, it is speculated that the ligand-binding groove of the binding protein may act as a chaperone that holds the incompletely folded foreign protein.

**BOX 3.13**

a more oxidizing environment and contains enzymatic systems that catalyze the formation and isomerization of disulfide bonds, in contrast to the cytosol of *E. coli*, which is very strongly reducing. It also contains several proteins that function both as chaperones and peptidyl prolyl isomerases, although the classical chaperones requiring ATP, such as Hsp60 and Hsp70, are absent. Thus, it is a common observation that foreign proteins, especially secreted proteins of mammalian origin such as hormones, form inclusion bodies much less frequently when they are secreted into the periplasmic space.

## AN EXAMPLE: PRODUCTION OF CHYMOSIN (RENNIN) IN *E. COLI*

Chymosin is the major protease produced in the fourth stomach (abomasum) of calves. Its production is limited to the few weeks during which the calves are nourished by milk. Chymosin is synthesized in the mucosal cells as preprochymosin (containing the "pre" signal sequence, the "pro" sequence removed at the time of activation, and the mature chymosin sequence). The signal sequence of 16 amino acid residues is removed, the protein is secreted as prochymosin (molecular weight 41,000), and this inactive zymogen becomes converted under acidic conditions into the active enzyme chymosin (molecular weight 35,600) by autocatalytic cleavage of the N-terminal "pro" sequence of 27 amino acid residues.

Chymosin is an aspartyl protease. It coagulates milk very efficiently through the limited hydrolysis of $\kappa$-casein and is used extensively in the manufacture of cheese. Because the production of cheese has increased rapidly in recent decades and the supply of suckling calves has declined, the availability of chymosin or chymosin substitutes has become an important issue in the dairy industry.

One major solution has been the commercialization of fungal enzymes from *Mucor* and *Endothia* as substitutes for chymosin. These enzymes are less expensive, but they do not quite attain the high coagulation/proteolysis ratio of calf chymosin, and this results in subtle but real differences in the flavor of the cheese. A more satisfactory solution, therefore, would be to produce chymosin by cloning, if it can be done in a cost-effective manner.

Several laboratories succeeded in cloning chymosin cDNA in the early 1980s. In every case, the original template was mRNA from the mucosa of the calf abomasum. cDNA was prepared from this mRNA, in some cases after size fractionation in order to further enrich for (pre)prochymosin mRNA. In the primary cloning step, the cDNA was inserted into *E. coli* plasmid vectors, and the recombinant plasmids were screened, for example, by probe hybridization (see Figure 3.18).

The next step was the cloning in a suitable expression vector. In the calf abomasum, chymosin is made as a preproprotein. Researchers had to decide in which form it should be expressed in *E. coli*. No attempt was made to express chymosin in its mature, processed form because the production of such an active protease in the *E. coli* cytoplasm was expected to be harmful to host cells. Nor, in the initial efforts, was an attempt made to express the entire preprochymosin sequence because of concern that the eukaryotic signal sequence might not lead to efficient secretion in *E. coli*. In several laboratories, therefore, prochymosin was chosen as the form to be expressed.

Two methods were used. In one, the prochymosin sequence was fused to the N-terminal portion of LacZ or TrpE, and the protein was expressed as the fusion protein. In this case, the prokaryotic promoters and the RBS present in front of these highly expressed prokaryotic genes were used to initiate transcription and translation efficiently. In another approach, the sequence coding for prochymosin was inserted directly behind a sequence containing a suitable prokaryotic promoter, RBS, and ATG codon. Some adjustment of distance between the RBS and ATG, as well as of the actual base sequence, was needed to optimize the expression in this case. Both approaches led to the production of prochymosin at a level corresponding to up to 5% of total *E. coli* protein.

The overproduced prochymosin, however, accumulated in a denatured form as inclusion bodies; in retrospect, this is not surprising as prochymosin contains several disulfide bonds. When attempts were made to purify the inclusion bodies and to renature prochymosin from this material by the procedure already described (pages 120), the yield of active prochymosin was disappointingly low, owing primarily to difficulties in the renaturation step. The increased production cost that would result might be tolerated if the product were a human therapeutic compound, but for agricultural products such as prochymosin, the cost was clearly prohibitive. Attempt to express prochymosin in secretion vectors also resulted in failure because parts of prochymosin apparently folded rapidly to prevent its secretion. Prochymosin has since been expressed more efficiently in yeasts (see below).

## PRODUCTION OF PROTEINS IN YEAST

We have already described the cloning of foreign genes in bacteria, mostly in *E. coli*. In passing, we touched on the difficulties encountered when bacteria are used to clone and express genes from eukaryotes. For example, many eukaryotic proteins normally undergo one or more posttranslational modifications that are important to their functions or stability. Yeast has often been referred to as a model eukaryote, and in this section, we show how yeast cells are able to carry out many of the posttranslational modifications necessary to produce accurately synthesized proteins using the genes or cDNA of higher organisms.

Glycosylation – the addition of oligosaccharide units to a protein – is one of the most important posttranslational modifications that occur to the gene products of eukaryotic cells (Box 3.14). Indeed, most secreted eukaryotic proteins are glycosylated. Glycosylation often helps ensure the correct folding of proteins and protects them from proteolytic enzymes. In some cases, specific receptors on animal cells recognize serum proteins whose N-linked oligosaccharides lack certain sugars and remove these proteins (usually "old" proteins) from circulation. Thus, the presence of the correct

**Posttranslational Modification of Eukaryotic Proteins**

Many eukaryotic proteins, especially secreted proteins (including hormones), are glycosylated. They acquire oligosaccharide substituents at asparagine residues via an *N*-glycosidic bond during the secretion process (see Box 3.15). In higher animals, these "*N*-linked" oligosaccharides are typically of the complex, branched type, containing *N*-acetylglucosamine, mannose, galactose, and sialic acid residues. Yeast glycoproteins characteristically carry oligosaccharides containing very large numbers of mannose residues. Other oligosaccharides can be linked to serine or threonine residues via an *O*-glycosidic bond. These "*O*-linked" oligosaccharides are generally less branched than the *N*-linked ones and typically contain *N*-acetylgalactosamine, galactose, and sialic acid.

In many eukaryotic proteins, the amino group of the N-terminal amino acid residue is modified by acylation – that is, by the formation of an acyl amide linkage. *N*-acetylation interferes with recognition of the protein by the intracellular proteolytic degradation machinery and thus preserves the proteins for a longer period within the animal or human body. Another characteristic modification of the N-terminal residue is *N*-myristylation, which adds the 14-carbon, saturated fatty acid known as myristic acid onto the amino group. The myristylated proteins can bind to membranes at the fatty acid, thus becoming peripheral membrane proteins. Similar targeting of certain other proteins occurs by the covalent attachment of palmitic acid, a 16-carbon, saturated fatty acid to the sulfhydryl groups of internal (not N-terminal) cysteine residues.

Redrawn based on artwork from the first edition (1995), published by W.H. Freeman.

**BOX 3.14**

**Protein-Secretion Pathways in Prokaryotic and Eukaryotic Cells**

In prokaryotes, secretory proteins are made with an N-terminal signal sequence and are secreted via the SecYEG(DF) protein complex found in the plasma membrane. SecA protein is thought to help bring the signal sequence to the export machinery. The signal sequence is cleaved when the junction between it and the mature sequence appears on the outer side of the cytoplasmic membrane (see Figure A).

A

Cytoplasm
N
+
SecA
N
SecA
Sec D, E, F, Y
N
N
Cleavage
Periplasmic space

SECRETION IN *E. coli*

Cytoplasm
N
+
N
SRP
N
N
SRP Receptor
Cleavage
ER lumen
Glyco-sylation

SECRETION IN EUKARYOTIC CELL

Secreted proteins made by eukaryotic cells also have an N-terminal signal sequence. However, the signal sequence is recognized by a complex structure, the signal-recognition particle (SRP), which contains six proteins held together by a small RNA. SRP then binds to the membrane-associated SRP receptor, thus guiding the nascent protein to the export apparatus located specifically within the membrane of the *rough endoplasmic reticulum*. One of the proteins in SRP also arrests translation until this "docking" at the export apparatus takes place, thus preventing the protein's misfolding in the cytosolic environment. The protein passes across the membrane, presumably in an extended form, and enters the lumen of the rough endoplasmic reticulum. The signal sequence is split off soon after a partial translocation of the protein across the membrane. The environment in the lumen is less reducing than the cytosol, and folding of the protein is often followed by the formation of disulfide bonds. Because the creation of disulfide bonds between "wrong" pairs of cysteine residues might produce a misfolded protein, the lumen contains disulfide isomerase, which splits and reforms disulfide bonds so as to allow the protein to reach the native conformation (see Box 3.11).

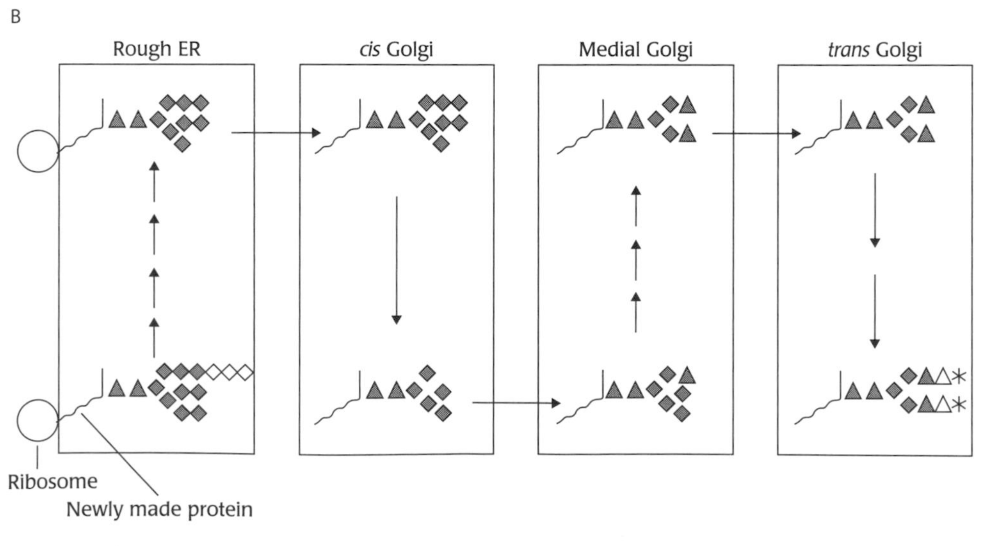


Symbols: ▲, *N*-acetylglucosamine; ◆, mannose; ◇, glucose; △, galactose; ✳, sialic acid.

Even while the polypeptide is being extruded through the membrane, some sites within the secreted protein become glycosylated. Figure B shows the formation of a complex type of *N*-linked oligosaccharide of the simplest structure in animal cells (there can be many variations on the details of the pathway). Within the endoplasmic reticulum, a "core" oligosaccharide containing two proximal *N*-acetylglucosamine residues, nine mannose residues, and three glucose residues is attached to an appropriate site on the protein; subsequently, all of the glucose residues and one mannose residue are "trimmed off." The glycoprotein is then transported, via small membrane vesicles, into the Golgi apparatus, another complex, membrane-bounded organelle: First, it enters *cis* Golgi vesicles, where three more of the mannose residues are removed. In the next compartment, the lumen of the medial Golgi vesicles, more mannose residues are trimmed off, and two *N*-acetylglucosamine residues are added on. Finally, in the *trans* Golgi compartment, two galactose residues are added to the *N*-acetylglucosamine residues, and sialic acid residues are added onto the galactose residues. The completed glycoprotein is then secreted from the cell by the fusion of glycoprotein-containing vesicles with the plasma membrane.

**BOX 3.15**

oligosaccharides is very important in producing recombinant human proteins that work well and last for a long time *in vivo*. Glycosylation and other modifications described in Box 3.14 do not occur if eukaryotic genes are expressed in bacteria such as *E. coli*.

In eukaryotic cells, secretory proteins are synthesized by ribosomes associated with the membrane of the endoplasmic reticulum and are translocated across that membrane cotranslationally by a mechanism involving a *signal-recognition particle*. On entering the lumen of the endoplasmic reticulum, these proteins are immediately glycosylated (Box 3.15). By contrast, the prokaryotes' homologs of signal-recognition particles do not play a major role in the export of proteins but are involved in the insertion of cytoplasmic membrane proteins, and bacterial secretory proteins are almost never glycosylated.

Because glycosylation is of such importance in eukaryotes, it follows that eukaryotic microbes, such as yeasts, may be better hosts for the production of proteins of higher eukaryotes. Yeast cells do export many proteins using the endoplasmic reticulum–Golgi pathway, apparently with the participation of a signal-recognition particle, and glycosylate those proteins in the process. Yeast cells also carry out posttranslational *N*-acetylation and myristylation of proteins (see Box 3.14).

One might even expect that yeasts, as eukaryotes, would be able to carry out the correct splicing of nascent RNA transcripts of mammalian genes. However, yeasts contain few introns and therefore may fail to process mammalian intron sequences. The safest procedure when expression of a mammalian protein is desired is to utilize an intron-free gene – that is, to generate a cDNA copy of the mature mRNA for the gene of interest and use that as a template.

Yeasts can be grown to very high densities in simple, inexpensive media. More important, the components and metabolic products of yeast cells are not toxic to humans (remember that LPS, an integral component of the *E. coli* outer membrane, is a very toxic molecule also known as endotoxin). In the following pages, we discuss how foreign DNA is expressed in yeast cells (most often *S. cerevisiae*, or baker's yeast).

## INTRODUCTION OF DNA INTO YEAST CELLS

DNA can be introduced into bacteria in a variety of ways. With yeasts, however, transformation is the only practical means of introducing DNA. In one method, the yeast cell wall is removed by enzyme digestion and the resulting "spheroplasts" (cells bounded essentially only by the cytoplasmic membrane) are incubated with DNA in the presence of $Ca^{2+}$ and polyethyleneglycol. Both $Ca^{2+}$ and polyethyleneglycol are agents that stimulate the membrane fusion process and thereby enhance fusion between spheroplasts. Possibly DNA is taken up by yeast cells in the process of spheroplast fusion, but it is not yet clear whether the fusion is necessary for this uptake to occur. In another method, intact yeast cells (with cell wall in place) are treated with $Li^{+}$ ions and then incubated with DNA and polyethyleneglycol. The mechanism of DNA uptake remains obscure in this case, too. A third method is to apply transient high voltages to a suspension of cells. This process, called *electroporation*, creates transient holes in the walls and membranes, as described earlier in this chapter.

## YEAST CLONING VECTORS

Several types of cloning vectors are used to manipulate recombinant DNA constructs in yeast. Most are "shuttle" vectors: vectors that can multiply in yeast as well as in *E. coli*. The reason shuttle vectors are preferred is that the basic recombinant DNA manipulations are more easily carried out in *E. coli*, but the resulting DNA constructs must be transferred to yeast to take advantage of the superior properties of this host, such as the expression of glycosylated proteins. Shuttle vectors can be moved between the two hosts because they also contain the origin of replication recognized by *E. coli* and selection markers useful in *E. coli*, as well as features that enable them to survive in yeast cells. There are five major types of yeast cloning vectors: yeast integrative plasmids (YIps), yeast replicating plasmids (YRps), yeast episomal plasmids (YEps), yeast centromeric plasmids (YCps), and yeast artificial chromosomes (YACs).

### Yeast Integrative Plasmids

YIps (Figure 3.24) are essentially bacterial plasmid vectors with an added marker that makes possible their genetic selection in yeast. As we have seen already, antibiotic resistance genes are commonly used as selection markers in bacterial vectors. However, not many antibiotics are effective against yeasts. Thus, selection procedures in yeast are commonly designed to utilize a host strain that is defective in the biosynthesis of amino acids, purines, or pyrimidines and a vector that contains a yeast gene for the missing function. Some commonly used nutritional markers are the yeast genes *LEU2* (a gene involved in leucine biosynthesis), *URA3* (a gene involved in uracil biosynthesis), and *HIS3* (a gene involved in histidine biosynthesis). For example, the *leu2* host strain (in yeast genetics, a mutated and, hence, usually functionally defective allele is denoted in lower-case letters) will not grow in a minimal medium – that is, a medium containing only the requisite minerals

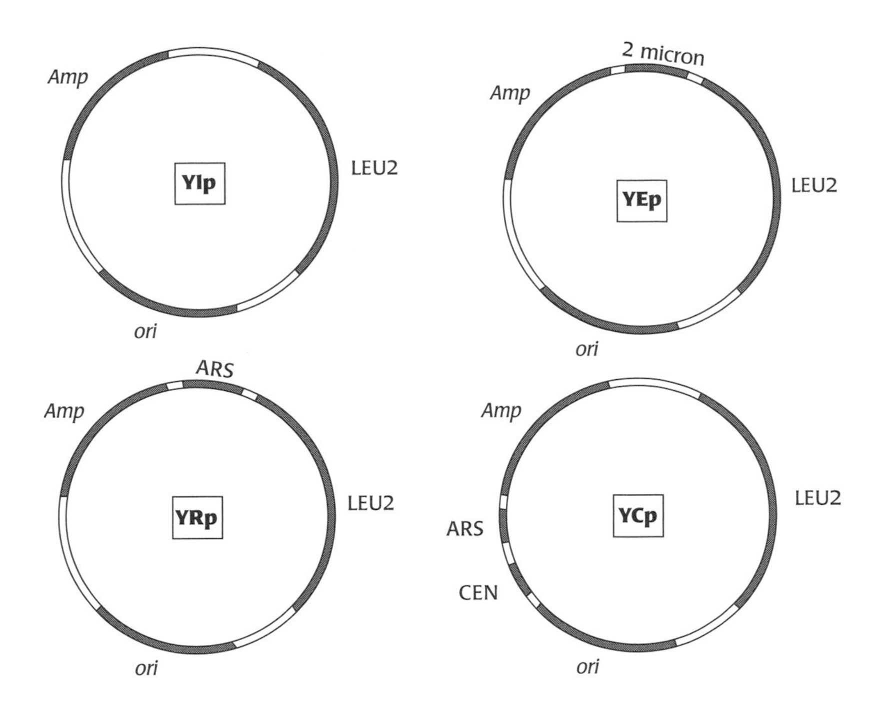


**FIGURE 3.24**

Plasmid vectors useful for cloning in yeast. Examples of four types of vectors are shown. Abbreviations: *ori, E. coli* origin of replication; Amp, ampicillin resistance gene for selection in *E. coli*; *LEU2*, a gene involved in leucine biosynthesis, for selection in yeast. Regions controlling replication and segregation in yeast, such as ARS, CEN, and 2-$\mu$m plasmid sequence, are described in the text.

Redrawn based on artwork from the first edition (1995), published by W.H. Freeman.

and a carbon source – but the same strain harboring an *LEU2*-containing plasmid will grow because it is able to synthesize leucine.

YIps lack an origin for replication that can be recognized by the yeast DNA synthesis machinery. Therefore, they can be maintained in yeast cells only when they become integrated into a yeast chromosome (usually by homologous recombination at the site of the yeast marker gene or one of the other yeast sequences present in the vector). Once integrated, they are inherited quite stably as a part of the yeast genome. However, integration is a rare event, so the frequency of transformation with plasmids of this type is extremely low (one to 100 transformants/$\mu$g DNA compared with the 100,000 transformants/$\mu$g that can be obtained with *E. coli*). The frequency of integration can be enhanced somewhat by cutting the plasmid within the region of yeast homology, a procedure that promotes homologous recombination to some degree (Box 3.16). Another drawback of these plasmids is their low copy number. Usually only one copy at most is integrated in one haploid yeast cell, effectively limiting the level of expression of the cloned gene. One way to circumvent this problem is to design the plasmid to integrate into genes that exist in multiple copies in the yeast chromosomes. For example, there are more than 100 copies of the genes coding for rRNA in a yeast cell, so multiple integrations into these sites could create a cell genome with many copies of the cloned genes. Alternatively, one can use as the selectable marker on the YIp vector a gene that has to exist in a large number of copies for the yeast to survive under certain conditions. For example, if the plasmid contains the *CUP1* gene, which codes for metallothionein, a protein that protects yeast cells by binding to heavy metals, yeast cells will survive in a medium containing $Cu^{2+}$ only when a large number of copies of the *CUP1* have been integrated into the genome – that is, when the gene has become "amplified." YIps are quite useful in spite of their typically low copy

**Homologous Recombination Process**

Homologous recombination begins with alignment of the homologous regions in two parental DNA duplexes. This is followed by "nicking" (single-stranded cleavage) of one parent-DNA helix and generation of single-stranded "whiskers" (step 1). A whisker wanders into the other duplex and forms Watson–Crick pairs with the complementary strand of the other DNA duplex (step 2). Finally, the end of the whisker is joined covalently to one of the strands of the other duplex, completing the process of crossing over (step 3). This mechanism requires an initial cut in one or both strands. Consequently, using plasmids that are already cut in the region of homology increases the frequency of recombination. More comprehensive schemes for the entire recombination pathway have been proposed [Orr-Weaver, T. L, Szostak, J. W., and Rothstein, R. J. (1981). Yeast transformation: a model system for the study of recombination. *Proceedings of the National Academy of Sciences U.S.A.*, 78, 6354–6358].

Redrawn based on artwork from the first edition (1995), published by W.H. Freeman.

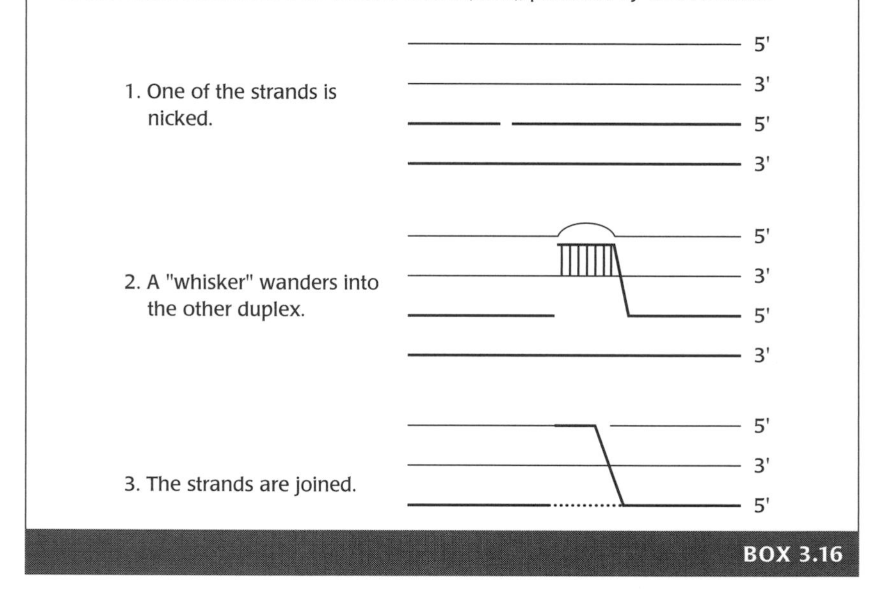


BOX 3.16

number because plasmid stability can often become a major problem with other kinds of yeast vectors (see below).

### Yeast Replicating Plasmids

In addition to selection markers useful in yeast, YRps (Figure 3.24) contain an origin of replication derived from the yeast chromosome and termed *ARS* (autonomously replicating sequence). With this origin, the plasmids can replicate without having to be integrated into the chromosome. However, yeast cells divide unequally by budding. In the process, only a disproportionately small fraction of the plasmids that were present in the mother cell are partitioned off into the buds, and many of the progeny cells are likely to lack plasmids entirely. Thus, YRp plasmids are lost rapidly unless constant selection pressures are applied. Consequently, they are not very useful for the reproducible expression of cloned genes.

### Yeast Episomal Plasmids

Some strains of *S. cerevisiae* contain an endogenous, autonomously replicating, high-copy-number plasmid called 2-$\mu$m plasmid. The origin of this plasmid is added to YIps to produce YEps (Figure 3.24), which can exist in high copy numbers (30 to 50 copies/cell). (An *episome* is a genetic element that can exist either free – as a plasmid – or as a part of the cellular chromosome.) Like YRps, YEps are poorly segregated into daughter cells, but they are maintained more stably because of their higher copy number. If the entire 2-$\mu$m DNA (6.3 kb) is inserted into a YIp plasmid and introduced into yeast cells that lack an endogenous 2-$\mu$m plasmid, copy numbers in excess of 200 per cell can be achieved under certain conditions. Plasmids of this type are obviously most suitable when high-level expression of a foreign gene is desired. One vector of this type, pJDB219, contains an intact *LEU2* gene but not its promoter. Because of the lack of promoter, this construction, called *leu2-d*, does not produce the full-scale expression of the LEU2 protein. Nevertheless, an extremely low level of expression does occur, presumably because of the nonspecific binding of the RNA polymerase or a very weak "readthrough" from upstream genes. This low-level expression produces a detectable phenotype because even a small amount of the enzyme is enough to produce some leucine. But because the amount of the enzyme produced by a single plasmid is far from sufficient, the plasmid-carrying cells cannot grow at a reasonable rate in minimal medium unless the plasmid is present in very large numbers (200 to 300 per cell) in order to complement the completely defective *leu2*$^{-}$ allele of the host. In other words, this plasmid is designed so that its presence in high copy numbers will be favored.

### Yeast Centromeric Plasmids

YCps (Figure 3.24) are YRps, or sometimes YEps, in which the sequence of a yeast centromere has been inserted. The centromeric sequence allows these plasmids to behave like regular chromosomes during the mitotic cell division, so YCps are faithfully distributed to daughter cells and are highly stable even without maintenance by selection. However, the "chromosomelike" behavior of these plasmids also means that their copy number is kept very low (one to three per haploid cell). This is a potential disadvantage when the plasmids are used for the expression of cloned genes, although the expression can be increased by the use of highly inducible promoters.

### Yeast Artificial Chromosomes

YACs are linear plasmids containing an ARS, a centromeric sequence, and, most important, a telomere (Box 3.17) at each end (Figure 3.25). These features allow the plasmids to behave exactly like chromosomes. Because the plasmid is linear, there is no limit to the amount of foreign DNA that can be cloned into it. This is the most important feature of YACs. Animal genes contain many introns and can exceed 100 kb in size. Such genes cannot be cloned in a single vector except in YACs (and BACs, which we

**Telomeres**

The ends of linear DNA duplexes, such as chromosomes and YACs, cannot be replicated faithfully because DNA synthesis always occurs in the 5′-to-3′ direction: One of the strands is thus replicated in short segments (Okazaki fragments) formed by the elongation of RNA primers (rectangles). When the RNA primer that is complementary to the very end of the DNA is degraded, there is no mechanism for synthesizing DNA to replace it. Consequently, such linear DNAs become shorter with each replication cycle. Eukaryotic chromosomes solve the problem by having repeated oligonucleotide sequences (telomeres) at their ends (e.g., telomeres in human chromosomes have the sequence $[TTAGGG]_n$). When telomeres become too short, they are elongated by an enzymatic mechanism that does not require a template.

Redrawn based on artwork from the first edition (1995), published by W.H. Freeman.

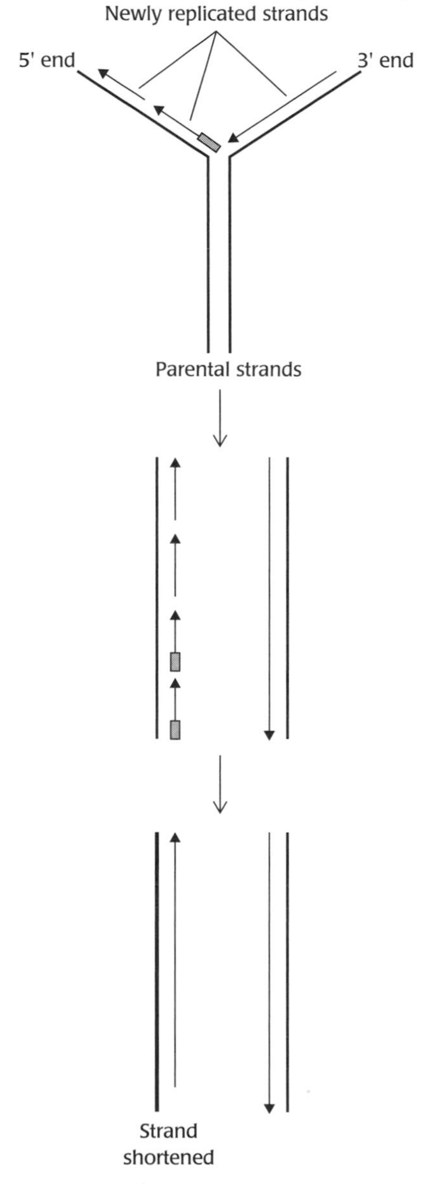


BOX 3.17

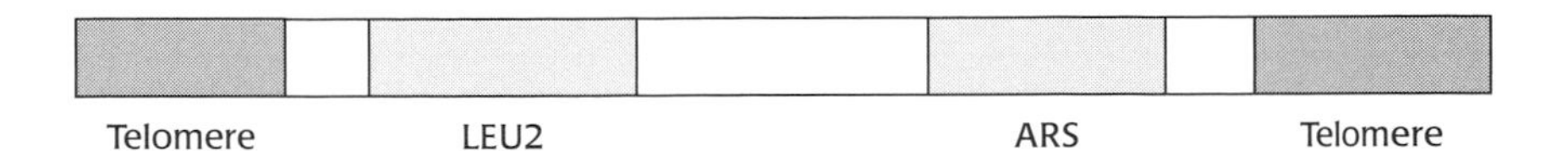


FIGURE 3.25

An example of a YAC vector. *LEU2* gene is needed for selection in yeast, ARS for replication in yeast, and telomeres for stability in yeast.

Redrawn based on artwork from the first edition (1995), published by W.H. Freeman.

discussed earlier in this chapter). YACs, however, are not the first choice when the main objective is a high level of expression of foreign genes.

## ENHANCING THE EXPRESSION OF FOREIGN GENES IN YEAST

There are several points to consider when designing a system for expressing foreign-gene products in yeast cells.

### Plasmid Copy Number

A high-copy-number plasmid of the YEp class is the best choice for maximal expression of any cloned gene. However, the expression of foreign proteins is often toxic for the yeast cells. Probably one of the main reasons is that some foreign proteins are likely to misfold in the cytoplasm, sequestering many of the chaperone molecules needed for the correct folding and functioning of the yeast's own proteins. In this situation, low-copy-number YEps and, paradoxically, even YIps may produce higher sustainable yields than high-copy-number plasmids.

Another problem that may be important in commercial production runs is the instability of some of the plasmids. In commercial fermentation, the organism must last through a far higher number of generations than is usually necessary in a small, laboratory-scale experiment. Thus, even a moderate degree of plasmid instability can cause a major problem.

### Promoter Sequence

Because promoters of foreign origin are unlikely to be expressed efficiently in yeast cells, the coding sequence of a foreign gene is usually inserted behind a strong yeast promoter. Yeast promoters are quite different from bacterial promoters. Although both contain AT-rich recognition sequences for RNA polymerase (typically TATATAA for yeast, in contrast to the TATAAT consensus sequence [see Box 3.9] for *E. coli*), the "TATA sequence" in yeast is located much further upstream (40 to 120 bases) from the mRNA initiation site than it is in *E. coli*, in which the "TATAAT," or Pribnow box, is typically located only 10 bases upstream of the transcription initiation site. In addition, yeast promoters usually require an upstream activator sequence (UAS), an enhancer-like sequence located very far upstream (100 to 1000 bases) from the transcription initiation site. Because of the location of the UAS, most yeast expression vectors contain a long, native "promoter sequence" (typically around 1 kb). Two frequently used promoters are the upstream sequences for an alcohol dehydrogenase gene (*ADH1*) and for a triose phosphate dehydrogenase gene (*TDH3*). *ADH1* was thought to be expressed constitutively at a high level, and its use was popular at one time. However, we now know that this particular isozyme of alcohol dehydrogenase becomes repressed when the culture reaches a high density, so its use has fallen off. (In contrast, another isozyme of alcohol dehydrogenase, *ADH2*, becomes derepressed when glucose in the medium becomes exhausted. The promoter for this enzyme is often used as a regulatable promoter, as we shall see.)

If the expression of the foreign protein inhibits the growth of the yeast cells, it becomes necessary to use regulatable promoters and to initiate expression of the foreign genes only when the culture has reached a high density. For example, the genes involved in galactose catabolism, *GAL1*, *GAL7*, and *GAL10*, have been extensively used as sources of regulatable promoters for cloned genes because they are repressed in the presence of glucose but are induced by the addition of galactose to the medium. The regulation of these genes involves the binding of a positive activator, *GAL4*, to upstream sequences of *GAL1*, *GAL7*, and *GAL10*. Thus, if the recombinant DNA containing the latter genes exists in multiple copies in a cell and *GAL4* is expressed from a single copy of the gene on the chromosome, the *GAL4* protein in the cell might become exhausted by binding before all the recombinant genes are activated. However, this limitation can be removed if *GAL4* is also introduced into the vector so that multiple copies of *GAL4* are present in a cell. A drawback of this system is that it tends to increase the expression of the cloned gene even in the absence of galactose, so it is dangerous if the product is toxic to yeast cells. Other regulatable promoters that have been used include ADH2 (alcohol dehydrogenase regulated by ethanol and glucose) and PHO5 (acid phosphatase, regulated by phosphate). Another attractive regulatable promoter is the one for CUP-1, which codes for metallothionein, a $Cu^{2+}$-binding protein, and which is induced by the addition of metal ions such as $Cu^{2+}$ or $Zn^{2+}$ to the medium. Several systems that are induced by elevated temperatures have been used successfully in the laboratory, but it may be difficult to get the temperature to change fast enough in a large fermentation tank.

Several hybrid promoters have also been used. These contain (1) a UAS from a regulatable promoter for controlling the level of expression of the gene and (2) the TATA box region from a strong, constitutive promoter for increasing the maximal level of expression. For example, a hybrid promoter containing the UAS sequence of ADH2 and the downstream sequences (containing the TATA box) from the TDH3 promoter has been effective in producing some foreign proteins at levels sometimes exceeding 10% of the total yeast protein.

### Transcription Termination and Polyadenylation of mRNA

In higher animals, termination of transcription usually occurs very far downstream from the coding sequence. Following termination, the nascent RNA transcript is cleaved at or near the cleavage signal, AAUAAA, present hundreds of nucleotides upstream from the transcript's 3′-terminus. This newly exposed 3′-terminus is then polyadenylated – that is, a stretch of A is added. In yeast, this processing follows a rather different pattern, with polyadenylation apparently occurring quite close to the 3′-end of the transcript. Unfortunately, the precise structure of a yeast terminator sequence is still not very clear. Because of this uncertainty, when recombinant DNA is constructed for gene expression in yeast, a large segment of "terminator" sequence, taken from the downstream sequence of yeast genes whose transcription is terminated efficiently, is usually placed downstream of the gene to be expressed.

#### Stability of mRNA

Yeast mRNAs differ greatly in their stability. The sequences that determine their degradation rates have been located in the 3′ untranslated regions and the coding regions of mRNA, but this knowledge is difficult to use for increasing the stability of foreign-gene transcripts in yeast.

If the gene is not followed by an effective terminator sequence, this usually produces unstable mRNA, presumably because it lacks the proper 3′-end that could become protected by polyadenylation. Repeated experiments have shown that such a situation drastically decreases the yield of foreign proteins in yeast. Thus, it is desirable to clone an efficient yeast terminator sequence downstream from the coding sequence of the foreign gene to be expressed, as described above.

#### Recognition of the AUG Initiation Codon

Efficient synthesis of mRNA, though necessary, is not sufficient in itself to ensure high-level production of a protein. Another basic requirement is efficient translation, for which the correct AUG codon must be readily recognized by initiation factors and by the ribosome machinery. In bacteria, this involves pairing of the Shine–Dalgarno sequence with the complementary sequence in 16S rRNA. There is no equivalent recognition sequence in eukaryotes, but the efficiency of translation initiation is known to depend on the sequences surrounding the AUG codon (such surrounding sequences are often called *context*). Analysis of gene sequences in yeast has shown that the consensus sequence (see Box 3.9) of the context is AxxAUGG (this is called Kozak's rule).

If the 5′ untranslated segment of the mRNA tends to form base-paired loops, translation can be inhibited quite significantly. G occurs infrequently (taking about 5% of the positions) among the 20 to 40 bases immediately preceding the AUG codon; the presence of a large number of G residues in this region is known to inhibit translation initiation. These facts should be taken into consideration when designing the part of the DNA sequence that codes for the 5′-terminal portion of the mRNA.

#### Elongation of the Polypeptide

Foreign genes often contain codons that are rarely used by yeast, and this may slow the translation process. Strings of rare codons occurring close together are especially detrimental. To enhance the expression of foreign genes in such cases, codons that yeast prefers have been substituted for those rarely used by yeast, usually by the site-directed mutagenesis procedure. The preferred yeast codons can be determined by analyzing the codons used in endogenous genes that are continuously expressed at high levels, such as genes for yeast glycolytic enzymes.

### Folding of the Foreign Protein

Many foreign proteins have been expressed at a high level in yeast cells and have been shown to fold correctly. For example, the hepatitis B virus core protein, the P-28–1 protective antigen of the schistosome, and human superoxide dismutase have all been expressed to a level that corresponds to 20% to 40% of total yeast protein, and yet the proteins do not appear to misfold or to form intracellular aggregates. This is in a striking contrast to the nearly ubiquitous formation of aggregates or inclusion bodies when foreign proteins are expressed in the cytoplasm of bacteria (see the earlier part of this chapter). Although the formation of such aggregates has been reported in yeast, they are not found nearly so frequently. This could be the result of the presence in the yeast cell cytoplasm of many kinds of molecular chaperones (the so-called heat-shock proteins).

### Proteolysis

Many proteins in eukaryotic cells are subject to degradation by the ubiquitin pathway. These proteins have at their N-terminus certain amino acids that are recognized by a small protein, called ubiquitin, that tags them for proteolytic degradation. All eukaryotic proteins are translated with methionine at the N-terminus, but subsequent removal of the N-terminal amino acid residues may expose one of the "destabilizing" amino acids and lead to destruction of the protein. If this problem exists in a cloning situation, one way of solving it is to alter the N-terminal amino acid sequence. Another recourse is to fuse the protein to the N-terminal segment of another protein, preferably of yeast origin, that is known not to be degraded by this pathway.

### Glycosylation

Animal proteins secreted through the endoplasmic reticulum–Golgi pathway (see Box 3.15) are usually glycosylated in the process. As we have noted, this may help the proteins fold correctly and make them more resistant to proteases. Such posttranslational modifications do occur to foreign proteins cloned in yeast cells (if the proteins successfully enter the secretion pathway; see also below), but the yeast system can add only the high-mannose type of oligosaccharides, not the complex type (see Figure B in Box 3.15) most common in the glycoproteins of higher animals. Sometimes this may affect the folding and protease sensitivity of the protein and, more important, the half-life of the protein *in vivo*. However, genes involved in the production of mammalian-type complex oligosaccharides have been successfully expressed in *Pichia pastoris*, and one can now produce glycoproteins with complex type side chains in yeasts.

## EXAMPLE: HEPATITIS B VIRUS SURFACE ANTIGEN

The commercial production of the hepatitis B virus surface antigen (HBsAg) in yeast, a process that led to the first recombinant DNA vaccine licensed in the United States (Chapter 5), illustrates several of the features we have discussed.

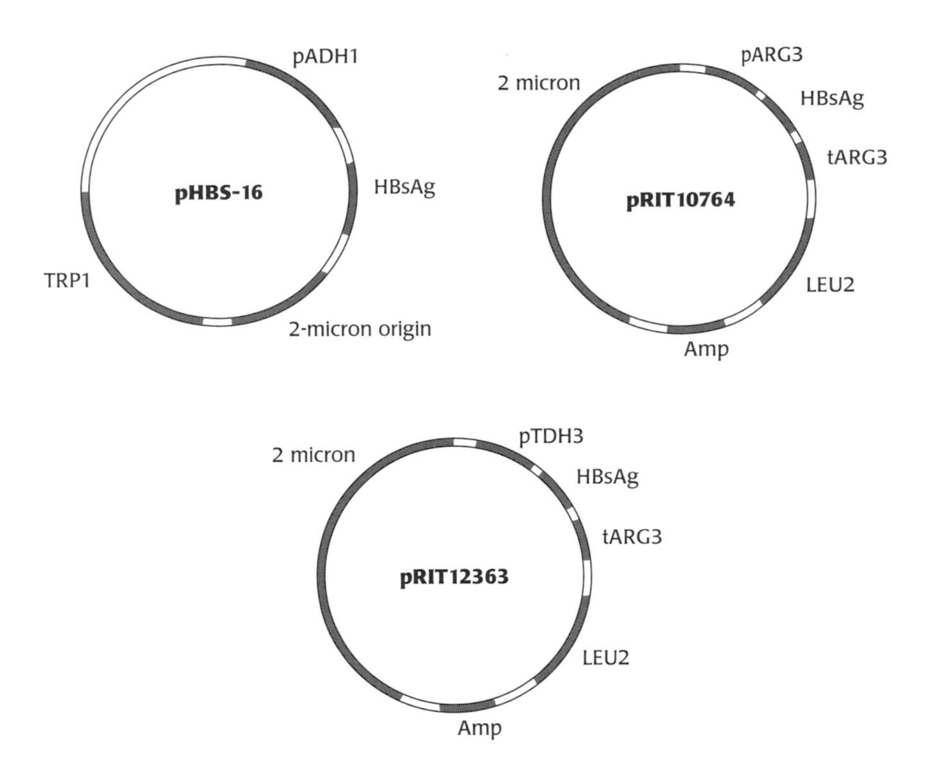


**FIGURE 3.26**

Recombinant plasmids used for the production of HBsAg. The "first-generation" plasmids include pHBS16 and pRIT10764. These were further developed for commercial use by Merck, Sharpe & Dohme and Smith Kline–RIT, respectively. pRIT12363 is an improved expression plasmid said to be in use at Smith Kline–RIT. Here, *p* indicates a promoter sequence, and *t* denotes a terminator sequence. *TRP1* in pHBS16 is a yeast gene coding for an enzyme of the tryptophan biosynthetic pathway and serves as the selective marker in yeast.

Redrawn based on artwork from the first edition (1995), published by W.H. Freeman.

HBsAg is a major component of the envelope of the hepatitis B virus, and immunization with this protein was known to confer good protection against viral infection (such a substance is called a protective antigen). The coding sequence for this 226-residue protein was identified on the virus genome, and it was successfully inserted into YEp-type yeast cloning vectors in several laboratories in the early 1980s (Figure 3.26). Remarkably, HBsAg folded correctly in yeast and became assembled in the form of empty envelopes, or "22-nm particles," making the subsequent purification somewhat easier. Several years later, the production of HBsAg was commercialized by two companies, Merck, Sharpe & Dohme in the United States and Smith Kline–RIT in Belgium.

*S. cerevisiae* strains transformed with the first-generation recombinant plasmids produced only small amounts of HBsAg. For example, pHBS-16 (see Figure 3.26), the first plasmid reported to produce HBsAg in yeast, made no more than 25 $\mu$g of HBsAg per liter of culture. The subsequent development that led from this plasmid to establishment of the commercial production process at Merck, Sharpe & Dohme is unfortunately not documented in detail in the open literature. However, some of the improvements carried out at Smith Kline–RIT have been documented, so we can get a glimpse of what they entailed.

**Plasmid Copy Number.** YEp vectors appear to be the most suitable for high-level expression because of their high copy numbers, and they were used in the production of HBsAg. As we have said, it is possible to increase the copy number of the plasmids by replacing LEU2 with the promoterless *leu2-d* so that only cells containing hundreds of copies of the plasmid can make enough leucine to survive. The Smith Kline–RIT group tried such an "improved" vector for HBsAg production but found that the cells rapidly lost the capacity to make HBsAg. It seems likely that the cells were losing the portion of the plasmid that coded for HBsAg. After all, when HBsAg is constitutively expressed (see below), a high level of this foreign protein is likely to be deleterious to the growth of the host cell. Therefore, progeny cells that inherit the *leu2-d*–containing part of the plasmid (which is essential for growth) but fail to inherit the gene for HBsAg are more likely to flourish. This example shows that one cannot blindly apply methods that are supposed to work better without testing and taking numerous factors into account. Production of foreign proteins is rarely neutral for the host, and one should always be alert to their possible toxic effects.

**Promoter Sequence.** In the first-generation plasmids pHBS16 (Merck) and pRIT10764 (Smith Kline–RIT), the promoter sequences came from the alcohol dehydrogenase (*ADH1*) and ornithine carbamoyl-transferase (*ARG3*) genes, respectively (see Figure 3.26). In the improved production strains used at both Merck and Smith Kline–RIT, the promoter comes from the gene for glyceraldehyde 3-phosphate dehydrogenase (*TDH3*). The TDH3 promoter is especially powerful, as one might have predicted from the fact that the dehydrogenase expressed from this promoter constitutes 5% of the total yeast protein. Clearly, use of the TDH3 promoter was advantageous.

As mentioned above, ADH1 was later found to become repressed toward the end of the exponential growth phase, so TDH3 remained preferable. In pRIT10764, the scientists chose a host strain with a leaky mutation in arginine biosynthesis so that the cells would be starved for arginine and so that the expression of *ARG3*, a gene involved in arginine synthesis, could be sustained at a high level (for regulation of amino acid biosynthetic genes; see Chapter 9). However, the paucity of arginine slowed the growth of the culture, again creating a less favorable situation for commercial fermentation. In these cases, there are rational explanations why the use of TDH3 promoter was preferable, but we must emphasize that in general, it is difficult to predict the levels of expression of foreign genes from the levels of expression of the endogenous yeast genes. There are many reasons why foreign genes may not be expressed as efficiently as host genes: instability of the mRNA, possible effects of the untranslated 5′ sequences of mRNA on the efficiency of translation initiation (see the next section), and the possibility that the coding sequences of the yeast genes contained enhancerlike sequences that were absent in the cloned foreign gene.

**Transcription Termination and Polyadenylation of mRNA.** Of the first-generation plasmids pHBS16 and pRIT10764, the latter was reported to produce a higher yield of HBsAg – about 200 $\mu$g/L of culture compared to the reported yield of less than 25 $\mu$g/L for pHBS16. Although much of this difference could be due to trivial factors such as the different quantitation methods used in different laboratories, there is an obvious difference between the two plasmids that could have contributed significantly to the higher yield of

HBsAg in strains containing pRIT10764. In this plasmid, the HBsAg sequence is followed by a terminator sequence taken from the downstream sequence of the *ARG3* gene, whereas no special terminator sequence is present in pHBS16.

**Recognition of the AUG Initiation Codon.** Another difference between pHBS16 and pRIT10764 is the relative content of G residues directly upstream of the AUG codon. We have noted that large numbers of G residues in this region inhibit the initiation of translation. Of 25 bases in this region in pHBS16, nine are G residues (36%), far more than the proportion found in native yeast promoter sequences. In contrast, pRIT10764, which produced a higher reported yield, contains only three G residues, well within the range found in native yeast promoters.

**Glycosylation, Folding, and Acetylation.** HBsAg made in human cells is *N*-glycosylated. This suggests that it is exported to the cell surface via the endoplasmic reticulum–Golgi pathway, in spite of the fact that there is no typical, cleaved, signal sequence at its N-terminus. When HBsAg is made in yeast cells, it is not glycosylated, and the protein accumulates in the cytoplasm without entering the endoplasmic–Golgi pathway. Perhaps the cloned sequence is incomplete. In the hepatitis B virus, the HBsAg sequence is preceded by an upstream "preS" sequence. Transcription in human cells may start at the preS sequence, which may contain the export signal. (Analysis of RNA transcripts is difficult with hepatitis B virus, because it cannot be grown in cultured cells.) When the HBsAg sequence is cloned and expressed together with the upstream extension, the product *is* glycosylated in yeast, a result that is consistent with the hypothesis that the preS sequence contains the export signal.

Despite the lack of glycosylation, HBsAg obviously folds correctly. This and the assembly of the protein into 22-nm particles presumably are important in achieving the desired overproduction of the antigen; if HBsAg were folded incorrectly to produce inclusion bodies in the cytoplasm, this would tie up foldases and chaperones that are needed for the folding of essential proteins of yeast, thereby interfering with the growth of the host cells.

The N-terminus of HBsAg becomes acetylated when produced in human cells; in yeast, at least a fraction of the HBsAg molecules become acetylated.

**Fermentation Conditions.** Some seemingly minor improvements in the fermentation conditions can have major effects on the yield. With the Smith Kline–RIT strains, the initial recombinant plasmid pRIT10764 was reported to produce HBsAg to a level of 0.06% of total yeast protein. Two years later, investigators in the same company reported a yield of 0.4% with the identical plasmid – an improvement presumably caused by a fine-tuning of culture conditions. However, the use of the ARG3 promoter still limited the growth of yeast cells to about 1 g/L. In pRIT12363, which was used for the production strain, use of the TDH3 promoter increased expression to about 1% of the yeast cell protein. However, the major improvement seems to have resulted from the fact that whereas pRIT10764 necessitated the use of leaky arginine biosynthesis mutants as the host, with pRIT12363, prototrophic strains could be used as the host, resulting in a much higher final density of yeast cells: about 60 to 70 g/L.

## EXPRESSION OF FOREIGN-GENE PRODUCTS IN A SECRETED FORM

As with bacterial hosts, it is advantageous in many ways if the yeast cell secretes the foreign-gene products into the medium. First, because *S. cerevisiae* does not naturally produce many extracellular proteins, purification of the products is much simpler: One does not have to start from a mixture containing thousands of other cytoplasmic proteins. Second, the secreted protein goes through the endoplasmic reticulum–Golgi pathway, where disulfide bonds – and hence, a stable protein – may be formed under optimal conditions with the help of protein disulfide isomerase, which is present in the lumen of the endoplasmic reticulum. Indeed, $\alpha$-interferon secreted by yeast cells has been shown to have disulfide bonds at the same positions as $\alpha$-interferon made by human cells. In contrast, when the same protein is made in the cytoplasm, a large fraction of it appears to become misfolded. Third, the proteins may become glycosylated during their passage through the endoplasmic reticulum and Golgi apparatus. Fourth, hormone precursors may be made into mature products by processing proteases during their secretion by yeast.

Proteins are brought into the endoplasmic reticulum–Golgi pathway when the components of the secretory pathway recognize their signal sequence (see Box 3.15). It is possible to design a recombinant plasmid so that the protein will enter this pathway, by fusing the DNA coding for an effective signal sequence to the coding sequence for the protein. Secretion vectors, which already contain DNA segments coding for the signal sequence, are useful in producing such recombinant plasmids. Signal sequences for secreted invertase (SUC2) and for secreted acid phosphatase (PHO5) have been used in this way and have resulted in the successful secretion of several animal proteins of interest. In some cases, however, a large fraction of the secreted foreign protein remains trapped within the yeast cell wall. This is reminiscent of certain secreted yeast proteins that do not seem to become freely dispersed in the culture medium. We can release these proteins into the medium by digestion of the cell wall, so it is clear that they are not anchored to the plasma membrane; rather, they appear to be trapped in the space between the cytoplasmic membrane and the cell wall. This problem has led to the exploration of yeast mechanisms that produce the genuine secretion of peptides into the surrounding medium.

In *S. cerevisiae*, one such system produces and excretes the mating factor $\alpha$, a 13-residue peptide. The immediate product of its structural gene, MF$\alpha$1, is a 165-residue polypeptide containing an N-terminal signal sequence and four copies of the $\alpha$-factor sequence. The $\alpha$-factor sequences are separated by a spacer with the sequence Lys-Arg-(Glu-Ala)$_n$, ($n$ = 2 or 3) (Figure 3.27). After cleavage of the signal sequence in the lumen of the endoplasmic reticulum, the polypeptide undergoes further proteolytic processing in the later

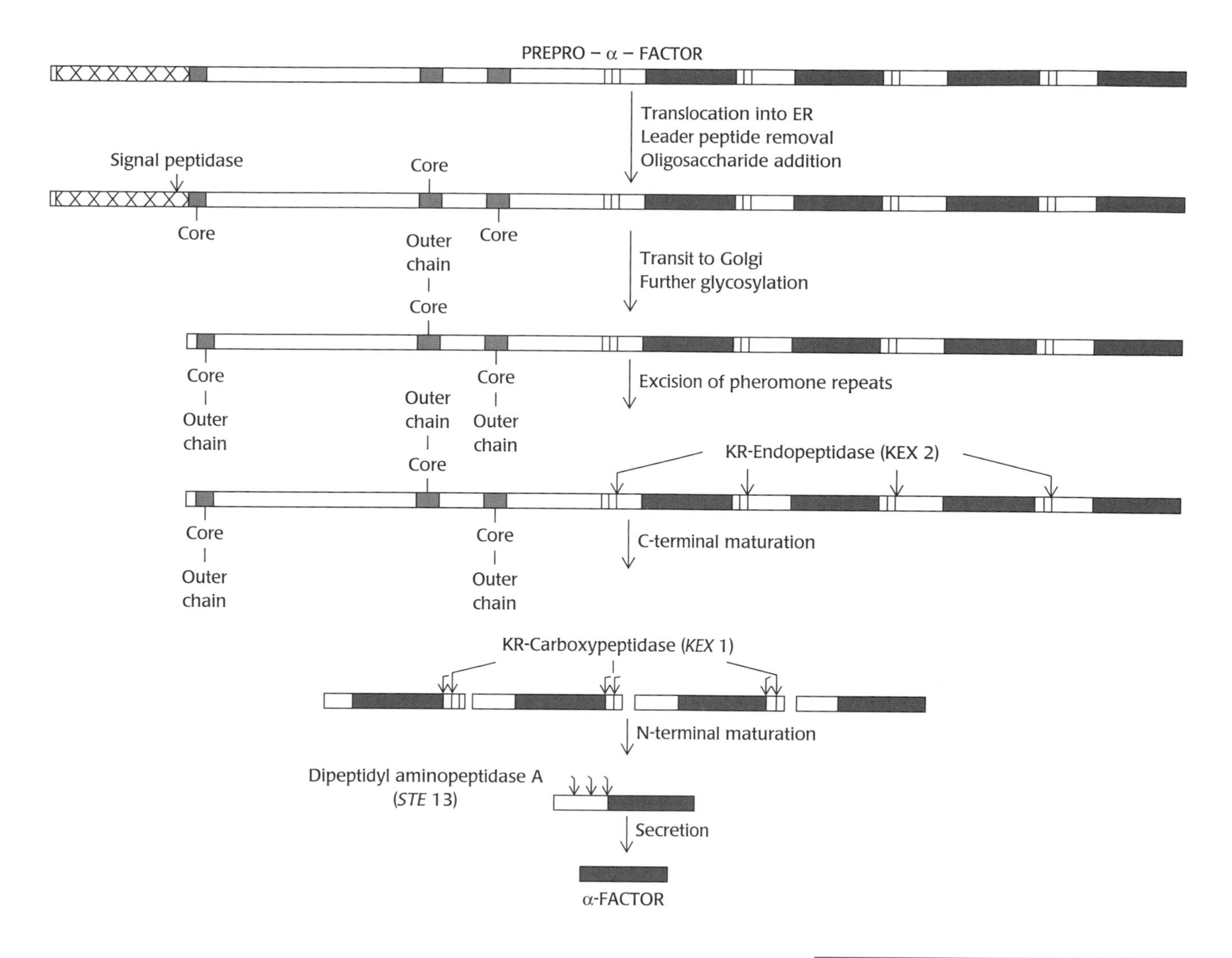


**FIGURE 3.27**

The processing of α-factor within the secretion pathway. [From Fuller, R. S., Sterne, R. E., and Thorner, J. (1988). Enzymes required for yeast prohormone processing. *Annual Review of Physiology*, 50, 345–362; with permission from the Annual Reviews, Inc.]

stages of the secretion process. First, KEX2 protease cleaves the bond after the Lys-Arg sequence. Then the peptide is shortened from both ends, KEX1 carboxypeptidase removing the Arg and Lys residues from the C-termini and STE13 dipeptidyl aminopeptidase removing Glu-Ala units from the N-termini. This complex processing scheme may prove useful to biotechnologists because it may afford them some flexibility in the design of fusion joints. Genes coding for animal and plant proteins have been fused to the N-terminal "prepro" portion of the *MFαl* gene, and successful secretion of products was observed in many cases. Furthermore, this method has now become a standard approach in the secretion of massive amounts of proteins by the use of "nonconventional" yeast species, and secretion of human proinsulin up to the level of 1.5 g/L has been reported.

In yeast, the Asn-linked core oligosaccharides that are attached to the *N*-glycosylation sites of foreign proteins sometimes become extended into large outer chains, producing high-mannose-type oligosaccharides. These enormous oligosaccharides may impair the proper folding and functioning of the animal-derived proteins. It may therefore be advantageous to use mutants, such as *mnn9*, that are defective in the addition of outer-chain mannose residues. For example, human $\alpha_1$-antitrypsin, secreted from a *S. cerevisiae mnn9* mutant, carries three N-linked oligosaccharides similar in size to those attached to the human protein. Finally, there is recent progress in efforts to produce mammalian-type glycosylation in yeasts (p. 136).

One general problem with yeast secretion systems is low yield. However, screening of mutagenized yeast cells that contain secretion plasmids has produced high-secretion mutants. In one case, the combination of two mutations produced a strain that secreted 80% of the prochymosin synthesized. An alternative strategy is the use of nonconventional yeast species that are highly efficient in protein secretion (see below).

## EXPRESSION OF PROCHYMOSIN IN YEAST

As described earlier in this chapter, the expression of calf prochymosin in the cytoplasm of *E. coli* resulted in the formation of inclusion bodies. It was thought that this problem might be overcome by expression of the protein in yeast cells because inclusion bodies are formed less frequently in yeast. The prochymosin gene was cloned into several YEp-type expression plasmids behind effective yeast promoters. However, the accumulated product was largely insoluble when overproduced.

Better results were expected with the yeast secretion vectors because the protein would then be secreted through the endoplasmic reticulum–Golgi pathway, which is similar to that in animal cells. The recombinant plasmid that was tested contained (1) a strong yeast promoter, such as the one for the phosphoglycerate kinase gene, (2) the DNA sequence coding for the signal sequence and several of the N-terminal amino acid residues of the mature invertase, a secreted yeast protein, and (3) the sequence for prochymosin fused to the invertase fragment. These YEp-type plasmids directed the secretion of prochymosin, but the fraction secreted was quite low, usually less than 5%. Mutagenesis and screening of the host strain have refined the system to the point where up to 80% of the synthesized fused protein is secreted from the yeast cell, as described above. However, the reported yield is still rather low – around 1 mg/g of total yeast protein – even in the best combinations of host with plasmid. A possible explanation is that *S. cerevisiae* normally secretes only a very small fraction of its cellular proteins across its cytoplasmic membrane; in wild-type strains, secreted invertase corresponds to much less than 0.1% of the total cellular protein. For this reason, the recent trend has been to explore other, more secretion-competent species of yeast.

In one experiment, a recombinant plasmid was made in which the sequence coding for prochymosin was placed between a strong LAC4 promoter and the LAC4 terminator sequence from *Kluyveromyces lactis*, a lactose-utilizing yeast species. When this plasmid was linearized and integrated into the *Kluyveromyces* genome, there was only a low-level expression of prochymosin. But most of the prochymosin was secreted into the medium, even though the cloned DNA lacked the sequence coding for the signal sequence. When the prochymosin gene was cloned together with the sequence coding for its own signal sequence, the prochymosin production increased 50- to 70-fold, and 95% of the product appeared in the medium in a correctly processed form. The yield was reported as about 100 enzyme units/ml, which corresponds roughly to 1 g/L, or about 10% of the total cellular protein. Other yeast species that have been shown to produce (and secrete) foreign proteins at a higher level than *S. cerevisiae* include *Pichia pastoris* and *Hansenula polymorpha*, both methylotrophic yeasts (yeasts capable of using methanol as the carbon source), and *Yarrowia lipolytica*, which can grow on alkanes. *P. pastoris*, for example, produces HBsAg to a level of about 50% of the total cellular protein.

There are also non-yeast fungal species that are known to secrete very large amounts of proteins; for example, *Trichoderma reesei* and *Aspergillus awamori* naturally secrete more than 20 g of protein per liter. These species were hypothesized to be even more proficient in catalyzing the export of large amounts of foreign proteins. Initial yields, obtained after cloning of the prochymosin gene in an expression vector, were not exceptional, but several optimization steps increased the yield significantly. These procedures included inactivation of the fungal gene that encodes a prochymosin-inactivating protease and fusion of the prochymosin sequence to the 3′-end of a complete sequence coding for a glucoamylase, an enzyme secreted in very large amounts by *A. awamori*. Such modifications increased the yield to the range of 100 mg/L. Finally, random mutagenesis and screening of the host *A. awamori* strain increased the yield to about 1 g/L, apparently a level that would make production commercially profitable. Importantly, the high-secretion mutant strain selected on the basis of prochymosin production also secretes other foreign proteins at a higher efficiency.

With a variety of tools for solving the production problem – chief among them the approaches described above – several laboratories are now attempting to modify, by site-directed mutagenesis, the structure of the prochymosin molecule itself. Recently, modifying the residues surrounding the glycosylation site to improve glycosylation efficiency resulted in the doubling of yield of prochymosin secreted from *A. awamori*.

## SUMMARY

Some proteins and peptides of therapeutic value are difficult to purify in sufficient amounts from their human and animal sources. Recombinant DNA methods have had a revolutionary impact in the production of these compounds. Once the DNA sequences coding for these proteins and peptides have been cloned and amplified in microorganisms, the latter can continue to function as living factories for the inexpensive production of such compounds. Bacteria, especially *E. coli*, are used extensively as the host microorganism. Segments of "foreign DNA" coding for these products are first obtained either by cutting the genomic DNA or by synthesizing a DNA sequence (cDNA) complementary to the mRNA with reverse transcriptase. Such segments must first be inserted into vector DNA, which contains information that makes possible its replication in the bacterial host. In addition to plasmids, which are widely used as general-purpose cloning vectors, there are several types of vectors for cloning in bacteria. λ Phage vectors, cosmids, and BAC vectors are useful for the cloning of large segments of DNA, and single-strand DNA phage vectors are especially well suited for phage display technology that allows the isolation of mutants producing proteins with desired properties. The recombinant DNA – that is, the vector DNA containing the foreign DNA insert – is then introduced into the host cell by transformation or by injection from phagelike particles after it has been packaged into phage heads. The clone that contains the desired gene sequence is identified, sometimes among a vast majority of clones not containing this sequence, by using either the DNA sequence itself or the protein product of the gene as the marker. In some cases, however, PCR enables one to bypass all the steps of primary cloning and screening by direct amplification of the DNA sequence *in vitro*.

Regardless of the source, the sequence coding for the desired product can then be inserted into expression vectors to maximize the synthesis of the product in bacteria. The overproduction of foreign proteins in bacteria, however, frequently results in misfolding and aggregation of these proteins. Several strategies for avoiding aggregation are available, but none of them appears to be universally applicable to all proteins. However, aggregation does not necessarily mean a total failure, because protein aggregates can be easily purified, totally denatured, and then renatured under controlled conditions.

*S. cerevisiae* and other yeast species have considerable potential as host organisms for the production of foreign proteins, especially proteins of animal origin. Many different vectors are available, and most are shuttle vectors, which allow the recombinant DNA manipulations to be conveniently carried out in *E. coli* before the final recombinant product is introduced into yeast.

One major advantage of expression in yeasts is that foreign proteins appear to have less tendency to become misfolded in yeast than in bacterial hosts, partly because yeast cells presumably contain more efficient chaperones and foldases. Furthermore, proteins can become glycosylated in yeast cells if they can be introduced into the endoplasmic reticulum–Golgi apparatus protein-secretion pathway. Glycosylation not only facilitates the correct folding of some proteins but also makes them less susceptible to degradation in the animal body, thus prolonging their half-life when they are administered as therapeutic agents. Because *S. cerevisiae* is not well-equipped to secrete a large amount of proteins, nonconventional yeast species such as *P. pastoris* and *K. lactis* are increasingly used as hosts of secretion vectors, often making use of the "prepro" sequence of *S. cerevisiae* mating factor $\alpha$ precursor.

HBsAg and prochymosin are two proteins of animal origin that have been successfully produced in yeasts. HBsAg did not enter the secretion pathway and was not glycosylated, however, apparently because the cloned DNA fragment lacked the segment coding for the signal sequence. Nevertheless, it was folded correctly and assembled into a structure resembling the envelope of the virus. When the cDNA for prochymosin was expressed in *E. coli*, it produced inclusion bodies that were difficult to renature. In contrast, when it was expressed from a secretion vector in an *S. cerevisiae* host, the protein entered the endoplasmic reticulum–Golgi pathway, was folded correctly and glycosylated, and was secreted, although the yield remained low. When non-*Saccharomyces* yeasts and non-yeast fungi that physiologically secrete very large amounts of proteins were used as hosts, commercially acceptable yields of secreted prochymosin were achieved.

## SELECTED REFERENCES


### General References on Recombinant DNA Methods

Primrose, S. B., and Twyman, R. M. (2006). *Principles of Gene Manipulation and Genomics*, 7th Edition, Oxford, UK: Blackwell Science.

Sambrook, J., and Russell, D. W. (2001). *Molecular Cloning: A Laboratory Manual*, 3rd Edition, Cold Spring Harbor, NY: Cold Spring Harbor Laboratory Press.

### Vectors

Balbas, P., Soberon, X., Merino, E., Zurita, M., Lomeli, H., Valle, F., Flores, N., and Bolivar, F. (1986). Plasmid vector pBR322 and its special-purpose derivatives – a review. *Gene*, 50, 3–40.

Casali, N., and Preston, A. (eds.) (2003). *E. coli Plasmid Vectors: Methods and Applications* (Vol. 235, Methods in Molecular Biology), Clifton NJ: Humana Press.

Shizuya, H., Birren, B., Kim, U.-J., Mancino, V., Slepak, T., Tachiri, Y., and Simon, M. (1992). Cloning and stable maintenance of 300-kilobase-pair fragments of human DNA in *Escherichia coli* using an F-factor-based vector. *Proceedings of the National Academy of Sciences U.S.A.*, 89, 8794–8797.

Kehoe, J. W., and Kay, B. K. (2005). Filamentous phage display in the new millennium. *Chemical Reviews*, 105, 4056–4072.

Lipovsek, D., and Plückthun, A. (2004). In-vitro protein evolution by ribosome display and mRNA display. *Journal of Immunological Methods*, 290, 51–67.

### PCR

Shamputa, I. C., Rigouts, L., and Portaels, F. (2004). Molecular genetic methods for diagnosis and antibiotic resistance detection of mycobacteria from clinical specimens. *APMIS*, 112, 728–752.

### Expression of Cloned Genes

Makrides, S. C. (1996). Strategies for achieving high-level expression of genes in *Escherichia coli*. *Microbiological Reviews*, 60, 512–538.

Baneyx, F. (1999). Recombinant protein expression in *Escherichia coli*. *Current Opinion in Biotechnology*, 10, 411–421.

Baneyx, F. (ed.) (2004). *Protein Expression Technologies: Current Status and Future Trends*, Norfolk, U.K.: Horizon Bioscience.

### Proteolysis

Enfors, S.-O. (1992). Control *of in vivo* proteolysis in the production of recombinant proteins. *Trends in Biotechnology*, 10, 310–315.

### Protein Folding, Foldases, and Molecular Chaperones

Baneyx, F., and Mujacic, M. (2004). Recombinant protein folding and misfolding in *Escherichia coli*. *Nature Biotechnology*, 22, 1399–1408.

Thomas, J. G., Ayling, A., and Baneyx, F. (1997). Molecular chaperones, folding catalysts, and the recovery of active recombinant proteins from *E. coli:* to fold or refold. *Applied Biochemistry and Biotechnology*, 66, 197–238.

Schmid, F. X. (2002). Prolyl isomerases. *Advances in Protein Chemistry*, 59, 243–282.

Bader, M. W., and Bardwell, J. C. A. (2002). Catalysis of disulfide bond formation and isomerization in *Escherichia coli. Advances in Protein Chemistry*, 59, 283–291.

Hartl, F. U., and Hayer-Hartl, M. (2002). Molecular chaperones in the cytosol: from nascent chain to folded protein. *Science*, 295, 1852–1858.

Kapust, R. B., and Waugh, D. S. (1999). *Escherichia coli* maltose-binding protein is uncommonly effective at promoting the solubility of polypeptides to which it is fused. *Protein Science*, 8, 1668–1674.

Terpe, K. (2003). Overview of tag protein fusions: from molecular and biochemical fundamentals to commercial systems. *Applied Microbiology and Biotechnology*, 60, 523–533.

### Protein Secretion

Georgiou, G., and Segatori, L. (2005). Preparative expression of secreted proteins in bacteria: status report and prospects. *Current Opinion in Biotechnology*, 16, 538–545.

Mergulhão, F. J. M., Summers, D. K., and Monteiro, G. A. (2005). Recombinant protein secretion in *Escherichia coli. Biotechnology Advances*, 23, 177–202.

Miot, M., and Betton, J.-M. (2004). Protein quality control in the bacterial periplasm. *Microbial Cell Factories*, 3, 4.

### Prochymosin

Beppu, T. (1988). Production of chymosin (rennin) by recombinant DNA technology. In *Recombinant DNA and Bacterial Fermentation*, J. A. Thomson (ed.), pp. 11–21, Boca Raton, FL: CRC Press.

### Cloning in Yeast

Guthrie, C., and Fink, G. R. (2002). *Guide to Yeast Genetics and Molecular and Cell Biology, Parts B and C* (Methods in Enzymology, volumes 350 and 351), New York: Academic Press.

Goffeau, A., Barrell, B. G., Bussey, H., et al. (1996). Life with 6000 genes. *Science*, 274, 546–567.

Kumar, A., and Snyder, M. (2001). Emerging technologies in yeast genomics, *Nature Reviews Genetics*, 2, 302–312.

Spencer, J. F. T., Ragout de Spencer, A. L., and Laluce, C. (2002). Non-conventional yeasts. *Applied Microbiology and Biotechnology*, 58, 147–156.

Liti, G., and Louis, E. J. (2005) Yeast evolution and comparative genomics. *Annual Review of Microbiology*, 59, 135–153.

Cereghino, G. P. L., Cereghino, J. L., Ilgen, C., and Cregg, J. M. (2002). Production of recombinant proteins in fermenter cultures of the yeast *Pichia pastoris. Current Opinion in Biotechnology*, 13, 329–332.

Gerngross, T. U. (2004). Advances in the production of human therapeutic proteins in yeasts and filamentous fungi. *Nature Biotechnology*, 22, 1409–1414.

Li, H., Sethuraman, N., Stadheim, T. A., et al. (2006). Optimization of humanized IgGs in glyco-engineered *Pichia pastoris.* Nature Biotechnology, 24, 210–215.

Kjeldsen, T. (2000). Yeast secretory expression of insulin precursors. *Applied Microbiology and Biotechnology*, 54, 277–286.

Mohanty, A. K., Mukhopadhyay, U. K., Grover, S., and Batish, V. K. (1999). Bovine chymosin: Production by rDNA technology and application to cheese manufacture. *Biotechnology Advances*, 17, 205–217.

van den Brink, H. M., Petersen, S. G., Rahbek-Nielsen, H., Hellmuth, K., and Harboe, M. (2006). Increased production of chymosin by glycosylation. *Journal of Biotechnology*, 125, 304–310.

# FOUR

# The World of "Omics": Genomics, Transcriptomics, Proteomics, and Metabolomics

## GENOMICS

### SEQUENCING OF GENOMES

As mentioned in Chapter 3, we now know the complete nucleotide sequences of genomes of many organisms. The availability of this large amount of data at an unprecedented scale now allows us, and indeed forces us, to think "globally," that is, on the scale of whole organisms, or even an assemblage of organisms, rather than of individual genes and enzymes. Here we describe very briefly how the genome sequences are determined.

Genome sequencing of viruses began in the late 1970s. The basic technique involved, the random sequencing of fragments by the Sanger dideoxy termination method, was proposed and applied successfully by Fred Sanger and associates to the complete sequencing of bacteriophage DNAs, notably that of phage λ in 1980 (see Figure 4.1 for the principle of the random shotgun method).

Historically, the first attempt to obtain a complete genome sequence of a *cellular* organism was geared toward *Escherichia coli*, the best studied organism outside of humans. This project started in 1989 and used a "directed" approach. Because a fairly detailed genetic map of *E. coli* was available thanks to the efforts of bacterial geneticists, it was possible to first produce a set of λ-based clones, each containing up to 20 kb DNA, with overlapping ends. The sequencing from here on represented the shotgun phase. The inserted segments in the λ vector were then randomly cut into much smaller fragments of a few kilobases, they were cloned into an M13 vector and were sequenced (for sequencing reactions, see Box 4.1), and the sequences were assembled by looking for overlaps. The ends of larger inserts in a plasmid vector, when sequenced, were useful also in positioning some raw sequences that overlap with them (Figure 4.1, where the "source DNA" corresponds to the 20-kb inserts in the λ vector). This two-stage approach (also called "clone-by-clone shotgun") was taken because it was thought that the shotgun approach could not be used for larger segments of DNA or for an entire genome.

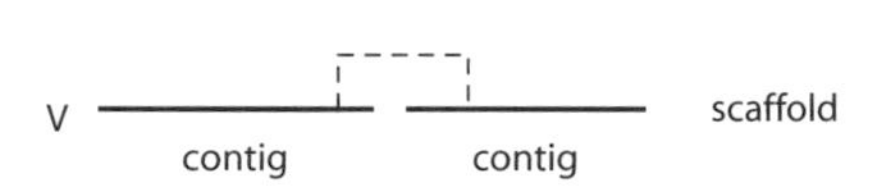

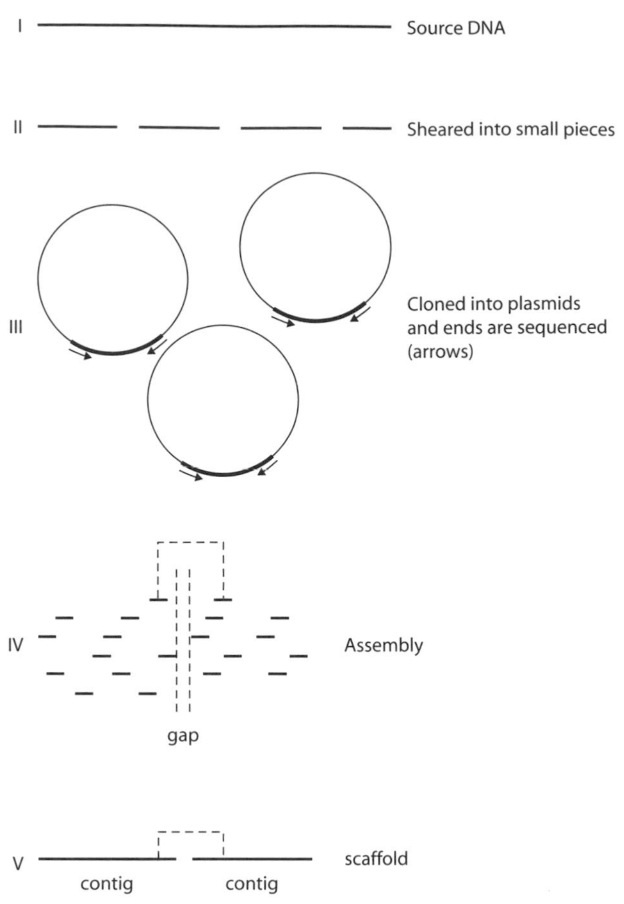

**FIGURE 4.1**

Principle of shotgun sequencing. The source DNA (step I) is sheared into random fragments of about 2 to 5 kb (step II), and they are cloned into plasmid vectors (step III). The ends of the inserts (*thick line*) are sequenced by using the "universal primers" (*arrows*) (which correspond to the vector sequences at the vector/insert borders). The sequences thus obtained are aligned by looking for overlaps and are assembled into contiguous stretches or contigs (steps IV and V). This usually leaves gaps, but contigs can still be connected by taking advantage of two segments that came from the same insert (connected by a *dotted line* in IV). These contigs with gaps are then called scaffolds (step V).

It was a surprise, therefore, that J. Craig Venter and his associates at The Institute of Genomic Research successfully used the random shotgun approach for the entire 1.8-megabase (Mb) genome of *Haemophilus influenzae* without first directly cloning and ordering large inserts (Figure 4.1, where the "source DNA" corresponds to the entire bacterial genome in this case) and published the results in 1995, two years before the publication of the *E. coli* genome sequence. The determination of *H. influenzae* genome sequence involved the use of 24,000 "reads," each with the average length of about 400 bases or slightly longer. Thus, the total sequences used for assembly were almost 12 Mb long, meaning that each segment of the genome was covered more than six times by the sequencing reaction (this is called "depth of coverage" and will be mentioned in the "Metagenomics" section later in this chapter). The assembly required a newly developed computer program.

For many of us, the crowning glory of the genome sequencing projects was that of the human genome – at about 3 Gb, more than 1600 times larger than that of *H. influenzae*. The International Human Genome Sequencing Consortium used the clone-by-clone shotgun approach, and the "first draft" was published in 2001. This approach was taken because the presence of very large amounts of repeated sequences in the mammalian genome was predicted to make the assembly of random sequences difficult. Thus, in the first stage, 100- to 200-kb segments of DNA were cloned into the BAC vector and were correlated with the physical map containing known genetic markers and other sites. A set of BAC constructs with overlapping ends was then chosen, and each insert was fragmented randomly into about 2- to 5-kb pieces, which were usually cloned into plasmid vectors, and the ends of the insert were sequenced (Figure 4.1, where the "source DNA" here corresponds to the large inserts in the BAC vector, or sometimes their fragments cloned into other vectors, such as cosmids). The first draft contained about 150,000 gaps, covering about 10% of the genome. Subsequent refinement published in 2004, however, reduced the number of gaps to only 341 and corrected numerous minor errors.

**Sequencing DNA**

Almost all of the sequencing is currently carried out by using the Sanger dideoxy chain termination method (Box 3.7). During the early years, the DNA chains that were synthesized were separated on polyacrylamide gel electrophoresis. The speed of analysis, which is most important in the shotgun sequencing, was greatly improved when the gel was eliminated by separation methods using capillary electrophoresis. (Capillary electrophoresis employs polymeric matrices, most frequently linear polyacrylamide that is then cross-linked. It is not just a buffer solution). Use of fluorescent dyes, rather than radioactive isotopes, as markers also improved the speed of detection and the accuracy of quantitation. Constant improvements in methodology have resulted in an instrument that is said to be able to sequence 2.8 million bases/day; this can be contrasted with the fact that in 1995, when the sequence of the first genome of a cellular organism, *H. influenzae*, was completed, the capacity was only somewhat above 1000 bases/day.

As mentioned in Chapter 3, both M13-based vectors, producing a single-stranded DNA, and conventional plasmid vectors, producing a double-stranded DNA, are used for sequencing. In both cases, the primer for sequencing comes from the vector sequence right next to the vector-cloned DNA junction, sometimes called a universal primer. The M13-based vectors are said to produce cleaner "reads," whereas plasmid vectors give the advantage of providing the sequences of two ends of the cloned fragments, a feature that is very useful in the assembly of the sequences to "contigs" or contiguous segments.

**BOX 4.1**

Interestingly, Venter and associates proposed that this sequencing of truly massive scale can also be achieved by the whole genome shotgun approach within a shorter time. Indeed, by using a large number of sequencing machines with a large capacity, 27 million raw sequences (each 500 to 750 bases long) were obtained and were assembled in just a few years. It is important to note, though, that the approach involved the cloning and sequencing of ends of not only small (about 2-kb) pieces but also much larger (10- or 50-kb) pieces so that the raw sequences could be connected into "scaffolds" (see Figure 4.1) in an unambiguous manner.

The merits and demerits of these two approaches have been hotly debated. The directed approach was successful in sequencing some large genomes, including *Saccharomyces cerevisiae* (16 Mb; 1996), the model plant *Arabidopsis thaliana* (115 Mb; 2000), the worm *Caenorhabditis elegans* (97 Mb; 1998), and rice (390 Mb; finished sequence in 2005). The whole genome shotgun sequencing was the basic approach in the sequencing of the small genomes of many prokaryotes, but it was also used successfully for the puffer fish genome (365 Mb; 2002), chicken genome (1000 Mb; 2004), and others, and most importantly, the human genome as seen above. However, these two approaches are not mutually exclusive. In fact, it is becoming a general consensus that a hybrid approach may be the most effective. Here a library containing large inserts – for example, the BAC library – is constructed, and the inserts are subjected to shotgun sequencing. At the same time, a large library containing random short segments is also made from the whole genome, and the inserts are sequenced. Out of the latter, only the sequences that partially

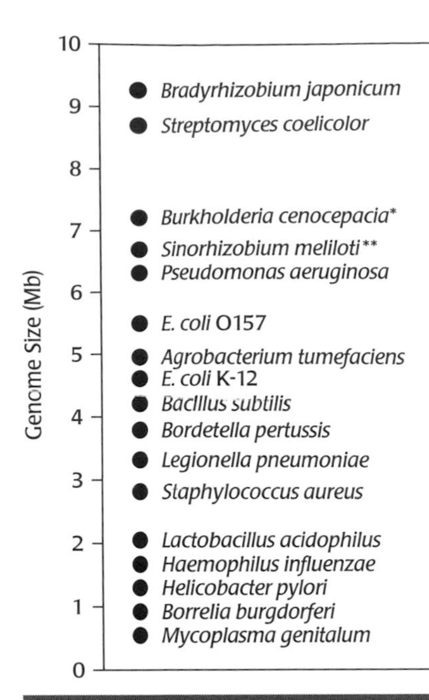


FIGURE 4.2

Genome sizes of some bacteria, derived from the finished genome sequences.
*Genomic DNA is distributed in three chromosomes.
**This size includes DNA in two large plasmids.

overlap with the "reads" from a BAC construct are retained and used for assembly. In this way, the enormous complexity of the fragments from the whole genome can be reduced, the assembly becomes much easier, and the probability of making errors is greatly reduced. This hybrid approach was used for the mouse genome (~3000 Mb; 2002), and rat genome (~3000 Mb; 2004). The sequencing of the *Drosophila melanogaster* genome (120 Mb; 2000) involved the use of clone-by-clone information at the finishing stage and thus can be considered a version of the hybrid approach.

## A GLIMPSE AT COMPARATIVE GENOMICS

Now that genomes of more than 300 cellular organisms have been sequenced, the first item on the agenda should be to compare these sequences.

### Prokaryotic Genomes

There is a large variation in the size of bacterial genomes (Figure 4.2). Compared with the *H. influenzae* genome (1.8 Mb), some obligate parasites, such as *Chlamydia trachomatis* (1.0 Mb) and *Mycoplasma genitalum* (0.6 Mb), have smaller genomes. On the other hand, the genome of *E. coli* (4.6 Mb) is much larger, and even larger genomes are found in *Pseudomonas aeruginosa* (6.3 Mb), *Streptomyces coelicolor* (8.7 Mb), and *Bradyrhizobium japonicum* (9.1 Mb). (With rare exceptions, most archeal genomes have a rather small size, between 1.5 and 3 Mb). In the prokaryotic genome, most of the DNA represents coding sequences or genes, unlike in higher animals and plants. So this large difference leads us to two questions. The first is, What is the minimal set of genes that allow simple cellular organism to grow and replicate?

Here we have to classify genes, and the proteins coded by these genes, on the basis of homology. Homology means descent from a common ancestor, and this is usually predicted by comparing sequences using computer programs such as BLAST. We distinguish two kinds of homologs. Homologous proteins in two different organisms, where they developed independently of each other, are called *orthologs*. In contrast, homologous proteins that exist in the same species, presumably created originally by gene duplication and often differentiated later in terms of functions, are called *paralogs*. Comparison of the genomes of *M. genitalum* (coding for 468 proteins) and *H. influenzae* in terms of orthologs led to the definition of a "minimal gene set" of about 300 genes. (The list was expanded to 382 genes in 2006 by experimental gene disruption using transposons [see Box 3.8]). This set includes genes needed for DNA replication and repair, transcription, and translation (including chaperones) and for biosynthesis of nucleotides, coenzymes, and lipids, as well as the glycolytic pathway and the $F_1F_0$ATPase. However, because *Mycoplasma*, which does not have peptidoglycan, was used as the basis, the list does not contain genes needed for peptidoglycan synthesis, nor does it contain genes involved in the biosynthesis of amino acids or purines and pyrimidines. The set of genes needed for an independent survival of a bacterium in an environment is usually thought to contain about 1500 genes.

The second question regarding different genome sizes is, What do larger genomes contain in addition to the minimal set just mentioned? Comparison of the genome of *H. influenzae*, which basically resides in only one environment – the upper respiratory tract of animals – with the much larger genome of *E. coli* already indicates that the latter organism, which must face "feast-or-famine" existence in the intestinal tract and must survive (at least for a short period) in natural waters, contains many more genes (often paralogs) needed for adaptation in different types of environment. *E. coli* also needs genes for complex regulatory responses for this adaptation. The same theme is found over and over. Thus, *P. aeruginosa*, which is basically a resident of soil and water but can also cause severe infection in humans, has a larger genome filled with a more complex array of genes. This tendency reaches its zenith in *S. coelicolor*, a soil bacterium that produces several antibiotics and undergoes differentiation into aerial spores. This genome is equipped for very complex regulatory responses, as seen, for example, from the record number of sigma factors (that control transcription specificity at the most basic level) – 55 – in comparison with *E. coli*, which produces only seven. Genes for many enzymes occur as a set of multiple paralogs, each gene known or suspected to be expressed only under certain conditions. For example, five paralogs of *fabH*, coding for the first enzyme of fatty acid biosynthesis, are found in *S. coelicolor*. One is in the main fatty acid synthesis operon and is essential. Three are in gene clusters involved in the biosynthesis of antibiotics and probably a polyunsaturated fatty acid and are predicted to function for these specialized pathways without interfering with the major "housekeeping" metabolism of the cell.

Another major mechanism that increases the prokaryotic genome size is the addition, presumably through horizontal transfer from other organisms, of large genomic islands. In *Salmonella*, "pathogenicity islands" 1 through 5 are located at the 63-, 31-, 82-, 92-, and 20-min positions on the circular map (with a circumference of 100 min). They range in size from several kilobases to 40 kb, and they code for proteins involved in pathogenesis, such as type III secretion systems that inject toxic proteins directly into mammalian cells. These islands are strikingly different from the endogenous sequence of the genome – for example, in GC content – a fact that suggests their "foreign" origin. The function of islands is not limited to pathogenicity. For example, the chromosome of *B. japonicum*, an $N_2$-fixing symbiont of the soybean root, contains a very large (610-kb) "symbiosis island" of much lower GC content that apparently codes for many functions needed for symbiosis. The analysis of prokaryote genomes then shows that the genome often has a mosaic origin, and horizontal transfer has occurred many times during evolution.

How do the genomes of organisms of limited habitat become smaller? Again, genomes give a clue. *Salmonella typhi* is a pathogen specializing in infecting humans and cannot cause major diseases in other animals. Its genome is quite similar to that of *Salmonella typhimurium*, which can infect many animal species. However, in *S. typhi*, more than 200 genes have been converted into pseudogenes and therefore are nonfunctional. An even more extreme example of gene decay is the situation found in the genome of *Mycobacterium leprae*, the causative organism for leprosy. Here, only 50% of the genome contains protein-coding genes, and 27% is occupied by 1116 pseudogenes that have functional orthologs in *Mycobacterium tuberculosis*. A large fraction of the remainder (23%) presumably corresponds to remnants of genes that were altered beyond recognition. The *M. leprae* genome indeed appears to be in the midst of contraction, because its size, 3.3 Mb, is significantly smaller than that of *M. tuberculosis* (4.4 Mb). Inactivation occurred even in genes coding for the central energy pathway, such as the respiratory electron transport, and it seems to explain the observation that this organism grows only in leprosy patients, in armadillos, or in mouse foot pads, and even there grows exceedingly slowly, with an estimated doubling time of two weeks.

## Comparative Genomics in Higher Animals

The size of eukaryotic genomes shows a wide range, from 12 Mb in *S. cerevisiae*, to 97 Mb in the worm *C. elegans* and 120 Mb in the fly *D. melanogaster*, and finally to about 3000 Mb in humans, rats, and mice. The genome size can be even larger: some protozoa are thought to have genomes of up to 600 Gb and some plants haploid genomes of up to 125 Gb. (These values were determined by scanning the stained nuclei and not by sequencing.) Sequencing confirmed what was suspected for some time, that the much larger genome size of mammals in comparison to worm and fly is basically caused by the presence of much larger amounts of repeat sequences. In the human genome, more than 50% is occupied by the repeats, most of which are interspersed repeats. Such repeats comprise only 3% and 6.5% of the genomes in fly and worm, respectively. The majority of these interspersed repeats in the human genome are transposons. As described in Chapter 3 (Box 3.8), a transposon is a piece of selfish DNA that codes for genes allowing random insertion of its own sequence into a genome. Unlike transposons in bacteria, these transposons are "retrotransposons" that insert themselves by using a reverse transcriptase, with RNA as an intermediate. They do not contain antibiotic resistance genes. "Alu element," a short (300-nucleotide) segment so named because it contains the recognition sequence (AGCT) for the restriction enzyme Alu, is present in more than 1 million copies in the human genome.

One of the questions that attracted attention in the sequencing of the human genome was the number of protein-coding genes. The predicted number was as high as 150,000, and such predictions fueled the hope that there will be literally hundreds of thousands of hitherto unknown, potential targets for pharmaceutical companies. However, the results of Venter and associates showed only 26,500 genes, plus 12,000 "computationally derived"

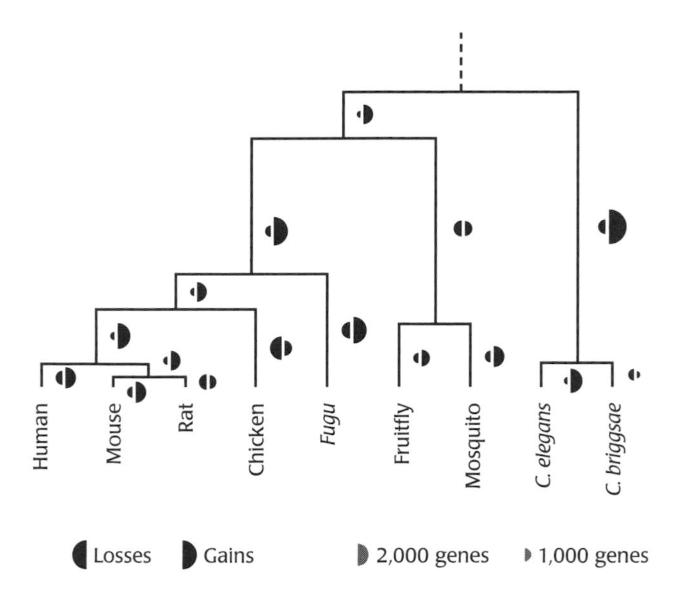


**FIGURE 4.3**

Estimates of loss and gain in genes during evolution of various groups of animals. Only groups of orthologous genes were considered, and the map was drawn assuming parsimony and no horizontal transfer of genes. The *left and right half-circles* indicate the gain and loss of genes during the development of each branch, and the areas of the *half-circles* are proportional to the changes in the number of genes. *Caenorhabditis briggsae* is a worm species distantly related to *C. elegans*. [After International Chicken Genome Sequencing Consortium (2004). *Nature*, 432, 695–716; with permission.]

genes. The draft sequence of the International Human Genome Sequencing Consortium predicted between 30,000 and 40,000 protein-coding genes, but this was decreased to only 25,000 to 30,000 in the finished sequence in 2004. These numbers are similar to those reported for rat and mouse, but for many it was disturbing to admit that humans contain only a few more genes than the lowly worm, *C. elegans* (predicted to contain around 19,000 genes). However, comparison of orthologs in various genomes indicates that some genes are lost and others are gained in the evolution of any branch of higher organisms (Figure 4.3).

Are the mammalian genes different in structure from the genes of the worm or the fly? Indeed they are, and this explains why it was so difficult to get a reasonable estimate of the number of genes in the human genome. Exons are usually very short, with an average length coding for just 50 amino acids. Introns separating the exons are much longer in humans (average length >3300 bp) than in worm (average length 267 bp, with a pronounced peak at 47 bp). This increases the length of the entire gene (exons plus introns), a factor that contributes to the larger size of the human genome. Yet coding regions of genes are estimated to occupy only 1.2% of the human genome sequence.

There are some features in human protein-coding genes that suggest more complex functions than those in worms and flies. First, alternative transcription and alternative splicing, producing different proteins from the same gene, is more common in the expression of human genes. Second, regulation of expression is thought to be more complex in humans. Finally, the human proteins often combine more domains in novel ways than are found in other organisms. For example, a trypsinlike serine protease domain occurs with one other domain in the same protein in yeast, occurs with five other domains in worms, but occurs with 18 different domains in humans.

## METAGENOMICS

A fundamental operation in biotechnology is the expression of a foreign gene coding for a useful product in a suitable host organism, such as *E. coli* or yeast. In Chapter 3, we described the cloning of such genes from living organisms. However, because what is cloned is a piece of DNA, it may come directly from environmental sources. This approach of *direct cloning* is extremely valuable because most of the microorganisms in the environment have not yet been cultured, and the use of samples (for example soil samples) from a diverse range of environments is likely to supply genes that code for proteins with widely different properties. Thus, it has now become a standard approach in the industry to maintain such samples for the direct cloning of useful genes, as we will see in Chapter 11.

We can extend this approach, and instead of cloning just a gene or a gene cluster, we may hope to reconstruct the entire genomes of yet-to-be-cultured organisms from DNA samples in the environment. We will then be dealing not with the genome of a single cultured organism, but with genomes of the entire community of microorganisms, many of which may be currently uncultured. This is what is often called *metagenomics*.

One example is the study by Jillian Banfield and associates of the biofilm from acid mine drainage. In an abandoned mine, the large surface of pyrite ($FeS_2$) becomes exposed to air and water, causing large-scale leaching of metal through oxidation by Fe(II)-oxidizing bacteria. As detailed in Chapter 14, this activity produces much sulfuric acid, and at the site studied, the pH of the effluent was 0.83 and the temperature was 42°C. Under these extreme conditions, the composition of the microbial community is expected to be rather simple, and this aided in the analysis. A random insert library was used in sequencing, generating 76-Mb sequences. The sequences were divided mostly by GC content into "bins," and assembly led to near-complete construction of the genomes of *Leptospirillum* group II (2.23 Mb, a eubacterium, high GC) and *Ferroplasma* (1.82 Mb, an archeon, low GC), as well as partial coverage sequences of two other organisms. Prediction of metabolic pathways suggested that *Leptospirillum* group II can fix carbon, and *Ferroplasma* presumably uses the organic compounds coming out of the former organism. $N_2$-fixing genes were not present in these two organisms but were found in *Leptospirillum* group III, which was a minority component of the community. These examples show that it is possible to reconstruct genomes and predict the biochemical interactions between the component organisms without cultivation, at least when the community is relatively simple.

A much more complex assembly of microorganisms was analyzed by the group of J. Craig Venter with the surface water from the Sargasso Sea. In this study, already described in Chapter 1, microorganisms of 0.1 to 3.0 $\mu$m in size were collected by filtration, and the mixed DNA samples were randomly split into 2- to 6-kb fragments, cloned, and sequenced from both ends. Raw sequence reads had the total size of 1.36 Gb. Note the massive scale of this effort, as the random shotgun sequencing of the entire human genome, three orders of magnitude larger than the average prokaryotic genome, required "reads" only 10 times larger (14.9 Gb). In spite of the fact that between a few thousand and several tens of thousands of different organisms were present in the sample, the genomes of the few, most abundant organisms could be assembled by first putting the sequences into "bins" by using the depth of coverage (because fragments from more abundant organisms tend to become covered or sequenced multiple times), the frequency of consecutive nucleotides, and similarity to known sequences. This pilot study showed that because of the complexity of the community, we need more extensive sequencing in order to assemble the genomes of even the predominant organisms. Yet it is important in showing the extent of complexity, the nature of the most abundant organisms (surprisingly, relatives of *Burkholderia*, a plant and animal pathogen, and *Shewanella*, known as an inhabitant of polluted waters but not of open ocean, were some of the most abundant), and the reliability of this method over the usual approach of PCR amplification of 16S rRNA genes.

## TRANSCRIPTOMICS

The presence of a gene in a genome does not obviously mean that it is expressed to produce mRNA and subsequently a protein. Furthermore, regulation of metabolism often involves alteration of expression of various genes, and therefore it is important to know their expression level in order to understand the physiology of the cell. Before the genome sequences became available, gene expression had to be examined on a one-by-one basis. The techniques used included Northern blot, which involves annealing of fractionated mRNA with the cognate pieces of DNA, and reverse transcriptase–PCR, in which mRNA is used as a template to produce cDNA, which is then amplified by PCR. These methods are still used when we already know the mechanism of regulation in precise detail and when we want to know the expression level of the genes involved. However, when the regulation mechanism is not known entirely, examining the expression levels of only a few selected genes out of many thousands becomes pure guesswork and has led to many wrong conclusions.

When genome sequences became available, methods were developed in 1995 to "print" fragments of many genes onto a glass or plastic slide and to examine the binding of mRNA to these *microarrays*. The invention of this microarray or "chip" technology allowed us, for the first time, to examine the gene expression in an entire cell, for example, on an unbiased, "global" scale.

The affinity of the mRNA to a cognate piece of DNA fragment is obviously affected by the length, base composition, and other characteristics of the probe, and the binding itself cannot be used as the measured marker. However, we are usually interested in *changes* in transcription patterns, and comparison can be made by labeling the mRNA populations from two samples (query and reference) by different fluorescent dyes. In 1997, this methodology was successfully used to identify changing patterns of gene expression in yeast cells undergoing physiological adaptation and has produced a

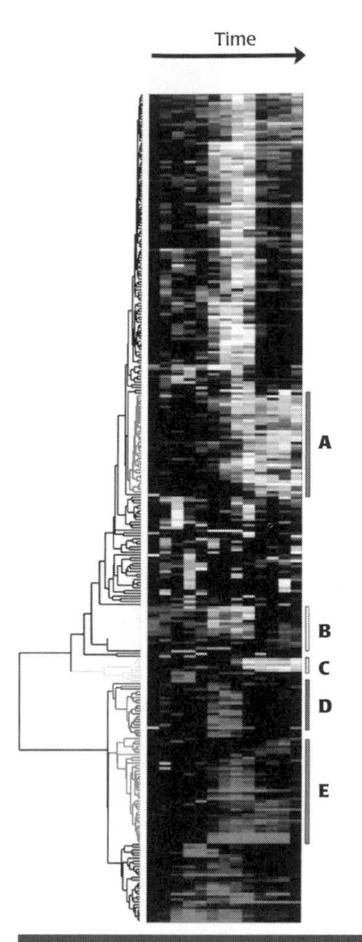


FIGURE 4.4

Expression levels of fibroblast genes upon stimulation with serum. Fibroblasts were kept deprived of serum for 48 hours and were then exposed to serum. Each column (from *left* to *right*) shows the transcriptome pattern after 0, 15 min, 30 min, 1 hour, 2 hours, 3 hours, 4 hours, 8 hours, 12 hours, 16 hours, 20 hours, and 24 hours. The rightmost column is a control with an unsynchronized sample. Each row shows the expression levels of more than 8000 human genes, clustered by the pattern of changes in expression. Group A, involved in cholesterol synthesis, becomes down-regulated (as shown in white, corresponding to the green in the original figure), whereas group E genes, involved in wound healing and tissue remodeling, become strongly up-regulated (as shown in gray, corresponding to the red in the original figure). [From Eisen M. B., Spellman P. T., Brown P. O., and Botstein D. (1998). Cluster analysis and display of genome-wide expression patterns. ***Proceedings of the National Academy of Sciences U.S.A.***, 95, 14863–14868; with permission.]

revolutionary change in the way we approach the molecular basis of the biology of the cell. Its impact may be seen by the fact that more than 10,000 papers using this technology were published in less than 10 years since its introduction. An example is shown in Figure 4.4.

Two methods are used to construct the microarrays. In one approach, usually 100- to 300-bp–long fragments of genes are amplified by PCR and deposited on glass slides. In another, slides containing much shorter (about 20- to 25-nucleotide) pieces of DNA are fabricated by Affymetrix by synthesizing each chain on the chip. In both cases, efforts are made to decrease the false-positive signals. Important features of the microarray method are that it generates data of an enormous size that biologists have not encountered earlier and that the data contain significant amounts of statistical variation as well as "noise." Thus, it is extremely important for the data to be treated in a manner that is statistically correct (Box 4.2).

What types of results were obtained by studies of these "transcriptomes" or "gene expression profiles"? The list will be too large because it covers almost every branch of biology. One of the areas investigated most intensely is the gene transcription pattern in human diseases such as cancer. In a study of human breast cancer in 2000, many clusters of genes were found to be overexpressed in different patients. In 2002, a group of scientists tried to define the "signature" genes, whose expression could serve as the prognostic marker in breast cancer. RNA was extracted from biopsy samples from primary breast cancer patients with no known metastasis to lymph nodes at the time of diagnosis. Because the purpose was to examine variations in individual tumors rather than to find variations in cancer in relation to normal tissues, a mixture of all samples was used as the reference. Use of a microarray containing 25,000 human genes showed that there were significant differences in expression in about 5000 genes in the 78 tumor tissues studied. For each of these 5000 genes, the correlation coefficient was calculated between the expression level of that gene and the prognosis (whether there were distant metastases within five years). Most of the genes showed only insignificant correlation and were discarded from further analysis. However, 231 genes showed significant correlation. These potential marker genes were used in increasing numbers to find out if a group of them could successfully predict the outcome of the disease, and it was found that the use of 70 genes with the highest correlation coefficient was enough for this purpose. This method could classify about 90% of the poor-prognosis group accurately so that they could have been directed to "adjuvant" therapy, such as chemotherapy or radiation therapy; whereas the patients in the good-prognosis group could have escaped such therapy, which puts a heavy burden on the patient's body. The prediction procedure was validated by using a series of 295 patients and, more recently, in an international multicenter study in 2006.

Another focus of transcription pattern studies has been the effort, mostly by pharmaceutical companies, to find a new target for a drug. Although transcriptomes can show only that the gene is transcribed, by using a bioinformatics approach, the functions of many genes may be predicted. If certain organs or pathological tissues, such as cancer cells, overexpress any of the genes, they may be attractive candidates for selective therapy. In fact, gene expression array analysis of multiple sclerosis lesions showed that hitherto unsuspected genes were up-regulated, potential targets of intervention. Pharmaceutical companies were thus the first to adopt transcriptomics, and the expectations were that a great many new targets would be discovered in a record short time, followed by hundreds of new drugs. In spite of a very large amount of work invested in this effort, however, it appears that a flood of new drugs has not yet been developed in this manner, with the approval of drugs for entirely new targets staying at a steady rate of a few per year. Possibly this is because most of the newly discovered targets are not really "drugable" targets – such as G protein–coupled receptors, serine/threonine kinases, tyrosine kinases, transcription factors, nuclear receptors, serine proteases, or ion channels, for which pharmaceutical companies already know how to develop low molecular weight drugs. Some scientists blame the poor correlation between the steady-state expression levels of mRNA, measured by transcriptomics, and the level of proteins that are produced. In any case, gene profiling is a very useful tool in predicting the types of drug action and its efficacy, as the drugs of one type, say, opioids, generate their own characteristic pattern of up- and down-regulation of various genes. (As a related example, the effect of serum on the expression of fibroblast genes is shown in Figure 4.4.) Furthermore, a similar technique is used in predicting the toxicity of the drugs in development. In the future, it may even be possible to predict the specific response of an individual to various drugs and thus to tailor the use of drugs to each individual, the ultimate goal of pharmacogenomics.

**Treatment of Microarray Data**

The raw data obtained by a microarray analysis must be treated carefully in order to yield meaningful and reliable pieces of information. First, the data must be *normalized*. At the simplest level, this is to correct for the different amounts of RNA from the two sources (query and reference), different efficiencies in the labeling and fluorescent detection of the RNA molecules, and so on. However, in addition to this simple normalization, the microarrays require more complex, sophisticated normalization, as the ratio of the two signals (query over reference) usually shows a systematic deviation dependent on the intensity of the signal. Lowess (locally weighted linear regression) normalization is the method most often employed for this purpose. These procedures for the treatment of data are described lucidly by Quackenbush.[1]

After normalization, we are still left with data on the expression level of many thousands of genes under many different conditions, for example, at different stages of cell cycle, in hundreds of patients, and so on. It will be nearly impossible to see any pattern in this ocean of unorganized data. Thus, "clustering" of genes that show similar patterns is indispensable in the further analysis of array data. Various methods have been described for this purpose[2,3]

1. Quackenbush, J. (2002). Microarray data normalization and transformation. ***Nature Genetics***, 32, 496–501.

2. Eisen, M. B., Spellman, P. T., Brown, P. O., and Botstein, D. (1998). Cluster analysis and display of genome-wide expression patterns. ***Proceedings of the National Academy of Sciences U. S.A***., 95, 14863–14868.

3. Tamayo, P., Slonim, D., Mesirov, J., et al. (1999). Interpreting patterns of gene expression with self-organizing maps: methods and application to hematopoietic differentiation. ***Proceedings of the National Academy of Sciences U. S.A***., 96, 2907–2912.

BOX 4.2

In the area of basic research, expression profiling is now essential in the effort to understand the molecular mechanisms of regulatory processes, as mentioned already. In addition, because genes will be simultaneously induced or repressed under a given condition if their products work together, the global gene expression profile may give hints to the "unknown" function of some proteins. In an extensive study of 276 deletion mutants and 13 mutants of essential genes (under the regulatable promoter) of *S. cerevisiae*

**Tiling arrays**

In the usual arrays, oligonucleotide probes for any given gene come from widely separated parts of the gene. In contrast, in the tiling arrays, they are "tiled" so that consecutive probes overlap each other. If we assume that each probe contains 25 nucleotides, the first spot in the five-nucleotide interval array will contain the sequence from nucleotides 1 through 25, the second spot nucleotides 6 through 30, and so forth.

BOX 4.3

in 2000, many expression patterns were defined, and in fact the functions of eight hitherto uncharacterized open reading frames were predicted on this basis. Recent studies also showed that in higher animals, alternative splicing of pre-mRNA, use of alternative promoters, and use of alternative polyadenylation sites generate many mRNA species from one single gene; such a mechanism for the generation of multiple mRNA species may be used quite extensively and often appears to occur in an organ- or tissue-specific manner.

One important development in biology, aided in many cases by transcriptomic studies, is our realization that the untranslated portion of the genome plays major roles in cell physiology. Even in bacteria, small untranslated RNAs (more than 50 are now known to exist in *E. coli*) play important roles in regulating the translation of many messages. In higher animals and plants, small interfering RNA (siRNA) is known to be generated from double-stranded RNA by a complex called Dicer and to inhibit gene expression through degradation of mRNA (see Box 6.3), most likely as a defense mechanism against double-stranded RNA viruses. In higher organisms, it is now known that hundreds of microRNA species, which are similar in size to siRNA (18 to 24 nucleotides long), play major roles by inhibiting mRNA translation and by inducing the degradation of mRNA. microRNA species are initially transcribed as a larger primary transcript, which folds upon itself to produce a double-stranded RNA structure that in turn is processed by nucleases including Dicer to produce the final short single-stranded form. Expression of some species of microRNA was known to become altered in cancer. microRNA levels are not detected in the conventional DNA arrays. However, recent examination of the levels of more than 200 microRNA species, using specially designed arrays and beads, showed very widespread changes in their levels in human cancer, suggesting that they play important roles in the development of cancer. The importance of the noncoding DNA region is not limited to the production of microRNA. A recent study of the transcription of 10 selected human chromosomes, using tiling arrays (Box 4.3) spaced at five-nucleotide intervals, showed that about 9% of the probes hybridized with some transcripts. The surprising finding was that even when polyadenylated cytosolic RNA fractions were used, more than half of them did not come from known genes (exons) but came from elsewhere. Some came from introns, but more came from intervening regions of chromosome, hitherto considered as "junk DNA." The fraction of such transcripts coming from an intervening region is even higher among cytosolic, nonpolyadenylated RNA, reaching 48% of the total. It is not known what function these novel transcripts are performing, but this analysis shows both the power of array analysis and the extent of our ignorance on how the genome of higher organisms really functions.

## PROTEOMICS

As described in the previous section, many of the efforts to examine the expression of the many genes in the genome have been done by quantitating their transcription, that is, by using the transcriptomics approach. However, mRNA must be translated, and translational control adds an important step to the expression of genetic information. Furthermore, especially in higher eukaryotes, many proteins are processed mainly by proteolytic cleavage, and various posttranslational modification steps also intervene. These steps are thought to increase the complexity of the "proteome" (the ensemble of proteins that are expressed in a given organism but sometimes also in a given tissue, organelle, etc.) The surprising finding that the human genome contains only 25,000 to 30,000 protein-coding genes prompted scientists to focus on the discrepancy between this number and up to a million or more proteins (and peptides) that are estimated by some scientists to be present in the human body. Some of this discrepancy may be the result of the coding sequences that were overlooked in the analysis of the genome. Many undoubtedly come from alternative transcription initiation, alternative splicing, and the fact that a single nascent polypeptide gives rise to many proteins through processing and modifications.

These considerations led to the realization that we need proteomic analysis in order to understand how an organism functions. However, proteins are not self-replicating and do not anneal to nucleic acids. Thus, it is not easy to construct an array or a chip that can be used to quantitate the expression of thousands of proteins. We first discuss the technique used for examining the expression of many proteins simultaneously and then discuss the application of proteomic approach in the analysis of protein-to-protein interactions.

## PROTEOMICS AND MASS SPECTROMETRY

Proteomic analysis requires first the separation of thousands of proteins or their fragments and then their identification. It is impossible to identify this many proteins and their fragments by using the traditional approaches, such as reactivity with specific antibodies or functional properties. Thus, the advances in mass spectrometry, now almost the universal approach in identification, were the major factor that made proteomics possible. Previously, mass spectrometry, which measures the mass-to-charge ratio of ions in a vacuum (Box 4.4), required the "hard" ionization methods that fragmented the sample molecule. Mass spectrometry became extremely useful in the study of proteins and peptides when the two "soft ionization" methods – matrix-assisted laser desorption/ionization (MALDI) and electrospray ionization (ESI) (Box 4.4) – were developed. These methods made possible the analysis of (unfragmented) peptides and proteins at high sensitivity and great precision. The significance of these advances may also be seen from the fact that two of the Nobel prizes in chemistry in 2002 were awarded to scientists who contributed to the development of these methods.

For the separation of proteins and peptides, two major approaches are used. In one, proteins are separated by two-dimensional polyacrylamide gel electrophoresis, usually by isoelectric points in the first dimension, and by molecular weight in the second dimension using sodium dodecylsulfate

**Introduction to Mass Spectrometry**

Mass spectrometry measures the mass-to-charge ratio (*m/z*) of ions. Thus, organic compounds must be first converted to volatile ions. Two soft ionization methods, which were developed only in recent years and which produce stable ions usually in protonated forms, have been instrumental in the application of mass spectroscopy to protein identification (and sometimes quantitation). In the MALDI technique, a peptide (typically a fragment of a protein generated by tryptic digestion) is mixed with an excess of light-absorbing organic compound (matrix), such as 3,5-dihydroxybenzoic acid, and a co-crystal is produced. This is then irradiated with a short pulse of a laser beam, and the sudden generation of heat then releases the peptide as an ion into the gas phase. In the ESI technique, a peptide dissolved in solvent (such as 50% acetonitrile) is pushed out of the end of a needle at high voltage. The positively charged droplets become smaller as a result of the evaporation of the solvent, and the electrostatic repulsion between positive charges increases, finally resulting in the creation of a positively charged molecular ion, which then enters the mass spectrometric analyzer. ESI can be coupled easily to liquid chromatography separation, and this has been a big advantage in "gel-free" analysis of proteomes.

As analyzers, the pulsed nature of MALDI makes the TOF analyzer ideal, as this measures the time it requires for an ion to travel the fixed distance to the detector under acceleration by the electrical potential gradient. For ESI, quadrupole analyzers, which determine the *m/z* by varying the voltage across the ion stream and thus affect the path of ions, have been in widespread use. ESI is often used for tandem mass spectrometry in order to obtain structural information on the peptides, not just the *m/z* information. For this purpose, the quadrupole ion trap analyzer, which can accumulate the desired ions in the first step, is very useful as the accumulated ions can then be fragmented by collision-induced dissociation (CID; through collision with neutral gas molecules, for example), and then the fragmentation pattern can be examined in the second analyzer.

**BOX 4.4**

(SDS)-polyacrylamide gel electrophoresis (Figure 4.5). The protein spots are stained, and in this way they can be roughly quantitated. Each of the protein spots can then be digested with trypsin, and the fragments are analyzed by mass spectrometry. Commonly, MALDI is the ionization method used, and time-of-flight (TOF) detection (Box 4.4) is used. The molecular weights of various fragments are compared with their predicted sizes, obtained from the genomic database, and this results in the identification of the protein. The TOF analyzer is sufficient for this setup, because the identification relies on the comparison of sizes of multiple fragments from a single protein. The weakness of this method is that only the highly expressed proteins can be analyzed in this way because the resolution of the two-dimensional gel is limited and poorly expressed proteins may be obscured by the spots of strongly expressed ones.

In the second, "gel-free" approach, the mixture of proteins is digested first with trypsin, and the resultant, enormously complex mixture of peptides is separated by liquid column chromatography. Because a single column is unlikely to resolve this mixture, multidimensional approaches are used. For example, a column may contain an ion exchange resin in its upper part and a reverse phase matrix in its lower part, and it can be eluted by pulses of salt solutions followed each time by gradients of a water-organic solvent mixture, such as acetonitrile. The advantage of this method is that the effluent from the column can be directly introduced into a mass spectrometer using ESI (Box 4.4), avoiding the cutting out of a gel spot, subsequent digestion, and preparation of the solid sample for MALDI. It is thus a method suitable for very large-scale analysis. The preferred detector is a tandem mass spectrometer (MS/MS), in which the isolated ion can then be analyzed further after collision-induced fragmentation (Box 4.4). Such a setup is necessary because the size of a peptide alone is not sufficient for identification as the initial mixture may contain many different peptides of approximately the same size; thus, the fragmentation pattern of each peptide, determined in the second step, becomes important. Another advantage of this approach is the sensitivity that is achieved because the entire sample

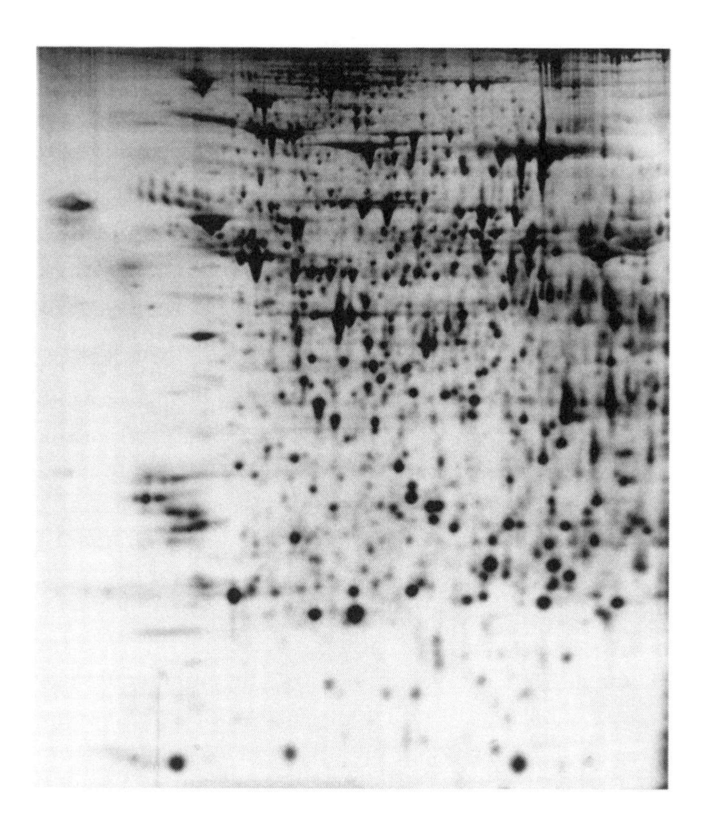

**FIGURE 4.5**

Two-dimensional separation of liver proteins. First, separation in the horizontal direction, based on isoelectric points, was achieved by isoelectric focusing, then separation based on molecular weights was carried out in the vertical direction by SDS-polyacrylamide electrophoresis. [From Cutler P. (2003). Protein arrays: the current state-of-the-art. ***Proteomics***, 3, 3–18; with permission.]

can be introduced into the mass spectrometer. A sample containing about 50 proteins was successfully analyzed by using a total of 0.2 $\mu$g protein by this method.

Following are a few examples. Samuel Miller and associates wanted to examine the expression levels of proteins in *P. aeruginosa*, a human pathogen that plays a major role in cystic fibrosis airways. They examined the effect of low $Mg^{2+}$ concentration in the medium and also strains isolated from cystic fibrosis patients. They took the second approach mentioned above, liquid chromatography of trypsin digests followed by ESI ionization and MS/MS analysis. Mass spectrometry is inherently poor in terms of quantitation. Thus, they labeled the two sets of proteins by using either a light or heavy version of the isotope-coded affinity tag (ICAT) that reacts covalently with the sulfhydryl group of cysteine residues. Because ICAT contains a biotin moiety, only the labeled peptides can be pulled out by using an avidin column (see Box 3.10), thus simplifying the analysis. They found that, among 1337 proteins thus examined, the expression of 145 proteins was affected by low $Mg^{2+}$, and the strains from cystic fibrosis patients had similar changes.

The second example is the analysis of protein expression in the malaria parasite *Plasmodium falciparum*. This parasite undergoes extensive morphological and biochemical changes in its development from the sporozoite stage (present in the salivary gland of mosquitoes), through the merozoite stage (invasive to red blood cells) to trophozoite (the form multiplying in erythrocytes), and finally gametocytes (the sexual stage; see Figure 5.16).

Samples of the parasites that could be collected were very small for some stages and were heavily contaminated by either human or mosquito proteins. This precluded gene expression analysis with the DNA arrays as it requires a few micrograms of mRNA; proteins are usually more abundant by a couple of orders of magnitude. Two-dimensional capillary chromatography coupled with MS/MS was used, and the study resulted in showing that many proteins are expressed in a stage-specific manner, a finding that will be most useful in the development of vaccines and therapeutic agents.

## PROTEIN INTERACTIONS

Many proteins function by interacting with other proteins. For example, the RNA polymerase II preinitiation complex in yeast is thought to contain at least 68 different proteins. Myriads of signaling pathways in eukaryotic (and also prokaryotic) cells involve cascades of protein-to-protein interactions. If a protein of unknown function can be shown to interact with proteins of known function, this will be a major step toward the understanding of the function of the former. Besides, some scientists argue that the complexity in higher animals is not the result of a large number of genes, but is caused by the increased complexity in protein-to-protein interactions.

For all these reasons, the study of protein-to-protein interaction is extremely important. A classical strategy for such a study is the use of a yeast two-hybrid system. Here the two proteins to be studied are produced as fusions to two domains of an activator of gene transcription in yeast. If the two proteins associate with each other, the two domains of the activator come together and the transcription of the indicator gene ensues. Although examining the 36 million combinations of the 6000 genes in yeast in this manner may sound like an impossible task, scientists overcame the difficulty by creating pools of dozens of genes. However, there are potential problems with this approach. First, because we are dealing with a transcriptional activator, the protein-to-protein interaction must occur in the nucleus, and we cannot use the method for cytoplasmic membrane proteins, for example. Second, it is difficult to deal with higher-level interactions involving more than two proteins. Finally, because we are dealing with fusion proteins, there is always a possibility that we may have interfered with the proper folding of the protein by making fusions.

Mass spectrometric protein identification provides an ideal approach to the study of protein interaction. An early example is the identification of about two dozen proteins that are found in spliceosomes, which splice out introns from the primary transcript in eukaryotes. On a larger scale, tags were added to the C-terminus of more than 1500 proteins in yeast, out of which close to 1200 were expressed at a reasonable level. Pulling out tagged proteins showed that at least 230 protein complexes exist in yeast. When the component proteins in these complexes were identified by mass spectrometry, most of them were found to contain one or more proteins of unknown function. As described already, this is a major step toward understanding the function of the latter proteins.

One potential problem with the purification of protein complexes is that some of the proteins might be contaminants, not the intrinsic components of the complexes. However, an approach that uses the differential labeling with light and heavy ICAT was proposed, and this should work when the formation of complexes can be controlled experimentally.

## PROTEIN ARRAYS

Can we make protein arrays, as we can make gene arrays? Indeed this is possible. In one study, 5800 open reading frames in the yeast genome were cloned and expressed as a fusion with GST and a hexahistidine tag (see Box 3.10). The proteins were purified on glutathione-agarose and attached to glass slides coated with nickel, using the affinity of the hexahistidine tag. The utility of such a "chip" was demonstrated when it was treated with a solution of biotinylated calmodulin. When the binding of the biotin tag was detected with a fluorescence-labeled streptavidin, 39 proteins were shown to bind calmodulin, out of which only six were the already known calmodulin-binding proteins (Figure 4.6). Similar approaches should be possible with many reagents, including drug candidates, and protein arrays may one day

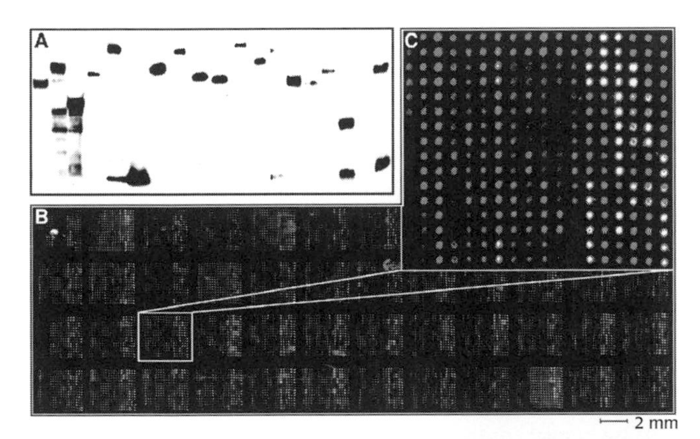


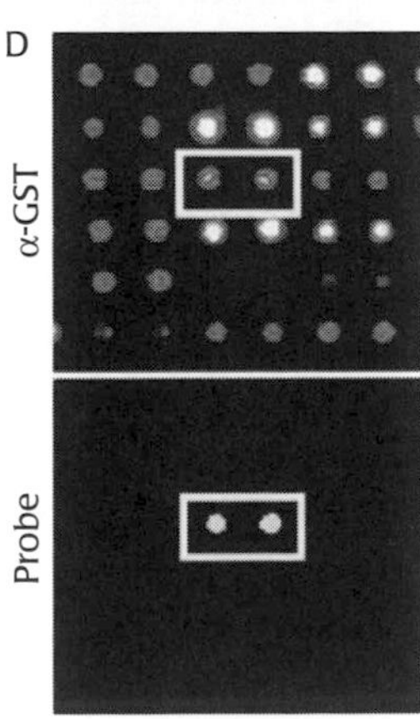


**FIGURE 4.6**

Use of a protein array for identification of proteins that interact with calmodulin. Five thousand eight hundred yeast genes were expressed as fusions with hexahistidine-tagged GST and were fixed onto Ni-coated slides. (**A**) Most proteins are indeed expressed as stable fusions, as the proteins were stained well with anti-GST antibody. (**B**) The protein array, when reacted with anti-GST antibody, indicates that all spots have enough proteins. (**C**) Enlargement of one small block. (**D**) The array was stained with fluorescent streptavidin, which reacts specifically with the biotinylated calmodulin that became bound to specific proteins in the array (probe). [From Zhu H., Bilgin M., Bangham R., et al. (2001). Global analysis of protein activities using proteome chips. ***Science***, 293, 2101–2105; with permission.]

become the standard approach for searching for proteins with specific functions on a global scale.

## METABOLOMICS AND SYSTEMS BIOLOGY

Once we learn the expression levels of proteins at the global level, it becomes important to know the levels of many metabolites, which represent the integrated results of the activities of these proteins, including enzymes and regulators. Such an analysis at the global scale – *metabolomics* – became possible in recent years, thanks to technical developments in mass spectroscopy and nuclear magnetic resonance (NMR). Because a very large amount of data is generated, its mathematical analysis becomes very important. Metabolomics on its own holds much promise in the area of clinical diagnosis; for example, an inexpensive, noninvasive NMR analysis of ingredients of patient sera was reported to diagnose coronary artery disease, although a later study found problems with the statistical treatment of data. In this case also, the visual inspection of NMR spectra showed no difference between patients and healthy people, and mathematical analysis was obligatory. However, from the perspective of biotechnology, metabolomics becomes significant only when it is combined with transcriptomics and proteomics. In this way, we are trying to integrate all the "omics" information together, in the direction of *systems biology*, where we are trying to consider all components of cells, tissues, and so on, from the level of genes up to the level of metabolites, as an integrated whole.

Metabolomics, for example, supplies an exceptionally sensitive indicator of phenotypes, showing that metabolism becomes altered in mutant strains of yeast, in which conventional tests showed no alteration of phenotype. Metabolomics was also embraced eagerly by scientists working on plants, as plants produce a very large number of compounds.

Here we give two examples in which a metabolomic approach was used in microorganisms in areas relevant to biotechnology. In *Corynebacterium glutamicum*, which produces large amounts of the amino acid lysine (Chapter 9), lysine begins to be secreted abruptly when the growth rate begins to decrease at the end of the exponential phase of culture. Metabolomic analysis showed that there is a rapid change in the flux of metabolites at this time, yet hardly any change is noted in the gene transcription pattern, except in the down-regulation of glucose-6-phosphate dehydrogenase. Although the data did not show us how the switching of metabolism occurs, they serve as a starting point for future studies. In *Aspergillus terreus*, the producer of the cholesterol-lowering drug lovastatin (Chapter 10), about a dozen genetically defined strains were constructed and lovastatin production and the gene expression profile were determined. Genes, whose expression levels were correlated with lovastatin production, were then modified to increase their expression levels to produce a strain improved in terms of drug production. Although the number of measured metabolites was very small (and thus this study hardly qualifies as a metabolomic study), this approach suggests that the combination of metabolomics with transcriptomics may be fruitful in the future.

## SUMMARY

There has been an explosive growth in our knowledge of DNA sequences, most prominently those of the complete genomes of hundreds of microorganisms and many plants and animals, including humans. This knowledge has changed our way of doing science in a profound way. We now think "globally" of transcription and translation of thousands of genes as well as the functioning of these gene products, rather than those of a few arbitrarily selected sets. This chapter gives some glimpses of this new science, the world of "omics."

To start, we have to obtain the sequence of genomes. The large genomes of higher plants and animals, containing many repeated sequences, were initially thought to be difficult to sequence by a random shotgun approach. However, this approach was successfully used in sequencing the human genome. Especially for sequencing of the environmental DNA samples (metagenomics), there is no alternative. Currently, the favored method combines the random shotgun sequencing of whole genomes with the stepwise, clone-by-clone approach. Comparative genomics has already produced many interesting insights into the evolution of genomes.

The ability to examine the transcription levels of thousands of genes at once using DNA chips created a revolutionary change in cell biology. Now we can ask which genes are expressed or repressed not only in pathological tissues, but also in cells treated with candidate drugs. The "transcriptome" study also led us to the realization that untranslated RNA (including but not limited to microRNA) plays a very important role in the regulatory processes.

What ultimately functions in cells are usually proteins, not mRNA. Because translational control plays a large role in some cells, we need identification and quantitation of thousands of cellular proteins (proteomics), which were made possible by the rapid progress in mass spectrometry instrumentation. Proteins can be separated in two-dimensional polyacrylamide gels, followed by the analysis of each protein spot. Alternatively, a mixture of thousands of proteins can be cleaved by proteases such as trypsin, and the enormously complex mixture of fragments can be separated by "two-dimensional" liquid chromatography prior to analysis by MS/MS. The unprecedented sensitivity of this method was successfully utilized in the analysis of proteomes of the malaria parasite at various stages of its life cycle. Protein arrays are useful, for example, in the analysis of protein-to-protein interactions on a global scale.

Finally, the quantitative analysis of metabolic intermediates on a global scale (metabolomics) may become integrated with other types of "omic" analysis to give us the complete picture of cells, tissues, or organs – as a whole, the goal of "systems biology."

## SELECTED REFERENCES


### General

Sensen, C. W. (ed.) (2005). *Handbook of Genome Research*, Volumes I and II, Weinheim, Germany: Wiley-VCH.

### Genome Sequencing

Green, E. D. (2001). Strategies for the systematic sequencing of complex genomes. *Nature Reviews Genetics*, 2, 573–583.

Fleischmann, R. D., Adams, M. D., White, O., et al. (1995). Whole-genome random sequencing and assembly of *Haemophilus influenzae* Rd. *Science*, 269, 496–512.

Blattner, F. R., Plunket III, G., Bloch, C. A., et al. (1997). The complete genome sequence of *Escherichia coli* K-12. *Science*, 277, 1453–1462.

International Human Genome Sequencing Consortium (2001). Initial sequencing and analysis of the human genome. *Nature*, 409, 860–933.

International Human Genome Sequencing Consortium (2004). Finishing the euchromatic sequence of the human genome. *Nature*, 431, 931–945.

Venter, J. C., Adams, M. D., Myers, E. W., et al. (2001). The sequence of the human genome. *Science*, 291, 1304–1351.

### Prokaryotic Genomes

Mushegian, A. R., and Koonin, E. V. (1996). A minimal gene set for cellular life derived by comparison of complete bacterial genomes. *Proceedings of the National Academy of Sciences U.S.A.*, 93, 10268–10273.

Glass, J. I., Assad-Garcia, N., Alperovich, N., et al. (2006). Essential genes of a minimal bacterium. *Proceedings of the National Academy of Sciences U.S.A.*, 103, 425–430.

Bentley, S. D., Chater, K. F., Cerdeño-Tárraga, A.-M., et al. (2002). Complete genome sequence of the model actinomycete *Streptomyces coelicolor* A3(2). *Nature*, 417, 141–147.

Dobrindt, U., Hochhut, B., Hentschel, U., and Hacker, J. (2004). Genomic islands in pathogenic and environmental microorganisms. *Nature Reviews Microbiology*, 2, 414–424.

Cole, S. T., Eiglmeier, K., Parkhill, J., et al. (2001). Massive gene decay in the leprosy bacillus. *Nature*, 409, 1007–1011.

### Metagenomics

Riesenfeld, C. S., Schloss, P. D., and Handelsman, J. (2004). Metagenomics: genomic analysis of microbial communities. *Annual Reviews of Genetics*, 38, 525–552.

Handelsman, J. (2004). Metagenomics: application of genomics to uncultured microorganisms. *Microbiology and Molecular Biology Reviews*, 68, 669–685.

Allen, E. E., and Banfield, J. F. (2005). Community genomics in microbial ecology and evolution. *Nature Reviews Microbiology* 3, 489–498.

### Transcriptomics

Stoughton, R. B. (2004). Applications of DNA microarray in biology. *Annual Review of Biochemistry*, 74, 53–82.

DeRisi, J. L., Iyer, V. R., and Brown, P. O. (1997). Exploring the metabolic and genetic control of gene expression on a genomic scale. *Science*, 278, 680–686.

van't Veer, L., Dai, H., van de Vijver, M. J., et al. (2002). Gene expression profiling predicts clinical outcome of breast cancer. *Nature*, 415, 530–536.

van de Vijver, M. J., He, Y. D., van't Veer, L., et al. (2002). A gene-expression signature as a predictor of survival in breast cancer. *The New England Journal of Medicine*, 347, 1999–2009.

Foekens, J. A., Atkins, D., Zhang, Y., et al. (2006). Multicenter validation of a gene expression-based prognostic signature in lymph node-negative primary breast cancer. *Journal of Clinical Oncology*, 24, 1665–1671.

Dechering, K. J. (2005). The transcriptome's drugable frequenters. *Drug Discovery Today*, 10, 857–864.

Lock, C., Hermans, G., Pedotti, R., et al. (2002). Gene microarray analysis of multiple sclerosis lesions yields new targets validated in autoimmune encephalomyelitis. *Nature Medicine*, 8, 500–508.

Gunther, E. C., Stone, D. J., Gerwien, R. W., Bento, P., and Heyes, M. P. (2000). Prediction of clinical drug efficacy by classification of drug-induced genomic expression profiles *in vitro*. *Proceedings of the National Academy of Sciences U.S.A.*, 100, 9608–9613.

Steiner, G., Suter, L., Boess, F., et al. (2004). Discriminating different classes of toxicants by transcript profiling. *Environmental Health Perspectives*, 112, 1236–1248.

Weinshilboum, R., and Wang, L. (2004). Pharmacogenomics: bench to bedside. *Nature Reviews Drug Discovery*, 3, 739–748.

Hughes, T. R., Marton, M. J., Jones, A. R., et al. (2000). Functional discovery via a compendium of expression profiles. *Cell*, 102, 109–126.

Soares, L. M. M., and Valcárcel, J. (2006). The expanding transcriptome: the genome as the 'Book of Sand.' *The EMBO Journal*, 25, 923–931.

Lu, J., Getz, G., Miska, E. A., et al. (2005). MicroRNA expression profiles classify human cancers. *Nature*, 435, 834–838.

Volinia, S., Calin, G. A., Liu, C.-G., et al. (2006). A microRNA expression signature of human solid tumors defines cancer gene targets. *Proceedings of the National Academy of Sciences U.S.A.*, 103, 2257–2261.

Cheng, J., Kapranov, P., Drenkow, J., et al. (2005). Transcriptional maps of 10 human chromosomes at 5-nucleotide resolution. *Science*, 308, 1149–1154.

Willingham, A. T., and Gingeras, T. R. (2006). TUF love for "junk" DNA. *Cell*, 125, 1215–1220.

### Proteomics and Mass Spectrometry

Speicher, D. W. (ed.) (2004). *Proteome analysis: Interpreting the genome*. Amsterdam: Elsevier.

Domon, B., and Aebersold, R. (2006). Mass spectrometry and protein analysis. *Science*, 312, 212–217.

Baldwin, M. A. (2005). Mass spectrometers for the analysis of biomolecules. *Methods in Enzymology*, 402, 3–48.

Yates, J. R. (2004). Mass spectral analysis in proteomics. *Annual Review of Biophysics and Biomolecular Structure*, 33, 297–316.

Ong, S.-E., and Mann, M. (2005). Mass spectrometry-based proteomics turns quantitative. *Nature Chemical Biology*, 1, 252–262.

Link, A. J., Eng, J., Schieltz, D. M., Carmack, E., Mize, G. J., Morris, D. R., Garvik, B. M., and Yates, J. R., III. (1999). Direct analysis of protein complexes using mass spectrometry. *Nature Biotechnology*, 17, 676–682.

Guina, T., Purvine, S. O., Yi, E. C., Eng, J., Goodlett, D. R., Aebersold, R., and Miller, S. I. (2003). Quantitative proteomic analysis indicates increased synthesis of a quinolone by *Pseudomonas aeruginosa* isolates from cystic fibrosis airways. *Proceedings of the National Academy of Sciences U.S.A.*, 100, 2771–2776.

Florens, L., Washburn, M. P., Raine, J. D., et al. (2002). A proteomic view of the *Plasmodium falciparum* life cycle. *Nature*, 419, 520–526.

### Protein Interactions

Figeys, D. (2003). Novel approaches to map protein interactions. *Current Opinion in Biotechnology*, 14, 119–125.

Borch, J., Jørgensen, T. J. D., and Roepstorff, P. (2005). Mass spectrometric analysis of protein interactions. *Current Opinion in Chemical Biology*, 9, 509–516.

Gavin, A.-C., Bösche, M., Krause, R., et al. (2002). Functional organization of the yeast proteome by systematic analysis of protein complexes. *Nature*, 415, 141–147.

Butland, G., Peregrin-Alvarez, J.M., Li, G. et al. (2005) Interaction network containing conserved and essential protein complexes in *Escherichia coli*. *Nature*, 433, 531–537.

**Protein Arrays**

Cutler, P. (2003). Protein arrays: the current state-of-the-art. *Proteomics*, 3, 3–18.

Zhu, H., Bilgin, M., Bangham, R. et al. (2001). Global analysis of protein activities using proteome chips. *Science*, 293, 2101–2105.

**Metabolomics**

Griffin, J. L. (2006). The Cinderella story of metabolic profiling: does metabolomics get to go to the functional genomics ball? *Philosophical Transactions of the Royal Society, Series B*, 361, 147–161.

van der Werf, M. J., Jellema, R. H., and Hankemeier, T. (2005). Microbial metabolomics: replacing trial-and-error by the unbiased selection and ranking of targets. *Journal of Industrial Microbiology and Biotechnology*, 32, 234–252.

Weckwerth, W. (2003). Metabolomics in systems biology. *Annual Review of Plant Biology*, 54, 669–689.

Vemuri, G. N., and Aristidou, A. A. (2005). Metabolic engineering in the -omics era: elucidating and modulating regulatory networks. *Microbiology and Molecular Biology Reviews*, 69, 197–216.

Krömer, J. O., Sorgenfrei, O., Klopprogge, K., et al. (2004). In-depth profiling of lysine-producing *Corynebacterium glutamicum* by combined analysis of the transcriptome, metabolome, and fluxome. *Journal of Bacteriology*, 186, 1769–1784.

Askenazi, M., Driggers, E. M., Holtzman, D. A., et al. (2003). Integrating transcriptional and metabolite profiles to direct the engineering of lovastatin-producing fungal strains. *Nature Biotechnology*, 21, 150–156.

# FIVE

# Recombinant And Synthetic Vaccines

In developing countries, infectious diseases still cause 30% to 50% of all deaths. Effective chemotherapeutic agents simply do not exist for many of the diseases that plague these regions, and many of the agents that do exist are far too costly for much of the population to afford. Vaccines thus have become the most important tool for fighting infectious diseases in those parts of the world.

The situation is very different in developed countries, where infectious diseases account for only 4% to 8% of all deaths. This is not to say, however, that vaccines are not important in those parts of the world. The low rate of infectious diseases in industrialized nations is in fact largely the result of the widespread use of vaccination (Figure 5.1). In addition to the well-known example of the smallpox vaccine, which has succeeded in eradicating the disease completely, other vaccines have brought dramatic decreases in the incidence of numerous grave diseases. For example, at the beginning of the twentieth century, diphtheria (caused by the bacterium *Corynebacterium diphtheriae*) infected about 3000 children yearly out of every million in developed countries. Because diphtheria targets young children in particular, this incidence corresponds to several percent of children of the susceptible age, and nearly one tenth of the infected children died. Now, thanks to a mass immunization program, diphtheria incidence in the United States is less than 0.2 per million, a decrease of more than a thousandfold. The effect of immunization was illustrated dramatically by the epidemics of diphtheria that occurred in the Baltic countries after the collapse of the Soviet Union, when enforcement of public health policies lapsed and many young children went unvaccinated or received poor-quality vaccines. Another example is furnished by poliomyelitis (caused by an RNA-containing virus). As recently as 1955, the U.S. and Canadian incidence of polio was 200 per million of the population. However, the development of vaccines has decreased polio cases by more than 4000-fold, to less than 0.05 per million in recent years. Similar rapid decreases in incidence occurred for measles and rubella (German measles) after introduction of those vaccines in the 1950s and 1960s. Currently, the U.S. government recommends that all children be treated with 11 vaccines (Table 5.1).

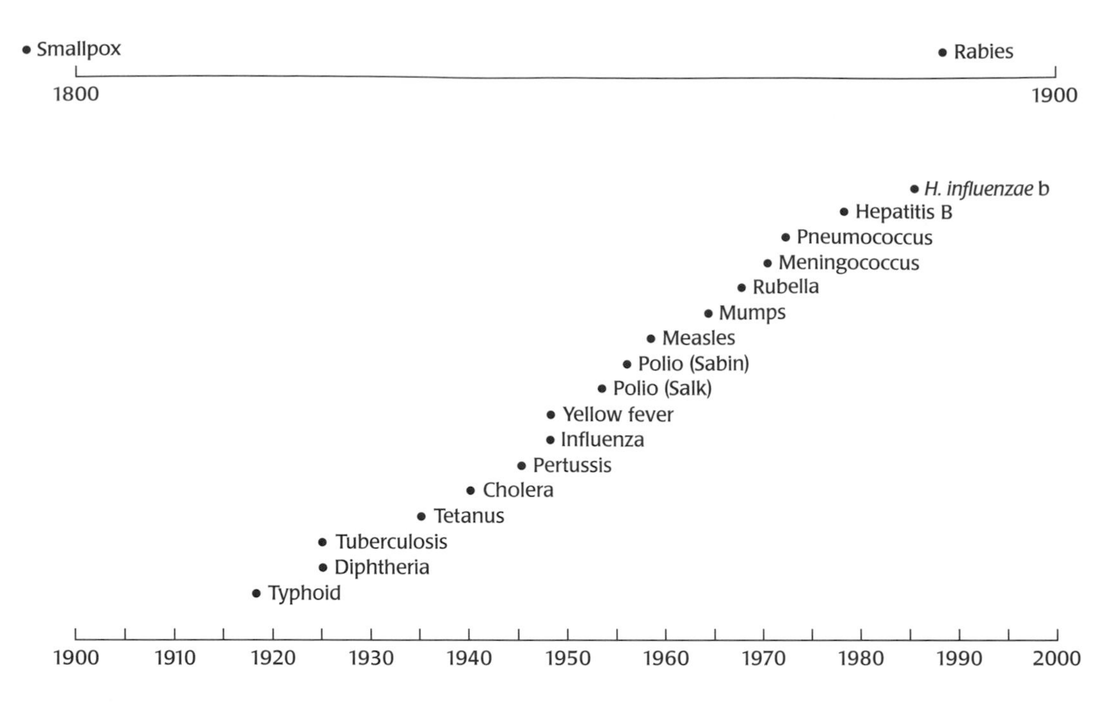


**FIGURE 5.1**

Vaccines introduced since Jenner's discovery of the smallpox vaccine. Among the newest are pneumococcal conjugate vaccine, varicella vaccine, and the hepatitis B surface antigen vaccine. [From Warren, K. S. (1983). New scientific opportunities and old obstacles in vaccine development. *Proceedings of the National Academy of Sciences U.S.A.*, 83, 9275–9277; with permission.]

Apart from their effectiveness, a second reason that vaccines remain important in developed countries is cost-efficiency. Vaccination is much less costly than treating people who are already sick. Not only can modern antibiotics and other chemotherapeutic agents be very expensive, but the cost of morbidity itself, both in lost productivity and increased allocation of resources to health care, can also be very high.

Finally, vaccines continue to play an important role in veterinary medicine, especially because cost pressures usually mean that farm animals are kept in tight quarters, a practice that enormously increases the chances of cross-infection.

The traditional methods of vaccine production are still used to manufacture many important vaccines. However, some vaccines produced in these ways have important problems. New methods, using recombinant DNA techniques and synthetic organic chemistry, have provided superior alternatives or substitutes, including an entirely new class of vaccine. These methods may also be used to develop vaccines against diseases for which traditional vaccines do not exist.

## PROBLEMS WITH TRADITIONAL VACCINES

Traditional vaccines are of two types, live and killed. Most *live vaccines* consist of attenuated (weakened) viral or bacterial strains, usually obtained by totally empirical procedures, such as prolonged storage or cultivation under suboptimal conditions. *Killed vaccines* are either killed whole cells of bacteria or inactivated toxin proteins, which are called *toxoids*. Many traditional vaccines are quite effective, but new vaccines, and new techniques for producing vaccines, are desperately needed. For many important diseases, vaccines have not yet been developed (Table 5.2). Moreover, certain of the traditional vaccines are not sufficiently effective or are not entirely safe.

**TABLE 5.1 Vaccines recommended for all children by the centers for disease control and prevention, U.S. department of health and human services (2002)**

| Vaccine | Constituent |
|---|---|
| Attenuated live pathogen: | |
| Measles | Attenuated live virus |
| Mumps | Attenuated live virus |
| Rubella (German measles) | Attenuated live virus |
| Varicella (chickenpox) | Attenuated live virus |
| Inactivated whole pathogen: | |
| Polio | Inactivated virus |
| Modified component of the pathogen: | |
| Diphtheria | Toxoid |
| Tetanus | Toxoid |
| Pertussis | Acellular vaccine containing toxoid and other proteins |
| *Haemophilus influenzae* type b | Capsular polysaccharide conjugated to carrier protein |
| *Streptococcus pneumoniae* | Capsular polysaccharides conjugated to carrier protein |
| Recombinant DNA–derived subunit vaccine: | |
| Hepatitis B | Surface antigen produced in yeast cells |

Foremost among the problems encountered with traditional live vaccines is the danger of reversion to the virulent state. For instance, the oral (Sabin) vaccine was thought to be generally safe and was used as the primary means of vaccination against polio in the United States and Europe since the mid-1960s. However, when the nucleotide sequences of the vaccine strains became available, they were found to be quite similar to those of the parent virulent strains, with one of the vaccine strains showing only two nucleotide substitutions. Mutant strains with such slight alterations do revert from time to time, and indeed the use of Sabin vaccine produced an estimated one case of poliomyelitis (VAPP, for vaccine-associated paralytic poliomyelitis) for every 520,000 administrations of the first dose. As shown in Figure 5.2, poliomyelitis from infection by wild-type virus has been essentially eradicated in the United States since around 1981, and all subsequent new cases were caused by the vaccination. In view of this situation, the U.S. Department of Health and Human Services recommended in 2000 that all childhood vaccinations against polio now be done with the inactivated polio vaccine (which is similar to the vaccines used before the advent of the live polio vaccine; see Table 5.1).

Another danger is that the viruses used in traditional live vaccines have to be grown in tissue culture cells, which poses the risk of introducing hidden viruses from those host cells. In one well-known case, a cell line used for propagation of the polio vaccine was found to contain a virus capable of producing tumors in experimental animals. Still another drawback is that even attenuated pathogens can produce severe diseases in individuals with immune system deficiencies. This could be a serious problem in developing countries, where many malnourished children suffer from such deficiencies.

The chief problem with the traditional killed vaccines is that they themselves can cause severe reactions. For example, the "whole-cell" vaccine for pertussis consists of whole killed cells of Gram-negative bacteria. Such preparations contain the principal component of the bacteria's outer membrane, lipopolysaccharide (LPS), also called endotoxin. Even very small amounts of endotoxin may elicit a strong toxic response. In sensitive animals, such as rabbits, endotoxin in amounts as low as 1 ng/kg of body weight can produce a measurable increase in body temperature – and the consequences of administering endotoxin are not limited to fever. Crude killed-cell preparations usually contain other toxic materials as well. Because of widespread

**TABLE 5.2 Examples of diseases for which effective vaccines are not yet available**

| Disease | Pathogen | New cases worldwide[a] (millions/year) | Deaths worldwide[a] (thousands/year) |
|---|---|---|---|
| AIDS | Virus | 8.4 | 2800 |
| Diarrheal diseases | Usually bacteria | 4500 | 1800 |
| Tuberculosis[b] | Bacterium | 7.6 | 1600 |
| Malaria | Protozoon | 408 | 1300 |
| Hepatitis C | Virus | 0.7 | 54 |
| Leishmaniasis | Protozoon | | 51 |
| Trypanosomiasis | Protozoon | | 48 |
| Schistosomiasis | Trematodes | | 15 |
| Chagas disease | Protozoon | 0.2 | 15 |

[a] These numbers are estimates by the World Health Organization for the year 2002, obtained from WHO Statistical Information Systems (WHOSIS).
[b] The only available attenuated live vaccine for tuberculosis, BCG, is generally thought to be ineffective in adults, and its efficacy in children is also disputed.

fear of side effects caused by such killed–bacterial cell preparations, many governments have had to change the status of pertussis vaccination of infants from compulsory (or highly recommended) to voluntary. A second problem is the direct risk run by the workers who cultivate dangerous pathogens in large amounts to manufacture the vaccines. A third is the possibility that the organism or toxin in the vaccine may not be completely killed or inactivated. The killing or inactivation procedure is usually a mild one, designed to inactivate the organism or toxin without destroying its ability to produce specific immunity. In several widely publicized cases, mass infections and toxic effects killed many who were inoculated, because a viral vaccine accidentally contained living viruses and a toxoid-based vaccine contained incompletely inactivated toxins. A final problem is that production of sufficient quantities of an infectious agent is not always possible or affordable. For example, to grow malaria parasites on a large scale using human blood cells would be prohibitively expensive and accompanied by a significant risk of introducing contaminating viruses into the vaccine produced. Another example is presented by the hepatitis B virus, which cannot be grown in tissue culture cells.

## IMPACT OF BIOTECHNOLOGY ON VACCINE DEVELOPMENT

Developments in biotechnology have led to the production of new kinds of vaccines. Some of these are directed at new targets; others are simply more effective or safer than traditional vaccines.

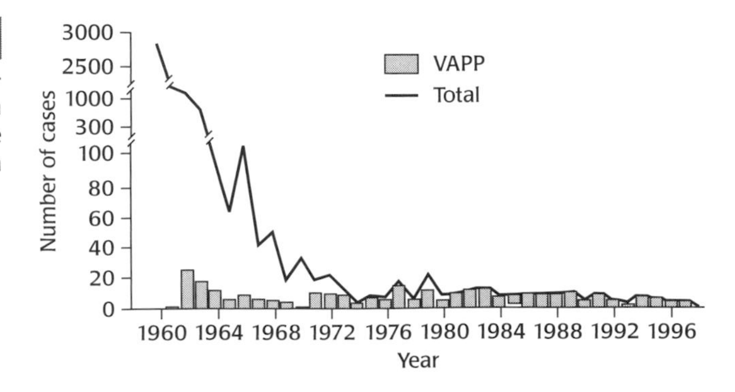


**FIGURE 5.2**

Incidence of all poliomyelitis cases and vaccine-associated paralytic poliomyelitis (VAPP) in the United States. [From Centers for Disease Control (2000). Poliomyelitis prevention in the United States. ***MMWR***, 49 (no. RR-5).]

Traditional vaccines are commonly made from intact pathogenic organisms or incompletely purified products of such organisms. In contrast, the new vaccines are based on purified components or products of pathogenic organisms. Once researchers have identified a *molecule* that can produce specific immunity – that is, a molecule that can be used as a *protective antigen* (Box 5.1) – among the thousands of components in a pathogenic microorganism, either a traditional purification strategy or recombinant DNA methods can be used to produce immunogenic quantities of the antigen. The latter approach is obviously more desirable than the traditional one because the antigen is produced in a safe, nonpathogenic organism such as *Escherichia coli* or yeast. Moreover, recombinant DNA technology allows vaccines to be produced even when the pathogens are difficult or impossible to cultivate. Because the new vaccines contain only one or some of the molecules found in the original pathogen, they are often called *subunit vaccines*. The following subsections contrast the development of acellular pertussis vaccine and conjugate polysaccharide vaccines, which are subunit vaccines produced mostly by nonrecombinant DNA methods, with the creation of the hepatitis B subunit vaccine, which was developed through recombinant DNA technology.

**Antigens**

An antigen is any molecule that elicits a specific immune response, either (1) the production of antibody proteins that have binding sites complementary to the antigen or (2) the proliferation of lymphocytes (T effector cells) that have specific surface receptors complementary to the antigen. In a narrower sense, an antigen is any molecule that binds to these complementary sites; such a molecule may be called an immunogen if it also elicits the immune response (the production or proliferation) described above.

An antigen is described as protective if the immune response it elicits in an organism protects the animal from later infection by the pathogen containing the antigen.

**BOX 5.1**

## ACELLULAR PERTUSSIS VACCINE

Pertussis, or whooping cough, is a childhood disease that, before the introduction of vaccine, accounted for 270,000 cases of illness resulting in 10,000 deaths annually in the United States. The World Health Organization (WHO) estimates that even now 45 million cases occur annually worldwide, with 400,000 deaths. The whole-cell vaccine decreased the number of cases dramatically in developed countries (Figure 5.3). However, because this vaccine consists of heat-killed, whole Gram-negative cells (of the causative organism, *Bordetella pertussis*) along with chemically inactivated culture supernatants containing many toxic components, it frequently causes adverse reactions.

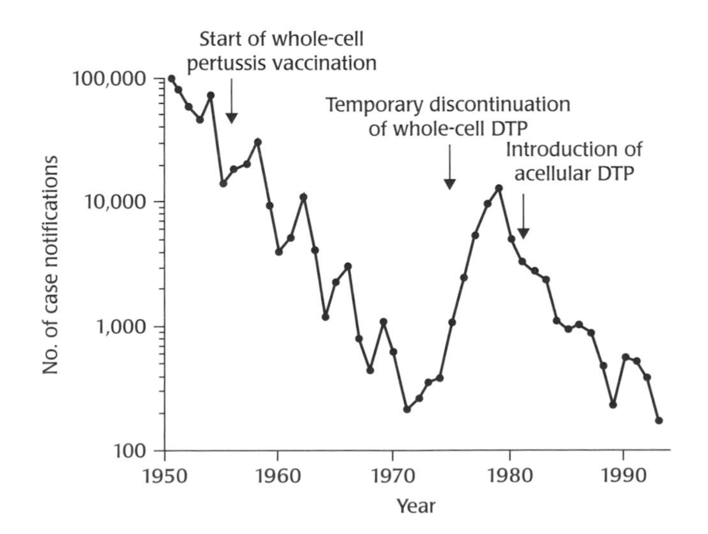


**FIGURE 5.3**

Incidence of reported cases of pertussis in Japan. [From Aoyama T. (1996). Acellular pertussis vaccines developed in Japan and their application for disease control. ***Journal of Infectious Diseases***, 174(Suppl 3), S264–269; with permission from the University of Chicago Press.]

Common ones include fever, local redness, and swelling (which occur in nearly 50% of infants who receive injections). During the 1970s, claims that the vaccine caused acute brain damage and sudden infant death led to a drastic decrease in the rate of immunization in Japan and Sweden, which in turn caused rapid increases in the occurrence of pertussis in these countries (Figure 5.3). In response, Japanese scientists developed acellular vaccines, which contain chemically inactivated, purified pertussis toxin, accompanied by a few purified proteins of *B. pertussis* that are thought to function as protective antigens. Similar formulations are now licensed in many countries, including the United States, and are widely used (see Table 5.1). These preparations are effective, produce fewer adverse effects, and have attained public acceptance, as seen from the decrease in pertussis cases in Japan in recent years (Figure 5.3).

Although preferable to the whole-cell vaccine, these acellular vaccines, developed before the advent of recombinant DNA technology, are in no way perfect. For example, although fever and local swelling are less frequent than with whole-cell vaccine, they still do occur. Chiron now produces a recombinant DNA–derived pertussis toxin vaccine, licensed in Europe, that has been inactivated by the introduction of two specific alterations in its amino acid sequence. Because the changes induced are so precise, they can be counted on to destroy the toxic activity without altering the overall protein conformation, which is needed to generate immunity. Such preparations are likely to be more effective at immunization than any nonspecifically inactivated protein toxin could be – say, one inactivated by treatment with formaldehyde, in which many molecules would be extensively altered in their conformation and thus would not generate immunity against the toxin.

## CONJUGATE POLYSACCHARIDE VACCINES

Before effective vaccines were made available, *Haemophilus influenzae* type b produced about 800 cases of "invasive disease" per 100,000 population in all age groups and 150 cases in children under 5 yearly in the United States, and *Streptococcus pneumoniae* produced about 200 cases per 100,000 population per year. These organisms are the leading causes of bacterial infection in young children, often leading to invasive infections such as meningitis, pneumonia, and bacteremia. In both, the protective antigen is the polysaccharide capsule. Polysaccharides can produce effective immunity in adults, but not in infants; however, when the purified polysaccharides are covalently linked to a "carrier" protein, the resulting conjugates function as very effective vaccines in infants (because the carrier protein can supply the T cell epitopes that are absent in the pure polysaccharide; see "Mechanisms for Producing Immunity" below). In the United States, the first of conjugate *H. influenzae* vaccines was licensed for immunization of infants in 1990, and a conjugate pneumococcal vaccine was licensed in 2000. The *H. influenzae* type b vaccine was phenomenally successful in the United States, decreasing the incidence of invasive infection from the prevaccination era figure cited

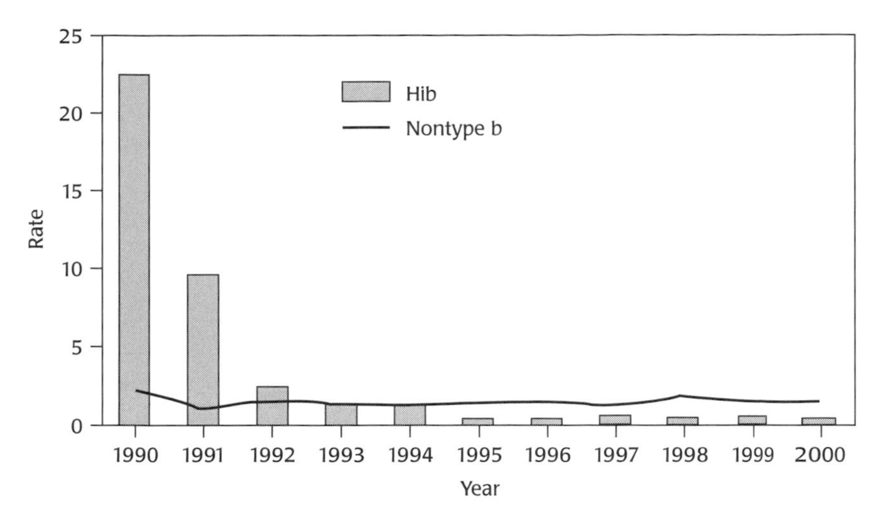


**FIGURE 5.4**

Incidence of invasive *Haemophilus influenzae* disease in U.S. children under 5 per 100,000 population. The *vertical bars* show the incidence of invasive infections caused by type b *H. influenzae,* and the *horizontal line* shows the incidence of those caused by other types. [From the Centers for Disease Control and Prevention (2002). Progress toward elimination of *Haemophilus influenzae* type b invasive disease among infants and children – United States, 1998–2000. ***MMWR***, 51, 234–237.]

above for children under 5 to only 0.3 cases per 100,000 per year, a decrease of more than 99% (Figure 5.4).

Some of these conjugate vaccines use a genetically inactivated version of diphtheria toxin called CRM197 as the protein component. Such genetically altered toxins, produced by recombinant DNA methods, preferably in "safe" host organisms, may one day be used in place of the chemically inactivated toxoids that are currently in use and that are often impure. Currently used diphtheria toxoid frequently produces mild adverse reactions, presumably because of its lack of purity (it is only about 60% pure); moreover, there is always a danger that it has not been completely inactivated.

## A RECOMBINANT SUBUNIT VACCINE FOR HEPATITIS B

The hepatitis B virus, transmitted through contaminated needles and sexual contact, infects an estimated 200,000 Americans every year. Of the 20,000 who subsequently become carriers, one in five dies of cirrhosis of the liver and one in 20 develops liver cancer. Surprisingly, the virus does not grow in tissue culture cells and until recently was available only from the plasma of carriers. Vaccines were made either by purifying the viral surface antigen or by inactivating living virus through chemical treatment (e.g., with formaldehyde). This source of the virus was quite limited, however, and use of the killed vaccine always carried the risk that not all the particles were inactivated. Fortunately, the surface antigen (the surface glycoprotein) of the virus was known to be an effective vaccine. The first step in producing a subunit vaccine was therefore to clone the gene for this protein from the viral genome.

The hepatitis B virus genome consists of largely double-stranded DNA (Figure 5.5) that codes for a core protein as well as for the major surface protein (S protein); most of the protein subunits making up the viral envelope are S protein molecules (226 residues). The DNA coding for the S protein was inserted into a YEp plasmid vector behind an effective yeast promoter and before a terminator (the construction of the first-generation recombinant

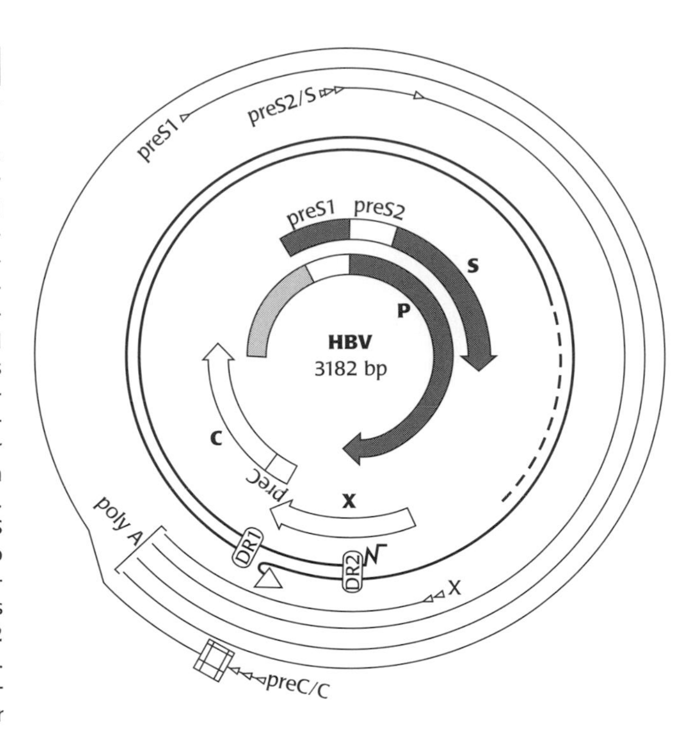


**FIGURE 5.5**

The genome of the hepatitis B virus. The figure shows the partially double-stranded genome (*inner ring*) and RNA transcripts (*outer arcs*). Open reading frames for four proteins – P, X, C, and S (with PreS2 and PreS1) – are also shown in the center. Protein P is the DNA polymerase that synthesizes one strand of DNA by reverse transcription of the longest mRNA (preC/C in the figure) and then synthesizes the other strand by using the just-synthesized DNA strand as the template. The precise function of protein X, which is present in minute quantities, is not known. Protein C is the major component of the inner capsid, and protein S is the major surface (envelope) protein. Some translation products cover only the S region, producing a protein of 226 amino acid residues, the hepatitis B surface antigen (HBsAg), but other translation products also cover PreS2 or both PreS1 and PreS2 regions, producing larger protein products. These latter, less abundant products containing PreS2 and PreS1-PreS2 are minor surface proteins of the virion. Although hepatitis B virus cannot infect tissue culture cells, its DNA can be introduced into cells by transfection, allowing analysis of the transcription pattern of the genome. [From Nassal, M., and Schaller, H. (1993) Hepatitis B virus replication. ***Trends in Microbiology***, 1, 221–228.

plasmid is shown in Figure 5.6; for plasmids made later, see Figure 3.26). Presumably, one reason the yeast was chosen as the host was the expectation that it would glycosylate the envelope protein (see Chapter 3). It did not do so, but the protein seemed to have folded properly nevertheless; it self-assembled into a form that resembled an empty virus envelope 22 nm in diameter and nearly indistinguishable from those found in the plasma of patients (Figure 5.7). (Maneuvers that increased the yield of the recombinant protein are discussed in Chapter 3, on page 136.) This yeast-produced vaccine, although lacking the oligosaccharides, was as effective as the vaccine derived from human plasma, and in 1986, it became the first recombinant DNA–based vaccine licensed for use in the United States. With earlier methodologies, about 40 L of infected human serum were required to produce a single dose of hepatitis B vaccine; now we can obtain many doses of the recombinant vaccine from the same volume of yeast culture.

Although the original hepatitis B recombinant vaccine was highly successful, there was room for improvement. A newer generation of vaccines is now produced using DNA that codes for PreS2 and PreS1 regions (see Figure 5.5) in addition to the S protein, because these N-terminal parts of the protein appear to help in the buildup of immunity. The plasmid also contains a promoter effective in animal cells and is introduced into a mammalian cell line (often a Chinese hamster ovary line). Under these conditions, the translation occurs on ribosomes attached to the endoplasmic reticulum, so that the protein products (some including PreS2 and PreS1 domains) are exported through the natural secretion pathway of the endoplasmic reticulum–Golgi apparatus, are glycosylated in the normal manner, and enter the medium as empty vesicles. Some studies suggest that these vaccines can produce

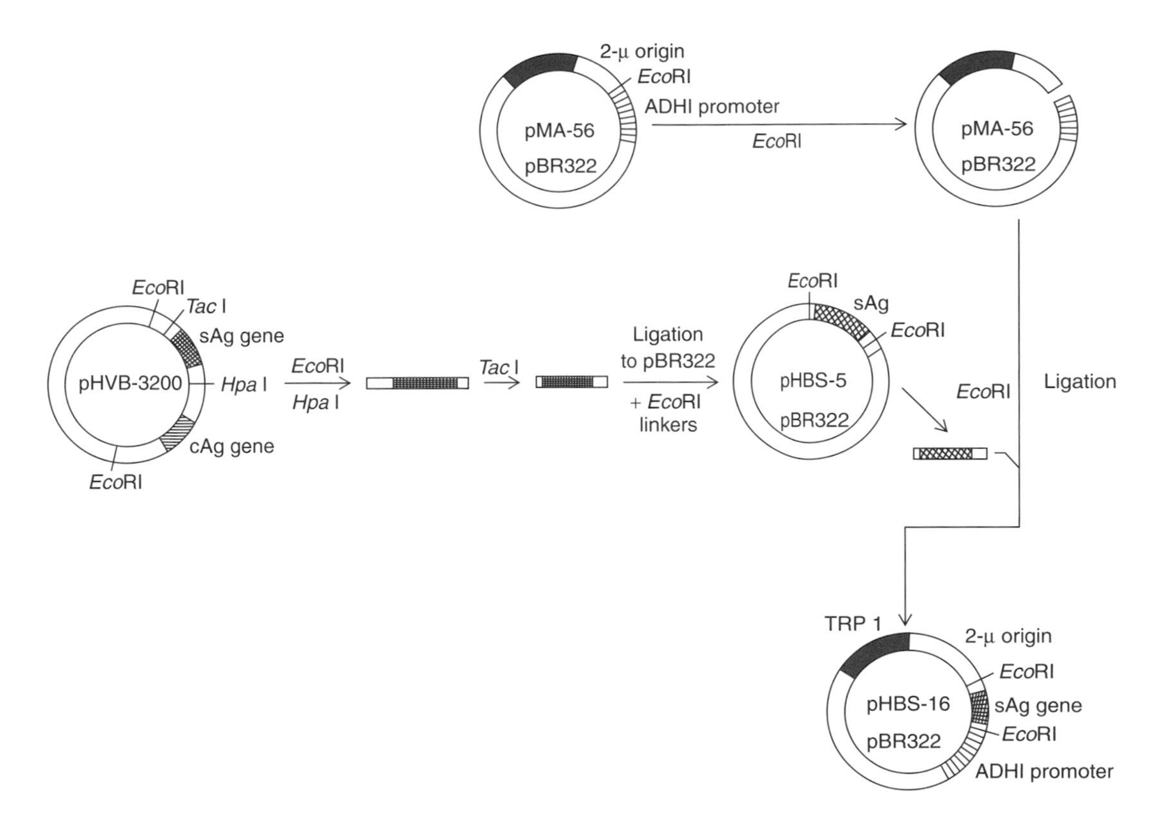


immunity even in the 5% to 10% of the population that does not respond to the older vaccines produced in yeast.

FIGURE 5.6

Construction of a plasmid expressing HBsAg in yeast. From a plasmid (pHVB-3200) that contains both the surface antigen (*sAg*) gene and the core antigen (*cAg*) gene of the hepatitis B virus, a clone containing only the *sAg* gene, pHBS-5, was constructed. The *sAg* gene was then inserted behind an alcohol dehydrogenase (ADH1) promoter in plasmid pMA-56 to produce pHBS-16. Note that the final plasmid contains not only the replication origin functional in *E. coli* (from plasmid pBR322) but also the sequence that enables the plasmid to replicate in yeast cells (a 2-$\mu$ plasmid origin from pMA-56). [From Valenzuela, P., et al. (1982). Synthesis and assembly of hepatitis B virus surface antigen particles in yeast. ***Nature***, 298, 347–350; with permission.]

## POTENTIAL PROBLEMS OF RECOMBINANT SUBUNIT VACCINES

Recombinant DNA subunit vaccines, when they are effective, have many advantages over traditional vaccines. They can be produced easily, safely, and inexpensively and are devoid of all the extraneous components of the pathogen that may cause undesirable side effects. Furthermore, there is absolutely no possibility that a living pathogen will be present in the subunit vaccine.

If subunit vaccines produced by recombinant DNA technology are really so effective and advantageous, why have they not yet replaced the traditional vaccines? A major reason is economic rather than scientific. When a traditional vaccine – chemically inactivated diphtheria toxoid, for example – is effective and causes no *major* adverse reactions, no vaccine manufacturer will be interested in spending the necessary funds to develop and test a recombinant DNA–derived substitute, even though "genetically inactivated" toxins are known to be safer and usually more effective. There is also a scientific reason: the recombinant DNA approach is still hampered by certain technical problems. Some genes have a low level of expression. Some proteins fold improperly in a nonmammalian host, or when they are produced in unusually large amounts, or because they require posttranslational processing (Chapter 3). Many viral proteins, including proteins from the viruses

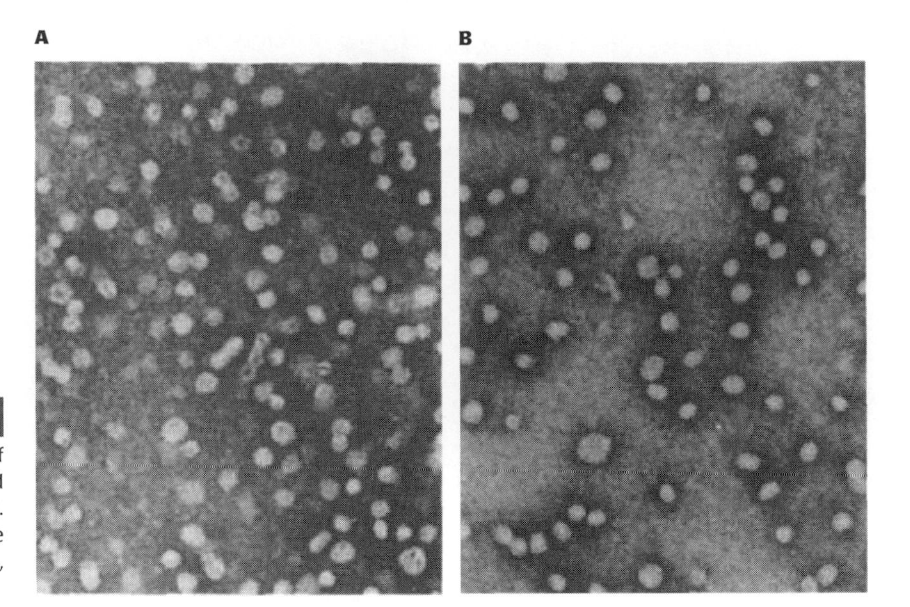


FIGURE 5.7

Negatively stained electron micrographs of (**A**) plasma-derived and (**B**) yeast-derived HBsAg vaccines. [From Hilleman, M. R. (1988) Hepatitis B and AIDS and the promise for their control by vaccines. ***Vaccine***, 6, 175–179; with permission.]

that cause fowl plague, vesicular stomatitis, and herpes viruses, *have* been expressed successfully in *E. coli*; unfortunately, improper folding apparently prevents the production of many others.

Cultured mammalian cells have been used to express genes for protective antigens in the hope that in these cells the products would be properly modified and folded. This method requires cell lines that multiply indefinitely, as tumor cells do. Indeed, cells of many of these lines are known to induce tumors when injected into appropriate hosts. In order to prevent the introduction of any tumor-causing DNA into a vaccine when such a system is being used, all of the host cell DNA must be removed from the vaccine, and this can be a difficult process. (It is facilitated in the new generation of hepatitis B vaccine mentioned above by the fact that the particle of HBsAg is secreted into the medium.)

With many subunit vaccines, an even bigger problem than incorrect folding is that the immunity produced is weak and of short duration. In fact, the recombinant DNA–based vaccine for *Borrelia burgdorferi* (Lyme disease), containing a surface protein of this spirochete, was the only new recombinant DNA–based subunit vaccine licensed in the United States since the first approval of the recombinant HBsAg vaccine in 1986, and it was taken off the market after a few years by the manufacturer, probably because of its limited efficacy. (However, in 2006 FDA approved a vaccine for several types of human papilloma virus, which causes cervical cancer. This vaccine is made by expressing the cloned gene for the virus capsid antigen in yeast. The protein spontaneously forms a spherical virus-like particle (just like the HBsAg), which is used as the vaccine). Presumably our immune systems, which have evolved to react against natural pathogens and thus respond well to the traditional vaccines, which are similar to the pathogens, often respond only feebly to subunit vaccines, which are radically different from the natural pathogens. Thus, whereas a detailed knowledge of the mechanisms of immunity was not necessary for the development of whole-cell vaccines, researchers will need to acquire a more thorough understanding of immune defenses if they wish to find ways of improving the performance of subunit vaccines. A brief overview of the current understanding of those mechanisms is provided in the section that follows.

## MECHANISMS FOR PRODUCING IMMUNITY

In vertebrates, the first lines of defense against pathogenic microorganisms are nonspecific. An infecting organism may be killed by the antimicrobial substances in tissues or ingested by macrophages in tissues or by polymorphonuclear leucocytes migrating into infected tissues from the bloodstream. Most infections are presumably arrested at this stage. In recent years, the discovery of Toll-like receptors on phagocytic cells (a topic revisited in a later part of this chapter) has increased our understanding of this "innate immunity." Only when the pathogens survive this initial defense are the body's specific immune responses activated.

### PRODUCTION OF SPECIFIC ANTIBODIES

In many cases, immunity is acquired through the production of *antibodies*, proteins with binding sites complementary to the structure of the immunizing foreign antigen (Figure 5.8). Many vaccines act by stimulating the synthesis of antibodies that bind to various components of the vaccines and the pathogens. Often these antibodies bind to the protein toxins produced by pathogens, inactivating (neutralizing) the toxins as a result. This is the way the diphtheria and tetanus vaccines work. Protein toxins secreted by diphtheria and tetanus bacteria cause the major symptoms of those diseases, but immunization with inactivated toxin vaccines stimulates the synthesis of antibodies that bind to the toxins and neutralize them. Even when toxins do not play a major role in the development of a disease, antibodies may still be effective in preventing it; when the antibodies bind to the surface of the invading pathogen, they are recognized by the phagocytic cells, which then ingest and kill the invader (Figure 5.9). This function of the antibody is called *opsonization*. The bound antibodies have other important effects as well: one is the initiation of the *complement cascade*, a series of reactions involving many serum proteins that leads to the migration of phagocytes out of the bloodstream; another is the direct killing of Gram-negative bacterial invaders without the involvement of phagocytosis (see Figure 5.9).

All antigens (see Box 5.1), including vaccines, stimulate antibody production through a process called *clonal selection*, in which the antigen first binds to an antibody on the surface of a particular lymphocyte (B cell), one of a preexisting collection of lymphocyte types that each produce a different antibody (Figure 5.10). The antigen binds to a given antibody because the latter has a combining site complementary to some portion of the antigen's structure. The antigen–antibody binding on the surface of a B cell stimulates

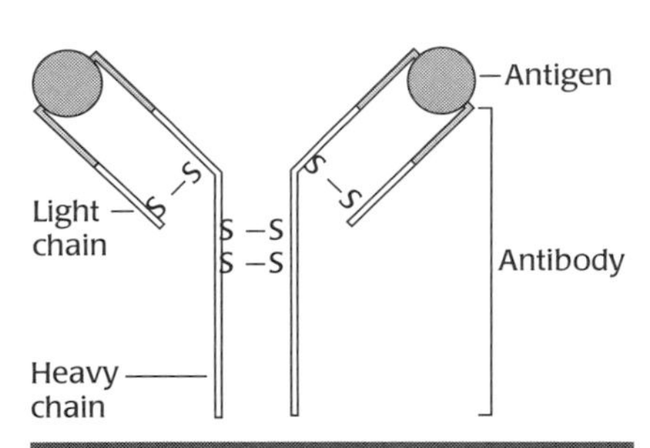


**FIGURE 5.8**

A schematic structure of an antibody of the immunoglobulin G (IgG) type. This type of antibody is composed of two heavy chains (the longer polypeptides shown in the center) and two light chains (the shorter polypeptides shown on the sides), linked via disulfide bridges. The two antigen-binding sites on the IgG molecule consist of the N-terminal ends of heavy and light chains, regions where there is much variation in amino acid sequence among antibody molecules (so-called *hypervariable regions*, shaded in the figure).

Redrawn based on artwork from the first edition (1995), published by W.H. Freeman.

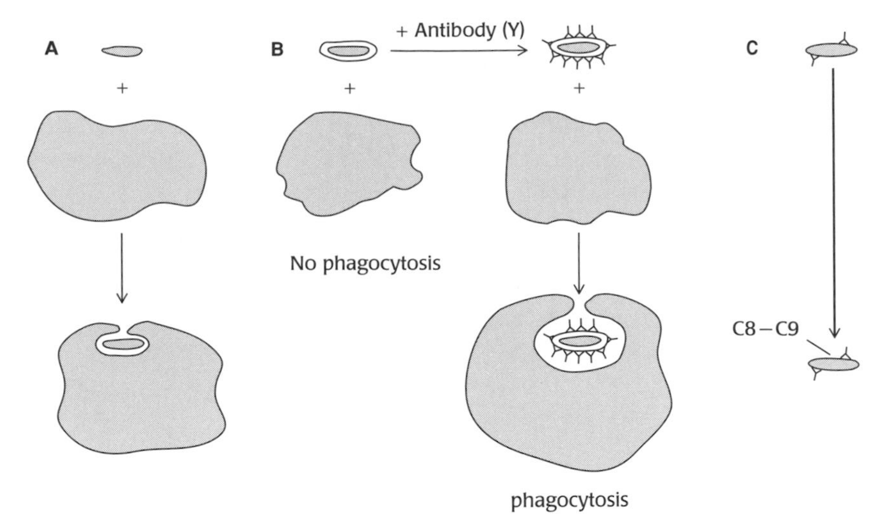


**FIGURE 5.9**

Killing of invading microorganisms. **(A)** If the invading microorganisms have a cell surface that can be easily recognized as "foreign" (e.g., a less hydrophilic surface than the host cells), they are nonspecifically ingested by phagocytes such as macrophages and polymorphonuclear leucocytes and usually killed off. Furthermore, molecules commonly present on the surface of invading organisms, such as LPS and peptidoglycan, are recognized by Toll-like receptors (see Figure 5.11) on the macrophage surface, and this recognition leads to the secretion of various cytokines (proteins that affect other cells) and finally to an inflammatory response, including the migration and activation of phagocytic cells, such as macrophages. **(B)** Most successful pathogens have hydrophilic surface structures (e.g., capsules) that enable them to evade the nonspecific phagocytosis. However, when antibodies bind to the surface of the pathogens, phagocytic cells recognize the Fc portion of the antibody and succeed in ingesting and killing the invading pathogen, that is, opsonization occurs. (Fc is the nonspecific domain of the antibody composed of the C-termini of heavy chains; in Figure 5.8, it corresponds to the bottom part of the structure.) **(C)** If the invading pathogen is a Gram-negative bacterium, it can be killed without phagocytosis. Antibody binding to the surface activates a series of reactions in the complement pathway, and the final components of this pathway, C8 and C9, form a *membrane attack complex* that inserts into the membranes of the pathogen and kills them directly.

Redrawn based on artwork from the first edition (1995), published by W.H. Freeman.

that line of B cells to proliferate, and the resulting clones differentiate into plasma cells that secrete large amounts of the specific antibody (see Figure 5.10). The result is immunization through vaccination.

It is important to note that antibodies bind to only a small portion of the macromolecular antigen. The antigen-binding site of an antibody can accommodate structures of $20 \times 30$ Å only – that is, structures containing 18 to 20 amino acids if the antigen is an $\alpha$-helical protein. Thus, the antibody actually binds to the *epitope*, a small portion of the antigen that determines the specificity of the particular antibody.

Although an effective vaccine stimulates antibody production, the body generally does not continue producing the antibodies indefinitely. Some vaccines, however, do produce a long-term – even lifelong – immunity. In these situations, a successful clonal selection of immune cells has left behind a small number of "memory cells" that persist and can respond immediately if the organism is challenged again by the same antigen (pathogen). In the optimum scenario, a vaccine will induce the persistence of both B and $T_H$ memory cells (see below) to produce an effective secondary, or *anamnestic*, response.

## CELL-MEDIATED IMMUNITY

The production of antibodies is not the only specific mechanism vertebrates use to fight invading pathogens. Antibodies have no effect against invaders that live *inside* host cells, because they cannot enter the cytoplasm by diffusing across the plasma membrane. When such invasion occurs, cellular immunity is the next line of defense, and this response involves another class of lymphocytes, T cells. A pathogen-infected cell expresses certain antigens, such as viral proteins, on its surface, and these are recognized by receptors on cytotoxic, or CD8, T cells. Recognition leads to the selection and proliferation of antigen-specific T cells, through a clonal mechanism similar to the one outlined for the selective propagation of specific B cells in Figure 5.10. Such activated T cells kill the target cell by direct contact with it. In

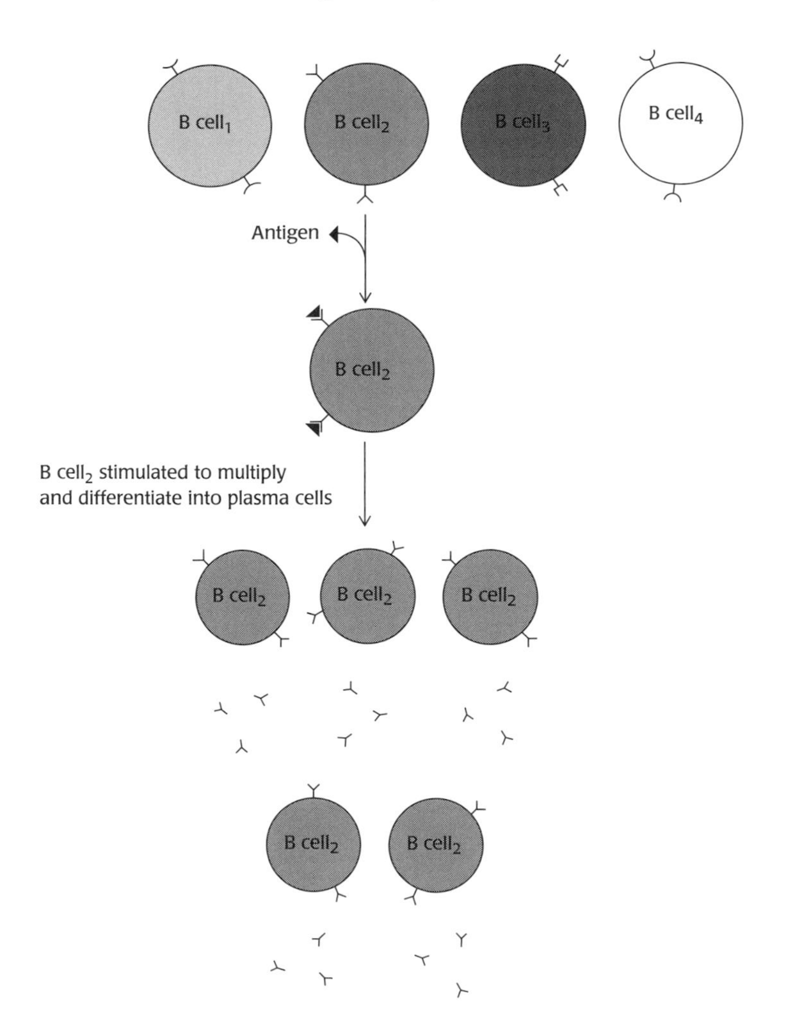


**FIGURE 5.10**

Clonal selection of B cells producing a specific antibody. Antigen molecules encounter a spectrum of B cells producing different kinds of antibodies. Only the B cell whose antibody binds the antigen (in this case, B $cell_2$) multiplies and differentiates eventually into a "clone" of plasma cells producing one particular kind of antibody.

Redrawn based on artwork from the first edition (1995), published by W.H. Freeman.

contrast, many bacterial and eukaryotic pathogens, especially intracellular pathogens such as *Mycobacterium tuberculosis* and *Leishmania*, are killed within macrophages. One way macrophages are activated is by interaction with a certain class of T helper cells, $T_H1$ cells, which also stimulate the local inflammatory response (see below).

Cellular immunity, especially the kind mediated by CD8 cells, may also be important in the body's early detection of malignant tumor cells. Tumor cells usually express abnormal antigens on their surface. The CD8 cells recognize the new antigens and destroy the cells carrying them. This *immune surveillance* is thought to eliminate most tumor cells that arise in the body.

## THE ROLE OF ANTIGEN-PRESENTING CELLS

One important feature of the recognition of antigenic epitopes by T cells of all types, including CD8 T cells – and in fact a process required for T cell stimulation – is that T cells cannot recognize antigen molecules until those molecules have been partially degraded or "processed" by other cells, specifically "antigen-presenting cells," or APCs. The classes of APC that are important at the time of initial infection include "dendritic cells" (so named because they

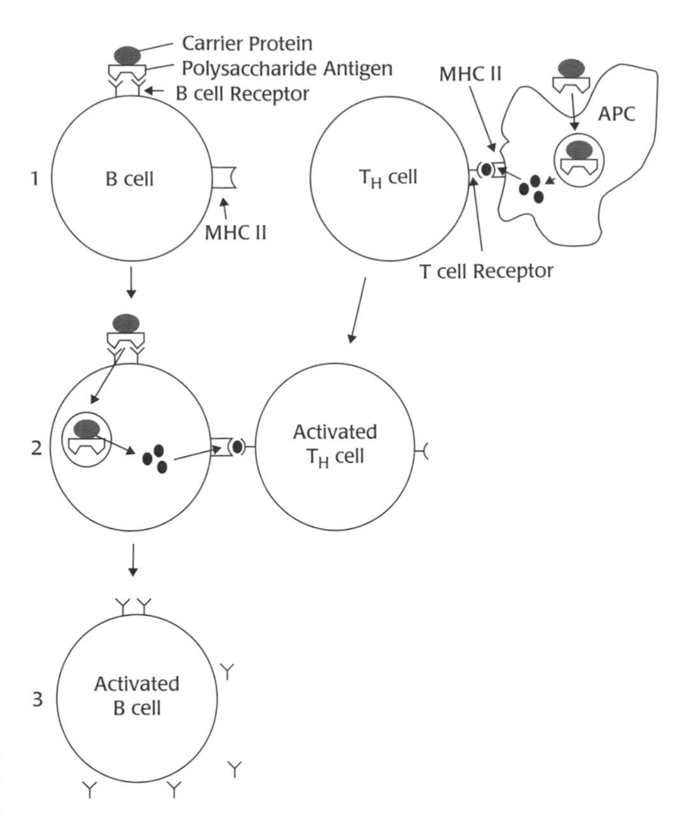


**FIGURE 5.11**

T cell/B cell cooperation in the production of antibody. The vaccine contains a polysaccharide antigen whose B cell epitopes (shown as a *polygon*) are conjugated to a carrier protein (shown as a *black ellipse*). The vaccine enters an APC (a dendritic cell or macrophage), where the carrier protein is degraded into peptides, and these are subsequently presented in a complex with MHC II molecules on the APC surface (stage 1, *right*). In a population of T helper cells, a small number will have a receptor that fits the presented peptide (i.e., the T cell epitope shown as small black dots) and will therefore be activated. The activation of B cells requires two signals. First, the antigen will bind to those rare B cells that happen to produce correct antibodies (called "B cell receptors" when they are located on the cell surface), specifically antibodies that fit the polysaccharide containing the B cell epitopes. Cross-linking of two B cell receptors by the antigen serves as the first signal (stage 1, *left*). B cells internalize the antigen, degrade the protein part of it, and then present the peptides on MHC II molecules on the B cell surface. Recognition of the peptide (T cell epitope) by the activated T cells (stage 2) serves as the second signal and leads to the activation of B cells, which then multiply and secrete antibodies (stage 3).

have many branchlike protrusions) and macrophages, although B cells also act as APCs in the process of T cell/B cell cooperation discussed below (see Figure 5.11). A requirement for proper "presentation," or the T cell will not recognize it, is that the fragment of the antigen (epitope) must be embedded in a special class of proteins called major histocompatibility complexes (MHCs, called human leukocyte antigens [HLAs] in humans). Histocompatibility complexes differ from one individual to the next and enable immune cells to distinguish the body's own cells from foreign cells.

This process of antigen presentation is the mechanism by which the immune response becomes directed toward a specific pathway. When proteins of viruses are released into the cytosol of host cells, they are digested by giant proteolytic complexes called proteasomes, and the resulting fragments, complexed with class I MHC, are then presented on the APC surface. The cytotoxic CD8 T cells interact exclusively with MHC class I, thus becoming activated and inducing cell-mediated immunity. In contrast, soluble toxins are phagocytized into acidic vesicles in APCs and then processed by the vesicular proteases, after which their fragments, complexed with class II MHC, are presented on the APC surface. Similarly, pathogen proteins phagocytized into the same acidic vesicles of macrophages are also presented (after digestion) on the macrophage surface in a complex with class II MHC. These antigens are recognized by another subclass of T cells, CD4 T cells (also called T helper cells), which cause various responses, including antibody production and activation of phagocytes.

## T CELL/B CELL COOPERATION

The mechanism of clonal selection of B cells is even more complex than that shown in Figure 5.10. The receptors on the surface of CD4 T cells (T helper, or $T_H$ cells), antibody-like proteins called T cell receptors, bind specifically to a part of an antigen, and the $T_H$ cells thus activated are essential for the activation of B-cells (Figure 5.11).

The various subclasses of T cells perform very different functions. Those that collaborate with B cells in the antibody response are more commonly a subclass of $T_H$ cells called $T_H2$ cells. For an effective antibody response to occur, the antigen molecules must bind to the receptor on the $T_H2$ cell surface (Figure 5.11). Usually, the part of the antigen recognized by the antibody on the B cells (the B cell epitope) is different from the part of the antigen recognized by the T cells (the T cell epitope). Effective production of antibodies occurs only if the vaccine contains both B cell and T cell epitopes in close proximity.

## $T_H1/T_H2$ DICHOTOMY

Another concept that has influenced the thinking of immunologists in recent years is that of $T_H1/T_H2$ dichotomy. According to this idea, certain antigens administered in certain ways activate a subclass of T helper cell called $T_H1$, which secretes interferon-gamma, a cytokine, whereas other antigens activate $T_H2$, which characteristically secretes interleukin 4. These two types of T cells have often been described as controlling cell-mediated and humoral (antibody-mediated) responses, respectively, an oversimplification that "has been the source of considerable confusion," in the words of one expert. It is now generally accepted that $T_H1$ stimulation results in local inflammatory responses, including the activation of macrophages, and at the same time, production of complement-fixing and opsonizing antibodies. The $T_H2$ pathway is believed to produce the subclass of IgG antibody that functions in the neutralization of toxins and also to produce immunoglobulin E (IgE) antibody (Box 5.2), important in allergy because it interacts with mast cells, leading to the release of histamine and serotonin. The $T_H2$ mediation often occurs in response to antigens that are found outside the epithelial barrier and penetrate the barrier only occasionally, such as most allergens or those that are present in large parasites (e.g., worms). This pathway also activates eosinophil leukocytes, whose granules contain proteins that are toxic to these parasites.

Because they produce such different responses, it will be important to be able to manipulate these two pathways pharmacologically. For example, to alleviate the symptoms of allergy patients, we might want to favor the $T_H1$ response to a given allergen. (In fact, we already know that a certain class of small synthetic drugs, imidazoquinolines, stimulate the $T_H1$ pathway, as seen in Figure 5.12).

If recognition of T cell epitopes by both $T_H1$ and $T_H2$ cells requires class II MHC on the APC surface, how does our body decide between generating a $T_H1$ response and a $T_H2$ response? Recent studies suggest that signaling

**Immunoglobulin Isotypes**

Antibodies (immunoglobulins) occur in various types, or isotypes, that differ in the structure of the constant region of their heavy chains (see Figure 5.8). The isotypes are IgG, IgA, IgD, IgM, and IgE. Among them, IgA and IgM form oligomers that bind tightly to antigens with repetitive epitopes. IgG is mainly responsible for opsonization. Some IgG subclasses and IgM play a major role in the complement cascade. IgA is responsible for mucosal immunity, and IgE causes allergic reactions by binding to mast cells. In humans, IgD exists only in a membrane-bound form but plays a role in the maturation of IgM-secreting B cells.

BOX 5.2

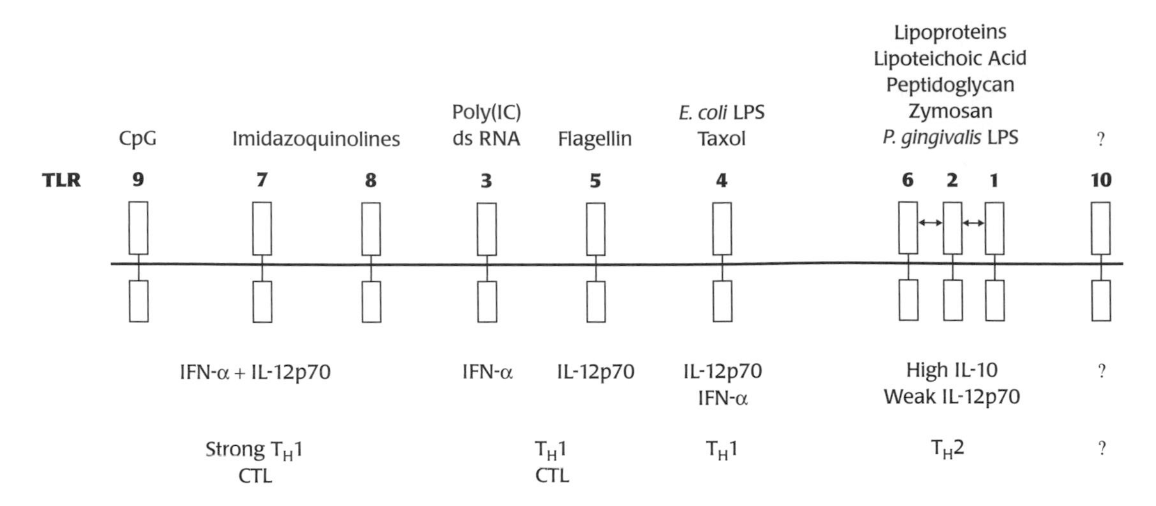


**FIGURE 5.12**

Toll-like receptors on human cells and the ligands that bind to these receptors. As shown, LPS from most sources stimulates TLR-4, and the unmethylated CpG dinucleotide sequences, abundant in bacterial DNA, stimulate TLR-9, with both events leading to the secretion of interferon-alfa (IFN-α) and interleukin 12 (IL-12p70), and to the stimulation of the $T_H1$ pathway as well as the activation of cytotoxic lymphocytes (CTLs). TLR-7 (and TLR-8 in humans) responds to synthetic imidazoquinoline compounds, although their natural ligands are not known. In contrast, lipoproteins, lipoteichoic acid, and *Porphyromonas gingivalis* LPS stimulate TLR-2, which works as a complex either with TLR-1 or TLR-6 (see the *small double-headed horizontal arrows*); this results in the secretion of inflammation-suppressing cytokine interleukin 10 (IL-10) and also in the stimulation of $T_H2$ response. [This is a simplified version of Figure 2 in Pulendran B. (2004). Modulating vaccine responses with dendritic cells and Toll-like receptors. ***Immunological Reviews***, 199, 227–250.]

through "Toll-like receptors" (so called because they are related to a receptor class known as Toll in *Drosophila*) may play a large part. Toll-like receptors are found on the surface of both macrophages and dendritic cells, and so far, 10 different Toll-like receptors are known to bind to various components commonly found in the cells of pathogens (Figure 5.12). The binding of these components (e.g., LPS to Toll-like receptor 4 or lipoteichoic acid to Toll-like receptor 2) produces signaling cascades that stimulate adaptive, specific immune responses that eventually favor either a $T_H1$ response or a $T_H2$ response. Much effort is under way to discover and use tools that would enable us to control $T_H1$ versus $T_H2$ responses. Indeed, CpG DNA and imidazoquinolines are actively pursued in order to produce a strong $T_H1$ response to the antigens in vaccines.

## IMPROVING THE EFFECTIVENESS OF SUBUNIT VACCINES

What they have learned about the mechanisms of protective immunity has enabled scientists to devise a number of strategies for improving the efficacy of subunit vaccines.

### STRATEGIES FOR ADMINISTERING ANTIGEN

As we have seen, the production of antibody in response to a vaccine requires the presence of both B cell and T cell epitopes in the vaccine. With the subunit protein vaccines, this is usually not a problem because a protective antigen, being a large protein, usually contains both. However, polysaccharides used in vaccination do present a problem because they cannot contain T cell epitopes. They can generate antibodies in adults through the action of T-independent B cells, but that action does not occur in infants. This is the reason capsular polysaccharides of *H. influenzae* and *S. pneumoniae* had to be attached to carrier proteins to produce conjugate vaccines that can immunize infants.

Our immune response mechanisms have become so specialized over the course of evolution that they are able to mount an effective immune response against real pathogens, yet they do not accidentally launch an attack against similar-looking antigens derived from our own tissues. The recognition of "pathogen-associated molecular patterns" plays a large role in this discriminative ability. Cells that participate in both innate and adaptive immunity, including APCs, use Toll-like receptors (Figure 5.12) to recognize common pathogen components such as LPS, peptidoglycan, and CpG DNA. The fact that the strongest immune responses generally occur when the antigen molecules are present in a concentrated form, as on the surface of a virus particle or a bacterial cell, suggests that such concentrated arrangements facilitate phagocytosis/pinocytosis by APCs and cause cross-linking between B cell receptors, the first signal for B cell activation (see Figure 5.11). Adjuvants are molecules that stimulate the immune response in a nonspecific manner, usually either because they preserve the local high concentration of immunogens or because they bind to and activate Toll-like receptors.

A pure subunit vaccine almost always produces a weaker response than does the whole pathogen, because the former lacks the typical pathogen-associated molecular pattern just mentioned. The surface antigens of the hepatitis B and human papilloma viruses were a lucky exception to this rule, because they assembled into particles of a size similar to that of the empty virus particles themselves, with the same antigens exposed at high concentrations on their surface as on the virus. Thus, the hepatitis and human papilloma virus vaccines are excellent at mimicking the natural virus and at satisfying at least one of the conditions noted above for provoking a strong immune response, an ability that is not achieved with most other subunit vaccines.

To produce a concentrated array of the proteins, various methods have been devised of fixing a large number of antigenic protein molecules on the surface of a particulate carrier. For example, practically all of the subunit vaccines (whether nonrecombinant or recombinant in type) contain, as adjuvants, insoluble aluminum salts, which not only maintain a locally high concentration of the immunogen but also may present the antigen as a high-concentration array by adsorbing antigen proteins to their surface. One new adjuvant licensed in the United States is an oil-in-water emulsion containing squalene and some detergents, called MF59. Presumably, amphiphilic immunogens become concentrated at the oil/water interface. This adjuvant is used in some commercial influenza vaccines, which are subunit vaccines containing largely proteins on virus surface, produced by the traditional nonrecombinant method. A similar oil-in-water adjuvant is depicted in Figure 5.13.

Another approach to enhancing the effect of a subunit antigen takes advantage of the self-assembling feature of the hepatitis B surface antigen, fusing the important parts of other subunit antigens to this protein by the recombinant DNA technique.

Adjuvants that act by interacting with Toll-like receptors are also used. In one approach, more than 130 synthetic analogs of the natural adjuvant

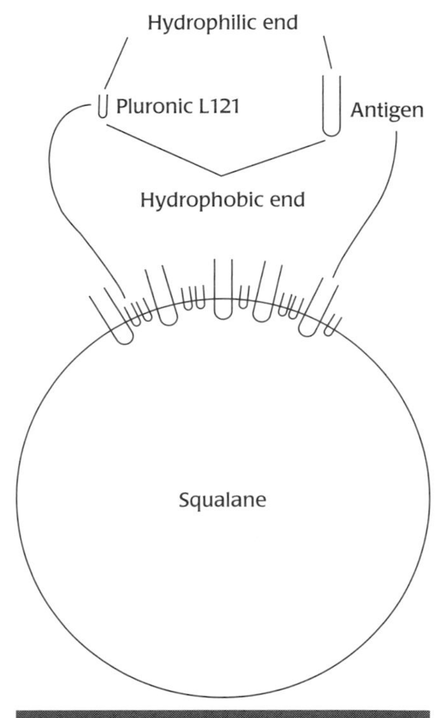


**FIGURE 5.13**

Antigenic protein molecules embedded in the squalane–Pluronic L121 system. Pluronic L121, being an amphiphilic molecule, inserts its hydrophobic ends into the surface of a droplet of squalane, a completely hydrophobic compound. With the hydrophilic ends of Pluronic L121 protruding all over the surface of the complex, the entire structure becomes stabilized. Antigen molecules also insert themselves partly into the surface of this complex, thereby achieving a dense two-dimensional array that favors recognition by immune cells. [Modified from Allison, A. C., and Byars, N. E. (1987) Vaccine technology: adjuvants for increased efficacy. ***Bio/Technology***, 5, 1041–1045.]

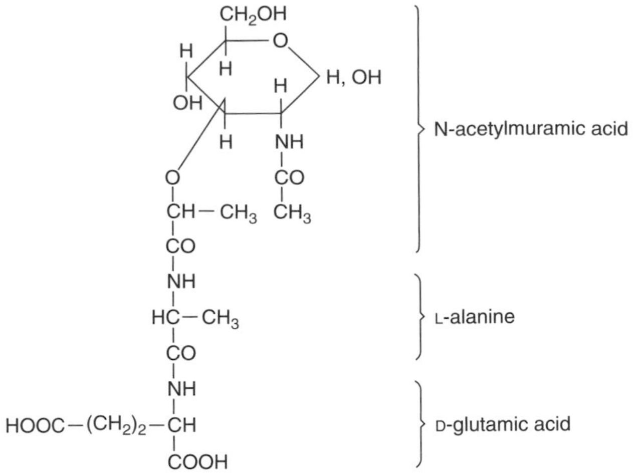

**FIGURE 5.14**

Structure of muramyldipeptide.

muramyldipeptide, or MDP (Figure 5.14), a fragment of the bacterial peptidoglycan, were tested, and one compound, in which the L-alanine residue of MDP was replaced by L-threonine, was found to be a potent stimulator of the immune response without the unwanted side effects of MDP. When this MDP analog was used in combination with several viral antigens produced by recombinant DNA methods and dispersed on the surface of the hydrophobic microsphere of squalane, a very effective vaccination was obtained in animal models. Another type of adjuvant that functions by interacting with Toll-like receptors is monophosphoryl lipid A and its derivatives. These compounds resemble LPS (or its lipid portion, lipid A) in structure but are not nearly as toxic.

## USE OF LIVE, ATTENUATED VECTORS

In some cases, the conditions required for a strong immune response can be met by incorporating the subunit antigens into live, attenuated viruses or bacteria. Such strategies have both advantages and disadvantages.

Some highly effective traditional vaccines are live vaccines (see Table 5.1). Because they present the antigens to the body's defense mechanisms in a "natural" manner (i.e., in a concentrated form, often accompanied by molecules that act as effective adjuvants), they very often confer stronger immunity, sometimes for a longer period, than killed vaccines. Furthermore, live vaccines can usually be administered in smaller dosages, because they may, to a limited extent, be able to multiply within the host. Another advantage of some live vaccines is that they do not have to be administered parenterally (by injection).

Introducing the gene for a protective antigen into a live vector creates a recombinant DNA vaccine that has all these advantages. In addition, such vaccines tend to be much less expensive than the subunit vaccines discussed previously, because there is no need for production and purification of the antigenic protein in a manufacturing plant. Remember, however, that the live vectors present the dangers we noted before, such as the chance they will revert to virulence or that they will act as virulent strains in hosts with weakened immune systems.

### Viral Vectors

The vaccinia (related to cowpox) vector seems to be the most promising candidate for a viral vector, as it is reasonably safe and its large genome can accommodate a fair amount of foreign DNA. The large vaccinia virus DNA is difficult to manipulate *in vitro*, but a series of clever techniques has been devised to overcome this obstacle. In a typical situation, the foreign genes are cloned into short stretches of vaccinia DNA in conventional plasmids, with *E. coli* as the host. The plasmid DNA is then isolated and introduced into mammalian cells that are simultaneously infected with vaccinia virus. The foreign DNA inserts into the vaccinia DNA through homologous recombination (Figure 5.15).

Approaches of this type were used to produce vaccines against rabies, hepatitis B, influenza, Friend murine leukemia, herpes simplex, and other diseases. Many proved highly efficacious in animal experiments, and some have been tested in field trials. Recombinant vaccinia virus containing the gene for a glycoprotein of rabies virus has been administered (hidden in bait) to wild animals and has virtually eradicated rabies in most of Western Europe. This is a significant accomplishment, because the live, attenuated rabies vaccine was known to cause disease in some species of wild animals and was also known to revert to the virulent state.

The vaccinia vector thus shows much promise. Perhaps as many as a dozen or more genes for foreign proteins can be inserted into its genome,

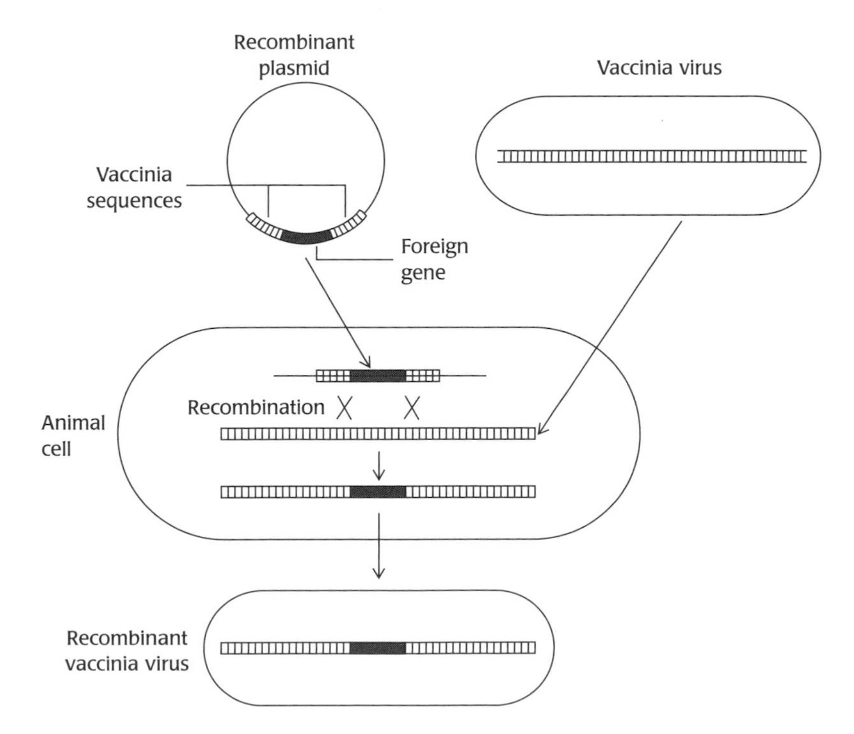


**FIGURE 5.15**

Cloning of foreign DNA into vaccinia virus DNA. The piece of foreign DNA is first cloned between vaccinia sequences in a plasmid. The plasmid is then introduced into cultured animal cells that are simultaneously infected with vaccinia virus. Homologous recombination between viral DNA and the vaccinia sequences in the plasmid leads to the creation of a hybrid vaccinia DNA, which when encapsulated becomes a recombinant vaccinia virus particle.

Redrawn based on artwork from the first edition (1995), published by W.H. Freeman.

suggesting the possibility (as yet theoretical) of producing a vaccine that in a single administration would provide immunity against many different diseases. Although the unmodified vaccinia virus is known to produce severe side effects on occasion, a strain (Ankara) with many extensive mutations and deletions is available and may be presumed to be a safe vector.

### Bacterial Vectors

Until recently, the only efficient protection against bacterial pathogens that cause infections of the gastrointestinal tract has been to generate a localized mucosal immunity by oral administration of attenuated bacteria. Killed vaccines have not been particularly effective. For example, parenteral administration of killed *Salmonella* vaccines generates only moderate immunity against typhoid fever, and the toxicity of the endotoxin (LPS) very often causes significant side effects. Now, however, oral vaccination with live, attenuated strains is being developed and appears to be more effective and to cause less severe side effects in both animals and people.

Several types of *Salmonella* mutants have been tested for this purpose. One class lacks enzymes for the synthesis of aromatic compounds, including *p*-aminobenzoic acid. This mutation prevents multiplication of the bacteria in animal tissues, because salmonella, like other bacteria, has to synthesize an essential cofactor, folic acid, from *p*-aminobenzoic acid and cannot utilize the prefabricated folic acid found in animals. Another class has deletions in the genes for adenylcyclase and cyclic adenosine monophosphate (cAMP)-binding protein; these deletions make the mutant avirulent, presumably because cAMP-dependent regulation controls the biosynthesis of various proteins the bacteria need, especially under conditions of starvation and stress. Another class lacks the enzyme for galactose synthesis, uridine diphosphate (UDP)-galactose 4-epimerase (*galE*). Because galactose is a major constituent of the LPS of many salmonella serotypes, including *Salmonella typhi* and *Salmonella typhimurium*, these mutants cannot synthesize the complete LPS required for virulence. However, they can utilize the small amounts of galactose available in the host tissues to build small numbers of complete LPS molecules, a feature that is thought to make them just capable of proliferating slowly in the host and providing a very effective immunity. A *galE* mutant strain of *S. typhi*, Ty21a, was made by chemical mutagenesis and has been studied extensively in field tests. It appears to be a safe vaccine without side effects. In the first field trial in Egypt, it was reported to be very effective, but a subsequent trial in Chile yielded less convincing results.

Some problems still complicate the use of these live vaccine strains. For example, a double mutant of *S. typhi*, lacking the ability to synthesize *p*-aminobenzoic acid as well as purines, was very safe to use but rather weak in eliciting antibody response, presumably because the nutritional deficiencies were too effective at blocking bacterial growth. On the other hand, when *S. typhi* strains with only the *galE* gene defect were constructed using recombinant DNA methods, they remained highly virulent in humans, indicating that the lack of virulence of Ty21a was the result of unknown mutations most likely introduced in the course of heavy chemical mutagenesis. This discovery leaves Ty21a open to the various criticisms that were marshaled against the traditional live vaccines.

As yet, no attenuated vaccine strains have been produced that are suitable for all applications, in spite of their effectiveness at stimulating local mucosal immunity. Current efforts are aimed at using these strains as vectors – that is, adding protective antigens of other pathogens to them. These include the antigens of *Shigella*, a relative of *E. coli*, that cause diarrhea, and even those of streptococci that are implicated in the generation of dental caries.

## FRAGMENTS OF ANTIGEN SUBUNIT USED AS SYNTHETIC PEPTIDE VACCINES

Subunit vaccines use only one or a small number of macromolecular components from the pathogenic organism. Because only small parts (the epitopes) of these macromolecules are needed for binding to the antibody or to the T cell receptor, this approach may be extended even further. In many cases, researchers are eliciting immunization responses with nothing more than the small peptide corresponding to the epitope. The peptide is first attached to a macromolecular "carrier" protein and then administered to animals.

A *peptide vaccine* has several advantages. The most prominent is that peptides can be made by chemical synthesis and thus do not require the purification steps necessary for the production of recombinant DNA-based subunit vaccines. Such purification is often difficult and expensive. Consequently, peptide vaccines tend to be less expensive, purer, and more stable than protein-containing subunit vaccines.

### IDENTIFYING THE EPITOPE

The first step in producing a peptide vaccine is to identify the antibody-binding epitopes on the surface of the antigenic protein. In itself, the identification of an epitope is not so difficult; the difficulty is in selecting the *correct* epitope that, when used as a vaccine, will protect the vaccinated human or animal against attack by the pathogenic organism.

The antigenic proteins of some virus species vary so much from strain to strain that infection by one does not necessarily produce immunity against the others. The influenza virus is a notorious example of this phenomenon. The proteins on its surface undergo such rapid variation that people infected by the virus one year have little immunity against the next year's epidemic. The coat proteins of the foot-and-mouth disease virus (FMDV), an animal pathogen, show similar variation. Fortunately, the natural variation itself often provides a clue to the location of epitopes in such cases: analyses of nucleic acid sequences usually pinpoint several regions of high variability on the antigenic protein molecule. Studies have shown that these regions produce variants differentiated by the body's immune response, an indication that they react with antibodies and are able to stimulate the proliferation of a particular line of B cells. In other words, these regions are *immunogenic epitopes*.

Another strategy for identifying epitopes uses *in vitro* creation of a special kind of antibody. *In vivo*, antibody diversity results from the random joining of many genes coding for different segments of the antibody polypeptide chains. Thus, before an animal is ever exposed to an antigen, its body already contains more than a million different kinds of B cells, many of which produce antibodies that can bind to multiple different epitopes of any particular antigen with different degrees of affinity. Any B cell will be stimulated to mature and divide when an antigen is introduced that binds to it. Thus, the antibodies produced in response to one kind of antigen in an immunized human or animal are actually a heterogeneous mixture, coming from many independent clones of antibody-producing cells; that is, the antibodies are "polyclonal." When polyclonal antibodies encounter an antigen, different ones will bind to different parts of the antigen molecule, making the identification of epitopes arduous and complex.

In the laboratory, however, individual B cell clones can be "immortalized" by fusion with a tumor cell line (so that those particular clones can be cultured indefinitely). From each clone can then be isolated a homogeneous population of antibodies, called *monoclonal antibodies*, that bind with uniform affinity to only one epitope. Thus, each monoclonal antibody may be used to identify the molecular structure of a site (epitope) that the antibody recognizes. When antibodies are generated using proteins as immunogens, a large fraction of those antibodies (usually 50% or more) bind well only when the proteins are intact and properly folded. These are antibodies whose corresponding epitopes are three-dimensional sites formed by the juxtaposition of regions that are widely separated from each other in the protein's primary structure. Such sites are often called *assembled topographic sites* or *discontinuous epitopes*. In contrast, some monoclonal antibodies – those that define *continuous epitopes* – bind well to certain continuous fragments of the protein that was used in the immunization. When a continuous epitope is identified on a pathogen, a synthetic peptide can be made to correspond to it. The hope is that when humans or animals are immunized with the synthetic peptide, they will generate antibodies that bind to that epitope and protect them against the pathogen.

## PREDICTING EPITOPES FROM PRIMARY STRUCTURE

Predictive strategies have been developed to make the search for continuous epitopes more efficient. Some pinpoint promising regions of primary sequence as presumptive epitopes, and others predict a peptide's immunogenicity.

It is often possible to recognize, on the basis of the primary structure alone, *short sequences* that act as good epitopes and are also immunogenic. Because epitopes must be on the surface of the protein in order to combine with the antigen-binding sites of the antibody, they must occur in exposed, hydrophilic regions of the protein. Hence, *hydrophilicity plots*, which identify such regions along the amino acid sequence, are popular tools for prediction.

Although hydrophilic and exposed segments tend to be good epitopes, experience has shown that many of them are not very immunogenic when administered as short peptides. A plausible reason for this is rapid fluctuation in the peptides' three-dimensional conformations. Short peptides usually do not assume stable conformations in water, whereas the corresponding short segments within the parent protein *are* likely to exist in a distinct conformation, stabilized by interaction with other parts of the protein. Thus, most of the antibodies generated in response to the short peptides do not bind to the corresponding segment of the protein. Researchers surmise that to be effective, a continuous epitope segment must have high mobility (flexibility) in the protein so that it can fit into the antigen-binding site of the antibody. Indeed, when the immunogenicity of peptides corresponding to various parts of the TMV protein was tested, highly immunogenic regions were shown by X-ray crystallography to correspond to flexible parts of the protein.

The correlation between immunogenicity and segmental mobility breaks down as the peptides get longer (14 to 20 residues). Surprisingly, longer peptides that are more immunogenic and produce antibodies with stronger affinity appear to correspond to protein regions with a stable secondary structure. It is likely that a significant fraction of these peptides assume definite structures in water. A case of successful production of an experimental vaccine using a peptide of this type is described below.

## FMDV VACCINE: AN EXPERIMENTAL PEPTIDE VACCINE

In economic terms, foot-and-mouth disease is the most important disease of farm animals. It afflicts cattle, sheep, goat, and swine populations around the world except in North America and Australia. Fear of the spread of FMDV is one of the major reasons why the U. S. Department of Agriculture forbids the importation of uncooked meat products from various parts of the world.

Currently, veterinarians use a traditional vaccine containing killed virus particles to inoculate against the disease. However, this vaccine becomes inactivated rapidly if it is not kept at refrigerator temperature. Presumably, this is why vaccination has not reduced the incidence of the disease drastically, except in Western Europe. (In Europe, many countries have stopped vaccination because they have been disease-free for some years; an explosive outbreak that occurred in England in 2001 thanks to this decision is fresh in our memory.) Furthermore, killed or inactivated vaccines may be dangerous, as we noted earlier, because some of the viruses may survive the inactivation process. This indeed occurred with one batch of the FMDV vaccine, causing an outbreak of the disease in Western Europe in 1981.

FMDV produces four major capsid proteins. Because the blood sera of animals who have survived foot-and-mouth disease contain antibodies that react with the capsid protein VP1, the first attempt to produce a subunit vaccine focused on producing VP1 with recombinant DNA techniques. The researchers learned that although VP1 can be overproduced in *E. coli*, it cannot, unfortunately, be recovered in its native conformation.

Subsequent efforts were directed at developing potential peptide vaccines. Using the approach described above, researchers discovered several regions of high variability in the VP1 protein and identified them as potential epitopes. When peptide 141–160 was synthesized, coupled to a large carrier protein, and used to vaccinate guinea pigs, high titers of antibodies were produced that successfully inactivated the virus particles and conferred very effective immunity against the disease. It should be noted that this peptide is fairly long and comes from a region that is likely to assume a stable secondary structure.

Unfortunately, there are still obstacles to overcome before the FMDV peptide vaccine is ready for use in the field. First, although the vaccine works extremely well in guinea pigs, it is only marginally effective in cattle. The different responses in different animal species were originally thought to be the result of a problem of recognition by T cells (this is discussed below). Second, in order to be of practical value, the vaccine will have to be made more powerful. Perhaps this can be achieved by arranging the antigenic molecules in closely spaced arrays or by improving the adjuvant. In one study, cloning of the DNA sequence that codes for the VP1 sequence 140–161 into the gene for hepatitis B virus core antigen dramatically improved the vaccine's potency, presumably because hepatitis core antigen, like hepatitis surface antigen, self-assembles into a particulate structure.

Perhaps the most serious problem – and one that is likely to plague other peptide vaccines as well – is the danger that mutant viruses will thrive in the vaccinated host. The antibodies generated in response to peptide vaccines are all directed at a single epitope and may not bind well to viruses with altered sequences in that region of the protein.

## RECRUITING THE ASSISTANCE OF T HELPER CELLS

As described earlier, antibody production requires the presence within the antigen of both B cell and T cell epitopes (see Figure 5.11). Most large antigenic proteins contain both epitopes. However, when a peptide vaccine has been constructed by identification and cloning of the B cell epitope only, there is no guarantee that the peptide will also contain a T cell epitope. It is true that the peptide is usually conjugated to a large carrier protein that supplies T cell epitopes, but the response to T cell epitopes varies greatly from one individual to another. Possibly because the T cell receptor also has to recognize a specific histocompatibility antigen (see Figure 5.11), the range of antigens it is able to recognize is rather limited. That is, T cell receptors in any given individual, with a given range of histocompatibility antigens, can recognize only a subset of T cell epitopes. This is not a problem when the entire pathogenic organism is being used in vaccination, because in any pathogen there are many proteins, each containing at least several T cell epitopes, and any individual host will respond well to at least one of them. However, when a single peptide is used in immunization, the immune response depends greatly on the makeup of the host's histocompatibility antigen, and a given individual may not respond significantly to the T cell epitope in the vaccine. A similar situation may explain the often pronounced differences with which various animal species react to a given peptide vaccine. (Recall that the experimental FMDV vaccine elicited an excellent response in guinea pigs but not in cattle.)

These considerations led to the design of an artificial "promiscuous" T cell epitope that functions well in different animals. When the FMDV VP1 B cell epitope was fused to this synthetic peptide, the vaccine was very effective in swine. However, another trial with cattle under more severe challenge conditions showed a disappointing result.

### PEPTIDES FOR GENERATING CELLULAR IMMUNITY

Although many problems are encountered in using peptides to stimulate B cells, peptides have nevertheless proved to be excellent instruments for generating cellular immunity through the selection of T cell clones. This is mainly because in T cell selection, the antigen "presented" to the T cells has already been processed (i.e., proteolytically cleaved); therefore, the relevant structures are short peptide strings, and most T cell epitopes are continuous ones (unlike the numerous B cell epitopes that are of the assembled type).

## DNA VACCINES

In a 1990 experiment, the injection of 100 $\mu$g of naked plasmid DNA into mice resulted in the detectable expression of foreign genes cloned downstream from a strong promoter active in animal cells. This research, which suggested much potential for gene therapy, was followed in a few years by reports that similar injection of plasmid DNA including genes coding for antigens led to the production of immunity in mice. Because DNA vaccines are easy to modify in the laboratory, these findings seemed to promise that even small laboratories and small companies could produce effective new vaccines. Most steps necessary for the production of recombinant protein vaccines, such as ensuring the good expression and correct folding of the protein, protein purification, and removal of the potentially toxic contaminants, would not be necessary with DNA vaccines, because the antigen gene could be expressed within the APCs. Furthermore, the expression of the antigenic protein within the cytoplasm of APCs meant that the immune response would be tilted toward the production of cytotoxic CD8 T cells, a result that is difficult to achieve with conventional vaccines.

Since then, an enormous amount of research has gone into the development of DNA vaccines that are effective in humans. Limited success has been achieved, indicated by increases in antibody titer or T cell proliferation, but the general consensus seems to be that DNA vaccines are usually not potent enough for use in humans. Current efforts are focused on increasing the potency by improving the method of introduction of DNA ("electroporation," use of adjuvants, etc.) and by combining the initial administration of DNA vaccine with subsequent "booster" shots containing recombinant viral constructs (see "Viral Vectors" above). Even when followed by such boosters, however, the antimalarial DNA vaccine was found to have no effect in the recent clinical trial (see "Malaria" below).

Why were DNA vaccines so effective in mice but so disappointing in humans? A major reason appears to be the dosage. The initial 1990 gene expression study in mice used 100 $\mu$g/mouse and showed that the expression level of the foreign gene goes down tenfold if only 10 $\mu$g DNA is injected. Thus 100 $\mu$g DNA/mouse appears to be the minimum dosage required, and subsequent DNA vaccine studies in mice all used that amount or even more per animal. On the basis of human body weight, which is nearly 10,000-fold higher than that of a laboratory mouse, this dosage would translate roughly into 1 g DNA/person. It is impossible to inject this much DNA into humans, and besides, manufacturing such a large amount would be quite expensive. Thus human trials have so far used between 1 and 3 mg DNA/person, a dosage that is obviously far from adequate. Novel approaches seem necessary if effective DNA vaccines are ever to be available for human use.

In addition to the issue of dosage, there is a very low yet theoretically possible chance that the foreign plasmid DNA may become incorporated into the human chromosome. Experiments using cultured animal cells showed that such an outcome should be extremely rare. Nevertheless, any successful vaccine will be administered to millions, even billions, of people, and thus even the unlikely adverse effects must be taken seriously.

## VACCINES IN DEVELOPMENT

Some pathogens have developed extraordinarily "clever" weapons against the immune defenses of the host, adding greatly to the challenge of producing effective vaccines. Even greater difficulties are encountered in the attempt to produce vaccines against nontraditional targets, such as cancer. We now present a brief example or two from each of these developing areas.

### VACCINES FOR "HIT-AND-STAY" VIRUSES

Viruses causing chronic infections, such as the AIDS virus, present a particular set of problems for vaccine development. For example, whereas viruses that cause acute infections usually generate rapid immune responses, including the generation of antibodies and activated CD8 T cells, which contribute to the recovery of the infected patients, the viruses that cause chronic infections (e.g., HIV and hepatitis C virus) rarely produce strong immune responses in the course of disease and are thought to have mechanisms that enable them to evade the immune system. Nevertheless, passive immunization of macaques with monoclonal antibodies against simian immunodeficiency virus, a relative of HIV, showed that the animals were strongly protected against infection. Thus, if vaccines could be developed that would mount a strong immune response, they might be able to provide protection. Much effort is being expended to develop such prophylactic vaccines. With HIV, the major envelope protein has been tested as subunit vaccines and peptide vaccines, and its gene has been cloned into live viral vectors, such as vaccinia virus. Although these trials sometimes generated neutralizing antibodies or cytotoxic T cells, they were not protective against viruses in the population because of the tremendous variation in antigens produced by the RNA genome of the virus. A similar problem plagues the effort to develop vaccines for hepatitis C, which is also an RNA virus (in contrast to hepatitis B virus, which is a double-stranded DNA virus).

## MALARIA

Malaria is a major problem in the tropical and subtropical parts of the world. Although its worldwide impact is notoriously difficult to determine, WHO estimated that 1.3 million deaths occurred from malaria in 2002 (Table 5.2). Furthermore, species of *Plasmodium*, the causative organism, are becoming increasingly resistant to the most effective therapeutic agent, chloroquine. It is thus most important to develop effective vaccines against this disease. Given the difficulty and high cost of growing malaria parasites, subunit vaccines produced by recombinant DNA technology would seem to be a preferable approach to prevention. Many laboratories around the world have been working intensively to produce such vaccines, but the results are still far from satisfactory.

Clearly, one of the major problems is the complexity of *Plasmodium*. Its life cycle includes at least three completely separate stages within its vertebrate host (Figure 5.16), and at each stage it has a completely distinct antigenic makeup. As is well known, malaria infection begins with the bite of an *Anopheles* mosquito. The organism that enters the bloodstream through this route is in the *sporozoite* stage. Sporozoites enter liver cells and multiply there. After one to two weeks, the organisms are released into the bloodstream again, this time as *merozoites*. Many of the symptoms of malaria are the result of merozoites infecting red blood cells, multiplying, and being released after two or three days to reinfect fresh red blood cells (see Figure 5.16). A small fraction of merozoites eventually develop into sexual forms called *gametocytes*. When these enter the gut of a mosquito (i.e., when a mosquito bites and ingests the blood of a malaria-infected host), they develop and mate, producing *ookinetes* that eventually produce sporozoites to begin the cycle again. Vaccine development has targeted each of the stages that occur in the human host.

A vaccine for preventing infection by malaria parasites must produce an immunity directed against sporozoites. Thus, most of the efforts so far have used sporozoite antigens. Following the standard approach to producing subunit vaccines, scientists immunized an experimental host with killed whole sporozoites and then identified the antigen to which most of the resulting antibodies were directed. In this manner, the *circumsporozoite* (CS) protein, a protein component on the surface of the sporozoite, was identified

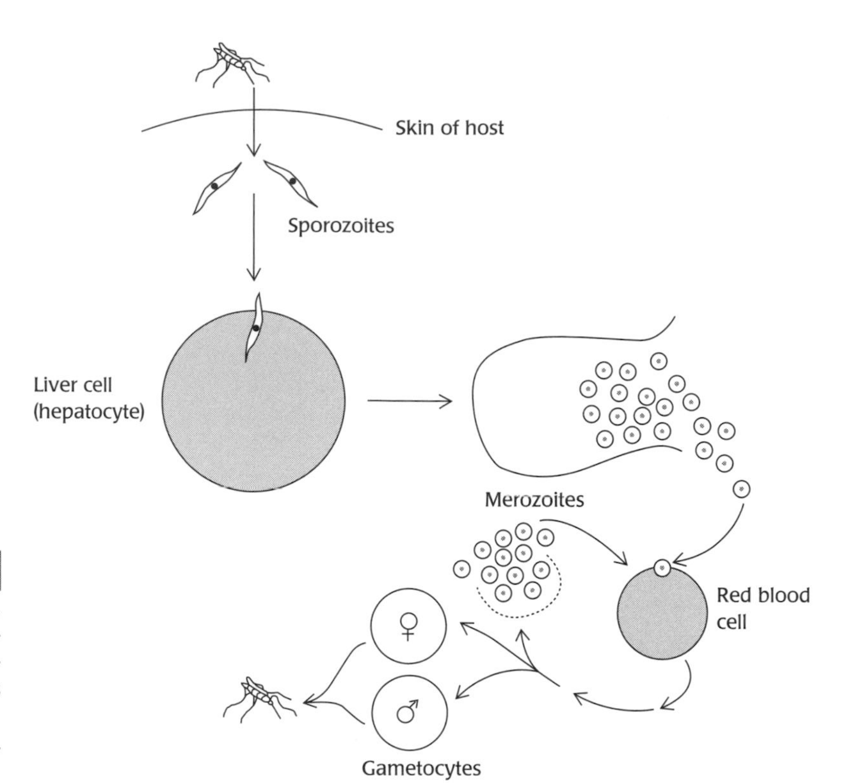


**FIGURE 5.16**

A simplified life cycle of the malaria parasite. In addition to the three stages shown (sporozoite, merozoite, and gametocyte), the parasite multiplying within the red blood cell is often said to be in the *trophozoite* stage.

Redrawn based on artwork from the first edition (1995), published by W.H. Freeman.

as the predominant antigen. The central part of this protein consists of many repetitions (37 times in one strain and 43 times in another strain of *Plasmodium falciparum*) of the tetrapeptide Asn-Ala-Asn-Pro, with a few repetitions of a similar tetrapeptide, Asn-Val-Asp-Pro (Figure 5.17). Because a high fraction of the antibodies produced upon immunization with the killed sporozoites are directed against this region, the first subunit vaccines consisted of a synthetic peptide containing these repeats conjugated to a protein carrier. The human subjects who received these vaccines developed good antibody titers, but only a small fraction of this group was protected against the disease.

Later studies showed that the repetitive region did not contain any T cell epitopes. As we noted earlier, for an effective immune response to occur, the antigen must contain both B cell and T cell epitopes. The predominant epitope recognized by $T_H$ cells is located in the nonrepeating, C-terminal region of CS protein. When subunit vaccines that presumably contained this T cell epitope were made by expressing the recombinant DNA in yeast, the immune response in experimental animals was superior.

Currently, the candidate malaria vaccine that is in the most developed stage is a vaccine called RTS,S/AS02A. To produce this vaccine, a polypeptide corresponding to residues 207 to 395 of the CS protein (containing both the repetitive domain and the T cell epitope domain) is fused to the HBsAg. The fusion construct is then expressed along with the unmodified HBsAg

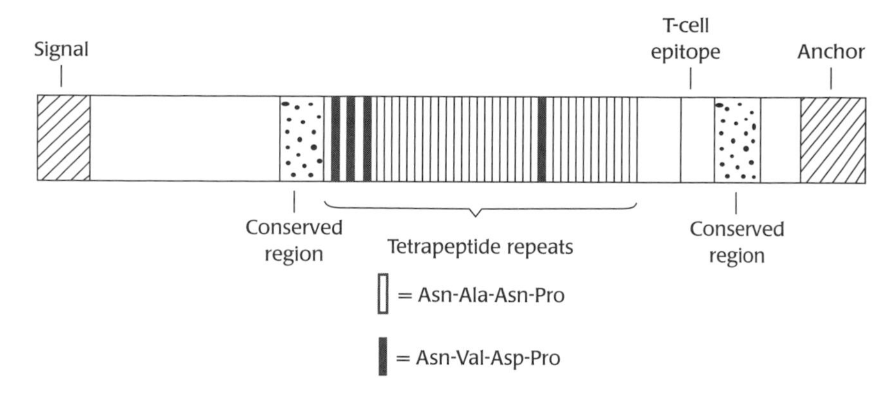


**FIGURE 5.17**

Schematic structure of CS antigen from a strain of *Plasmodium falciparum*. The N-terminus of the protein (at the *left* in this drawing) begins with a signal sequence for export, and the C-terminus (at the *right*) ends with a hydrophobic sequence that presumably anchors the protein at the surface of the cell membrane. The middle portion of the protein consists of tetrapeptide repeats (37 repeats of Asn-Ala-Asn-Pro and four repeats of Asn-Val-Asp-Pro). The T cell epitope is located on the C-terminal side of the repeat region. [Based on Kemp, D. J., Coppel, R. L., and Anders, R. F. (1987). Repetitive proteins and genes of malaria. ***Annual Review of Microbiology***, 41, 181–208.

so that the malaria antigen is presented as a surface array on the vesicles composed of the HBsAg. To boost the immune response further, this preparation is presented in an oil-in-water emulsion containing an adjuvant, monophosphoryl lipid A, mentioned earlier in this chapter. Following a field trial in Gambia that showed partial protection in adults, a large-scale double-blind trial in young children in Mozambique showed that the prevalence for *P. falciparum* infection was lowered by 37% and the prevalence for severe malaria by 58%. This is not total protection, but it certainly shows that vaccination is feasible for malaria.

A major problem for such a recombinant polypeptide vaccine is likely to be its high cost. In this respect, synthetic peptide or DNA vaccines would be much more attractive. However, a recent field trial of a DNA vaccine with a "booster" scheme using a recombinant vaccinia virus (see "DNA Vaccines" above) showed almost no statistically significant protection.

Alternatively, some research has targeted the merozoite stage, in which the most extensive proliferation of malarial parasites occurs. Among natural populations of humans, this proliferation is inhibited in individuals who are heterozygous for a mutant form of hemoglobin, sickle cell hemoglobin. (Its association with malaria resistance is assumed to be the reason why this mutant allele is so common – occurring in as much as 40% of the population in some parts of Africa – even though it causes a severe disease, sickle cell anemia, in people who are homozygous for it.) If a vaccine could produce significant levels of antibody directed at merozoites, the merozoites could be attacked during their transition from one red blood cell to the other, and the symptoms of malaria might be alleviated. (Antibodies cannot cross the cell membrane, so they cannot attack parasites multiplying within the red blood cells.) Scientists in Bogota, Colombia, synthesized a 45-residue peptide containing potential epitopes from three merozoite protein antigens and then polymerized this peptide into a macromolecule by means of disulfide bonds. Unfortunately, this vaccine, which appeared quite promising in the initial experiment, was not effective in a rigorous field trial in Africa.

Finally, there is the problem of frequent antigenic variation, an example of which is the variability of the aforementioned predominant T cell epitope of the CS antigen. Some vaccines may show good efficacy in laboratory experiments that use known challenge strains propagated in the laboratory but may produce disappointing results in the field, where antigen variation occurs rapidly.

## THERAPEUTIC VACCINES FOR AUTOIMMUNE DISEASES AND CANCER

The newest frontier of vaccine science is the effort to produce therapeutic vaccines for noninfectious diseases. Many human diseases – for example, type 1 diabetes, multiple sclerosis, and several forms of arthritis – are thought to be autoimmune diseases, in which patients produce antibodies or cytotoxic T cells targeted against the components of their own body. Because immune responses in mammals can take various different forms, it is thought that a proper immunization with antigens could divert a dangerous immune response into one that is less harmful. In most cases, this would involve a shift from a $T_H1$ response to a $T_H2$ response.

Multiple sclerosis in humans is a very complex disease, but the animal model is known to involve an autoimmune reaction to myelin basic protein, a component of myelin, the multilayered membranous sheath surrounding the neuron. A random copolymer of glutamic acid, lysine, alanine, and tyrosine in a predefined ratio, called glatrimer, has been in clinical use to retard the progress of multiple sclerosis and is thought to regulate the direction taken by the immune response by mimicking the epitopes in myelin basic protein. Intense research is ongoing to develop similar therapeutic vaccination for type 1 diabetes and other diseases, and in animal models the shift from $T_H1$ to $T_H2$ response has been achieved in several cases. However, $T_H2$ response also generates IgE antibody, which is involved in allergic reactions; therefore, repeated administration of antigens could generate a sudden, severe, and perhaps fatal reaction ("anaphylactic reaction"), which could be a serious problem in human therapy.

Cancer cells often produce characteristic new antigen molecules on their surface. Targeting therapy at these molecules should produce far better results than the usual chemotherapy that targets both cancer cells and normal cells. In fact, the utility of immunological approaches has already been demonstrated by the now widespread use of monoclonal antibodies – such as tratuzumab (Herceptin), used for certain types of breast cancer – directed against the molecules on cancer cell surfaces. Although immune mechanisms against such tumor-specific antigens are usually suppressed in cancer patients, artificial stimulation may lead to the production of significant immune responses. Extensive research is under way in the attempt to produce therapeutic vaccines for cancer and has already yielded solid results in the area of melanoma therapy. Extracts of human melanoma cells injected in combination with powerful adjuvants (composed of fragments of mycobacterial cell wall and monophosphoryl lipid A) have been studied since 1988. A recent study has shown that the response to this therapeutic vaccine varies depending on the MHC (or more correctly HLA) types the patients expressed. In patients expressing HLA types 2 or C3, the vaccine therapy was clearly effective, resulting in a five-year, relapse-free survival rate of 83%, in contrast to the rate of 59% seen in the control group. These results give us hope that immune therapy will prove to be an effective treatment for (and possibly prophylaxis against) malignant tumors.

## SUMMARY

Traditional vaccines consist of either live organisms with attenuated virulence, killed organisms, or inactivated toxins. Usually, only minor genetic alterations distinguish virulent organisms from the attenuated vaccine strains, which therefore have the potential to revert to the virulent state. With killed or inactivated vaccines, there is always a danger that their inactivation might have been incomplete. Furthermore, they often produce undesirable side effects because they are likely to contain extraneous toxic components that are not needed to produce the protective immunity.

These and other shortcomings of traditional vaccines can be avoided by the use of subunit vaccines, which contain only the immunity-conferring "protective antigen," most often a single protein component of the pathogenic organism. Such an antigen may be produced from the pathogens by the traditional technology, as with acellular pertussis vaccine and conjugate polysaccharide vaccines of *H. influenzae* and pneumococci. It may also be produced safely and inexpensively by introducing, into harmless organisms such as *E. coli* or yeast, recombinant DNA molecules containing the appropriate gene. Recombinant hepatitis B vaccine, composed of the major capsid protein of the virus, is a successful commercial product of the latter class and is now used widely.

The subunit vaccines, however, are not always as effective as the traditional vaccines because the vertebrate immune system is optimized to recognize features of the whole pathogen, such as the presence of multiple copies of the same antigen on its surface and the presence of certain components, such as LPS or peptidoglycan, that occur commonly in foreign, invading microorganisms but not in host cells. Therefore, to obtain a better immune response with a subunit vaccine may require an administration strategy that causes the vaccine to mimic the appearance of antigen in the whole pathogen or that combines the vaccine with chemicals that act as adjuvants. In some cases, attenuated live pathogens, which can elicit a natural immune response, are used as an effective vector to carry the vaccine subunit. (Thus, vaccinia virus carrying the gene for the capsid glycoprotein of rabies virus is now successfully used to immunize wild animals against rabies). It is especially difficult to develop subunit vaccines for pathogens that go through complex life cycles (e.g., malaria parasites) or evade the normal immune response (e.g., viruses causing chronic infections). Because subunit vaccines present antigens to animals and humans in an artificial context, the production of effective subunit vaccines requires much more in-depth knowledge of the microbes and of the functions of immune cells than does the production of traditional vaccines. Researchers were unable to produce a successful malaria vaccine, for example, until their understanding of immunology led them to combine the antigen with very strong adjuvants.

Because the parts of the antigenic protein that are recognized by antibodies are very small, it is also possible to generate immunity by injecting small peptides (usually conjugated to a carrier protein) instead of the whole subunit or antigen. Peptide vaccines are attractive because pure, stable preparations can be produced in large quantities by chemical synthesis. A more ambitious strategy is the creation of DNA vaccines. These are plasmids that contain the gene for the protective antigen behind a promoter able to drive an efficient expression in vertebrate cells. Although DNA vaccines were shown to be effective in mice, human trials have rarely produced significant protection, presumably because in humans, it is difficult to achieve the dosage level that is needed for strong immunity.

Much of vaccine research is currently focused on developing therapeutic vaccines targeted at human diseases other than infection. These include autoimmune diseases and cancer, and some encouraging results have been obtained.

## SELECTED REFERENCES


### General

Plotkin, S. A., and Orenstein, W. A. (eds.). (2004). *Vaccines*, 4th Edition, Philadelphia: W. B. Saunders.

Plotkin, S. A. (2005). Vaccines: past, present, and future. *Nature Medicine*, 11(Suppl.), S5–S11.

### Subunit Vaccines Made with Traditional Technology

Sato, Y., and Sato, H. (1999). Development of acellular pertussis vaccines. *Biologicals*, 27, 61–69.

Decker, M. D., and Edwards, K. M. (2000). Acellular pertussis vaccines. *Pediatric Clinics of North America*, 47, 309–335.

Robbins, J. B., Schneerson, R., Trollfors, B., Sato, H., Sato, Y., Rappuoli, R., and Keith, J. M. (2005). The diphtheria and pertussis components of diphtheria-tetanus-toxoids-pertussis vaccine should be genetically inactivated mutant toxins. *Journal of Infectious Diseases*, 191, 81–88.

Mäkelä, P. H., and Käyhty, H. (2002). Evolution of conjugate vaccines. *Expert Review of Vaccines*, 1, 399–410.

### Subunit Vaccines Made with Recombinant DNA Technology

Valenzuela, P., Medina, A., Rutter, W. J., Ammerer, G., and Hall, B. D. (1982). Synthesis and assembly of hepatitis B virus surface antigen particles in yeast. *Nature*, 298, 347–350.

Michel, M.-L., Sobczak, E., Malpièce, Y., Tiollais, P., and Streeck, R. E. (1985). Expression of amplified hepatitis B virus surface antigen genes in Chinese hamster ovary cells. *Bio/Technology*, 3, 561–566.

Shouval, D. (2003). Hepatitis B vaccines. *Journal of Hepatology*, 39, S70–S76.

Roden, R., and Wu, T.C. (2006). How will HPV vaccines affect cervical cancer? *Nature Reviews Cancer*, 6, 753–763.

### Techniques for Delivery of Vaccines

Abbas, A. K., Murphy, K. M., and Sher, A. (1996). Functional diversity of helper T lymphocytes. *Nature*, 383, 787–793.

Petrovsky, N., and Aguilar, J. C. (2004). Vaccine adjuvants: current state and future trends. *Immunology and Cell Biology*, 82, 488–496.
Stills, H. F. (2005). Adjuvants and antibody production: dispelling the myths associated with Freund's complete and other adjuvants. *ILAR Journal*, 46, 280–293.
Persing, D. H. (2002). Taking toll: lipid A mimetics as adjuvants and immunomodulators. *Trends in Microbiology*, 10(Suppl.), S32–S37.
Krieg, A. M. (2006). Therapeutic potential of Toll-like receptor 9 activation. *Nature Reviews Drug Discovery*, 5, 471–484.

**Live, Attenuated Vectors**

Brochier, B., et al. (1991). Large-scale eradication of rabies using recombinant vaccinia-rabies vaccine. *Nature*, 354, 520–522.
Antoine, G., Sceiflinger, F., Dorner, F., and Falkner, F. G. (1998). The complete genomic sequence of the modified vaccinia Ankara strain: comparison with other orthopoxviruses. *Virology*, 244, 365–396.
Drexler, I., Staib, C., and Sutter, G. (2004). Modified vaccinia virus Ankara as antigen delivery system: how can we best use its potential? *Current Opinion in Biotechnology*, 15, 506–512.
Kochi, S. K., Killeen, K. P., and Ryan, U. S. (2003). Advances in the development of bacterial vector technology. *Expert Review of Vaccines*, 2, 31–43.

**Synthetic Peptide Vaccines**

Sobrino, F., et al. (2001). Foot-and-mouth disease virus: a long known virus, but a current threat. *Veterinary Research*, 32, 1–30.
Rodriguez, L. L., Barrera, J., Kramer, E., Lubroth, J., Brown, F., and Golde, W. T. (2003). A synthetic peptide containing the consensus sequence of the G-H loop region of foot-and-mouth disease virus type-O VP1 and a promiscuous T-helper epitope induces peptide-specific antibodies but fails to protect cattle against viral challenge. *Vaccine*, 21, 3751–3756.
Celada, F., and Sercarz, E. E. (1988). Preferential pairing of T-B specificities in the same antigen: The concept of directional help. *Vaccine*, 6, 94–98.
Davis, M. M., and Bjorkman, P. J. (1988). T-cell antigen receptor genes and T-cell recognition. *Nature*, 334:395–402.

**Mechanism of Immune Response**

Goldsby, R. A., Kindt, T. J., Osborne, B. A., and Kuby, J. (2003). *Immunology*, 5th Edition, New York: W. H. Freeman.
Janeway, C. A., Jr., Travers, P., Walport, M., and Shlomchik, M. J. (2005). *Immunobiology*, 5th Edition, New York: Garland Publishing.
Takeda, K., Kaisho, T., and Akira, S. (2003). Toll-like receptors. *Annual Review of Immunology*, 21, 335–376.
Netea, M. G., van der Meer, J. W. M., Sutmuller, R. P., Adema, G. J., and Kullberg, B.-J. (2005). From the Th1/Th2 paradigm towards a Toll-like receptor/T-helper bias. *Antimicrobial Agents and Chemotherapy*, 49, 3991–3996.
Holmgren, J., and Czerkinsky, C. (2005). Mucosal immunity and vaccines. *Nature Medicine*, 11, S45–S53.

**DNA Vaccines**

Wolff, J. A., et al. (1990). Direct gene transfer into mouse muscle in vivo. *Science*, 247, 1465–1468.
Donnelly, J., Wahren, B., and Liu, M. A. (2005). DNA vaccines: progress and challenges. *Journal of Immunology*, 175, 633–639.
Moorthy, V. S., et al. (2004). A randomised, double-blind, controlled vaccine efficacy trial of DNA/MVA ME-TRAP against malaria infection in Gambian adults. *PLoS Medicine*, 1, 128–136.

**Vaccines in Development**

Berzofsky, J. A., Ahlers, J. D., Janik, J., Morris, J., Oh, S.-K., Terabe, M., and Belyakov, I. M. (2004). Progress on new vaccine strategies against chronic viral infections. *Journal of Clinical Investigation*, 114, 450–462.

Good, M. F. (2005). Vaccine-induced immunity to malaria parasites and the need for novel strategies. *Trends in Parasitology*, 21, 29–34.

Balou, W. R., et al. (2004). Update on the clinical development of candidate malaria vaccines. *American Journal of Tropical Medicine and Hygiene*, 71 (Suppl. 2), 239–247.

Alonso, P. L., et al. (2004). Efficacy of the RTS,S/AS02A vaccine against *Plasmodium falciparum* infection and disease in young African children: randomised controlled trial. *Lancet*, 364, 1411–1420.

Malkin, E., Dubovsky, F., and Moree, M. (2006) Progress towards the development of malaria vaccines. *Trends in Parasitology*, 22, 292–295.

Deen, J. L., and Clemens, J. D. (2006). Issues in the design and implementation of vaccine trials in less developed countries. *Nature Reviews Drug Discovery*, 5, 932–940.

Kooij, T. W. A., Janse, C. J., and Waters, A. P. (2006). *Plasmodium* post-genomics: better the bug you know? *Nature Reviews Microbiology*, 4, 344–359.

Hohlfeld, R., and Wekerle, H. (2004). Autoimmune concepts of multiple sclerosis as a basis for selective immunotherapy: from pipe dreams to (therapeutic) pipelines. *Proceedings of the National Academy of Sciences U.S.A.*, 101, 14599–14606.

Finn, O. J. (2003). Cancer vaccines: between the idea and the reality. *Nature Reviews Immunology*, 3, 630–641.

Sondak, V. K., and Sosman, J. A. (2003). Results of clinical trials with an allogeneic melanoma tumor cell lysate vaccine: Melacine. *Seminars in Cancer Biology*, 13, 409–415.

Stevenson, F. K., et al. (2004). DNA vaccines to attack cancer. *Proceedings of the National Academy of Sciences U.S.A.*, 101, 14646–14652.

Banchereau, J., and Palucka, A. K. (2005). Dendritic cells as therapeutic vaccines against cancer. *Nature Reviews Immunology*, 5, 296–306.

Lollini, P. L., Cavallo, F., and Nanni, P., et al. (2006). Vaccines for tumor prevention. *Nature Reviews Cancer*, 6, 204–216.

SIX

# Plant–Microbe Interactions

Humans need food in order to survive, and most of the food in the modern world is the product of agriculture. In 1798, Thomas Robert Malthus published the famous essay in which he argued that the human population increases geometrically yet food production can increase only arithmetically. What he could not predict at that time was the contribution of science to the increased production of food. As Malthus foretold, the world population has increased at an almost alarming rate. It took slightly more than 100 years to double from the 1.25 billion in Malthus's day to 2.5 billion in 1950, but the next doubling, to 5 billion, was achieved in less than 40 years, as seen in Figure 6.1. However, the yield of major food crops per unit area (represented by wheat in Figure 6.1) has increased at an even steeper rate, tripling in slightly more than 40 years. One of the major contributing factors to this increase has been the development of high-yielding varieties of crops, for example, semi-dwarf varieties of wheat and rice, which direct a larger portion of their energy to the production of seeds (grains) than to plant growth; this development, which occurred in the 1960s and 1970s, is often called the "Green Revolution." Thanks to this increase in yield, the world production of food (represented by cereals in Figure 6.1) could more than keep pace with the increase in population, in spite of the steadily decreasing total land area devoted to agricultural production. (The huge problems of malnutrition and hunger seen in the developing world in spite of all this are mostly the consequence of unequal distribution of food.)

Agriculture represents a very large fraction of the global economy, and yet a precise estimate of its monetary value is notoriously difficult to make. Table 6.1 provides an estimate of sorts, but for many reasons (including scarcity of information), the figures are inexact. The values in the table are based on international import prices, but these prices are likely to be different from domestic prices, because the quality of a nation's exported items may be different from that of items earmarked for domestic consumption and also because prices may be affected by government regulations on exports. Nevertheless, even these imprecise estimates demonstrate that agriculture is one of humankind's major economic activities. Any improvement in agriculture

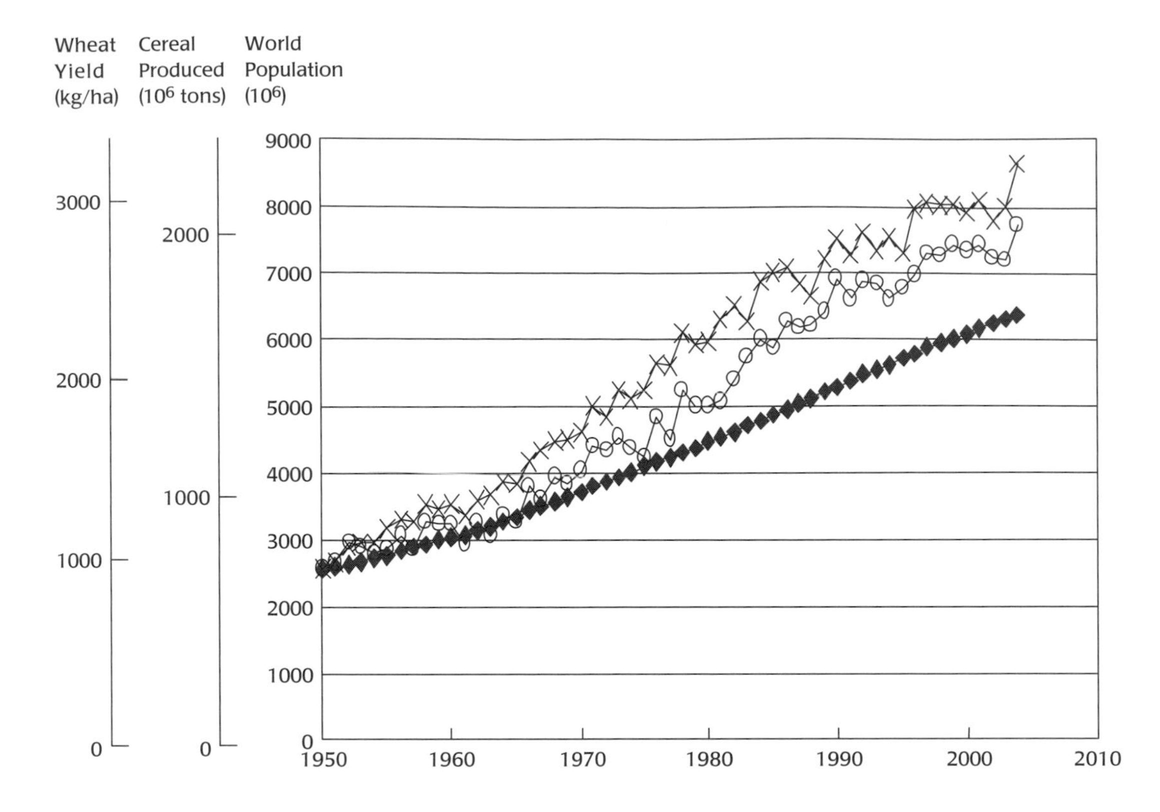


**FIGURE 6.1**

Increases in world population and food production. World population (*diamonds*) continues to increase. However, agricultural production per area (*x's*), in this case wheat) as well as the total cereal production in the world (*o's*) have increased at rates surpassing that of the population. The population estimate is from the U.S. Census Bureau. The production figures are from World Crop and Livestock Statistics 1948 to 1985; FAO (Food and Agriculture Organization of the United Nations) for 1950 to 1960; and from the FAO website (http://www.fao.org) for 1961 to 2003.

can therefore have a major economic impact. In this chapter, we describe the types of improvement that may be possible through the methods of biotechnology.

## USE OF SYMBIONTS

As we shall see, much of the current agricultural research effort is directed at introducing potentially beneficial foreign genes into plant stocks. However, because there are many symbiotic bacteria normally associated with specific organs of various plants, a simpler plan might be to modify such bacteria and then use their interactions with the plants to introduce the modified traits at the appropriate locations. This alternative approach is technically easier, because the engineering of bacterial DNA through recombinant DNA methods is now routine, whereas the manipulation of plant DNA has yet to be perfected.

### PROTECTION OF PLANTS FROM FROST DAMAGE USING ENGINEERED SYMBIOTIC BACTERIA

One of the earliest examples of successful modification of a symbiotic bacterium was performed in *Pseudomonas syringae*, which is found at high concentrations on the leaves of many plants. Many strains of this bacterium produce an ice nucleation protein that is apparently located on the surface of the bacterial cell, and the presence of this protein causes the formation of ice at temperatures only a little below 0°C, inflicting significant frost damage on important crop plants and so facilitating the subsequent invasion of plant tissues by the bacteria.

Steven Lindow and his associates first cloned the ice nucleation gene of *P. syringae* by using cosmid vectors. *Escherichia coli* cells containing the recombinant DNA were screened for nucleation of ice formation at −9°C. Then a deletion was made in the gene by a recombinant DNA method, and the deletion was put back into the *P. syringae* chromosome by transformation followed by a homologous recombination process. The resulting $Ice^-$ strain was identical to the parent $Ice^+$ strain in all other properties, and heavy application of the mutant onto the leaves of strawberries, for example, led to a colonization that competed successfully with the wild-type bacteria, protecting the plants from frost damage. More recently, a wild-type strain of *Pseudomonas fluorescens* that competes well against various ice-nucleating organisms has been commercially produced for protection of plants.

**TABLE 6.1 Estimated worldwide values of key agricultural products (2003)**

| | Production per year ($10^6$ tons) | Unit price ($/kg) | Estimated value ($10^9$ $) |
|---|---|---|---|
| Cereals | 2075 | 0.167 | 346 |
| Root crops[a,b] | 679 | 0.164 | 111 |
| Vegetables[c,d] | 842 | 0.64 | 539 |
| Fruits[e] | 626 | 0.77 | 482 |
| Oil crops[f] | 330 | 0.26 | 89 |
| Sugar (raw) | 146 | 0.25 | 37 |
| Coffee (green) | 7.2 | 1.24 | 8.9 |
| Cocoa beans | 3.3 | 1.93 | 6.4 |
| Tea | 3.2 | 2 | 6.4 |
| Vegetable fibers[g] | 23 | 1.05 | 24 |
| Tobacco | 6.2 | 7.4 | 46 |
| Rubber | 7.4 | 1 | 7.4 |
| Meat | 253 | 2.03 | 514 |
| Milk | 600 | 0.51 | 306 |
| Eggs | 56 | 1.3 | 73 |

*Source*: The production figures are from the FAO (Food and Agriculture Organization of the United Nations) website (http://www.fao.org). The unit prices were obtained from the same website and represent average international import prices.

[a] Potato makes up about half of this category (by weight).

[b] Unit price was calculated by assuming the price of non-potato root crops to be 50% that of potato.

[c] Vegetable production in small fields is not included in many countries' statistics, although it is estimated to correspond to as much as 40% of the total production in some cases. This category also includes melons.

[d] Statistics on many individual categories of vegetables are available in FAO yearbooks, but they still comprise only half of the total vegetable production. Because of the extreme diversity and complexity of this category, our unit price – the weighted average of the unit prices of tomatoes, watermelons, cabbages, onions, cucumbers, eggplants, cantaloupe melons, carrots, chilies, and lettuce (the 10 vegetables produced in largest amounts) – is a great oversimplification.

[e] Apples, bananas, citrus fruits, and grapes account for about two thirds of the total fruit production. The unit price was calculated by averaging the prices of these items.

[f] Soybean comprises the major part of this category.

[g] 75% of this category is cotton; about 20% is jute.

## USE OF NITROGEN-FIXING BACTERIA TO IMPROVE CROP YIELDS

Another focus of current research – one that has a potentially much wider impact – is the process of nitrogen fixation. All animals and plants and most bacteria depend on the availability in their environment of some form of "combined nitrogen" or "reactive nitrogen" – nitrate ($NO_3^-$), ammonia ($NH_3$), or nitrogen-containing organic compounds such as amino acids. The huge amounts of $N_2$ that exist in the atmosphere are unavailable to the biological world except through the process of nitrogen fixation.

Nitrogen has been "fixed," or converted into combined nitrogen in the form of fertilizers (mostly ammonium salts), by industrial processes (e.g., the Haber–Bosch process) since the beginning of the twentieth century. Because

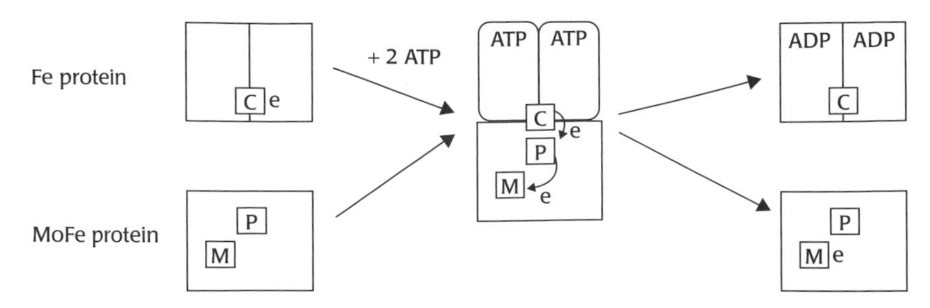


**FIGURE 6.2**

Enzymatic nitrogen fixation. Recent x-ray crystallographic studies of Mo nitrogenase show that the Fe protein, which exists as a homodimer, binds two ATP molecules. This produces a conformational alteration that leads to the binding of the MoFe protein by the Fe protein. The ATP binding also pushes the 4Fe–4S redox center (shown as *C*) in the Fe protein closer to the 8Fe–7S center (shown as *P*) in the MoFe protein in order to facilitate the reduction of the latter. This process is indicated by the movement of one electron (*e*) from the 4Fe–4S center of the Fe protein to the 8Fe–7S center of the MoFe protein. The electron then travels to the Mo–7Fe–9S center (*M*) of the MoFe protein, where the reduction of nitrogen takes place. ATP hydrolysis then brings about the dissociation of the MoFe protein and the Fe protein, and the latter is reduced by ferredoxin or flavodoxin before beginning the next cycle. In the reduction of $N_2$ (and of the accompanying two protons), a minimum of eight such cycles are required. Recent studies with model compounds suggest that the reduction occurs by a stepwise addition of electrons and protons to the substrate.

$N_2$ is an exceptionally stable compound, its conversion to $NH_3$ requires very extreme conditions – for example, temperatures around 500°C and pressures exceeding 200 atm. Thus, the manufacture of chemical fertilizers consumes a significant portion of the energy expended globally. In addition, a considerable amount of the fertilizer applied to fields is washed into streams, ponds, and eventually the ocean, polluting the water and promoting the growth of unwanted microalgae and other microorganisms.

In contrast, the *biological* process of nitrogen fixation, carried out by a small number of prokaryotic species, does not require the consumption of fossil fuels or electricity, and because it produces no more nitrogen than is needed in a given environment (because the expression of relevant genes is repressed by excess ammonia and nitrate), it does not produce pollution. The fostering of biological nitrogen fixation has therefore been an important goal for biotechnology.

The conversion of nitrogen and hydrogen molecules into $NH_3$ is thermodynamically favored, but the biological fixation of nitrogen is complex and consumes a large number of ATP molecules, because the enzymes involved in it must overcome the huge activation energy barrier. Two enzymes are required: an MoFe protein (also called component I or nitrogenase) and an Fe protein (also called component II or nitrogenase reductase). After the Fe protein is reduced by a strong biological reductant (ferredoxin or flavodoxin), ATP molecules are hydrolyzed to accomplish the reduction of the MoFe protein, which is followed by the reduction of $N_2$ to two molecules of $NH_3$ (Figure 6.2). The overall equation for the nitrogenase reaction is:

$$N_2 + 16\,ATP + 8\,e^- + 8\,H^+ \rightarrow 2\,NH_3 + 16\,ADP + 16\,Pi + H_2.$$

Nitrogen fixation is a strongly reductive reaction, and the enzymes involved are usually irreversibly inactivated when they are exposed to oxygen. This oxygen sensitivity is important in understanding the biology of the $N_2$-fixing microorganisms, described below.

The ability to fix nitrogen is found in scattered members of the Bacteria and Archaea (but not in eukaryotes). Some of the nitrogen-fixing genera are only very distantly related to others, an observation that suggests that this function was probably transferred "laterally" – that is, between different organisms – during evolution. Several groups fix nitrogen as free-living organisms. Among those, *Clostridium* and *Klebsiella* fix nitrogen only under anaerobic conditions, an observation that is consistent with the oxygen sensitivity of the enzymes. Other free-living bacteria, however, can fix nitrogen even under aerobic conditions, because each of these organisms has developed a complex machinery for protecting the nitrogen-fixing apparatus from oxygen. Thus, the cyanobacteria, which carry out oxygen-evolving photosynthesis, perform nitrogen fixation only in specialized cells called heterocysts, which do not produce oxygen. *Azotobacter* consumes oxygen at an extremely high rate, which seems to protect its nitrogen-fixing machinery. Another group of bacteria fix nitrogen only when they are in a symbiotic relationship with plants. The best-studied of the symbiotic nitrogen fixers is the group that used to be called *Rhizobium* (today, many species have been transferred to genus *Sinorhizobium*, *Mesorhizobium*, or *Bradyrhizobium*, but here all the species will be described by the general name, rhizobia), which invades the root tissues of leguminous plants, such as alfalfa, pea, clover, and soybean, and lives in intracellular vacuoles, where it differentiates into a form called "bacteroid." The bacteroids are usually much larger and sometimes have more complex and irregular shapes (e.g., a Y-shape) than the vegetative cells. Above all, bacteroids carry out nitrogen fixation, which the vegetative cells cannot do. The bacteroids also carry out oxidation of energy sources supplied by the plant, thus depleting the free oxygen level and so creating favorable conditions for nitrogen fixation. The vacuoles are also filled with an oxygen-binding protein, leghemoglobin, produced by the plants. This protein is thought to facilitate the transport of oxygen to the bacteroids.

The organization of genes involved in nitrogen fixation was first elucidated in *Klebsiella*. Remarkably, in that genus a very large number of genes are organized into a single *nif* gene cluster. This finding led, in the early 1970s, to the idea that cloning this cluster and putting the clones into desired crop plants might produce plant stocks that did not require chemical fertilizers – a possibility that if realized would revolutionize agriculture. Of course, the situation is much more complicated. If nitrogen-fixation machinery were produced in plant cells that were not also supplied with the necessary protective mechanisms, it would rapidly be inactivated by oxygen. Furthermore, the large amounts of ATP needed for the process must also be supplied. Leguminous plants have evolved together with rhizobia and contribute heavily to the successful symbiosis by expressing more than 20 genes specifically for that purpose. One of the contributions of these host plants is to supply a steady stream of compounds, such as dicarboxylic acids, that serve as the energy source for the bacteria. In short, simply introducing *nif* genes does not bestow on a plant the ability to fix nitrogen.

These considerations led scientists to try more modest approaches in their attempts to improve symbiotic $N_2$-fixing bacteria. These efforts will be aided by the knowledge of complete genome sequences of *Sinorhizobium meliloti* (an alfalfa symbiont), *Mesorhizobium loti* (a clover symbiont), and *Bradyrhizobium japonicum* (a soybean symbiont), which range in size from 6.7 Mb distributed in three replicons (*S. meliloti*) to 9.1 Mb in a single chromosome (*B. japonicum*). The large genome sizes presumably reflect the complex life cycle of these symbionts.

One possible target for improvement is the rate of nitrogen fixation. In these bacteria, at least eight electrons are needed for the reduction of $N_2$, a process that should theoretically require only six electrons. The remaining two electrons (and probably many more in root nodules) are used for the reduction of protons to produce $H_2$. Some of the rhizobia produce "uptake hydrogenase" to reutilize $H_2$ by oxidizing it with $O_2$, thereby regenerating ATP. Mutants lacking this enzyme are indeed less efficient in $N_2$ fixation. Overexpression of uptake hydrogenase is thus expected to increase the efficiency of $N_2$ fixation. Although the hydrogenase is a very complex enzyme whose production and assembly requires nearly 20 genes, the genes are clustered together, and a transposon containing all the known genes has been used successfully to bring hydrogenase activity to strains originally lacking the enzyme.

Another area with potential for improvement is the host–bacterium interaction. Although the interaction between rhizobia and their hosts is extremely complex, many of the relevant genes have been identified. A member of the rhizobia not only recognizes a given plant as a host but also induces a whole set of reactions in it, causing the root hair to curl, an infection thread to form, and the thread to develop into a membrane that envelops the bacterium. It also induces the plant to secrete leghemoglobin, filling the space around the bacterium, and to supply constantly a large amount of an energy source (such as dicarboxylic acids) to the bacterium. For example, the *nodD* bacterial gene product responds to specific flavonoid compounds produced by plants and activates the other genes involved in nodulation. By altering the sequence of *nodD*, it has been possible to change (and sometimes broaden) the host specificity of a given strain. In a later step of the nodulation process, the *nodH* and *nodQ* gene products of rhizobia synthesize a low molecular weight signaling molecule that is recognized by a specific host plant, which then responds with curling of the root hair and so on. Substituting genes from a different species of rhizobia for these genes resulted in successful alteration of the host range.

These results are impressive, but usually an "improved" strain performs rather poorly under field conditions because it is not competitive in the natural soil. In fields where leguminous plants are grown on a regular basis, the soils tend to contain a wealth of rhizobia strains that are especially well suited to surviving in that particular environment, even though their $N_2$-fixing efficiency may not rival that of the newly engineered strain. Studies have shown that when rhizobial strains that supposedly fix $N_2$ more efficiently are introduced into such fields, they rarely survive the pressures of competing with the indigenous strains. Any hope of introducing a genetically engineered rhizobial strain hinges on the production of better survivors and better colonizers – unfortunately, an area where our knowledge is as yet incomplete.

The effort to expand the host range of rhizobia to include non-leguminous plants is ongoing. Many pessimistic views have been expressed, especially concerning the effort to find a symbiotic $N_2$ fixer for rice and wheat. However, researchers found that in Egypt, where clover and rice have been rotated in cultivation since antiquity, specially adapted strains of *Rhizobium leguminosarum* occur in close association with rice roots, and inoculation of seeds with this strain increases the rice yield by nearly 50% in the absence of any chemical fertilizer. Most of the bacteria appear to be attached to the roots, rather than to form nodules, and the growth enhancement seems to result from the production of plant hormones by *Rhizobium* cells. Another $N_2$-fixing bacterium, *Klebsiella pneumoniae*, was found to enter the roots of wheat and contribute to its growth by $N_2$ fixation. These results give us renewed hopes for this line of research.

Another approach is to use the "associative" $N_2$ fixers, whose symbiotic relationship with plants is much less intimate. *Azospirillum* species, for example, which grow in association with important monocotyledonous crop plants such as sugar cane, associate with these host plants only loosely, most of the time by colonizing the surface of the roots. There is a price to pay for the looseness of the interaction, however. Because the plants cannot supply nutrients rapidly to the bacteria under such conditions, the efficiency of $N_2$ fixation in such a system cannot be very high.

Finally, it is important to ask whether all these efforts to improve nitrogen fixation are worthwhile. In the immediate present, the answer is probably no, because chemical fertilizers are quite inexpensive. However, in the long run, the effort is important for the preservation of ecological balance on Earth. As shown in Figure 6.3, human activities now affect the global cycling of elements, which used to be conducted almost entirely by nonhuman organisms. The emission of $CO_2$ and oxidized sulfur compounds into the air have increased at an alarming rate in recent years and have created problems such as global warming and acid rain. The manufacture and use of chemical fertilizers also increased dramatically in the latter half of the twentieth century. In the absence of human activity, biological nitrogen fixation by $N_2$-fixing prokaryotes is estimated to convert about $100 \times 10^6$ tons of nitrogen per year in the terrestrial environment. ($N_2$ fixation in the oceans is difficult to estimate and remains a controversial topic.) In comparison with this, humans are currently applying more than $80 \times 10^6$ tons of nitrogen in the form of

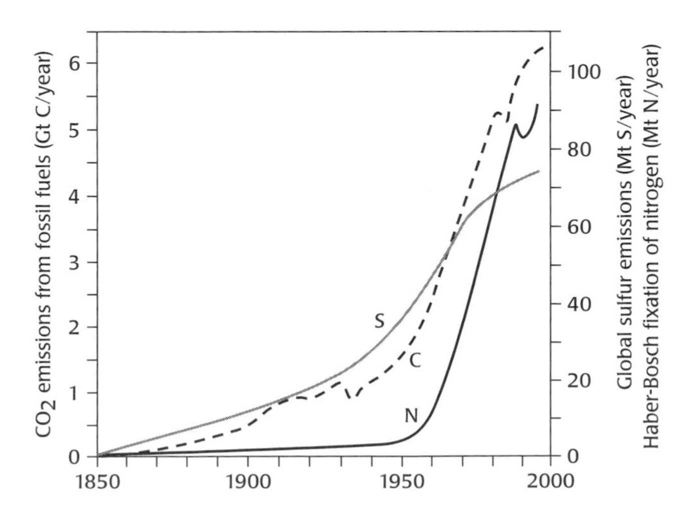


**FIGURE 6.3**

Human interference in the global cycling of the elements S, C, and N. [From Smil, V. (2001). *Enriching the Earth: Fritz Haber, Carl Bosch, and the Transformation of World Food Production*. Cambridge, MA: MIT Press, Figure 9. 1, p. 179; with permission.]

chemical fertilizers and adding about 40 × $10^6$ tons of combined nitrogen through the cultivation of crops containing $N_2$-fixing symbionts, as well as 20 × $10^6$ tons of reactive nitrogen through the burning of fossil fuels. Thus, the human activity in the nitrogen cycle now surpasses the flow due to natural processes, and the human contribution will keep increasing rapidly because the growing world population must be fed. Because about half the chemical fertilizers escape into the air and water and because nearly all the reactive nitrogen created from fossil fuels escapes into the air and water, these huge amounts of combined nitrogen create enormous environmental problems, such as the greenhouse gas effect of nitrogen-containing compounds and eutrophication of the coastal waters followed by depletion of $O_2$, resulting in the creation of "dead zones" in bodies of water. Because biological $N_2$ fixation does not produce such excesses, feeding crops through this route rather than the massive application of chemical fertilizers is the much-preferred approach for the future.

## PRODUCTION OF TRANSGENIC PLANTS

Improved plant stocks have been produced during the long history of humankind by the extremely slow and laborious process of selection and crossing of random mutations. The introduction of recombinant DNA methods brought about a revolutionary change in this process by making it possible to insert desired genes of "foreign" origin into plants. The impact of this revolution may be seen from the fact that as of 2004, more than 80 million hectares of cropland are planted with these "transgenic," or genetically modified, crops (the explosive growth in the adoption of such plant stocks during the last decade is seen in Figure 6.4) and that more than one half (56%) of the soybean cropland worldwide is planted with transgenic stocks.

Inserting cloned genes with desirable traits into plants is not a trivial matter. Plant cells are surrounded by a thick and rigid cell wall, and DNA cannot usually be brought into them unless the cell wall is first removed. Even when

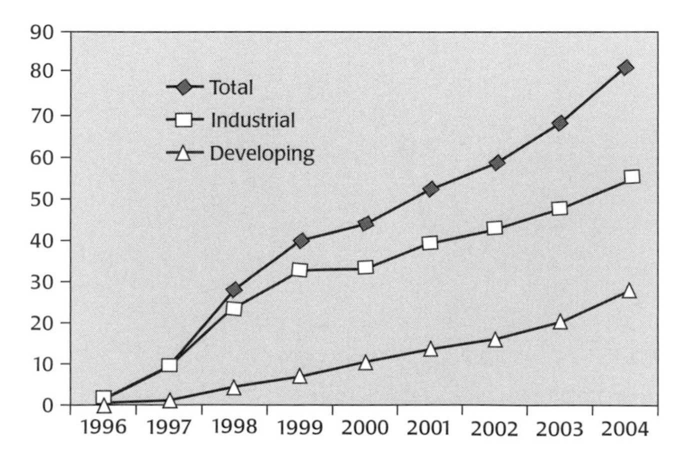


FIGURE 6.4

Total world cropland areas (in million hectares) planted with transgenic stocks. [From ISAAA (International Service for the Acquisition of Agri-biotech Applications, www.isaaa.org) Brief 32–2004; with permission.]

pieces of DNA are successfully brought into the plant cell, there are obstacles to success because such pieces of DNA are not expected to be replicated in succeeding generations. The standard strategy for overcoming this difficulty is to put the cloned DNA segment into plasmids, so the plasmids can be replicated indefinitely in the host cell. Lower eukaryotes, such as yeast, sometimes contain plasmids, and these have been used in the construction of shuttle vectors. However, most plant cells are not known to contain any plasmid DNA. Alternatively, the cloned DNA can survive in the host cell if it becomes integrated into the host chromosome, but there is no guarantee that such a process will occur with a high frequency.

For the production of transgenic plants, therefore, the crucial steps are efficiently introducing the cloned genetic material into the plant cell nucleus, then facilitating the integration of the cloned gene into the plant chromosome. It is interesting that the best method for doing this uses a system that already exists in nature, the system by which the plant-pathogenic bacterium *Agrobacterium tumefaciens* injects a portion of its plasmid DNA into plants and inserts it into the plant genome. (This underscores again the practical importance of studying the "natural history" of bacteria–plant interactions.)

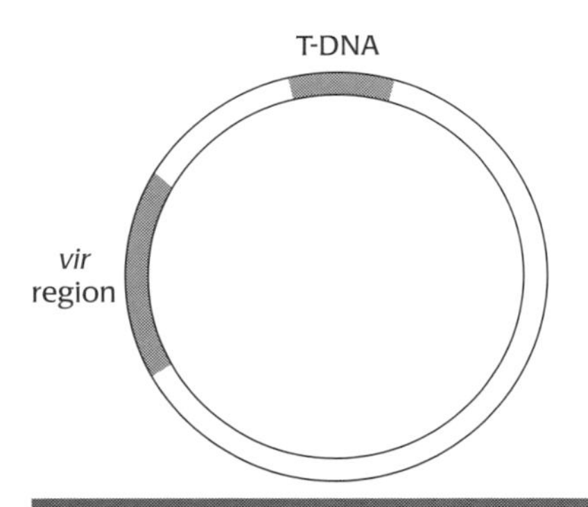


**FIGURE 6.5**

The structure of the Ti plasmid, with its *vir* region and T-DNA. In addition, the plasmid contains an origin of replication and several genes that assist in the colonization of host plants, such as the genes for the enzymes that degrade octopines and nopalines.

Redrawn based on artwork from the first edition (1995), published by W.H. Freeman.

## INTRODUCTION OF CLONED GENES INTO PLANTS BY THE USE OF *A. TUMEFACIENS*

*A. tumefaciens* is a Gram-negative bacterium that causes uncontrolled multiplication of "transformed," "tumorlike" cells – the disease known as crown gall – in host plants. Interestingly, rRNA homology has shown *A. tumefaciens* to be closely related to *Rhizobium*, but *A. tumefaciens* also contains a large (200-kb or even larger) "tumor-inducing" plasmid, the Ti plasmid (Figure 6.5). A small region of the plasmid, the *vir* (virulence) region, contains about two dozen genes that are involved in the infection of plants and in the transfer of a small part of the plasmid, T-DNA, into plant cells (Figure 6.5). Recent years have seen impressive progress in the understanding of this very complex process.

The genes of the *vir* region (Figure 6.6) become activated in response to substances exuded by injured plant tissues. These substances thus serve as

**FIGURE 6.6**

The organization and functions of the virulence region of the Ti plasmid. The *large open arrows* indicate the direction of transcription. *C* and *M* denote the cytoplasmic and membrane locations of the gene products, respectively. [Based on Zambryski, P. (1988). Basic processes underlying *Agrobacterium*-mediated DNA transfer to plant cells. *Annual Review of Genetics*, 22, 1–30, and Kuldau, G. A., et al. (1990). The *virB* operon of *Agrobacterium tumefaciens* pTiCSS encodes 11 open reading frames. *Molecular and General Genetics*, 221, 256–266.]

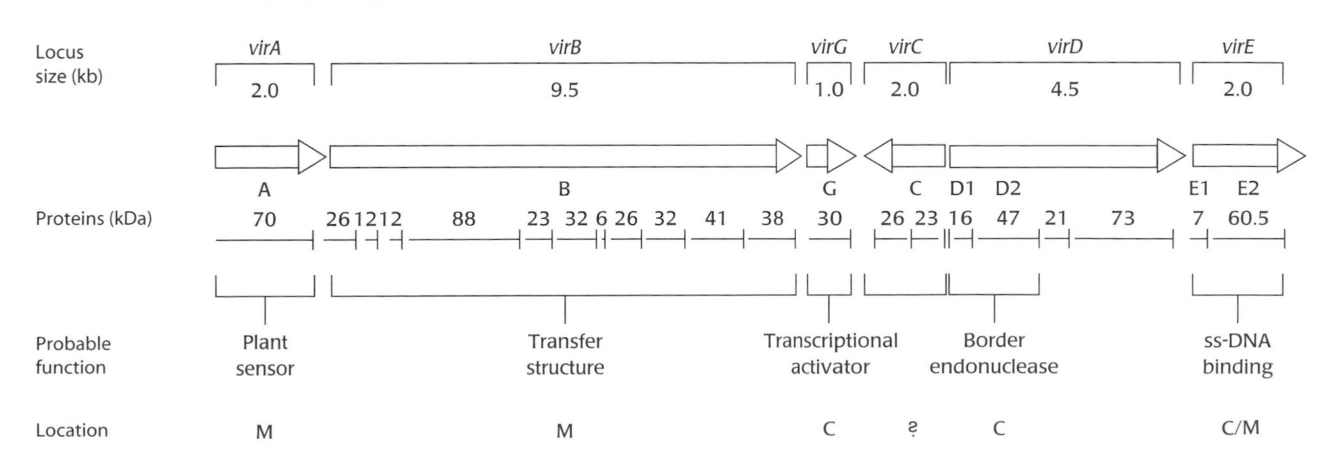

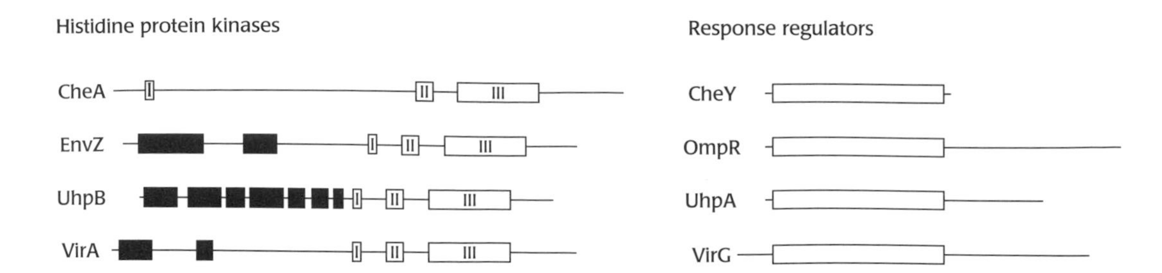


**FIGURE 6.7**

Two-component regulatory systems of bacteria. A system is composed of two proteins, a histidine kinase and a response regulator. All histidine kinases are related, sharing three regions of homology (shown as *I*, *II*, and *III*), and all response regulators have certain similar sequences in their N-terminal region (shown as a *box*). In many cases, the histidine kinase is a membrane protein (the putative transmembrane domain is shown as a *black box*) and senses certain factors in the external environment – for example, osmotic pressure (EnvZ), the presence of hexose phosphate (UhpB), or the presence of plant injury signals such as acetosyringone (VirA). Activation of the histidine kinase results in the phosphorylation of a conserved histidine residue in region I of these proteins, after which the phosphate is transferred to a conserved aspartate residue on the response regulator protein. This activates the response regulator, which typically affects the transcription rates of relevant genes (OmpR, phosphorylated by EnvZ, regulates porin genes; UhpA, phosphorylated by UhpB, regulates the expression of hexose phosphate transport genes; and VirG, phosphorylated by VirA, stimulates the transcription of *vir* genes). Some systems, however, have cytosolic histidine kinases – for example, CheA that functions in chemotaxis by phosphorylating CheY, which acts to determine the sense of flagellar rotation. [For reviews on two-component systems, see Parkinson, J. S., and Kofoid, E. C. (1992). Communication modules in bacterial signaling proteins. *Annual Review of Genetics*, 26, 71–112, and Chang, C., and Stewart, R. C. (1998). The two-component system. Regulation of diverse signaling pathways in prokaryotes and eukaryotes. *Plant Physiology*, 117, 723–731.]

"signals" telling an *A. tumefaciens* cell that it is next to a plant that is wounded and vulnerable to invasion. The proteins that recognize and respond to these signals, VirA and VirG, belong to a large family of prokaryotic regulatory proteins, often called two-component systems, that enable various organisms to respond adaptively to changes in environmental conditions, such as osmolarity, availability of nitrogen sources, and presence of chemoattractants (Figure 6.7). Such two-component systems typically contain a sensor protein located in the cytoplasmic membrane. The presence of signal molecules activates the protein kinase function of this sensor, leading to phosphorylation of the second component, the response regulator or transducer protein. This phosphorylation activates the response regulator, and it can, for example, activate the transcription of pertinent genes.

In the VirA–VirG system, VirA, a membrane protein, apparently acts as the sensor. It is activated synergistically by the presence of phenolic compounds such as acetosyringone (Figure 6.8) leaking out of damaged plant tissues and by the presence of D-glucose, D-galactose, L-arabinose, or other sugars commonly found in plants. The sugars bind to a binding protein in the periplasm, then the binding protein interacts with the periplasmic domain of VirA. Acetosyringone apparently interacts directly with VirA. The activated VirA phosphorylates its own cytoplasmic domain, and the phosphate group is transferred to VirG, which is in the cytoplasm. The phosphorylated VirG then binds to the promoter regions of other *vir* genes and activates their transcription.

The next stage in the process is the nicking of the Ti DNA at specific points by "border nucleases" encoded by the *virD1* and *virD2* genes. A single-stranded DNA fragment of about 22 kb, called the T-strand, is released by an unwinding reaction, and at the same time, a replacement strand is synthesized (Figure 6.9). The VirD2 protein remains attached to the 5′-end of the T-strand and is thought to function as a "pilot" that leads the DNA into the plant cell. This is an orderly transfer that begins with the "right" border. The process itself – the nicking of the double-stranded DNA and the unwinding and injection of the single-stranded fragment – is remarkably similar to the events that occur in bacterial conjugation, and both phenomena are thought to have a common evolutionary origin.

The injection is catalyzed by the products of 11 *virB* genes and the *virD4* gene. These genes are now known to be homologs of the type IV secretion system (machinery for extracellular secretion of proteins in Gram-negative bacteria is described in Box 6.1), which not only exports but injects proteins into animal cells in pathogens such as *Bordetella pertussis* and *Helicobacter pylori* and presumed DNA-protein complexes into another bacterial cell in the plasmid-mediated conjugation process. Indeed, the F-factor system catalyzing the transfer of both plasmid and chromosomal DNA in *E. coli* (see Chapter 3) carries out this process through the type IV secretion system. VirB1 serves as a glycosidase that produces a hole in peptidoglycan, making possible the assembly and function of the export apparatus. Three proteins (VirB4, VirB11, and VirD4) appear to be ATPases, which coordinately energize the transport process. VirB6, VirB8, and VirB10 appear to form an export channel that reaches the inner surface of the outer membrane. This membrane is also punctured by an oligomeric assembly of VirB9, which resembles other "secretin" proteins producing outer membrane channels for protein export in type II and III secretion systems. This apparatus secretes and assembles a special conjugational pilus composed of VirB2 and VirB5. Although the pilus brings *A. tumefaciens* cells close to the plant cell, it is not absolutely required for the injection of VirD2–T-DNA complex.

After it is injected into a plant cell, the T-strand presumably becomes associated with the binding protein for single-stranded-DNA, VirE2, which is also injected by *A. tumefaciens*. Sequences close to the carboxy terminus of both VirD2 and VirE2 proteins serve as "nuclear translocation signals" and facilitate the entry of the T-DNA complex into the cell nucleus through nuclear pores. A complementary strand has to be synthesized at some point, and the double-stranded product, the "T-DNA," must become integrated into the plant genome. These latter processes remain largely unidentified, although one plant protein that interacts with VirE2 was found also to interact with histones.

Once the T-DNA has been integrated into one of the plant chromosomes, various genes in the fragment begin to be expressed. These genes code for the synthesis of opines and of plant hormones. The opines (amino acid derivatives that can be used only by fellow *Agrobacterium* cells; see Figure 6.10) encourage the invasion of the plant by more *Agrobacterium* cells, and the hormones (auxin: indoleacetic acid; cytokinin: $N^6$-isopentenyladenine) stimulate the division and growth of plant cells, producing the characteristic tumor or gall.

The *Agrobacterium* Ti system seems almost tailor-made for exploitation by biotechnologists. The transfer of T-DNA is determined essentially by the 25-bp "border" repeats. The DNA between those repeats is transferred and integrated regardless of its sequence. Thus, the insertion of extraneous genetic material into the middle of the T-DNA segment of the plasmid has no negative impact on the transfer and integration of that segment into a plant cell.

Unfortunately, the Ti plasmid is too large to be conveniently manipulated *in vitro*. Therefore, the piece of foreign DNA to be cloned is almost always introduced into a smaller vector plasmid first. Then one of two strategies is used to transfer the cloned sequence into plants.

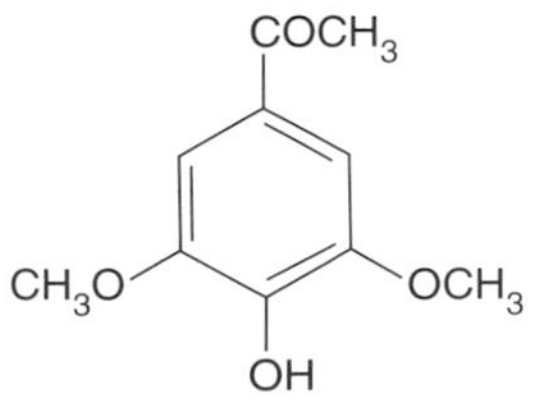

$COCH_3$ $CH_3O$ $OCH_3$ OH

**FIGURE 6.8**

Structure of acetosyringone.

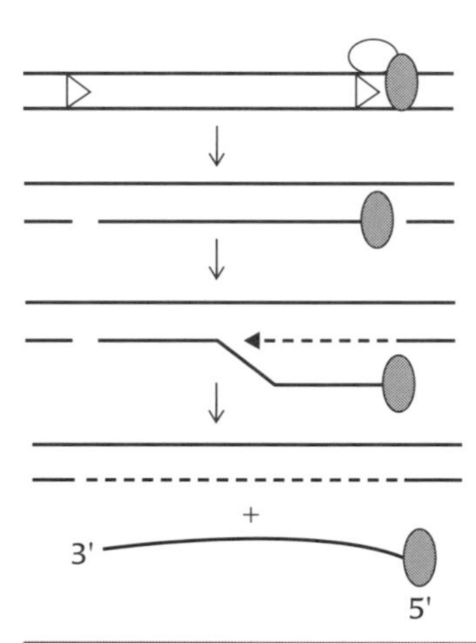


**FIGURE 6.9**

Nicking and transfer of T-DNA. The *open arrowheads* in the top diagram represent the 25-bp direct repeat border sequences that are recognized by VirD1 (with probable assistance from VirC1) and are cleaved by the specific endonuclease VirD2. The VirD2 protein (*gray ellipse*) remains attached to the 5′-terminus of the cleaved single-stranded T-DNA fragment. Synthesis of the replacement strand (*broken line*) results in the release of the T-DNA as the single strand (T-strand) that is then injected into plant cells.

Redrawn based on artwork from the first edition (1995), published by W.H. Freeman.

**Secretion of Proteins into Extracellular Space in Gram-Negative Bacteria**

The Gram-negative bacterial cell is surrounded not only by the cytoplasmic membrane but also the outer membrane. Secretion across the cytoplasmic membrane into the periplasmic space usually occurs through the SecYEG(DF) complex (see Box 3.15). (However, more recently, biotechnologists have been exploring the possible use of the "twin arginine transport" [TAT] pathway, which is attractive because it appears to export a folded protein molecule.) In contrast, secreting proteins across both membranes into the extracellular space requires complex machineries. As shown in the figure below, this may occur by utilizing one of the four pathways, types I through IV secretion systems.

In the simplest system, type I secretion system, used for the export of some colicins and hemolysins, the ATP-energized transporter of the ABC (ATP-binding cassette) class in the inner membrane exports the protein, which then crosses the outer membrane through the channel of TolC protein. The complex is held together by a periplasmic protein, a member of the Membrane Fusion Protein family.

In the type II secretion system, proteins usually cross the inner membrane by the classical SecYEG system, then get transported across the outer membrane through a channel made by an oligomer of D protein. More than a dozen proteins are involved in producing this second-step extrusion process across the outer membrane, presumably energized by ATP hydrolysis by the E protein. Because of the involvement of the SecYEG complex, this pathway has sometimes been called the "Main Terminal Branch" of the "General Secretory Pathway." However, this nomenclature is criticized[1] now that it is known that proteins exported by the Tat pathway are also secreted by the type II system, and that some "autotransporter" proteins are secreted via SecYEG, then by an autocatalytic mechanism not involving the type II system (this system is sometimes called type V secretion system).

The type III secretion system uses a complex assembly of proteins morphologically resembling the basal portion of bacterial flagella, containing a needlelike projection. The proteins are exported without using the SecYEG system. This system is used by many pathogens to inject (sometimes toxic) proteins directly into the cytosol of animal and plant host cells.

The type IV secretion system is also complex. The best-studied system is the VirB system of *Agrobacterium tumefaciens* (shown), which injects protein-coated DNA into plant cells. Homologous systems are involved in the secretion of toxin in the human pathogen *Bordetella pertussis* and in the transfer of plasmid DNA between bacterial cells. Pertussis toxin is made with a signal sequence, and thus the possibility remains that it crosses the inner membrane via the SecYEG system, similar to the situation in the type II system.

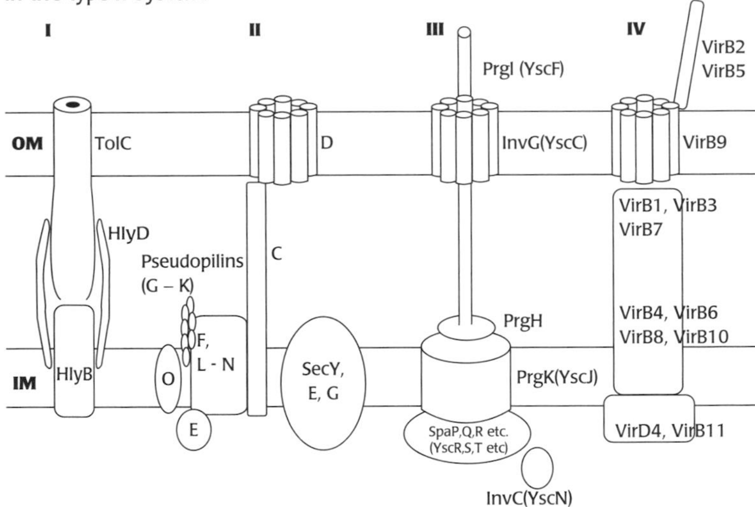


Type I through IV secretion systems of Gram-negative bacteria. The type I system shown here as an example uses the hemolysin secretion system of *E. coli*. The type II system is exemplified by the pullulanase secretion system of *Klebsiella oxytoca*. It was proposed to convert the gene symbols in that system, *pulE*, *pulO*, and so on, into "general secretory pathway" symbols, such as *gspE*, *gspO*, and so on.[2] Here we show only the last letters of the symbols. For type III secretion systems, the nomenclature of the *Shigella* proteins (Prg, Inv, Spa) are shown, with the *Yersinia* protein nomenclature in parentheses. The type IV secretion system is shown with the *A. tumefaciens* protein nomenclature. [This figure is based on Figure 7 of Nikaido, H. (2003). Molecular basis of bacterial outer membrane permeability revisited. *Microbiology and Molecular Biology Reviews*, 67, 593–56.]

[1]Desvaux, M., Parham, N. J., Scott-Tucker, A., and Henderson, I. R. (2004). The general secretory pathway: a general misnomer? ***Trends in Microbiology***, 12, 306–309.

[2]Francetic, O., and Pugsley, A. P. (1996). The cryptic general secretory pathway (gsp) operon of *Escherichia coli* K-12 encodes functional proteins. ***Journal of Bacteriology***, 178, 3544–3549.

**BOX 6.1**

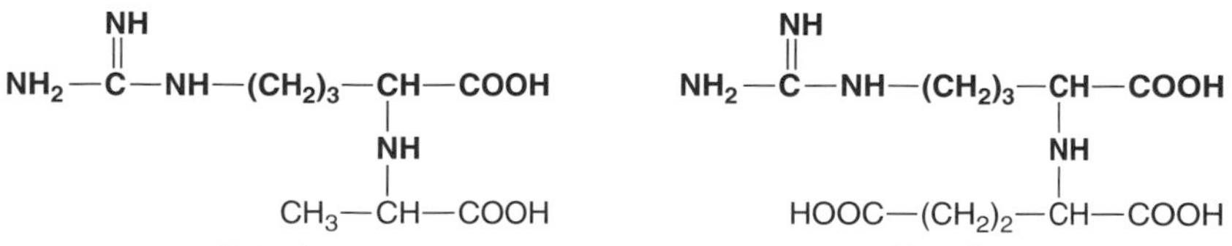

$$NH_2-\overset{\overset{\displaystyle NH}{\|}}{C}-NH-(CH_2)_3-\underset{\underset{\displaystyle CH_3-CH-COOH}{|}}{\overset{}{CH}}-COOH$$

(Octopine: the CH bears an NH group linked to $CH_3-CH-COOH$)

Octopine

$$NH_2-\overset{\overset{\displaystyle NH}{\|}}{C}-NH-(CH_2)_3-CH-COOH$$

(Nopaline: the CH bears an NH group linked to $HOOC-(CH_2)_2-CH-COOH$)

Nopaline

**FIGURE 6.10**

Structure of the opines, octopine, and nopaline. These compounds are made of arginine residues (*bold letters*) joined to an acidic compound. *Agrobacterium* also produces families of similar compounds in which the arginine residue is replaced by other basic amino acids.

### Use of a Cointegrate Intermediate

This is the method that was developed first and was used more in the early days. The foreign gene is inserted into a small vector, such as a derivative of pBR322 (often used for cloning in *E. coli*; see Chapter 3), by the usual *in vitro* methods of recombinant DNA technology. Then *A. tumefaciens* cells that contain modified (non–tumor-producing) Ti plasmids are transformed with this recombinant plasmid. The modified Ti plasmids also contain a stretch of pBR322 sequence between the left and right borders of their T-DNA region. Because of this homology with the pBR322 replicon, recombination can take place to generate a cointegrate plasmid containing the entire sequence of the smaller plasmid between its left and right T-DNA borders (Figure 6.11). If the population is selected for the presence of a marker gene, such as a drug resistance gene on the pBR322 plasmid, only the cells that contain the cointegrate survive the selection. The cointegrate then injects the sequence between the two borders containing the foreign gene into plants.

The cointegrate method has some drawbacks, however. The piece of DNA that is injected is relatively large and contains much extraneous information, so it is difficult to control the gene transfer process with precision. Often, only portions of this DNA become integrated into the plant genome. Furthermore, antibiotic resistance "marker" genes may get transferred into the Ti plasmid by mechanisms other than the homologous recombination, and thus there is no guarantee that the donor *Agrobacterium* cell contains the desired cointegrate. Finally, ascertaining the structure of the cointegrate by

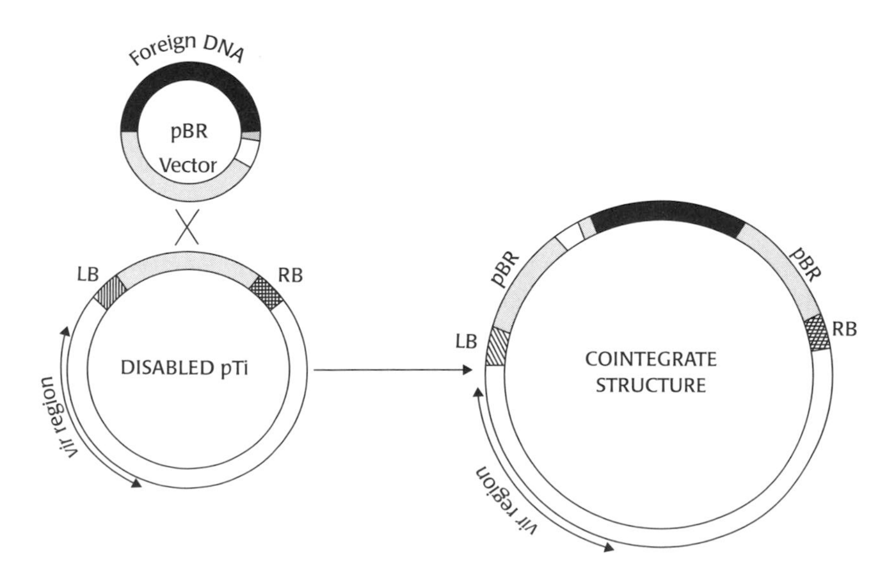


**FIGURE 6.11**

Transfer of T-DNA via the formation of a plasmid cointegrate. The pBR plasmid contains a marker selectable in *A. tumefaciens*, such as an appropriate antibiotic resistance marker (*open box*). The foreign DNA is cloned in the pBR plasmid, and the composite plasmid is introduced into *A. tumefaciens*, which already contains a Ti plasmid that has been "disarmed" by removing genes responsible for tumor production and has been further modified by incorporating a small fragment of the pBR sequence in the middle of the T-DNA. The origin of replication in the pBR plasmid functions well in *E. coli* but not at all in *A. tumefaciens*. Thus, the antibiotic selection enriches only the cells that contain cointegrates, which are formed by the homologous recombination, utilizing the homologous pBR sequences, between the Ti plasmid and the pBR plasmid. Because the cointegrate contains all the *vir* genes, the foreign gene will be transferred into plants as a part of the modified T-DNA. [From Lurquin, P. F. (1987). Foreign gene expression in plant cells. *Progress in Nucleic Acid Research*, 34, 143–188; with permission.]

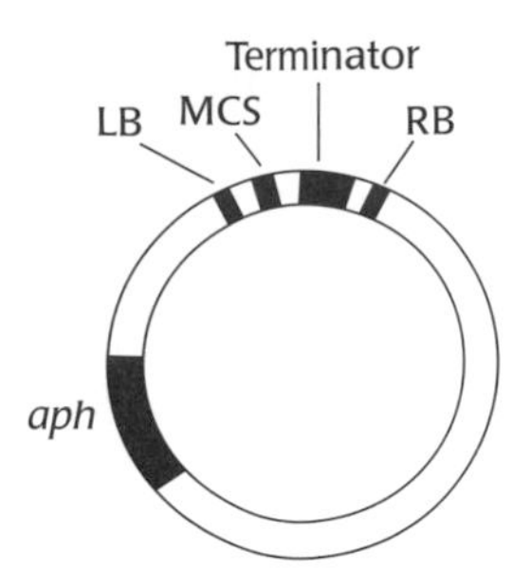

**FIGURE 6.12**

Transfer of T-DNA by the binary vector system. The foreign DNA is cloned into the middle of T-DNA in the plasmid, shown at the top, which contains one or more origins of replication that function in both *E. coli* and *A. tumefaciens*. In the particular vector shown, the original T-DNA sequence has been replaced by a multiple cloning site (MCS), followed by a terminator sequence functional in plants, located between the left border (LB) and right border (RB) sequences. The plasmid also contains an antibiotic resistance gene (in this case, an aminoglycoside phosphoryltransferase, *aph*, to make possible selection of the plasmid-containing cells on aminoglycoside-containing plates). This composite plasmid is introduced into *A. tumefaciens* cells that contain another plasmid, shown at the bottom. This larger plasmid, a derivative of the Ti plasmid, contains only the *vir* region and the origin of replication and is totally devoid of the T-DNA region. Because the two plasmids have no common sequences, there is no recombination and cointegrate formation. Nevertheless, the products of the *vir* genes on the larger plasmid can mediate the transfer, into plants, of the T-DNA sequence of the other plasmid.

Redrawn based on artwork from the first edition (1995), published by W.H. Freeman.

the usual *in vitro* methods is difficult because of its large size. Consideration of these points prompted the development of the binary plasmid approach.

### Use of Binary Vectors

This method takes advantage of the fact that the *vir* genes on one plasmid can catalyze the excision and transfer of a T-DNA sequence located on another plasmid; that is, these genes can act in *trans*. The binary plasmid approach consists of cloning the DNA fragment of interest into the T-DNA sequence of a plasmid vector with a broad host range that is capable of replicating in both *E. coli* and *A. tumefaciens*. The plasmid DNA is then introduced into *A. tumefaciens* cells that contain a "disarmed" Ti plasmid with *vir* genes but no T-DNA sequence. The *vir* genes of the disarmed plasmid effect the transfer of the T-DNA from the other plasmid without the formation of a cointegrate intermediate (Figure 6.12). In this method, only the piece of DNA that had been inserted between the left and right borders of the smaller plasmid is transferred into plants, allowing more precise control of the process.

Both of these methods require several modifications in the Ti plasmid. In the Ti plasmid used for the cointegrate method, tumor-producing genes are either inactivated or removed; otherwise, transformed plant cells would become tumors, not healthy "transgenic plants." In addition, a pBR sequence is inserted into the T-DNA region, as shown in Figure 6.11. In the Ti plasmid used for the binary plasmid strategy, the entire T-DNA region is removed (see Figure 6.12); otherwise, T-DNA from the Ti plasmid would compete with the injection of T-DNA from the smaller plasmid.

### General Considerations

Regardless of which strategy is used, the site of insertion of the foreign gene is usually sandwiched between a promoter sequence that functions effectively in plants and a terminator sequence. The 35S protein promoter from the cauliflower mosaic virus (CaMV) was popular in the early experiments because of its reliably high expression in a variety of plants. However, CaMV promoter drives the expression of foreign genes in any plant tissue, a situation that may be unnecessary or unwanted. In more recent attempts, promoters that drive tissue-specific expressions were also used. For example, the promoter for a rice seed protein is expected to drive expression of cloned genes only in rice grains; therefore, it was used for the expression of introduced protein genes for the modification of amino acid content in rice. A popular terminator sequence is the one for nopaline synthetase. (An effective terminator ensures that the 3′-ends of mRNA are processed and polyadenylated so that the mRNA achieves a reasonable degree of stability [see Chapter 3].)

In addition, the region to be introduced into plants must contain a good marker so that plant cells that have received and integrated the foreign genes can be recognized easily. $\beta$-Glucuronidase is a marker whose activity can be detected readily in plant tissue. Even more useful are markers that enable one to select, not just screen, for plant cells with integrated T-DNA. Currently, the most popular marker of this kind is neomycin (aminoglycoside) phosphotransferase II gene (*nptII*; see Table 10.1), which effectively inactivates aminoglycoside antibiotics that have detectable activity against plant cells, such as neomycin and kanamycin.

It is the common practice to use "explants," that is, plant tissues and cells removed from the whole plant and placed in growth medium, as recipients of DNA. For example, tobacco leaves can be cut into small pieces, mixed with an *Agrobacterium* donor strain containing binary plasmids, and incubated for a few days to allow the transfer of DNA. The explants are removed from the bulk of the medium containing most of the bacteria and transferred to a solid medium containing a mixture of auxin and cytokinin in order to stimulate the division and growth of plant cells. The medium must also contain one antibiotic (e.g., a $\beta$-lactam) to kill the remaining bacterial cells and another antibiotic (e.g., kanamycin) to select for only the plant cells that have received the DNA segment containing the antibiotic marker (such as *nptII*) as well as the foreign gene to be introduced. In many plant species, whole plants can be regenerated from single explant cells, and we end up with transgenic plants in this manner. (However, much crossing and selection at the whole-plant level are needed, as described below.)

Methods using explants take time and with many plant species, it is difficult to obtain suitable explants with a high probability of regenerating healthy whole plants. Thus, it is of interest that methods using whole plants ("in planta" transformation) are being developed by using a model plant, *Arabidopsis thaliana* (Box 6.2). In one of these methods, the flowering plant is dipped in a suspension of *Agrobacterium* and the bacterial penetration into tissues is enhanced by the application of a vacuum. A fraction of seeds from such plants contain introduced DNA segment and can generate transgenic plants.

The introduced piece of DNA between the T-DNA ends becomes integrated into plant chromosomal DNA by the process of illegitimate recombination. This process does not involve any specificity, so the location of integration is essentially random. Furthermore, often several copies of the introduced DNA are inserted into one location, although the copy number is usually smaller than that obtained by the direct introduction method discussed below. Because of this imprecise nature of integration, it is necessary to start from many (usually hundreds) of transformant lines and select the one that produces a strong and stable expression of the foreign gene. It is also necessary that unwanted alterations in the plant genome, a frequent by-product of transformation, are eliminated by careful backcrossing.

A puzzling and frustrating observation was frequently made with the immediate products of transformation. The plant cells, which seemed to express the foreign gene product at high levels initially, often showed rapidly decreasing levels of expression after a few days. Also, when the introduced gene had a homolog within the plant genome, the expression of both the introduced and endogenous genes was simultaneously decreased. The studies of these phenomena, together with the observations made with animal

**Arabidopsis**

*Arabidopsis* is a small cruciferous plant that has become the favored species for genetic and recombinant DNA studies in plants, somewhat like *E. coli* among bacteria. It offers many advantages, including a very small genome size (70,000 kilobase pairs [kbp], only five times the size of the yeast genome and only 10% of the size of the genome of typical crop plants), an exceptionally low content of repetitive DNA, and a rapid reproduction cycle (seeds can be obtained in six weeks after germination). In addition, it is self-fertile, so mutant strains can be maintained easily, and it is susceptible to *Agrobacterium* transformation, so genetic material can be introduced by recombinant DNA methodology. For a review, see Estelle, M. A., and Somerville, C. R. (1986). The mutants of *Arabidopsis*. *Trends in Genetics*, 2, 89–93, and Meyerowitz, E. M. (1989). *Arabidopsis*, a useful weed. *Cell*, 56, 263–269.

BOX 6.2

systems, led to the discovery of a complex and important mechanism called gene silencing (see Box 6.3). Gene silencing occurs at both the transcriptional level and the posttranscriptional level. With the foreign genes (sometimes called transgenes), the mRNA generated is often recognized as "aberrant" by the plant cell, possibly because the 5′-end capping or 3′-end polyadenylation is incomplete. This will generate double-stranded RNA via the action of cellular RNA-dependent RNA polymerase (see Box 6.3) and start the process of silencing. Thus, it is most important to use proper promoters and terminators. In another situation, integration of two copies of transferred DNA in head-to-head (or tail-to-tail) fashion generates RNA that could fold upon itself, and this double-stranded RNA will again produce silencing. Thus, it is desirable to use conditions that favor the integration of single copies of foreign DNA. In addition, one can introduce viral suppressors of gene silencing, which was shown in one example to increase the expression of foreign genes by up to 50-fold.

## DIRECT INTRODUCTION OF CLONED GENES INTO PLANTS

In nature, *Agrobacterium* does not infect monocotyledonous plants (monocots), the subclass of plants that includes all the cereal crops. Thus extensive modification of the protocol (supplying acetosyringone to induce *vir* gene expression, use of embryonic cells as the target, etc.), which required many years, was needed to produce *Agrobacterium*-mediated transformation of cereal plants on a reliable basis. This and other difficulties have led to a number of attempts to introduce DNA directly into plant cells.

- *Incubation of plant protoplasts with DNA.* Protoplasts take up DNA in the presence of polyethyleneglycol (Chapter 3). When the protoplasts are made from embryogenic cells, it is often possible to regenerate the transgenic plants.
- *Introduction of DNA into protoplasts by electroporation.* Transient application of high electric voltage across a protoplast membrane will produce large pores through which DNA in the medium diffuses spontaneously or possibly electrophoretically (Chapter 3 ).
- *Bombardment of plant cells with DNA-coated microprojectiles.* Pellets of microscopic size (usually gold or tungsten particles) are coated with solutions of DNA and are literally "shot" into tissues and cells. This method, sometimes called biolistics, has many advantages and has become the predominant approach for direct gene transfer. Its advantages include the following. (a) There is no species barrier, so that DNA can be introduced into any plant, including cereal crops and forest tree species. (b) With the *Agrobacterium* system, the final transgenic plant is usually constructed by crossing the easily transformed variety with the elite cultivars. In contrast, particle bombardment may be applied directly to any cultivar of choice. In an extreme scenario, pollen (haploid) can be bombarded, then grown up, and finally made into a diploid plant by the use of colchicine, for example. (c) For introducing cloned foreign genes into chloroplasts, particle bombardment is the only

**Gene Silencing**

Gene silencing (repression of gene expression) attracted attention in the studies of transgenic plants because of the difficulty in expressing foreign genes at high levels. At almost the same time, researchers using antisense RNA to inhibit the translation of mRNA in nematodes (*Caenorhabditis elegans*) discovered that sense RNA was as effective as the antisense RNA in inhibiting the expression of the corresponding gene and that double-stranded RNA was at least 10 times more active in gene silencing than the sense or antisense RNA. Such RNAi (RNA interference) preparations are now used widely to selectively block expression of many genes in various eukaryotes.

In plants also, double-stranded RNA plays a central role in gene silencing. As shown in the figure below, short pieces of RNA called siRNA (short interfering RNA) play a major role in degrading the mRNA, the mRNA often arising from the transcription of introduced foreign genes.

Posttranscriptional gene silencing is thought to have evolved as a means of plant defense against invading viruses and transposons. Double-stranded RNA viruses introduce double-stranded RNA directly into plant cells, and double-stranded RNA is often created after the infection of viruses containing DNA or single-stranded RNA. Retrotransposons, which are very abundant in many plant species (they comprise as much as 70% of the nuclear DNA in maize), resemble the genome of a single-stranded RNA virus and are thought to have originated from invaders of plants. They may produce double-stranded RNA or at least DNA–RNA complex during their transposition event. In order to combat this defensive strategy by plants, viruses often produce proteins that inhibit (or "suppress") the posttranscriptional silencing mechanism. These suppressor proteins are useful weapons for biotechnologists, as described in the text.

Related regulatory mechanisms involving double-stranded RNA molecules play other important roles, as described in the legend to the figure in this box. In addition, similar mechanisms are also hypothesized to play a major role in phenomena such as the silencing of one of the X chromosomes in females and genomic imprinting (a process whereby copies of selected genes on chromosomes coming from one of the parents are silenced).

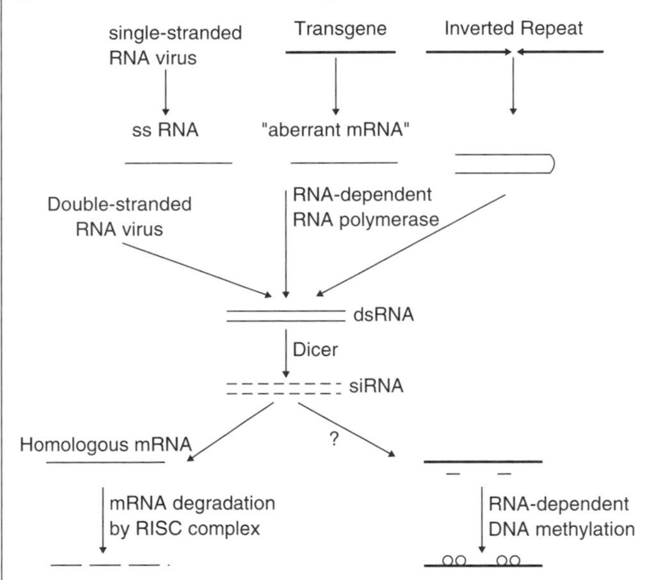


Posttranscriptional gene silencing in plants. Double-stranded RNA (dsRNA) may be introduced directly as the component of double-stranded RNA virus, or made by RNA-dependent RNA polymerase from "aberrant" mRNA (which may be missing the proper modification at the 5′- or 3′-end, or may simply be overproduced) or from the genomic RNA of a single-stranded RNA virus, or from the transcription products of DNA sequences containing head-to-head (or tail-to-tail) tandem repeats (as often occurs when transgenes are introduced into plant chromosomes). The dsRNA is then cut into short (usually 21 to 24 bases long) pieces by a homolog of Dicer, a protein originally identified in *Drosophila*. These siRNA molecules then anneal with the homologous target mRNA, and this leads to the degradation of mRNA by a complex called RISC (RNA-induced silencing complex).

Regulation involving these double-stranded RNA molecules, however, is often more complex. In some systems, such RNA molecules interact with the chromosomal DNA, causing the methylation and the consequent inhibition of gene transcription (transcriptional gene silencing). In other systems, transcription of short, incomplete inverted repeat sequences in the chromosome, followed by cleavage by Dicer-like proteins, creates short dsRNA called microRNA or miRNA, which are involved in endogenous regulatory networks.

**BOX 6.3**

practical choice because the *Agrobacterium* system introduces T-DNA into the nucleus. Modification of the chloroplast genome is attractive because the large number of chloroplasts per cell and the large copy number of DNA within a single chloroplast produce a very high copy number per cell (and presumably a very high level of expression) for the introduced gene, because there seems to be no gene silencing in chloroplasts, and because the introduced gene (and marker genes) will not be disseminated to neighboring fields as chloroplasts are virtually absent in pollens.

There are, however, some drawbacks to the particle bombardment method. The most important one is that the cloned DNA segment often is not integrated intact into plant chromosome, or is integrated as multiple contiguous stretches (often containing several dozen copies). The latter point is of particular disadvantage in terms of potential gene silencing. The integration also tends to produce chromosomal rearrangements. Another limitation is the rather low frequency of successful introduction and integration of the foreign DNA.

## EXAMPLES OF TRANSGENIC PLANTS

The *Agrobacterium* system is an excellent system for introducing foreign genes into the chromosomes of plants. The transfer into intact plant cells occurs at high frequency, the T-DNA is usually integrated into the plant chromosomes at high frequencies without undergoing structural alterations, and the cells that have received the T-DNA can be selected easily by using antibiotic resistance, such as the neomycin resistance marker. Finally, the transgenic plants produced in this manner tend to be stable. Consequently, many of the existing transgenic plants with potentially desirable traits have been obtained by this method.

In the simplest case, the desirable traits arise in transgenic plants through the continuous expression of the foreign genes. Such transgenic plants have been commercialized and adopted very rapidly by farmers in many parts of the world during the last decade, as seen in Figure 6.4. We described already that more than one half (56%) of the cropland planted with soybeans worldwide is used for the cultivation of transgenic soybeans. For other crop plants, transgenic varieties now occupy 28%, 19%, and 14% of the cropland used for growing cotton, canola, and corn, respectively. Transgenic rice is expected to be approved in China in the next few years, and this is expected to increase the fraction of transgenic crops very much. Below we discuss representative classes of important transgenic crop plants.

### Herbicide-Resistant Plants

The advantages of making crop plants resistant to herbicides are obvious, because a large fraction of efforts in agriculture is devoted to the control of weeds. Although there is a fear that such resistance may eventually increase the use of herbicide chemicals, there are also reasons to expect that these

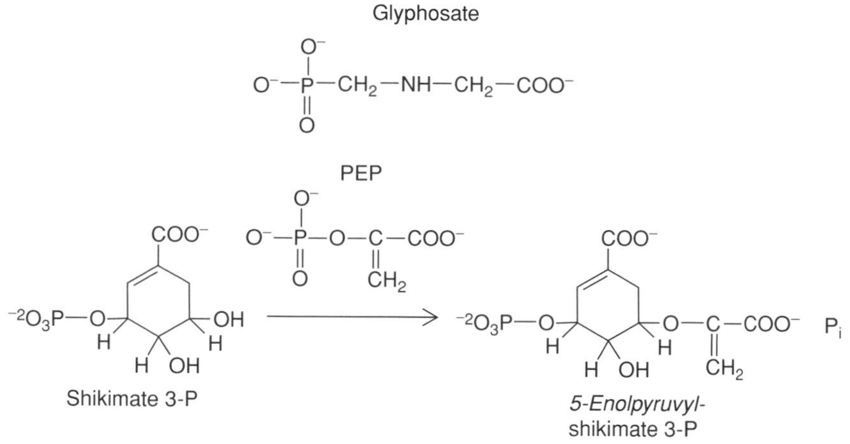

Glyphosate

$O^- - P(=O)(O^-) - CH_2 - NH - CH_2 - COO^-$

PEP

$O^- - P(=O)(O^-) - O - C(=CH_2) - COO^-$

Shikimate 3-P $\longrightarrow$ 5-*Enolpyruvyl*-shikimate 3-P + $P_i$

**FIGURE 6.13**

Glyphosate and its mode of action. Glyphosate, acting as an analog of phospho*enol*pyruvate, inhibits the formation of 5-*enol*pyruvylshikimate 3-phosphate, a precursor of aromatic amino acids.

transgenic plants will promote the use of safer, more biodegradable herbicides, perhaps in smaller amounts. One example involves the herbicide glyphosate, which inhibits 5-*enol*pyruvylshikimate 3-phosphate synthase – an enzyme involved in the biosynthesis of aromatic amino acids – by acting as a structural analog of phospho*enol*pyruvate (PEP; Figure 6.13).

This enzyme was purified from crop plants and sequenced, and DNA probes corresponding to its amino acid sequence were synthesized. These probes were used to isolate cDNA for the enzyme from the cDNA library of a plant cell line known to overproduce 5-*enol*pyruvylshikimate 3-phosphate synthase. The cDNA was then cloned behind the strong CaMV 35S promoter and ahead of the nopaline synthetase terminator, and the gene complex was introduced into plant cells (e.g., petunia) via a disarmed Ti plasmid vector. The transgenic plants produced a much higher level of the target enzyme and thus were significantly more resistant to glyphosate (Figure 6.14). These results were encouraging because glyphosate has very low toxicity to animals and is rapidly degraded in soil.

An obvious improvement to this strategy is to use DNA coding for a glyphosate-resistant mutant enzyme for the construction of transgenic plants. Early experiments using the genes from glyphosate-resistant mutant

**FIGURE 6.14**

Production of glyphosate-resistant plants. The gene for 5-*enol*pyruvylshikimate 3-phosphate synthase, cloned behind a strong promoter, was introduced into petunia plants as described in the text. Three weeks after these transgenic petunia plants (top) and unaltered control plants (bottom) were sprayed with Roundup (a pesticide containing glyphosate), the control plants were dead but the transgenic, resistant plants were completely healthy. [From Shah, D. M., et al. (1986). Engineering herbicide tolerance in transgenic plants. *Science*, 233, 478–481; with permission.]

strains of bacteria were not encouraging because the enzyme produced lacked the N-terminal sequence that guided its transport into chloroplasts ("chloroplast transit peptide"). The synthesis of aromatic amino acids occurs mainly in chloroplasts, and the plant enzyme used in the earlier experiments contained this sequence and was therefore effectively moved into chloroplasts. The final commercial products were thus made by introducing a DNA containing a strengthened CaMV 35S promoter, a sequence coding for a chloroplast transit peptide, a sequence coding for a bacterial glyphosate-resistant enzyme, and a nopaline terminator. The gene coding for the enzyme came from a wild-type strain of *Agrobacterium*, and it binds the substrate with an affinity similar to the plant enzyme, whereas its affinity to glyphosate is 5000-fold lower. It is interesting that none of the mutants of normally glyphosate-susceptible enzymes from bacteria or plants displayed resistance even remotely comparable to this naturally occurring enzyme. Herbicide-resistant soybean was planted on nearly 48 million hectares in 2004, accounting for nearly 60% of all the cropland worldwide planted with transgenic plants. Among these, glyphosate-resistant plants presumably represent the overwhelming majority.

Altering the levels or nature of the enzymes targeted by herbicides is thus an effective approach to producing herbicide-resistant plants. However, this approach has several limitations. First, theoretically it cannot produce an absolute level of resistance because greater amounts of herbicide will still inhibit an overproduced or less sensitive target enzyme. Second, the overproduction of an endogenous enzyme may produce unexpected and undesirable results. For example, glyphosate-resistant soybean is known to suffer, in hot weather, from splitting of the stems, which is thought to be caused by the higher lignin content of these plants. The overproduced enzyme catalyzes the synthesis of aromatic compounds, which are predominant building blocks in lignin (see Chapter 12). Third, the overproduction of one enzyme in a complex pathway may disturb the balance in such a pathway and may lead to the slowing of growth. Considerations such as these have prompted the exploration of alternative methods for developing herbicide resistance in plants.

One technique for producing resistant transgenic plants is the introduction of genes coding for herbicide-detoxifying enzymes. The development of glyphosate-resistant plants through this approach is described on p. 410. Another important example involves phosphinothricin, an analog of glutamic acid that inhibits glutamine synthetase (Figure 6.15).

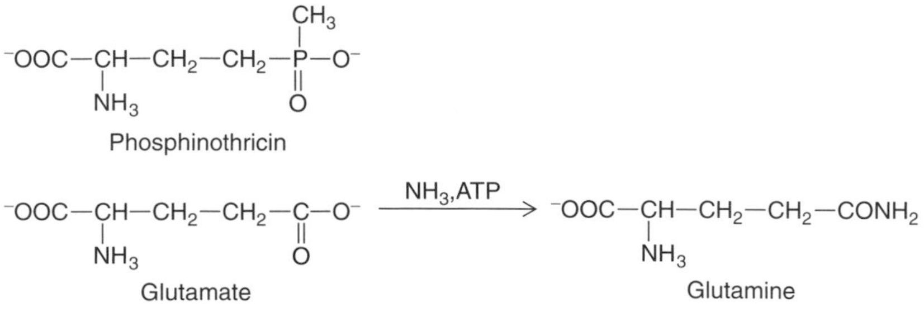

$$^{-}OOC-CH(NH_3)-CH_2-CH_2-P(CH_3)(=O)-O^{-}$$

Phosphinothricin

$$^{-}OOC-CH(NH_3)-CH_2-CH_2-C(=O)-O^{-} \xrightarrow{NH_3,ATP} {}^{-}OOC-CH(NH_3)-CH_2-CH_2-CONH_2$$

Glutamate — Glutamine

**FIGURE 6.15**

Phosphinothricin and its mode of action. Phosphinothricin is an analog of glutamate. It binds to glutamine synthetase to inhibit the synthesis of glutamine. Bialaphos, an antibiotic produced by a *Streptomyces* species, is a tripeptide (phosphinothricinyl-alanyl-alanine) that is converted into phosphinothricin by plant peptidases.

Phosphinothricin was originally discovered as the active moiety of an antibiotic, bialaphos, produced by a *Streptomyces* strain. As described in Chapter 10, antibiotic-secreting microorganisms frequently produce enzymes that detoxify the antibiotic to protect themselves. The bialaphos-producing *Streptomyces* indeed produced an enzyme that inactivated both bialaphos and phosphinothricin by acetylation. The gene coding for this enzyme was cloned from the *Streptomyces*, put behind the CaMV 35S promoter, and introduced into crop plants, such as potato. The transgenic plants showed strong resistance to phosphinothricin. Currently, transgenic maize, canola, soybean, and other crops produced by a similar procedure are available commercially. Phosphinothricin-resistant rice has been developed and was planted experimentally in Texas in 2004.

The issues of environmental impact of genetically engineered plants will be discussed more extensively in the next section, but the possibility of cross-pollination with the same or related species was a real concern with the glyphosate- and phosphinothricin-resistant plants. When pollens do not travel for a long distance (as with maize), this is not a problem; however, with canola, the pollen was shown to travel as far as a mile and to cross-pollinate other plants. Thus, planting of herbicide-resistant canola poses serious problems for "organic" farmers in the neighboring field. However, the transfer of herbicide resistance genes to a related species appears to be relatively rare. One may be able to avoid these problems by introducing the herbicide tolerance genes into chloroplasts, because pollens do not contain chloroplasts.

### Insect-Resistant Plants

*Bacillus thuringiensis* is used in the biological control of caterpillars because its sporulating cells contain toxic proteins (see Chapter 7). Initially, the entire gene for one of the toxic proteins (Cry1A) was cloned and transferred to plants, but the expression levels were extremely low. In the next stage, only the portion of the gene coding for the N-terminal toxin fragment (see Chapter 7) of about 650 amino acids was cloned between promoters and terminators that are effective in plants and was introduced into plant cells via a Ti plasmid vector. This improved the expression levels and produced plants that are toxic to the more sensitive insect species, such as tobacco hornworm (*Manduca sexta*). Still, the expression levels of the toxin protein were too low to kill the more toxin-resistant species of insects, such as corn earworm or beet armyworm. Use of "improved" CaMV 35S promoter, with a tandemly repeated promoter sequence (see Chapter 7), still did not increase the level of mRNA, although it increased 10-fold the expression of endogenous plant genes. Thus, it was important that Monsanto scientists altered extensively the coding sequence of the toxin gene in 1991, and succeeded in increasing the expression level of the truncated toxin almost 100-fold. They changed the third positions of codons from A/T to G/C to increase the GC content from 37% to 49% without altering the amino acid sequence of the product. In this process, 17 out of 18 potential polyadenylation signal sequences (AATAAA or

AATAAT) were eliminated as well as all 13 of the ATTTA sequences reported to destabilize mRNA in animals. When plants were provided with this modified gene (placed behind the CaMV promoter with a duplicated enhancer region), they showed impressive resistance to common lepidopteran insects that damage unmodified plants. Subsequent transgenic plants all incorporated similar modification of the coding sequences for *B. thuringiensis* toxins. (Detailed case studies of examples are provided in Chapter 7.)

Although the modification of the toxin coding sequence was experimentally successful, the mechanism behind the increased toxin production is not entirely clear. Efforts to eliminate only the ATTTA sequence or rare codons have produced rather contradictory results, but it appears that the extensive modification of the coding sequence acts mostly by stabilizing the mRNA. A factor that was not known at the time of these studies is gene silencing. In fact, even with the fully modified gene, many of the transformed plant cells produced almost no toxin, and in these cases posttranscriptional gene silencing is likely to have been involved.

The safety issues and environmental impacts of the toxin-containing crops have received much attention. The major issues are described in Chapter 7. There is so much data on the total absence of human and vertebrate toxicity of *B. thuringiensis* toxins spanning many decades of their use as sprays that there is no problem on this issue. However, it is not advisable to generate such a large number of copies of neomycin phosphotransferase (*nptII*) gene, so future generations of transgenic crops should be made without such drug resistance marker genes. The issue of allergenicity was highlighted by the fate of the StarLink corn. This variety, producing Cry9C toxin, was marketed as being appropriate only for animal feed, because its allergenicity had not been tested and because it was more stable under acidic conditions (mimicking the contents of the human stomach) than the toxins produced in other insect-resistant strains. When traces of the StarLink corn were detected in human food, this led to much public uproar and the recall of various corn products. It is still unsettled whether the Cry9C protein is more allergenic than other Cry toxins, but this episode shows that serious attention must be paid to the issue of allergenicity of foreign proteins expressed in food crops.

Some seeds contain high concentrations of protease inhibitors, which are thought to interfere with the digestive process in insects. The trypsin inhibitor gene was cloned from a variety of African cowpea that is resistant to a number of insects. Its transfer to tobacco plants has resulted in good resistance to a wide variety of leaf-eating insects even under field conditions. In this case, there was no problem with the expression of the protein. It reached levels as high as 1% of the total plant protein, presumably because the cloned gene is of plant origin. Similarly, lectins (proteins that bind to carbohydrates) are often present in high concentrations in seeds and inhibit insects ingesting them. Thus, some of these lectins were also expressed in plants. However, none of the transgenic plants expressing protease inhibitors or lectins has been commercialized so far, primarily because these plants are protected only to a modest degree.

### Virus-Resistant Plants

The method most frequently used for producing virus-resistant plants sprang from the observation that, oftentimes, plants infected by a nearly avirulent virus are thereafter resistant to superinfection by a related, highly virulent one. Various observations suggested that the cross-resistance between viruses may occur because of the presence of virus coat proteins in the previously infected plant cells. Indeed, transgenic plants, whose genomes contain the introduced tobacco mosaic virus (TMV) coat protein gene, show resistance to TMV infection. However, the hypothesis that the presence of coat protein confers resistance had to be abandoned when it was discovered that even mutated coat protein genes, which are not translated, are fully effective in protection. It is thus currently assumed that the resistance caused by the introduction of coat protein genes into plants is mostly a result of gene silencing (see Box 6.3). Most plant viruses are positive-strand RNA viruses. When the coat protein gene is transcribed in plants, the ensuing mRNA is somehow recognized as "aberrant," and this recognition presumably results in the production of double-stranded RNA, which is cleaved to generate short pieces of siRNA. The siRNA in turn will initiate the degradation of homologous single-stranded RNA molecules, including that of viral RNA.

Transgenic plants containing viral coat protein genes have been commercialized as virus-resistant stocks of summer squash and zucchini. These stocks have not been adopted by farmers as enthusiastically as the insect-resistant or herbicide-resistant varieties, presumably because it is difficult to make the plants resistant to all of the many existing viruses. However, the transgenic papaya stock resistant to papaya ringspot virus, constructed by introducing the coat protein gene of the virus by particle bombardment, was extremely successful and saved the papaya industry on the Big Island of Hawaii from extinction. This success is related to the presence of essentially only one viral strain that caused disease in papaya in the given location.

### Plants Resistant to Fungi and Bacteria

Fungal (and to a lesser extent, bacterial) diseases are estimated to result in an annual loss of between $10 billion and $33 billion per year in the United States. To minimize this damage, $700 million is spent every year for fungicides alone. Within the last two centuries, Irish famine, caused by potato blight (a fungal disease), is still remembered. Thus, the development of stocks resistant to attack by fungi and bacteria would be most valuable. Recent progress in our understanding of the microbe–plant interaction gives us some hope that a rational approach may be possible.

Plant pathologists knew for a long time that the outcome of microbial infection is decided by the interaction between the Avr (for "avirulent") gene product of the pathogen and the R (for "resistance") gene product of the plant. The Avr proteins correspond to many diverse but essential proteins of the invader, and one class is now known to be injected into plant cells by the

**Box 6.4. Phytoalexins**

Unlike higher animals, plants cannot produce specific antibodies to fight invading microorganisms. Many plants instead produce, as a response to microbial infection, low molecular weight secondary metabolites – phytoalexins – that inhibit the growth of invading microorganisms. The structure of phytoalexins is specific to the producing plant species. Here we show two examples, phaseolin (an isoflavonoid compound produced by green beans) and rishitin (a norsesquiterpene produced by potatoes).

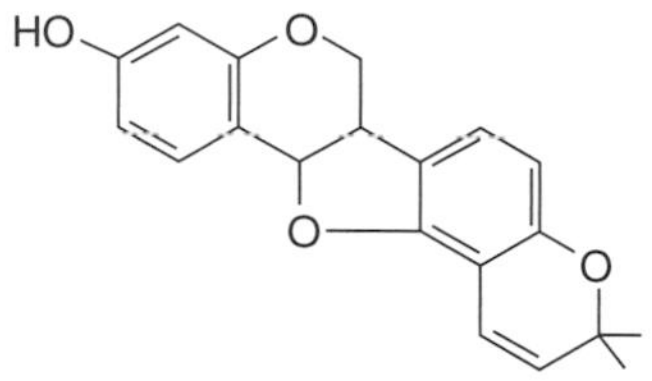

Phaseollin

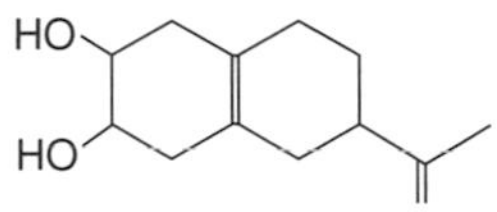

Rishitin

BOX 6.4

type III secretion pathway (see Box 6.1). When it is recognized specifically by the cognate R protein of the plant, a local "hypersensitive response" ensues, which sends signals throughout the plant to increase the level of defense and make it eventually survive the attack. When either the Avr or R protein is missing, systemic diseases result. A remarkable conclusion from recent studies is that some of the R proteins have an overall domain structure reminiscent of the Toll-like receptors of animal cells (see Chapter 5), which are used to recognize less specific features of pathogens, such as the presence of LPS. Thus, plants and animals appear to use proteins of the same type for recognition of the components of invading pathogens. In any case, crop plants can probably be made resistant to additional pathogens by "arming" them with the introduction of genes for additional R proteins that would recognize them. This approach was successful, at least under laboratory conditions. However, production of commercially useful levels of resistance would require identification, cloning, and introduction of many different R genes, a feat that is not so easy.

Scientists have tried to create crop plants with resistance to pathogens of a broader range. Recognition of pathogens either through the Avr–R interaction or through different mechanisms usually results in the release of signaling molecules, such as salicylate. This in turn produces a cascade of reactions, eventually resulting in the increased expression of an array of proteins needed for plant defense, as well as the synthesis of other systemic signaling molecules, such as phytoalexins (Box 6.4). Earlier attempts were aimed at the overexpression of such defense proteins. However, such approaches usually resulted in plants compromised in growth. Several attempts at overproducing the central regulatory protein in this cascade, NPR1, however, produced plants that are broadly resistant to a number of fungi and bacteria yet are not compromised in their growth. The latter property probably resulted from the fact that the activation of the defensive cascade did not occur until the plants came into contact with invaders.

In one clever approach, the *avr* gene from a pathogen is placed under a "pathogen-inducible promoter" and is then cloned into the crop plant genome. When the plant is invaded by any pathogen that would activate this promoter, the Avr protein will be produced, which will interact with its cognate R protein to activate the defense response. The difficulty in this approach is to find a proper promoter sequence, but this task may have become easier now thanks to modern array technology (see Chapter 4). In summary, great advances in our knowledge of molecular interaction pathways give us hope that a broad-range resistance to pathogenic microbes may be achieved with the transgenic approaches.

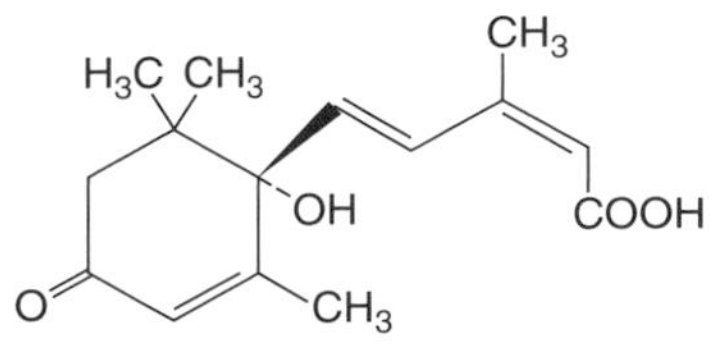

**FIGURE 6.16**

Abscisic acid. This plant signaling molecule is produced mainly when plants are stressed as a result of dehydration.

### Stress-Tolerant Plants

Plants have to survive different kinds of stresses, such as cold weather, hot weather, and drought conditions. Salt (or drought) tolerance has attracted much attention because high salinity (much of it caused by years of irrigation) limits crop yield in 30% of cropland, exemplified by the Central Valley of California. The strategy used for this survival is extremely complex. First, plants produce a signaling molecule (abscisic acid; Figure 6.16) that activates a cascade of regulatory proteins, resulting in the expression of many effector proteins. Second, some of these proteins produce compatible osmolytes, such as quaternary amino compounds (e.g., glycine betaine; Figure 6.17) and sugars and sugar alcohols (e.g., ononitol; Figure 6.18), which maintain the high osmolarity in the cytosol without disturbing the structure of proteins. Finally, $Na^+$ ions are sequestered away from the cytosol into vacuoles by the use of the $Na^+/H^+$ antiporter.

Because of this complexity of salt tolerance response, much of the work done so far utilized the simple approach to produce compatible osmolytes through the introduction of foreign genes. These studies have produced only limited success because the level of osmolyte overproduction was usually low (see the example of trehalose-producing transgenic rice discussed in Chapter 2) and the protection was only moderate. However, recently, transgenic carrots were created by introducing the gene coding for betaine aldehyde dehydrogenase (see Figure 6.17) into chloroplasts. Because of the strong amplification effects obtained by chloroplast cloning, the plants produced glycine betaine at a much higher level and could grow even in the presence of 400 mM NaCl. These are very promising results indeed, although it is still unclear whether we can achieve the ultimate tolerance without the participation of many other factors involved in the natural reaction of salt-resistant plants.

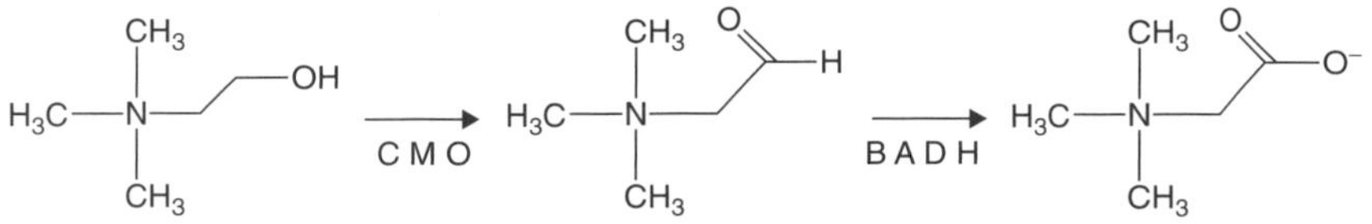

**FIGURE 6.17**

Biosynthesis of glycine betaine in plants. The synthesis starts from choline (left), which is converted by choline monooxygenase (CMO) into betaine aldehyde (center), which in turn is converted by betaine aldehyde dehydrogenase (BADH) into glycine betaine (right). *E. coli* uses a similar pathway, except that the first step is catalyzed by a conventional $NAD^+$-linked dehydrogenase. *Arthrobacter*, a Gram-positive bacterium, converts choline in one step into glycine betaine by choline oxidase.

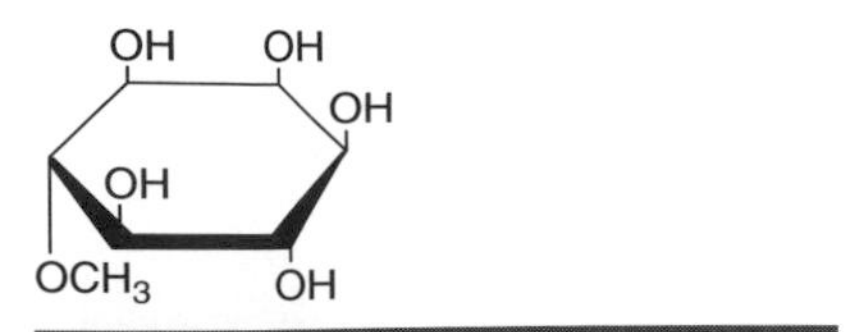


FIGURE 6.18
D-ononitol (1D-4-methyl-*myo*inositol).

## Improvement of Quality and Yield of the Products

In the examples we have discussed, the main improvement in crop plants occurred mostly in properties that were of importance to farmers. More recently, emphasis has been placed on the improvement of properties such as the quality of the product, some of which would be of more immediate interest to consumers. One example that attracted much public attention is the creation of "golden rice" that contains $\beta$-carotene, the precursor of vitamin A, by the collaborative efforts of Swiss and German university scientists. They introduced three genes, coding for phytoene synthase, phytoene desaturase, and lycopene $\beta$-cyclase, respectively, from two plasmids into rice, using the *Agrobacterium* Ti system. The first and last genes came from plants and the second gene from a bacterial species, because this enzyme can catalyze a succession of desaturation steps normally catalyzed by two separate enzymes in higher eukaryotes. Two of the genes were put behind a rice promoter producing endosperm-specific expression, and all the genes contained sequences coding for "chloroplast transit peptides," which induce the transport of expressed proteins into chloroplasts. It is difficult to introduce more than one gene into transgenic plants, and it took eight years for this effort to succeed. The final product yielded $\beta$-carotene–containing rice, which is expected to benefit children in some parts of the world who are currently suffering from vitamin A deficiency that some believe is causing the death of up to 1 million per year. Although the golden rice was created in 1999 in the laboratory, it had to clear many hurdles as a genetically modified food crop, and it is estimated that it will not reach farmers until 2009. In a similar strategy, transgenic rice expressing the iron-binding protein ferritin has been generated to produce rice grains enriched in iron and zinc.

Cereals serve not only as a source of energy but also as the main source of protein (amino acids) in many developing countries. Cereal proteins, however, are quite low in the content of a few essential amino acids, including lysine. For this reason, lysine is often added to both human and animal food (see Chapter 9). It would be even better if one could manufacture cereal plants producing seeds containing more of these amino acids. Storage proteins, however, must go through complex export and sorting pathways that begin with secretion into the lumen of endoplasmic reticulum, followed by a series of covalent modifications and oligomeric assembly process, finally ending up by sequestration in vacuole-like structures. Most of these compartments contain proteases, and it is expected that the storage proteins will be degraded if their sequence is altered by the introduction of more lysine, for example. For this reason, the improvement of the amino acid composition of cereal proteins has been very difficult. In one successful experiment, soybean glycinin cDNA was introduced into rice plants behind the rice glutelin promoter. Both of these proteins are the components of the seeds and belong to the same "11S globulin" family, although the lysine content of glycinin (5%) is about twice that of glutelin (2.5%). Presumably because glycinin is exported and sorted through the same pathway as glutelin, rice grains containing some glycinin were obtained. However, glycinin

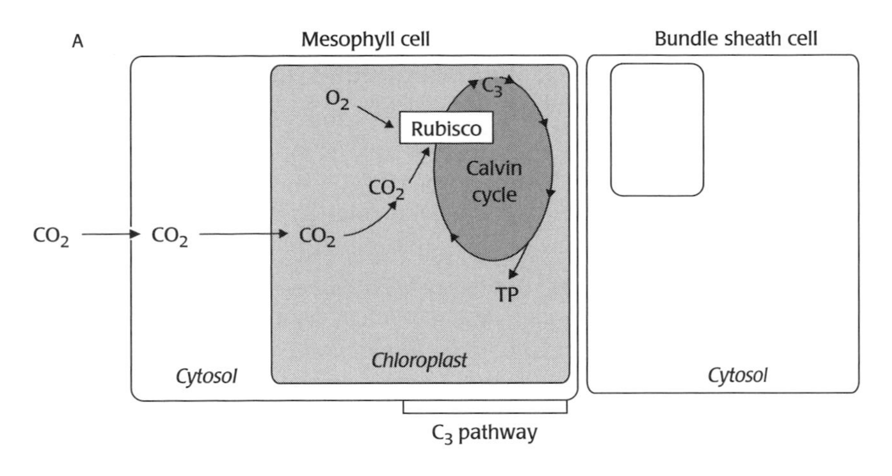


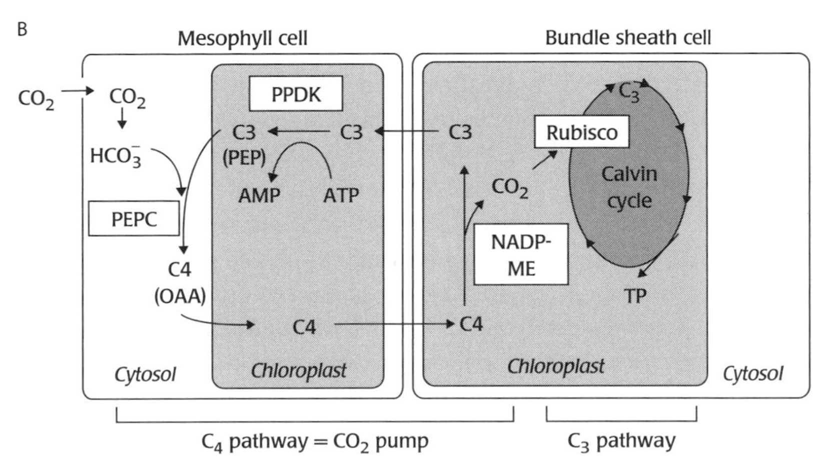


**FIGURE 6.19**

Schematic representation of photosynthetic pathways in (**A**) $C_3$ and (**B**) $C_4$ plants. In the $C_3$ plants, $CO_2$ is fixed by RuBisCo to generate the product, triose phosphate (TP). In the $C_4$ plants, in contrast, $CO_2$ is first fixed by PEP carboxylase (PEPC) to generate oxaloacetate (OAA), which is then converted into another C4 acid (malate) that then enters a bundle sheath cell. Malate releases $CO_2$ by malic enzyme (ME) with the participation of $NADP^+$, building up a high concentration of $CO_2$ locally. The other product, pyruvate (C3) is brought back to the mesophyll cell, where it regenerates PEP through the action of pyruvate orthophosphate dikinase (PPDK). [From Miyao M. (2003). Molecular evolution and genetic engineering of $C_4$ photosynthetic enzymes. *Journal of Experimental Botany*, 54, 179–189; copyright © 2003 Oxford University Press, with permission.]

represented only 5% of the total protein of transgenic rice even in the best stock obtained, and in terms of lysine content of the whole grain, this hardly made a difference. In an alternative strategy, a gene for tRNA that would translate the UGA stop codon as lysine was introduced into rice plants. The lysine content of the rice grain was increased, but the extent of increase was very small, from 4.27% in the untransformed rice to 4.55% in the best transformant.

There have been efforts to increase the yield of crops. The yield is obviously related to the efficiency of photosynthesis. In many plants, including rice, potato, and wheat – called $C_3$ plants – $CO_2$ fixation occurs at the stage of ribulose-1,5-bisphosphate carboxylase (RuBisCo). In some plants, including maize – called $C_4$ plants – the $CO_2$ fixation first occurs by using PEP carboxylase, and the $C_4$ product enters a neighboring bundle sheath cell, where it regenerates $CO_2$, which is then finally fixed by RuBisCo. This two-step process is energetically more expensive, but it allows the plant to circumvent the inhibition of RuBisCo by $O_2$, a $CO_2$ analog, because it can generate a much higher local concentration of the real substrate, $CO_2$, in the area where RuBisCo is functioning (Figure 6.19). After much effort, the maize PEP carboxylase was strongly overexpressed in rice, but the desired increase in yield was not obtained. The difference between $C_4$ and $C_3$ plants involves not only the activity of PEP carboxylase but also the levels and regulation of other enzymes and the cellular structures. Thus, the goal of a higher yield might be quite difficult to obtain with the currently available technology.

All the examples of transgenic plants discussed so far were created by bringing in genes from other organisms into plants. However, modification of endogenous plant genes is also possible. One area in which success has already been attained through this means is the modification of flower color. The biosynthetic pathways of flower pigments are known in detail, so it has been possible to inactivate or overexpress endogenous genes (as well as bring in genes from different species of plants) to produce “unnatural” colors that dramatically change a flower’s appearance.

Another activity of interest is the effort to delay the softening of fruits such as tomatoes. Slowing the softening process would obviously lengthen the storage life of fruits and facilitate their transportation. Evidence suggested that hydrolysis of polygalacturonate, a polysaccharide component of the fruit cell wall, by the enzyme polygalacturonase was involved in the softening of tomatoes. The gene coding for this enzyme was cloned, and it was put behind the strong CaMV 35S promoter in an *inverted* orientation. This construct was then introduced into tomato plants via the *Agrobacterium* Ti system. The resulting transgenic plants produced much lower levels of polygalacturonase because the “antisense mRNA” produced by the reading of the gene in reverse orientation interfered with translation of the normal mRNA or caused the posttranscriptional silencing by annealing to it. One version of these engineered tomatoes was marketed in the United States as the first genetically modified crop approved by the FDA, but it was not a commercial success. Some claim that the particular variety of tomato used for this construction was not attractive.

There are many plant traits that might one day be modified to our benefit. The list is endless, but plant size and flowering season would be obvious ones. Considering that the Green Revolution owed much to the development of dwarf stocks of crop plants, the alteration of plant size will be important. If we could make it possible to harvest a given crop twice or three times a year, instead of just once, by the modification of flowering season, this would lead to very significant increases in crop yields per unit area. Most of these traits are governed by multiple genes, whose products interact in complex ways. However, with the availability of array methods for elucidating transcription and translation levels, we may at least get closer to these goals in the not-too-distant future.

## SUMMARY

Many species of microorganisms interact with plants either as symbionts or as pathogens, and scientists have taken advantage of this intimate relationship in their efforts to improve agricultural production through biotechnology. For example, plant-pathogenic bacteria, inactivated through the deliberate deletion of genes essential for pathogenesis (such as the ice nucleation gene) have been used successfully as competitors against natural pathogens. Similar efforts are being made to improve the properties – for example, the host range – of beneficial symbionts such as nitrogen-fixing bacteria. In most cases, however, the focus of improvement has been the genetic constitution of a given plant – that is, the production of transgenic plants. However, even these cases have exploited the natural capacity of the plant pathogen *A. tumefaciens* to introduce a portion of its plasmid DNA, T-DNA, into plant cells. The introduction of exogenous genes from other plants or microorganisms in this manner has resulted in the production of plants that are resistant to herbicides, insect pests, or viruses and that are now planted on a vast scale, totaling more than 80 million hectares worldwide. In the laboratory, some success also has been obtained in the creation of transgenic plants that are broadly pathogen resistant or salt tolerant. The next generation of transgenic plants will place more emphasis on the improvement of the quality of an agricultural product, and rice plants producing grains of higher nutritional values have already been obtained in the laboratory. Fruits and vegetables with an improved shelf life and flowers with new and unexpected colors have also been successfully produced. Even the production of crops at higher yields may not be out of reach. A better understanding of the regulation of plant genes, however, is essential before these goals can be achieved.

## SELECTED REFERENCES


### General

Slater, A., Scott, N., and Fowler, M. (2003). *Plant Biotechnology: The Genetic Manipulation of Plants*. New York: Oxford University Press.

Jauhar, P. P. (2006). Modern biotechnology as an integral supplement to conventional plant breeding: The prospects and challenges. *Crop Science*, 46, 1841–1859.

### Use of Symbiotic Microorganisms

Lindow, S. E., and Leveau, J. H. J. (2002). Phyllosphere microbiology. *Current Opinion in Biotechnology*, 13, 238–243.

Gage, D. J. (2004). Infection and invasion of roots by symbiotic, nitrogen-fixing rhizobia during nodulation of temperate legumes. *Microbiology and Molecular Biology Reviews*, 68, 280–300.

Maier, R. J., and Triplett, E. W. (1996). Toward more productive, efficient, and competitive nitrogen-fixing symbiotic bacteria. *Critical Reviews in Plant Science*, 15, 191–234.

Yanni, Y. G., et al. (2001). The beneficial plant growth-promoting association of *Rhizobium leguminosarum* bv. trifolii with rice roots. *Australian Journal of Plant Physiology*, 28, 845–870.

Dobbelaere, S., Vanderleyden, J., and Okon, Y. (2003). Plant growth-promoting effects of diazotrophs in the rhizosphere. *Critical Reviews in Plant Sciences*, 22, 107–149.

Iniguez, A. L., Dong, Y., and Triplett, E. W. (2004). Nitrogen fixation in wheat provided by *Klebsiella pneumoniae* 342. *Molecular Plant-Microbe Interactions*, 17, 1078–1085.

Galloway, J. N., Schlesinger, W. H., Levy, H., II, Michaels, A., and Schnoor, J. L. (1995). Nitrogen fixation: anthropogenic enhancement-environmental response. *Global Biogeochemical Cycles*, 9, 235–252.

#### Agrobacterium Ti System

Christie, P. J. (2004). Type IV secretion: the *Agrobacterium* VirB/D4 and related conjugation systems. *Biochimica Biophysica Acta*, 1694, 219–234.

Zupan, J., Muth, T. R., Draper, O., and Zambryski, P. (2000). The transfer of DNA from *A. tumefaciens* into plants: a feast of fundamental insights. *Plant Journal*, 23, 11–28.

Cascales, E., and Christie, P. J. (2004). Definition of a bacterial type IV secretion pathway for a DNA substrate. *Science*, 304, 1170–1173.

Gelvin, S. B. (2003). *Agrobacterium*-mediated plant transformation: the biology behind the "gene-jockeying" tool. *Microbiology and Molecular Biology Reviews*, 67, 16–37.

Bent, A. F. (2000). *Arabidopsis* in plant transformation. Uses, mechanisms, and prospects for transformation of other species. *Plant Physiology*, 124, 1540–1547.

Broothaerts, W., Mitchell, H. J., Weir, B., et al. (2005). Gene transfer to plants by diverse species of bacteria. *Nature*, 433, 629–633.

#### Direct Gene Transfer

Taylor, N. J., and Fauquet, C. M. (2002). Microparticle bombardment as a tool in plant science and agricultural biotechnology. *DNA Cell Biology*, 21, 963–977.

Maliga, P. (2004). Plastid transformation in higher plants. *Annual Review of Plant Biology*, 55, 289–313.

#### DNA Integration into Plant Genome

Somers, D. A., and Mararevitch, I. (2004). Transgene integration in plants: poking or patching holes in promiscuous genomes? *Current Opinion in Biotechnology*, 15, 126–131.

Koohli, A., Twyman, R. M., Abranches, R., Wegel, E., Stoger, E., and Christou, P. (2003). Transgene integration, organization and interaction in plants. *Plant Molecular Biology*, 52, 247–258.

#### Gene Silencing

Baulcombe, D. (2004). RNA silencing in plants. *Nature*, 431, 356–363.

Waterhouse, P. J., Wang, M.-B., and Lough, T. (2001). Gene silencing as an adaptive defence against viruses. *Nature*, 411, 831–842.

Voinnet, O., Rivas, S., Mestre, P., and Baulcombe, D. (2003). An enhanced transient expression system in plants based on suppression of gene silencing by the p19 protein of tomato bushy stunt virus. *Plant Journal*, 33, 949–956.

#### Herbicide-Resistant Plants

CaJacob, C. A., Feng, P. C. C., Heck, G. R., Alibhai, M. F., Sammons, R. D., and Padgette, S. R. (2004). Engineering resistance to herbicides. In *Handbook of Plant Biotechnology*, Volume 1, P. Christou and H. Klee (eds.), pp. 333–372, Chichester, U.K.: John Wiley & Sons.

Légère, A. (2005). Risks and consequences of gene flow from herbicide-resistant crops: canola (*Brassica napus* L) as a case study. *Pest Management Science*, 61, 292–300.

#### Insect-Resistant Plants

Perlak, F. J., Fuchs, R. L., Dean, D. A., McPherson, S. L., and Fischhoff, D. A. (1991). Modification of the coding sequence enhances plant expression of insect control protein genes. *Proceedings of the National Academy of Sciences U.S.A.*, 88, 3324–3328.

Diehn, S. H., De Rocher, E. J., and Green, P. J. (1996). Problems that can limit the expression of foreign genes in plants: lessons to be learned from B.t. toxin genes. In *Genetic Engineering*, New York: Plenum Press, Volume 18, J. K. Setlow (ed.), pp. 83–99.

Prieto-Samsónov, D. L., Vásquez-Padrón, R. I., Ayra-Pardo, C., González-Cabrera, J., and de la Riva, G. A. (1997). *Bacillus thuringiensis*: from biodiversity to biotechnology. *Journal of Industrial Microbiology and Biotechnology*, 19, 202–219.

Bernstein, J. A., Bernstein, I. L., Bucchini, L., Goldman, L. R., Hamilton, R. G., Lehrer, S., Rubin, C., and Sampson, H. A. (2003). Clinical and laboratory investigation of allergy to genetically modified foods. *Environmental Health Perspectives*, 111, 1114–1121.

Murdock, L. L., and Shade, R. E. (2002). Lectins and protease inhibitors as plant defenses against insects. *Journal of Agricultural and Food Chemistry*, 50, 6605–6611.

### Virus-Resistant Plants

Gonsalves, D. (1998). Control of papaya ringspot virus in papaya: a case study. *Annual Review of Phytopathology*, 36, 415–437.

### Plants Resistant to Fungi and Bacteria

Dangle, J. L., and Jones, J. D. G. (2001). Plant pathogens and integrated defence responses to infection. *Nature*, 411, 526–533.

Campbell, M. A., Fitzgerald, H. A., and Ronald, P. C. (2002). Engineering pathogen resistance in crop plants. *Transgenic Research*, 11, 599–613.

Gurr, S. J., and Rushton, P. J. (2005). Engineering plants with increased disease resistance: what are we going to express? *Trends in Biotechnology*, 23, 275–282.

### Stress-Tolerant Plants

Zhang, J. Z., Creelman, R. A., and Zhu, J.-K. (2004). From laboratory to field. Using information from *Arabidopsis* to engineer salt, cold, and drought tolerance in crops. *Plant Physiology*, 135, 615–621.

Flowers, T. J. (2004). Improving crop salt tolerance. *Journal of Experimental Botany*, 55, 307–319.

Kumar, S., Dhingra, A., and Daniel, H. (2004). Plastid-expressed betaine aldehyde dehydrogenase gene in carrot cultured cells, roots, and leaves confers enhanced salt tolerance. *Plant Physiology*, 136, 2843–2854.

Umezawa, T., Fujita, M., Fujita, Y. , et al. (2006). Enginnering drought tolerance in plants: discovering and tailoring genes to unlock the future. *Current Opinion in Biotechnology*, 17, 113–122.

### Improvement of Quality and Yield of the Products

Ye, X., Al-Babili, S., Klöti, A., Zhang, J., Lucca, P., Beyer, P., and Potrykus, I. (2000). Engineering the provitamin A ($\beta$-carotene) biosynthetic pathway into (carotenoid-free) rice endosperm. *Science*, 287, 303–305.

Bajaj, S., and Mohanty, A. (2005). Recent advances in rice biotechnology – towards genetically superior transgenic rice. *Plant Biotechnology Journal*, 3, 275–307.

Katsube, T., Kurisaka, N., Ogawa, M. Maruyama, N., Ohtsuka, R., Utsumi, S., and Takaiwa, F. (1999). Accumulation of soybean glycinin and its assembly with the glutelins in rice. *Plant Physiology*, 120, 1063–1073.

Poletti, S., Gruissem, W., and Sautter, C. (2004). The nutritional fortification of cereals. *Current Opinion in Biotechnology*, 15, 162–165.

Miyao, M. (2003). Molecular evolution and genetic engineering of C4 photosynthetic enzymes. *Journal of Experimental Botany*, 54, 179–189.

van Camp, W. (2005). Yield enhancement genes: seeds for growth. *Current Opinion in Biotechnolology*, 16, 147–153.

SEVEN

# *Bacillus thuringiensis* (*Bt*) Toxins: Microbial Insecticides

The concerted effect of the exponentially increasing costs of insecticide development, the dwindling rate of commercialization of new materials, and the demonstration of cross or multiple resistance to new classes of insecticides almost before they are fully commercialized makes pest resistance the greatest single problem facing applied entomology. The only reasonable hope of delaying or avoiding pest resistance lies in integrated pest management programs that decrease the frequency and intensity of genetic selection by reduced reliance upon insecticides and alternatively rely upon multiple interventions in insect population control by natural enemies, insect diseases, cultural manipulations, and host-plant resistance.

– Metcalf, R. L. (1980). Changing role of insecticides in crop protection. *Annual Review Entomology*, 25, 219–256.

The competition for crops between humans and insects is as old as agriculture, but chemical warfare against insects has a much shorter history. Farmers began to use chemical substances to control pests in the mid-1800s. Not surprisingly, the development of insecticides paralleled the development of chemistry: early insecticides were in the main inorganic and organic arsenic compounds, followed by organochlorine compounds, organophosphates, carbamates, pyrethroids, and formamidines, many of which are in use today. In 2001, global sales of chemical insecticides included more than 1.23 million pounds of active ingredients and reached about $9.1 billion a year.

There are disadvantages to relying exclusively on chemical pesticides. Foremost is that widespread use of single-chemical compounds confers a selective evolutionary advantage on the progeny of pests that have acquired resistance to the substances. For example, housefly strains (*Musca domestica*) worldwide have developed resistance to virtually every insecticide used against them. A second problem is that some pesticides affect non-target species, with disastrous results. Unintentional elimination of desirable predator insects has resulted in explosive multiplication of secondary pests. A third concern is the environmental persistence and toxicity of many pesticides, which have led to the abandonment of many chemical pesticides and increased the costs of developing new and safer ones. Cumulatively, these disadvantages provide a strong incentive to find alternative ways of controlling pests.

Like all living things, insects are susceptible to infection by pathogenic microorganisms (bacteria, fungi, and protozoa) and viruses. Many of these biological agents have a narrow host range and consequently do not cause random destruction of beneficial insects and are not toxic to vertebrates. In spite of this very attractive feature, microbial pest control agents represent less than 1% of total insecticide sales.

*Bacillus thuringiensis* has been used for pest control since the 1920s and still accounts for over 90% of the miniscule share of the insecticide market attributable to biological control agents. In addition, since 1996, transgenic crop plants (primarily soybeans, corn, and cotton) that express *B. thuringiensis* insecticidal proteins have been widely adopted. In 2002, more than 35 million acres of these transgenic cotton and corn crops were planted worldwide. The widespread and increasing use of *B. thuringiensis* and of its toxins has made this soil bacterium and its entomocidal proteins the subjects of intense, multidisciplinary examination. This chapter attempts a synthesis of the current knowledge gained through these studies.

## *BACILLUS THURINGIENSIS*

The discovery of *B. thuringiensis* is credited to Shigetane Ishiwata. In Japan in 1901, he isolated the organism responsible for flacherie, a disease of silkworm (*Bombyx mori*) larvae and named it *Bacillus sotto*. (*Sotto* is a Japanese word roughly equivalent to *limp* in English. Larvae dying of this disease become soft and flaccid and eventually turn black.) A similar bacillus was isolated in 1911 by Ernst Berliner from diseased larvae of the Mediterranean flour moth (*Anagasta kühniella*). Berliner named his isolate *B. thuringiensis* after the province of Thüringen, where the discovery was made. Numerous strains of *B. thuringiensis* have been described since that time, and each has its own distinct spectrum of pathogenic effects on host insects.

Up to 1976, only *B. thuringiensis* strains pathogenic to Lepidoptera (butterflies and moths) were known. These strains showed poor larvicidal activity against blackflies, mosquitoes, and beetles. Subsequent surveys, however, have led to the isolation of strains pathogenic to dipteran (flies, midges, and mosquitoes) and coleopteran (beetles) pests. These pathogens are neither rare nor difficult to isolate. The greater than 60-year gap between the discovery of the lepidopteran pathotype and that of the dipteran and coleopteran pathotypes was solely the result of the lack of a strong incentive to search. This incentive finally came from the eventual recognition of an urgent need to find new biological agents to control disease-causing insect pests. Joel Margalit has provided a vivid account of the discovery of the first dipteran pathotype, *B. thuringiensis* var. *israelensis*:

> In 1975–76, Drs. Tahori and Margalit conducted a survey in Israel for biocontrol agents against mosquitoes. During this survey (August 1976) the senior author of this paper came across a small pond in a dried-out riverbed in the north central Negev Desert, near Kibbutz Zeelim. This mosquito breeding site, 15 × 60 m, with a maximum depth of 30 cm, contained brackish water with an approximate salinity of 900 mg Cl/liter and a heavy load of decomposing organic material. A very dense population of exclusively *Culex pipiens* [a common species of mosquitoes] complex dead and dying larvae was found as a "thick carpet" on the surface in an epizootic situation [an epizootic is a disease that affects many animals of one kind at the same time; corresponding to *epidemic* as applied to diseases of humans]. In addition, pupae and sunk adults attempting to emerge from their pupal cases were floating on the surface.
>
> A sample collected from the edge of the pool, containing dead and decomposing larvae, water and silty mud was taken to the laboratory and refrigerated. Bacteria were isolated from this sample in the lab, in association with Mr. L. H. Goldberg, and purified to single colonies. Thus, from a single colony designated ONR 60A, were derived all known cultures of *B.t.i.* now in use.
>
> – Margalit, J., and Dean, D. (1985). The Story of *Bacillus thuringiensis* var. *israelensis* (*B.t.i.*). *Journal of the American Mosquito Control Association*, 1, 1–7.

The potential practical importance of *B. thuringiensis* var. *israelensis* was recognized immediately. Bloodsucking dipteran insects, such as mosquitoes and blackflies, transmit a broad spectrum of animal diseases. The blood-borne pathogens they carry include viruses, bacteria, protozoa, and helminths. Mosquitoes, for example, are the vectors that spread the protozoan that causes malaria, with its annual incidence of 200 to 300 million cases. Insecticidal preparations from *B. thuringiensis* strains toxic to mosquitoes and blackflies are now used successfully as biological control agents in virtually every country where such pests are a severe problem.

*B. thuringiensis* var. *tenebrionis*, a pathotype effective against the larvae of Coleoptera was described in 1983. This strain can infect the Colorado potato beetle, the most damaging pest of potatoes in Europe and North America. Populations of this beetle can reach a density of hundreds of insects on a single plant. It has developed resistance to many chemical insecticides and is extremely difficult to control. Fortunately, among 850 strains of *B. thuringiensis* isolated from a wide variety of locations in the United States in 1988, 55 were active against Coleoptera.

*B. thuringiensis* is a Gram-positive soil bacterium that can grow either by digesting organic matter derived from dead organisms (*saprophytic metabolism*) or by colonization within living insects (*parasitic metabolism*). The bacterium has been isolated from the dust inside silkworm-rearing houses in Japan and from the soils surrounding them. Outbreaks of infection with *B. thuringiensis* are found in insectaries rearing pink bollworm and among larvae inhabiting grain-storage bins. Although *B. thuringiensis* is commonly found inside insects, epizootics, such as the one described by Margalit, are rare in nature. The organism appears to have a low capacity to spread through insect populations. One study confined healthy larvae of *A. kühniella* and *Pieris brassicae* with larvae of the same species that had been infected with *B. thuringiensis*. Most of the diseased larvae died within a few days, and their carcasses were left in the cages. Nevertheless, none of the initially healthy *A. kühniella* and only three of 180 *P. brassicae* larvae developed *B. thuringiensis* infections. Thus, *B. thuringiensis* acts more like a chemical insecticide than as an infectious agent.

*B. thuringiensis* strains currently are classified into different serotypes or varieties (subspecies) on the basis of their flagellar antigens. This taxonomy is important in identifying the particular strains, but it has little value for predicting the specificity and potency of the insecticidal proteins (see below) produced by a particular strain.

Strains of *B. thuringiensis* are also classified into six pathotypes, on the basis of their insecticidal range: (a) lepidopteran-specific (e.g., var. *berliner*); (b) dipteran-specific (e.g., var. *israelensis*); (c) coleopteran-specific (e.g., var. *tenebrionis*); (d) active against both Lepidoptera and Diptera (e.g., var.*aizawai*); (e) active against both Lepidoptera and Coleoptera (var. *thuringiensis*); and (f) no known toxicity in insects (e.g., var. *dakota*). Even within each of these pathotypes, the various strains differ markedly with respect to potency and to specificity against different insects.

## CRYSTALLINE INCLUSION BODIES

The early studies on flacherie, in 1915, found that only those *B. thuringiensis* cultures that had undergone sporulation were toxic to silkworm larvae. The importance of this observation was not appreciated for over 40 years. It seems obvious now that the toxic agent must therefore be some molecular substance produced specifically during the sporulation process. Once this connection was made in the 1950s, the toxic substance was not hard to identify. Unlike almost all other *Bacillus* species, *B. thuringiensis* produces parasporal crystalline inclusion bodies during sporulation, and these are readily visible with a light microscope. The stages in the sporulation of *B. thuringiensis* var. *kurstaki* are shown schematically in Figure 7.1. Approximately eight hours into the sporulation process, a large bipyramidal crystal and a smaller cuboidal crystalline inclusion develop within the vegetative cell. Chemical analysis showed the inclusion bodies to consist of proteins that exhibit highly specific insecticidal activity. Actively growing cells lack the crystalline inclusions and thus are not toxic to insects.

## MECHANISM OF ACTION OF A *B. THURINGIENSIS* INSECTICIDE

The commercial insecticide Dipel™ illustrates how a *B. thuringiensis* insecticide works. This particular product is a dry powder consisting of sporulated cells of *B. thuringiensis* var. *kurstaki* that is applied to vegetation by dusting. The active ingredients, which are the large protein-containing crystalline inclusions and the spores, are ingested by larvae consuming the treated leaves. The inclusions contain five different insecticidal crystal proteins.

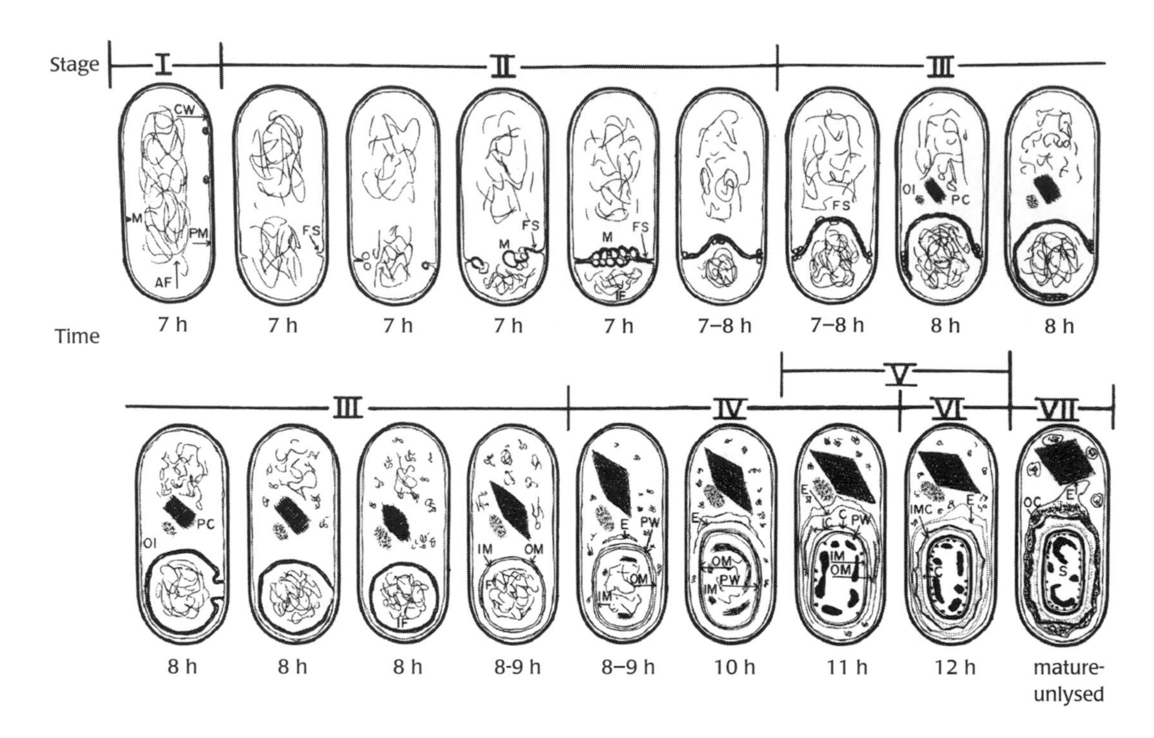


FIGURE 7.1

Diagrammatic scheme of sporulation in *Bacillus thuringiensis*. Abbreviations: M, mesosome; CW, cell wall; PM, plasma membrane; AF, axial filament; FS, forespore septum; IF, incipient forespore; OI, ovoid inclusion; PC, parasporal crystal; F, forespore; IM, inner membrane; OM, outer membrane; PW, primordial cell wall; E, exosporium; LC, lamellar spore coat; C, cortex; UC, undercoat; OC, outer spore coat; S, mature spore in an unlysed mother cell. [Reproduced with permission from Bechtel, D. B., and Bulla, L. A., Jr. (1976). Electron microscopic study of sporulation and parasporal crystal formation in *Bacillus thuringiensis*. *Journal of Bacteriology*, 127, 1472–1481.]

The crystals consist of inactive protoxin molecules known as δ-endotoxins. After the larva's alkaline midgut juices have dissolved the crystals, larval gut proteases cleave the protoxins, thus generating the active protein toxins. Many insects have a peritrophic membrane – a sleevelike, noncellular semipermeable membrane – lining the midgut region and separating the contents of the gut lumen from the digestive epithelial cells of the midgut wall. The mature protein toxins diffuse through the peritrophic membrane, bind to specific receptors on the plasma membrane of larval gut epithelial cells, and insert into the membrane to form cation-conducting pores, 10 to 20 Å in diameter, thereby making the cells permeable to ions and protons. The influx of water that accompanies the entrance of ions into the intestinal cells causes them to swell and lyse (Figures. 7.2 and 7.3). The loss of ion regulation also causes paralysis of the muscles of the gut and mouth parts. As a result, feeding stops soon after ingestion of the crystals.

In addition to the destruction and paralysis caused by the crystalline inclusions, vegetative *B. thuringiensis* bacteria germinating from the spores enter the larval hemolymph through the damaged gut epithelium and multiply. The resulting bacteremia promotes an intoxication process leading to death within one to three days. Some of the individual steps in the above sequence are explored in greater detail later in this chapter.

## MULTIPLE δ-ENDOTOXINS WITH DIFFERING SPECIFICITIES IN A SINGLE *B. THURINGIENSIS* STRAIN

Some *B. thuringiensis* strains produce only one δ-endotoxin; others produce several δ-endotoxins having different specificities. Most current

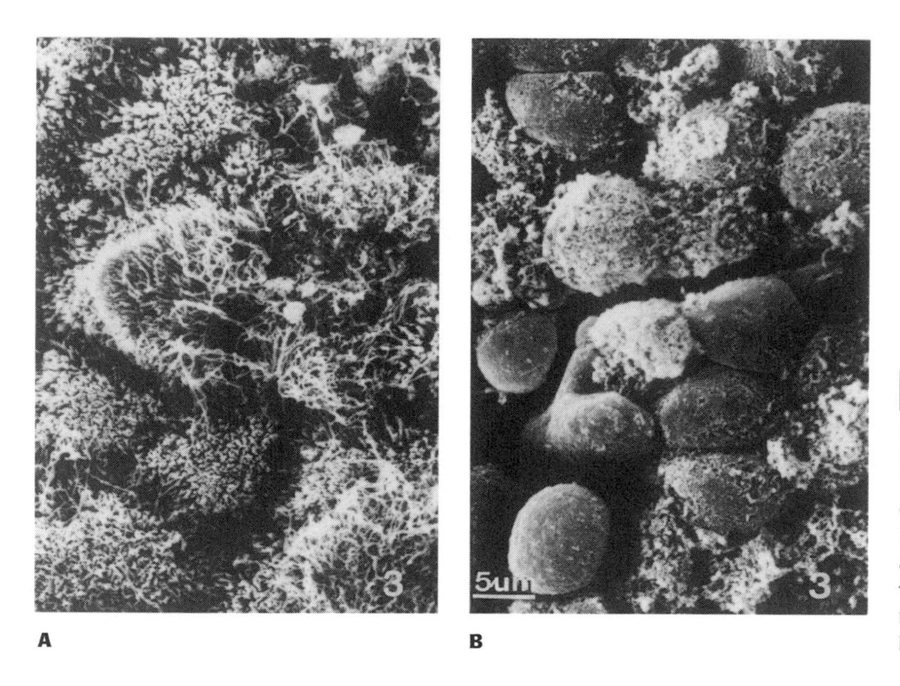


FIGURE 7.2

(**A**) Scanning electron micrograph of a healthy midgut epithelium of a large white butterfly (*Pieris brassicae*) larva. (**B**) Scanning electron micrograph of the midgut epithelium of a larva fed 5 μg of δ-endotoxin and dissected 15 min after endotoxin ingestion. Note the resulting disappearance of microvilli from the epithelium. (Courtesy of Dr. Peter Lüthy.)

preparations for the control of caterpillars, such as Dipel™, use *B. thuringiensis* var. *kurstaki* strain HD-1. This strain produces two crystalline inclusions, a large bipyramidal structure and a small cuboidal one that is often located at one apex of the bipyramidal crystal. The bipyramidal crystal contains several protoxin proteins of 135 to 145 kDa that have insecticidal activity against Lepidoptera. The cuboidal crystal contains a single protoxin of 65 kDa with activity against both Lepidoptera and Diptera.

The gene for the 135-kDa protein resides on a 67-kbp plasmid, whereas genes encoding 140-kDa protoxins and the gene for the 65-kDa protein reside on a 174-kbp plasmid. The 67-kbp plasmid is readily transmitted from strain HD-1 to other *B. thuringiensis* strains. The 174-kbp plasmid is not self-transmissible.

## NOMENCLATURE FOR THE *B. THURINGIENSIS* CRYSTAL PROTEINS

The crystal protein genes of *B. thuringiensis* fall into numerous classes, whether on the basis of the protein sequences they encode or the resulting proteins' insecticidal spectra. In many cases, the two methods of classification yield discordant results. On the basis of insecticidal specificity, for example, genes for crystal (Cry) proteins may be classed as Lepidoptera specific, Diptera specific, Lepidoptera and Diptera specific, Coleoptera specific, and Lepidoptera and Coleoptera specific. The classifications based on protein-sequence homology, however, are not rigorously predictive of these activities. In *B. thuringiensis* var. *israelensis*, one crystal constituent is a small protein (designated CytA) that exhibits cytolytic activity against a variety of cells from invertebrates and vertebrates, but it is totally unrelated in sequence to the *cry* genes. The pathotype of a given *B. thuringiensis*

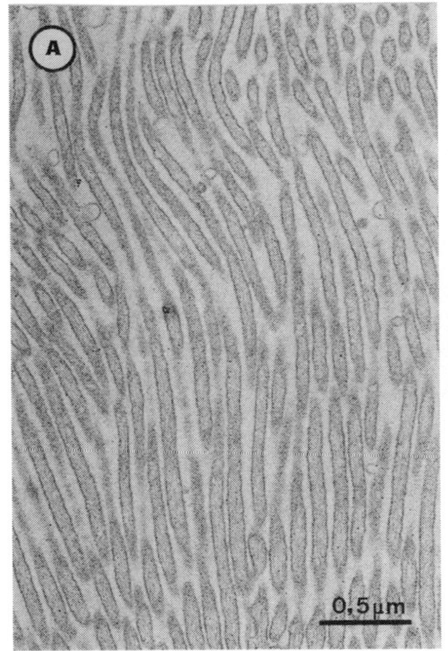


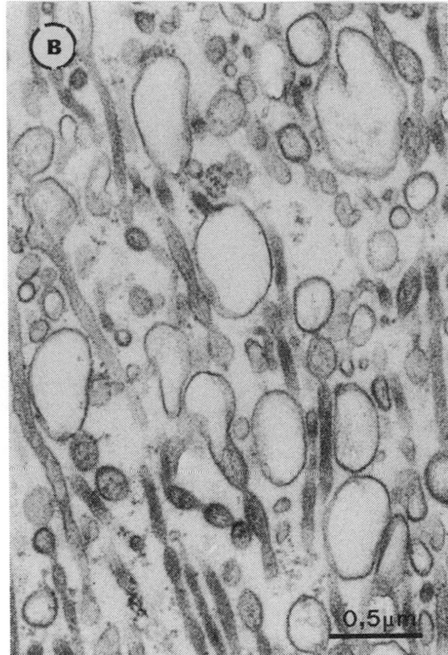


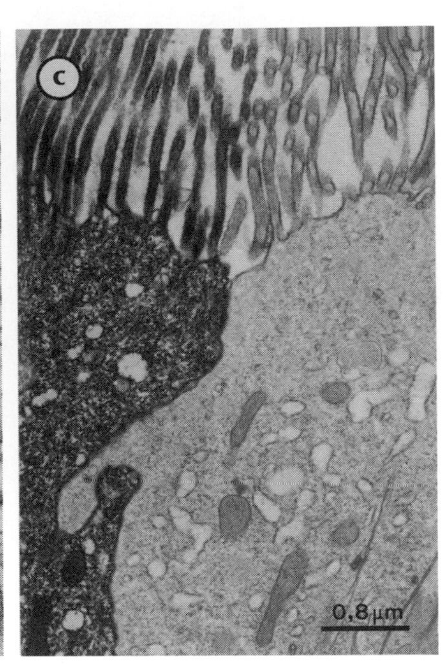


**FIGURE 7.3**

(**A**) Intact microvilli of a columnar cell from the gut epithelium of *Pieris brassicae* (control). (**B**) Appearance of microvilli 10 min after $\delta$-endotoxin ingestion. (**C**) Within 10 min of ingestion of $\delta$-endotoxin, the cells of the midgut epithelium begin to lose the ability to control permeability and thus become permeable to the indicator stain, ruthenium red. The cell on the left has taken up the stain, whereas the cell on the right, which still retains control of permeability, does not. [Reproduced with permission from Lüthy, P., and Ebersold, H. R. (1981). *Bacillus thuringiensis* delta-endotoxin: histopathology and molecular mode of action. In *Pathogenesis of Invertebrate Microbial Diseases*, E. W. Davidson (ed.), p. 244, Totowa, NJ: Allanheld, Osmun Publishers.]

strain is a reflection of the particular endotoxin gene or genes the strain expresses.

The nomenclature now adopted for the *B. thuringiensis* insecticidal proteins is entirely sequence based. It is used to classify the crystal protein genes of *B. thuringiensis* strains into two families: (1) A Cry protein is a parasporal inclusion (crystal) protein that shows a demonstrable toxic effect on a target organism, or any protein that has obvious sequence similarity to a known Cry protein. (2) A Cyt protein is a parasporal inclusion (crystal) protein that exhibits hemolytic (cytolytic) activity, or any protein that has obvious sequence similarity to a known Cyt protein.

The nomenclature for Cry and Cyt proteins is based entirely on phylogenetic trees constructed from a multiple alignment and distance matrix of full-length toxin sequences for each of these two families of proteins. As shown in Figure 7.4, the nomenclature reflects the relationship between Cry proteins based on the branching pattern of the tree and classified according to boundaries set at three levels of percent amino acid sequence identity – 45%, 76%, and 95%. The designation given to a particular protein, such as Cry1Ab or Cry1Hb, depends on the location of the node where the protein enters the tree relative to these boundaries.

## *B. THURINGIENSIS* $\beta$-EXOTOXIN

During the active phase of vegetative growth, certain varieties of *B. thuringiensis* produce a low molecular weight heat-stable toxin called $\beta$-exotoxin. This toxin has a nucleotide-like structure (Figure 7.5) and inhibits the activity of DNA-dependent RNA polymerase of both bacterial and mammalian cells. Its potential to control the Colorado potato beetle has been

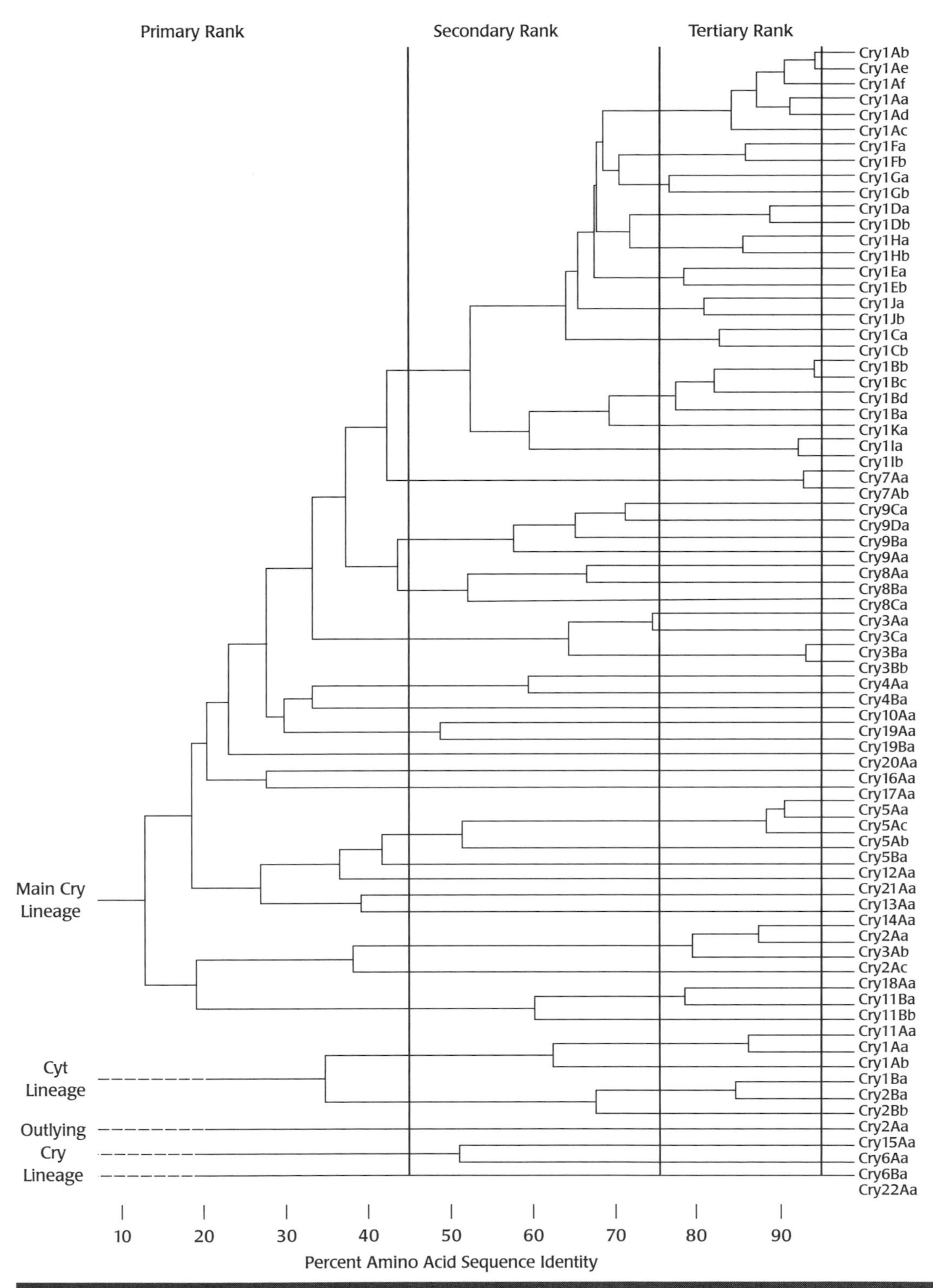


**FIGURE 7.4**

A phylogram based on similarity of amino acid sequences in Cry and Cyt proteins. This phylogenetic tree was constructed from a multiple alignment and distance matrix of the full-length toxin sequences as described in the source reference. The *vertical bars* mark the four levels of nomenclature ranks. The lower four Cyt lineages share a low percentage of identical residues and no conserved sequence blocks with the Cry proteins and are regarded as a separate family. [Reproduced with permission from Crickmore, N., et al. (2004). Revision of the nomenclature for the *Bacillus thuringiensis* pesticidal crystal proteins. *Microbiology and Molecular Biology Reviews*, 62, 807–813.]

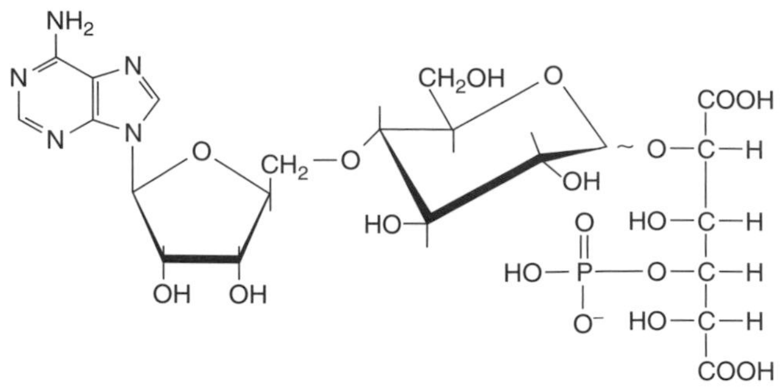

FIGURE 7.5

$\beta$-Exotoxin of *B. thuringiensis*.

examined in the United States, but its similarity to a nucleotide, its teratogenic effects on insects, and its toxicity when injected into mammals have argued against its use as a biological control agent. Currently, the use of *B. thuringiensis* preparations containing $\beta$-exotoxin is forbidden in North America and Western Europe. In Eastern Europe and some parts of Africa $\beta$-exotoxin preparations) have been used effectively to control fly larvae in piggeries, latrine and compost toilets at insecticidal doses that do not affect vertebrates.

## MECHANISM OF $\delta$-ENDOTOXIN ACTION

To give a more detailed description of the mechanism of *B. thuringiensis* insecticide action, we divide the mechanism into four stages.

### Stage 1: Proteolysis of Protoxins in the Insect Gut to Generate Active Toxin Fragments

The molecular weights of the protoxins range from about 70 to 145 kDa. Regardless of the protoxin's molecular weight, however, or its ultimate insecticidal specificity, proteolysis in the gut generates active toxin fragments of similar size, 60 to 70 kDa. Deletion mapping of the different gene types for various 130- to 135-kDa protoxins has shown that in each case the mature toxin resides within the amino-terminal 62- to 70-kDa portion of the protoxin. Depending on the particular protoxin, the amino-terminal border of the active fragment ranges from residue 29 to 39, and the carboxyl-terminal border from residue 607 to 677 (Figure 7.6). The *cry3Aa* gene, encoding a Coleoptera-specific protoxin, directs the synthesis of a 72-kDa protein, which is converted into a 66-kDa toxin by spore-associated proteases that remove 57 amino-terminal residues (Figure 7.6). The *cry3Aa* gene is homologous to the toxin-encoding domain of the genes specifying the larger 130- to 135-kDa protoxins but lacks a region corresponding to the 3′ portion of these genes. Deletion analysis confirmed these conclusions, showing that any truncation of the *cry3Aa* gene at its 3′-end leads to loss of toxic activity.

The 3′ portions of the large *cry* genes, encoding the part of the protoxin sequence starting at about residue 700, show a high degree of sequence homology. This finding suggests that the carboxyl-terminal portion of such protoxins may play a role in the formation of the crystalline protoxin inclusions.

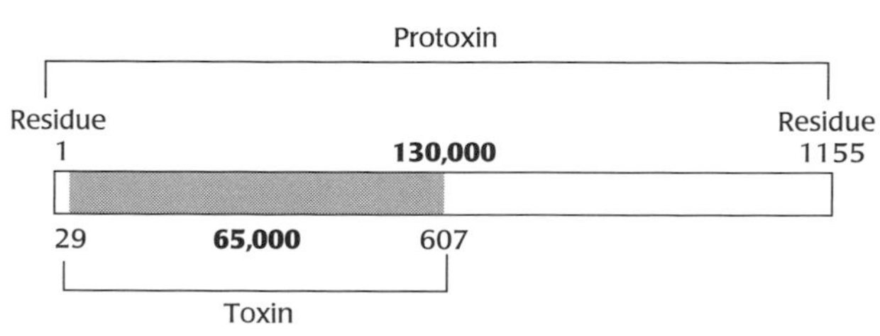


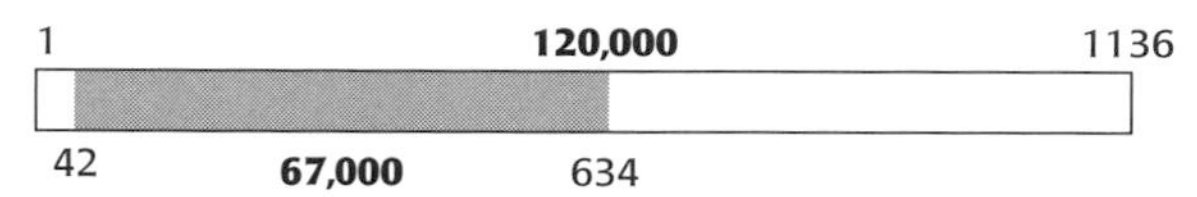


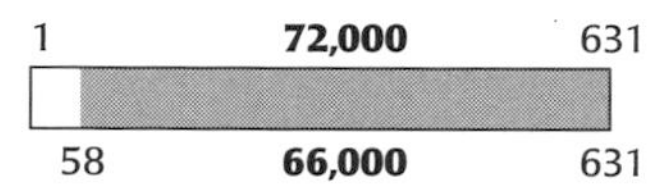


**FIGURE 7.6**

Structural features of *B. thuringiensis* insecticidal crystal proteins. Boldface numbers above the *bars* give the molecular weights of the protoxins; those below the *bars* give the molecular weights of the toxins.

Redrawn based on artwork from the first edition (1995), published by W.H. Freeman.

The need for dissolution of the crystalline inclusions in the insect midgut followed by proteolysis before the mature Cry protein can be generated focuses attention on variations in the midgut pH and in the nature of the proteolytic enzymes in various orders of insects and various families within these orders. The midgut pH and the specificity of the proteolytic enzymes present are likely to be important determinants of the selective toxicity of particular Cry protoxins. The pH of the contents in the midgut of Lepidoptera is highly alkaline, ranging from pH 9.5 to 10.5, and the proteases present are trypsins and chymotrypsins with pH optima at the prevailing midgut pH. In most Coleoptera, the pH range of the midgut contents is 5.5 to 8.0 and the proteases present are aspartic and cysteine proteases, enzymes with very different sequence specificity for polypeptide cleavage from trypsins and chymotrypsins. The pH of the midgut contents of Diptera varies widely.

### Stage 2: Binding of Cry Toxins to Specific Receptors on Midgut Epithelial Cells

The toxicity and specificity of *B. thuringiensis* Cry toxins correlate with their high-affinity binding to receptors on midgut epithelial cells. Specific classes of glycoproteins and glycolipids have been identified as the receptors on the surface of these epithelial cells.

In *Manduca sexta* (the tobacco hornworm) and *B. mori*, specific cell adhesion molecules, cadherins, act as Cry toxin receptors in the gut epithelium. Cadherins are members of a large family of calcium-dependent transmembrane glycoproteins that mediate cell-to-cell adhesion. Binding of Cry toxins to the cadherin receptors initiates disruption of the epithelium and severe damage to the entire midgut tissue. It is of special interest that in *M. sexta*, the particular cadherin that serves as the receptor for Cry1A toxins is specifically expressed in the larval stage of the insect's life cycle and is not present in any other stage of its life. In *Heliothis virescens*, retrotransposon-mediated disruption of a specific cadherin gene results in resistance to the Cry1Ac

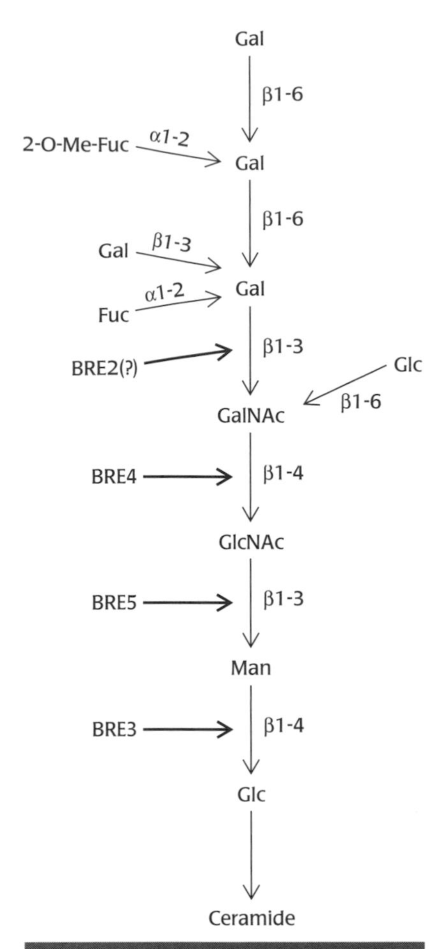


**FIGURE 7.7**

Structure of a *Caenorhabditis elegans* glycosphingolipid receptor for Cry5B. All of the *C. elegans* glycosphingolipids that bind Cry5B share a common core oligosaccharide structure: GalNAc($\beta$1–4)GlcNAc($\beta$1–3)Man($\beta$1–4) Glc. The glycosidic linkages proposed to be catalyzed by the BRE enzymes are indicated by arrows in bold face. Abbreviations: Glc, glucose; GlcNAc, *N*-acetylglucosamine; Gal, galactose; Fuc, fucose; Man, mannose; 2-*O*-Me-Fuc, 2-*O*-methylfucose. [Griffitts, J. S., et al. (2005). Glycolipids as receptors for *Bacillus thuringiensis* crystal toxin. *Science*, 307, 922–925.

toxin. These observations show that toxin interactions with specific midgut epithelial cadherins have an essential role in determining the host range of entomopathogenic *B. thuringiensis* strains.

Another class of receptors for Cry1Ac and several other Cry toxins found on the midgut epithelium surface in *M. sexta* is a ubiquitous midgut protease, aminopeptidase N (APN), a 120-kDa glycoprotein with a glycosylphosphatidylinositol anchor. An *N*-acetyl-D-galactosamine moiety forms a part of the site recognized by the Cry toxin and contributes in a major way to its high-affinity binding.

Certain of the *B. thuringiensis* Cry proteins target nematodes. Elegant studies with the nematode *Caenorhabditis elegans* have shown that in this host organism, glycosphingolipids function as specific receptors for Cry14A and Cry5B. Cry14A targets nematodes and insects, and Cry5B targets nematodes. The minimum portion of the glycosphingolipid that is essential to its function as a receptor is the core tetrasaccharide (*N*-acetylgalactosamine $\beta$1-4 *N*-acetylglucosamine $\beta$1-3 mannose $\beta$1-4 glucose) linked to ceramide. This core is a specific carbohydrate signature in invertebrates, conserved in nematodes and insects but absent in vertebrates. The Cry proteins bind the intact receptor (Figure 7.7) much more tightly than they bind the minimum core.

The findings with *C. elegans* suggested that glycolipids may also function as receptors in insects. Indeed, Cry1Aa, Cry1Ab, and Cry1Ac all bind specifically the same glycolipids extracted from the midguts of *M. sexta*. These results support the proposition that specific glycolipids act as receptors for Cry toxins in insect cells. Why would there be two kinds of receptors, glycoprotein and glycolipid, in the mechanism of action of these toxins? The favored hypothesis is that glycolipid and glycoprotein receptors both play a role, either sequentially or simultaneously, in positioning or clustering Cry proteins on the outer face of the cytoplasmic membrane in a manner that facilitates the insertion of the toxins into the bilayer.

### Stage 3: Formation of Transmembrane Pores

Binding of the Cry toxin to the receptors leads to two outcomes: (1) The toxin is localized at the external surface of the cytoplasmic membrane of the epithelial cells of the midgut brush border membrane, and (2) the binding induces a conformation change in the toxin which is necessary for membrane insertion. The initial interaction of the toxin with the receptor is reversible, but the subsequent membrane insertion step is irreversible. Upon insertion into the membrane, the toxin oligomerizes and forms transmembrane cation-selective pores. The pores allow equilibration of cation concentrations across the membrane. The entry of sodium ions with an accompanying influx of water leads to swelling and eventual rupture of the cell.

### Stage 4: Bacteremia

The vegetative cells of *B. thuringiensis* that germinate from the ingested spores are able to enter the hemolymph through the damaged epithelial cell layer but, apparently, do not multiply there. Bacteremia, combined with starvation, leads to the death of the larva. Surprisingly, studies with the gypsy moth have shown that *B. thuringiensis* does not kill the larvae in the absence of the indigenous *Enterobacter* sp. midgut bacteria. The *Enterobacter* sp. achieve a high population in the hemolymph, whereas *B. thuringiensis* appeared to die in the hemolymph. In summary, *B. thuringiensis*-induced mortality depends on the indigenous enteric bacteria.

## STRUCTURE–FUNCTION RELATIONSHIPS IN THE INSECTICIDAL CRYSTAL PROTEINS

Known Cry toxin crystal structures include Coleoptera-specific Cry3Aa and Cry3Bb1, Lepidoptera-specific Cry1Aa and Cry1Ac, and Lepidoptera- and Diptera-specific Cry2Aa. All of these structures, apart from the characteristics that give them their insect specificities, share a common topology made up of three domains (Figure 7.8). The great value of sequence and structure

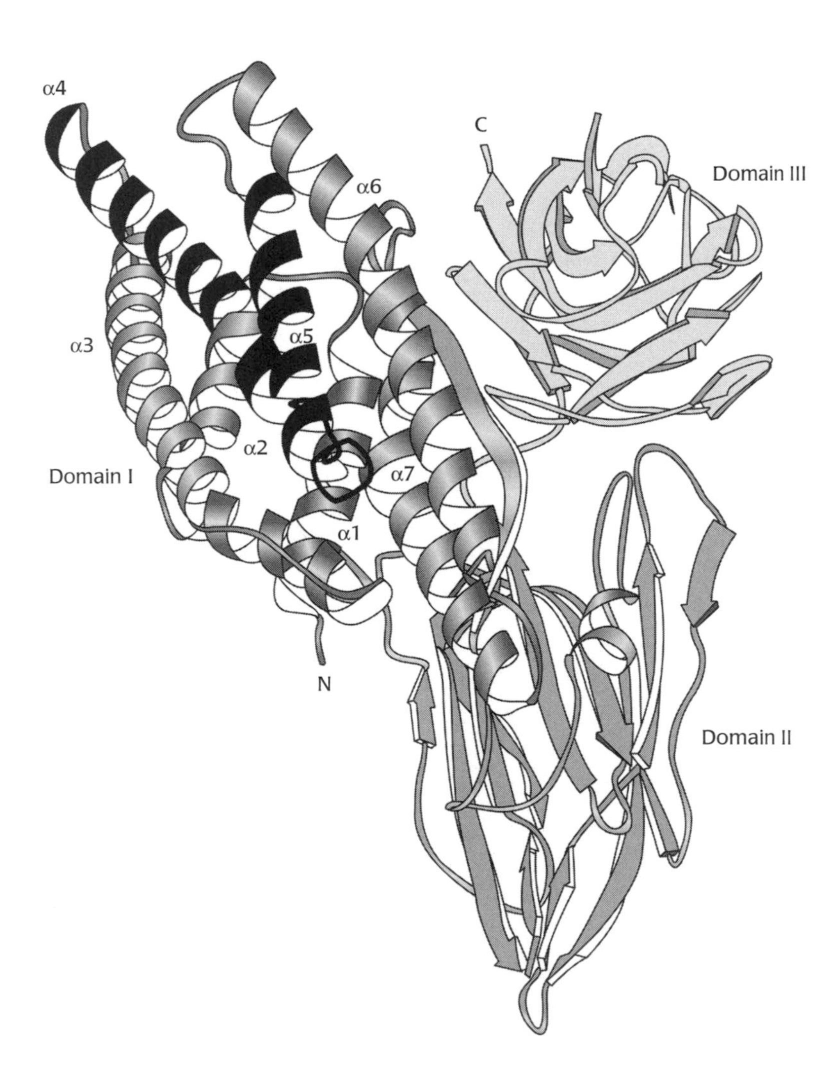


**FIGURE 7.8**

Ribbon diagram of the structure of the 644-residue activated form of a Cry3A δ-endotoxin (Protein Data Base code 1dlc). This protein, which is toxic to the Colorado potato beetle, is representative of the three-domain Cry proteins. *Dark shading* indicates the putative transmembrane region of domain I. [Reproduced with permission from Parker, M. W., and Feil, S. C. (2005). Pore-forming protein toxins: from structure to function. *Progress in Biophysics and Molecular Biology*, 88, 91–142.]

databases is illustrated by reviewing what such data have revealed about the roles of each of these domains.

The amino-terminal domain (domain I in Figure 7.8), a seven-helix bundle with long amphipathic helices, forms part of the transmembrane pore. Two of the helices are sufficiently long to span a 30-Å thickness of a typical bilayer membrane. It is proposed that formation of the pore involves oligomerization of several toxin molecules. Domain I shares many structural similarities with other pore-forming bacterial toxins, such as hemolysin E and colicins Ia and N. Toxins with mutations in domain I bind receptors but frequently fail to insert in the membrane.

Domain II (Figure 7.8) makes a major contribution to receptor binding and consequently to the toxin's insect specificity. This domain consists of the three antiparallel $\beta$-sheets that form a $\beta$-prism remarkably similar to the structure seen in three unrelated carbohydrate-binding proteins – jacalin and Mpa, which are lectins, and vitelline, with 64%, 65%, and 75% overlap of equivalent residues, respectively. Determination of the structure of jacalin in complex with its carbohydrate ligand showed that the ligand was bound to the exposed loops at a location corresponding to the apex of domain II. Cry toxins with mutations in these loops showed large changes in the kinetics of binding to insect midgut brush border membrane vesicles. This finding led to the inference that insecticidal specificity is determined by the carbohydrate ligand specificity of the domain II lectin fold.

The carboxyl-terminal domain III is a sandwich of two twisted antiparallel sheets (Figure 7.8). This domain has a close structural similarity to the cellulose-binding domain from *Cellulomonas fimi* $\beta$-1,4-glucanase C (with 75% overlap of similar residues) and to the structures of several other carbohydrate-binding proteins. Domain III is believed to bind to *N*-acetylgalactosamine moieties attached at *O*-glycosylation sites on APN. If so, the insecticidal specificity of Cry toxins may be the outcome of interaction of two different lectinlike domains with distinct receptors. It remains to be seen whether these domains function independently or cooperatively.

## STUDY OF BACTERIAL TOXIN–TARGET HOST INTERACTION BY GENE TRANSCRIPTION PROFILING

The soil bacterium *B. thuringiensis* lives in close proximity to many species of nematodes that inhabit the soil and feed on bacteria. This suggests that *B. thuringiensis* toxins may have evolved as a way to protect the bacteria from nematodes. One of the most extensively studied and well understood of all organisms is the nematode *C. elegans*, which naturally feeds on bacteria.

Screening of the effect of $\delta$-endotoxins belonging to the three-domain Cry protein family on *C. elegans* revealed that four of these proteins, Cry5B, Cry6A, Cry14A, and Cry 21A, each showed a high level of toxicity. Their effects on progeny production furnished a measure of their relative toxicities, as brood size appears to reflect the health of the mother's intestine. By this assay, Cry14A was judged the most toxic, with a 50% inhibition (or $IC_{50}$) of 16 ng/$\mu$l, and Cry6A had the lowest toxicity, with an $IC_{50}$ of 230 ng/$\mu$l.

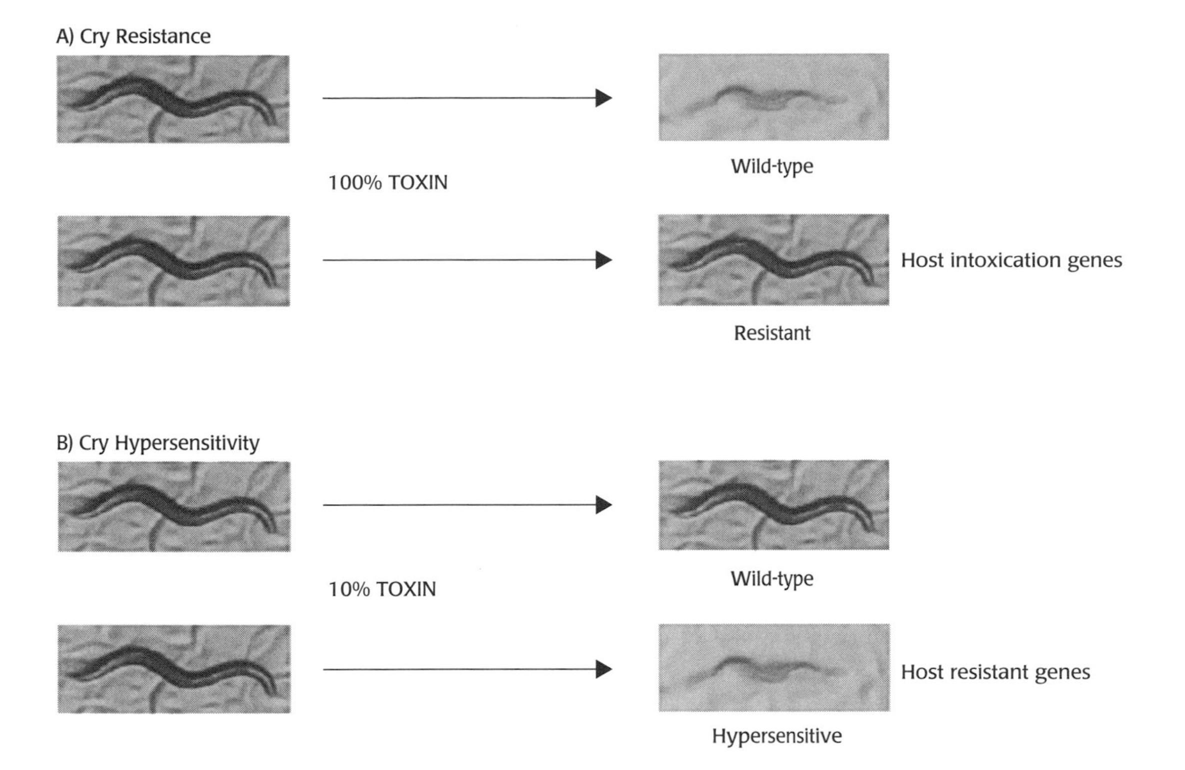


**FIGURE 7.9**

An approach to understanding the interactions between Cry toxins and the nematode *C. elegans*. **(A)** Screening for host resistance. Wild-type worms fed *E. coli* in which 100% of the bacteria express Cry protein become intoxicated, as indicated by their small size and pale color, whereas worms with a resistant mutation do not. Identification of the mutant alleles in the resistant worms will reveal host components required for Cry toxicity. **(B)** Screening for host hypersensitivity. Wild-type worms fed a mixture of *E. coli* in which only 10% of the bacteria produce the Cry protein do not become intoxicated. Hypersensitive worms have increased susceptibility to the Cry protein and do become intoxicated on this lower dose. Identification of alleles that confer a hypersensitive phenotype will reveal host components that are required for defense against the toxin. [Reproduced with permission from Huffman, D. L., Bischof, L. J., Griffitts, J. S., and Aroian, R. V. (2004). Pore worms: using *Caenorhabditis elegans* to study how bacterial toxins interact with their target host. *International Journal of Medical Microbiology*, 293, 599–607.]

The complete sequence of the *C. elegans* genome is known, and microarrays can be prepared for whole-genome gene transcription profiling under various conditions. Appropriate experiments can be designed that lead to identification of host factors that promote pathogenesis and of those that protect the host.

In one such study, a recombinant *Escherichia coli* strain was engineered to express Cry5B. *C. elegans* feeding on this recombinant *E. coli* strain showed signs of intoxication, whereas feeding on wild-type *E. coli* produced no ill effects. Researchers controlled the dose of toxin by mixing recombinant and wild-type *E. coli* cells in desired proportions.

Subsequently, ethylmethanesulfonate-induced mutants of *C. elegans* were screened to identify those that had acquired resistance to Cry5B. Genes required for intoxication were detected by the screening method illustrated in Figure 7.9. Mutants in four genes, *bre-2* to *bre-5*, detected by this screen were shown to be resistant to Cry5B. In contrast to wild-type *C. elegans*, these mutants were unable to absorb Cry5B into their intestinal cells. Further analysis showed that the genes encode glycosyltransferases that function in a single genetic pathway required for intoxication, the pathway that leads to the biosynthesis of the common oligosaccharide core within complex ceramide-linked oligosaccharides (Figure 7.7). This research provides a solid foundation for the conclusion discussed earlier that these glycolipids function in nematodes, and in at least some insects, as receptors of Cry toxins.

Genes that contribute to host resistance can be detected by looking for mutants that show hypersensitivity to the toxin. Such an assay is illustrated in

Figure 7.9B. In this experiment, *C. elegans* was fed a mixture of wild-type and recombinant *E. coli* cells so that the dose of toxin ingested by the worm was decreased 10-fold. At this dose, as the figure shows, wild-type *C. elegans* is not affected but hypersensitive mutants are intoxicated. Identification of the alleles that confer hypersensitive phenotypes reveals the host components that contribute to defense against the toxin. Gene-transcription profiling experiments demonstrate the complexity of the situation by showing that over 5% of the *C. elegans* genome is transcriptionally regulated in response to Cry5B.

## *Bt* CYTOLYTIC TOXINS

As noted previously, the δ-endotoxins make up two multigenic families, *cry* and *cyt* (Figure 7.4). The Cyt proteins are produced by *B. thuringiensis* subsp. *israelensis* and a few other subspecies. Whereas the Cry proteins are predominantly toxic to Lepidoptera and Coleoptera, the Cyt proteins are toxic *in vivo* to members of the Diptera, such as the larvae of *Aedes* and *Anopheles* mosquitoes, which transmit dengue fever and malaria, respectively, and to blackflies, the vectors for river blindness.

The Cyt proteins are much shorter polypeptides than the Cry proteins and show no amino acid sequence homology with them. The structure of Cyt2Aa, a 245–amino acid δ-endotoxin found in the parasporal inclusions of *B. thuringiensis* subsp. *kyushuensis*, is shown in Figure 7.10. Unlike the three-domain Cry proteins, Cyt2Aa is an $\alpha/\beta$ protein with a three-layer core. Amino acid–sequence comparisons support the view that the structure of Cyt2Aa is fairly representative of that of the Cyt toxin family in general.

The receptors for Cyt δ-endotoxins are phospholipids. The nature of a given phospholipid's polar head group and the need for an unsaturated fatty acyl chain at the *syn*-2 position determine whether or not the toxin will bind to that phospholipid. Phosphatidylcholine, phosphatidylethanolamine, and sphingomyelin all bind to the toxin. The lipids of Diptera are much richer in phosphatidylethanolamine and unsaturated fatty acids than are those of other insects. Mutational analysis indicates that the loops at the bottom of the molecule (in the orientation shown in Figure 7.10) are the part of the protein responsible for toxicity and lipid binding.

Alternative hypotheses concerning the manner in which the Cyt toxins permeabilize cell membranes include the pore model, which postulates that upon interaction with the outer lipid leaflet of the membrane, Cyt2Aa undergoes a conformational change wherein the helix pair on the left side of the molecule, as represented in Figure 7.10, pulls away from the sheet to lie on the membrane surface, while the sheet region rearranges and associates with sheet regions of other membrane-bound toxin molecules to form an oligomeric transmembrane pore. The alternative model envisages a detergentlike action of Cyt2Aa in which Cyt2Aa aggregates bound to the membrane surface cause large nonspecific defects in lipid packing through which intracellular molecules can leak out. There is experimental support for each of the models described here but as yet no unequivocal proof of either.

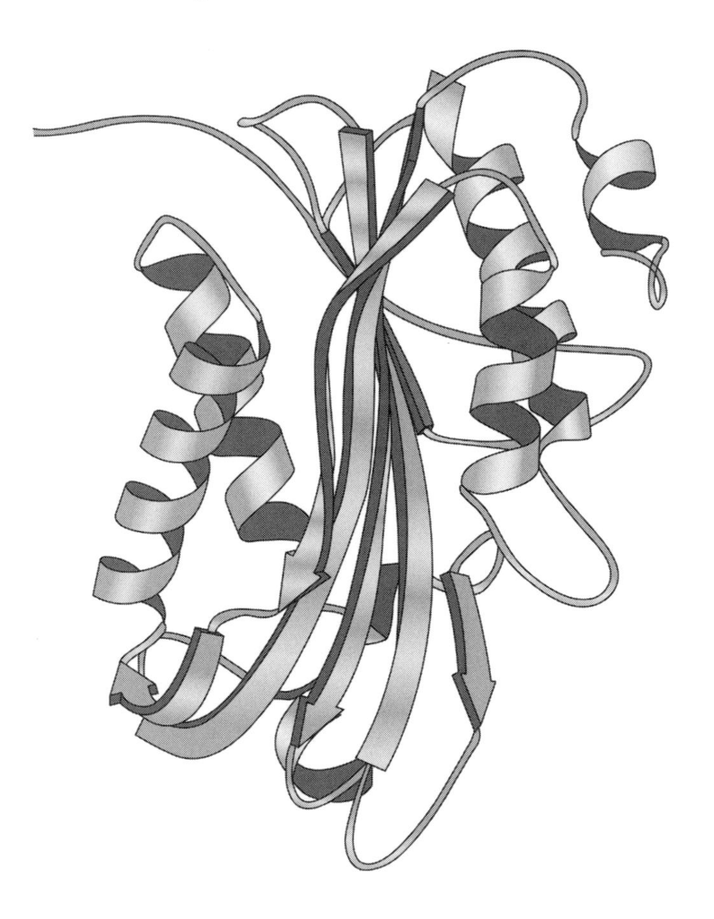

**FIGURE 7.10**

Ribbon diagram of the structure of Cyt2Aa1, a cytolytic *Bacillus thuringiensis* insecticidal δ-endotoxin (Protein Data Base code 1cby). [Reference: de Maagd, R.A., Bravo, A., Berry, C., Crickmore, N., and Schnepf, H.E. (2003). Structure, diversity and evolution of protein toxins from spore-forming entomopathogenic bacteria. *Annual Review of Genetics*, 37, 409–433.

## *B. THURINGIENSIS* SUBSP. *ISRAELENSIS* AS AN INSECTICIDE

The parasporal inclusions of *B. thuringiensis* subsp. *israelensis* (*Bti*) contain four proteins that are toxic to the larvae of Diptera. The Cry4A (128 kDa), Cry4B (134 kDa), and Cry11A (72 kDa) δ-endotoxins are members of the three-domain Cry protein family. The fourth protein is the cytolytic toxin Cyt1Aa1 (27 kDa). The simultaneous presence of these proteins in the larvae results in a much higher toxicity than the individual proteins would have alone. The insecticidal crystal inclusions derived from *Bti* are used for larvicidal treatment of the breeding grounds of the dipteran vectors of malaria (mosquitoes) and of river blindness (blackflies; see Box 2.2 ). *Bti* is grown on a large scale for the production of these crystal inclusions. In contrast to the toxin preparations containing only Cry proteins along with spores (such as Dipel, described above), the addition of *Bti* spores does not cause a significant increase in mortality, so spores are not included in commercial *Bti* preparations.

Mosquito breeding habitats include floodwaters, standing ponds, salt marshes, rice fields, irrigation and roadside ditches, and even the small amounts of water held in tree crotches and flowerpot saucers. Worldwide, between 700,000 and 2.7 million people die from malaria each year. Over 75% of them are African children. *Bti* preparations can be applied to the

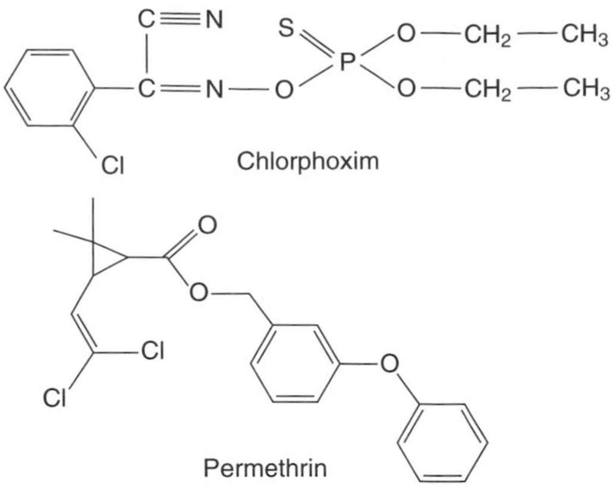

Chlorphoxim

Permethrin

**FIGURE 7.11**

Structures of the insecticides chlorphoxim and permethrin.

habitats as granules or spread aerially as a powder and are very widely used, particularly in Africa.

A centerpiece of the Onchocerciasis Control Program of WHO in 11 West African countries is the control of the river-breeding sites of blackflies. As part of this program, an annual integrated pest management strategy utilizes *Bti* during the dry season, when river flows are low, to minimize the amounts needed. The organophosphate insecticide chlorphoxim is then used for about eight weeks at the start of the wet season, followed by use of the neurotoxic insecticide permethrin for six-week periods when water levels are high (Figure 7.11). This annual treatment alternation regime prevents the development of resistance.

## INSECT-RESISTANT TRANSGENIC CROPS

More than 30 million acres around the world are growing crops engineered to carry *Bt* insecticidal genes. Corn, cotton, and potatoes are currently the major *Bt* crops, but they will be joined by *Bt* rice, which is soon to be introduced in China and India. Table 7.1 lists some of the important pests that feed on these crops and that are susceptible to particular *Bt* Cry toxins. The history of genetically modified (GM) *Bt* crops offers a valuable perspective on the difficulties of assessing the benefits and risks of large-scale introduction of recombinant plants carrying genes new to plant genomes.

### DEVELOPMENT OF INSECT-RESISTANT PLANT LINES

Three different methods have been employed to introduce foreign DNA into plant cells: (1) protoplast electroporation, (2) bombardment of plant cells with particles coated with DNA encoding the intended insert, and (3) transformation with various "disarmed" and modified *Agrobacterium tumefaciens* Ti plasmids. (Disarmed Ti plasmids lack the tumor-inducing genes of *A. tumefaciens*, as explained in the detailed discussion of *A. tumefaciens*–mediated transformation presented in Chapter 6.) *Bt* crops (Table 7.2) have been generated from cell lines transformed by methods 2 and 3.

**TABLE 7.1 Main target insect pests of corn, cotton, and potatoes, susceptible to particular *B. thuringiensis* Cry toxins**

| Crop | Common name of pest | Scientific name of pest |
|---|---|---|
| Corn | Black cutworm | *Agrotis ipsilon* (Hufnagel) |
| *(Zea mays)* | Corn earworm | *Helicoverpa zea* (Boddie) |
| Cotton | Common stalk borer | *Papaipema nebris* (Guen.) |
| *(Gossypium hirsutum)* | European corn borer | *Ostrinia nubilalis* (Huebner) |
| | Fall armyworm | *Spodoptera frugiperda* (J. E. Smith) |
| Potato | Southern corn stalk borer | *Diatracea crambidoides* (Grote) |
| *(Solanum tuberosum)* | Southwestern corn borer | *Diatracea grandiosella* (Dyar) |
| | Cotton bollworm | *Helicoverpa zea* (Boddie) |
| | Pink bollworm | *Pectinophora gossypiella* (Saunders) |
| | Tobacco budworm | *Heliothis virescens* (Fabricius) |
| | Colorado potato beetle | *Leptinotarsa decemlineata* (Say) |

*Source*: U. S. Environmental Protection Agency (2001). Biopesticides Registration Action Document – *Bacillus thuringiensis* Plant-Incorporated Protectants. http://www.epa.gov/oppbppd1/biopesticides/pips/bt_brad.htm.

The following descriptions of GM insect-resistant and herbicide-tolerant cotton and corn plants may appear, on superficial consideration, to be presented in undue detail. However, it is precisely these details – concerning the makeup of the vectors used to generate the transformed plant lines, and of the resulting foreign DNA introduced into particular crops and maintained stably there – that have given rise to much of the opposition to the introduction of such GM crops. Thus, they are worth discussing here.

Much that is currently known about all the genomes studied to date is applied to the creation of vectors for genetic modification. Thus, the DNA fragments assembled into vectors for the transformation of plants are drawn from diverse bacterial genomes, plasmids, viruses, and plants. As the knowledge base grows, the choice of vector components will also evolve, and so will the number and kinds of traits chosen for introduction as transgenes. As a result, many important methodological details will certainly change. Nevertheless, the broad principles informing the methods by which GM crops are generated will likely remain important for a long time to come.

## DEVELOPMENT AND CHARACTERIZATION OF GM COTTON LINE 531 EXPRESSING *cry1AC*

In 1992, Monsanto initiated the field-testing of a GM cotton resistant to key lepidopteran insect pests. This cotton line, designated Monsanto Technology LLC Bollgard Cotton Event 531, has since given rise to widely grown commercial cotton lines and, as illustrated below, has been used in traditional crosses to generate new GM cotton lines that are both insect resistant and herbicide tolerant.

Development of line 531 was initiated by transformation of cotton tissue with a binary plasmid vector incorporating a disarmed and modified

**TABLE 7.2 Some crops for use in human food and animal feed that express *B. thuringiensis (Bt)* insecticidal proteins**

| Crop | Protein | Source | Intended effect |
|---|---|---|---|
| Cotton | Cry1Ac | *Bt* subsp. *kurstaki* | Resistance to cotton bollworm, pink bollworm, tobacco budworm, and European corn borer |
| Cotton | Cry1Ab | *Bt* subsp. *kurstaki* | Resistance to European corn borer |
| Cotton | Cry2Ab and Cry1Ac | *Bt* subsp. *kumamotoensis* | Resistance to lepidopteran insects |
| Corn[a] | Cry9C | *Bt* subsp. *tolworthi* | Resistance to certain lepidopteran insects |
| Corn | Cry1F | *Bt* subsp. *aizawai* | Resistance to certain lepidopteran insects |
| Corn | Cry3Bb1 | *Bt* subsp. *kumamotoensis* | Resistance to coleopteran insects, including corn rootworm |
| Corn | Cry34Ab1 and Cry35Ab1 | *Bt* strain PS149B1 | Resistance to coleopteran insects |
| Potato | Cry3A | *Bt* subsp. *tenebrionis* | Resistance to Colorado potato beetle |

[a] For use in animal feed only.
*Source*: http://cfsan.fda.gov/_lrd/biocon.html.

*A. tumefaciens* Ti plasmid. Figure 7.12 shows a map of this vector, PV-GHBK04; the role of its components is detailed below, proceeding counterclockwise from the top of the figure.

*Ori322/rop* region. The engineering and amplification of PV-GHBK04 is performed in *E. coli*. *Ori322* is derived from *E. coli* plasmid pBR322. Plasmids containing *ori322* can replicate autonomously in *E. coli*. *Rop* encodes a small protein that is involved in the regulation of plasmid replication initiation and hence plasmid number. This region also contains *oriT*, which is necessary for conjugal plasmid transfer from *E. coli* to *A. tumefaciens*.

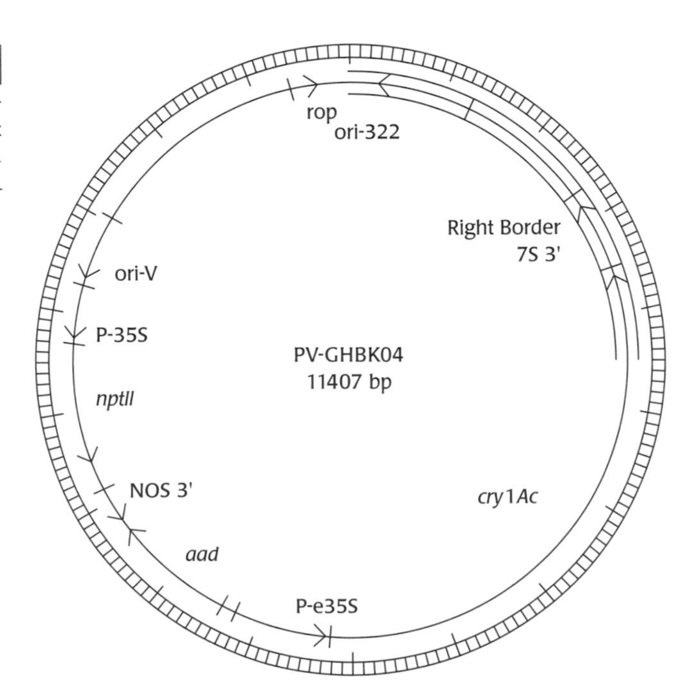


**FIGURE 7.12**

Map of plasmid vector PV-GHBK04. [Monsanto Company (2002). Safety assessment of Bollgard Cotton Event 531. http://www.monsanto.com/monsanto/sci_tech/product_safety/bollgard/es.pdf (accessed 07.17.05).]

*OriV*. *OriV* is derived from the broad recipient–range plasmid RK2. Plasmids containing *oriV* can replicate autonomously in *A. tumefaciens*.

P-35S, *nptII*, and NOS3′. The next three elements, P-35S, *nptII*, and NOS3′, constitute the *nptII* gene expression cassette. The P-35S sequence is the 35S promoter region from cauliflower mosaic virus (CaMV). *NptII* encodes neomycin phosphotransferase type II, which confers kanamycin resistance. The *nptII* gene originates from *E. coli* transposon Tn5. *NptII* is used to select recombinant plant cells expected also to contain the gene of interest, in this case, *cry1Ac*. NOS3′ is the 3′ untranslated region of the nopaline synthase (NOS) gene from *A. tumefaciens*. This sequence terminates transcription and induces polyadenylation of the messenger RNA.

*Aad*. The *aad* gene, derived from *Staphylococcus aureus*, encodes 3″(9)-*O*-aminoglycoside adenylyl transferase. This enzyme confers resistance to the antibiotics spectinomycin and streptomycin. The *aad* gene is under the control of a bacterial promoter. It is included in PV-GHBK04, so the *A. tumefaciens* bacteria that contain the plasmid can be identified by their ability to grow on media containing spectinomycin or streptomycin.

P-e35S, *cry1Ac*, and 7S 3′. The next three elements, P-e35S, modified *cry1Ac*, and 7S 3′, constitute the *cry1Ac* gene expression cassette. The P-e35S sequence is the 35S promoter region with a duplicated enhancer, derived from CaMV. The differences in sequence between the modified Cry1Ac protein and the wild-type protein from *B. thuringiensis* var. *kurstaki* are confined to the amino-terminus and were introduced to enhance its expression level in plants. The *cry1Ac* gene has a higher percentage of A-T nucleotide pairs compared with plant DNA, which is higher in G-C pairs. In modifying *cry1Ac*, the substitution of A-T pairs with G-C pairs was done in such a manner as to minimize changes in the amino acid sequence of Cry1Ac. Overall, the level of amino acid sequence homology of the modified protein with the wild-type protein is 99.4%. The modified Cry1Ac shows the same insecticidal activity and specificity as the wild-type protein. 7S 3′ is the 3′ untranslated region from the $\alpha$ subunit of the soybean $\beta$-conglycinin gene. This sequence terminates the transcription of *cry1Ac* and induces polyadenylation of the mRNA.

Right border. The DNA sequence containing the 24-bp *right border* sequence of nopaline-type T-DNA derived from Ti plasmid pTiT37 serves as the initiation point for the transfer of T-DNA from *A. tumefaciens* to the plant genome.

To create line 531, researchers introduced the T-DNA region of plasmid vector PV-GHBK04 into the hypocotyls of cotton cultivar Coker 512 using the *A. tumefaciens*–mediated transformation system. Transformed plants were selected by culturing in media containing kanamycin.

T-DNA integrates into the plant genome through illegitimate recombination mechanisms in which no homology with plant DNA sequences is required. Thus, *Agrobacterium*-mediated DNA integration may result in complex integration patterns, including directed and inverted repeats. The T-DNA may insert as a single copy or as repeated and multiple insertions, and the multiple insertions may occur in linked or unlinked sites.

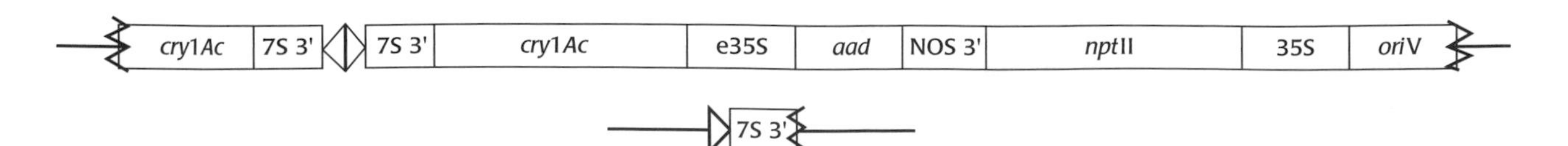


FIGURE 7.13

Map of inserted DNA segments in the genome of GM cotton line 531 generated by *A. tumefaciens*–mediated transformation with vector PV-GHBK04 (see Figure 7.12). A detailed discussion is provided in the text.

At this time, it is not possible to predict either the site of integration of a particular T-DNA construct into the plant genome or the stability of the inserted DNA. Because chromosomal rearrangements may occur at the site of insertion into plant DNA, a GM plant must undergo comprehensive genome analysis, and the GM line intended for commercialization must be shown to be substantially equivalent to the traditionally cultivated non-GM varieties.

The genome characterization of line 531 illustrates some of the complexities of *Agrobacterium*-mediated transformation noted above. The line 531 genome contained two DNA inserts (Figure 7.13). A large insert contained single copies of the full-length *cry1Ac* gene, the *nptII* gene, and the *aad* antibiotic resistance gene. This T-DNA insert also contained an 892-bp portion of the 3′-end of the *cry1Ac* gene fused to the 3′ transcriptional termination sequence. The latter segment of DNA is at the 5′-end of the insert, contiguous with and in reverse orientation to the full-length *cry1Ac* gene cassette, and does not contain a promoter. The second T-DNA insert contained a 242-bp portion of the 7 S 3′-polyadenylation sequence from the terminus of the *cry1Ac* gene. Line 531 expresses active Cry1Ac and NPTII. As noted above, the *aad* gene is under the control of a bacterial promoter and, as expected, is not expressed in cotton.

The *cry1Ac* gene in Bollgard Cotton Event 531 has a stable Mendelian inheritance pattern. Crosses to other cotton varieties show consistent transfer of the functional insert from generation to generation. In short, the *cry1Ac* gene is stably integrated in the cotton genome. Analyses with seed obtained from multisite trials over eight years showed similar levels of Cry1Ac and NPTII proteins.

Safety assessment of GM crops requires that the chemical composition of the GM plant lie within the natural variability range for plants produced by conventional breeding (except for the presence of the deliberately added modifications, such as the Cry1Ac and NPTII proteins in the GM cotton line described above). Comprehensive comparison of line 531 with other cotton varieties revealed no significant differences in the content of protein, lipid, carbohydrate, ash, or moisture; fatty acid profile; amino acid composition; or caloric value. The levels of several cyclopropenoid fatty acids, gossypol, aflatoxin, and $\alpha$-tocopherol were similar to those in the parental cotton variety. Cyclopropenoid fatty acids inhibit the desaturation of stearic to oleic acid and are thus undesirable components of human food and animal feed. Gossypol is a terpenoid aldehyde present in cottonseed and toxic to humans and animals. Aflatoxins are potent toxins and carcinogens in animals and are believed to act as carcinogens in humans. With the exception of its resistance to lepidopteran insects, line 531 met the requirement of "substantial equivalence" and was commercialized in 1996.

## DEVELOPMENT AND CHARACTERIZATION OF GM COTTON LINE 15985 EXPRESSING *cry1AC* AND *cry2AC*

Insect-protected Bollgard II cotton line 15985, containing both Cry1Ac and Cry2Ac, has broader resistance to lepidopteran insect pests than does line 531. Cry1Ac exhibits insecticidal activity toward major pest insects that damage cotton: tobacco budworm, pink bollworm, and cotton bollworm. Cry2Ac is also active against these insects and in addition is active against fall armyworm, beet armyworm, and the soybean looper, insects that show little sensitivity to Cry1Ac.

The first step in the generation of line 15985 was to perform a traditional cross between cotton variety DP50 and line 531 to produce DP50B that stably inherited and expressed Cry1Ac and NPTII.

A plasmid vector, PV-GHBK11 (Figure 7.14), was engineered to contain a *cry2Ab* expression cassette and a *uidA* expression cassette. *UidA*, derived from *E. coli* plasmid pUC19, encodes the marker enzyme $\beta$-glucuronidase (GUS), which can be detected in transformed plant cells by histochemical staining. The vector was produced in large amounts in *E. coli* and then digested with the restriction enzyme KpnI to produce PV-GHBK11L (Figure 7.14). This DNA segment, which carries the *uidA* and the *cry2Ab* gene cassettes, was purified, precipitated onto gold particles, and introduced into the meristems of the recipient cotton variety DP50B by particle acceleration. After staining, nontransformed tissue was gradually removed, which promoted the growth of meristems containing the introduced DNA. The seeds from the resulting GUS-positive plants were screened for the production of the Cry2Ab protein, leading to selection of the transformant referred to as

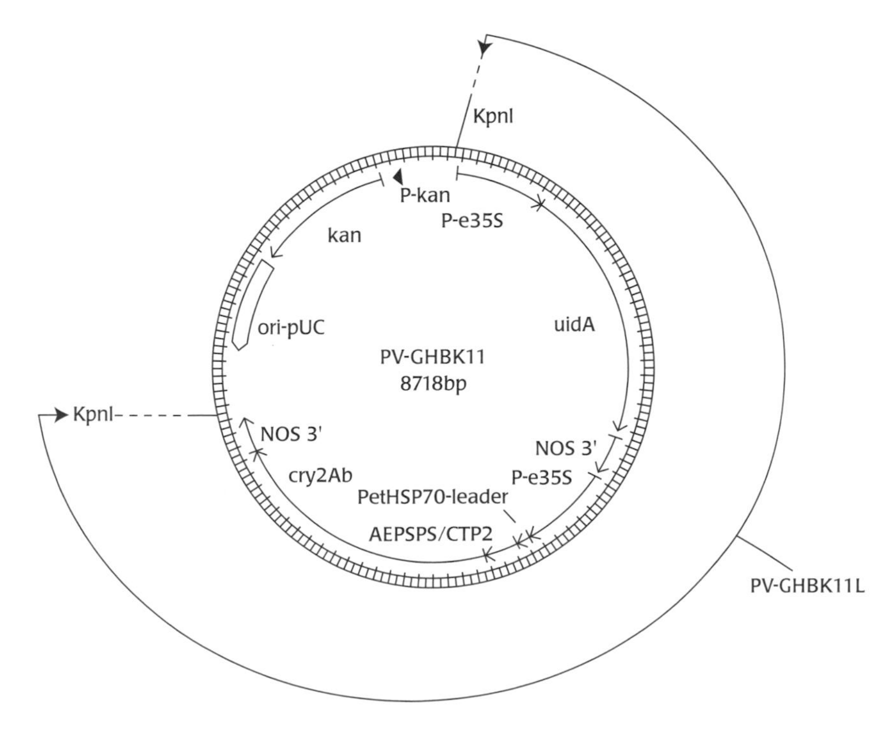


FIGURE 7.14

Map of plasmid vector PV-GHBK11 used in the engineering of GM cotton line 15985. A detailed discussion is provided in the text. [Ministry of the Environment. Japan Biosafety Clearing House. http://www.bch.biodic.go.jp/download/en_lmo/15985_1445enRi.pdf (accessed 07.17.05).]

Insect-Protected Bollgard II cotton line 15985 for commercial development. The inserts in this transformant showed a stable Mendelian inheritance pattern. Line 15985 expresses Cry1Ab, NPTII, AAD, Cry2Ab, and GUS. When analyzed in the manner described above for line 531, line 15985 met the requirements of "substantial equivalence," with the exception of its resistance to lepidopteran insects.

## INSECT-PROTECTED AND HERBICIDE-RESISTANT BOLLGARD II COTTON LINE 15985X1445

A traditional cross performed between line 15985 and line 1445 produced the widely used insect-protected and herbicide-resistant Bollgard II cotton line 15985x1445. GM cotton line 1445, which was produced by *Agrobacterium*-mediated transformation of cotton variety Coker 312, expresses the following introduced genes: *Cp4-epsps* (which encodes CP4-*enol*pyruvylshikimate-3-phosphate synthase from *A. tumefaciens* CP4), *aad*, and *nptII*. The transgenic DNA in line 1445 is stably inherited.

5-Enolpyruvylshikimate-3-phosphate synthase (EPSPS) is one of the enzymes in the shikimate biosynthetic pathway for aromatic amino acids (Phe, Tyr, and Trp) and is located in chloroplasts or plastids. This enzyme is specifically inhibited by glyphosate, the active ingredient in the nonselective herbicide Roundup Consequently, plants treated with glyphosate die. *Cp4 epsps* gene confers resistance to Roundup because glyphosate does not effectively inhibit the *Agrobacterium* CP4 EPSPS. Plants that synthesize as much as 40 times as much EPSPS compared with normal controls show no change in the content of aromatic amino acids. This finding indicates that EPSPS is not a rate-determining enzyme in the shikimate pathway.

In sum, Bollgard II cotton line 15985x1445 expresses the insecticidal proteins Cry1Ac and Cry2ab, the herbicide-resistant enzyme CP4 EPSPS, and the enzymes NPTII (that confers resistance to neomycin and kanamycin), AAD (that confers resistance to streptomycin and spectinomycin), and $\beta$-D-glucuronidase. All these proteins are absent in non-GM cotton.

## GENERATION AND CHARACTERIZATION OF AN HERBICIDE- AND INSECT-RESISTANT CORN LINE EXPRESSING 5-ENOLPYRUVYLSHIKIMATE-3-PHOSPHATE SYNTHASE AND *Bt CRY3BB1*

As a final example, we consider an insect-protected, glyphosate-tolerant corn line, MON 88017, generated by much the same methodologies as described above for the GM cotton lines. A noteworthy feature of corn line MON 88017 is the absence of introduced antibiotic resistance genes in the GM plants.

The *Bt* Cry3Bb1 corn line MON 88017 was generated by *A. tumefaciens*–mediated transformation of corn cells with a disarmed *A. tumefaciens* Ti plasmid, PV-ZMIR39 (Table 7.3). The DNA sequences to be inserted into the plant genome were assembled as a continuous sequence between the left

**TABLE 7.3 The DNA sequences assembled within the T-DNA region of the plasmid vector PV-ZMIR39 utilized in the *Agrobacterium tumefaciens*–mediated transformation of corn**[a]

| DNA sequence | Description |
| --- | --- |
| Left border | Left border sequence from an octopine Ti plasmid, essential for transfer of T-DNA |
| p-ract1 | Rice actin gene promoter to drive *cp4 epsps* expression |
| rac1 intron | Rice actin gene first intron, enhances *cp4 epsps* expression |
| *CTP2* | Sequence encoding a chloroplast transit peptide from *Arabidopsis thaliana*, targets protein expression to the chloroplast |
| CP4 *epsps* | Coding sequence for the CP4 EPSPS[b] protein from *Agrobacterium* sp. strain CP4 |
| NOS 3′ | 3′ untranslated region of the nopaline synthase (NOS) coding sequence, terminates transcription and directs polyadenylation |
| p-e35S | Promoter with duplicated enhancer region from CaMV |
| wt CAB leader | 5′ untranslated leader from the wheat chlorophyll a/b-binding protein |
| ract 1 intron | Rice actin gene first intron, enhances *cry3Bb1* expression *cry3Bb1* |
| *Cry 3Bb1* | Coding sequence for a synthetic variant of the protein from *B. thuringiensis* subsp. *kumamotoensis* |
| tahsp17 3′ | 3′ untranslated region of wheat heat-shock protein 17.3, terminates transcription and directs polyadenylation |
| Right border | Right border sequence from a nopaline Ti plasmid, essential for transfer of T-DNA |

[a] *Agrobacterium tumefaciens* binary transformation vector PV-ZMIR39 is "disarmed" (see text). This plasmid carries the transgenes for insertion into the plant genome between consensus T-border sequences. The plasmid backbone of PV-ZMIR39 contains a bacterial selectable marker gene *aad* that encodes 3″(9)-*O*-aminoglycoside transferase and confers resistance to spectinomycin and streptomycin antibiotics. The presence of this gene facilitates cloning and maintenance of the transformation vector in bacterial hosts. It has been reported that no sequences from the plasmid backbone of PV-ZMIR39 integrate into the transformed corn line.

[b] The *epsps* gene was originally isolated from the soil bacterium *Agrobacterium* sp. strain CP4. This gene was modified to encode a version of the enzyme 5-*enol*pyruvylshikimate-3-phosphate synthase that is resistant to inhibition by the herbicide glyphosate (*N*-phosphonomethylglycine; Roundup).

*Source*: U. S. Food and Drug Administration. Office of Food Additive Safety. (2005). Biotechnology Consultation Note to the File BNF No. 000097, January 5, 2005. Subject: *Bacillus thuringiensis* Cry3Bb1 corn line MON 88017. http://www.cfsan.fda.gov/.

and right border sequences. This region includes genes for two proteins, specifically, *c4 epsps*, which is the gene for CP4 5-epsps, and *cry3Bb1*, which is the gene for Cry3Bb1, a coleopteran-specific insecticidal protein.

The EPSPS gene was isolated from *Agrobacterium* species strain CP4, and a synthetic version was generated. That synthetic *c4 epsps* gene is fused at the 5′-end to the region coding for the chloroplast transit peptide from *Arabidopsis thaliana* EPSPS, and the codon usage of the EPSPS coding region has been modified to enhance expression of the *c4 epsps* gene in plants. The expression of *c4 epsps* in the plant is modulated by a cassette of noncoding DNA regulatory elements fused to the 5′-end of the gene, consisting of the 5′ region of the rice actin 1 gene (p-ract1-ract1 intron). At the 3′-end, the gene is fused to a sequence derived from the 3′ nontranslated region of the nopaline synthase gene (*nos*) that terminates transcription and directs polyadenylation of the messenger RNA.

The synthetic CP4 EPSPS is resistant to glyphosate, and the plant stably transformed with the gene for this enzyme is tolerant to Roundup. Consequently, the synthetic CP4 EPSPS is used as a selectable marker for transformed plant cells.

The *cry3Bb1* gene was isolated from *B. thuringiensis* subsp. *kumamotoensis* strain EG4691. A synthetic version of the coding sequence of *cry3Bb1* was generated wherein the codon usage has been modified to enhance expression in corn. The expression of *c4 cry3Bb1* in corn is modulated by noncoding DNA regulatory elements fused to the gene at the 5′-end: a promoter from CaMV, the 5′ untranslated leader sequence from the wheat chlorophyll *a*/*b*-binding protein, and the first intron from the rice actin gene. At the 3′-end, the gene is fused to a sequence derived from the 3′ nontranslated region of wheat heat-shock protein 17.3 that terminates transcription and directs polyadenylation of the messenger RNA (Table 7.3). The Cry3Bb1 δ-endotoxin protects corn from the larvae of the corn rootworm (*Diabrotica* spp.).

Transformation of callus material from a select corn line with *A. tumefaciens* carrying plasmid PV-ZMIR39 led to the creation of the GM corn line MON 88017. The backbone of PV-ZMIR39 carries a bacterial selectable marker gene, *aad* (see above). The presence of this gene facilitates cloning and maintenance of the transformation vector in bacterial hosts. Because only the sequences (T-DNA) that lie between the left and right borders of the Ti plasmid are transferred, it is expected that no sequences from the plasmid backbone of PV-ZMIR39 should integrate into the transformed corn line. Southern blot analysis of MON 88017 genomic DNA indicated the integration of a single, intact copy of the T-DNA sequence carrying both the *cp4 epsps* and the *cry3Bb1* cassette, but sequences (including those from *aad*) from the backbone of the PV-ZMIR39 vector were not detected.

Plant cells are totipotent. A single cell from any part of a plant can divide and give rise to a complete plant. The first generation of GM corn plants was obtained by growing new plants from the transformed cells. These plants were then bred with other high-quality plants of the same variety. The resulting elite lines of hybrids were tested for the genetic stability of the *cry3Bb1* and *cp4 epsps* genes across 10 generations by Southern blot and segregation analysis. The studies showed that the integration of the T-DNA was stable and inherited in a Mendelian pattern.

Corn is grown worldwide as a source of food and animal feed. Cornstarch is an important source of sugar and ethanol. An exhaustive comparison of grain from the GM corn line MON 88017 with that from non-GM corn showed no significant difference in the content of fiber or in mineral, amino acid, fatty acids, secondary metabolite, or vitamin composition. Thus, corn line MON 88017 met the requirement of "substantial equivalence" to nonrecombinant corn currently consumed in human food and animal feed.

## CRY PROTEIN TISSUE EXPRESSION LEVELS IN CORN

The GM corn lines express the *Bt* transgene in all of their tissues. According to Table 7.4, which provides the expression levels for several lines, nearly 90% of the *Bt* protein is in the leaf tissue. In one set of field data for Bt11 corn expressing Cry1Ab, an acre of corn that produced 83,300 pounds of fresh tissue also produced 0.57 pounds of *Bt* toxin per acre. For a rough comparison, 1994 estimates of the amount of chemical insecticide applied

**TABLE 7.4 Cry protein tissue expression in various GM corn lines**

| Active ingredient | Leaf, ng/mg | Root, ng/mg | Pollen | Seed, ng/mg | Whole plant, ng/mg |
|---|---|---|---|---|---|
| Cry1AB *Bt*11 (006444) | 3.3 | 2.2–37/ protein | <90/g | 1.4 dry wt (kernel) | – |
| Cry1Ab MON 810 (006430) | 10.34 | – | <90/g dry wt | 0.19–0.39 (grain) | 4.65 |
| Cry 1F (006481) | 56.6–148.9 total protein | – | 31–33 ng/mg | 71.2–114.8 total protein | 803.2–1572.7 total protein |
| Cry1Ac (006445) | 2.04 | – | 11.5 ng/g | 1.62 | – |
| Cry3A (006432) | 28.27 ng/mg | 0.39 (tuber) | – | – | 3.3 |

*Source*: U. S. Environmental Protection Agency (2001). Biopesticides Registration Action Document – *Bacillus thuringiensis* Plant-Incorporated Protectants, pp. IIA4–IIA5. http://www.epa.gov/oppbppd1/biopesticides/pips/bt_brad.htm.

per acre of corn were on the order of one pound. On a molar basis, such an amount of insecticide is more than 500-fold higher than the amount of *Bt* toxin in 0.57 pounds.

## BENEFIT AND RISK ASSESSMENT OF *Bt* CROPS

Insect resistance, conferred on plants by introduction of genes encoding various modified *Bt cry* genes, is the second most widely introduced trait in commercial GM crops after herbicide resistance. Although cotton and corn currently represent the great majority of *Bt* crop acreage, recent very successful trials of *Bt* rice suggest that this crop will be commercialized on a large scale in the near future.

In trials comparing regular with insect-resistant rice seed, the performance of the GM seed was very impressive. The data in Table 7.5 show that farmers growing *Bt* rice reduced pesticide use by over 80% while their rice yields increased by several percent. In the village trials in 2003 that provided part of the data summarized in Table 7.5, 10.9% of households reported that their health was adversely affected by pesticide use, whereas none of the adopters of the GM rice varieties reported adverse effects. Comparison of cotton crop yields in China before and after commercialization of *Bt* cotton varieties reveals a substantial decrease in crop loss due to pest damage and more than a 40% decrease in the use of pesticides (Table 7.6). These data are particularly impressive because at the time they describe, the *Bt* cotton had supplanted the non-GM varieties by less than 60%.

The insect control practices used in growing nontransgenic cotton and corn are known to have undesirable environmental effects. As a recent review pointed out, "In evaluating the use of *Bt* crops and their possible environmental damage, it is important to take into account the environmental damage caused by the use of pesticides in agriculture generally. It is argued that millions of birds and billions of insects, both harmful and beneficial (including pollinators and biological control agents), are killed each year in the United States alone as a result of pesticide use." The significant decrease in

**TABLE 7.5 Pesticide use and crop yields for adopters and nonadopters of insect-resistant GM rice[a] in preproduction trials in China, 2002–2003 (mean ± SD)[b]**

| Parameter | Adopters | Nonadopters |
|---|---|---|
| Pesticide spray (times) | 0.50 ± 0.81 | 3.70 ± 1.91 |
| Expenditure on pesticide (yuan/ha) | 31 ± 49 | 243 ± 185 |
| Pesticide use (kg/ha) | 2.0 ± 2.8 | 21.2 ± 15.6 |
| Pesticide spray labor (days/ha) | 0.73 ± 1.50 | 9.10 ± 7.73 |
| Rice yield (kg/ha) | 6364 ± 1294 | 6151 ± 1517 |

[a] Insect-resistant GM rice included two varieties, GM Xianyou 63 (with an introduced engineered *Bt* gene) and GM II-Youming 86 (with an introduced modified cowpea trypsin inhibitor gene, *cpTI*).
[b] Adopters used only insect-resistant GM seed, whereas nonadopters used non-GM rice seed. The trials were conducted in eight villages, of which seven used GM Xianyou 63 and one used GM II-Youming 86 rice. ha, hectare.
*Source*: Huang, J., Hu, R., and Pray, C. (2005). Insect-resistant GM rice in farmers' fields: assessing productivity and health effects in China. *Science*, 308, 688–690.

pesticide use made possible by the introduction of *Bt* crops is expected to be seen as a large positive trade-off against any adverse consequences that may yet be discovered. This view is not universally held, however. GM crops are regarded by many with distrust and prohibited by certain countries. We next explore some of the areas of concern.

## THE RELIABILITY OF "SUBSTANTIAL EQUIVALENCE"

Approval of a particular GM crop for use as human food or animal feed relies on the demonstration of substantial equivalence. Thus, if the GM food product is shown to be essentially equivalent in composition to its non-GM counterpart, then it can be considered as safe as its conventional equivalent. However, there have been concerns that the processes leading to the generation of transformed plant genomes result in undetected mutations that may lead to subtle but significant changes not detected by the compositional analyses. Mutations are introduced by tissue culture procedures; by the gene transfer methods, whether *Agrobacterium*-mediated or utilizing particle bombardment; and by the insertion of foreign DNA segments, both those that carry the transgene and others. Analysis of sites of T-DNA insertion shows that more than a third are within gene sequences. This may be an underestimate, because it is sometimes difficult on the basis of DNA sequence analysis alone to recognize whether the insert has disrupted such a sequence in the plant genome.

A partial response to this concern is that the development of transgenic plant varieties screens out clones of mutants that perform poorly. Here, in addition to selection for the introduced trait (e.g., insect resistance), the selection of clones relies on evaluations of plant growth, growth habit, yield, crop quality, and disease susceptibility. It is recognized that this screening

**TABLE 7.6 Major arthropod pests of China cotton and their impact on estimated crop loss before and after *Bt* cotton commercialization[a]**

| Pest complex | Estimated % crop loss, 1994 | Estimated % crop loss, 2001 |
|---|---|---|
| Cotton bollworm[b] | 6.60 | 2.47 |
| Cotton aphids | 1.03 | 0.52 |
| Pink bollworm | 0.85 | 0.27 |
| Spider mites | 0.63 | 0.59 |
| Mirids | 0.15 | 0.28 |
| Other pests | 0.29 | 0.24 |
| Total | 9.55 | 4.37 |

[a] *Bt* cotton represented 58% of the total cotton acreage of 4.8 million hectares in 2003. The exact fraction for 2001 was not given. The commercialized *Bt* cotton carried the following modified transgenes: *cry1A* or *cry1Ac*,or *cry1A* + *cpTI*.

[b] Cotton bollworm is the most important cotton pest in China. There was no difference in cotton bollworm resistance efficiency between *cry1A* cotton and *cry1A* + *cpTI* cotton. The number of pesticide applications per year for cotton bollworm decreased from 8.5 in 1994 to 3.7 in 2001.

*Source*: Data from Wu, K. M., and Guo, Y. Y. (2005). The evolution of cotton pest management practices in China. *Annual Review of Entomology*, 50, 31–52.

process focuses on agronomic performance and does not directly address concerns about potential health or environmental impacts.

## TRANSFER OF TRANSGENES FROM TRANSGENIC PLANTS TO WILD RELATIVES OR BACTERIA

A transgenic plant might breed with a wild relative and transfer a transgene that confers added fitness to the latter. A *cry* gene that confers ability to kill pests would be an example. Whereas this trait is valuable in the domesticated transgenic plant, the transformed wild plant might have a negative environmental impact by killing beneficial insects and soil organisms.

In a laboratory experiment, sunflowers were genetically engineered to possess a *Bt* gene that conferred protection against sunflower insect pests, and the transgenic plants were crossed with wild sunflowers. The *Bt* gene remained active in the hybrids and continued to be expressed when the hybrids were crossed with other wild sunflowers. The *Bt* gene contributed a significant fitness advantage to the progeny of these crosses. Its presence reduced insect predation, and the wild sunflower hybrids produced more flower heads and more seeds.

The presence of antibiotic resistance genes in GM plants raises a different kind of concern. GM cotton is used as animal feed. Rumen bacteria, oral streptococci, and soil organisms can all be transformed with naked DNA and express genes taken up in this manner. The insect-protected and herbicide-resistant Bollgard cotton lines, described in detail above, contain the *aad* gene. The spectinomycin and streptomycin resistance conferred by the *aad* gene is associated with integrons (see Chapter 10). Given the massive amount of transgenic plant material containing the *aad* gene and the multiple potential routes for lateral gene transfer, there is concern that through consecutive transfers, this gene will find its way to *Neisseria gonorrhoeae*. Spectinomycin is currently used for the treatment of *N. gonorrhoeae* infections resistant to other antibiotics. In humans, this organism causes diseases that include septic arthritis, endocarditis, and pelvic inflammatory disease, as well as serious infections of newborns.

There is now a general consensus that antibiotic resistance genes should not continue to be used in GM plants. Other selectable markers that do not pose health risks have become available.

### Role of *Bt* Toxins in Nature

The overwhelming majority of studies of *Bt* toxins have examined the toxicity of specific Cry proteins toward particular pest insects. In comparison, the ecological role of *Bt* proteins has received little critical attention. *B. thuringiensis* is a soil organism and as such does not come into contact with feeding lepidopteran or coleopteran larvae. Moreover, stability studies on *B. thuringiensis* insecticides that largely consist of the parasporal crystals and spores have shown that when sprayed on plant leaves, both of these components are rapidly inactivated by the ultraviolet radiation in direct sunlight. Consequently, the pest insects targeted by *Bt* sprays or by the *Bt* toxins expressed in GM crops are unlikely to be commonly represented among *B. thuringiensis* hosts in nature. This conclusion begs the question of the ecological role of *Bt* toxins. The fact that *Bt* toxins represent as much as 10% of the total protein in sporulating *B. thuringiensis* cells makes it highly probable that these proteins make an important contribution to the reproductive fitness of the bacteria.

*B. thuringiensis* shares its soil habitat with nematodes, one of the most diverse groups of soil invertebrates. Nematodes eat bacteria, and one might speculate that coevolution of *B. thuringiensis* strains and different nematodes led to the evolution of *Bt* toxins specific to particular nematodes in the course of which the bacterium turned from prey to predator. Strong support for this hypothesis comes from the studies of *Bt* toxins active in *C. elegans*, described earlier in this chapter. These considerations suggest that the potential long-term impact of persistent *Bt* toxins on soil arthropod populations and subsequent influence on the aboveground food web deserves careful attention.

Some studies do show that exposure to realistic levels of *Bt* toxins leads to certain changes in the soil biota, but thus far the results are inconclusive.

### Impact on Nontarget Organisms

The toxicity of *Bt* proteins that are very similar to those expressed by GM crops has been tested on nontarget taxa, including honeybee larvae and adults, ladybird beetles, green lacewings, earthworms, springtails, *Daphnia*, bobtail quail, and mice. No adverse effects were observed. This result was anticipated because the specific receptors for *Bt* toxins are not thought to occur in these organisms.

Surprisingly, the information on toxicity toward insects is limited. A report lists 376 lepidopteran species as feeding on *Zea mays* (corn). Of these, fewer than 15 are believed to have been tested for susceptibility to the *Bt* toxins expressed in GM corn. Nontarget species that depend on *Zea mays* as a host plant or that consume corn pollen may be important as plant pollinators for other plants. Negative impacts on the populations of such insect species may not be readily noticed.

## SUMMARY

Many chemical pesticides have an unfavorable environmental impact. Moreover, many pests have acquired resistance to widely used chemical pesticides. *Integrated pest management programs* decrease the frequency and intensity of genetic selection of resistant insect mutants by employing in concert different means of insect population control: insecticides, microbial or viral pathogens, other natural enemies, and resistance of the plant host, either natural or genetically engineered.

Many bacteria and viruses cause disease and death in insects. *B. thuringiensis* (*Bt*) is the most widely exploited of these organisms. *Bt* is a Gram-positive, aerobic, rod-shaped flagellated bacterium that synthesizes insecticidal parasporal crystals during sporulation. The parasporal crystal proteins ($\delta$-endotoxins) of *Bt* strains are classified on the basis of their amino acid sequences into two distinct families, Cry and Cyt (the latter are proteins with hemolytic activity). The $\delta$-endotoxins can represent up to 30% of the dry weight of sporulating cells.

Particular *Bt* strains produce toxins active against Lepidoptera (butterflies and moths), Diptera (flies, midges, and mosquitoes), or Coleoptera (beetles). Commercial *Bt* insecticide is a dry powder consisting of sporulated *Bt* cells. The active ingredients in this powder are the large protein-containing crystalline inclusions and the spores. The powder is applied to vegetation by dusting. Larvae consuming treated leaves ingest the proteinacous crystals and the spores. After ingestion by larvae, the $\delta$-endotoxins are processed to active toxins. The latter bind to specific receptors on the plasma membrane of larval gut epithelial cells, cause small pores to form in the membrane, and thereby destroy the permeability barrier to ions. The destruction of the larval gut epithelium leads to cessation of feeding and opens a path into the hemolymph for indigenous enteric bacteria in the larval gut. Bacteremia follows, and death ultimately ensues.

The parasporal inclusions of *B. thuringiensis* subsp. *israelensis* (*Bti*) contain four proteins that are toxic to the larvae of Diptera: three different Cry proteins and a cytolytic toxin. *Bti* is grown on a large scale to produce the crystal inclusions used for larvicidal treatment of the breeding grounds of the dipteran vectors of malaria (mosquitoes) and of river blindness (blackflies). In contrast to the toxin preparations containing exclusively Cry proteins, *Bti* spores do not cause a significant increase in mortality, and commercial *Bti* preparations are free of spores.

*Bt*, primarily a soil organism, does not come into contact with feeding lepidopteran or coleopteran larvae. It shares its soil habitat with nematodes, which eat bacteria, leading to speculation that coevolution of *Bt* strains and different nematodes has led to the evolution of *Bt* toxins specific to particular nematodes, with the bacterium turning from prey to predator. In support of this view, certain *Bt* Cry proteins have been shown to target nematodes.

More than 30 million acres around the world are planted with crops engineered to carry *Bt* insecticidal genes. *Cry* gene-based insect resistance is the second most widely introduced trait in commercial GM crops after herbicide resistance. Corn, cotton, and potatoes are currently the major *Bt* crops. To generate insect-resistant crop lines, plant cells are transformed with various "disarmed" and modified *A. tumefaciens* Ti plasmids engineered to introduce the transgenes of interest, or by bombardment of plant cells with particles coated with DNA encoding the intended insert. To be approved for commercialization, GM lines must be shown to be substantially equivalent to the traditionally cultivated non-GM varieties. GM crops have significantly reduced pesticide use and greatly decreased the incidence of illness caused by pesticides among farmers.

Concerns about GM crops expressing high levels of *Bt* toxins focus in part on the following potential negative consequences: toxicity to soil microorganisms, impact on nontarget insects, and undesirable transfer of transgenes from the GM plants to wild relatives or to bacteria. Of the GM plant lines in current large-scale use, some contain antibiotic resistance genes, including *aad*, which encodes 3″(9)-*O*-aminoglycoside transferase and confers resistance to spectinomycin and streptomycin. Transfer of this gene to certain pathogenic bacteria currently sensitive to these antibiotics, especially *N. gonorrhoeae*, would limit the options for the treatment of infections caused by such pathogens.

## SELECTED REFERENCES AND ON-LINE RESOURCES

### General

Steinhaus, E. A. (1975). *Disease in a Minor Chord*. Columbus: Ohio State University Press.

Broderick, N. A., Raffa, K. F., and Handelsman, J. (2006). Midgut bacteria required for *Bacillus thuringiensis* insecticidal activity. *Proceedings of the National Academy of Sciences U.S.A.* 103, 15196–15199

Gill, S. S., Cowles, E. A., and Pietrantonio, P. V. (1992). The mode of action of *Bacillus thuringiensis* endotoxins. *Annual Review of Entomology*, 37, 615–634.

Casida, J. E., and Quistad, G. B. (1998). Golden age of insecticide research: past, present, or future? *Annual Review of Entomology*, 43, 1–16.

Navon, A. (2000). *Bacillus thuringiensis* insecticides in crop protection – reality and prospects. *Crop Protection*, 19, 669–676.

### Nomenclature

Crickmore, N., et al. (1998). Revision of the nomenclature for the *Bacillus thuringiensis* pesticidal crystal proteins. *Microbiology and Molecular Biology Reviews*, 62, 807–813.

### Host Specificity of *B. thuringiensis* δ-Endotoxins

Terra, W. R., and Ferreira, C. (1994). Insect digestive enzymes: properties, compartmentalization and function. *Comparative Biochemistry and Physiology*, 109B, 1–62.

de Maagd, R. A., Bravo, A., and Crickmore, N. (2001). How *Bacillus thuringiensis* has evolved to colonize the insect world. *Trends in Genetics*, 117, 193–199.

Wei, J-Z., Hale, K., Carta, L., Platzer, E., Wong, C., Fag, S-C., and Aroian, R. V. (2003). *Bacillus thuringiensis* crystal proteins that target nematodes. *PNAS*, 100, 2760–2765.

### Insect and Nematode Receptors for Cry δ-Endotoxins

Knight, P. J. K., Crickmore, N., and Ellar, D. J. (1994). The receptor for *Bacillus thuringiensis* Cry1A(c) delta-endotoxin in the brush-border membrane of the lepidopteran *Manduca sexta* is aminopeptidase-N. *Molecular Microbiology*, 11, 429–436.

Stephens, E., Sugars, J., Maslen, S. L., Williams, D. H., Packman, L. C., and Ellar, D. J. (2004). The N-linked oligosaccharides of aminopeptidase N from *Manduca sexta*. Site localization and identification of novel N-glycan structures. *European Journal of Biochemistry*, 271, 4241–4258.

Griffitts, J. S., et al. (2005). Glycolipids as receptors for *Bacillus thuringiensis* crystal toxin. *Science*, 307, 922–925.

### Structure–Function Relationships in Cry Proteins

Morse, R. J., Yamamoto, T., and Stroud, R. M. (2001). Structure of Cry2Aa suggests an unexpected receptor binding epitope. *Structure*, 9, 409–417.

Galitsky, N., Cody, V., Wojtczak, A., Ghosh, D., Luft, J. R., Pangborn, W., and English, L. (2001). Structure of the insecticidal bacterial δ-endotoxin C3Bb1 of *Bacillus thuringiensis*. *Acta Crystallographica*, D57, 1101–1109.

Tuntitippawan, T., Boonserm, P., Katzenmeier, G., and Angsuthanasombat, C. (2005). Targeted mutagenesis of loop residues in the receptor-binding domain of the *Bacillus thuringiensis* Cry4Ba toxin affects larvicidal activity. *FEMS Microbiology Letters*, 242, 325–332.

Boonserm, P., Davis, P., Ellar, D. J., and Li, J. (2005). Crystal structure of the mosquito larvicidal toxin Cry4Ba and its biological implications. *Journal of Molecular Biology*, 348, 363–382.

Parker, M. W., and Feil, S. C. (2005). Pore forming protein toxins: from structure to function. *Progress in Biophysics & Molecular Biology*, 88, 91–143.

### Structure–Function Relationships in Cyt Proteins

Li, J., Koni, P. A., and Ellar, D. J. (1996). Structure of the mosquitocidal δ-endotoxin CytB from *Bacillus thuringiensis* sp. *kyushuensis* and implications for membrane pore formation. *Journal of Molecular Biology*, 257, 129–152.

Butko, P. (2003). Structure-function relationships in Cyt proteins. Cytolytic toxin Cyt1A and its mechanism of membrane damage: data and hypotheses. *Applied and Environmental Microbiology*, 69, 2415–2422.

### Resistance

Forcada, C., Alcácer, E., Garcerá, M. D., Tato, A., and Martinez, R. (1999). Resistance to *Bacillus thuringiensis* Cry1Ac toxin in three strains of *Heliothis virescens*: proteolytic and SEM study of the larval midgut. *Archives of Insect Biochemistry and Physiology*, 42, 51–63.

Loseva, O., Ibrahim, M., Candas, M., Koller, C. N., Bauer, L. S. and Bulla, L. A., Jr. (2002). Changes in protease activity and Cry3Aa toxin binding in the Colorado potato beetle: implications for insect resistance to *Bacillus thuringiensis* toxins. *Insect Biochemistry and Molecular Biology*, 32, 567–577.

Ferré, J., and Van Rie, J. (2002). Biochemistry and genetics of insect resistance to *Bacillus thuringiensis*. *Annual Review of Entomology*, 47, 501–533.

Griffitts, J. S., et al. (2003). Resistance to a bacterial toxin is mediated by removal of a conserved glycosylation pathway required for toxin-host interactions. *Journal of Biological Chemistry*, 278, 45594–45602.

### Insect-Resistant Transgenic Crops

Entwistle, P. F., Cory, J. S., Bailey, M. J., and Higgs, S. (eds.) (1993). *Bacillus thuringiensis, an Environmental Biopesticide: Theory and Practice*, New York: John Wiley & Sons.

U. S. Environmental Protection Agency (2001). *Biopesticides Registration Action Document – Bacillus thuringiensis Plant-Incorporated Protectants*. http://www.epa.gov/oppbppd1/biopesticides/pips/bt_brad.htm.

Metz, M. (ed.) (2003). *Bacillus thuringiensis: a Cornerstone of Modern Agriculture*, New York: Haworth Press. [Co-published as *Journal of New Seeds*, Volume 5, Numbers 1 and 2/3, 2003.]

Miki, B., and McHugh, S. (2004). Selectable marker genes in transgenic plants: applications, alternatives and biosafety. *Journal of Biotechnology*, 107, 193–232.

Zhao, J-Z., et al. (2005). Concurrent use of transgenic plants expressing a single and two *Bacillus thuringiensis* genes speeds insect adaptation to pyramided plants. *PNAS*, 102, 8426–8430.

### Benefit and Risk Assessment of *Bt* Crops

National Research Council (NRC) (2000). *Genetically-Modified Pest-Protected Plants: Science and Regulation*. Washington, DC: National Academy Press.

National Research Council (NRC) (2002). *Environmental Effects of Transgenic Plants: the Scope and Adequacy of Regulation*. Washington, DC: National Academy Press.

Dale, P. J., Clarke, B., and Fontes, E. M. G. (2002). Potential for the environmental impact of transgenic crops. *Nature Biotechnology*, 20, 567–574.

Cellini, F., et al. (2004). Unintended effects and their detection in genetically modified crops. *Food and Chemical Toxicology*, 42, 1089–1125.

Wu, K. M., and Guo, Y. Y. (2005). The evolution of cotton pest management practices in China. *Annual Review of Entomology*, 50, 31–52.

O'Callaghan, M., Glare, T. R., Burgess, E. P. J., and Malone, L. A. (2005). Effects of plants genetically modified for insect resistance on nontarget organisms. *Annual Review of Entomology*, 50, 271–292.

Huang, J., Hu, R., Rozelle, S., and Pray, C. (2005). Insect-resistant GM rice in farmers' fields: assessing productivity and health effects in China. *Science*, 308, 688–690.

EIGHT

# Microbial Polysaccharides and Polyesters

The use of green plants as industrial factories will potentially become an important component of "green chemistry" efforts. Realization of this technology will likely require metabolic engineering of multi-step pathways and significant use of plant primary metabolites.

– Slater, S., et al. (1999). Metabolic engineering of *Arabidopsis* and *Brassica* for poly(3-hydroxybutyrate-*co*-3-hydroxyvalerate) copolymer production. *Nature Biotechnology*, 17, 1011–1016.

This chapter deals with two classes of biopolymers: polysaccharides and polyesters. Polysaccharides include some of the most abundant carbon compounds in the biosphere, the plant polysaccharides, cellulose, and hemicelluloses (discussed in Chapter 12), as well as the much less abundant but useful algal polymers, such as agar and carrageenan. Bacteria and fungi also produce many different types of polysaccharides, some in amounts well in excess of 50% of cell dry weight. High molecular weight polyesters are produced exclusively by prokaryotes and for a long time were of interest only to students of microbial physiology.

Polysaccharides are used to modify the flow characteristics of fluids, to stabilize suspensions, to flocculate particles, to encapsulate materials, and to produce emulsions. Among many other examples is the use of polysaccharides as ion-exchange agents, as molecular sieves, and, in aqueous solution, as hosts for hydrophobic molecules. Polysaccharides are used in enhanced oil recovery and as drag-reducing agents for ships.

The discovery that many bacteria synthesize large amounts of *biodegradable* polyester polymers of high molecular weight, which can be used to manufacture plastics, has aroused considerable interest. There are hundreds of varieties of synthetic plastics; their uses are too many to enumerate. Current annual production of these materials in the United States alone exceeds 30 billion pounds. Plastics are manufactured from petrochemicals and decompose extremely slowly in the natural environment.

In this chapter, we discuss the structures of some of the microbial polysaccharides and polyesters, their biosynthesis and functions, and some of the uses of these polymers.

**TABLE 8.1 Major food and industrial polysaccharides in order of decreasing consumption**

| |
|---|
| Cornstarch and derivatives |
| Cellulose and derivatives |
| Guar gum and derivatives |
| Gum arabic |
| Xanthan[a] |
| Alginate |
| Pectin |
| Carrageenan |
| Locust bean gum |
| Ghatti |

[a] By weight, the consumption of xanthan represents less that 1% that of starches and derivatives.

## POLYSACCHARIDES

Microorganisms and plants produce polysaccharides of widely varying composition and structure. Cellulose, hemicelluloses, and starch from plants are the most abundant, important, and familiar of these biopolymers, but many other naturally occurring polysaccharides have interesting and useful properties. Agar, a mixture of polysaccharides extracted from marine red algae, has been manufactured in Japan since about 1760 and is one of the most effective gelling agents known. Gel formation occurs at agar concentrations as low as 0.04%. Dried agar suspended in water melts in the temperature range of 60°C to 90°C (the precise temperature varies with the algal source) and, at about 1% to 1.5% by weight, sets between 32°C and 39°C to form a firm gel. This large temperature hysteresis is a particularly valuable property of agar that it shares with few other polysaccharides. Among polysaccharides produced by bacteria and fungi, some have properties resembling those of agar whereas others have distinctive rheological (flow) properties valuable in certain pharmaceutical or industrial applications.

Only one microbial polysaccharide, xanthan, ranks among the 10 industrial polysaccharides utilized in the largest amount (Table 8.1). In terms of market value, it ranks high among fermentation products (Table 8.2). Much of this chapter is devoted to this biopolymer. Other microbial exopolysaccharides (Table 8.3) have found only modest commercial use, and only a few are produced on a large scale. However, the enormous range of polysaccharides synthesized by microorganisms has yet to be adequately explored.

### BACTERIAL POLYSACCHARIDES

The majority of bacterial species, under the proper culture conditions, secrete mucoid substances of high molecular weight. When these viscous materials remain associated with the cell, they are called variously *capsules*, *sheaths*, or *slime layers* (Figure 8.1). Whereas capsules and sheaths are well-defined layers external to the cell wall, slime accumulates in large quantities outside the cell wall and diffuses into the medium. The slime formers may produce such copious amounts of viscous slime when they are grown in a liquid medium so that the culture flask may be inverted and the culture will remain in place. Mucoid colonies grown on solid media usually have a moist, glistening surface.

Capsules have long been a subject of great interest because of their roles as important virulence factors in bacteria that cause invasive infections. Capsules protect bacteria from phagocytosis and impart resistance against bactericidal effects of serum. Frequently, pathogens cultured in the laboratory will spontaneously produce unencapsulated mutants; uniformly, these mutants are no longer pathogenic.

Most of the extracellular polymers produced by bacteria are polysaccharides, although a few bacteria produce capsules made up of polypeptides of D–amino acid residues. The precise structure of capsular polysaccharides of many pathogenic organisms is strain specific. Such individual "coding"

**TABLE 8.2 Estimated global market for fermentation products by category (in millions of U.S. dollars)**

|  |  |
|---|---|
| Crude antibiotics | 5000 |
| Amino acids | 3435 |
| Organic acids | 2321 |
| Enzymes | 2006 |
| Vitamins | 1013 |
| Xanthan | 335 |

*Source*: Business Communications Co., Inc. http://www.bccresearch.com/editors/RGA-103R.html (accessed on 07.16.05).

**TABLE 8.3 Bacterial and fungal polysaccharides with commercial uses**

| Polysaccharide | Organism | Composition |
|---|---|---|
| **Bacteria** | | |
| Xanthan | *Xanthomonas campestris* | A pentasaccharide repeating unit containing glucose, mannose, glucuronic acid, and acetyl and pyruvate substituents |
| Dextran | *Acetobacter* sp.<br>*Leuconostoc mesenteroides*<br>*Streptococcus mutans* | Polyglucose linked by $\alpha$-1,6-glycosidic bonds; some 1,2-, 1,3-, or 1,4-bonds are also present in some dextrans |
| Alginate | *Pseudomonas aeruginosa*<br>*Azotobacter vinelandii* | Blocks of $\beta$-1,4-linked D-mannuronic residues, blocks of $\alpha$-1,4-linked L-guluronic acid residues, and blocks with these uronic acids in either random or alternating order |
| Curdlan | *Alcaligenes faecalis* | $\beta$-1,3-glucan (polyglucose) |
| Gellan | *Pseudomonas elodea* | Partially *O*-acetylated polymer of glucose, rhamnose, and glucuronic acid |
| Cellulose | *Gluconacetobacter xylinus* | $\beta$-1,4-linked D-glucopyranose polymer |
| **Fungi** | | |
| Scleroglucan | *Sclerotium glutanicum* | Glucose units primarily $\beta$-1,3-linked with occasional $\beta$-1,6-glycosidic bonds |
| Pullulan | *Aureobasidium pullulans* | Glucose units primarily $\alpha$-1,4-linked with occasional $\alpha$-1,6-glycosidic bonds |

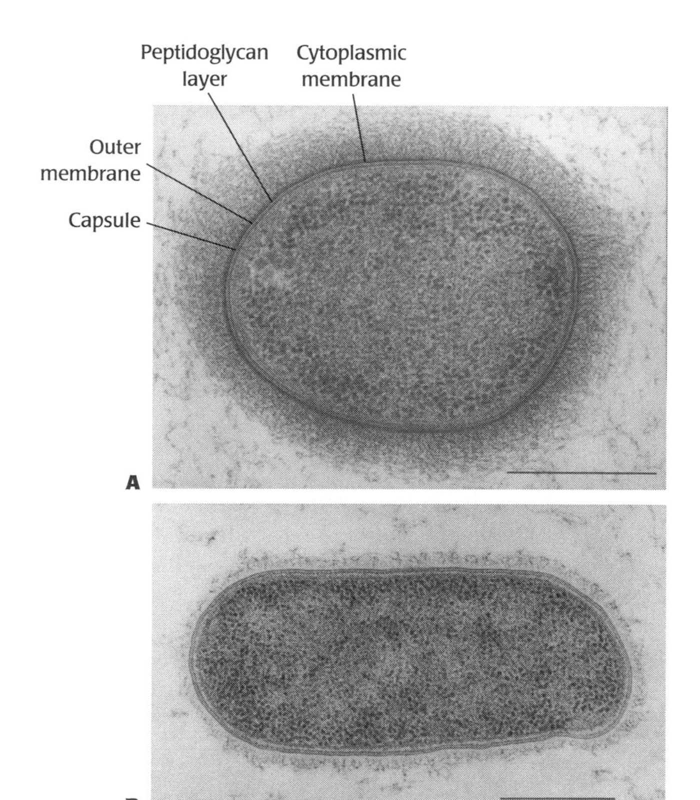


**FIGURE 8.1**

(*Upper panel*) Thin section of the bacterium *Klebsiella pneumoniae* showing the fine structure of the polysaccharide capsule. The capsule is approximately 160 nm thick and appears to be made up of two layers. The inner layer is formed by a palisade of thick bundles of fibers extending outward at right angles to the surface of the outer membrane. In the second layer, thin fibers spread from the ends of the bundles forming a fine network. (*Lower panel*) Thin section of the bacterium *E. coli* K1 showing a capsule much thinner than that of *K. pneumoniae*, but with a similar morphology. Staining was with uranyl acetate and lead citrate. *Bars*, 0.5 $\mu$m. [Reproduced with permission from Amako, K., Meno, K. Y., and Takade, A. (1988). Fine structure of the capsules of *Klebsiella pneumoniae* and *Escherichia coli* K1. *Journal of Bacteriology*, 170, 4960–4962.]

of its outermost layer protects the bacterial cell from being attacked by an immune system that may have developed antibodies in response to an earlier, perhaps closely related, invader.

In contrast, the structures of the extracellular polysaccharides of many slime-forming organisms are simpler in composition and less homogeneous in size. The molecular weight and even the composition of these extracellular polysaccharides may vary depending on the culture conditions.

## ASPECTS OF POLYSACCHARIDE STRUCTURE

The glycosidic bond between two monosaccharides is formed between the hydroxyl group on the anomeric carbon of one monosaccharide and any hydroxyl group on the other monosaccharide. Hence, the formation of a disaccharide from two identical hexopyranose ring structures of the D-series may result in 11 different isomers. In eight of the isomers, the glycosidic linkage is between C-1, in either the $\alpha$ or $\beta$ anomeric configuration, and C-2, C-3, C-4, or C-6 of the other pyranose residues. These are designated $\alpha$-D-(1→2) linkages, $\alpha$-D-(1→3), linkages, and so on, where $\alpha$ and $\beta$ refer to the anomeric configuration at C-1. The other three isomers are created by acetal formation between both C-1 atoms through the glycosidic oxygen atom in the $\alpha,\alpha$, $\alpha,\beta$, or $\beta,\beta$ configuration. A similar series of 11 isomers will result if the two identical hexopyranose residues are of the L-series. The number of possible isomers is even higher if furanose forms (five-membered rings) are included. When the two monosaccharides are not identical, the number of possible structures increases, because either carbohydrate residue can occupy the first or the second position – that is, can be the reducing or nonreducing residue. Each additional carbohydrate residue brings a large increase in the number of possible isomers. For complex polysaccharides, the number of theoretically possible structures is astronomical.

In reference to the structure of polysaccharides and of proteins, the terms *primary* and *secondary structure* have a similar meaning. The primary structure of a polysaccharide is the identity, sequence, linkage(s), and anomeric configurations of all its monosaccharide residues, and the nature and position of any other substituents. The individual sugar rings in a polysaccharide are essentially rigid. The secondary structure of the polysaccharide is thus determined by the relative orientations of the component monosaccharide residues about the glycosidic linkage. As illustrated in Figure 8.2, two rotational angles ($\phi$ and $\psi$) specify the glycosidic bond between two carbohydrate residues, except for (1→6) linked residues, in which specification of a third rotational angle, $\omega$, is required. The range of permitted values for the rotational angles $\phi$, $\psi$, and $\omega$ is limited by steric constraints, including the geometric relations within each residue of the polysaccharide and, to a lesser degree, the interactions between adjacent residues in the chain. Depending on the nature of such constraints, polysaccharides with a primary structure made up of identical repeating units exist in solution as relatively stiff extended or crumpled ribbons, loose helical coils, or [for (1→6)-linked polysaccharides] flexible "random" coils. For example, curdlan,

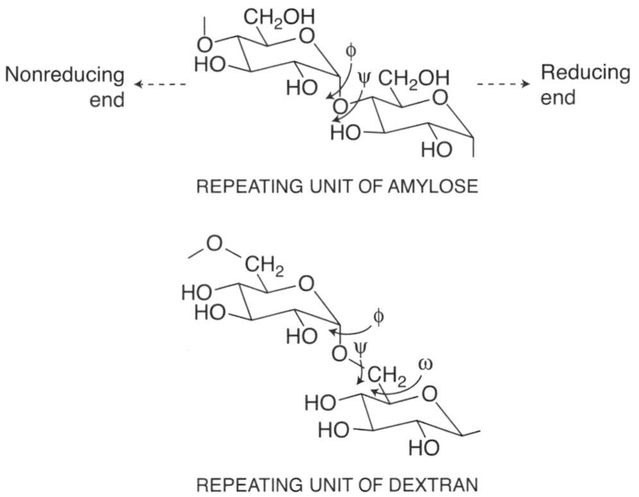

Nonreducing end

Reducing end

REPEATING UNIT OF AMYLOSE

REPEATING UNIT OF DEXTRAN

**FIGURE 8.2**

The relative orientations of adjacent carbohydrate residues joined by glycosidic linkage to a ring hydroxyl are defined by two torsion (rotational) angles, $\phi$, and $\psi$, except for (1→6)-linked polysaccharides, in which a third angle, $\omega$, specifies the rotation around the exocyclic carbon–carbon bond.

a (1→3)-$\beta$-D-glucan, exists in solution as a flexible helix with sevenfold symmetry. For charged polysaccharides, interaction with ions may be very important. Alginate, a block copolymer of (1→4)-$\beta$-D-mannuronic acid and (1→4)-$\alpha$-L-guluronic acid (Table 8.3) exists as a stiff random coil in solution, but, on binding $Ca^{2+}$ ions, the poly-L-guluronate sequences undergo dimerization as shown in Figure 8.3, a step leading to gel formation. Noncovalent aggregation of polysaccharide chains to form higher-order

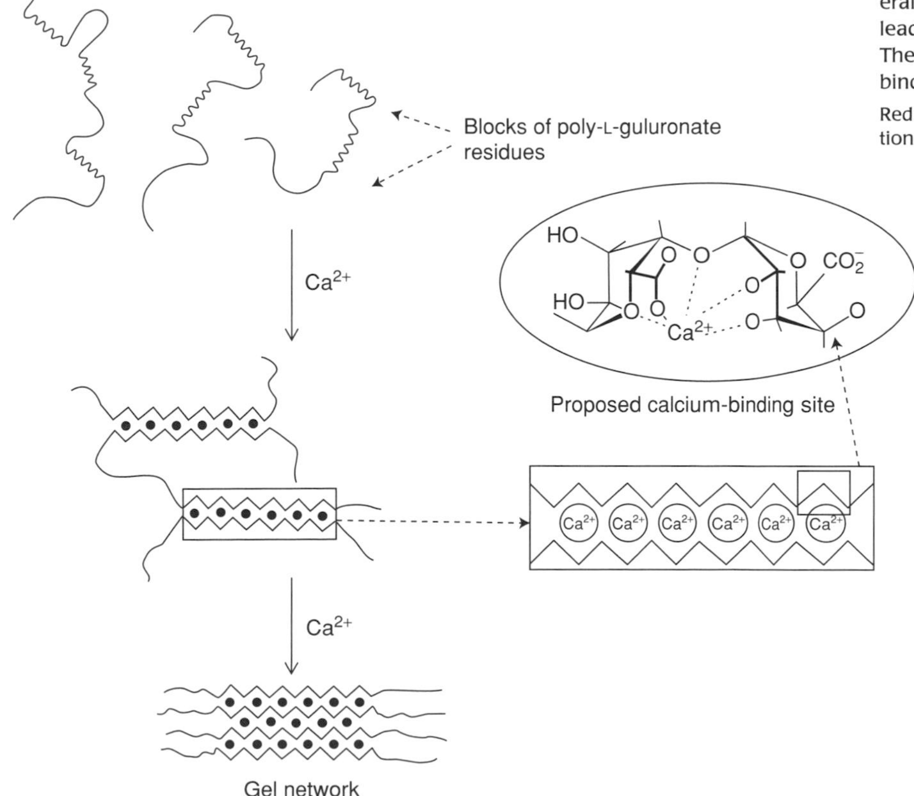


**FIGURE 8.3**

Interaction of $Ca^{2+}$ ions with the poly-L-guluronate sequences in alginate causes lateral association of these chain segments, leading to the formation of a gel network. The proposed structure of the calcium-binding site is shown on the right.

Redrawn based on artwork from the first edition (1995), published by W.H. Freeman.

structures is analogous to subunit aggregation and *quaternary structure* formation in proteins.

The wide range of physical properties seen among microbial polysaccharides is a direct consequence of the differences in the nature and proportion of their monosaccharide building blocks, of substituents (such as acetyl groups) on the monosaccharides, of the linkages between the building blocks, and, ultimately, of the stereochemical consequences of all of these factors. It follows that modulation of the composition of microbial polymers, through genetic modification or manipulation of culture conditions, may enhance their existing desirable properties or generate entirely new polymers with novel characteristics.

### ROLES OF MICROBIAL POLYSACCHARIDES IN NATURE

Organisms that produce extracellular polysaccharides have been isolated from a wide variety of environments. The functions of these polysaccharides are equally varied. The capsular polysaccharides that protect pathogenic microorganisms from immune system defenses were mentioned above. Capsules may also serve as physical barriers to infection by bacteriophages. Moreover, capsules and sheaths retain water and in some settings play an important role in preventing dehydration of the cells. A common function of extracellular polysaccharides of water and soil microorganisms is to bind cells to surfaces, such as soil particles or rocks, as well as to each other. Likewise, it has been proposed that the extracellular polysaccharides of some plant pathogens play a role in attaching the bacteria to the surfaces of host plants.

## XANTHAN GUM

Xanthan gum was discovered in the mid-1950s, during a systematic search for useful biopolymers. In a screen of its large microbial culture collection, the Northern Utilization Research and Development Division of the U.S. Department of Agriculture discovered that the bacterium *Xanthomonas campestris* (originally isolated from the rutabaga plant) produced a polysaccharide with potentially valuable physical properties. Substantial commercial production of xanthan gum began in 1964, and in 1969 the U.S. FDA authorized its use in food. Today, xanthan gum has numerous uses, in the food industry and elsewhere (Table 8.4). The total U.S. consumption of xanthan in 2005 is estimated at more than 80 million pounds.

### *XANTHOMONAS*: A PLANT PATHOGEN

The genus *Xanthomonas* belongs to the $\gamma$ subdivision of the proteobacteria and consists exclusively of plant pathogens. It is a relative of the genus *Pseudomonas*, a group of bacteria discussed elsewhere in this book in the context of the degradation of xenobiotics (Chapter 14).

**TABLE 8.4 Major industrial applications of xanthan gum**

| Application and concentration (% w/w) | Function |
|---|---|
| **Foods and beverages** | |
| Salad dressing (0.1–0.5) | Emulsion stabilizer |
| Dry mixes (0.05–0.2) | Facilitates dispersion in hot or cold water |
| Syrups, relishes, sauces, toppings (0.05–0.2) | Thickener; provides heat stability and uniform viscosity |
| Beverages (0.5–0.2) | Stabilizer |
| Frozen foods (0.05–0.2) | Improves freeze–thaw stability |
| **Pharmaceuticals and cosmetics** | |
| Creams and suspensions (0.1–1.0) | Emulsion stabilizer |
| Controlled-release tablets | Regulates disintegration rate |
| Shampoo, liquid soaps, toothpaste (0.2–1.0) | Improves flow properties and lather stability |
| **Agriculture** | |
| Additive in animal feeds and pesticide formulations (0.03–0.4) | Suspension stabilizer; increases spray cling and permanence |
| **Oil production** | |
| Oil drilling aid (0.1–0.4) | Thickener in drilling fluids that flush rock fragments away from drill bit |
| Enhanced oil recovery (0.05–0.2) | Facilitates oil displacement by increasing water viscosity |
| **Other** | |
| Textile printing and dyeing (0.2–0.5) | Forms temperature-stable foams for printing and finishing; acts as flow modifier for dyeing heavy fabrics |
| Ceramic glazes | Suspending agent |

w/w, weight to weight.
*Source*: Garcia-Ochoa, F., Santos, V. E., Casas, J. A., and Gómez, E. (2000). Xanthan gum: production, recovery, and properties. *Biotechnology Advances*, 18, 549–579.

Bacteria of the genus *Xanthomonas* are yellow-pigmented, motile, aerobic, Gram-negative rods. *X. campestris* pathovar (abbreviated "pv.") *campestris*, the producer of xanthan gum, causes black rot, one of the most serious diseases of plants in the genus *Brassica*, which includes vegetables such as cabbage, cauliflower, brussels sprouts, broccoli, rutabaga, and turnips. Seeds contaminated with bacteria spread the infection. As the seedling emerges and grows, the bacteria colonize the surface of the plant. In the case of *X. campestris* pv. *campestris*, the epiphytic bacteria enter the internal tissues through the *hydathodes*, structures on the leaf margin that allow the plant to excrete water. They then migrate through the vascular system and progressively cause chlorosis (destruction of chlorophyll), vein blackening (deposition of melaninlike pigments), and, ultimately, rotting.

Mutants of *X. campestris* with defects in xanthan biosynthesis show greatly reduced virulence in plants. Xanthan is but one of many contributory factors conferring virulence. The genome of *X. campestris* pv. *campestris* has been completely sequenced and a comprehensive insertional mutant library constructed using a transposon-based mutagenesis system. The analysis of the virulence of the mutants indicated that as many as 75 genes of a total of around 4200 contribute to the pathogenicity of *X. campestris*. Xanthan is

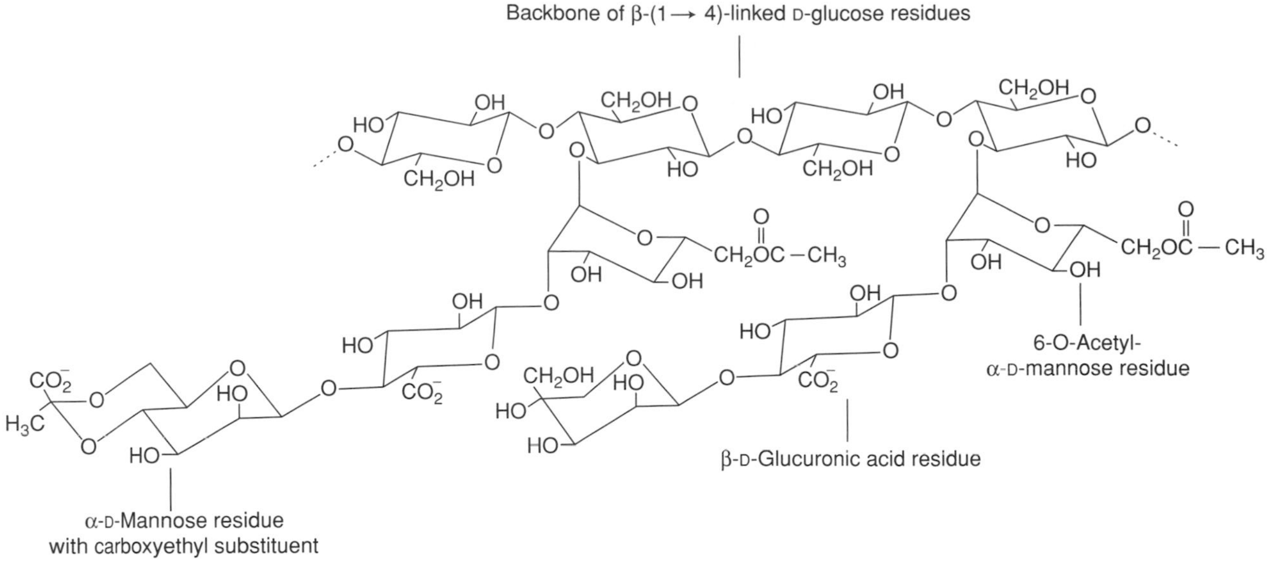

Backbone of β-(1→4)-linked D-glucose residues

6-O-Acetyl-α-D-mannose residue

β-D-Glucuronic acid residue

α-D-Mannose residue with carboxyethyl substituent

**FIGURE 8.4**

Primary structure of xanthan gum. Some of the terminal mannose residues are carboxyethylated, others are acetylated.

necessary for the assembly of *X. campestris* in liquid culture into biofilms. This has led to the suggestion that a role of xanthan in the disease process may be in the formation of biofilms that might protect the bacteria against desiccation, and against antimicrobial compounds produced by the plant. There is no evidence that *X. campestris* forms biofilms within plants.

## STRUCTURE OF XANTHAN GUM

The primary structure of xanthan gum is shown in Figure 8.4. In common with cellulose, the xanthan backbone consists of $\beta$-(1→4)-linked D-glucose; however, the 3-position of alternate glucose monomer units carries a trisaccharide side chain containing one glucuronic acid and two mannose residues. The nonterminal D-mannose unit carries an acetyl group at the 6-position, and a pyruvate is attached to the terminal D-mannose residue by a ketal linkage to the 4- and 6-positions. Occasionally, the terminal D-mannose carries a 6-*O*-acetyl group rather than pyruvate. The degree of acetylation and the pyruvate ketal content vary with culture conditions and from one strain of the microorganism to another.

In solution, xanthan forms a stiff right-handed double helix with fivefold symmetry and a pitch of 4.7 nm, with the two chains most likely running antiparallel to each other as shown in Figure 8.5. The trisaccharide branches are closely aligned with the polymer backbone. The double-helical molecules interact side-by-side to form body-centered lattices, which in turn form microcrystalline fibrils. Estimates of the apparent molecular weight of xanthan in solution range from 2 to 20 million. This wide spread in estimated values reflects the varying extent of intermolecular association between xanthan chains. In solution, xanthan chains transform from single- to double-helical structures in the presence of salts. As noted below, this transition is dependent on the shielding of the negative charges on xanthan by cations.

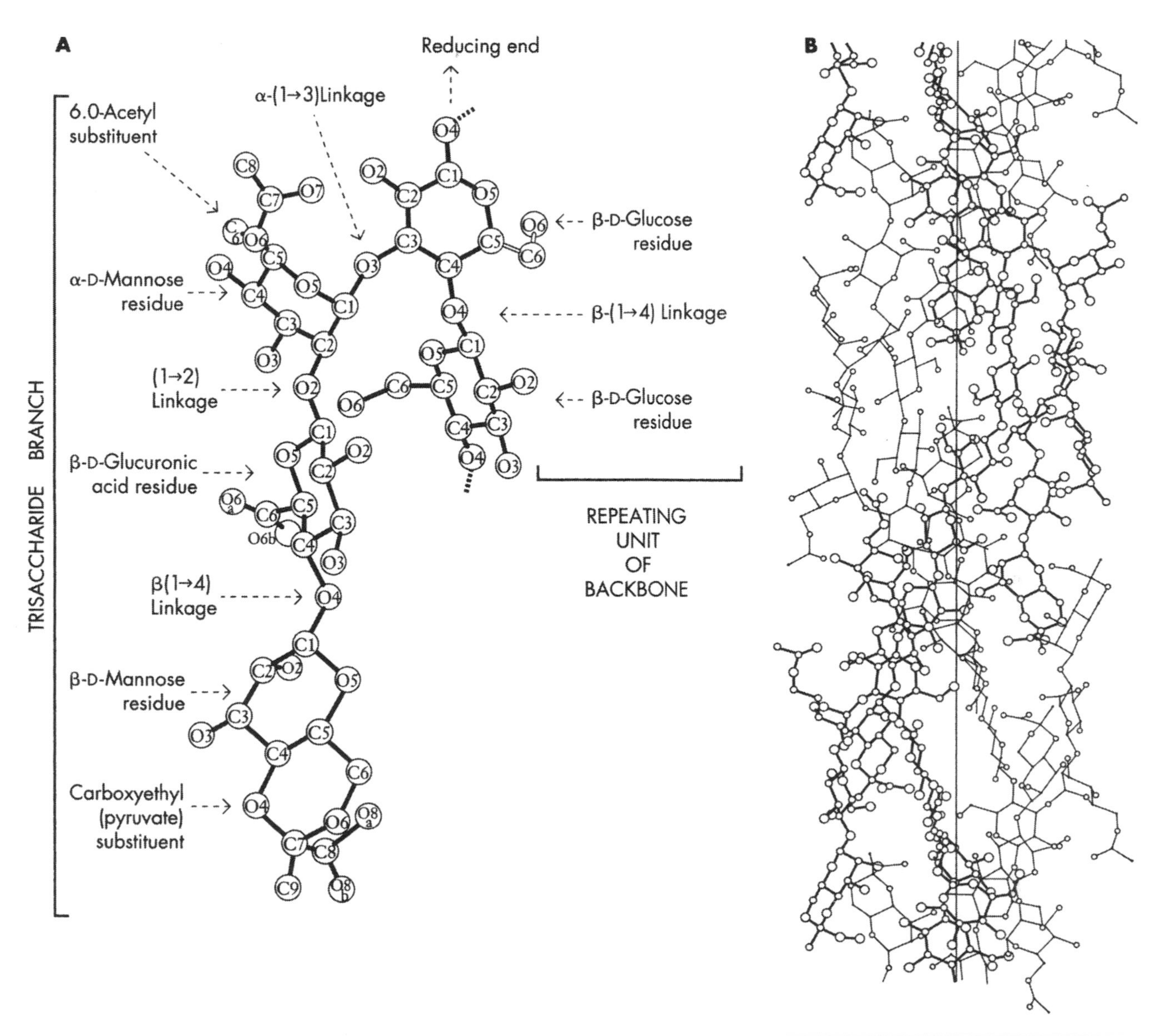


FIGURE 8.5

Structure of xanthan exopolysaccharide. (**A**) The conformation of the pentasaccharide repeating unit of xanthan. (**B**) The $5_1$ antiparallel xanthan double helices viewed perpendicular to the helix axis. [Based on Figures 3 and 4 in Okuyama, K., Arnott, S., Moorhouse, R., Walkinshaw, M. D., Atkins, E. D. T., and Wolf-Ullish, Ch. (1980). Fiber diffraction studies of bacterial polysaccharides. In *Fiber Diffraction Methods*, A. D. French and K. H. Gardner (eds.), ACS Symposium Series Volume 141, pp. 411–427, Washington, DC: American Chemical Society.]

## PROPERTIES OF XANTHAN GUM

Xanthan gum displays an extraordinary combination of physical properties that make this exopolysaccharide ideal for a wide range of applications (Table 8.4). Dilute xanthan solutions (0.5% to 1.5%, by weight) have a high viscosity, and when the concentration of salt in a solution is 0.1% or higher, the viscosity stays uniform over a temperature range of 0°C to 100°C. As a consequence, gravies and sauces containing xanthan maintain their thickness even at 80°C. The high viscosity of xanthan solutions is a result of the very stable two-chain helical, rigid, and rodlike structure of this branched exopolysaccharide (Figure 8.5). The small amounts of salt added to these solutions shield the negative charges on the trisaccharide branches and minimize electrostatic repulsion. Often salt is already present in the materials to which xanthan is added; if not, the amount of salt that must be added does

**Rheology** is the study of the deformation and flow of matter. It is that part of mechanics that deals with the relation between force and deformation in material bodies. The term *rheological behavior* is applied to materials that show nonlinear and time-dependent deformation in response to shear stress.

BOX 8.1

not present a problem in most products. In the complete absence of salt, the electrostatic repulsion destabilizes the two-chain structure sufficiently that the individual strands dissociate at elevated temperatures. Such separation can immediately be reversed by the addition of salt.

A particularly important property of xanthan is its rheological behavior (Box 8.1). In the absence of shear stress, a xanthan solution is viscous. When a shear stress is applied beyond a certain low minimum value, the viscosity decreases steeply with shear rate. The term *shear thinning* describes this behavior. Moreover, xanthan solutions show no hysteresis (Box 8.2): shear thinning and recovery are instantaneous. This property has considerable practical significance. For example, paints are formulated to be shear thinning: a xanthan-containing paint will maintain a high viscosity at low shear rates; it will not drip from the brush, and a film of the paint freshly applied to a vertical wall will not sag. However, a brushing motion applies a shear stress. Because shear thinning of xanthan solutions takes place at modest shear rates, the paint will thin and can be applied without undue exertion.

The shear-thinning and particle-suspending properties of xanthan solutions are particularly important in the inclusion of this polymer as a component of drilling muds used in drilling oil wells. Drilling muds suspend the sand particles from oil wells and carry them to the surface; they also act as lubricants for the drill bit. (At the tip of the drill bit, the shear rate is very high and the viscosity of xanthan solutions very low. The reverse situation holds in the drill shaft: the shear rate is low and the solution viscosity high.) Settling tests show that the suspending ability of xanthan gum surpasses that of any other polymer used in drilling fluids. Although xanthan gum is more expensive than guar gum, the plant polysaccharide most commonly used in the oil industry, the exceptional suspending ability of xanthan gum solutions at low polymer concentration favors its use where transportation costs are high.

The molecular explanation of the shear-thinning behavior of xanthan solutions is found in the way the two-chain xanthan molecules associate to form an entangled network of stiff molecules. This weak network is maintained by hydrogen-bonding interactions and entanglement between the trisaccharide side chains of the extended xanthan molecules. In the absence of shear or at low shear rates, this network accounts for the effective suspending properties exhibited by xanthan solutions. As shear rates increase, the side chains flatten against the backbone of the molecule and the network is progressively disrupted, with attendant decrease in viscosity.

**Hysteresis** is a phenomenon in which the relationship between two or more physical quantities depends on their prior history. More specifically, the response, *B*, takes on different values for an increasing input *A* than for a decreasing *A*. For example, when solid agar is heated, it does not dissolve until a temperature of 60°C or greater is reached. However, once dissolved, agar remains in solution until the temperature decreases to about 39°C or lower.

BOX 8.2

In the presence of 0.1% (or higher) salt, xanthan shows high solubility, uniform viscosity, and excellent chemical stability over a pH range from 1 to 13. This chemical stability is particularly impressive. It is reported that aqueous solutions of xanthan in 5% sulfuric or nitric acid, 25% phosphoric acid, or 5% sodium hydroxide are reasonably stable for several months at room temperature. The unusual chemical stability of xanthan at high temperature, or in strong acid or base, is a direct outcome of the protective effect of the trisaccharide branches. In the presence of salt, the branches of xanthan interact with the main chain and shield the labile glycosidic linkages in the cellulose backbone from hydrolytic cleavage.

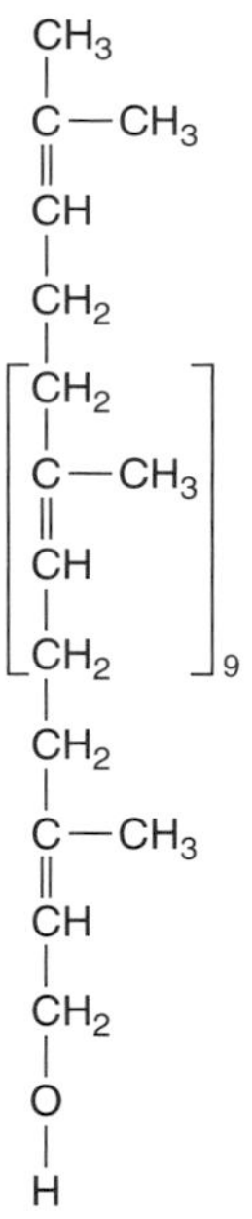

## BIOSYNTHESIS OF XANTHAN GUM

The biosynthesis of xanthan gum is similar to the biochemical pathways that produce such common bacterial extracellular polymers as peptidoglycan and lipopolysaccharide. In all of these cases, the challenge is to transfer a largely hydrophilic macromolecule across a lipid bilayer, the cell membrane. Microorganisms accomplish this task by utilizing a $C_{55}$ isoprenoid lipid carrier, bactoprenol (undecaprenyl alcohol; Figure 8.6), in the stepwise construction of the polysaccharide and its transfer across the cytoplasmic membrane. The assembly of the five-sugar repeating unit of xanthan by consecutive transfer of sugars from sugar nucleotide diphosphates is shown in detail in Figure 8.7. Each step is catalyzed by a specific glycosylase. The site-specific acetylation of the repeating unit is catalyzed by a specific acetylase with acetyl coenzyme A as acetyl donor. Pyruvoylation (carboxyethylation) of the terminal mannose is catalyzed by a specific ketalase with phospho*enol*pyruvate as cosubstrate. The acetylation and carboxyethylation reactions need not be consecutive. The completed xanthan building block, still linked through a pyrophosphoryl linkage to the bactoprenol, is transferred to another lipid-linked repeating unit by "tail-to-head" polymerization. The carrier bactoprenol pyrophosphate, released from the growing xanthan molecule in the addition step, is hydrolyzed to the monophosphate (reaction 9), contributing favorably to the energetics of the polymerization reaction (reaction 8), and regenerating lipid carrier for the synthesis of another repeating unit. The lipid carrier attached to the growing xanthan chain remains behind when the polysaccharide is ultimately released into the medium.

The enzymes required for the stepwise assembly of the pentasaccharide units attached to the polyisoprenyl phosphate carrier, the subsequent acetylation and ketalation of the mannose units, and the polymerization and export steps, are encoded in a 16-kb region, the 12-gene *gumBCDEFGHIJKLM* operon (Figure 8.8). This region of the *X. campestris* genome is sufficient to confer the ability to synthesize and export xanthan. An extracellular polysaccharide minus mutant of *Sphingomonas* species, a member of the $\alpha$-proteobacteria, was used as a recipient for the *X. campestris* strain B1459 *gumBCDEFGHIJKLM* cluster. The transformed strain secreted large amounts of xanthan gum largely indistinguishable from that produced by the donor strain.

$$CH_3-C(CH_3)=CH-CH_2-[CH_2-C(CH_3)=CH-CH_2]_9-CH_2-C(CH_3)=CH-CH_2-O-H$$

**FIGURE 8.6**
Bactoprenol (undecaprenyl alcohol).

## PRODUCTION OF XANTHAN GUM

The yield of microbial polymers is frequently strongly influenced by the medium composition and the nature of the growth-limiting nutrient. Optimization of culture conditions is performed by growing cells continuously in a chemostat, in which the cell growth is limited by a selected nutrient.

**1.** Bactoprenol-**P** + UDP-Glc → Bactoprenol-**P**-**P**-Glc + UMP

**2.** Bactoprenol-**P**-**P**-Glc + UDP-Glc → Bactoprenol-**P**-**P**-Glc-(4←1)-β-D-Glc + UDP

**3.** Bactoprenol-**P**-**P**-Glc-(4←1)-β-D-Glc + GDP-Man → Bactoprenol-**P**-**P**-Glc-(4←1)-β-D-Glc + GDF
3
↑
1
α-D-Man

**4.** Bactoprenol-**P**-**P**-Glc-(4←1)-β-D-Glc + UDP-GlcUA → Bactoprenol-**P**-**P**-Glc-(4←1)-β-D-Glc + UDP
3 ↑ 1 α-D-Man
3 ↑ 1 β-D-GlcUA-(1→2)-α-D-Man

**5.** Bactoprenol-**P**-**P**-Glc-(4←1)-β-D-Glc + GDP-Man → Bactoprenol-**P**-**P**-Glc-(4←1)-β-D-Glc + GDF
3 ↑ 1 β-D-GlcUA-(1→ 2)-α-D-Man
3 ↑ 1 β-D-Man-(1→4)-β-D-GlcUA-(1→ 2)-α-D-Man

**6.** Bactoprenol-**P**-**P**-Glc-(4←1)-β-D-Glc + Acetyl-CoA → Bactoprenol-**P**-**P**-Glc-(4←1)-β-D-Glc + CoA
3 ↑ 1 β-D-Man-(1→4)-β-D-GlcUA-(1→2)-α-D-Man
3 ↑ 1 β-D-Man-(1→4)-β-D-GlcUA-(1→2)-α-D-Man-6-O

**7.** Bactoprenol-**P**-**P**-Glc-(4←1)-β-D-Glc + PEP → Bactoprenol-**P**-**P**-Glc-(4←1)-β-D-Glc + **P**
3 ↑ 1 β-D-Man-(1→4)-β-D-GlcUA-(1→2)-α-D-Man-6-OAc
3 ↑ 1 β-D-Man-(1→4)-β-D-GlcUA-(1→2)-α-D-Man-6-OAc
4 6
C
$H_3C$ COOH

**8.** Bactoprenol-**P**-**P**-[Glc-(4←1)-β-D-Glc-. . .]$_n$ + Bactoprenol-**P**-**P**-Glc-(4←1)-β-D-Glc
3 ↑ 1 β-D-Man-(1→4)-β-D-GlcUA-(1→2)-α-D-Man-6-OAc
4 6
C
$H_3C$ COOH
3 ↑ 1 β-D-Man-(1→4)-β-D-GlcUA-(1→2)-α-D-Man-6-OAc
4 6
C
$H_3C$ COOH
↓
Bactoprenol-**P**-**P**-[Glc-(4←1)-β-D-Glc-. . .]$_{n+1}$ + Bactoprenol-**P**-**P**
3 ↑ 1 β-D-Man-(1→4)-β-D-GlcUA-(1→2)-α-D-Man-6-OAc
4 6
C
$H_3C$ COOH

**9.** Bactoprenol-**P**-**P** → Bactoprenol-**P** + **P**

**FIGURE 8.7**

Biosynthetic pathway for the formation of bactoprenol-linked xanthan polymer. Steps 1 through 5 are catalyzed by specific glycosyltransferases; step 6 is catalyzed by an acetylase and step 7 by a ketalase. Step 8 is catalyzed by a polymerase. Abbreviations: Glc, glucose; Man, mannose; GlcA, glucuronic acid; Ac, acetyl; P, inorganic phosphate; UDP-Glc, uridine-5′-diphosphoglucose; GDP-Man, guanosine-5′-diphosphomannose; PEP, phospho*enol*pyruvate.

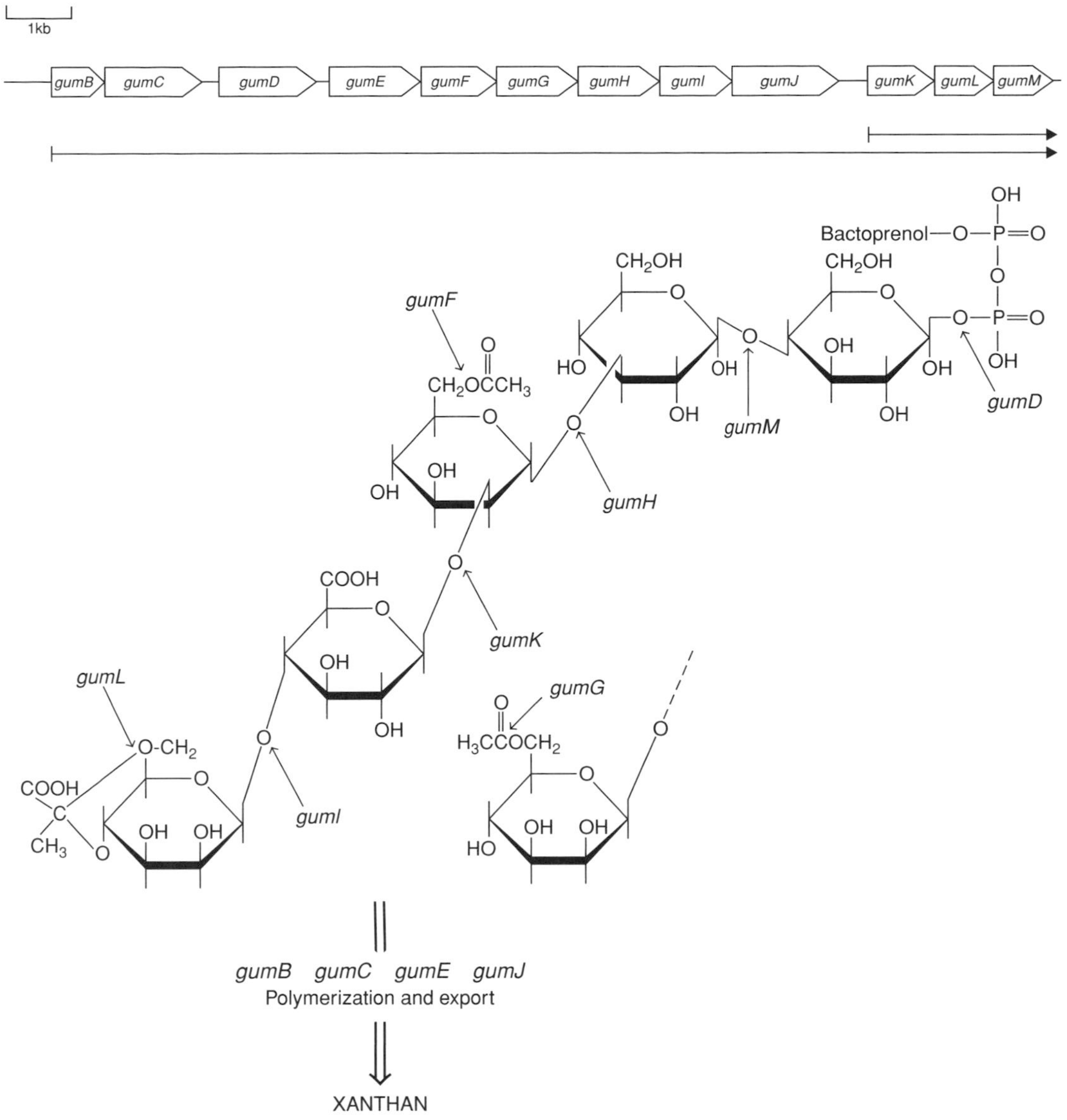

**FIGURE 8.8**

Genetic map of the *Xanthomonas campestris* pv. *campestris* gum operon that encodes the proteins required for the biosynthesis and export of xanthan and a diagram showing the bonds (or steps in polymerization and export) that require these gene functions. GumD, GumM, GumH, GumK, and GumI catalyze reactions 1 through 5 presented in Figure 8.7. GumL catalyzes the pyruvyl addition to the external mannose. GumF catalyzes the acetylation of the internal $\alpha$-mannose and GumG the acetylation of the external $\beta$-mannose. The synthesis of the xanthan polymer and its export requires the functions encoded by *gumB, gumC, gumE,* and *gumJ*.

Chemostat experiments (Box 8.3), performed at a constant dilution rate, showed that the yield of xanthan from glucose in *X. campestris* is strongly influenced by the choice of the growth-limiting nutrient, with the highest yield seen under conditions of nitrogen limitation. Temperature is an important variable. Cell growth is most rapid between 24°C and 27°C, whereas the yield of xanthan is highest between 30°C and 33°C. Commercial fermentations are run at about 28°C. Xanthan production invariably lags behind cell growth (Figure 8.9). It is possible that this lag reflects competition for bactoprenol between three synthetic pathways – those for peptidoglycan, lipopolysaccharide, and xanthan – during the period of exponential cell growth.

Xanthan gum is produced commercially by submerged aerobic batch fermentation of a pure *X. campestris* culture, using glucose, sucrose, or starch

The chemostat is an apparatus for the continuous culture of microorganisms. Sterile medium is fed into the culture vessel at a constant rate. The volume of the culture is kept constant by removing cells and spent medium at the same rate at which fresh medium is added. The chemostat controls the growth rate of the culture by limiting the supply of one essential nutrient while providing other essential nutrients in excess.

The dilution rate is the volume of fresh medium added per hour and is expressed as a fraction of the volume of the vessel.

**BOX 8.3**

as a carbon source in a simple medium. Acids are produced during the fermentation, and the pH is maintained near neutrality by periodic addition of sodium hydroxide. This adjustment of the pH is necessary because xanthan gum production ceases below pH 5. In the production of food-grade gum, fermentation is stopped when the carbohydrate in the medium is exhausted, and the broth is pasteurized at near-boiling temperature. Isopropyl alcohol is added to the broth to precipitate the xanthan gum, and then the precipitate is dried and milled.

Solid-state fermentation on very low–cost substrates can give xanthan yields comparable to those obtained in submerged aerobic batch fermentation (up to 30 $g\,l^{-1}$), as shown by laboratory-scale studies with food industry waste products. The latter included spent malt grains, apple pomace, and citrus peels. In another approach to using a waste product as substrate, construction of a genetically modified lactose-utilizing *X. campestris* strain permitted the use of whey as a cheap fermentation medium.

## MODIFICATION OF XANTHAN STRUCTURE

Xanthans with altered structure have been obtained in several different ways. To begin with, different strains of *X. campestris* synthesize xanthans with widely differing numbers of acetyl and pyruvyl groups. Modifications in the culture media and environmental conditions also lead to xanthans that vary widely with respect to these substituents. Insertional mutagenesis with transposons (see Box 3.8) has been used to generate mutant strains of *X. campestris* that produce xanthan polymers with truncated side branches.

### Variation in Acetyl and Carboxyethyl (Pyruvate) Group Content

The composition of the carbohydrate structure of xanthan is unaffected by changes in the carbon source supplied to *X. campestris*. However, the modification of the branch sugars is strongly influenced by the choice of carbon

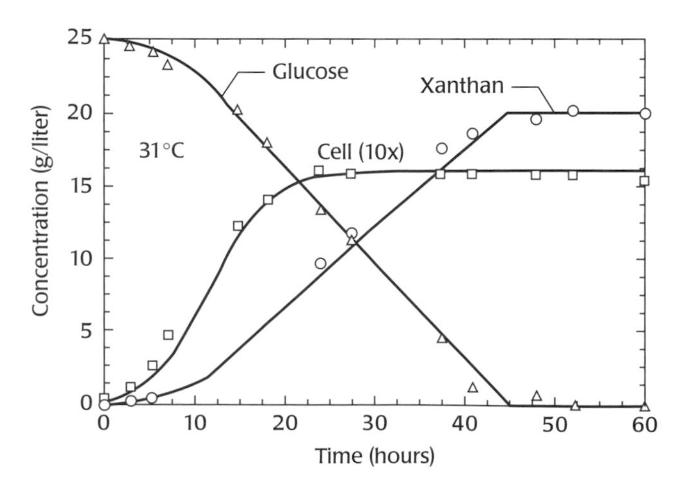


**FIGURE 8.9**

Consumption of glucose, cell growth, and xanthan gum formation in a submerged aerobic batch fermentation of *X. campestris*. [Based on Shu, C-H., and Yang, S-T. (1990). Effects of temperature on cell growth and xanthan production in batch cultures of *Xanthomonas campestris*. *Biotechnology and Bioengineering*, 35, 454–468, Figure 1 (a).]

substrate. For example, carboxyethyl groups represent 8% of the dry weight of xanthan when glycerol is used as the carbon source but only 5% when the carbon source is alanine; with glucose as substrate, pyruvate represents 7.1%, and with pyruvate as substrate 10.6%.

The acetyl and carboxyethyl groups of xanthan influence the conformational properties and the association behavior of xanthan in solution. The acetyl groups stabilize the ordered conformation of xanthan, which can be induced by specifically increasing the ionic strength of dilute aqueous solutions (by adding calcium ions). Carboxyethyl groups have been shown to have a strong destabilizing effect on the ordered conformation. This is most likely a result of an unfavorable electrostatic contribution due to repulsion between these negatively charged groups. The optical rotation of xanthan solutions depends in part on the xanthan's conformation. This dependence can be exploited to follow the transition between the ordered and random coil conformations of xanthan in solution and to assess the effect of the acetyl and carboxyethyl substituents on the stability of the ordered conformation. The midpoint temperature ($T_m$) for the optical rotation change in deionized water for xanthan containing, by weight, 4.5% of acetyl groups and 4.4% of carboxyethyl groups is 44°C, but rises to 54.5°C for polymer containing 7.7% acetyl and 1.7% carboxyethyl groups.

Thus, xanthan gum provides an excellent example of how the interplay between structure and physical and chemical properties can endow a particular exopolysaccharide with considerable industrial value. Similar accounts could be given of many other microbial exopolysaccharides but at this time would not be justified by the relatively modest demand for these materials. However, studies of the exopolysaccharides listed in Table 8.2 and, to a lesser extent, of other microbial exopolysaccharides have provided extensive information concerning the structure–property relationships in such polymers. This information coupled with the ability to genetically engineer microbial strains so that they produce high yields of modified polysaccharides will lead to new biopolymers with valuable properties.

## POLYESTERS

Bacteria differ from one another in the type of *reserve material* they accumulate when they are grown with unbalanced supplies of nutrients. The nature of the reserve material depends upon the genotype of the organism and the kind of limitation. If the carbon-to-nitrogen ratio is high, and if nitrogen, phosphorus, or oxygen are limiting, many bacteria will accumulate glycogen and/or aliphatic polyesters, polyhydroxyalkanoates (PHAs), in amounts up to 80% or more of their cellular dry weight.

### OCCURRENCE OF POLYHYDROXYALKANOATES IN NATURE

In 1926, M. Lemoigne described the characterization, in *Bacillus megaterium*, of the first compound of this class, poly-(*R*)-(3-hydroxybutyrate)

(P[3HB]):

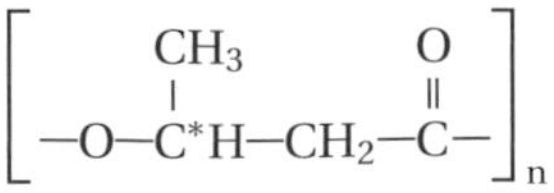

$$\left[ -O-\overset{\overset{CH_3}{|}}{C^*}H-CH_2-\overset{\overset{O}{||}}{C}- \right]_n$$

The asterisk indicates an asymmetric center.

By the late 1950s, the presence of P(3HB) was noted within the cells of many Gram-negative bacteria and substantial evidence had accumulated that it functioned as a reserve of carbon and energy. Within 15 years thereafter, indications were detected of considerable chemical complexity in naturally occurring PHAs. In 1974, examination of PHAs in extracts of activated sewage sludge showed that in addition to 3HB, the PHA monomers included 3-hydroxyvalerate (3HV) as a major component and 3-hydroxyhexanoate (3HHx) as a minor one. In 1983, analysis of PHA from marine sediments revealed 3HB and 3HV as the major components, and 11 other hydroxyalkanoate (HA) monomers. Another 1983 study showed that growth of *Pseudomonas oleovorans* on 50% volume to volume water-*n*-octane led to the formation of intracellular inclusions consisting of 3-hydroxyoctanoate [3HO] building blocks. Further analysis showed the presence of a small amount of 3HHx, but no 3HB.

PHAs from different bacteria, extracted by means of organic solvents, have molecular weights of up to $2 \times 10^6$, corresponding to a degree of polymerization of about 20,000. The polymer accumulates in highly refractive, discrete granules, typically 0.2 to 0.5 $\mu$m in diameter, in the cytoplasm of more than 100 prokaryote genera, with each granule containing several thousand polymer chains. These include aerobic and anaerobic heterotrophic bacteria (e.g., *Azotobacter beijerinckii*, *Zoogloea ramigera*, and *Clostridium butylicum*); many methylotrophs; chemolithotrophs, such as *Ralstonia eutropha*; aerobic and anaerobic phototrophs (*Chlorogloea fritschii*, *Rhodospirillum rubrum*, *Chromatium okenii*); and archaea (e.g., *Haloferax mediterranei*). The maximal levels of PHA (as percentages of dry weight) range widely from one organism to another, from 25% in the nitrogen-fixing aerobe *Azotobacter* grown on glucose, to 70% in the methylotroph *Methylocystis* grown on methane, to as high as 80% in the phototrophic anaerobe *Rhodobacter* grown on acetate. In sum, more than 125 different hydroxyalkanoate units, derived either from products of cell metabolism or originating from different precursor substrates added to the growth medium are present in various biosynthetic PHAs. The structures of a few of these units are shown in Figure 8.10.

As an interesting aside, until the mid-1980s, PHAs were believed to serve exclusively as stores of carbon and energy in bacteria. Short-chain P(3HB) made up of 130- to 170-monomer units is now known to be a ubiquitous constituent of both prokaryotic and eukaryotic cell membranes (both animal and plant), in which it seems to participate in the transport of polyanionic salts (particularly polyphosphate), transformation, and calcium signaling. This soluble low molecular mass polymer represents less than 0.001% of cellular dry weight.

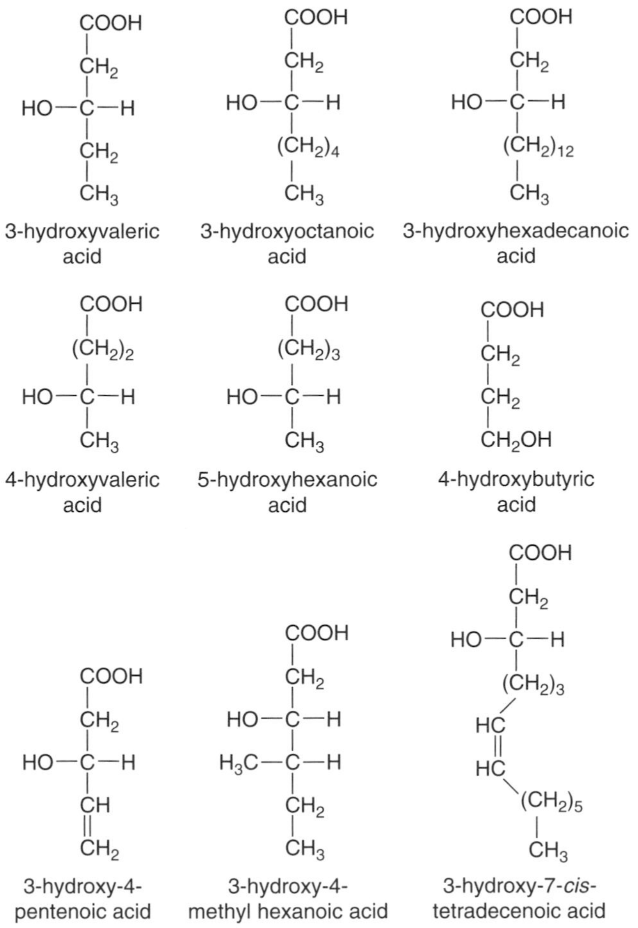

3-hydroxyvaleric acid

3-hydroxyoctanoic acid

3-hydroxyhexadecanoic acid

4-hydroxyvaleric acid

5-hydroxyhexanoic acid

4-hydroxybutyric acid

3-hydroxy-4-pentenoic acid

3-hydroxy-4-methyl hexanoic acid

3-hydroxy-7-*cis*-tetradecenoic acid

**FIGURE 8.10**

The diverse biosynthetic polyhydroxyalkanoate monomers found in prokaryotic PHAs illustrate the low substrate specificity of PHA synthases. [Steinbüchel, A., and Valentin, H. E. (1995). Diversity of bacterial polyhydroxyalkanoic acids. *FEMS Microbiology Letters*, 128, 219–228.]

## BIOCHEMICAL BASIS OF PHA DIVERSITY

Different microorganisms produce PHA copolymers containing different monomers in varying proportions. Careful analysis shows that pure homopolymeric PHAs are the exception. Novel polymeric PHAs with desired properties can be produced by the addition to the culture medium of organic compounds that can be taken up by cells and converted to (*R*)-hydroxyalkanoic acids. Even the ester backbone of the polymer can be varied. *R. eutropha* is able to utilize 3-mercaptopropionyl monomers (derived from added 3-mercaptopropionic acid) to form poly(3-hydroxybutyrate-*co*-3-mercaptopropionate), with monomers linked through oxygen ester or thioester linkages.

Here are the main factors that influence the composition of the PHA formed by a particular microorganism from endogenous substrates or from organic compounds added to the medium:

(a) *The substrate specificity range of the PHA synthase (PhaC)*. PHA synthases from different prokaryotes vary with respect to subunit composition and substrate specificity. These key enzymes in PHA biosynthesis use the CoA thioesters of the (*R*)-isomers of hydroxyalkanoic acids as

**TABLE 8.5 PHA synthases: genes and biochemical properties**

| Organism | Structural gene(s) | CoA thioester substrate specificity |
|---|---|---|
| *Ralstonia eutropha* | *phaC* (1767 bp) | $3HA_{SCL}$ |
| *Pseudomonas aeruginosa* | *phaC1* (1677 bp); *phaC2* (1680 bp) | $3HA_{MCL}$ |
| *Chromatium vinosum* | *phaC* (1063 bp) and *phaE* (1074 bp) encoding subunits of a single PHA synthase | $3HA_{SCL}$ |
| *Thiocapsa pfennigii* | *phaC* (1074 bp) and *phaE* (1104 bp) encoding subunits of a single PHA synthase | $3HA_{SCL}$ and $3HA_{MCL}$ |

*Source*: Rehm, B. H. A., and Steinbüchel, A. (1999). Biochemical and genetic analysis of PHA synthases and other proteins required for PHA synthesis. *International Journal of Biological Macromolecules*, 25, 3–19.

substrates and catalyze their polymerization into PHAs with the concomitant release of CoA. The ability of various prokaryotes to synthesize PHAs varying widely in the structure of their building blocks results from the distinctive, yet broad, substrate specificities of the various types of PHA synthases. Type I synthases preferentially utilize CoA thioesters of various monomers (designated $HA_{SCL}$) that contain only three to five carbon atoms. Type II synthases preferentially utilize CoA thioesters of monomers (designated $3HA_{MCL}$) with at least five carbon atoms. Type III synthases have a broad substrate range comprising CoA thioesters of $3HA_{SCL}$ and $3HA_{MCL}$ (Table 8.5). Monomers differing in the position of the hydroxyl-bearing carbon, such as 4-HA or 5-HA, can also be used as substrates by the various PHA synthases.

**(b)** *The supply of precursors and the repertoire of metabolic reactions available to an organism.* The endogenous precursors to the monomers incorporated into PHAs are intermediates in central biosynthetic and energy-producing pathways of the cell. As shown in Figures 8.11 and 8.12, PHA biosynthesis may compete with the tricarboxylic acid cycle and fatty acid biosynthesis for acetyl-CoA and with fatty acid biosynthesis or degradation for 3-hydroxyacylalkanoates of varying chain length.

The biosynthesis of poly(3-hydroxybutyrate-*co*-3-hydroxyvalerate; P(3HB-*co*-3HV)) in *R. eutropha* exemplifies the contribution of the repertoire of metabolic reactions available to an organism. *R. eutropha* produces P(3HB) when grown under appropriate conditions with glucose as the sole carbon source. However, when in addition to glucose, propionic acid is added in controlled amounts, a random P(3HB-*co*-3HV) copolymer is formed, containing a predictable fraction of randomly distributed 3HV units. This outcome depends on the presence of two different 3-ketothiolases in *R. eutropha*. One, encoded by the *phbA* gene has a high specificity for acetyl-CoA. The other, encoded by the *bktB* gene, has a higher specificity for propionyl-CoA. BktB 3-ketothiolase efficiently catalyzes the condensation of propionyl-CoA and acetyl-CoA to form 3-ketovaleryl-CoA. A single acetoacetyl-CoA reductase (PhbB) catalyzes the reduction of 3-ketovaleryl-CoA

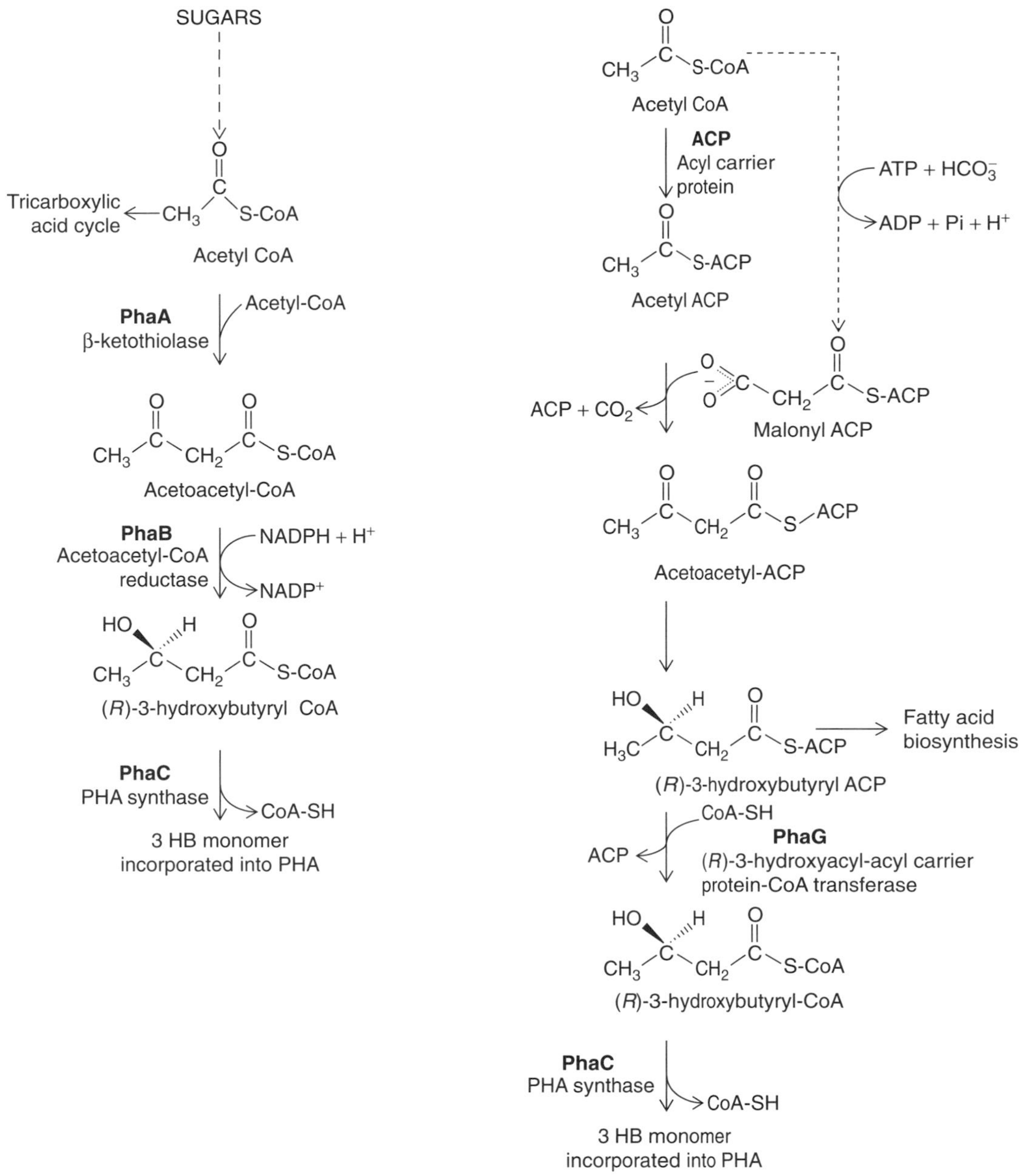

**FIGURE 8.11**

Utilization of acetyl-CoA from sugar catabolism and of intermediates from the fatty acid biosynthetic pathway in the biosynthesis of hydroxyalkanoate monomers for incorporation into PHAs.

and 3-acetoacetyl-CoA to 3-hydroxyvaleryl-CoA and 3-hydroxybutyryl-CoA, respectively. The subsequent polymerization to form P(3HB-*co*-3HV) is catalyzed by PHA synthase (PhbC). The genes encoding the enzymes for PHA biosynthesis in *R. eutropha* were first described and named *phbA*, *phbB*, and *phbC* in recognition of their roles in poly-3-hydroxybutyrate biosynthesis. The corresponding enzymes in other organisms are designated *phaA*, *phaB*, and *phaC*, respectively.

Several pseudomonads accumulate PHA containing 3-hydroxydecanoate as the predominant monomer when grown on various carbon sources, such as fructose, glycerol, acetate, or lactate. These strains possess a

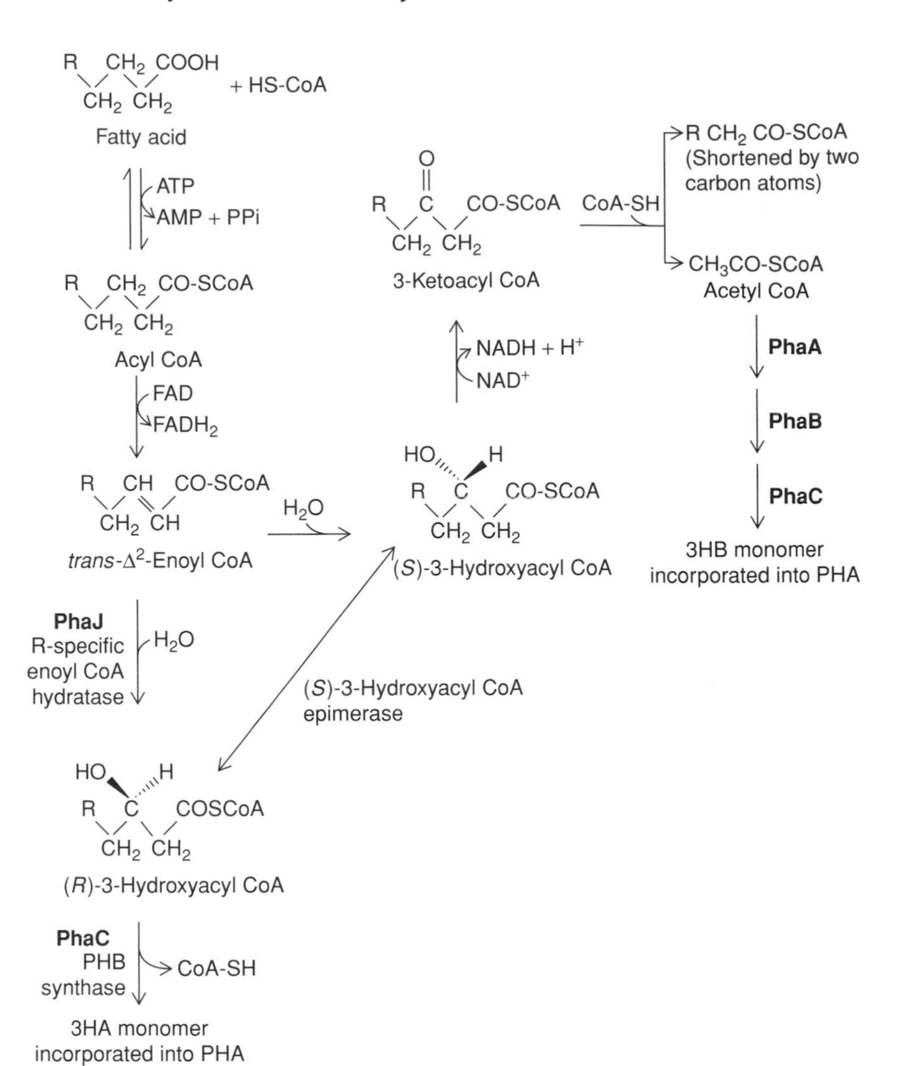


**FIGURE 8.12**

Utilization of intermediates from the fatty acid degradation pathway in the biosynthesis of hydroxyalkanoate monomers for incorporation into PHAs. Details of the pathway catalyzed by PhaA, PhaB, and PhaC for the conversion of acetyl-CoA to the 3HB monomer in PHA are shown in Figure 8.11.

3-hydroxyacyl-acyl carrier protein-CoA transferase, encoded by the *phaG* gene, and derive the 3-hydroxyacyl monomers from the *de novo* pathway of fatty acid biosynthesis (Figure 8.11). This enzyme converts the (*R*)-3-hydroxyacyl-acyl carrier protein (ACP) intermediates to (*R*)-3-hydroxyacyl-CoA. The monomer composition of the resulting PHAs reflects the preference of the pseudomonad PHA synthase for longer-chain substrates.

When alkanoic acids are added to the medium and taken up by the bacteria, they are converted to CoA esters preparatory to their entry into the fatty acid $\beta$-oxidation pathway. This pathway generates (*S*)-3-hydroxyacyl-CoA intermediates. Conversion to the *R* isomer is necessary for these compounds to be utilized as substrates by the PHA synthase. Organisms such *Pseudomonas oleovorans* have epimerases capable of interconverting the *S* and *R* isomers of 3-hydroxyacyl-CoA (Figure 8.12). Other organisms have an *R*-specific enoyl-CoA hydratase (PhaJ; Figure 8.12) that converts *trans*-$\Delta^2$-enoyl-CoA directly to (*R*)-3-hydroxyacyl-CoA.

## BIOSYNTHESIS OF COPOLYESTERS FROM SINGLE SUBSTRATES

*R. eutropha* requires two substrates – glucose and propionate – to synthesize 3HB-*co*-3HV copolymers. Other microorganisms synthesize copolymers with varying ratios of 3HB to 3HV when grown on a single substrate. This is illustrated in Table 8.6 for *Rhodococcus ruber* grown on different carbon sources. The 3HV content of the polymer ranges from above 99% for *R. ruber* grown on valerate (pentanoate) to about 65% for cells grown on malate.

**TABLE 8.6 Composition and yield of 3HB-co-3HV copolymers formed by *Rhodococcus ruber* grown on different substrates**

| Substrate | Molar ratio 3HV:3HB | Yield of PHA as % of cell dry weight |
|---|---|---|
| Malate | 1.3 | 7.1 |
| Acetate | 2.5 | 40.4 |
| Pyruvate | 3.6 | 9.0 |
| Glucose | 3.8 | 31.1 |
| Lactate | 5.1 | 32.2 |
| Succinate | 12.0 | 7.1 |

*Source*: Anderson, A. J., Williams, R. D., Taidi, B., Dawes, E. A., and Ewing, D. F. (1992). Studies on copolyester synthesis by *Rhodococcus ruber* and factors influencing the molecular mass of polyhydroxybutyrate accumulated by *Methylobacterium extorquens* and *Alcaligenes eutrophus*. *FEMS Microbiology Letters*, 103, 93–102.

## MANIPULATING GROWTH CONDITIONS TO PRODUCE NOVEL BACTERIAL POLYESTERS: AN EXAMPLE

In nature, *P. oleovorans* utilizes *n*-alkanes and *n*-alkanoic acids as sole carbon sources for growth. Under nutrient-limiting conditions, but with excess carbon source, *P. oleovorans* forms large amounts of poly(3-hydroxyalkanoates) from these substrates. *P. oleovorans* can use the "unnatural" substrate, 5-phenylvalerate (Figure 8.13), as the only carbon source. With this substrate in excess under conditions of nutrient deprivation, *P. oleovorans* forms a pure polyester, poly(3-hydroxy-5-phenylpentanoate) (Figure 8.13).

## BIOSYNTHESIS OF NOVEL BACTERIAL POLYESTERS THROUGH COMETABOLISM

A phenomenon known as *cometabolism* also adds greatly to the variety of copolymers that bacteria are able to synthesize. Cometabolism was first reported in 1960, in studies of methane-oxidizing bacteria. When these bacteria metabolize a growth-supporting hydrocarbon substrate, such as methane, they can also incorporate and oxidize other hydrocarbons, even though those hydrocarbons would not be utilized if they were present alone. We will encounter this phenomenon again in Chapter 14 in the context of environmental degradation of foreign organic compounds.

Examples of cometabolism are seen in PHA metabolism in *P. oleovorans*. When *P. oleovorans* grows on a hydrocarbon substrate, which itself can supply monomer units for PHA formation, it is able to incorporate either a non–polymer-producing substrate or a non–growth-producing substrate into the polymer. Thus, these bacteria can form polyesters from compounds not found in nature. The proportion of each building block in the copolymers produced by cometabolism is determined by the ratio of the starting materials in the culture medium. Table 8.7 shows data for a copolymer of nonanoic acid (a natural substrate) and 11-cyanoundecanoic acid (able to support growth of *P. oleovorans* but not polymer synthesis on its own).

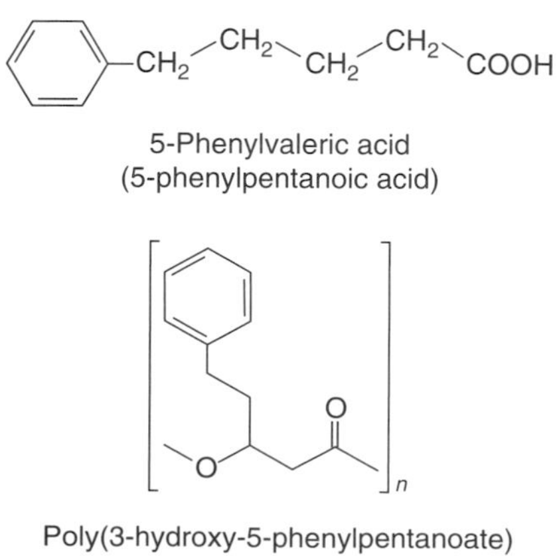

$C_6H_5-CH_2-CH_2-CH_2-CH_2-COOH$

5-Phenylvaleric acid
(5-phenylpentanoic acid)

Poly(3-hydroxy-5-phenylpentanoate)

**FIGURE 8.13**

Structures of 5-phenylvalerate and poly(3-hydroxy-5-phenylpentanoate).

## INTRACELLULAR AND EXTRACELLULAR BIODEGRADATION OF POLY(3-HYDROXYALKANOATES)

When the carbon sources in the growth medium are exhausted, P(3HB) is degraded by an intracellular P(3HB) depolymerase (PhaZ) to the monomeric hydroxy acids. The depolymerase is an *exo*-type hydrolase.

**TABLE 8.7 Composition and yield of copolymers produced by *P. oleovorans* grown with mixtures of nonanoic acid and 11-cyanoundecanoic acid**

| Molar ratio of nonanoic to 11-cyanoundecanoic acid in growth medium | Yield of copolymers, % of biomass | Mole % in copolymer of units derived from 11-cyanoundecanoic acid[a] |
|---|---|---|
| 1:1 | 19.6 | 32 |
| 7:5 | 30.5 | 25 |
| 2:1 | 36.3 | 17 |

[a] Monomers, derived from 11-cyanoundecanoic acid and incorporated into the polyester, are 9-cyano-3-hydroxynonanoate and 7-cyano 3 hydroxyheptanoate.
*Source*: Lenz, R. W., Kim, Y. B., and Fuller, R. C. (1992). Production of unusual bacterial polyesters by *Pseudomonas oleovorans* through cometabolism. *FEMS Microbiology Letters*, 103, 207–214.

D-(−)-Hydroxybutyric acid is then oxidized to acetoacetate by a nicotinamide adenine dinucleotide (NAD)- specific dehydrogenase. The acetoacetate is converted to acetoacetyl-CoA by acetoacetyl-CoA synthetase. Thus acetoacetyl-CoA is an intermediate common to both the biosynthesis and degradation of P3(HB). The intracellular depolymerases involved in the degradation of other PHAs are homologous to that for P(3HB).

The ability to degrade extracellular PHAs is widely distributed among microorganisms, including representatives of nearly a hundred genera of fungi. PHA-degrading microorganisms are ubiquitous in terrestrial and aquatic ecosystems. P(3HB)-degrading bacteria are particularly common. The microorganisms that are able to degrade extracellular PHAs either secrete specific PHA depolymerases or display them on their surface. These depolymerases hydrolyze PHAs by surface erosion to monomers and/or oligomers. These are subsequently biodegraded further to water and either methane or carbon dioxide.

## POLYHYDROXYALKANOATES AS BIODEGRADABLE THERMOPLASTICS OR ELASTOMERS

PHAs with aliphatic monomers of three to five carbons are thermoplastics. A thermoplastic material softens when heated and hardens again when it cools. In practical terms, for such polymers the transition temperature from hard to soft is high. The heating–cooling cycle may be repeated many times. Familiar thermoplastic materials, such as polyethylene, polypropylene, polyvinyl chloride, polystyrene, polycarbonate, and nylon, are used in the manufacture of hundreds of different types of plastic objects used in everyday life. PHAs with medium-length carbon side chains, such as poly(3-hydroxyoctanoate-*co*-3-hydroxyhexanoate) are elastomers. They are amorphous, flexible polymers, relatively soft and deformable. Below their glass transition temperature they become rigid and brittle. Their behavior resembles that of rubber. PHAs with even longer side chains behave like waxes. The ability to make different homopolymers and copolymers that vary in the nature and proportion of their building blocks has much potential practical importance.

Poly(3HB) shows similarities in its molecular structure and physical properties to polypropylene, a widely used polymer. Polypropylene is used for packaging, rope, wire insulation, pipe and fittings, bottles, and appliance parts. Both polymers are *isotactic*; that is, in each polymer, the methyl group attached to the backbone is present in a single configuration throughout the chain. The physical properties of poly(3HB) and of polypropylene and their degradability are compared in Table 8.8. Polypropylene is highly resistant to biodegradation, whereas poly(3HB) is ultimately completely degraded in a variety of environments. Even though polypropylene and P(3HB) have similar melting points, P(3HB) is difficult to process because it decomposes at about 10°C above its melting point of about 177°C. Moreover, the extension to break (<10%) for P(3HB) is very much lower than that for polypropylene (40%) (Table 8.8). Consequently, P(3HB) is a stiffer and more brittle plastic material than polypropylene.

Plastics composed of 3HB-*co*-3HV copolymers have more favorable characteristics. The physical properties of 3HB-*co*-3HV copolymers are very sensitive to the mole percentage of 3HV. As this increases (Figure 8.14), the decomposition temperature remains unchanged but the melting temperature drops progressively. The percent crystallinity of the plastic decreases from over 80% for P(3HB) to under 30% at a mole fraction of 25% 3HV. Likewise, the copolymers show severalfold improvement in flexibility and toughness.

**TABLE 8.8 Comparison of some properties of polypropylene and poly(3-hydroxybutyrate)**

| Property | Polypropylene | P(3HB) |
|---|---|---|
| Molecular weight | $(2.2–7) \times 10^5$ | $(1–8) \times 10^5$ |
| Melting point (°C) | 171–186 | 171–182 |
| Glass-transition temperature (°C) | −10 | 4 |
| Crystallinity (%) | 65–70 | 65–80 |
| Density ($g/cm^{-3}$) | 0.905–0.94 | 1.23–1.25 |
| Flexural modulus (Gpa)[a] | 1.7 | 3.5–4.0 |
| Tensile strength (MPa)[b] | 39 | 40 |
| Extension at break (%) | 40 | Good |
| Resistance to ultraviolet radiation | Poor | Good |
| Biodegradability | Very poor | Excellent |

[a] Flexural modulus is a measure of the elastic stiffness of a material. It relates the strain to the applied stress. GPa = $10^9$ pascals or newtons/$m^2$.

[b] Tensile strength of a material is a measure of the maximum stress a material can withstand. MPa = $10^6$ pascals.

*Source*: Brandl, H., Gross, R. A., Lenz, R. W., and Fuller, R. C., (1990). Plastics from bacteria and for bacteria: poly($\beta$-hydroxyalkanoates) as natural, biocompatible, and biodegradable polyesters. *Advances in Biochemical Engineering/Biotechnology*, 41, 77–93, Table 6.

Poly(3HB-*co*-3HV) was introduced on the market by Imperial Chemical Industries in the 1990s under the trade name of Biopol as a biodegradable alternative to polyethylene. Biopol was produced by fermentation in *R. eutropha*, whose complete genome sequence is known. Initial products made from Biopol were containers for shampoos and cosmetics and bicycle helmets made from a mixture of Biopol fibers and cellulose fibers. These products were water resistant and impermeable to oxygen. In simulated landfill experiments, over a 19-week period, buried Biopol bottles showed a weight loss of approximately 30% when

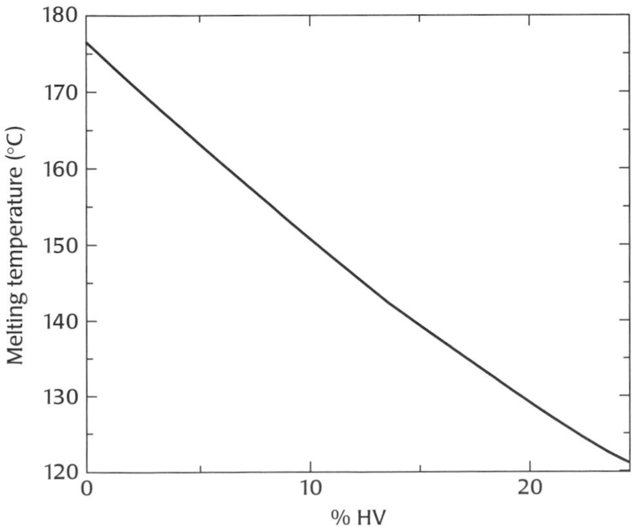

**FIGURE 8.14**

Influence of the 3HV content on the melting point of 3HB-co-3HV copolymers. [Data from Luzier, W. D. (1992). Materials derived from biomass/biodegradable materials. *Proceedings of the National Academy of Sciences U.S.A.*, 89, 839–842.]

**TABLE 8.9 Rate of degradation of random copolymers of 3-hydroxybutyrate and 3-hydroxyvalerate in various environments**

| Environment | Weeks required for 100% weight loss |
|---|---|
| Anaerobic sewage | 6 |
| Estuarine sediment | 40 |
| Aerobic sewage | 60 |
| Soil | 75 |
| Seawater | 350 |

*Source*: Luzier, W. D. (1992). Materials derived from biomass/biodegradable materials. *Proceedings of the National Academy of Sciences U.S.A.*, 89, 839–842.

oxygen was present and approximately 80% under anoxygenic conditions. These findings were encouraging because oxygen is very low or absent in many landfills (also see Table 8.9). However, the price of Biopol was about 10-fold higher than that of polyethylene derived from petrochemicals, and production ceased within a few years.

PHAs have many demonstrated applications in the manufacture of packaging containers, bottles, wrapping films, bags, and the like. They have important medical uses as well, serving as surgical pins, staples, wound dressings, bone replacements and plates, and carriers for long-term release of medicines (Table 8.10). The controllable biodegradability of these polymers is very important for many of these applications.

It is estimated that about 40% of the 75 billion pounds of plastics produced worldwide is discarded into landfills. Hundreds of thousands of tons of plastics are discarded into waterways and the oceans. Polyethylene and polypropylene are less dense than water and are very resistant to degradation. When discarded in rivers or the oceans, objects made of these materials float and persist in the environment for long periods of time. In contrast, the much higher density of PHAs ensures that objects made from these materials sink to the bottom sediment layers, where they will degrade.

More than 270 million tons of oil and gas are consumed annually worldwide in manufacturing plastics. The prospect of producing biodegradable substitute products derived from renewable resources is very attractive from the perspective of green chemistry. This prospect has motivated an intense effort to develop processes to replace conventional plastics with PHAs produced either in bacteria or in crop plants.

**TABLE 8.10 Applications of affordable biodegradable PHA polymers**

- Disposable utensils and dishes for use in the food industry
- Plastic wrap
- Moisture barrier films
- Coatings for paper products
- Components of fabrics for the textile industry
- Slow-release formulations for pesticides and fertilizers
- Medical devices such as bone screws, pins, surgical sutures, stents, patches, controlled drug delivery devices[a]

[a] For a discussion of the advantages of medical devices made from poly(4HB), see Martin, D. P., Skraly, F., and Williams, S. F. Polyhydroxyalkanoate compositions having controlled degradation rates. U.S. Patent 6,878,758, issued April 12, 2005. These materials show little acute inflammatory reaction or adverse tissue reactions. The degradation rates of the polymer can be adjusted for use as a resorbable wound closure material.

## GENETIC ENGINEERING OF PLANTS FOR THE PRODUCTION OF POLY(3-HYDROXYALKANOATES)

To compete in price with petroleum-derived plastics such as polyethylene, bacterial polyesters need to be produced inexpensively in massive amounts. Bacterial biomass is much more expensive to produce than plant biomass. To appreciate the potential of plants as producers of polymers, one need only consider starch. The yield from a field of potatoes is about 20,000 kg of starch per hectare (1 hectare = 10,000 $m^2$).

As discussed in detail above, a great diversity of PHAs, with more than 125 different monomers, are accessible through microbial fermentation. Such diversity is not attainable in plants, in which a much more limited selection of monomers can be derived through diversion of intermediates in normal plant metabolism.

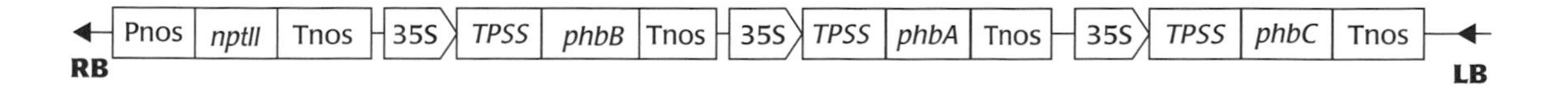


**FIGURE 8.15**

T-DNA of a binary vector used to transform *A. thaliana* plants. Abbreviations: LB and RB, left and right T-DNA borders, respectively; Pnos and Tnos, promoter and terminator region of the *A. tumefaciens* nopaline synthase gene; *nptII*, neomycin phosphotransferase gene; 35S, cauliflower mosaic virus (CaMV) 35S promoter; *TPSS*, encodes the transit peptide of the small subunit of *Pisum sativum* ribulose bisphosphate carboxylase, required for transfer of the protein to which it is fused into the chloroplast; *phbA*, *phbB*, and *phbC*, 3-ketothiolase, acetoacetyl-CoA reductase, and PHB synthase gene of *R. eutropha*, respectively. [Bohmert, K., et al. (2000). Transgenic *Arabidopsis* plants can accumulate polyhydroxybutyrate up to 4% of their fresh weight. *Planta*, 211, 841–845.]

### Poly(3-Hydroxybutyrate) Expression in Transgenic *Arabidopsis thaliana*

Beginning with a report on P(3HB) in 1994, various transgenic lines of *A. thaliana* have served as the models for the study of P(3HB) synthesis in plants. The most successful of these studies is described in detail here.

The T-DNA carrying the three *R. eutropha* genes (*phbA*, *phbB*, and *phbC*) encoding the pathway of P(3HB) formation from acetyl CoA was introduced into *A. thaliana* by *A. tumefaciens*–mediated transfer. The gene constructs were designed for import of the protein into the plastid (Figure 8.15). The *nptII* gene (conferring kanamycin resistance) on the T-DNA served as the selectable marker. Transformed seeds were selected from plants grown on a kanamycin-containing medium and used as the source of transgenic lines. The content of the 3-hydroxybutyrate monomer of the leaf material obtained directly from each of these lines was quantified by a mass spectrometry–based assay.

Line 6, which accumulated the highest amount of P(3HB), approximately 40% per gram dry weight of leaf material, showed stunted growth and did not produce any seeds. Second-generation plants, lines 1, 2, and 3, with P(3HB) level unchanged from the first generation, produced 3%, 5%, and 28% of P(3HB) per gram dry weight of leaf material. Whereas these lines were fertile, all three showed growth retardation and slight chlorosis in their leaves. Transmission electron microscopy of the mature leaves of the transgenic *Arabidopsis* plants showed tightly packed P(3HB) granules in the stroma of the plastids.

The P(3HB) biosynthetic proteins were targeted to the plastid in anticipation that the biosynthesis of the polymer would take place at the expense of the very productive fatty acid biosynthesis pathways in that organelle. However, detailed profiling of metabolites in the transgenic lines showed no changes in the composition or quantity of fatty acids, but significant decreases in the levels of fumarate and isocitrate, suggesting that depletion of the acetyl-CoA pool was leading to reduction in tricarboxylic acid cycle activity. This would provide a plausible explanation for the growth retardation of the P(3HB)-producing lines.

Appropriately modified constructs of the bacterial P(3HB) biosynthetic genes used in *A. thaliana* were also targeted to the plastid of corn. It was found that the plastids of leaf mesophyll cells contained very few polyhydroxybutyrate granules, whereas plastids of the bundle sheath cells associated with the vascular tissue were filled with granules (Figure 8.16). This observation indicates that the amounts of acetyl-CoA available for PHB synthesis differ widely among plant cell types.

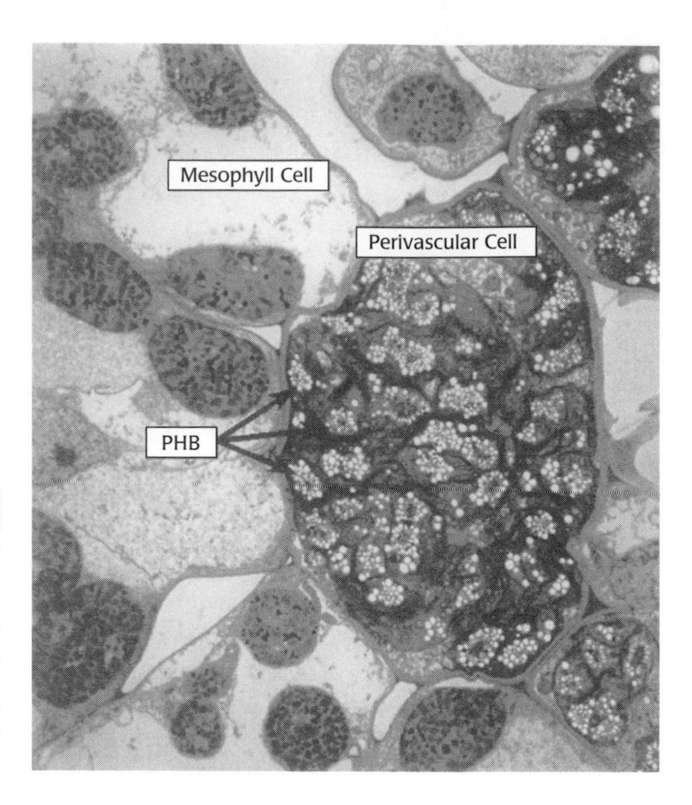


**FIGURE 8.16**

Accumulation of poly(3-hydroxybutyrate) (PHB) inclusions in transgenic corn expressing the biosynthetic pathway in plastids. Note the much greater abundance of PHB inclusions in the plastids of the perivascular cell compartment compared with the surrounding mesophyll cells. [Photograph from Kenneth J. Gruys, Monsanto Company (Davis, CA).]

### Poly(3-Hydroxybutyrate-Co-3-Hydroxyvalerate) Expression in Transgenic *A. thaliana* and *Brassica*

P(3HB) plastics are too brittle and stiff for use in the majority of the applications in Table 8.10. However, the commercialization of Biopol illustrated that products manufactured with P(3HV-*co*-3HV) met the flexibility and impact resistance demanded of commodity plastics. As described earlier, production of P(3HV-*co*-3HV) by fermentation in *R. eutropha* was achieved by adding propionate to the growth medium to provide the substrate for the formation of propionyl-CoA, a precursor of valeryl-CoA. Replication of this approach in plants requires engineering a pathway to generate adequate amounts of propionyl-CoA. The necessary metabolic engineering, performed in *A. thaliana* and in *Brassica napus* (oilseed rape), introduced a "shunt" into the branched amino acid biosynthetic pathway to generate propionyl-CoA.

The plants were transformed with four separate genes: *ilvA*, *btkB*, *phbB*, and *phbC*. The *ilvA* gene encodes threonine deaminase and was derived from *E. coli*. In *E. coli*, threonine deaminase catalyzes the first committed step in the biosynthesis of isoleucine from threonine. In plants, aspartate is the anabolic precursor of threonine and isoleucine.

In the engineered pathway, threonine deaminase converts threonine to 2-ketobutyrate, which is then converted to propionyl-CoA by the plant pyruvate dehydrogenase complex (Figure 8.17). As a footnote, at a 2-ketoacid concentration of 1.5 mM, the specific activity of the pyruvate dehydrogenase complex toward 2-ketobutyrate is 10-fold lower than that toward pyruvate.

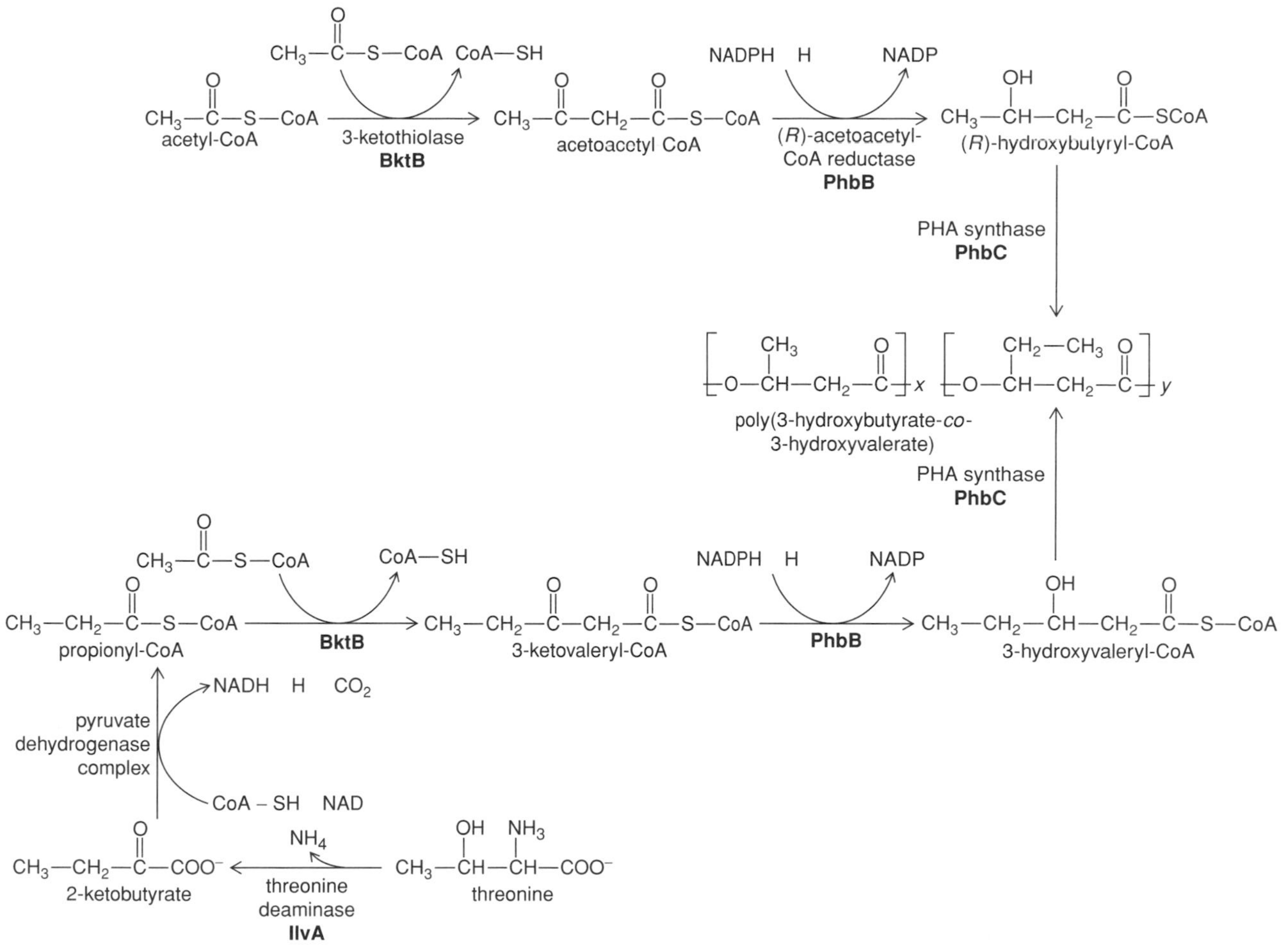

acetyl-CoA
3-ketothiolase
**BktB**
acetoacetyl CoA
(*R*)-acetoacetyl-CoA reductase
**PhbB**
(*R*)-hydroxybutyryl-CoA
PHA synthase
**PhbC**
poly(3-hydroxybutyrate-*co*-3-hydroxyvalerate)
PHA synthase
**PhbC**
propionyl-CoA
**BktB**
3-ketovaleryl-CoA
**PhbB**
3-hydroxyvaleryl-CoA
pyruvate dehydrogenase complex
2-ketobutyrate
threonine deaminase
**IlvA**
threonine

**FIGURE 8.17**

Pathway for the synthesis of 3HB-*co*-3HV in the chloroplasts of *A. thaliana* and the seeds of *Brassica napus* introduced into these plants by transformation with appropriate constructs of *E. coli* threonine deaminase gene (*ilvA*), of the *R. eutropha* genes, *phbA*, *phbB*, and *phbC*, encoding the 3-ketothiolase, acetoacetyl-CoA reductase, and PHB synthase genes, respectively. [Slater, S., et al. (1999). Metabolic engineering of *Arabidopsis* and *Brassica* for poly(3-hydroxybutyrate-*co*-3-hydroxyvalerate) copolymer production. *Nature Biotechnology*, 17, 1011–1016.]

A mutant form of *ilvA*, with much decreased sensitivity to feedback inhibition by isoleucine, was overexpressed in the transformed plants. BtkB, PhbB, and PhbC then catalyzed the remaining reactions needed to convert propionyl-CoA and acetyl-CoA to P(3HB-*co*-3HV). As discussed earlier, the BtkB 3-ketothiolase has a high affinity for both propionyl-CoA and acetyl-CoA and efficiently synthesizes both 3-ketovaleryl-CoA and acetoacetyl-CoA (Figure 8.17).

In *A. thaliana*, BtkB, PhbB, and PhbC were targeted to the chloroplast. For polymer production in *B. napus* seeds, the enzymes were targeted to the leucoplast. In the latter case, all four genes were expressed from a single vector and driven by a promoter with seed-specific expression. Analysis of a number of transgenic *A. thaliana* lines produced P(3HB-*co*-3HV). The concentration of the copolymer ranged from 0.08% to 0.84% per dry weight of material in shoots with a copolyester content of 3-hydroxyvalerate of 4 to 17 mol%. The 3-hydroxyvalerate content was highest in the lines accumulating the lowest amounts of P(3HB-*co*-3HV). In *B. napus*, copolymer accumulated to 1.5% per dry weight of seeds with 3 mol% 3-hydroxyvalerate. No deleterious phenotypic characteristics were consistently seen in the transgenic lines. However, these data were not conclusive. Nuclear magnetic resonance analyses showed that both in the transgenic *A. thaliana* and in the *B. napus* lines, the 3-hydroxybutyrate and 3-hydroxyvalerate monomers were randomly distributed within the copolymer.

## OBSTACLES TO THE COMMERCIAL PRODUCTION OF PLASTICS IN PLANTS

The results of studies such as those described above are indeed promising. Certain of the transformed plant lines produced P(3HB-*co*-3HV) copolymers, albeit in low yield, with monomer ratios similar to those of the corresponding commercialized Biopol copolymers produced by bacterial fermentation. Moreover, it is hoped that PHA production ultimately could take place in crop plants without compromising the products used for food. For example, in transgenic canola, the process would be designed to allow extraction of the oil prior to extraction and purification of the PHA. In corn, the PHA production would be targeted to the leaves and stem. This would allow harvesting of the corn grain. The stover (the stalks and leaves) would then be harvested for PHA production. However, there are still significant challenges to be met before such goals are realized.

Viable commercial production requires production of copolyester of the right composition in plants in a yield of at least 15% dry weight. Current yields are too low. Presumably, the yields can be improved by more elaborate metabolic engineering and by the use of transgenes encoding enzymes optimized by “directed evolution.” Significant success in the latter regard has been achieved with PHA synthase. Improved methods for the control of expression of multiple foreign genes would also contribute to raising yields.

Altered plant phenotypes, such as stunting and leaf chlorosis, have been observed in plants producing higher levels of PHA. These indicate that massive diversion of primary metabolites to PHA production may adversely affect normal plant functions. On another note, transgene stability in the transformed plant lines needs to be ensured.

The Achilles’ heel of the entire effort may be in the large overall energy requirement for the production of pure PHA in plants. There are major incentives perceived for producing PHAs in plants: the anticipated saving of fossil fuels, the resulting decrease in carbon dioxide emissions, and the biodegradability of plastic products made of PHAs. Environmental life cycle comparisons of PHA production from renewable carbon resources by bacterial fermentation provide grounds for modest optimism on these counts. A partnership was formed in 2004 between Procter and Gamble Co. and Kaneka Corporation to market products made of P(3HB-*co*-3HHx), and a small German company, Biomer, produces several tons of P(3HB) annually. Both of these polymers are made by bacterial fermentation. However, the results of a life cycle assessment study (Box 8.4) in transgenic corn indicated that the production of PHA showed no advantages over production of polyethylene or polystyrene from petrochemicals with respect to energy consumption or with respect to environmental impacts of the production process. The

**Life cycle assessment** is the analysis of the impacts – for example, energy, environmental, or economic impacts – of a system over its complete lifetime from creation to destruction, sometimes including the lifetimes of key constituents and components.

*Source*: U.S. Department of Energy Office of Science, Pacific Northwest Laboratory, http://energytrends.npl.gov/glosi_m.htm.

**BOX 8.4**

Monsanto Company, which had made a large-scale multiyear commitment to producing PHAs in transgenic plants, terminated this program in 1998.

## SUMMARY

Plants and algae are the sources of the most commonly used polysaccharides, such as starch, cellulose, guar gum, gum arabic, agar, and carrageenan. Bacteria and fungi also produce polysaccharides – under appropriate conditions in amounts well in excess of 50% of cell dry weight. Of the many microbial polysaccharides, which vary widely in composition, structure, and properties, only xanthan is in widespread use. Xanthan is produced by a plant pathogenic bacterium, *X. campestris*. The xanthan backbone consists of $\beta$-(1→4)-linked D-glucose, where the 3-position of alternate glucose monomer units carries a trisaccharide side chain containing one glucuronic acid and two mannose residues. The D-mannose units carry acetyl and carboxyethyl substituents. In solution, xanthan forms a stiff right-handed double helix with fivefold symmetry, with the two chains running antiparallel to each other. Xanthan has extraordinary chemical stability. Aqueous solutions of xanthan in strong acid or base are stable for several months at room temperature. The high viscosity of dilute xanthan solutions stays uniform over a temperature range of 0°C to 100°C. However, xanthan displays unusual flow behavior. In the absence of shear stress, a xanthan solution is viscous. When a shear stress is applied beyond a certain low minimum value, the viscosity decreases sharply with shear rate. Because of this unique behavior, xanthan is used widely as a stabilizer, thickener, or gelling or suspending agent in many foods, as a suspending agent in paints, and as a water-thickening polymer in drilling muds (used to displace oil in enhanced oil recovery).

Many bacteria, when grown with unbalanced supplies of nutrients, accumulate massive amounts of polyhydroxyalkanoate polymer granules. Both homopolyesters and/or copolyesters are formed depending on the nutrients supplied in the growth medium. Cometabolism allows formation of copolymers that include unnatural building blocks. For example, when *P. oleovorans* is grown on nonanoic acid (a natural substrate) and 11-cyanoundecanoic acid (a carbon source able to support growth of *P. oleovorans* but not polymer synthesis on its own), a copolyester accumulates whose yield and composition are determined by the molar ratio of nonanoic to 11-cyanoundecanoic acid in the medium.

Some of the microbial polyhydroxyalkanoates are thermoplastic; others are elastomers. Thermoplastic polymers melt when heated to a certain temperature, but harden again as they cool. The biodegradable polyhydroxyalkanoates are potential replacements for familiar nonbiodegradable thermoplastic materials such as polyethylene, polypropylene, polyvinyl chloride, polystyrene, polycarbonate, and nylon, used in the manufacture of a multitude of plastic objects. Under appropriate culture conditions, bacterial strains produce very high levels of various polyhydroxyalkanoate homopolymers and copolymers, some of which have been commercialized on a small scale. The future of these polyesters depends on the ability to produce these materials cheaply in massive amounts. Poly(3-hydroxybutyrate) and poly(3-hydroxybutyrate-*co*-3-hydroxyvalerate) have been successfully expressed in transgenic crop plants, including canola, corn, and soybean. However, life cycle assessments show that at this time the energy cost and environmental impact of producing these materials in crop plants are greater than those of producing polyethylene or polystyrene from petrochemicals.

## REFERENCES

### General

Robyt, J. F. (1998). *Essentials of Carbohydrate Chemistry*, New York: Springer-Verlag.

Turner, N., and Johnson, M. (eds.) (2004). *Low Environmental Impact Polymers*, Ontario, Canada: ChemTech Publishing, Inc.

Gerngross, T. U., and Slater, S. C. (2000). How green are green plastics? *Scientific American*, 282, 36–41.

Moire, L., Rezzonico, E., and Poirier, Y. (2003). Synthesis of novel biomaterials in plants. *Journal of Plant Physiology*, 160, 831–839.

Scheller, J., and Conrad, U. (2005). Plant-based material, protein and biodegradable plastic. *Current Opinion in Plant Biology*, 8, 188–196.

### Microbial Polysaccharides

Sutherland, I. W. (1999). Microbial polysaccharide products. *Biotechnology and Genetic Engineering Reviews*, 16, 217–229.

Sutherland, I. W. (2002). A sticky business. Microbial polysaccharides: current products and future trends. *Microbiology Today*, 29, 70–71.

Dumitriu, S. (ed.) (2005). *Polysaccharides: structural diversity and functional versatility*, 2nd Edition, New York: Marcel-Dekker.

### The Genus *Xanthomonas*

Starr, M. P. (1981). The genus *Xanthomonas*. In *The Prokaryotes*, Volume 1, M. P. Starr, H. Stolp, H. G. Trüper, A. R. Balows, and H. G. Schlegel (eds.), pp. 742–763, Berlin: Springer-Verlag.

Palleroni, N. J. (1985). Biology of *Pseudomonas* and *Xanthomonas*. In *Biology of Industrial Microorganisms*, A. L. Demain and N. A. Solomon (eds.), pp. 27–56, Benjamin/Cummings.

Swings, J. G., and Civerolo, E. L. (1993) *Xanthomonas*, London: Chapman and Hall.

Qian, W., et al. (2005). Comparative and functional genomic analyses of the pathogenicity of phytopathogen *Xanthomonas campestris* pv. *campestris*. *Genome Research*, 15, 757–767.

### Xanthan Gum

Okuyama, K., Arnott, S., Moorhouse, R., Walkinshaw, M. D., Atkins, E. D. T., and Wolf-Ullish, Ch. (1980). Fiber diffraction studies of bacterial polysaccharides. In *Fiber Diffraction Methods*, A. D. French and K. H. Gardner (eds.), ACS Symposium Series Volume 141, pp. 411–427. Washington, DC: American Chemical Society.

Tait, M. I., Sutherland, I. W., and Clarke-Sturman, A. J. (1986). Effect of growth conditions on the production, composition and viscosity of *Xanthomonas campestris* exopolysaccharide. *Journal of General Microbiology*, 312, 1483–1492.

Garcia-Ochoa, F., Santos, V. E., Casas, J. A., and Goméz, E. (2000). Xanthan gum: production, recovery, and properties. *Biotechnology Advances*, 18, 549–579.

Camesano, T. A., and Wilkinson, K. J. (2001). Single molecule study of xanthan conformation using atomic force microscopy. *Biomacromolecules*, 2, 1184–1191.

Pollock, T. J., Mikolajczak, M., Yamazaki, M., Thome, L., and Armentrout, R. W. (1997). Production of xanthan gum by *Sphingomonas* bacteria carrying genes from *Xanthomonas campestris*. *Journal of Industrial Microbiology and Biotechnology*, 19, 92–97.

**Polyhydroxyalkanoates**

Lemoigne, M. (1926). Produits de déshydratation et de polymérisation de l'acide $\beta$ oxybutyrique. *Bulletin de Sociéte Chimique et Biologique*, 8, 770–782.

Brandl, H., Gross, R. A., Lenz, R. W., and Fuller, R. C. (1990). Plastics from bacteria and for bacteria: poly($\beta$-hydroxyalkanoates) as natural, biocompatible, and biodegradable polyesters. *Advances in Biochemical Engineering/Biotechnology*, 41, 77–93.

Huisman, G. W., Wonink, E., Mima, R., Kazemier, B., Terpstra, P., and Witholt, B. (1991). Metabolism of poly(3-hydroxyalkanoates) (PHAs) by *Pseudomonas oleovorans*. Identification and sequences of genes and function of the encoded proteins in the synthesis and degradation of PHA. *Journal of Biological Chemistry*, 266, 2191–2198.

Steinbüchel, A., and Valentin, H. E. (1995). Diversity of bacterial polyhydroxyalkanoic acids. *FEMS Microbiology Letters*, 128, 219–228.

Rehm, B. H. A., and Steinbüchel, A. (1999). Biochemical and genetic analysis of PHA synthases and other proteins required for PHA synthesis. *International Journal of Biological Macromolecules*, 25, 3–19.

Sudesh, K., Abe, H., and Doi, Y. (2000). Synthesis, structure and properties of polyhydroxyalkanoates: biological polyesters. *Progress in Polymer Science*, 25, 1503–1555.

Jendrossek, D. (2001). Microbial degradation of polyesters. *Advances in Biochemical Engineering and Biotechnology*, 71, 293–325.

Tsuge, T. (2002). Metabolic improvements and use of inexpensive carbon sources in microbial production of polyhydroxyalkanoates. *Journal of Bioscience and Bioengineering*, 94, 579–584.

Lütke-Eversloh, T., and Steinbüchel, A. (2003). Novel precursor substrates for polythioesters (PTE) and limits of PTE biosynthesis in *Ralstonia eutropha*. *FEMS Microbiology Letters*, 221, 191–196.

Pohlmann, A. et al. (2006). Genome sequence of the bioplastic producing "Knallgas" bacterium *Ralstonia eutropha* H16. *Nature Biotechnology*, 24, 1257–1262.

Stubbe, J., and Tian, J. (2003). Polyhydroxyalkanoate homeostasis: the role of the PHA synthase. *Natural Products Reports*, 20, 445–457.

Steinbüchel, A., and Lütke-Eversloh, T. (2003). Metabolic engineering and pathway construction for biotechnological production of relevant polyhydroxyalkanoates in microorganisms. *Biochemical Engineering Journal*, 16, 81–96.

Taguchi, S., and Doi, Y. (2004). Evolution of polyhydroxyalkanoate (PHA) production system by "enzyme evolution": successful case studies of directed evolution. *Macromolecular Bioscience*, 4, 145–156.

**Production of Polyhydroxyalkanoates in Plants**

Slater, S., et al. (1999). Metabolic engineering of *Arabidopsis* and *Brassica* for poly(3-hydroxybutyrate-*co*-3-hydroxyvalerate) copolymer production. *Nature Biotechnology*, 17, 1011–1016.

Bohmert, K., et al. (2000). Transgenic *Arabidopsis* plants can accumulate polyhydroxybutyrate up to 4% of their fresh weight. *Planta*, 211, 841–845.

Snell, K. D., and Peoples, O. P. (2002). Polyhydroxyalkanoate polymers and their production in transgenic plants. *Metabolic Engineering*, 4, 29–40.

Poirier, Y. (2002). Polyhydroxyalkanoate synthesis in plants as a tool for biotechnology and basic studies of lipid metabolism. *Progress in Lipid Research*, 41, 131–155.

Matsumoto, K., et al. (2005). Enhancement of poly(3-hydroxybutyrate-*co*-3-hydroxyvalerate) production in the transgenic *Arabidopsis thaliana* by the *in vitro* evolved highly active mutants of polyhydroxyalkanoate (PHA) synthase from *Aeromonas caviae*. *Biomacromolecules*, 6, 2126–2130.

**Energy Cost and Environmental Impacts of Polyhydroxyalkanoate Production in Bacteria and Transgenic Plants**

Gerngross, T. U. (1999). Can biotechnology move us towards a sustainable society? *Nature Biotechnology*, 17, 542–544.

Akiyama, M., Tsuge, T., and Doi, Y. (2003). Environmental life cycle comparison of polyhydroxyalkanoates produced from renewable carbon resources by bacteria and fermentation. *Polymer Degradation and Stability*, 80, 183–194.

Kurdikar, D., Fournet, L., Slater, S. C., Paster, M., Gruys, K. J., Gerngross, T. U., and Coulron, R. (2001). Greenhouse gas profile of a plastic material derived from a genetically modified plant. *Journal of Industrial Ecology*, 4, 107–122.

Kim, S., and Dale, B. E. (2005). Life cycle assessment study of biopolymers (polyhydroxyalkanoates) derived from no-tilled corn. *International Journal of Life Cycle Assessment*, 10, 200–210.

NINE

# Primary Metabolites: Organic Acids and Amino Acids

Microorganisms can function as efficient factories for the industrial scale production of the primary metabolites. Among these, ethanol will be discussed in Chapter 13. Other important primary metabolites currently produced by fermentation are listed in Table 9.1. Some organic acids and amino acids are seen to be the most important products in this category (excluding ethanol, of course).

## CITRIC ACID

About 1 billion pounds of citric acid are produced worldwide (Table 9.1) by fermentation with the wild-type strains of a fungus, *Aspergillus niger*. Citric acid is used, for example, as a flavoring agent in food and drinks and to prevent oxidation and rancidity of fats and oils. In the very efficient fermentation process, up to 80% of the source sugar is converted into citric acid. Because this process, in operation since 1916, has been studied "probably more than any other process in mold metabolism," according to one review, and could serve as a model for primary metabolite fermentation processes including glutamate fermentation (discussed later), we will examine this process in some detail.

Citric acid, unlike ethanol, is not one of the typical waste products of energy metabolism, and is usually degraded completely into $CO_2$ and water through the operation of the citric acid cycle. Thus, the efficient conversion of sugars, such as glucose and sucrose, into citric acid by *A. niger* is rather unexpected, especially because mitochondria of *A. niger* contain all the enzymes of the citric acid cycle. The citric acid production occurs predominantly in the stationary phase, and indeed requires several unusual conditions. (1) The medium should be strongly acidic, with the pH value between 1.6 and 2.2. (2) The sugar concentration should be very high (120 to 250 g/L). (3) The medium should be deficient in $Mn^{2+}$. (4) The medium should contain a high concentration of $NH_4^+$ ions.

Extensive studies showed that under these conditions, glucose is fermented rapidly by the glycolytic pathway. Normally, the key controlling step,

**TABLE 9.1 Estimated annual production of important primary metabolites of microorganisms worldwide**

| Metabolite | Production (tons) | Market value (million dollars) |
|---|---|---|
| AMINO ACIDS | | |
| L-Glutamate | 1,000,000 | 3000 |
| L-Lysine | 800,000 | 915 |
| L-Threonine | 20,000 | 100 |
| L-Aspartate | 13,000 | 198 |
| L-Isoleucine | 400 | 43 |
| NUCLEOTIDES | | |
| 5′-IMP + 5′-GMP | 2500 | 350 |
| ORGANIC ACIDS | | |
| Citric acid | 400,000 | 1400 |
| VITAMINS | | |
| $B_{12}$ | 3 | 100 |
| C[a] | 60,000 | 71 |
| Riboflavin | 2000 | 60 |

[a] Partially synthetic.
IMP, inosine monophosphate; GMP, guanosine monophosphate.
Based mainly on Demain, A. L. (2000). Small bugs, big business: the economic power of the microbe. *Biotechnology Advances*, 18, 499–514.

phosphofructokinase (PFK1), would be inhibited by citrate, to which the glycolytic product, pyruvate, is converted. However, this citrate inhibition is abolished by the presence of high concentrations of $NH_4^+$. Furthermore, very high concentrations of sugars lead to the increased level of fructose-2,6-bisphosphate, the most powerful activator of PFK1. These factors presumably contribute to increased flux through the glycolytic pathway, culminating in the runaway production of pyruvate.

Some of the pyruvate molecules enter mitochondria and are converted into acetyl-CoA by the normal process catalyzed by pyruvate dehydrogenase. Acetyl-CoA then condenses with oxaloacetate to produce citrate. However, the flux is far higher than what can be handled by the rest of the citric acid cycle, especially because the isocitrate dehydrogenase is inhibited by glycerol, which is generated as a part of the hyperosmotic response (see Box 9.1) to the high concentration of sugars. Thus, citrate is exported out of mitochondria via citrate/malate antiporter, concomitantly with the influx of malate, which is converted to oxaloacetate for the next round of citrate synthesis. The remainder of the pyruvate molecules generated through glycolysis is converted into oxaloacetate by pyruvate carboxylase, which is located in the cytosol in this group of fungi. The cytosolic oxaloacetate is then reduced to malate by cytosolic malate dehydrogenase, and malate enters mitochondria as described above. The conversion of pyruvate to malate ensures the continuous operation of the glycolytic pathway, at least partially, by regenerating adenosine diphosphate (ADP) and $NAD^+$ from ATP and NADH, which were generated during the conversion of sugars to pyruvate.

Until recently, however, there was almost nothing known about one area of this proposed scheme. That is the exit of citrate from the cytosol into the medium. Citrate, with its three negative charges, cannot cross membranes spontaneously, and its export into the medium is likely to be catalyzed by a transporter protein. Furthermore, intracellular concentration of citrate is several millimolars, and the "uphill" export into the medium, which could contain more than 0.5 M of citrate, is likely to be catalyzed by an active,

**Hyperosmotic and Hypoosmotic Responses of Microorganisms**

The cytosolic osmolarity of microorganisms must be maintained slightly above that of the environment so that the cytoplasm does not collapse and a small turgor pressure is constantly applied to the rigid cell wall. Thus, a microbial cell is like a balloon in a small cage. When the environment becomes hyperosmotic, microorganisms must respond by increasing the solute concentration in the cytoplasm in order to maintain the small difference in osmolarity. Large increases in mineral salt concentration, however, are often deleterious to the proper function of proteins and organelles. Thus in such situations, they accumulate "compatible solutes," which are less toxic to proteins. They include proline, betaine, trehalose, and glycerol. When the environment suddenly becomes hypoosmotic, small cytosolic solutes are released into the medium through the opening of stretch-activated channels and the activation of other transporters. Amino acid exporters in *Corynebacterium glutamicum*, described later in this chapter, apparently play this role when large amounts of peptides are taken up and their hydrolysis in cytoplasm suddenly increases the osmolarity of cytoplasm.

**BOX 9.1**

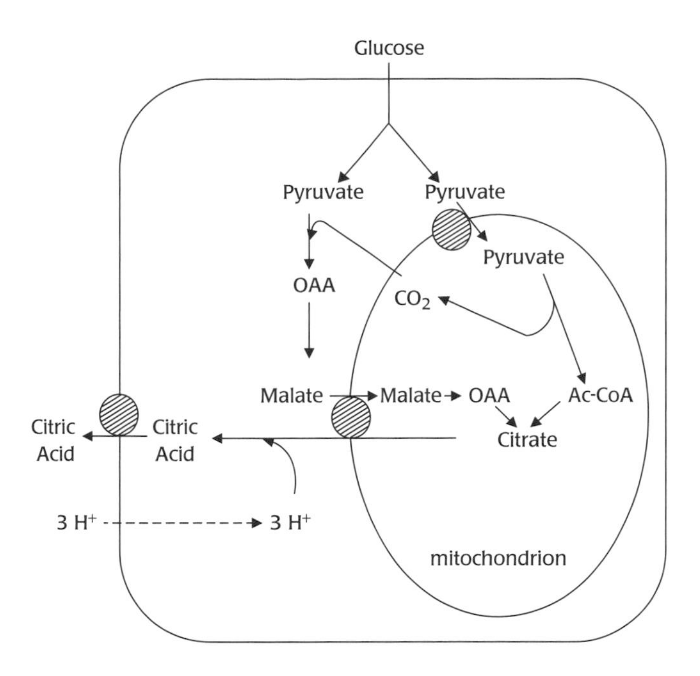


**FIGURE 9.1**

Metabolic pathways involved in the production of citric acid by *A. niger*. *Shaded circles* represent specific transport proteins. Thus, one of the pyruvate molecules generated from glucose via glycolysis enters the mitochondrion through the ubiquitous pyruvate/$H^+$ symporter and is converted into acetyl-CoA by pyruvate dehydrogenase. The other pyruvate molecule is converted into oxaloacetate (OAA) by the cytosolic pyruvate carboxylase. However, because there is no OAA transporter in the mitochondrion, OAA must be reduced to malate and then must enter the mitochondrion through a specific carrier, and OAA is regenerated inside the mitochondrion. The malate carrier is known to catalyze the exchange of malate with $\alpha$-ketoglutarate, but in this case it may well catalyze the exchange of malate with citrate generated within the mitochondrion. Citric acid leaves the cell via a specific carrier. It seems plausible that the carrier catalyzes the export of the protonated citric acid, thereby preventing the acidification of the cytosol from the protons that have leaked into the cytosol from a highly acidic medium. However, this part of the pathway is still a speculation, and a possibility exists that citrate$^{3-}$, for example, is exported without using any energy input (see text).

energy-consuming process. Indeed, export of citrate from *A. niger* was shown in 1997 to be inhibited by sodium azide and a proton conductor, carbonyl cyanide (*m*-chlorophenyl)hydrazone (CCCP), a result consistent with the presence of an active exporter protein. (Interestingly, cultivation of *A. niger* in $Mn^{2+}$-poor medium was necessary to activate this process, and this may be the major reason for the inhibitory effect of $Mn^{2+}$ for citrate production). If the export substrate is the protonated form, citric acid, perhaps its extrusion may decrease the proton concentration in the cytoplasm and thereby help the survival of *A. niger* in the strongly acidic environment, in which proton leakage from the medium will produce the acidification of the cytoplasm. In this speculative scheme (Figure 9.1), the production of citrate by this organism under strongly acidic conditions will therefore be a process that is physiologically significant. Alternatively, if the transporter catalyzes the flux of citrate$^{3-}$, the accumulation of citric acid in the medium can be explained as the result of conversion of most of citrate$^{3-}$ into uncharged citric acid; this flux might be facilitated by the outside-positive membrane potential, if such a potential exists.

## AMINO ACID: L-GLUTAMATE

### OVERVIEW

The industrial production of an amino acid, L-glutamate, dates back to 1908, when the Japanese agricultural chemist K. Ikeda discovered that L-glutamate was responsible for the characteristic taste, much appreciated in Japan, of foods cooked with dried kelp (konbu). For the first 50 years, monosodium L-glutamate (MSG) was manufactured by expensive chemical processes, based largely on the acid hydrolysis of proteins. This hydrolysis process was costly because glutamate had to be separated from all other amino acids in the hydrolyzate. Significant amounts of MSG were also made by chemical synthesis. This process was expensive too, because it produced a mixture of D- and L-glutamate that had to be resolved to eliminate the D-isomer, which is tasteless.

A revolutionary change was introduced in 1957, when scientists at Kyowa Hakko Co. discovered a soil bacterium that excreted large amounts of L-glutamate into the medium. Similar bacteria were soon isolated by several other companies, ushering in a new industrial technology: amino acid fermentation, the production of amino acids by microorganisms. Except for ethanol, some other organic solvents, and a few vitamins, glutamate was the first organic compound produced on an industrial scale by a microbial fermentation technique.

MSG is used as a flavor enhancer in very large amounts (Table 9.1). It was also the first amino acid to be produced by the fermentation process, using wild-type strains. Because amino acids are building blocks for the assembly of proteins, their biosynthesis is usually regulated tightly so that energy and carbon are not wasted for the synthesis of needless excesses of these compounds. Thus, the overproduction of glutamate by wild-type bacteria is surprising, and the mechanism, despite much study, is not yet totally clear.

MSG overproduction (or fermentation) requires *Corynebacterium glutamicum*. Various companies isolated glutamate-producing organisms named, for example, *Brevibacterium* and *Arthrobacter*, but it now appears that all these strains are close relatives of *C. glutamicum*. One striking observation made in the early days of development of glutamate fermentation technology was that all these strains were naturally auxotrophic for (requiring as growth factors) biotin, and the production of glutamate required starvation for biotin (Figure 9.2).

## NEED FOR BIOTIN STARVATION

Why is biotin starvation necessary for the excretion of glutamic acid? The major function of biotin is to serve as a prosthetic group in acetyl-CoA carboxylase, the first enzyme of fatty acid synthesis, so it was hypothesized that the excretion might be related to an increased general permeability of the cell membrane, caused by an insufficiency of fatty acids. This notion appeared to be supported by the finding that other treatments affecting the integrity of the cell membrane – for example, adding to the medium a detergent, Tween 60, or a low concentration of $\beta$-lactam antibiotics (the latter treatment strains the cell membrane by weakening the peptidoglycan cell wall) or limiting the amount of exogenously fed oleic acid in unsaturated fatty acid auxotrophs – also led to glutamic acid production, even in the presence of excess biotin.

These observations were also of practical importance because they suggested ways to stimulate glutamic acid excretion while feeding the bacteria with inexpensive carbon sources (such as molasses) that happen to be rich

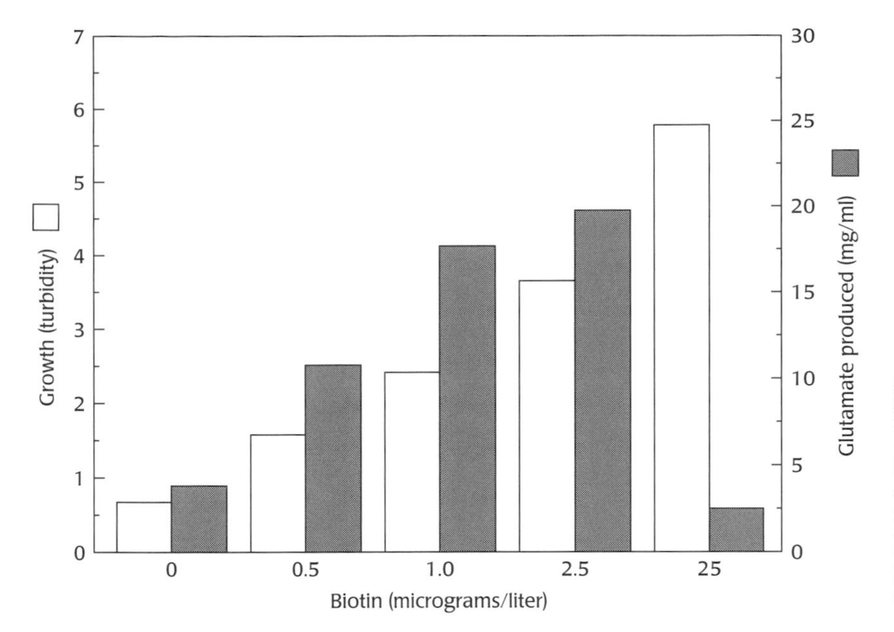


FIGURE 9.2

Influence of biotin concentration on the production of glutamic acid by *C. glutamicum*. Based on Tanaka, K., Iwasaki, T., and Kinoshita, S. (1960), Nihon Nogei Kagakukai Zasshi, 34, 593–600.

Redrawn based on artwork from the first edition (1995), published by W.H. Freeman.

in biotin. During the period when petroleum was inexpensive, petroleum hydrocarbons were also tried as a carbon source. When scientists found that unsaturated fatty acid auxotrophs do not produce glutamic acid when petroleum hydrocarbons are present (presumably because they are easily converted to unsaturated fatty acids by the organism), they tried an alternative way of weakening the cell membrane: They constructed conditional auxotrophs for glycerophosphate so that the synthesis of phospholipids from glycerophosphate became deficient under certain conditions. Such mutants also excreted glutamate when the supply of glycerophosphate became limiting.

The results described above seemed to support the hypothesis that a leaky cell membrane is responsible for the excretion of glutamic acid. Several important questions remained, however. Principally, how do the cells remain alive if the membrane becomes so leaky? In other words, what prevents the leakage of essential metabolites and ions? And why is this treatment needed only for the production of glutamic acid and not for that of other amino acids, described later in this chapter?

In the early 1990s, it was shown that *C. glutamicum* and its relatives produce specific exporters that function solely to excrete amino acids, such as lysine, that they also overproduce (see below). These "efflux transporters" are thought to be useful when the cytoplasm becomes flooded with amino acids, possibly as a consequence of the rapid uptake and intracellular hydrolysis of peptides from the medium. The intracellular glutamate concentration is unusually high among amino acids, around 200 mM, in contrast to those of other amino acids, which are usually around a few millimolars or less. The involvement of a specific glutamate efflux transporter presumably explains the export of only glutamate, if its activity is activated by alterations leading to the straining of the cell membrane. However, this glutamate exporter

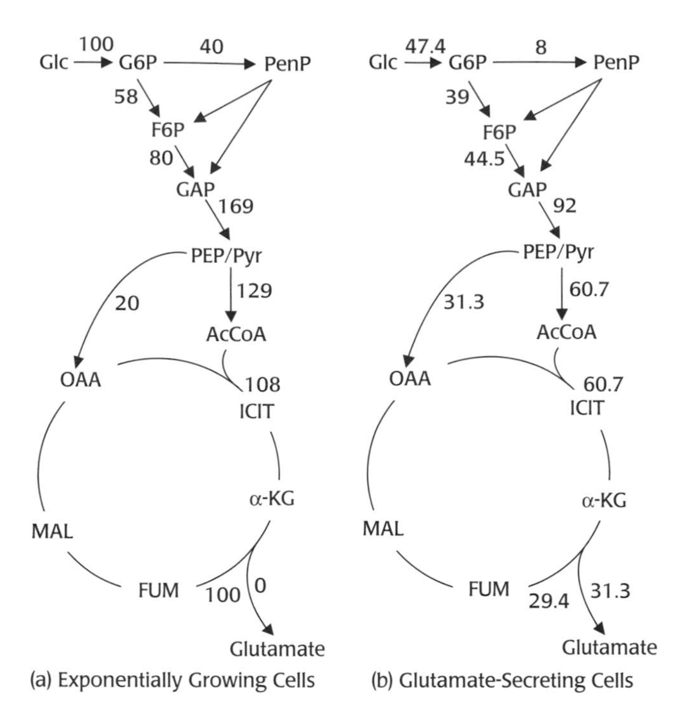


**FIGURE 9.3**

Carbon metabolism in *C. glutamicum*. In an older model, the citric acid cycle was thought to be truncated, with low or nonexistent $\alpha$-ketoglutarate dehydrogenase activity. However, more recent studies in which metabolic balancing was combined with $^{13}$C-NMR shows that the $\alpha$-ketoglutarate dehydrogenase remains quite active. (**A**) Flux values in *C. glutamicum* growing in biotin-rich medium and therefore not excreting glutamate. (**B**) Flux values in *C. glutamicum* growing in biotin-limited medium and excreting glutamate into the medium. The net flux values (in molar units) are shown as percentages of the flux of glucose in non–glutamate-excreting cells, here taken as 100. In glutamate-excreting cells, the growth is slowed (the influx of glucose is less than half that seen in the exponentially growing cells), the contribution of the pentose phosphate pathway is minimized, the production of oxaloacetate from pyruvate/PEP is greatly increased, and about half the $\alpha$-ketoglutarate is siphoned off as glutamate, producing a yield of 0.66 mole of glutamate per one mole of glucose taken up. At some steps, there are large bidirectional (exchange) fluxes, but they are not shown here for simplicity. In both cases, the contribution of the glyoxylate cycle was negligible. Abbreviations: G6P, glucose 6-phosphate; F6P, fructose 6-phosphate; PenP, pentose phosphate; GAP, glyceraldehyde 3-phosphate; PEP, phospho*enol*pyruvate; Pyr, pyruvate; AcCoA, acetyl-CoA; OAA, oxaloacetate; ICIT, isocitrate; $\alpha$-KG, $\alpha$-ketoglutarate, FUM, fumarate; MAL, malate. Flux values are based on Sonntag K., et al. (1995). *Applied Microbiology and Biotechnology*, 44, 489–495.

Redrawn based on artwork from the first edition (1995), published by W.H. Freeman.

still remains unidentified in spite of the knowledge of the complete genome sequence of *C. glutamicum*.

## IS THE CITRIC ACID CYCLE TRUNCATED?

Another factor that was thought to be important in the early days of glutamate fermentation was the presence of a truncated citric acid cycle. It was felt that the secretion of very large amounts of glutamate must involve unusual features in metabolic pathways, which lead to its extensive accumulation. The metabolic precursor of glutamic acid is $\alpha$-ketoglutaric acid, an intermediate in the citric acid cycle (Figure 9.3A). When the enzymes of the citric acid cycle were first examined, glutamic acid producers were reported to contain all the enzymes of the cycle except $\alpha$-ketoglutarate dehydrogenase. More recently, $\alpha$-ketoglutarate dehydrogenase was shown to be present. However, it is difficult to decide, from enzyme activities determined *in vitro*, how much substrate flux the enzyme is catalyzing in intact cells. In this connection, it is important that recent elucidation of detailed substrate flux (Box 9.2), determined by growing the organism with 1-$^{13}$C-glucose followed by the NMR-based determination of the labeling pattern of products, showed that the section of the cycle after the $\alpha$-ketoglutarate dehydrogenase step had full activity in nonsecreting cells (Figure 9.3A), and even in glutamate-secreting cells it still had about half the activity of the flux of the section preceding $\alpha$-ketoglutarate (Figure 9.3B); that is, the citric acid cycle is not truncated in *C. glutamicum*, and the change in the flow of substrates through the citric acid cycle, observed during glutamate secretion, may be a result, rather than the cause, of this process.

**Metabolic Flux Analysis**

Quantitative determination of metabolite flux through each step of the central metabolic pathway is essential in order to understand processes such as amino acid production and to design rational improvements for such processes (see Box 9.3). However, such a determination is difficult in intact cells. Earlier, activities of the enzymes and their kinetic properties, measured in cell extracts, were used to build models. However, often the enzymes are labile, and their activities measured *in vitro* have only a remote relation to the activities in intact cells. Another approach is Metabolic Control Analysis (MCA).[1] In this elegant mathematical approach, flux alterations caused by alterations in the amount of an enzyme are measured, and the determination of these "flux control coefficients" leads to the quantitative, dynamic model of the pathway. Because amplification of a gene is easily accomplished by cloning on plasmid vectors, this approach has been strongly advocated by some scientists. This method works when the segment of the pathway does not contain an enzyme that is regulated by the intermediate or product of the pathway. However, practically all important metabolic pathways contain such feedback-regulated enzymes, and in such cases, this approach fails, as shown by a simple computer simulation,[2] presumably because the mathematics involves the simple integration of partial derivatives. This problem has not been dealt with properly by the advocates of MCA, and the situation makes any MCA-derived models rather suspect.

The most reliable and useful models are generated, in the authors' view, by starting from Metabolite Balancing Analysis. Here one measures the consumption of the carbon and energy source – say, glucose – and the excretion and accumulation of products such as lactate, acetate, and amino acids in the proteins synthesized. One then tries to balance the uptake of glucose into the cell with the synthesis of all these products, using the knowledge of the central catabolic and anabolic pathways. However, presence of parallel pathways and bidirectional flow in many parts of metabolism make this simple analysis impossible. Thus, one combines the balancing analysis with the isotopic labeling patterns of various products by growing cells on, say, 1–$^{13}$C-glucose. For example, with regard to the parallel pathways, C-3 of pyruvate generated by glycolysis will be labeled with $^{13}$C at the 50% level, but C-3 of pyruvate coming from the pentose phosphate pathway will contain no $^{13}$C. The position and the degree of enrichment in $^{13}$C can be determined by NMR or mass spectroscopy, and the patterns in isotopic labeling in various products can then be simulated by using a computer. The final result is a model with precise flux figures for each step of the metabolism.[3,4]

1 Fell D. (1997). *Understanding the Control of Metabolism*, London: Portland Press.

2 Atkinson, D. E. (1990). An experimentalist's view of control analysis. In *Control of Metabolic Processes*, A. Cornish-Bowden and M. L. Cárdenas (eds.), pp. 413–429, New York: Plenum Press.

3 Sahm H., Eggeling, L., and de Graaf, A. A. (2000). Pathway analysis and metabolic engineering in *Corynebacterium glutamicum*. *Biological Chemistry*, 381, 899–910.

4 Weigert, W. (2001). $^{13}$C metabolic flux analysis. *Metabolic Engineering*, 3, 195–206.

BOX 9.2

In the simplest hypothesis, we may assume that the trigger of the whole process is the activation of the glutamate exporter by straining of the membrane. This will lower the cytosolic concentration of glutamate as discussed in the next section, shifting the steady-state fluxes around $\alpha$-ketoglutarate in the citric acid cycle. When only insufficient amounts of oxaloacetate are generated at the end of the cycle because of the siphoning off of a large amount of $\alpha$-ketoglutarate (Figure 9.3B), the replenishing (or anaplerotic) reactions, generating C4 compounds from C3 compounds (pyruvate and phospho*enol*pyruvate) become important. Indeed, the overproduction of pyruvate carboxylase, which catalyzes this reaction, was reported to improve the yield of glutamic acid.

Regardless of the nature of the trigger, glutamate fermentation results in the consumption of NADPH (generated by the isocitrate dehydrogenase step) by glutamate dehydrogenase (Figure 9.3A). In fact, glutamate fermentation is known to require oxygen but not very high concentrations of dissolved oxygen, which tend to favor the excretion of $\alpha$-ketoglutarate rather than glutamate. It is also obvious that the process demands high concentrations of $NH_3$. Oxygen tension and ammonia concentration are controlled precisely in the industrial process.

> **Metabolic Engineering**
>
> Metabolic engineering means in a narrow sense the application of recombinant DNA methods in order to alter the flow of metabolites in the cell, usually in order to increase the yield of a desired end product or to produce a novel product. Such approaches often require knowledge of the carbon flow in the central metabolic pathway (see Box 9.2), especially when the synthesis of the product consumes a large fraction of the cellular resources.
>
> BOX 9.3

## POSSIBLE PHYSIOLOGICAL ROLE OF THE GLUTAMATE PRODUCTION

Scientists used to assume that citric acid or glutamate fermentation occurred because of a complete block at relevant steps in general metabolism. Metabolic flux studies now show that such blocks are unlikely to exist and that massive excretion of primary metabolites may occur as a result of modest adjustments in the central metabolic pathway, as described already. This new knowledge of the immense flexibility of the major metabolic pathways is extremely important in the emerging field of metabolic engineering (see Box 9.3), which deals with the alterations of the fluxes of the major pathways, often required for the production of not only the usual primary metabolites but also novel products.

The processes leading to the production of primary metabolites, especially those involving fungi (such as citric acid fermentation), used to be – and sometimes still are – explained as "overflow metabolism." This is an old idea in which microbial cells are thought to consume excess carbon sources (the "overflow") when cell growth is limited by another nutrient – for example, by the paucity of divalent cations in the *A. niger* citric acid fermentation process. We now know, however, that catabolism is linked tightly to anabolism through many regulatory processes. Thus, the overproduction of metabolites is not simply an overflow phenomenon but must have physiological and ecological relevance. Such a physiological "purpose" of the glutamate excretion response is difficult to assess. However, *C. glutamicum* and its relatives, being natural biotin auxotrophs, are likely to become starved for biotin in their habitat, presumably soil. Blocked synthesis of membrane lipids will then cause mechanical strains in the cytoplasmic membrane because the intracellular osmotic pressure is usually much higher than the osmotic pressure of the extracellular medium. Perhaps the important point here is that the intracellular concentration of glutamate is about two, sometimes three, orders of magnitude higher than those of other amino acids. Efflux of glutamate, thus a major anionic component of the cytosol, can then be seen as a response designed to reduce the strain in the membrane by decreasing the osmotic pressure difference across the membrane (see Box 9.1). The realignment of metabolic fluxes caused by the excretion of glutamate and the decrease in growth rate will then lead to the characteristic flux pattern seen in Figure 9.3B. For *C. glutamicum*, in addition to relieving the membrane stress through the export of glutamate, the glutamate production process produces additional ATP through the oxidation of pyruvate to acetyl-CoA and in addition produces a much less acidic product (0.66 mole of glutamate per one mole of glucose) in comparison with pure fermentation (two moles of lactate per one mole of glucose), thus keeping the environment closer to the neutral pH.

The operation of the glutamate production system as a shifted yet balanced and interconnected central metabolic response is illustrated in the example of an ATP synthase mutant of *C. glutamicum*. In this strain that showed only 25% activity of ATP synthase (or $H^+$-ATPase), secretion of glutamic acid was totally abolished, when the biotin starvation conditions were moderate; instead, a large amount of lactic acid was excreted. It appears that with low ATP synthase activity, not enough ATP can be made from the NADH obtained in the altered flow through the citric acid cycle, so glycolysis (which generates ATP directly) becomes enhanced instead, resulting in the excretion of lactate.

It thus appears that *C. glutamicum* produces glutamate as a waste product of catabolism under a specific set of environmental conditions. This suggests that the enzyme responsible for the process, glutamate dehydrogenase, may not be regulated as tightly as are the obligatorily biosynthetic enzymes. Evidence that this is indeed the case is described below.

## REGULATION OF THE ENZYMES INVOLVED IN GLUTAMATE SYNTHESIS

In most microorganisms, the major pathway for glutamate synthesis is the GS-GOGAT pathway:

(1) L-Glutamate + $NH_3$ + ATP → L-glutamine + ADP + $P_i$

(GS: glutamine synthetase)

(2) L-Glutamine + $\alpha$-ketoglutarate + NADPH + $H^+$ → 2 L-glutamate + $NADP^+$

(GOGAT: glutamine oxoglutarate aminotransferase)

Sum (1) + (2):

$\alpha$-ketoglutarate + $NH_3$ + ATP + NADPH + $H^+$ → L-glutamate + ADP + $P_i$ + $NADP^+$.

Although this pathway consumes one ATP for every molecule of L-glutamate synthesized, it is preferred over the alternative pathway described below because it has a low Michaelis constant ($K_m$; much lower than 1 mM) for $NH_3$ and thus can efficiently utilize ammonia.

The second pathway is the glutamate dehydrogenase pathway:

$\alpha$-Ketoglutarate + $NH_3$ + NAD(P)H + $H^+$ → L-glutamate + $NAD(P)^+$ + $H_2O$.

Because of the rather high $K_m$ value (usually around 1 mM) for $NH_3$, this pathway becomes significant only when $NH_3$ concentrations are high. In most organisms, mutational inactivation of this enzyme does not make the mutants auxotrophic for glutamate, a finding that confirms the relatively minor role of this pathway for glutamate synthesis in most species. Because the GS-GOGAT pathway is so much more important, its enzymes are strongly regulated, as is typical of obligatorily biosynthetic enzymes; on the other hand, regulation of the glutamate dehydrogenase activity tends to be rather loose.

In *C. glutamicum*, the regulation of glutamate dehydrogenase is exceptionally loose; 50% inhibition of the enzyme requires the addition of 0.2 M glutamate to the cell extract, whereas similar inhibition of *Alkaligenes eutrophus* glutamate dehydrogenase requires only 20 mM.

These considerations suggest that the overproduction of glutamic acid in *C. glutamicum* is the result of a fortuitous combination of several factors. Over evolutionary time, the organisms of this group have developed a peculiar combination of enzymes, presumably equipping them to deal with ecological conditions perhaps related to those described above. It appears that the fermentation conditions used in the laboratory and in industry are precisely those that led to stimulation of the synthesis, and excretion, of glutamic acid.

## AMINO ACIDS OTHER THAN GLUTAMATE

### NEED FOR AMINO ACIDS OTHER THAN GLUTAMATE

Glutamate, as we have seen, is used widely as a flavor enhancer. But other amino acids are also needed in large amounts worldwide. The most important use of some of these amino acids is as an additive to human food and animal feed. This is because humans and most higher animals cannot synthesize lysine, threonine, tryptophan, phenylalanine, isoleucine, leucine, valine, methionine, and histidine – the "essential amino acids" – but must obtain them from the proteins in their diet. However, the less expensive, more abundant sources of food proteins, the seeds of crop plants (the main source of proteins for humans, especially in developing countries), are rather deficient in some of the essential amino acids, particularly lysine, methionine, and tryptophan (Table 9.2). The demand for such amino acids is particularly high in growing children and young animals. Yet the lysine content in wheat flour and corn proteins is only 21% and 27%, respectively, of that in pork protein (Table 9.2). The nutritional value of these cereal proteins can be increased significantly if seeds can be fortified with the deficient amino acid. Lysine is now produced at the rate of about 800,000 tons/year by fermentation, and its production is increasing at the rate of almost 10% per year. Lysine is added to both human food and animal feed. Similarly, L-threonine is produced by fermentation at the rate of 20,000 tons/year, and used mainly as a feed additive. Nutritionists evaluate food proteins by calculating the ratio between the weight the animal gains and the weight of the protein it is fed: the *protein efficiency ratio (PER)*. Corn proteins, because of their low lysine and tryptophan content (see Table 9.2), have a low PER of 0.85, but this value can be increased about threefold, to 2.55, by adding 0.4% lysine and 0.07% tryptophan to corn.

**TABLE 9.2 Levels of some essential amino acids in cereal and animal meat proteins[a]**

| | Cereal proteins | | | Meat proteins | |
|---|---|---|---|---|---|
| | Maize | Wheat flour | Sorghum | Beef | Pork |
| Lysine | 167 | 130 | 126 | 556 | 625 |
| Threonine | 225 | 168 | 189 | 287 | 319 |
| Tryptophan | 44 | 67 | 76 | 70 | 85 |
| Methionine | 120 | 91 | 145 | 169 | 188 |

[a] Amino acid content is expressed as milligrams per gram of protein nitrogen.
*Source*: Food and Agriculture Organization of the United Nations (1970). *Amino-Acid Content of Foods and Biological Data on Proteins. FAO Nutritional Study No. 24*, Rome, Italy: FAO.

## FERMENTATION WITH MUTANT STRAINS IS NEEDED FOR AMINO ACIDS OTHER THAN GLUTAMATE

Except glutamate, which is produced as an end product of an incomplete oxidation process in biotin-limited *C. glutamicum* as described above, other amino acids are end products of biosynthetic pathways. These amino acids are synthesized – at a net expense of energy – as building blocks of proteins. It is to the cells' advantage not to waste energy by producing more amino acids than they need. Consequently, amino acid production is usually effectively regulated.

Our understanding of such regulatory processes has come primarily through studies of *Escherichia coli*, which lives in the intestinal tract of higher vertebrates and thus leads a life of "feast or famine." It can predict neither the nature of the amino acids that will become available in its environment nor when they will arrive. Consequently, the regulatory mechanisms of *E. coli* are much more sophisticated than those of many soil microorganisms – for example, those whose environments are less complex and more stable.

In *E. coli*, the biosynthesis of most amino acids is regulated at two different levels: (1) by control of the *activity of preexisting enzymes*, and (2) by control of the *synthesis of new enzyme molecules*. The former governs the response of *E. coli* to the sudden appearance of amino acids in the environment. The biosynthesis of amino acids consumes a large amount of chemical energy. Thus, when a supply of exogenous amino acids becomes available, it is advantageous for the bacterium to shut down its own biosynthetic pathway and start utilizing the prefabricated amino acids. However, the cells already contain a full complement of the enzymes that participate in the formation of amino acids. Under these circumstances, the best response for the cell is to lower the activity of preexisting enzymes. This is usually achieved by *feedback inhibition*, a process in which an excess of end product, in this case an amino acid, inhibits the activity of the first enzyme of the biosynthetic pathway via an allosteric mechanism (Figure 9.4).

Feedback inhibition is an ideal mechanism for coping with rapid fluctuations in the supply of amino acids in the environment, but it is inefficient

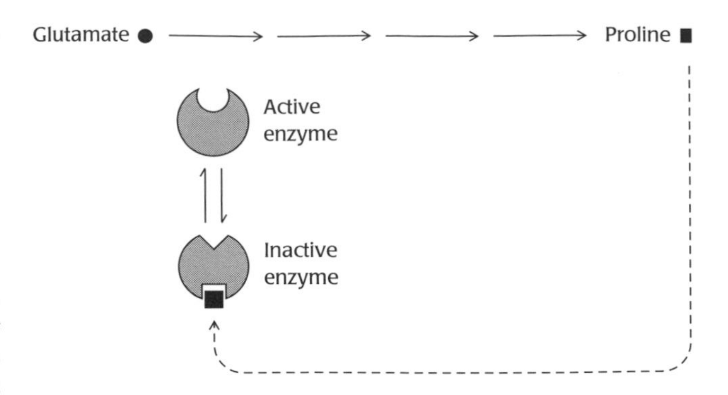


**FIGURE 9.4**

Regulation of proline biosynthesis by feedback inhibition. In this example, the biosynthesis of the amino acid proline is regulated by altering the activity of the first enzyme of the pathway, glutamate kinase. This enzyme is an allosteric enzyme, with a binding site for the substrate, glutamate, as well as a separate binding site for the inhibitor, proline. Whenever excess proline is present, it binds to the regulatory site of glutamate kinase, thereby converting the enzyme into an inactive form. This shuts down the endogenous proline synthesis instantaneously.

Redrawn based on artwork from the first edition (1995), published by W.H. Freeman.

for long-term adaptation. If a population of bacteria continues to live in an environment in which tryptophan, for example, is always available at high concentrations, it makes no sense for the bacteria to synthesize all the enzymes needed for tryptophan biosynthesis, using much energy in the process, and then to keep all of these enzymes from functioning by a system of feedback inhibition. It is much more economical to shut down the synthesis of the unneeded enzymes. Thus, regulation at the level of enzyme synthesis is also important, although it does not help the organism adjust to sudden changes in its environment.

Enzyme synthesis can be regulated by one of two mechanisms. In *repression*, the amino acid end product of the pathway binds to a specific repressor protein as a *corepressor*, altering its conformation. The unliganded repressor has no effect on enzyme synthesis, but the corepressor–repressor complex binds to an upstream sequence of the genes and operons coding for the biosynthesis of that particular amino acid, preventing transcription of the mRNA (Figure 9.5). *Attenuation*, the other mechanism, works by controlling the frequency of RNA chain termination during mRNA transcription. In this mechanism, the 5′-end of the mRNA can form one of the two alternative stem-loop (or hairpin) structures (Figure 9.6). When segments 3 and 4 form a hydrogen-bonded stem loop, that structure acts as a rho-independent termination signal for RNA polymerase (see Box 3.6), because its GC-rich stem

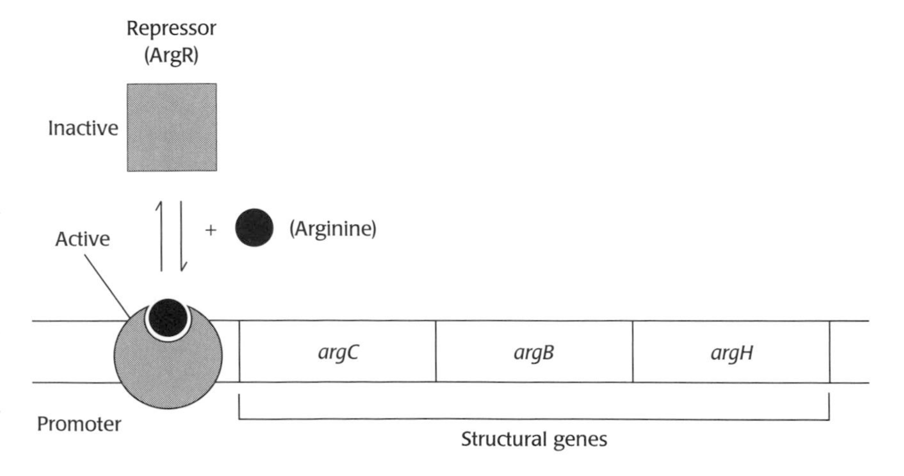


**FIGURE 9.5**

Regulation of amino acid biosynthesis by repression. In *E. coli* K-12, the transcription by RNA polymerase of genes for arginine biosynthesis (here *argC*, *argB*, and *argH*, which form an operon) is regulated by the repressor protein (ArgR). When no free arginine is present in the cytoplasm, ArgR does not bind to the upstream region of the operon, and transcription occurs. When arginine is present, it binds to ArgR and the liganded form of ArgR binds to the region of DNA just behind the promoter, presumably preventing the RNA polymerase from initiating the transcription process. Note that repression affects the synthesis of *all* the enzymes of the pathway (other genes of arginine synthesis are located elsewhere on the chromosome but are similarly regulated by ArgR). In contrast, feedback inhibition usually affects the activity of only the first enzyme of the pathway (or of the branch in a branched pathway).

Redrawn based on artwork from the first edition (1995), published by W.H. Freeman.

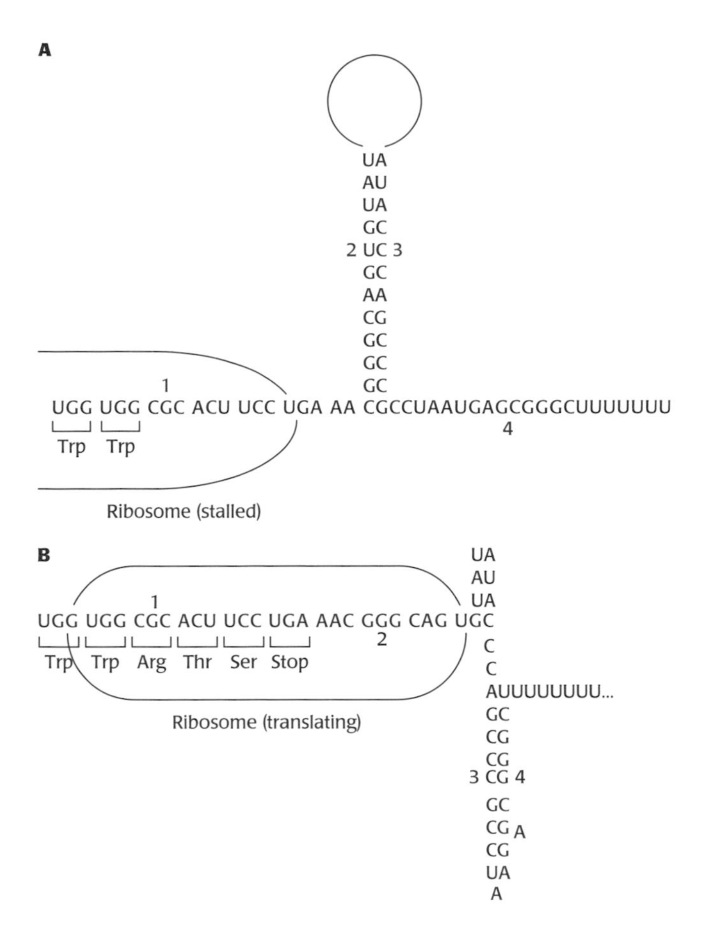


**FIGURE 9.6**

Regulation of amino acid biosynthesis by attenuation. In *E. coli* K-12, the levels of the enzymes responsible for tryptophan biosynthesis are regulated predominantly by attenuation. The 5′-terminal part of the mRNA for the *trp* operon can form two alternative looped structures, (**A**) one involving the pairing of regions 2 and 3, and (**B**) one involving the pairing of regions 3 and 4. Close to the beginning of this stretch of mRNA, there are two consecutive codons for tryptophan. When tryptophan (and therefore tryptophanyl-tRNA) is scarce, translation is stalled at the position of the tryptophan codons. In this state (**A**), region 1 is entirely covered by the ribosome, the 2-3 stem loop is formed, and transcription continues. When tryptophan is readily available, the ribosome moves past the tryptophan codons, inhibiting formation of the 2-3 stem loop, and structure (**B**) forms instead. The stem-loop structure created by the pairing of regions 3 and 4 is a typical rho-independent terminator, with a loop followed by stretches of U (see Box 3.6). In its presence, mRNA transcription is terminated, and tryptophan biosynthetic enzymes are not produced. [Yanofsky, C. (1981). Attenuation in the control of expression of bacterial operons. *Nature*, 289, 751–758.]

is followed by a continuous stretch of U residues. However, the stem loop formed when segment 2 becomes aligned with segment 3 does not act as a termination signal, because there is no stretch of U right after this hairpin. Attenuation works because segment 1 contains multiple codons for the particular amino acid whose synthesis is to be regulated – for example, tryptophan. If no tryptophanyl-tRNA molecules are present, then translation becomes stalled in segment 1. Segment 2 then forms a hairpin structure with segment 3, and transcription continues. In contrast, if tryptophanyl-tRNA is abundant, the continued translation inhibits the formation of a stem structure between segments 2 and 3, and the chain-terminating hairpin between 3 and 4 is formed instead, halting production of the enzymes needed to synthesize tryptophan.

In *E. coli*, the biosynthesis of many amino acids is regulated at the level of enzyme synthesis by both the repression and the attenuation mechanisms. This redundancy apparently expands the range of amino acid concentrations to which *E. coli* can respond. There is evidence that repression plays the predominant regulatory role when the amino acid concentrations in the medium are rather high and that attenuation becomes more important when the amino acids are scarce.

## PRODUCTION BY THE USE OF AUXOTROPHIC MUTANTS

The foregoing discussion of the multiple regulatory mechanisms in *E. coli* may give the impression that to create mutant strains that overproduce amino acids is a hopelessly complex task. This is true to a certain extent, as we will see. Much more practicable, however, is the creation of strains that overproduce biosynthetic intermediates of amino acids. All regulatory mechanisms – feedback inhibition, repression, and attenuation – require the end product in order to function. If one of the steps in amino acid biosynthesis is blocked, none of the regulatory machineries will work, and the cell will overproduce the intermediate that lies just before the blocked step. This type of mutant is termed *auxotrophic* (requiring growth factors, such as amino acids, purines, pyrimidines, or vitamins, for nutrition) because it cannot survive without the amino acid whose biosynthesis is blocked. These auxotrophic mutants are useful if the intermediate itself is a useful commodity or if it can be converted easily into the desired amino acid by a chemical process.

To isolate mutants that are defective in amino acid biosynthesis, one must create an environment in which they have the survival advantage. A classical technique is *penicillin selection*. Assume that the parent strain can synthesize all of the amino acids; that is, it is "prototrophic" and can grow in a mineral medium containing a single organic compound as its major source of cellular carbon and energy. In order to isolate mutants that are unable to synthesize L-arginine ("arginine auxotrophs"), one usually exposes the population of wild-type cells to mutagens so that the frequency of mutants will increase. Then the bacteria are grown in the same mineral medium with the carbon source, but this time it is fortified with L-arginine as well. When the population has reached a suitable density, the cells are harvested, washed, and transferred to a medium exactly like the last one except that it lacks L-arginine. When penicillin is added to this culture, only arginine auxotrophs survive. Penicillin kills growing bacteria because it inhibits the cross-linking of the newly synthesized cell walls and eventually causes lysis of the cells. However, it is totally without effect on nongrowing bacteria, which do not possess new peptidoglycan material. The prototrophic cells are killed because they are capable of growth in the arginine-free medium. The mutant cells, which are incapable of growth, survive. The arginine auxotrophs can then be recovered by plating the mixture on a medium containing L-arginine but no penicillin.

L-Ornithine, which has medicinal applications, is an example of a useful biosynthetic intermediate. Because L-arginine is made from L-citrulline, which in turn is derived from L-ornithine, L-ornithine is overproduced by mutants that are blocked in the arginine biosynthetic pathway at the step converting L-ornithine to L-citrulline. Mutants blocked at earlier or later steps are useless. To find L-ornithine–producing mutants, researchers isolated arginine auxotrophs and then selected those that were able to grow after the addition of citrulline to the medium but were unable to grow after the addition of ornithine. Such mutants are defective in ornithine transcarbamylase, which catalyzes the conversion of ornithine to citrulline, and they accumulate large amounts of ornithine because there is no arginine present to trigger negative regulation. During 1957, the first year of the modern era of amino acid fermentation, researchers isolated such a mutant belonging to the newly discovered glutamate producer, *C. glutamicum*, described earlier.

Auxotrophic mutations are also useful in the production of certain amino acids that are synthesized by branched pathways. If several amino acids – say, A, B, C, and D – are produced as end products of a branched pathway, the effect, in some organisms, of cutting off the branches leading to B, C, and D is often an increase in the production of A. This is primarily because the negative regulatory effects of B, C, and D are eliminated and secondarily because the flow of carbon to the cut-off branches is decreased. These cases will be considered in more detail in our discussion of regulatory mutants of branched pathways.

A major drawback to the use of auxotrophic mutants is that the amino acid that the auxotrophic strain cannot synthesize must be added to the medium. If too much is added, the excess exerts negative regulatory effects on the biosynthetic pathway, the very effect the mutant was designed to avoid. Thus, the mutant strains must be grown in a defined medium (such media tend to be expensive), and the required amino acids must be added in carefully controlled amounts (this process is called fed-batch fermentation). As a result, auxotrophic mutants are usually not the ideal strains for amino acid production.

## PRODUCTION OF AMINO ACIDS BY REGULATORY MUTANTS OF UNBRANCHED PATHWAYS

Mutants that are defective in the negative regulation of amino acid biosynthesis overproduce the amino acid. Unlike auxotrophic mutants, these regulatory mutants can be grown in inexpensive, complex media, and they do not require careful control of growth conditions.

In almost all cases, these regulatory mutants are isolated through the use of amino acid analogs that inhibit the growth of the parent strain in a minimal medium. It has been known for many years that some analogs of amino acids are toxic. Earlier, the prevailing view was that these analogs inhibited growth by becoming misincorporated into proteins, thereby producing nonfunctional proteins. This may well occur in some instances, but in a vast majority of cases, the major cause of inhibition appears to be that the analogs mimic the way the amino acid regulates its own production. Thus, an analog may bind to the allosteric site of the first enzyme of the synthetic pathway or may bind effectively to the repressor, in either case shutting off the synthesis of that particular amino acid. Growth is inhibited because the cells now become starved of that amino acid. Researchers take advantage of this phenomenon to select for regulatory mutants by synthesizing a wide variety of analogs of an amino acid, selecting those analogs that effectively inhibit growth of the wild-type strain in a minimal medium, and then selecting for mutants that are able to grow in the presence of these analogs. In many such

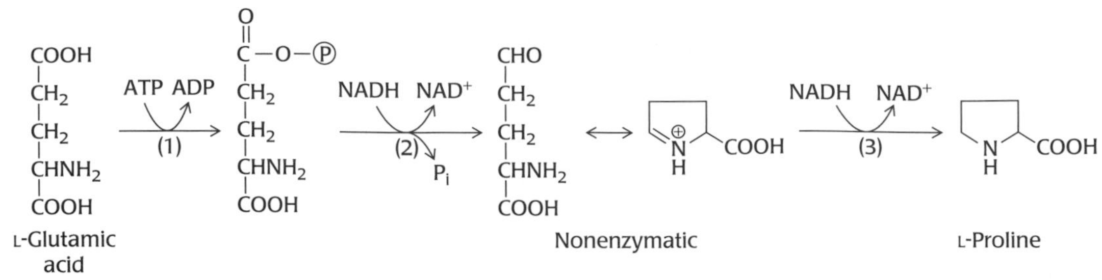

**FIGURE 9.7**

Proline biosynthesis pathway. The enzymes are (1) glutamate kinase, (2) glutamic $\gamma$-semialdehyde dehydrogenase, and (3) $\Delta^1$-pyrroline 5-carboxylate reductase.

mutants, either the repressor or the first enzyme of the pathway is altered, which is why the amino acid synthesis proceeds even in the presence of the analog and will do so in the presence of the amino acid itself.

This approach does not work well when a biosynthetic pathway is regulated at more than one level. Mutation is a rare event, so the probability of isolating a strain that has simultaneously acquired two desirable mutations is miniscule. This is a major concern because in *E. coli*, many pathways are regulated at *three* levels: feedback inhibition, repression, and attenuation. Fortunately, many aquatic and soil microorganisms regulate amino acid biosynthesis more simply than does *E. coli*, presumably because, unlike *E. coli*, they live in a more or less constant environment that is always poor in amino acids. In these organisms, the major function of the regulation of amino acid biosynthesis is not to enable the organism to adjust to a varying supply of the amino acids in the environment but to control the rate of amino acid biosynthesis so that it will meet the demands of the cell's protein synthesis machinery.

The foregoing points are well illustrated by certain mutant strains of a water and soil inhabitant, *Serratia marcescens*, that overproduce proline. Proline is the end product of an unbranched pathway (Figure 9.7). In *S. marcescens*, its synthesis is controlled almost entirely by feedback inhibition, making the system ideally suited to the use of amino acid analogs. A few of the proline analogs tested, such as thiazolidine-4-carboxylic acid and 3,4-dehydroproline (Figure 9.8), selected mutants in which the first enzyme of the pathway, glutamate kinase, was altered in such a way that it was no longer susceptible to feedback inhibition by proline. These mutants overproduced proline, as expected, but maximizing the production required the introduction of two additional changes. First, many wild-type strains of *S. marcescens* produce an enzyme that degrades proline. This enzyme enables the organism to use proline as a source of carbon and nitrogen and, in so doing, decreases the yield of proline by breaking some of it down after it has been synthesized. Thus, it was necessary to inactivate the gene coding for this enzyme, proline oxidase. Second, proline functions as an osmoprotectant solute. When bacterial cells are forced to grow in a high-osmolarity medium, they try to avoid the injurious effect of high salt concentration in the cytoplasm by accumulating proline (see Box 9.1), both by overproducing it and by transporting it from the medium. Because of this additional regulatory mechanism, the yield of proline increases when the bacteria are grown in high-salt media. With these two improvements, yields of 60 to 70 g/L of

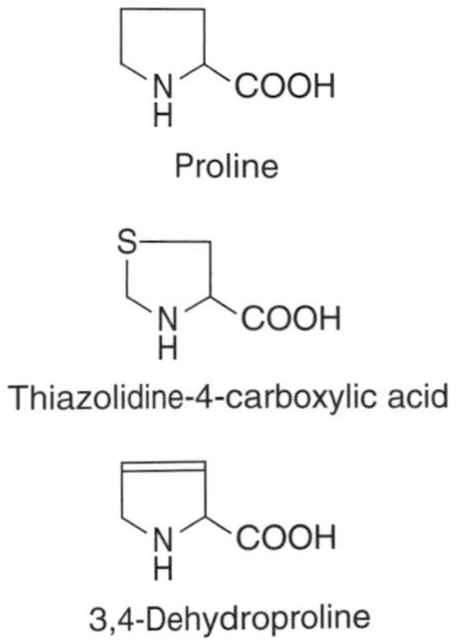

**FIGURE 9.8**

L-Proline and two toxic analogs.

**TABLE 9.3 Histidine biosynthesis in *Serratia* strains[a]**

| | 1 | 2 | 3 |
|---|---|---|---|
| | Extent of Repression (specific activity of histidinol dehydrogenase, units/mg protein) | Extent of Feedback Inhibition of the First Enzyme (inhibition of PRPP adenyltransferase by 10 mM histidine, %) | Histidine Production (g/L) |
| Sr41 (wild type) | 1.1 | 100 | 0 |
| Hd-16 (histidase$^-$) | 0.9 | 100 | 0 |
| 581 (2MH$^r$) | 1.6 | 0 | 0.8 |
| 142 (TRA$^r$) | 12.4 | 94 | 1.3 |
| 2604 (142 × 581) | 14.7 | 0 | 12.9 |

[a] This table shows the effectiveness of the two regulatory mechanisms of histidine biosynthesis in various strains of *S. marcescens.* In column **1**, the specific activity of one of the enzymes of the pathway is shown. The activity is low in the strains in which repression is inhibiting the expression of the genes, but it becomes higher in strains in which this regulatory mechanism has been altered. In column **2**, the efficiency of feedback control is indicated. In the wild-type strain, the first enzyme of the pathway, phosphoribosylpyrophosphate (PRPP) adenyltransferase, is completely inhibited by the end product, histidine, but in mutant 581 this regulatory mechanism is practically eliminated.

2MH, 2-methylhistidine; TRA, 1,2,4-triazole-3-alanine.

*Source*: Kisumi, M., Komatsubara, S., Sugiura, M., and Takagi, T. (1987). Transductional construction of amino-acid hyper-producing strains of *Serratia marcescens. Critical Reviews in Biotechnology*, 6, 233–252.

L-proline could be achieved. It is estimated that about 350 tons of L-proline are produced worldwide by fermentation each year.

When a pathway is regulated at more than one level, a production strain must contain a mutation at each level. Mutants selected with an analog are usually altered in only one of the regulatory mechanisms. However, mutations can be combined via the classical methods of bacterial genetics, especially in organisms such as *Serratia* that are phylogenetically related to *E. coli.* This principle is illustrated by the construction of a histidine-producing mutant of *S. marcescens*. Because *Serratia* uses histidine as a source of carbon and nitrogen, the first step was to create mutants in which the degradative enzyme, histidase, had been inactivated. Starting from this mutant, investigators then selected strains capable of growing in the presence of the toxic histidine analogs 2-methylhistidine and 1,2,4-triazole-3-alanine (Figure 9.9), respectively. As summarized in Table 9.3, one of the triazole-3-alanine (TRA)-resistant mutants, 142, overproduced enzymes of the histidine biosynthetic pathway (Figure 9.10), here exemplified by the much higher specific activity of one such enzyme, histidinol dehydrogenase, but the first enzyme of the pathway, phosphoribosyl phosphate adenyltransferase, was unaltered in its susceptibility to the feedback inhibition. In contrast, one of the 2-methylhistidine (2MH)-resistant mutants, 581, produced an adenyltransferase that is remarkably resistant to feedback inhibition by histidine. Because each mutant was altered in only one regulatory mechanism, neither mutant produced large amounts of histidine (see column 3 of Table 9.3). However, scientists at Tanabe, a pharmaceutical company in Japan, were able to combine the two regulatory mutations by using transduction. The transductant, 2604 (see Table 9.3), was desensitized in both

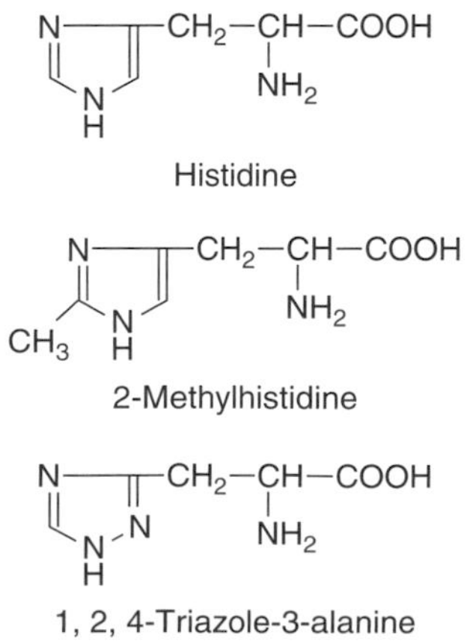

Histidine

2-Methylhistidine

1, 2, 4-Triazole-3-alanine

**FIGURE 9.9**

L-Histidine and two toxic analogs.

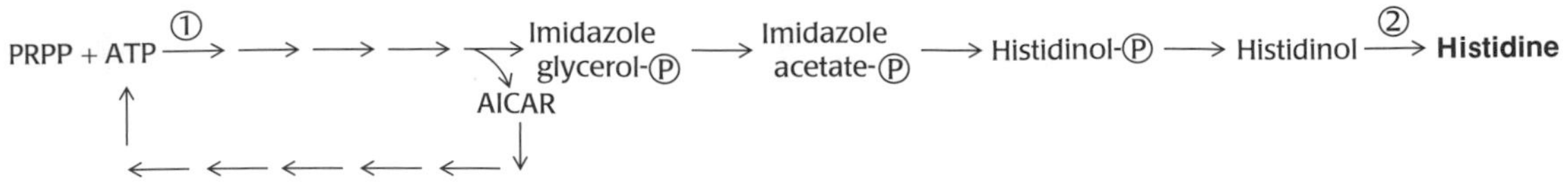

PRPP + ATP ① → → → → Imidazole glycerol-Ⓟ → Imidazole acetate-Ⓟ → Histidinol-Ⓟ → Histidinol ② → **Histidine**

AICAR

**FIGURE 9.10**

Histidine biosynthetic pathway. Steps 1 and 2 are catalyzed by PRPP adenyltransferase and histidinol dehydrogenase, respectively. Abbreviations: PRPP, 5-phosphoribosyl pyrophosphate; AICAR, 5-aminoimidazole-4-carboxamide ribonucleotide.

Redrawn based on artwork from the first edition (1995), published by W.H. Freeman.

repression and feedback inhibition processes, and, as expected, it produced large amounts of histidine.

Creating a system that overproduces metabolites is usually more complicated than creating one that overproduces primary gene products. The increased flux through metabolic pathways may become limited in unexpected ways. This is one of the reasons why straightforward applications of recombinant DNA technology have not always produced spectacular results in this field. In the foregoing example, strain 2604 and similar histidine-production strains developed adenine deficiency that is presumably related to the consumption of ATP in the first step of the histidine pathway. Although an adenine moiety is expected to be regenerated from 5-aminoimidazole-4-carboxamide ribonucleotide (AICAR; see Figure 9.10), this process probably cannot keep pace with the extremely rapid rate of histidine biosynthesis. Tanabe scientists solved the problem by putting the mutant strain through yet another selection cycle, this time using a toxic adenine analog, 6-methylpurine. The result was a resistant strain that overproduces adenine and hence is able to produce histidine at an increased level (23 g/L) without requiring the addition of adenine to the medium. Histidine is now produced at the rate of about 400 tons/year worldwide, by fermentation processes.

## PRODUCTION OF AMINO ACIDS BY REGULATORY MUTANTS OF BRANCHED PATHWAYS

Proline and histidine, which we have described, are the products of unbranched biosynthetic pathways, but many other amino acids are produced by branched pathways. Members of the aspartate family of amino acids (lysine, methionine, threonine, and isoleucine – all essential amino acids) are produced by one branched pathway, those of the pyruvate family (valine and leucine) by another branched pathway, and those belonging to the aromatic family (tryptophan, phenylalanine, and tyrosine) by still another. The regulation of these pathways is much more complex, because it requires a system in which an excess of one product does not accidentally shut off the entire pathway. The regulatory mechanisms in the branched pathways of *E. coli* are very complicated indeed. For example, in the aspartate pathway (Figure 9.11), each product inhibits and/or represses the first enzyme of the common pathway, aspartate kinase. However, so that one product does not shut down all of the kinase activity, three different isozymes (different enzymes that catalyze the same chemical reaction) of aspartate kinase are produced: one sensitive to lysine inhibition, one sensitive to threonine inhibition, and one repressed mainly by an excess of methionine. In addition, *E. coli* is able to adjust to the presence of an unbalanced mixture of the products of the pathway. For example, if it encounters an excess

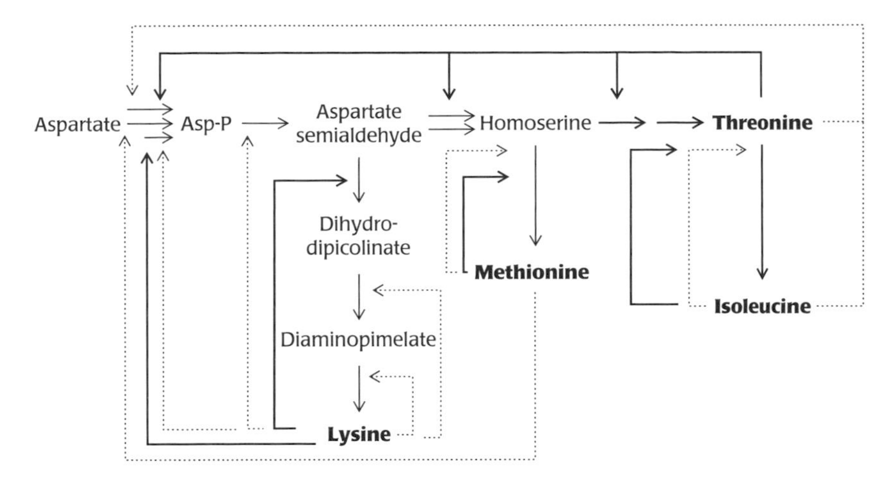


**FIGURE 9.11**

Regulation of synthesis of the aspartate family amino acids in *E. coli*. The pathways are shown in abbreviated form; for example, a *single arrow* from threonine to isoleucine represents five consecutive enzyme-catalyzed steps. *Continuous thick arrows* indicate regulation by feedback inhibition, and *dotted arrows* represent repression or attenuation regulation. Repression and attenuation affect all enzymes of a branch, but here only the effect on the first enzyme of the branch is shown. Step 1 is catalyzed by three isozymes of aspartate kinase. [Modified from Tosaka, O., and Takinami, K. (1986). Lysine. In *Biotechnology of Amino Acid Production*, Aida, K., et al. (eds.), pp. 152–172, Tokyo: Kodansha.]

of methionine but low levels of the other products (lysine, isoleucine, and threonine), regulating the aspartate kinase alone will not lead to the proper ratio of methionine to the other three amino acids. Thus, each amino acid usually controls the first enzyme of its own particular branch. Methionine and isoleucine regulate the pathway through both feedback inhibition and repression mechanisms, whereas lysine and threonine participate in more complex regulatory patterns.

Eliminating various types of controls in a system like this is a major endeavor. However, it is not impossible, especially given the highly developed state of techniques in bacterial genetics and in recombinant DNA manipulations when applied to *E. coli*. Indeed, an *E. coli* strain containing at least nine mutations and producing threonine almost at the level of 100 g/L has been reported.

An easier approach, however, would be the use of microorganisms that have much simpler regulatory mechanisms. Presumably these organisms do not often encounter mixtures of amino acids, sometimes of unbalanced composition, in their environment. If the organism usually lives in an environment that is poor in amino acids, the major function of the regulation of amino acid biosynthesis is to adjust its rate in response to the growth rate of the organism. Thus, there is no need to adjust the ratios of the various amino acids produced. When such an organism is producing excess lysine (e.g., because its growth has slowed), it is also likely to be producing excess methionine, threonine, and isoleucine. The major regulation of the branched pathway can then be achieved by a system in which only one or a few of the products inhibit the first common enzyme. This type of simple regulatory scheme is indeed found in the soil bacterium *C. glutamicum* and its relatives (Figure 9.12).

Lysine production was extensively studied in these organisms, because lysine is one of the amino acids that occur only in small amounts in cereal proteins, as mentioned earlier. The simplicity of the regulatory system in *C. glutamicum* makes it possible to obtain a relatively large amount of lysine simply by cutting off the branches leading to other amino acids. For example, a yield of 34 g lysine per liter of medium was obtained from a mutant that

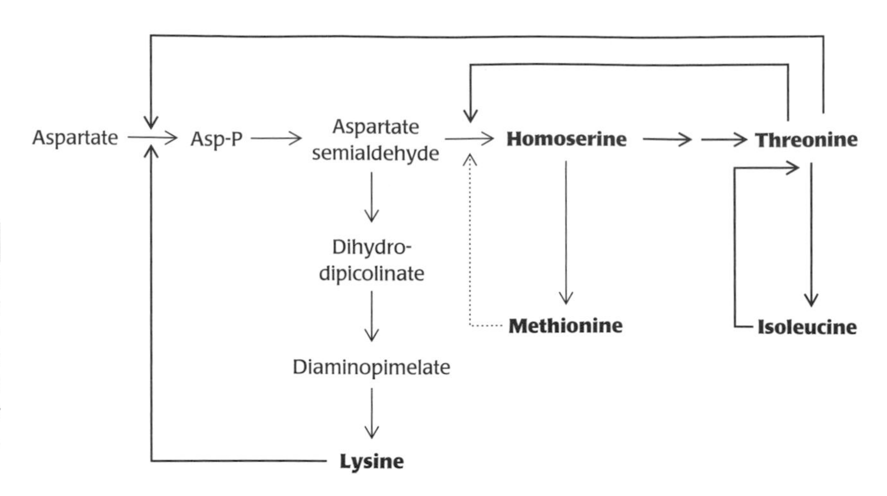


**FIGURE 9.12**

Regulation of synthesis of the aspartate family amino acids in *C. glutamicum. Thick continuous arrows* show regulation by feedback inhibition, and *dotted arrows* represent repression control. [After Tosaka, O., and Takinami, K. (1986). Lysine. In *Biotechnology of Amino Acid Production*, Aida, K., et al. (eds.), pp. 152–172, Tokyo: Kodansha.]

was defective in the branches leading to methionine and to threonine and isoleucine. The diversion of metabolites that would normally go to create those other amino acids contributed to the overproduction of lysine, but the primary reason it occurred is that efficient feedback inhibition of aspartate kinase requires both lysine and threonine (see Figure 9.12). However, lysine-producing auxotrophic mutants of this type must be fed methionine, threonine, and isoleucine continuously and in carefully measured amounts so that they are never present in large excess. Consequently, such mutants are not the desirable type for the commercial production of lysine.

Regulatory mutants of *C. glutamicum*, better suited for commercial production, were obtained by using the lysine analog *S*-aminoethylcysteine (AEC) (Figure 9.13). Like lysine, this analog inhibits the activity of aspartate kinase and hence inhibits the growth of wild-type bacteria. Thus, AEC-resistant mutants are likely to have an alteration in their aspartate kinase such that the altered allosteric, regulatory site of the enzyme has a lower affinity for AEC. These mutant enzymes are also likely to bind lysine with lower affinity and to be defective in the feedback regulation of lysine synthesis. A yield of 32 g lysine per liter was reported for such a mutant; additional fine-tuning that presumably modified the minor regulatory mechanisms increased the yield to 60 g/L. In such fermentation processes, more than 0.3 mole of lysine is obtained from one mole of glucose consumed.

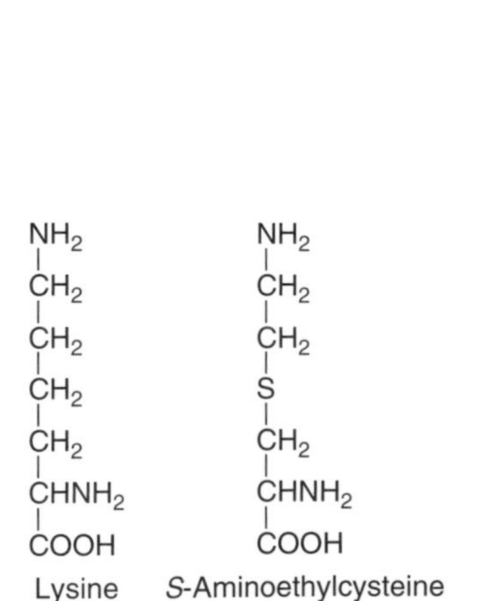


**FIGURE 9.13**

Structures of lysine and *S*-aminoethylcysteine.

The pathway for lysine synthesis in *C. glutamicum* has been studied in depth in recent years, and the flux of intermediates was determined by several methods, notably by combining metabolic flux balancing with the fate of 1-$^{13}$C carbon in glucose, analyzed by $^{13}$C-NMR (see Box 9.2). Another important finding is that lysine is exported into the medium by a specific export transporter, LysE. The sequencing of the *C. glutamicum* genome also supplied additional pieces of information. Because a large fraction of the central metabolic flux is diverted into the synthesis and excretion of lysine, many scientists feel that the best approach is the metabolic engineering method (see Box 9.3), in which one would apply recombinant DNA methods based on the detailed information on the metabolic flux. It was found that expanding an isolated step of the terminal pathway without regard to its effect on other parts of metabolism may have unexpected effects. For example, increasing the level of the first enzyme, aspartate kinase, by the amplification of the *lysC* gene, inhibited the growth of the organism presumably because so much of the flux of central metabolites was now siphoned off into the aspartate pathway, and this did not lead to the greatly increased production rate for lysine (Table 9.4). In contrast, *lysC* amplification combined with amplification of the genes coding for the first and the second enzymes in the lysine-specific branch (Figure 9.12) (*dapA* and *dapB*), significantly increased the lysine production (Table 9.4). Lysine synthesis starts from aspartate, which is made by the amination of oxaloacetate. Thus, increasing the production of oxaloacetate in the cell, by amplification of the pyruvate carboxylase gene (*pyc*) was attempted, and was indeed found to increase the final concentration of lysine in the medium by about 50%.

**TABLE 9.4 Effect of gene amplification on lysine production in *C. glutamicum***

| Strain and the genes introduced on plasmids[a] | Lysine produced (g/L) | | Growth ($OD_{562}$) |
|---|---|---|---|
| | at 40 hrs | at 72 hrs | |
| AJ11082 | 22.0 | 29.8 | 0.450 |
| +*lysC*$^{fbr}$ | 16.8 | 34.5 | 0.398 |
| +*lysC*$^{fbr}$ *dapA* | 19.7 | 36.5 | 0.360 |
| +*lysC*$^{fbr}$ *dapB* | 23.3 | 35.0 | 0.440 |
| +*lysC*$^{fbr}$ *dapA dapB* | 23.0 | 45.0 | 0.425 |

[a] Strain AJ11082 is an AEC-resistant mutant of *C. glutamicum* and produces feedback-resistant LysC protein (aspartate kinase). The feedback-resistant *lysC* gene (*lysC*$^{fbr}$) was cloned from this strain in a multicopy plasmid, and the plasmid was added to the strain in the row "+*lysC*$^{fbr}$." Similarly, the strains contained multicopy plasmids containing additional genes in other rows. $OD_{562}$, optical density at 562 nm.
Data from Otsuna, S., et al., patent WO 96/40934.

In the exploratory experiments discussed above, gene amplification was achieved by using recombinant plasmids. Such a strategy is not suitable for the industrial production of amino acids because the addition of antibiotics needed for the maintenance of plasmids is expensive and because the presence of antibiotic resistance genes is undesirable for the products used as food and feed additives. Use of amino acid/purine/pyrimidine biosynthesis genes as selective markers, as done for yeast plasmids (see Chapter 3), is not practical either because the inexpensive carbon source material used in the growth medium may contain such nutrients. A recent innovation is the use of *dlr* mutant host, defective in alanine racemase, and plasmid vectors containing the *dlr* gene as the selective marker. Here only plasmid-containing cells can synthesize D-alanine, needed for the construction of peptidoglycan cell wall, and therefore can survive. In this way, the use of antibiotic selection can be avoided.

In another approach, recombinant plasmids are used only to create alterations in the bacterial chromosome. Lysine-overproducing strains have been improved by the successive isolation of mutants with better properties. Since some of the mutants were isolated after generalized, non-specific mutagenesis, these multiple steps resulted in the inadvertent introduction of undesirable mutations that affected the "vigor" of the producing strain. Now that the complete genome of *C. glutamicum* is available, it became possible to introduce, into a vigorous wild-type parent, only the desired mutations that increase lysine production, such as those in aspartokinase (*lysC*), homoserine dehydrogenase (*hom*, catalyzing the first step in the threonine branch), and pyruvate carboxylase (*pyc*), generating a strain with the highest lysine-producing capacity. Improving the regeneration of a cofactor, NADPH, through the introduction of a mutant allele of the 6-phosphogluconate dehydrogenase gene resulted in further enhancement of the yield of lysine.

## AMINO ACID PRODUCTION WITH ENZYMES

Although many amino acids are produced by fermentation processes, it is sometimes advantageous to use an enzymatic process instead. A good example is the production of L-aspartic acid, which is used as a starting material for the chemical synthesis of the sweetener aspartame. L-Aspartic acid has been produced in Japan since 1969, primarily by exploiting the *E. coli* enzyme aspartase, which catalyzes the conversion of L-aspartate into fumarate and $NH_3$:

$$\text{L-aspartic acid} \rightarrow \text{fumaric acid} + NH_3.$$

Aspartase is a catabolic enzyme used by *E. coli* to degrade exogenous aspartic acid into readily utilizable sources of carbon and nitrogen for growth. Under physiological conditions, the reaction equilibrium lies far toward the right. More precisely, because the equilibrium constant ($K_{eq}$) is 20 mM, if the initial concentration of aspartate is 1 mM, almost 95% of it will be converted into fumarate and $NH_3$. Thus, at first the enzyme appears to be a poor choice for *synthesizing* aspartate. However, the law of mass action favors aspartate synthesis when the reactant concentrations are high – that is, under conditions that would be used for industrial production of an amino acid. Based on the $K_{eq}$ value cited above, we can calculate that if we start from 2 M fumarate and 2 M $NH_3$, the reaction will proceed far toward the left and will come to equilibrium when 90% of the reactants are converted into L-aspartate. Thus, aspartase, although it is a degradative enzyme, is ideal for the industrial synthesis of aspartate. In contrast, most enzymes that catalyze biosynthetic reactions in intact cells are not suitable for the industrial production of amino acids, either because they use complex, expensive substrates or because they require an expensive cofactor, such as NADPH or ATP.

When amino acids *can* bc producd by a simple enzyme reaction, such a process has many advantages.

1. The amino acids are produced in high concentrations (in the case of aspartate, about 1 M), making their recovery easy.
2. Because the solution contains only a few reactants, purification of the product is simple. In contrast, the supernatant in a fermentation run always contains some of the growth medium, the ingredients of which are usually quite complex and difficult to eliminate.
3. Compared with fermentation processes, the rate of production per unit volume is much higher. Fermentation requires huge tanks and much material for growth media, whereas enzymatic production can be done in a small space. Consequently, the required capital investment is much smaller.

Although enzymatic production can be carried out conveniently with soluble enzymes, it is much more efficient when the enzymes are immobilized by attachment to a supporting matrix. Separation and recovery of an immobilized enzyme are much simpler, so the enzyme can be reused economically. In addition, immobilization sometimes increases the stability of the enzyme. The immobilized enzyme preparations are usually packed into columns, a space- and time-saving format.

High aspartic acid yields have been achieved with columns of immobilized aspartase. In the early years, the aspartase was purified from *E. coli* and attached covalently to an insoluble matrix. It was soon discovered that the purification is unnecessary and that attaching intact *E. coli* cells to the matrix produces a very effective column. Because aspartase is an inducible enzyme, the *E. coli* must be grown in the presence of aspartate. Interestingly, the activity of the enzyme column increases nearly 10-fold during the first one to two days of incubation at 37°C; apparently this is because of the destruction of the membrane permeability barrier by autolysis. Even an intermediate method, in which *E. coli* cells were immobilized by entrapment in polyacrylamide gel, resulted in a cost reduction of 40% over the older batch method. This decrease was the result of the increased production rate we have cited and to the elimination of the cell-removal step associated with the batch process. In 1978, Tanabe scientists introduced a better immobilization medium, $\kappa$-carrageenan (a seaweed polysaccharide). Production with a properly treated column of this material is 170% higher than production with a polyacrylamide column. Moreover, the stability of the enzyme in this column is excellent; the half-life at 37°C is about two years. Adding a fresh mixture of fumarate and ammonia continuously to a small column 1 L in volume results in a yield of several kilograms per day of L-aspartate. To obtain yields of this magnitude by fermentation would require tanks of at least several hundred liters in volume.

Other amino acids, including L-alanine and L-lysine, have been produced successfully via enzymatic synthesis on a commercial scale. For example, L-alanine is produced by the enzymatic decarboxylation of L-aspartate, which can be generated by the process described above:

$$\text{L-aspartate} \rightarrow \text{L-alanine} + CO_2.$$

A wild-type strain of *Pseudomonas dacunhae* is used as the source of the decarboxylase, and a batch process employing immobilized whole cells is reported to produce a yield of 400 g/L.

Although L-lysine is usually produced by fermentation, it is also produced by the following ingenious enzymatic process:

$$\text{DL-}\alpha\text{-amino-}\varepsilon\text{-caprolactam} \xrightarrow{\text{hydrolase}} \text{L-lysine} + \text{D-}\alpha\text{-amino-}\varepsilon\text{-caprolactam}$$

$$\text{D-}\alpha\text{-amino-}\varepsilon\text{-caprolactam} \xrightarrow{\text{racemase}} \text{L-}\alpha\text{-amino-}\varepsilon\text{-caprolactam}.$$

This process uses the inexpensive starting material DL-aminocaprolactam, produced by organic synthesis, and exploits the stereospecificity of the hydrolase, which is found in the yeast *Cryptococcus laurentii*, to produce L-lysine. The remaining D-isomer of the caprolactam is brought back into the production pathway by a racemase found in the bacterium *Achromobacter obae*. Batch-wise production with intact cells of these organisms is reported to produce nearly complete conversion of 100 g/L of aminocaprolactam into L-lysine.

There was an attempt to produce L-phenylalanine by utilizing phenylalanine aminolyase, which under normal, physiological conditions would function in the direction of phenylalanine degradation:

$$\text{L-phenylalanine} \rightarrow \textit{trans}\text{-cinnamic acid} + NH_3.$$

As in the case of L-aspartic acid synthesis, the law of mass action favors the synthesis of phenylalanine when the substrates are present at high concentrations. Yields of 80% or more, with a final L-phenylalanine concentration of 50 to 60 g/L, have been produced in column reactors containing immobilized *Rhodotorula* yeast cells. Thus, there is no problem with the process itself. Nevertheless, it has not been a commercial success, because the starting material, *trans*-cinnamic acid, is almost as expensive as the product.

## SUMMARY

Some primary metabolites are not the typical waste products of energy metabolism, yet can be overproduced by wild-type strains under unusual conditions. Thus citric acid is produced by *Aspergillus niger* at around pH 2 in the presence of about 1 M sugar, and L-glutamate by *Corynebacterium glutamicum* under biotin starvation. Recent studies suggest that such overproduction phenomena can occur with moderate shift in the normal metabolic pathway, and need not involve total blocks in some metabolic steps.

Other amino acids, such as lysine, threonine, proline, are produced by regulatory mutants of bacteria. Such mutants may be selected by resistance to amino acid analogs, which normally act to repress the synthesis of amino acid biosynthetic pathway enzymes or to inhibit the first enzyme of the pathway. Genome sequence of *C. glutamicum* is contributing to the construction of robust overproduction strains. Finally, some amino acids, such as aspartate, can be produced by immobilized enzyme columns.

### SELECTED REFERENCES


#### General

Demain, A. L. (2000). Small bugs, big business: the economic power of the microbe. *Biotechnology Advances*, 18, 499–514.

#### Citric Acid

Karaffa, L., and Kubicek, C. P. (2003). *Aspergillus niger* citric acid accumulation: do we understand this well-working black box? *Applied Microbiology and Biotechnology*, 61, 189–196.

Roehr, M., Kubicek, C. P., and Kominek, J. (1996). Citric acid. In *Biotechnology*, 2nd Edition, H.-J. Rehm and G. Reed (eds.), Volume 9, pp. 307–345, Weinheim, Germany: Verlag Chemie.

Burgstaller, W. (2006). Thermodynamic boundary conditions suggest that a passive transport step suffices for citrate excretion in *Aspergillus* and *Penicillium*. *Microbiology*, 152, 887–893.

**Amino Acids (in General)**

Krämer, R. (2004). Production of amino acids: physiological and genetic approaches. *Food Biotechnology*, 18, 171–216.

Ikeda, M., and Nakagawa, S. (2003). The *Corynebacterium glutamicum* genome: features and impacts on biotechnological processes. *Applied Microbiology and Biotechnology*, 62, 99–109.

Kirchner, O., and Tauch, A. (2003). Tools for genetic engineering in the amino acid–producing bacterium *Corynebacterium glutamicum*. *Journal of Biotechnology*, 104, 287–299.

**Glutamic Acid**

Kinoshita, S. (1985). Glutamic acid bacteria. In *Biology of Industrial Microorganisms*, A. L. Demain and N. A. Solomon (eds.), pp. 115–142, Menlo Park, CA: Benjamin/Cummings.

Hoischen, C., and Krämer, R. (1989). Evidence for an efflux carrier system involved in the secretion of glutamate by *Corynebacterium glutamicum*. *Archives of Microbiology*, 151, 342–347.

Kimura, E. (2003). Metabolic engineering of glutamate production. *Advances in Biochemical Engineering/Biotechnology*, 79, 37–57.

Sonntag, K., Schwinde, J., de Graaf, A. A., Marx, A., Eikmanns, B. J., Wiechert, W., and Sahm, H. (1995). $^{13}C$ NMR studies of the fluxes in the central metabolism of *Corynebacterium glutamicum* during growth and overproduction of amino acid in batch cultures. *Applied Microbiology and Biotechnology*, 44, 489–495.

Sekine, H., Shimada, T., Hayashi, C., Ishiguro, A., Tomita, F., and Yokota, A. (2001). $H^+$–ATPase defect in *Corynebacterium glutamicum* abolishes glutamic acid production with enhancement of glucose consumption rate. *Applied Microbiology and Biotechnology*, 57, 534–540.

Eggeling, L., and Sahm, H. (2003). New ubiquitous translocators: amino acid export by *Corynebacterium glutamicum* and *Escherichia coli*. *Archives of Microbiology*, 180, 155–160.

Sauer, U., and Eikmanns, B. J. (2005). The PEP-pyruvate-oxaloacetate node as the switch point for carbon flux distribution in bacteria. *FEMS Microbiology Reviews*, 29, 765–794.

**Lysine and Threonine**

Debanov, V. G. (2003). The threonine story. *Advances in Biochemical Engineering/Biotechnology*, 79, 113–136.

de Graaf, A. A., Eggeling, L., and Sahm, H. (2001). Metabolic engineering for L-lysine production by *Corynebacterium glutamicum*. *Advances in Biochemical Engineering/Biotechnology*, 73, 9–29.

Pfefferle, W., Möckel, B., Bathe, B., and Marx, A. (2003). Biotechnological manufacture of lysine. *Advances in Biochemical Engineering/Biotechnology*, 79, 59–111.

Vrljic, M., Sahm, H., and Eggeling, L. (1996). A new type of transporter with a new type of cellular function: L-lysine export from *Corynebacterium glutamicum*. *Molecular Microbiology*, 22, 815–826.

Ohnishi, J. et al. (2002). A novel methodology employing *Corynebacterium glutamicum* genome information to generate a new L-lysine producing mutant. *Applied Microbiology and Biotechnology*, 58, 217–223.

Koffas, M., and Stephanopoulos, G. (2005). Strain improvement by metabolic engineering: lysine production as a case study for systems biology. *Current Opinion in Biotechnology*, 16, 361–366.

Wendisch, V. F., Bott, M., and Eikmanns, B. J. (2006). Metabolic engineering of *Escherichia coli* and *Corynebacterium glutamicum* for biotechnological production of organic acids and amino acids. *Current Opinion in Microbiology*, 9, 268–274.

TEN

# Secondary Metabolites: Antibiotics and More

The preceding chapter described the industrial production of some primary metabolites, including citric acid and amino acids. In contrast to these compounds, which are present in most living organisms and are produced by the ubiquitous major metabolic pathways, secondary metabolites are produced only by special groups of organisms through specialized pathways. Their chemical structure tends to be complex, and they are often produced only during the special growth phase, most often during the stationary phase. The most important of these secondary metabolites are the antibiotics.

Many science historians argue that among the many scientific discoveries of the twentieth century, that of the first antibiotic, penicillin (Figure 10.1), by Alexander Fleming (reported in 1928) is the discovery that had the largest impact on human life. The story is well known. Fleming is supposed to have kept a rather untidy laboratory and to have discovered, after returning from his vacation, that the bacterial colonies neighboring a contaminating mold colony were lysed on one of the Petri plates left on his bench. This story is often cited as an example of the importance of serendipity in science. This view, however, totally disregards the fact that Fleming dedicated his entire career to the search for natural products that lyse bacterial cells, in an effort to find agents that could be used in the treatment of bacterial infections. He had, in fact, discovered the enzyme lysozyme several years earlier but was disappointed that most human pathogens were intrinsically resistant to its lytic action. (Pathogenic bacteria presumably have been subjected to evolutionary selection for such resistance because lysozyme is present in most tissues and body fluids of higher animals.) Another important aspect of Fleming's discovery of penicillin is that he tried to isolate the active substance and to use it in the treatment of infections. Dozens of scientists before Fleming had published reports of either lysis or growth inhibition of bacteria caused by products of molds and other microorganisms. Yet nothing came out of these findings because the authors operated as classical "naturalists" reporting "curious" phenomena. The ultimate industrial-scale production and the clinical application of penicillin required, more than a decade later, great efforts by many outstanding chemists and a wartime collaboration

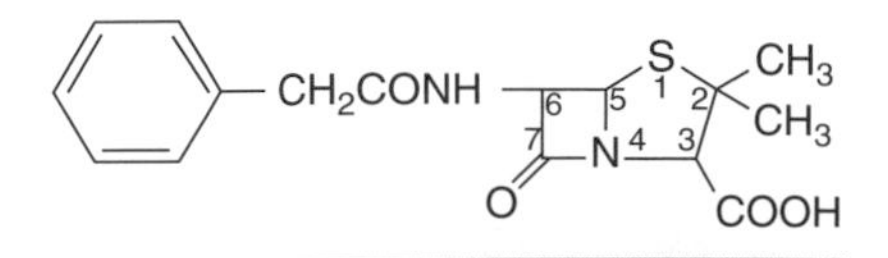


FIGURE 10.1

Penicillin G.

among pharmaceutical companies in the United States. Therefore, Fleming's early efforts are all the more impressive because he could already show the efficacy of the culture filtrate of his mold as a local therapeutic agent in an animal infection model.

Many different classes of antibiotics were discovered among secondary metabolites since then, as will be discussed below. However, it is not clear whether all of these natural antibiotics are synthesized by the producing organisms for the purpose of killing off their neighbors and competitors. Many of these compounds show only marginal antimicrobial activity and have to be altered by chemical modification to produce semisynthetic agents that can be used in therapy. Furthermore, some of them are produced only at very low levels. It is thus likely that many of the "natural antibiotics" are produced for other purposes, for example, as signaling molecules. This hypothesis fits with the observation that there are a great many secondary metabolites that do not show antimicrobial activity even with the very sensitive assays available. On the other hand, compounds like penicillin probably have served as an antibiotic even in natural microbial communities, because genes coding for resistance against $\beta$-lactams (including penicillins) are found on the chromosomes of many bacterial species.

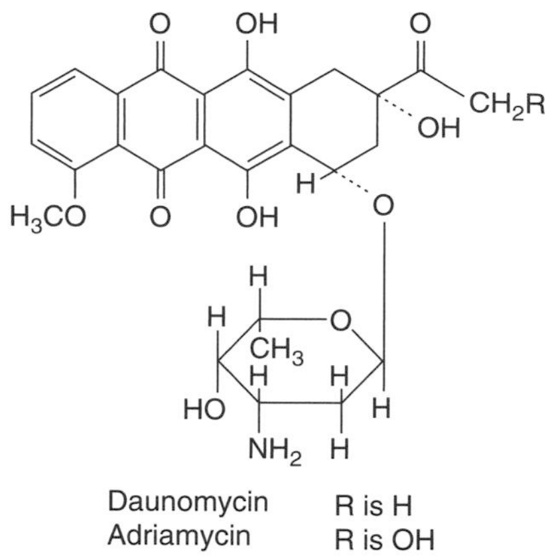

Daunomycin R is H
Adriamycin R is OH

FIGURE 10.2

Daunorubicin (Daunomycin) and Doxorubicin (Adriamycin).

## ACTIVITIES OF SECONDARY METABOLITES

### SECONDARY METABOLITES WITH USEFUL NONANTIBIOTIC ACTIVITIES

Because of the considerations described above, culture filtrates of microorganisms are screened not only for antimicrobial activity, but also for the presence of other activities, such as inhibition of particular animal enzymes or binding to specific receptors on animal cell surfaces. By this means, a number of compounds with interesting activity were discovered. A few examples will serve to illustrate the range of targets for which compounds have been found (see also Chapter 2). These compounds do not fit the traditional definition of *antibiotics*.

#### Antitumor Agents

Perhaps the most important agents in cancer chemotherapy are secondary metabolites of microorganisms belonging to the anthracycline family. These compounds, including doxorubicin and daunorubicin, contain a reduced naphthacene ring in which four benzene rings are fused together (Figure 10.2).

These compounds intercalate between the bases in double-helical DNA and are thought to act by inhibiting the topoisomerase reaction. Another family, including dactinomycin, contains three fused rings (Figure 10.3), also intercalates into double-stranded DNA, and inhibits both the transcription and synthesis of DNA.

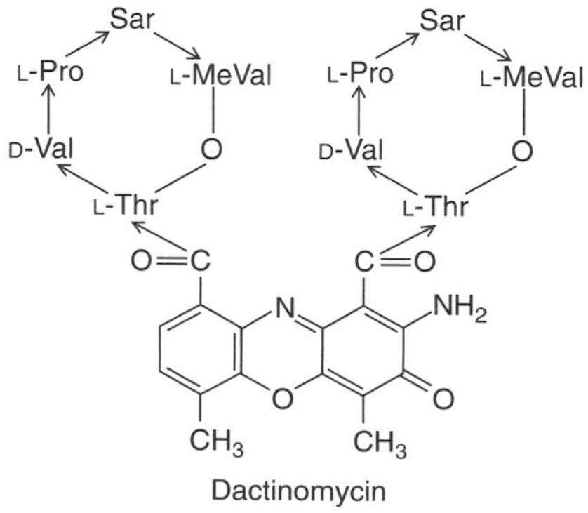

Dactinomycin

FIGURE 10.3

Dactinomycin. *Arrows* indicate the direction of peptide bonds (–CO–NH–). Abbreviations: Sar, sarcosine; MeVal, *N*-methylvaline.

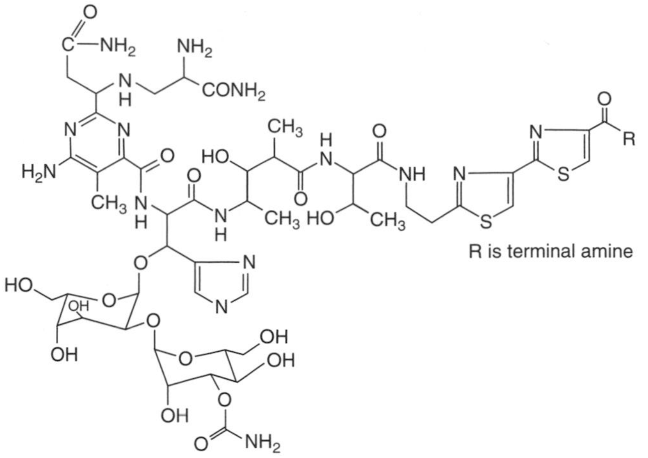

FIGURE 10.4
Bleomycin.

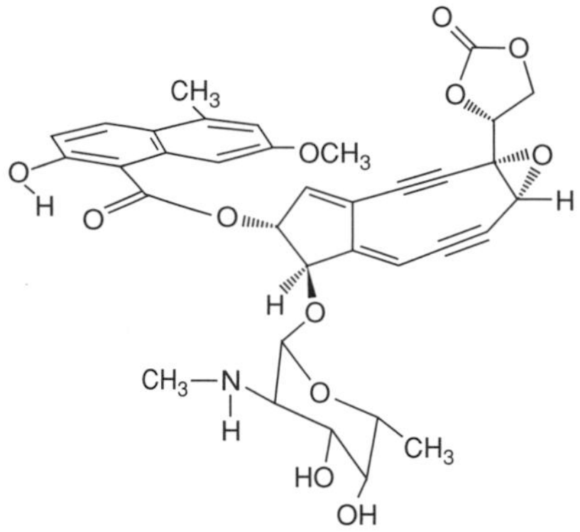

FIGURE 10.5
The chromophore of zinostatin.

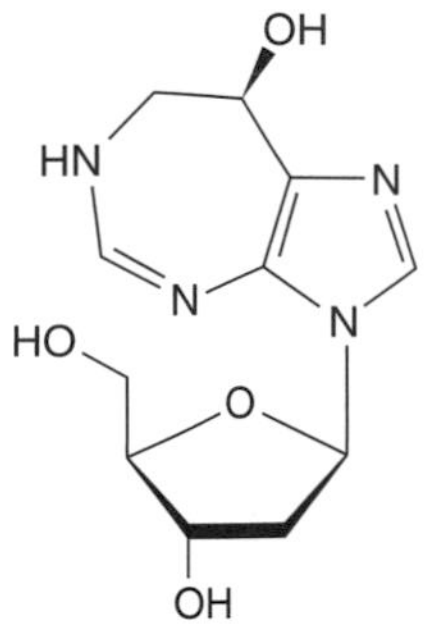

FIGURE 10.6
Pentostatin.

Bleomycin, which also has a high affinity for DNA, has a totally different structure and destroys DNA by generating oxygen free radicals (catalyzed by a ferrous ion that is coordinated by this antibiotic; Figure 10.4).

Finally, a fascinating compound, zinostatin (neocarzinostatin), licensed in Japan, is composed of a chromophore (Figure 10.5) that is protected by its binding to an apoprotein. When the chromophore leaves the protective protein and is reduced by sulfhydryl agents such as glutathione, it becomes converted to a free radical, and because its naphthalene ring system intercalates into double-stranded DNA, the free radical produces oxidation of the deoxyribose in DNA, followed by the cleavage of DNA strand(s). All of these compounds are products of *Streptomyces* species, the most important source of traditional antibiotics.

Some chemotherapeutic agents used for the treatment of tumors are totally synthetic. These include antimetabolites (e.g., methotrexate and fluorouracil), alkylating agents (e.g., cyclophosphamide), and DNA-cross-linking agents (e.g., cisplatin). However, even these classes include secondary metabolites of microorganisms – for example, pentostatin (a purine nucleoside analog produced by *Streptomyces* spp. [Figure 10.6]), streptozocin (an alkylating agent produced by *Streptomyces* spp. [Figure 10.7]), and mitomycin C (another *Streptomyces* product that becomes an alkylating agent after reduction in cells [Figure 10.8]). Some others are either plant products or synthetic compounds whose structures are based on these natural products – for example, the antitubulin drugs vinblastine, vincristine, and paclitaxel (taxol; see Chapter 2) and the DNA topoisomerase inhibitor etoposide.

## Protease/Peptidase Inhibitors

In the late 1960s, Hamao Umezawa had the foresight to screen for microbial secondary products that would inhibit proteases. Proteases are essential for many normal physiological functions (e.g., the blood-clotting cascade) but also are implicated in many pathological states (e.g., elastase in emphysema). Furthermore, proliferation of many animal viruses requires proteolytic processing of viral polyprotein precursors by virally encoded proteases (e.g., the aspartyl protease of the AIDS virus). A search for such protease inhibitors in culture filtrates of *Streptomyces* and related species led to the discovery of several potent inhibitors of various types of proteases. Some of these molecules, such as antipain, leupeptin, and pepstatin, are very useful reagents in the research laboratory (Figure 10.9). Some others may prove to be useful in human therapy. Bestatin (Figure 10.10), which inhibits aminopeptidase, is used for immunopotentiation in cancer patients in Japan. Another interesting example is A58365A isolated by scientists at Eli Lilly (see Box 10.1), a *Streptomyces* product that inhibits angiotensin-converting enzyme, a key enzyme in the renin–angiotensin system that controls blood pressure.

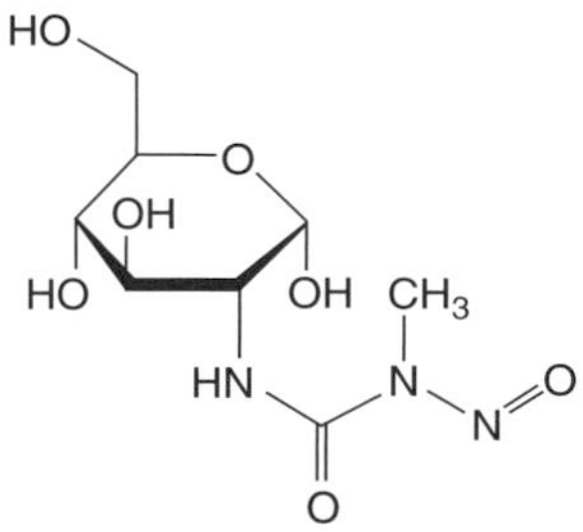

**FIGURE 10.7**

Streptozocin.

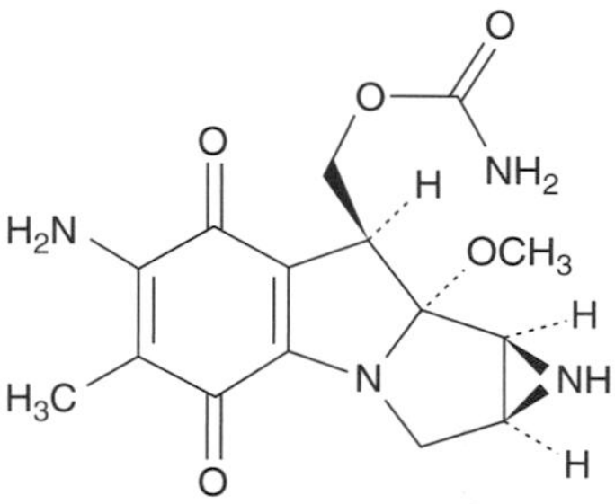

**FIGURE 10.8**

Mitomycin C.

### Inhibitor of Cholesterol Biosynthesis

Limiting dietary intake of foods rich in cholesterol lowers blood cholesterol levels significantly in some individuals, but in others high cholesterol levels result from elevated endogenous synthesis of cholesterol rather than from dietary intake. In the 1970s, scientists at Sankyo Co. in Tokyo isolated several novel products from cultures of *Penicillium* species that inhibited cholesterol biosynthesis by rat liver extract. Further study established that

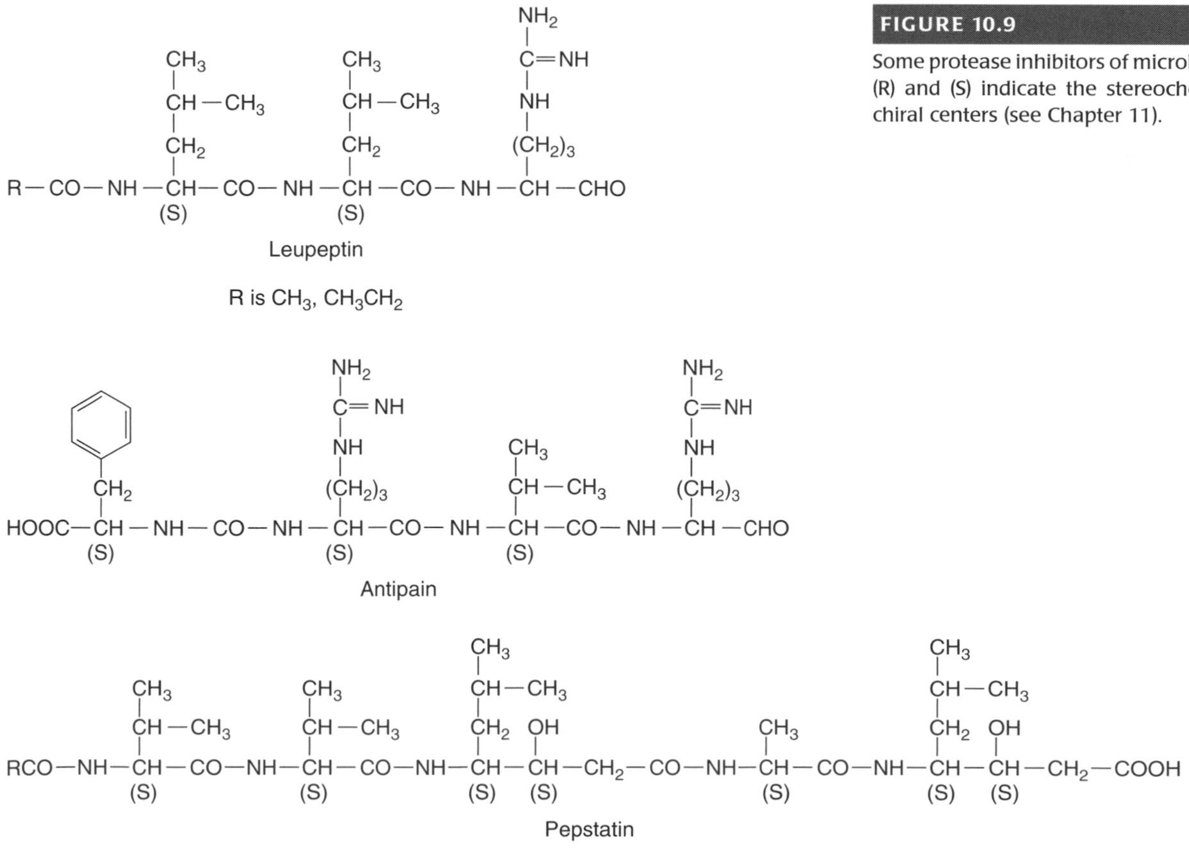

Leupeptin

R is $CH_3$, $CH_3CH_2$

Antipain

Pepstatin

**FIGURE 10.9**

Some protease inhibitors of microbial origin. (R) and (S) indicate the stereochemistry of chiral centers (see Chapter 11).

**Inhibitors of Angiotensin-Converting Enzyme (ACE)**

Angiotensin II, which is produced by the splitting of C-terminal dipeptide from angiotensin I, produces constriction of blood vessels and thus plays an important role in the regulation of blood pressure:

Asp-Arg-Val-Tyr-Ile-His-Pro-Phe-His-Leu (angiotensin I)
↓ ACE
Asp-Arg-Val-Tyr-Ile-His-Pro-Phe (angiotensin II) + His-Leu.

This fact was dramatically brought home by the observation that the snake venom of a Brazilian viper, whose bite produces the sudden lowering of blood pressure and collapse in its victims, contained a potent peptide inhibitor of the enzyme (ACE). When Squibb scientists identified the peptides, one of the most potent compounds had the sequence Pyr-Trp-Pro-Arg-Pro-Gln-Ile-Pro-Pro. ("Pyr" stands for pyroglutamic acid, or L-2-pyrrolidone-5-carboxylic acid). They tested a number of synthetic peptides and found that Glu-Lys-Trp-Ala-Pro was active as an inhibitor, but this peptide had to be administered by intravenous injection. They then designed smaller structural analogs of this pentapeptide inhibitor, beginning with the idea that ACE, being a carboxypeptidase that releases a dipeptide, must have three sites interacting with the substrate peptide, as shown below. They used proline at the C-terminus because inhibitory peptides all contained proline here. They added a succinyl group in front of proline because substituted succinic acid was shown to be an effective inhibitor of pancreatic carboxypeptidase. This initial compound (the lower compound in the scheme below) was improved by replacing the succinic acid moiety with the 3-mercaptopropionic acid group, culminating in the production of the first clinically effective ACE inhibitor for treatment of hypertension – captopril.

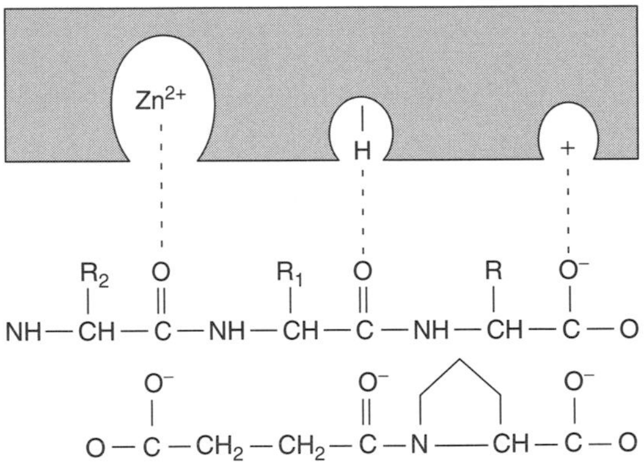

NH—CH—C—NH—CH—C—NH—CH—C—O

O—C—$CH_2$—$CH_2$—C—N—CH—C—O

It is instructive to compare the structure of these "rationally designed" drugs with the natural product with ACE inhibitory activity, A58365A below, which actually turned out to be a conformationally restricted analog of the succinyl proline shown above!

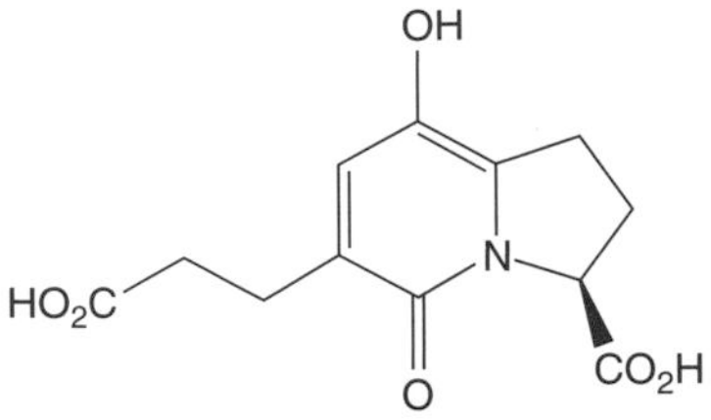

**BOX 10.1**

these compounds, including compactin (Figure 10.11), competitively inhibit the enzyme that catalyzes the first unique step of cholesterol synthesis, hydroxymethylglutaryl-CoA (HMG-CoA) reductase.

Compactin was also effective in lowering the serum cholesterol levels in experimental animals and in humans. In 1980, scientists at Merck, Sharpe & Dohme isolated a closely related compound, mevinolin, from culture filtrate of another fungus, *Aspergillus*. This compound is now successfully marketed as lovastatin. Several other semisynthetic or synthetic "statins" are now available, and these belong to the drugs with the highest market values; for example, atorvastatin (Lipitor) alone had sales of more than $12 billion worldwide in 2005.

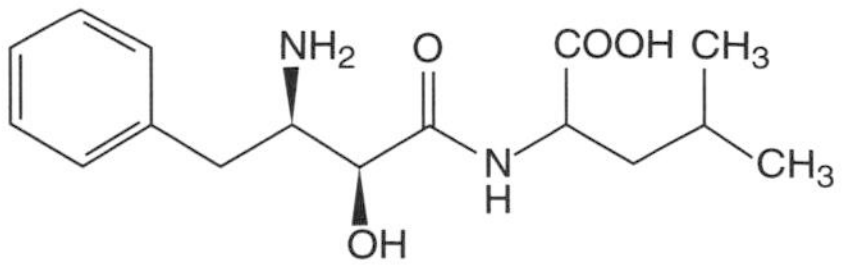

$NH_2$ O COOH $CH_3$ N H $CH_3$ OH

FIGURE 10.10

Bestatin.

### Some Other Inhibitors

Although not isolated originally by a screen for enzyme inhibition, some antibiotics turned out to be exceptionally strong and specific inhibitors of important enzyme reactions. Examples of such agents that have contributed enormously to experimental studies in biochemistry and cell biology include cerulenin (Figure 10.12; a fungal analog of a fatty acid derivative), which inhibits fatty acid synthesis in other organisms, and tunicamycin (Figure 10.13), which inhibits the transfer of the *N*-acetylglucosamine-1-phosphate moiety onto the lipid carrier in the biosynthesis of the asparagine-linked carbohydrate chains of glycoproteins.

### Immunosuppressants

Some microbial products act as powerful immunosuppressants. Treatment of patients with these compounds is therefore useful in decreasing the incidence of allograft rejection in organ transplantation operations. The best known among these compounds is cyclosporin A (Figure 10.14A), a cyclic peptide produced by a fungus. Interestingly, this compound was caught in the screening net originally because of its antifungal activity. Cyclosporin forms a complex with a peptidyl prolyl *cis–trans* isomerase, cyclophilin, in eukaryotic cytoplasm. The cyclosporin–cyclophilin complex binds to and inhibits the action of a phosphoprotein phosphatase known as calcineurin.

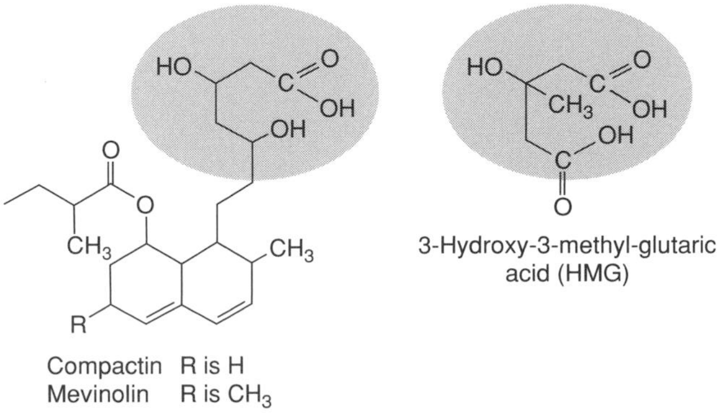

HO C=O OH OH O O $CH_3$ $CH_3$ R

Compactin R is H
Mevinolin R is $CH_3$

HO C=O $CH_3$ OH OH C O

3-Hydroxy-3-methyl-glutaric acid (HMG)

FIGURE 10.11

Compactin, mevinolin (Lovastatin), and 3-hydroxy-3-methylglutaric acid. The antibiotics are drawn in their acid rather than lactone form because lactones are hydrolyzed rapidly into acids in the human body. Thus, in the case of mevinolin, the structure would more accurately be called mevinolic acid. *Shaded areas* indicate regions of structural similarity.

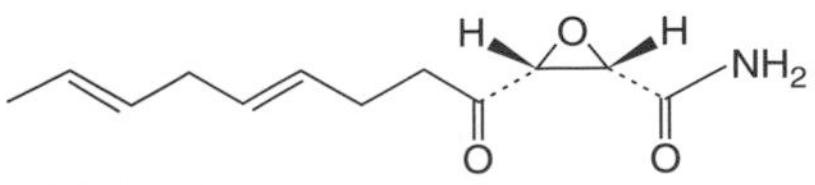

FIGURE 10.12

Cerulenin.

Inhibition of this phosphatase prevents a dephosphorylation event that is required for the nuclear entry of a transcription factor necessary for the expression of the autocrine lymphokine interleukin 2 in T cells, and in this manner prevents T cell activation.

More recently, rapamycin (now called sirolimus) was identified through its antifungal activity, and FK-506 (now called tacrolimus; Figure 10.14B) was isolated by screening microbial products for the activity to suppress interleukin 2 production. Both of these compounds are important agents in the clinic.

### Compounds Binding to Specific Receptors

An example of a compound binding to a specific receptor is asperlicin (Figure 10.15A), a product of *Aspergillus* species identified by Merck scientists. This compound binds strongly to the cholecystokinin receptor. Cholecystokinin (Figure 10.15B) is a peptide hormone, which was initially isolated from intestinal tissue and is known to stimulate gut motility and contraction of the gallbladder. The structure of asperlicin could have never been conceived as a "rational" analog of the peptide hormone.

## A BRIEF SURVEY OF ANTIBACTERIAL AGENTS

Most of the antibiotics that are now in commercial production are compounds active against bacteria. It is important to distinguish between the agents that act against Gram-positive bacteria only and those that are also active against Gram-negative bacteria.

The cells of Gram-negative organisms have the added protection of an outer membrane (see Figure 1.3). Small solutes (<1000 daltons) cross this permeability barrier primarily through the water-filled channels of special proteins called porins. The porin channels can be quite narrow, with a cross-section of only 0.7 × 1 nm in *Escherichia coli* and its relatives and with a diameter estimated to be only about 2 nm in the bacteria with the largest channels. For this reason, the diffusion of water-soluble antibacterial agents larger than 1000 daltons through the outer membrane is severely limited. Not surprisingly, the permeability of these water-filled channels to lipophilic solutes is also poor. Furthermore, the lipid bilayer region of the outer membrane has an unusually low permeability toward lipophilic molecules,

FIGURE 10.13

Tunicamycin.

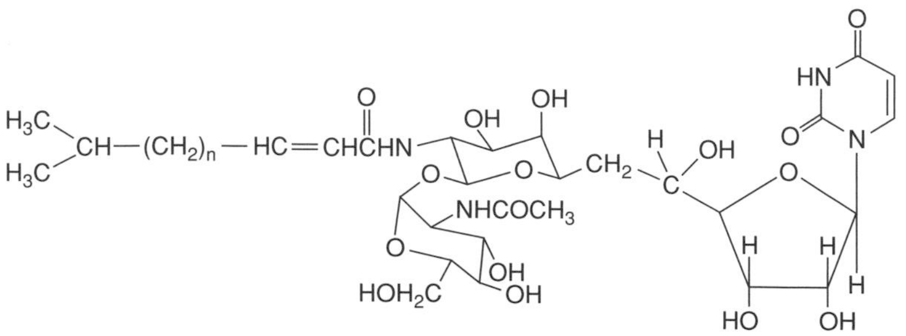

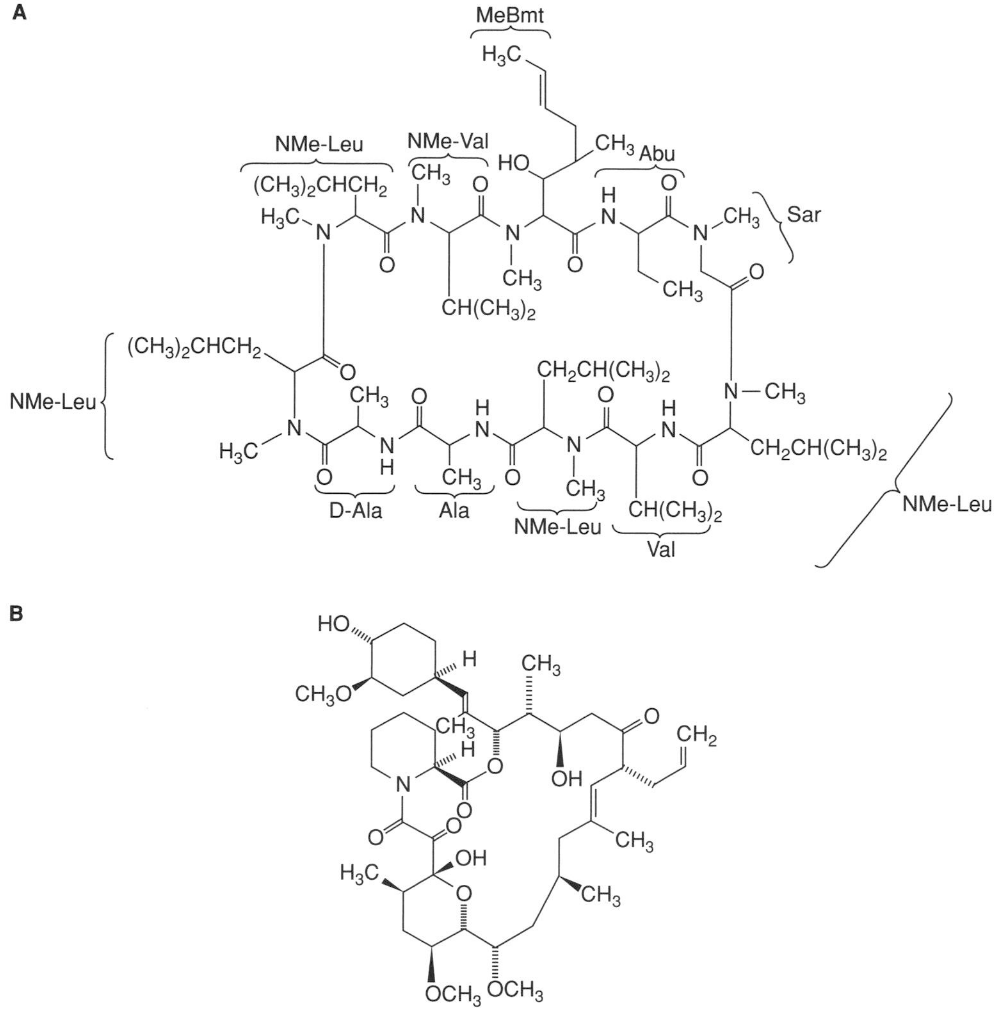

FIGURE 10.14

**(A)** Cyclosporin A. The new amino acid, MeBmt, is apparently essential for the biological activity as is the presence of numerous *N*-methyl groups, which inhibit hydrogen bonding between backbone residues and presumably help cyclosporin assume a particular conformation. **(B)** FK-506 (tacrolimus). Abbreviations: NMe, *N*-methyl; Sar, sarcosine; Abu, 2-aminobutyric acid; MeBmt, 4-(2-butenyl)-4,*N*-dimethyl-L-threonine.

apparently because its outer leaflet is composed exclusively of an unusual lipid molecule, the lipopolysaccharide (LPS). Consequently, the larger and more lipophilic antibiotics tend to have significant activity against Gram-positive bacteria only.

### β-Lactams

Penicillin (benzylpenicillin, penicillin G) (Figure 10.1) is the classical example of a β-lactam antibiotic. The β-lactams inhibit the synthesis of the eubacterial cell wall, which is composed of a polymer unique to the bacterial world, called peptidoglycan. Peptidoglycan is a network of polysaccharide (or glycan) chains composed of alternating *N*-acetylglucosamine and *N*-acetylmuramic acid residues (see also Figure 1.1). The polysaccharide chains are cross-linked to each other via short peptide chains, which include some D-amino acids and are attached to the *N*-acetylmuramic acid residues (Figure 10.16). This structure gives peptidoglycan exceptional chemical stability, mechanical strength, and rigidity.

A

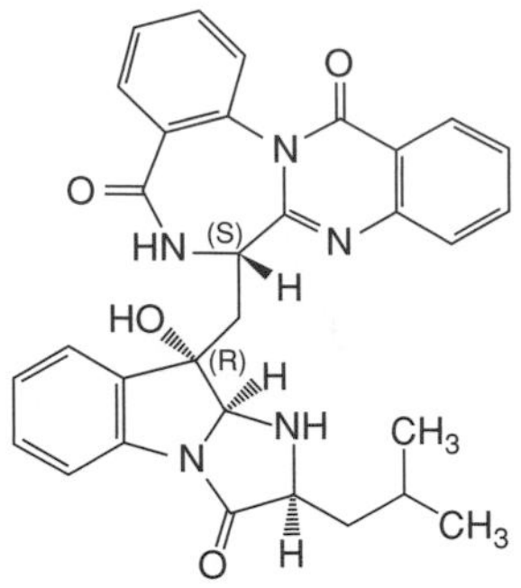

B

Lys—Ala—Pro—Ser—Gly—Arg—Met—Ser—Ile—Val—
Lys—Asn—Leu—Gln—Asn—Leu—Asp—Pro—Ser—His—
Arg—Ile—Ser—Asp—Arg—Asp—Tyr ($SO_3H$) —Met—Gly—Trp— Met—Asp—Phe—$NH_2$

**FIGURE 10.15**

(**A**) Asperlicin. (**B**) Human cholecystokinin.

The addition of new glycan chains to the cell wall in growing cells takes place by the cross-linking of the peptide side chain of a new unit to the preexisting peptidoglycan structure (Figure 10.17). The cross-linking reaction is catalyzed by a DD-transpeptidase. This enzyme cleaves the peptide bond between the two D-alanine residues in the side chain of a newly made glycan chain and transfers the glycan–peptidyl complex to the free amino group of the diamino acid residue in the side chain of the preexisting peptidoglycan, thereby cross-linking different chains. As first pointed out by Tipper and Strominger, the $\beta$-lactam ring system structurally resembles the D-alanyl-D-alanine of the nascent side chain (Figure 10.18). They demonstrated that the interaction between transpeptidase and penicillin generates a covalent penicilloyl enzyme, resulting in the irreversible inactivation of the enzyme. Thus, a $\beta$-lactam is an example of a "suicide inhibitor" because it interacts with its target enzyme like a substrate and then undergoes chemical reaction with the enzyme, thereby causing its permanent inactivation. Suicide

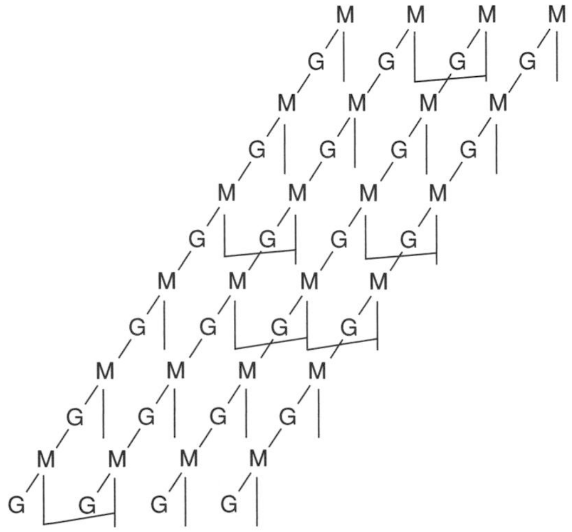

**FIGURE 10.16**

Structure of peptidoglycan from Gram-negative bacteria. Polysaccharide chains are made of alternating *N*-acetylmuramic acid units (M) and *N*-acetylglucosamine units (G). Short peptides (*vertical lines*) are attached to the *N*-acetylmuramic acid residues, and the polysaccharide chains are connected via peptide cross-links (*nearly horizontal lines*).

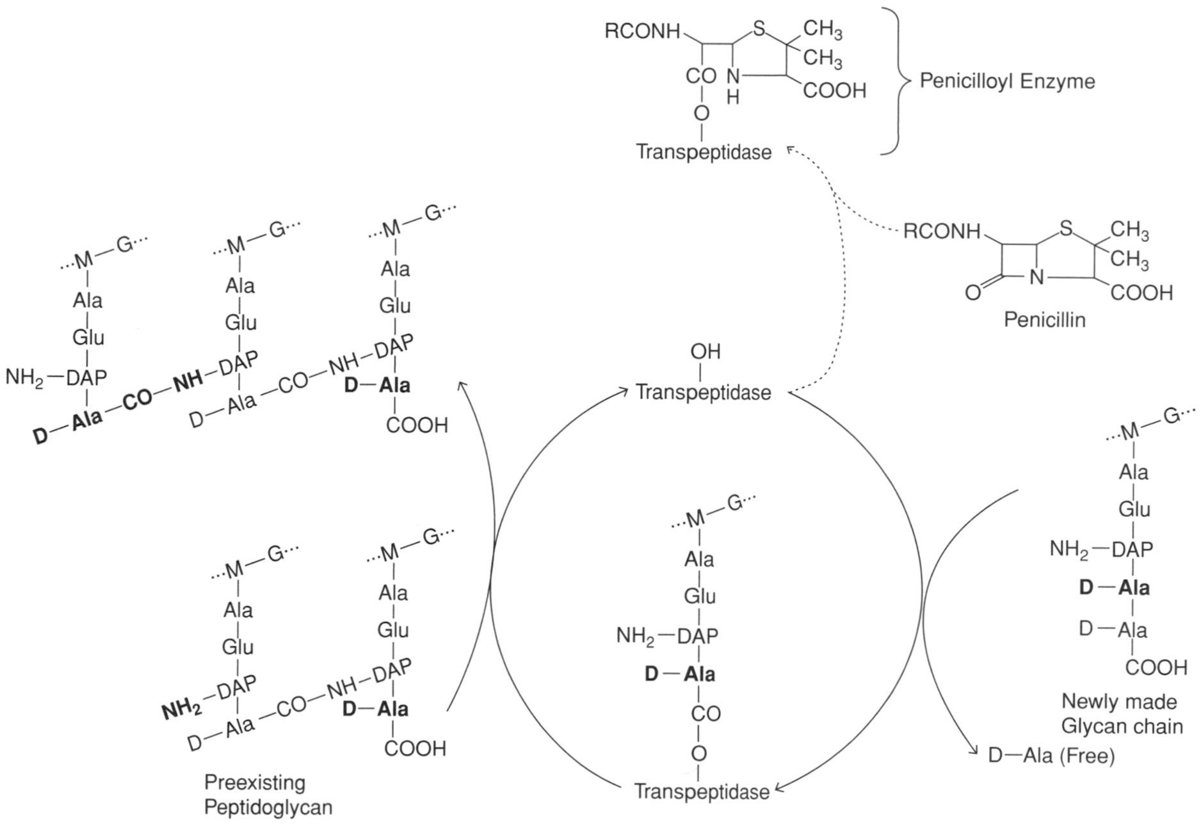

**FIGURE 10.17**

Addition of new material to the existing peptidoglycan and the mechanism of action of penicillin. The newly made, still un-cross-linked peptidoglycan (lower right) becomes covalently linked to transpeptidase (center bottom) by a reaction involving the splitting of the D-alanyl-D-alanine bond of the peptide. The new peptidoglycan material is then transferred onto the amino group of the diaminopimelic acid residue in a preexisting peptidoglycan complex (left bottom), regenerating the transpeptidase enzyme (center). Penicillin also produces a covalent complex with the transpeptidase (penicilloyl enzyme, top center), but this complex is stable and the enzyme becomes permanently inactivated (*broken arrows*).

Redrawn based on artwork from the first edition (1995), published by W.H. Freeman.

inhibitors are more desirable antimicrobial agents than are competitive inhibitors (such as sulfonamides) because complete inhibition of the target is achieved.

Penicillin G is very effective in killing most Gram-positive bacteria but is ineffective against most Gram-negative bacteria because of its lipophilicity. However, the benzyl side chain at the 6-position (Figure 10.1) can be chemically replaced with other substituents. Some of the resulting compounds show significant activity against Gram-negative bacteria and thus are "broad-spectrum" antibiotics (see below).

## Macrolides

The basic macrolide structure – seen, for example, in erythromycin (Figure 10.19) – is synthesized by bacterial species related to *Streptomyces* through head-to-tail condensations of several C3 units to produce a large lactone ring. Macrolides are therefore examples of polyketide compounds, whose synthesis is discussed later in this chapter (p. 374). These macrocyclic compounds are sufficiently large and hydrophobic that their inhibitory action is largely limited to Gram-positive bacteria. However, more recently, semisynthetic

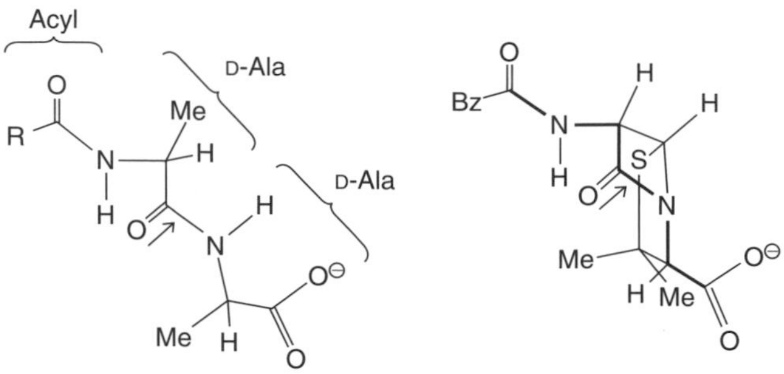

**FIGURE 10.18**

Structural similarity between the acyl-D-alanyl-D-alanine part of the peptidoglycan (left) – that is, the natural substrate of transpeptidase – and penicillin G (right). Bz in the penicillin structure denotes the benzyl moiety. The bonds in penicillin G that correspond to the peptide backbone in acyl-alanyl-alanine are drawn in *heavy lines*. The bonds that are cleaved by the peptidoglycan transpeptidase are indicated by *arrows*. [The drawing is based on Tipper, D. J., and Strominger, J. L. (1965) Mechanism of action of penicillins: a proposal based on their structural similarity to acyl-D-alanyl-D-alanine. *Proceedings of the National Academy of Sciences U.S.A.*, 54, 1133–1141, but it is turned around and simplified by representing the methyl group as Me.]

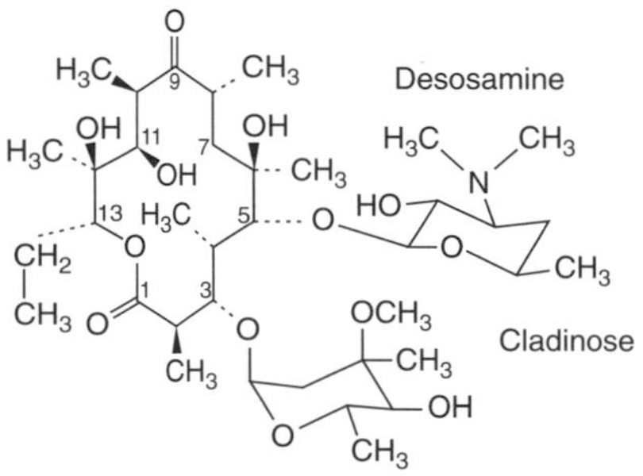

**FIGURE 10.19**

Structure of erythromycin A. The macrocyclic rings in most macrolides are substituted by sugar residues.

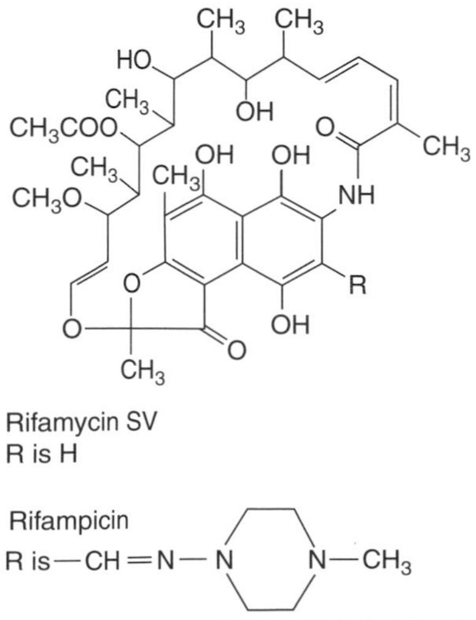

Rifamycin SV
R is H

Rifampicin
R is—CH=N—N N—$CH_3$

**FIGURE 10.20**

The structure of rifamycins.

macrolides (such as azithromycin) have been introduced that show significant anti–Gram-negative activity.

### Ansamycins

Ansamycins also have a macrocyclic structure, but the ring differs from that of the macrolides because it contains an aromatic chromophore as well as an amide (lactam) linkage (Figure 10.20). These compounds are isolated from *Amycolatopsis* species, which (like *Streptomyces*), are members of the actinomycete branch of eubacteria (see Chapter 1). Rifamycins inhibit prokaryotic RNA polymerase. Natural rifamycins, like rifamycin SV (Figure 10.20), have significant activity only against Gram-positive bacteria, as expected from their hydrophobic character and large size. When a positively charged group is introduced chemically to make the molecule more polar, as in rifampicin (Figure 10.20), some activity against Gram-negative bacteria becomes detectable. These compounds are important in combating a special class of Gram-positive pathogens including *Mycobacterium tuberculosis* (cause of tuberculosis) and *Mycobacterium leprae* (cause of leprosy).

### Tetracyclines

Tetracyclines, produced by several species of *Streptomyces*, contain a fused four-ring system (Figure 10.21). These compounds are inhibitors of prokaryotic protein synthesis. Because these agents are relatively hydrophilic due to the presence of several hydroxyl groups, an amide moiety, and a tertiary amine substituent, they can cross the outer membrane of Gram-negative bacteria efficiently through porin channels. Tetracyclines are thus broad-spectrum antibiotics active against both Gram-positive and Gram-negative bacteria. In the past, addition of low concentrations of tetracyclines to animal feed was a common practice all over the world because this treatment tended to increase the rate of weight gain and to decrease the incidence of infectious disease in livestock. However, in England there have been several epidemics in calves caused by *Salmonella* containing tetracycline-resistance plasmids that have later spread to the human population. In most European countries, it is now prohibited to give tetracycline (or other antibiotics that are commonly used for the treatment of human bacterial infections) to animals without veterinary prescription.

### Chloramphenicol

Chloramphenicol (Figure 10.22) was originally isolated from the filtrate of *Streptomyces venezuelae* cultures. However, because it is a small molecule with a simple structure, it is more economical to produce by chemical synthesis than by fermentation. Chloramphenicol inhibits prokaryotic protein synthesis. It penetrates through the outer membrane porin channels of Gram-negative bacteria easily because of its small size and is therefore another broad-spectrum antibiotic. However, chloramphenicol can enter eukaryotic cells and inhibit mitochondrial protein synthesis. Thus, use of chloramphenicol may prevent growth of rapidly proliferating cells and this may be related to at least one of its side effects, the suppression of bone marrow cells. Because of this and other side effects, chloramphenicol is not widely used today, except in the treatment of diseases in which the pathogen survives within human cells, such as infection by *Salmonella typhi* (typhoid fever).

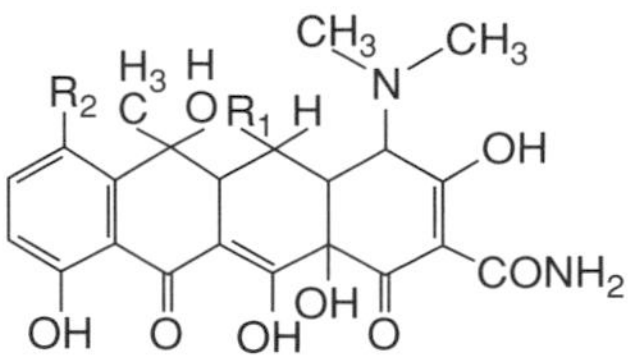

| | $R_1$ | $R_2$ |
|---|---|---|
| Tetracycline | H | H |
| Chlortetracycline | H | Cl |
| Oxytetracycline | OH | H |

**FIGURE 10.21**

The structure of tetracyclines.

### Peptide Antibiotics

Several peptide antibiotics (Figure 10.23) are now in commercial production. Many of these are produced by *Bacillus* species. These peptides contain unusual amino acids, such as D-amino acids, ornithine, and diaminobutyric acid, and have significant toxicity to humans when injected, and are thus useful only for topical applications. Some of these agents are used in Europe as feed additives because they are not used in human therapy (see the discussion above on tetracycline).

Many of these antibiotics are too bulky and hydrophobic to cross the outer membrane barrier of Gram-negative bacteria, and their efficacy is therefore limited to Gram-positive bacteria. An interesting exception is polymyxin, which has a two-step mode of attack on Gram-negative bacteria. This polycationic antibiotic apparently binds to the highly negatively charged LPS molecules in the outer membrane of many Gram-negative bacteria and disrupts the molecular organization of the outer membrane bilayer. Once the integrity of the outer membrane is destroyed, polymyxin can bind to and insert into the cytoplasmic membrane through its hydrophobic tail, thereby killing the bacterial cell by permeabilizing the plasma membrane. Polymyxin shows strong activity against organisms like *Pseudomonas aeruginosa*, which is resistant to almost all other antibiotics because of the unusually low permeability of its outer membrane. Hence, polymyxin still has its place in the treatment of life-threatening infections by *P. aeruginosa*, in spite of its significant toxicity.

Vancomycin, a glycopeptide antibiotic of 1449 daltons (Figure 10.24), acts only on Gram-positive bacteria through an unusual mechanism of binding to the D-Ala-D-Ala moiety of the peptidoglycan precursor. This has become an important drug for the treatment of infection caused by methicillin-resistant *Staphylococcus aureus* (MRSA). Daptomycin, a lipopeptide antibiotic (Figure 10.25), inserts into the cell membrane of Gram-positive bacteria (including MRSA) in the presence of $Ca^{2+}$, and kills them by producing leakage of

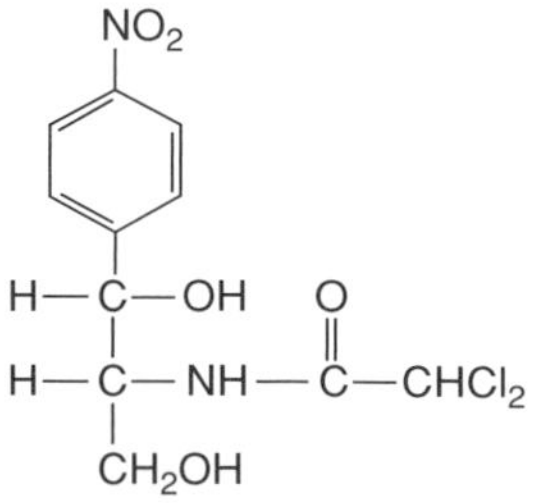

**FIGURE 10.22**

The structure of chloramphenicol.

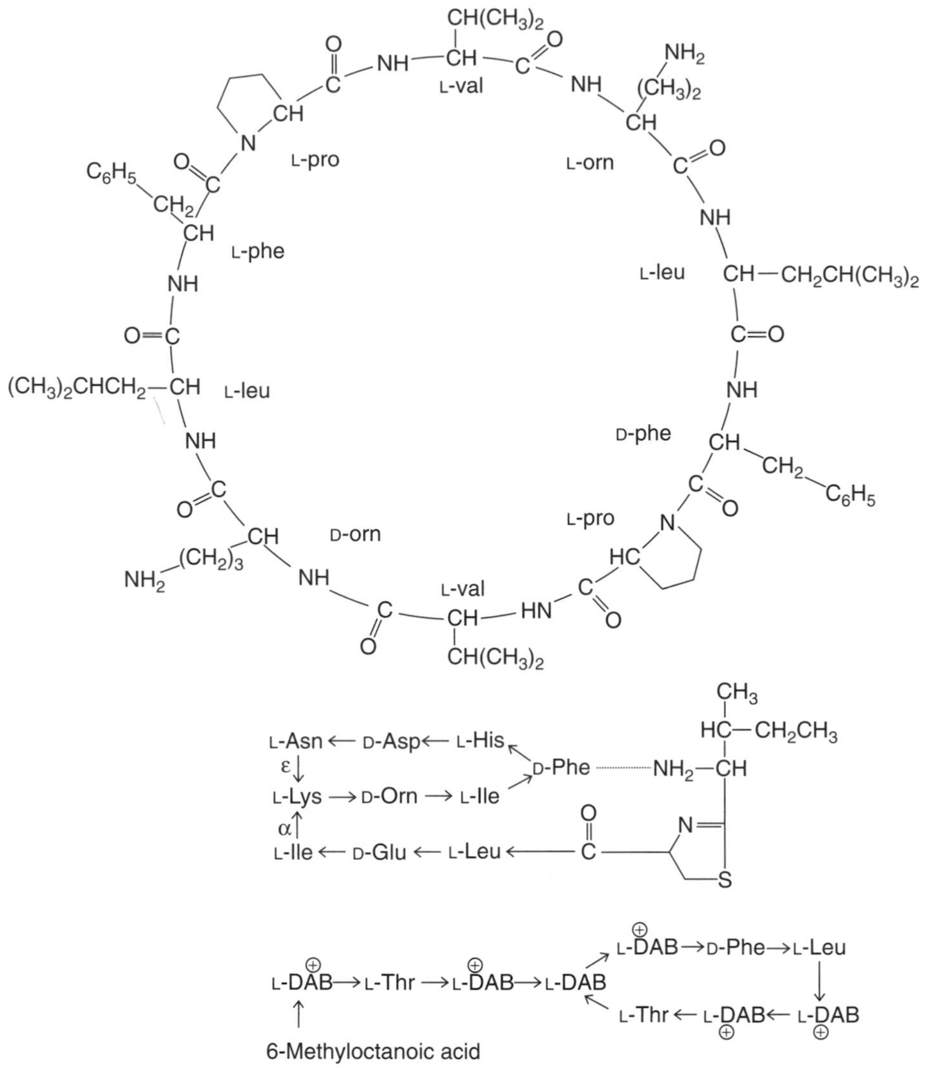

Gramicidin S

Bacitracin A

Polymyxin $B_1$

**FIGURE 10.23**

Some examples of peptide antibiotics. The *arrows* indicate the direction of peptide bonds (–CO→NH–). In the structure of polymyxin, DAB indicates 2,4-diaminobutyric acid. The positive charge due to the free 4-amino group of this residue is also indicated.

cytosolic ions. It is remarkable that this compound is tolerated by humans; the plasma membrane of higher animals does not contain much anionic lipids in its outer leaflet, and perhaps this prevents the $Ca^{2+}$-bridged interaction between the drug and the membrane.

## Aminoglycosides (Aminocyclitols)

Aminoglycosides, which are discussed in detail later in this chapter, consist of an aminocyclitol moiety (streptidine in streptomycin, 2-deoxystreptamine in many other aminoglycosides; see Figure 10.26) to which amino sugars are attached in various ways. All of these compounds are produced by the prokaryotes of the actinomycete line (*Streptomyces* in most cases, *Micromonospora* for gentamicin). These antibiotics are inhibitors of prokaryotic protein synthesis. Because they are quite hydrophilic and sufficiently small, aminoglycosides can cross the Gram-negative outer

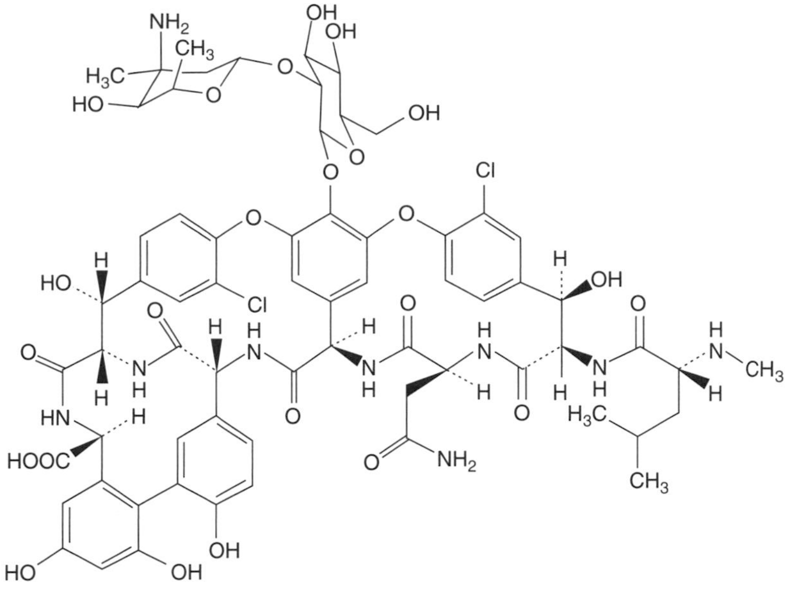

FIGURE 10.24
Structure of vancomycin.

membrane through the porin channels, and thus these compounds are equally effective against both Gram-positive and Gram-negative bacteria.

## A BRIEF SURVEY OF ANTIFUNGAL AGENTS

Because fungi are eukaryotes and contain the same type of machinery for protein and nucleic acid synthesis as the cells of higher animals, it is more difficult to achieve selective inhibition of fungal metabolism.

Polyoxin B has a structure reminiscent of UDP-*N*-acetylglucosamine and inhibits the synthesis of chitin, a polysaccharide found uniquely in fungal cell walls. It is used in agriculture as a fungicide.

Griseofulvin (Figure 10.27), produced by *Penicillium griseofulvum*, also inhibits fungi. It binds to proteins involved in the assembly of tubulin in

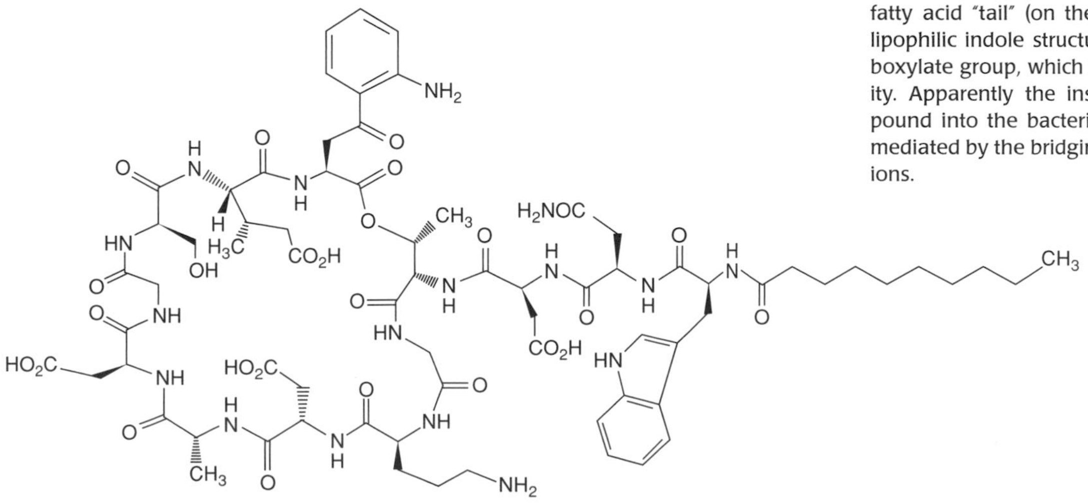

FIGURE 10.25
Structure of daptomycin. The main portion of the cyclic peptide structure is hydrophilic and polyanionic, but the portion next to the fatty acid "tail" (on the right) contains the lipophilic indole structure as well as a carboxylate group, which is essential for activity. Apparently the insertion of this compound into the bacterial cell membrane is mediated by the bridging action of the $Ca^{2+}$ ions.

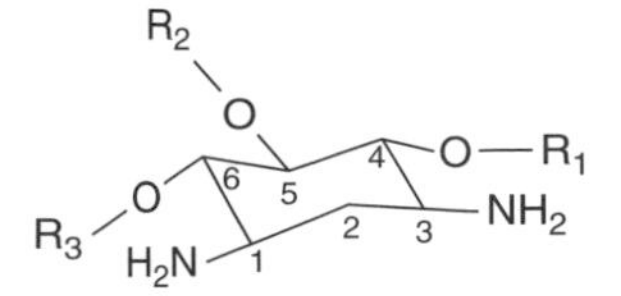

| Class | $R_1$ | $R_2$ | $R_3$ |
|---|---|---|---|
| I | sugar | H | sugar |
| II | sugar | sugar | H |
| III | sugar | H | H |
| IV | H | sugar | H |

FIGURE 10.26

A general structure of aminoglycosides containing 2-deoxystreptamine. The class with substituents at the 4- and 6-positions (class I) includes kanamycins, tobramycin, gentamicins, and sisomicin. The class with substituents at the 4- and 5-positions, here called class II, includes neomycins, paromomycins, lividomycins, and ribostamycin. Class III contains substituents only on the 4-position and includes apramycin. Class IV contains substituents only on the 5-position and includes destomycins and hygromycin B.

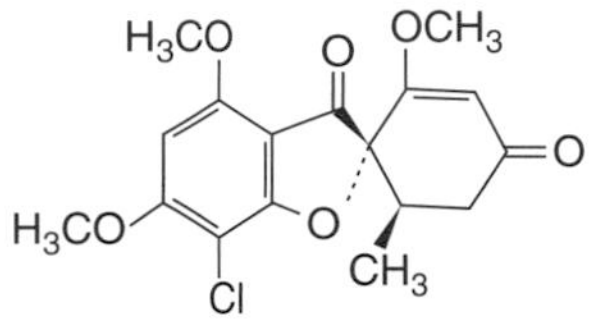

FIGURE 10.27

Griseofulvin.

microtubules and inhibits mitosis in fungi. It is used in human therapy as well as in agriculture.

Polyenes (Figure 10.28) have macrocyclic lactone structures similar to those in macrolides; however, the ring structure is larger, involving at least 26 atoms and at least three conjugated double bonds. Polyenes are produced by *Streptomyces* species and complex with sterols in the fungal cell membranes, thereby perturbing the membrane and increasing its nonspecific permeability. Because animal cell membranes also contain a sterol – cholesterol – polyenes generally have high levels of toxicity and therefore are used mostly in topical applications.

Several inhibitors of the biosynthesis of fungal cell wall polysaccharides have been identified among microbial products and developed into therapeutic compounds. Echinocandins, a type of lipopeptides, belong to this class (Figure 10.29). Another important class of antifungal agents is azoles, totally synthetic compounds that inhibit the synthesis of ergosterol in fungi.

## PRIMARY GOALS OF ANTIBIOTIC RESEARCH

The study and development of antibiotics certainly share some of the same aims as other areas of biotechnology. For example, it is always desirable to try to improve the yield of an antibiotic during fermentation and subsequent processing steps. However, much of the research on the antibiotics for human use has a very different focus. When we are dealing with the treatment of life-threatening infections, the cost of the drug is not the most important factor.

A very large fraction of antibiotic research is directed toward the development of new agents. New agents are needed because there are still microorganisms for which we do not yet possess truly effective agents with little or no side effects. These organisms include most fungi and viruses. Some bacteria, such as *P. aeruginosa*, also are intrinsically resistant to many antibiotics.

An even more vexing problem is the emergence of resistant strains among the organisms that were sensitive to antibiotics before the drugs became widely used. We will encounter numerous examples later in this chapter. This phenomenon tends to severely limit the useful life of any new antibiotic, requiring the pharmaceutical industry to come up with new compounds continually. The need is especially acute because of the following unfortunate situation. In any modern hospital, huge amounts of antibiotics are used

FIGURE 10.28

Filipin, an example of a polyene antibiotic.

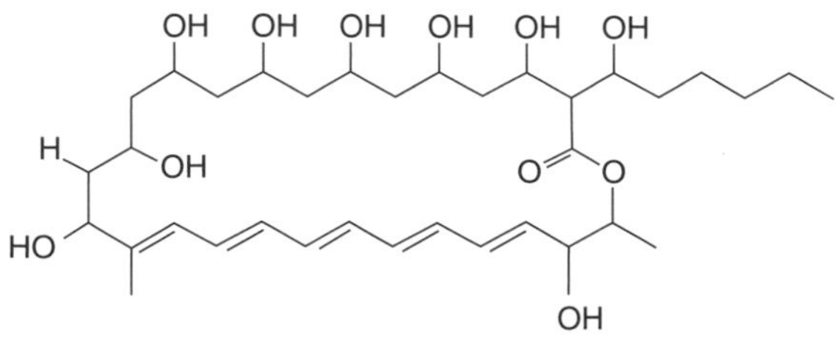

in the treatment as well as the prevention of infectious disease. As a result, the hospital environment becomes highly enriched for bacteria that are resistant to those antibiotics. At the same time, the immune and other defense mechanisms of the body are not functioning well in many hospitalized patients, who are thus especially vulnerable to hospital-acquired (nosocomial) infection by these resistant bacteria.

Although many different kinds of antibiotics have been developed, the same basic concepts and principles have governed the research in almost every case. For this reason, we will describe the steps in the development of only two classes of antibiotics: aminoglycosides and $\beta$-lactams. Readers who wish to learn the particulars of how other classes of antibiotics were developed should consult the references listed at the end of this chapter.

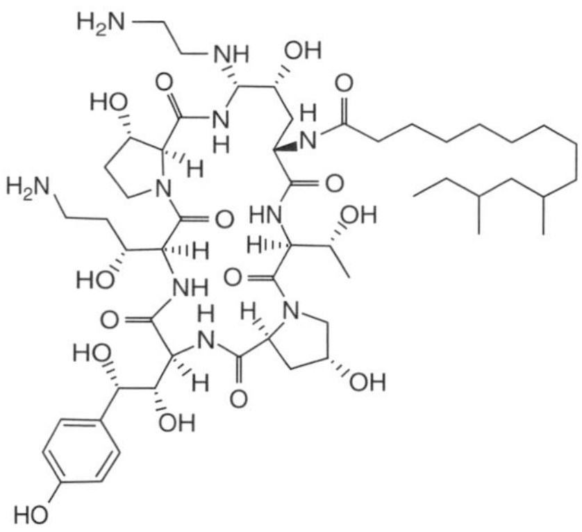

**FIGURE 10.29**

Caspofungin, an example of an echinocandin.

## DEVELOPMENT OF AMINOGLYCOSIDES

More than 150 naturally occurring aminoglycoside antibiotics have been isolated from culture filtrates of *Streptomyces* species and other members of the actinomycete line. These compounds characteristically contain an aminocyclitol moiety, which usually is either streptamine (as in streptomycin) or 2-deoxystreptamine (as in most other compounds; Figure 10.30).

Usually the aminocyclitol is further substituted by sugars, very often amino sugars. It appears that some of the enzymes involved in aminoglycoside biosynthesis do not have a very stringent substrate specificity, and it is common to find that a single strain produces a number of structurally related compounds. For example, one gentamicin-producing strain was found to produce more than 20 compounds, all related to each other in terms of structure, with only minor variations.

The history of the development of various aminoglycosides is very interesting, because it was influenced strongly by rational, synthetic approaches, yet at several points also was affected by the discovery of natural compounds of novel types that could not have been imagined, even by the most capable medicinal chemists. In order to follow this fascinating history, one has to have some understanding of the mechanism of aminoglycoside action and the mechanism of aminoglycoside resistance.

The aminoglycosides inhibit protein synthesis. The main target of action of aminoglycosides is the eubacterial – that is, the 70S – ribosome. Aminoglycoside molecules contain two or more amino groups, and their polycationic nature is important in their binding to the ribosome (or more correctly, the anionic 16S rRNA) as shown by x-ray crystallographic studies in recent years. Many of the antibiotics that inhibit protein synthesis (e.g., chloramphenicol, tetracycline, and erythromycin) are merely bacteriostatic; they stop the growth of bacteria but do not kill them. However, as already noted by Selman Waksman and coworkers in their first report on streptomycin in 1944, aminoglycosides are unusual among antibiotics in that they are clearly bactericidal. Once susceptible bacteria are exposed to aminoglycosides for several minutes, even extensive washing cannot revive them.

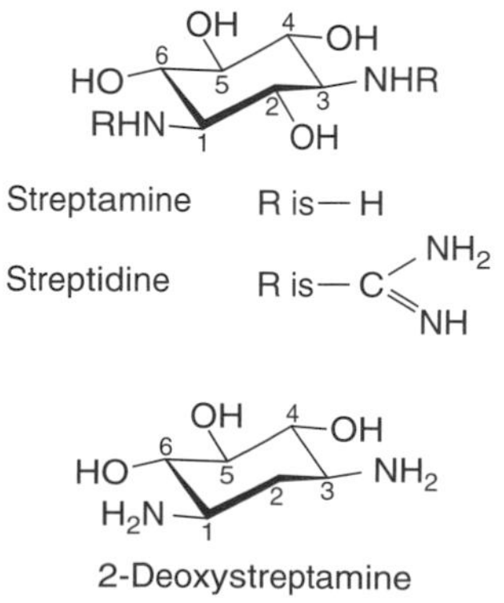

Streptamine R is—H

Streptidine R is—C(=NH)$NH_2$

2-Deoxystreptamine

**FIGURE 10.30**

Streptamine and 2-deoxystreptamine. In streptomycin, the two amino groups of streptamine are substituted by amidino groups, producing a cyclitol often called streptidine.

Apparently the irreversible nature of aminoglycoside action is related to their irreversible entry into the cells. A plausible scenario, proposed by Bernard D. Davis, is as follows. The small number of aminoglycoside molecules that enter the cell shortly after exposure to the antibiotic bind to the ribosomes and not only inhibit protein synthesis but also produce misreading and truncation of polypeptides. These abnormal polypeptides become inserted into membranes, and make them leaky. This is the irreversible step in aminoglycoside action. Because aminoglycosides are polycations and because there is a substantial membrane potential (interior negative) across the plasma membrane, large amounts of the drug are now "sucked in" through the leaky membrane by the interior-negative membrane potential. This massive influx will finally produce a complete cessation of protein synthesis.

In the laboratory, it is easy to isolate streptomycin-resistant mutants of *E. coli* by plating $10^9$ cells or more on streptomycin-containing media. Such mutants are usually altered in one of the proteins located in the decoding site, discussed in Box 10.2, of the small (30S) subunit of the ribosome. These mutants are thus altered in or near the target of streptomycin action. However, *E. coli* mutants of this type are practically never found in patients. Instead, among enteric bacteria such as *E. coli*, aminoglycoside-resistant strains isolated from patients almost always carry drug-resistance plasmids (R plasmids), which contain genes coding for the inactivation of aminoglycosides. The reason seems to be that in the intestine, so many bacteria coexist at such high population densities that there is a high probability of genetic exchange between them. (In contrast, streptomycin-resistant strains of *M. tuberculosis* are apparently ribosomal mutants.) The resistant *E. coli* inactivate the drug by enzymatically attaching various groups onto it and thus diminishing its polycationic character, needed both for entry into the cells and for interaction with the ribosomes. A group of enzymes called AACs (aminoglycoside acetyltransferases) transfer acetyl groups to amino groups, thereby decreasing the number of positive charges on the drug molecule. Another enzyme group, the APHs (aminoglycoside phosphoryltransferases), transfer phosphoryl groups to hydroxyl groups on the drug, thus adding negative charges and decreasing the drug's net positive charges. ANTs (aminoglycoside nucleotidyltransferases), some of which were once also called AADs (aminoglycoside adenylyltransferases), work in the same way, by adding nucleotide groups to hydroxyl groups on the drug. The way the kanamycin B molecule can be inactivated by various enzymes is illustrated in Figure 10.31.

The toxicity of aminoglycosides is probably associated with its polycationic nature. All aminoglycosides are somewhat toxic to humans and often damage the inner ear and kidney. Thus, the use of aminoglycosides has become limited to certain applications:

1. Streptomycin was the first drug that proved to be efficacious in the treatment of tuberculosis, caused by *M. tuberculosis*. Tuberculosis is a serious disease that claimed the lives of many young people through

**Molecular Mechanism of Aminoglycoside-Induced Misreading**

In the hypothesis proposed by B. D. Davis, misreading is the crucial element in the bactericidal action of aminoglycosides. The recent structural studies on ribosomes shed considerable light on this process. Protein synthesis by ribosomes is remarkably free of errors, in comparison with the prediction from the energy of pairing between the codon in mRNA and the anticodon in the tRNA. We now know that the short double-stranded RNA helix formed between the codon and anticodon interacts with three bases in the 16S rRNA – A1492, A1493, and G530 – producing a tight, stable structure (see color figure, part **B**). When there is a mismatch between the codon and the anticodon of tRNA, this interaction with the 16S rRNA is prevented and the mRNA–tRNA interaction becomes unstable, resulting in the release of the wrong amino-acyl-tRNA, a process that increases the fidelity in translation enormously. Now crystal structures have shown that aminoglycosides bind in an area very close to A1492, A1493, and G530 and shift the structure of the 16S rRNA to the one resembling the tight, stable one (see color figure, part **C**). This shift produces a stable complex even when there is a mismatch between codon and anticodon, strongly increasing the frequency of misreading.

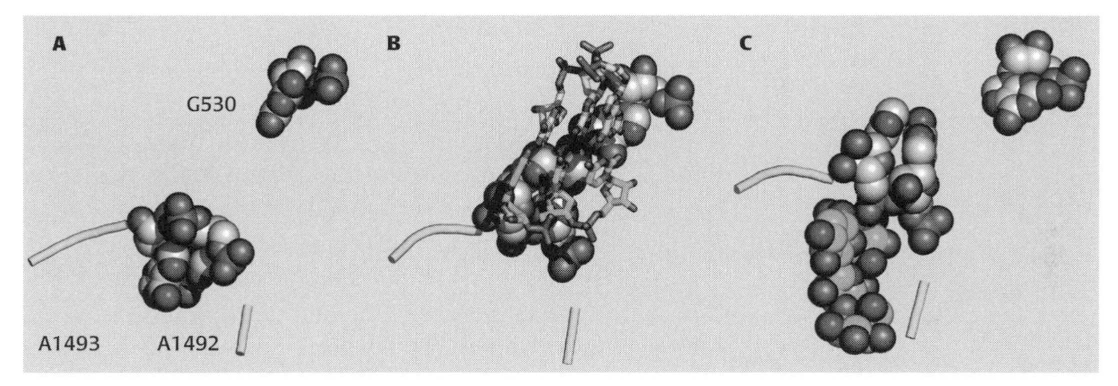


This figure shows the important parts of the A-site, the site on the 30S ribosomal subunit where aminoacyl-tRNA binds. In part **A**, the empty A-site is shown with the three nucleotide residues – G530, A1492, and A1493 of 16S rRNA (in CPK representation) – that play a prominent part in stabilizing the codon–anticodon pairs in part **B**. The loop at the bottom (painted in cyan) shows a small portion of the backbone of the 16S rRNA that contains A1492 and A1493. In part **B**, the codon (UUU for phenylalanine) of the mRNA (carbons as *green sticks*) and the anticodon (AAG) part of the phenylalanyl tRNA (carbons as *yellow sticks*) are shown. The codon and anticodon make Watson–Crick hydrogen-bonding interactions, as expected. This pairing, however, is not enough to explain the exceptional fidelity in translation, which is made essentially error-free because the codon–anticodon pairs are stabilized by the three nucleotides already shown in part **A**. Note especially that A1492 and A1493, which face toward the interior of the loop, now are flipped over and face the exterior. In part **C**, the structure of the complex between the 30S subunit and an aminoglycoside, paromomycin, is shown. Paromomycin inserts just underneath A1492 and A1495, causing these residues to flip outward, and creates a structure that resembles that containing the correct codon–anticodon pairs (**B**). This conformational alteration likely stabilizes the mRNA–tRNA complex, even when the tRNA does not contain the correct anticodon, strongly increasing the frequency of misreading. The figure was drawn with the PyMol program (DeLano Scientific, San Carlos, CA) using Protein Data Bank (PDB) files 1J5E, 1IBK, and 1IBM. For further details see Ogle J. M. (2001). Recognition of cognate transfer RNA by the 30S ribosomal subunit. *Science*, 292, 897–902. See the color page, this chapter.

**BOX 10.2**

the first half of the twentieth century, and a regimen including streptomycin used to be standard treatment until the early 1980s. However, more recently, rifampicin has replaced streptomycin because rifampicin can kill even those bacteria found in closed caseous lesions and because it can be taken orally.

**2.** Aminoglycosides are poorly absorbed from the intestinal tract when given orally. This means, however, that the aminoglycoside concentration *within* the intestinal tract remains high and that enteric bacteria are killed efficiently. They were thus used widely in the past in the treatment

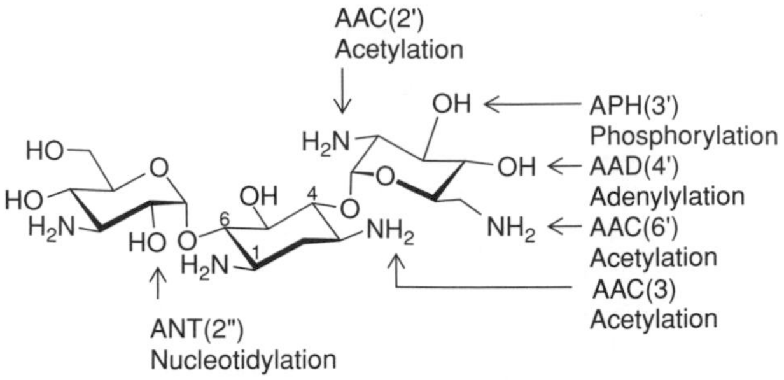

**FIGURE 10.31**

Kanamycin B can be inactivated in many different ways. The name of the enzyme and the nature of the modification reaction are shown. Note that the atoms in the sugar linked to the 4-position of the aminocyclitol (at the right end of the molecule) are numbered with primes and that the atoms on the 6-substituent sugar (at the left end of the molecule) are numbered with double primes. The enzymes include all those listed in Table 10.1 as being capable of inactivating kanamycin A, plus AAC(2′), which inactivates kanamycin B but not kanamycin A (because the latter lacks the 2′-amino group).



of gastrointestinal infections, especially in developing countries. Unfortunately, this use may have contributed heavily to the selection and dissemination of R plasmids.

**3.** Currently, the most important use of aminoglycosides is in the treatment of systemic infections caused by species of Gram-negative bacteria that are intrinsically resistant to most other antibiotics. Under these circumstances, we tolerate the inherent toxicity of these antibiotics because we do not have many other choices and because we are treating life-threatening infections. Aminoglycosides can cross even outer membranes of very low permeability, possibly because these polycations can bind to the anionic surface of the outer membrane, disorganize it, and thereby destroy the permeability barrier in a manner reminiscent of the action of polymyxin (page 335). The presence of inactivating enzymes, however, can make the drug totally ineffective, so one major thrust of the research in aminoglycoside development has been the search for compounds that are not inactivated significantly by commonly encountered enzymes.

## STREPTOMYCIN

Streptomycin (Figure 10.32) was discovered by Selman A. Waksman, a soil microbiologist interested in the antagonism between different groups of soil organisms. Waksman and coworkers were examining the products of *Streptomyces* species, a typical inhabitant of soil, and were looking for agents that were active against Gram-negative bacteria, for most of which penicillin G, the only commercially produced antibiotic at that time, was totally ineffective. (*Neisseria* spp. are susceptible to penicillin G, possibly because their porin channel favors anionic solutes, and thus constitute an exception among Gram-negative bacteria.) Their short note published in 1944 describing the discovery of streptomycin is impressive because they described not only the activity of streptomycin against both Gram-positive and

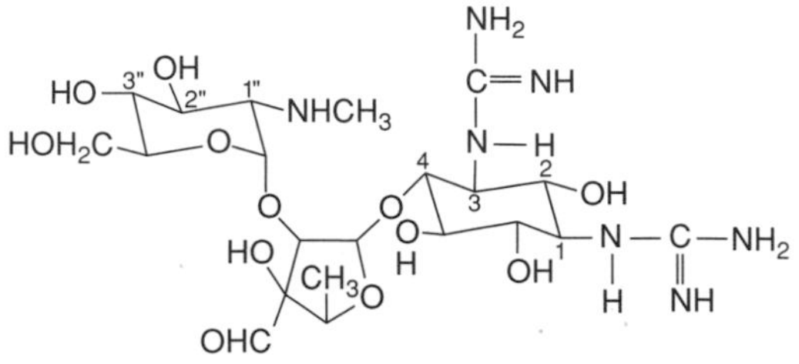

**FIGURE 10.32**

Streptomycin.

Gram-negative bacteria but also its bactericidal property and the fact that its production is affected greatly by the nature of the growth medium and the growth phase.

Although the toxicity of the compound prevented it from becoming the "penicillin for Gram-negative infections," the discovery of its activity against *M. tuberculosis* made this antibiotic extremely important. As mentioned earlier, streptomycin is no longer a first-line drug for the treatment of tuberculosis. However, because *M. tuberculosis* can develop resistance to first-line drugs such as rifampicin, streptomycin still remains an important agent for the treatment of cases caused by such resistant organisms.

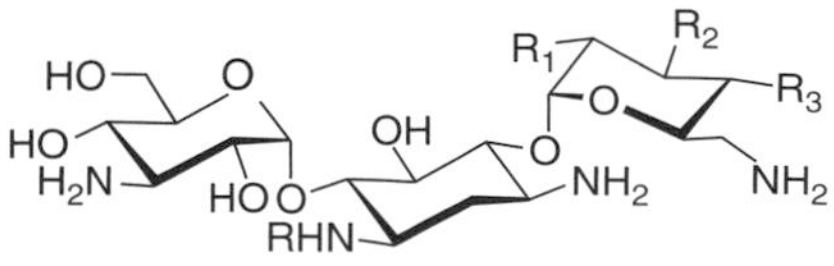

## DEVELOPMENT OF NEWER AMINOGLYCOSIDES

### Kanamycin

Kanamycin (Figure 10.33), the second commercially produced aminoglycoside, was discovered by Hamao Umezawa and coworkers in 1957 from the culture filtrate of another *Streptomyces* species. This was a timely discovery because kanamycin proved to be active on enteric bacteria that produce streptomycin-inactivating enzymes. When studies of plasmid-coded resistance to streptomycin showed that the antibiotic is inactivated through adenylylation or phosphorylation of its 3″-hydroxyl group, it became clear why the kanamycins remain fully active against strains that contain the streptomycin resistance plasmids. In streptomycin, the amino sugar is connected via another sugar to the 4-position of the aminocyclitol streptamine, but in kanamycins, the amino sugar is connected directly to the corresponding position of the aminocyclitol. Thus, in kanamycins, there is no group that corresponds to the 3″-hydroxyl group of streptomycin.

As mentioned earlier (see Figure 10.30), the aminocyclitol in kanamycin, 2-deoxystreptamine, is different from the one found in streptomycin, streptamine. 2-Deoxystreptamine is the aminocyclitol found in most other aminoglycosides either isolated or synthesized since the discovery of kanamycin. In kanamycin and most other compounds, 2-deoxystreptamine is substituted by amino sugars at both the 4- and 6-positions, although exceptions such as neomycin, ribostamycin, and butirosin (Figure 10.34), which are substituted at the 4- and 5-positions, exist.

Kanamycin proved to be significantly more active against most Gram-negative bacteria than streptomycin. However, it had poor activity against *P. aeruginosa*, partly because many strains of this organism contain enzymes, sometimes coded for by a chromosomal gene, that inactivate kanamycin.

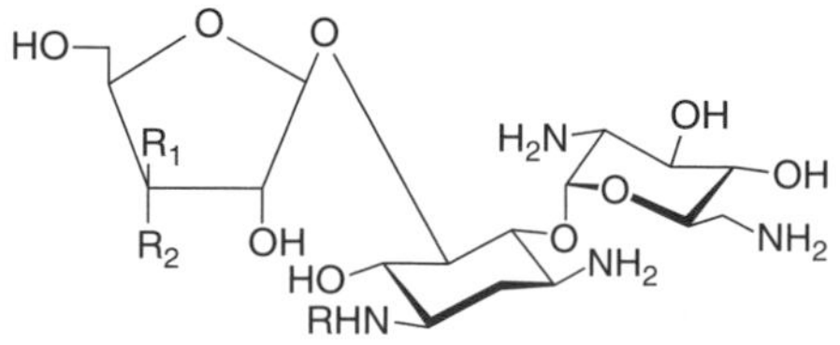

| Antibiotic | R | $R_1$ | $R_2$ | $R_3$ |
|---|---|---|---|---|
| Kanamycin A | H | OH | OH | OH |
| Kanamycin B | H | $NH_2$ | OH | OH |
| Tobramycin | H | $NH_2$ | H | OH |
| Dibekacin | H | $NH_2$ | H | H |
| Amikacin | HABA | OH | OH | OH |

FIGURE 10.33

Kanamycin A and its relatives. The molecule is drawn in such a way as to roughly represent its three-dimensional structure as determined by x-ray crystallography of the free drugs as well as the drug-16S rRNA complexes. Abbreviation: HABA, 2-hydroxy-4-aminobutyryl. [See for example, Bau, R., and Tsyba, I. (1999). Crystal structure of amikacin. *Tetrahedron* 55, 14839–14846 and Vicens, Q., and Westhof, E. (2002) Crystal Structure of a complex between the aminoglycoside tobramycin and an oligonucleotide containing the ribosomal decoding A site. *Chemistry & Biology* 9, 747–755.]

| | R | $R_1$ | $R_2$ |
|---|---|---|---|
| Ribostamycin | H | OH | H |
| Butirosin A | $NH_2$—$(CH_2)_2$—CHOH—CO | H | OH |

FIGURE 10.34

Ribostamycin and butirosin A.

### Semisynthetic Aminoglycosides: Dibekacin, Amikacin, and Netilmicin

In the 1960s, kanamycin was used extensively in Japan and elsewhere for the treatment of bacterial infections of the intestinal tract. This presumably has led to the spread of resistance plasmids with genes coding for kanamycin resistance. The occurrence of such plasmids increased dramatically in a short time in many countries. In one study of North American nurseries in 1971, for example, 70% of *E. coli* strains isolated were found to carry R plasmids coding for kanamycin resistance. This led to the development of aminoglycosides that would circumvent the mechanism of kanamycin inactivation, as detailed in this and the following sections. The successful development of these newer agents also resulted in the near disappearance of kanamycin from the clinic.

In Japan, H. Umezawa and coworkers found that the most frequent mechanism of such resistance was the phosphorylation of the drug at the 3′-position by APH(3′). Among kanamycin-resistant strains of Gram-positive bacteria, adenylylation at the 4′-position by AAD(4′) was often seen. Sumio Umezawa, a brother of H. Umezawa, then took the unprecedented "rational" approach of taking away the targets of these modification reactions, that is, the hydroxyl groups at 3′ and 4′. This was achieved by a chemical modification of the kanamycin molecule. The resultant compound, dibekacin (see Figure 10.33), retained the efficacy of kanamycin and was totally resistant to the kanamycin-modifying enzymes mentioned above, as expected; this agent has been used extensively, especially in Japan and neighboring countries such as South Korea and Taiwan.

A different approach for new aminoglycosides not modified by existing enzymes led to an excellent compound, amikacin (Figure 10.34). The scientists involved in amikacin development noted that a *Bacillus* species produces an aminoglycoside, butirosin A, which carries a 4-amino-2-hydroxybutyryl substituent on the 1-amino group of the 2-deoxystreptamine moiety of ribostamycin, a product of *Streptomyces* species. The presence of the unusually bulky substituent at the 1-position was a feature not seen before among products of *Streptomyces* and illustrates the benefit of going to phylogenetically distant organisms in search for compounds of novel structure. In any case, butirosin apparently was not modified by the APH(3′) enzyme that readily inactivated the ribostamycin. Taking hints from these data, scientists at Bristol-Myers Co. added the 4-amino-2-hydroxybutyryl substituent to the 1-amino group of kanamycin A, producing amikacin. The addition of this substituent had a remarkable effect. First, the modification of the 2″-hydroxyl group of kanamycin (by ANT(2″) enzyme) was abolished. This was not predicted from the behavior of butirosin because butirosin does not have a substituent sugar at the 6-position. However, in amikacin in aqueous solution, the position of the 2″-hydroxyl group is very close to that of the 1-amino group (Figure 10.35), and thus the bulky substituent on the latter group probably presents steric hindrance to the approach of the modifying enzyme. The second effect, making amikacin resistant to enzymes modifying the 3-amino group of the 2-deoxystreptamine residue, is also not difficult to understand because of the proximity of the modified 1-amino group. The

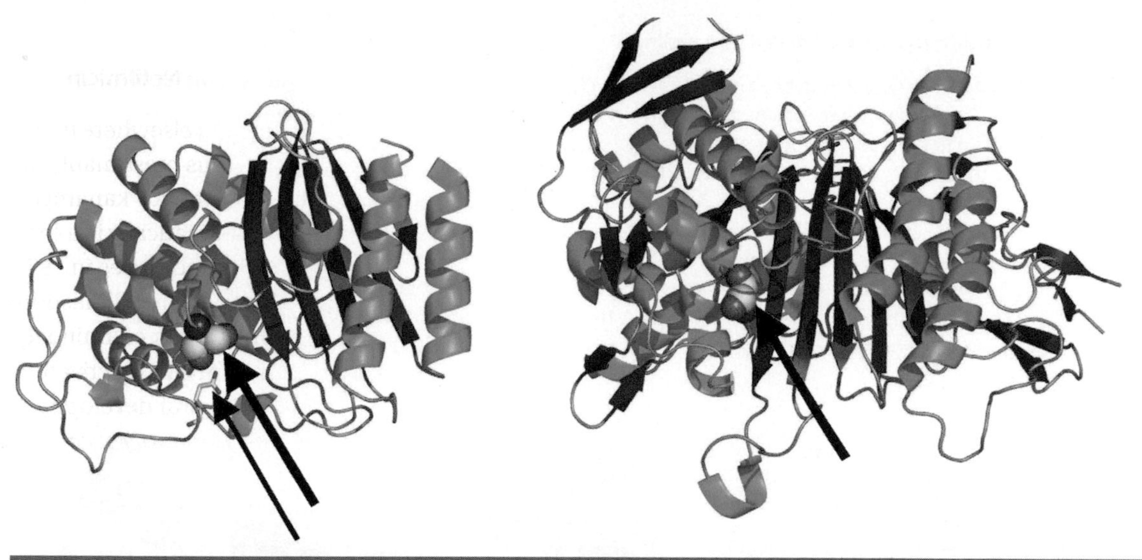

FIGURE 10.61

The folding patterns of a class A $\beta$-lactamase (TEM; left) and a penicillin-binding protein (PBP2x from *S. pneumoniae*; right). Note the very similar overall folding patterns. *Thick arrows* point to the active site serine residues (shown in the CPK model), and the *thin arrow* in the left panel shows the glutamic acid residue providing activated water molecules for the hydrolysis of the acyl-enzyme (shown in the stick model). This figure was drawn using PyMol based on PDB files 1BTL and 1QME.

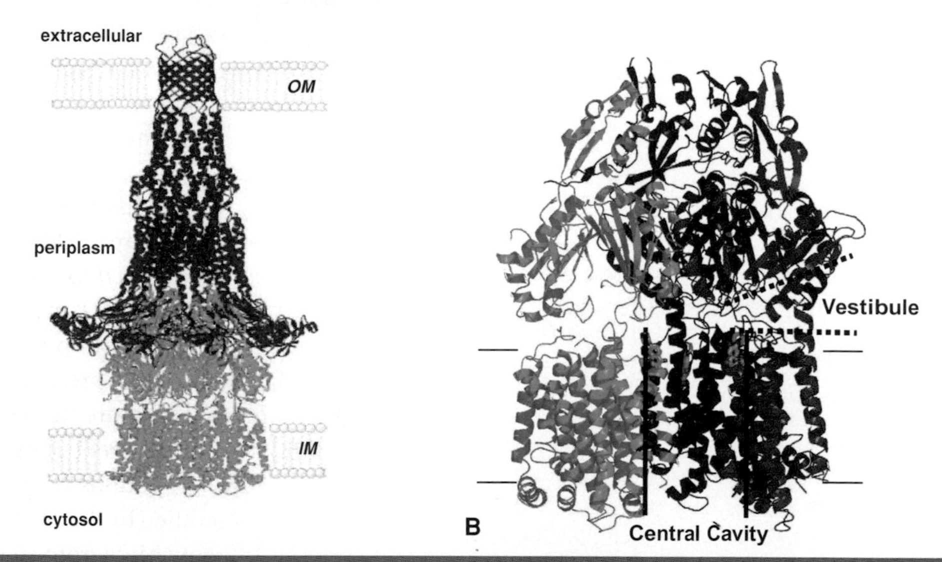


FIGURE 10.73

(**A**) The model of the AcrB-AcrA-TolC tripartite complex that exports drugs directly into the medium. The transmembrane domain of the AcrB pump trimer (green) is embedded in the cytoplasmic membrane, whereas its periplasmic domain is connected to the TolC channel (red) with a number of periplasmic AcrA linker proteins (blue). [From Eswaran, J., et al. (2004) Three's company: component structures bring a closer view of tripartite drug efflux pumps. *Current Opinion in Structural Biology*, 14, 741–747.] (**B**) AcrB trimer with the bound ciprofloxacin molecules. Each protomer is shown in cyan, mauve, and blue, and ciprofloxacin molecules are shown in the green stick models. The large central cavity is connected to the periplasm through vestibules between protomers. The proximal portion of the structure was cut away to reveal the presence of the vestibule, through which drug molecules are thought to be captured from the periplasm. The figure was drawn using PyMol with PDB coordinate 1OYE. Recent elucidation of X-ray structures of asymmetric trimers of AcrB (Murakami S., et al. (2006) Crystal structures of a multidrug transporter reveal a functionally rotating mechanism. *Nature*, 443, 173–179; Seeger M. A., et al. (2006) Structural asymmetry of AcrB trimer suggests a peristaltic pump mechanism. *Science* 313, 1295–1298) suggests how the substrate binds to a binding site within AcrB periplasmic domain and how it is exported into the TolC channel.

## Molecular Mechanism of Aminoglycoside-Induced Misreading

In the hypothesis proposed by B. D. Davis, misreading is the crucial element in the bactericidal action of aminoglycosides. The recent structural studies on ribosomes shed considerable light on this process. Protein synthesis by ribosomes is remarkably free of errors, in comparison with the prediction from the energy of pairing between the codon in mRNA and the anticodon in the tRNA. We now know that the short double-stranded RNA helix formed between the codon and anticodon interacts with three bases in the 16S rRNA – A1492, A1493, and G530 – producing a tight, stable structure (see color figure, part **B**). When there is a mismatch between the codon and the anticodon of tRNA, this interaction with the 16S rRNA is prevented and the mRNA–tRNA interaction becomes unstable, resulting in the release of the wrong amino-acyl-tRNA, a process that increases the fidelity in translation enormously. Now crystal structures have shown that aminoglycosides bind in an area very close to A1492, A1493, and G530 and shift the structure of the 16S rRNA to the one resembling the tight, stable one (see color figure, part **C**). This shift produces a stable complex even when there is a mismatch between codon and anticodon, strongly increasing the frequency of misreading.

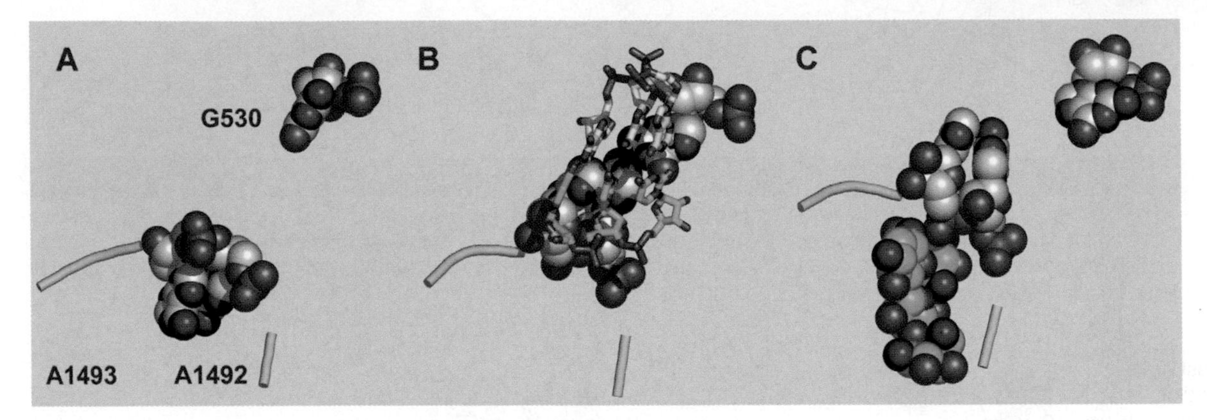


This figure shows the important parts of the A-site, the site on the 30S ribosomal subunit where aminoacyl-tRNA binds. In part **A**, the empty A-site is shown with the three nucleotide residues – G530, A1492, and A1493 of 16S rRNA (in CPK representation) – that play a prominent part in stabilizing the codon–anticodon pairs in part **B**. The loop at the bottom (painted in cyan) shows a small portion of the backbone of the 16S rRNA that contains A1492 and A1493. In part **B**, the codon (UUU for phenylalanine) of the mRNA (carbons as *green sticks*) and the anticodon (AAG) part of the phenylalanyl tRNA (carbons as *yellow sticks*) are shown. The codon and anticodon make Watson–Crick hydrogen-binding interactions, as expected. This pairing, however, is not enough to explain the exceptional fidelity in translation, which is made essentially error-free because the codon–anticodon pairs are stabilized by the three nucleotides already shown in part **A**. Note especially that A1492 and A1493, which face toward the interior of the loop, now are flipped over and face the exterior. In part **C**, the structure of the complex between the 30S subunit and an aminoglycoside, paromomycin, is shown. Paromomycin inserts just underneath A1492 and A1495, causing these residues to flip outward, and creates a structure that resembles that containing the correct codon–anticodon pairs (**B**). This conformational alteration likely stabilizes the mRNA–tRNA complex, even when the tRNA does not contain the correct anticodon, strongly increasing the frequency of misreading. The figure was drawn with the PyMol program (DeLano Scientific, San Carlos, CA) using Protein Data Bank (PDB) files 1J5E, 1IBK, and 1IBM. For further details see Ogle J. M. (2001). Recognition of cognate transfer RNA by the 30S ribosomal subunit. *Science*, 292, 897–902.

**BOX 10.2**

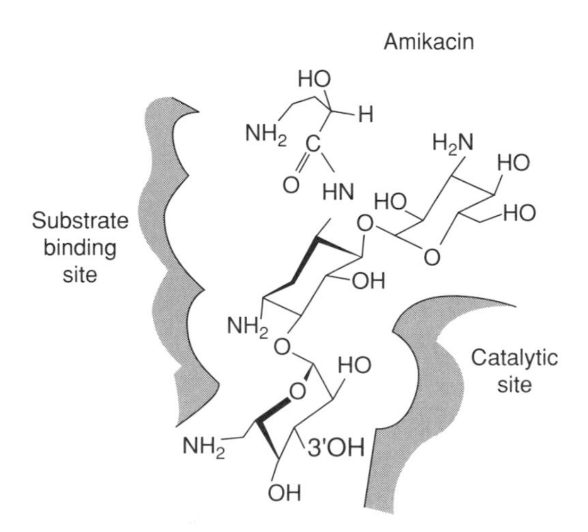


**FIGURE 10.35**

A possible mechanism for the stability of amikacin against enzymes that would modify the 3′-, 4′-, or 6′-position of the aminoglycoside. Kanamycin A binds to the APH(3′) enzyme by using the amino groups at the 1- and 3-positions of the 2-deoxystreptamine; thus, the catalytic site of APH(3′) comes to be positioned just above the 3′-OH group (above). In comparison, the binding of amikacin to the enzyme is skewed because the amino group of the 4-amino-2-hydroxybutyryl substituent is displaced in relation to the 1-amino group of the 2-deoxystreptamine. This skewed binding holds the 3′-OH group away from the catalytic site, thus making amikacin resistant to the APH(3′) enzyme.

Redrawn based on artwork from the first edition (1995), published by W.H. Freeman.

third effect, which made amikacin relatively resistant to enzymes modifying the groups at the 3′- and 6′-positions, was actually expected from the resistance of butirosin to APH(3′) but is somewhat difficult to explain by theoretical considerations because these groups are located very far away from the 1-amino substituent. One explanation is that the 4-amino group of the aminohydroxybutyl substituent and the 3-amino group of 2-deoxystreptamine may fit into the binding sites of the enzyme that normally bind the 1-amino and 3-amino groups of kanamycin, and this may move the potential modification sites far away from the active sites of the enzyme. In any case, amikacin and fortimicin A are the compounds that currently are resistant to probably the widest range of aminoglycoside-modifying enzymes (Table 10.1).

Because the continued use of antibiotics results in the spread of resistance factors that can inactivate an increasingly wide range of compounds, finding new compounds that are resistant to all these enzymes is becoming more and more difficult. It is thus unrealistic to expect to find a natural compound that, without further chemical modification, already has all the desirable properties. Thus, with more recent compounds, the role of the natural

**TABLE 10.1 Susceptibility of Aminoglycosides to Common Modifying Enzymes**

| | | Aminoglycoside | | | | | | |
|---|---|---|---|---|---|---|---|---|
| **Enzyme** | **Occurrence**[a] | **Streptomycin** | **Kanamycin A** | **Dibekacin** | **Tobramycin** | **Amikacin** | **Gentamicin $C_1$** | **Netilmicin** |
| AAC(3)-I | 11% | – | (–) | [–] | [–] | (–) | + | (–) |
| AAC(3)-II | 60% | – | (–) | + | + | (–) | + | + |
| AAC(3)-III | 4% | – | + | + | + | (–) | + | (–) |
| APH(3′)-I | 46% | – | + | – | – | [–] | – | – |
| ANT(4′)-I | [b] | – | + | +[c] | + | + | – | – |
| AAC(6′)-I | 60% | – | + | + | + | + | [–] | + |
| AAC(6′)-II | [d] | – | + | + | + | (–) | + | + |
| AAC(6′)-APH(2″) | [e] | – | + | + | + | + | + | + |
| ANT(2″)-I | 15% | – | + | + | + | (–) | + | (–) |
| ANT(3″)-I | [f] | + | – | – | – | – | – | – |
| APH(3″)-I | | + | – | – | – | – | – | – |

[a] This column shows the percentage of general Gram-negative bacterial strains (excluding *Serratia, Acinetobacter*, and *Pseudomonas*) that show the particular phenotype expected of the enzyme-carrying strain among recent clinical isolates showing resistance to at least one aminoglycoside (excluding streptomycin). Based on Shaw, K. J., et al. (1993) Molecular genetics of aminoglycoside resistance genes and familial relationships of the aminoglycoside-modifying enzymes. *Microbiological Reviews*, 57, 138–163.
[b] Present in 30% of resistant *S. aureus* strains.
[c] Although dibekacin lacks the 4′-OH group, the enzyme modifies the 4″-OH group instead.
[d] Present in 48% of resistant *P. aeruginosa* strains.
[e] This hybrid enzyme, made by fusing AAC(6′) to ANT(2″), is present in 99% of the resistant *S. aureus* strains.
[f] In one study, 59% of the strains that were resistant to other aminoglycosides were also resistant to streptomycin, and of those, 56% carried this gene.
Symbols: +, modification resulting in resistance; –, the target functional group is absent and the drug is not modified; (–) the target functional group is present but is not modified; [–], the drug is modified *in vitro*, but the rate (or affinity) is insufficient to produce significant resistance.

products has been either to serve as examples (as in the case of butirosin) or to serve as the starting materials for further chemical modification (as with kanamycin for the production of amikacin and dibekacin).

An example of this phenomenon is also seen in the development of netilmicin (Figure 10.36). The starting material in this case was sisomicin, which is produced not by *Streptomyces* but by another actinomycete, *Micromonospora*. This compound is unusual in that the amino sugar linked to the 4-position of the 2-deoxystreptamine is an unsaturated sugar and lacks the major targets of modification, such as the hydroxyl groups at the 3′- and 4′-positions. Thus, sisomicin is an excellent antibiotic and is produced commercially. However, the addition of a bulky substituent on the 1-amino group resulted in the blocking of another potential modification

**FIGURE 10.36**

Sisomicin and netilmicin.

Redrawn based on artwork from the first edition (1995), published by W.H. Freeman.

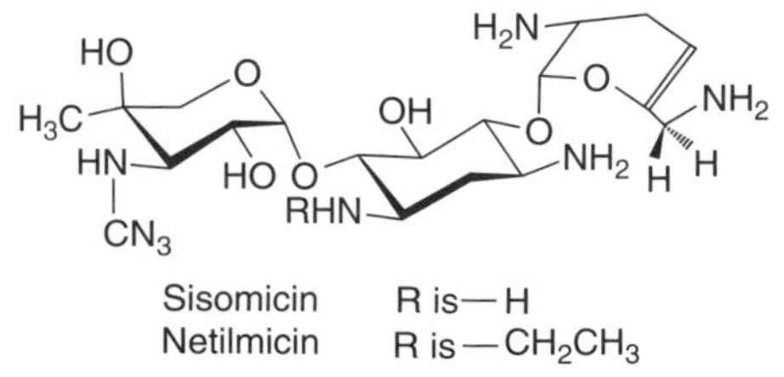

Sisomicin R is—H
Netilmicin R is—$CH_2CH_3$

site, the 2″-hydroxyl group, just as we have seen with amikacin, and produced netilmicin, which retains all the good properties of sisomicin and in addition is not inactivated by ANT(2″) or APH(2″).

### Tobramycin and Gentamicin C: Nature's Work as a Medicinal Chemist

The "rational" approach of eliminating the target sites of modification and introducing bulky substituents is not the only route for the production of drugs that would withstand the inactivating enzymes. At about the same time medicinal chemists were producing dibekacin and amikacin by chemical modification, screening for natural microbial products led to the isolation of a compound that showed good activity against many kanamycin-resistant Gram-negative bacteria. This compound, tobramycin, already lacks the 3′-hydroxyl group, one of the prime targets of modification of kanamycin (see Figure 10.33). Nature had long ago accomplished something that the best medicinal chemists were trying to do! As seen in Table 10.1, tobramycin is as inert to enzymatic modification as is dibekacin. It is also active against *P. aeruginosa,* in contrast to kanamycins that are inactivated by APH(3′). Because it is a natural product, it is less expensive than some of the semisynthetic compounds, and it is still used widely, especially in the United States.

Another natural product is gentamicin C (Figure 10.37). If we did not know that it is synthesized by *Micromonospora* species, we would surely believe it to be a semisynthetic compound produced by the medicinal chemists, because it lacks both the 3′- and 4′-hydroxyl groups Umezawa's

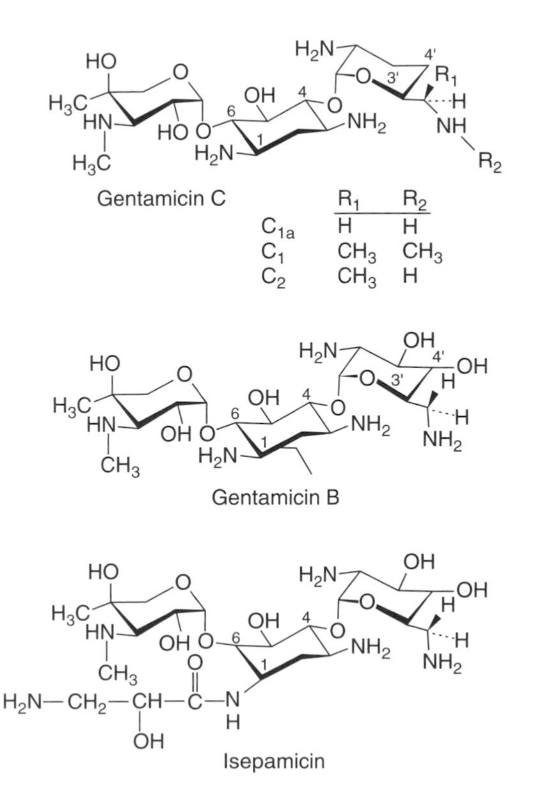


**FIGURE 10.37**
Gentamicins and isepamicin.

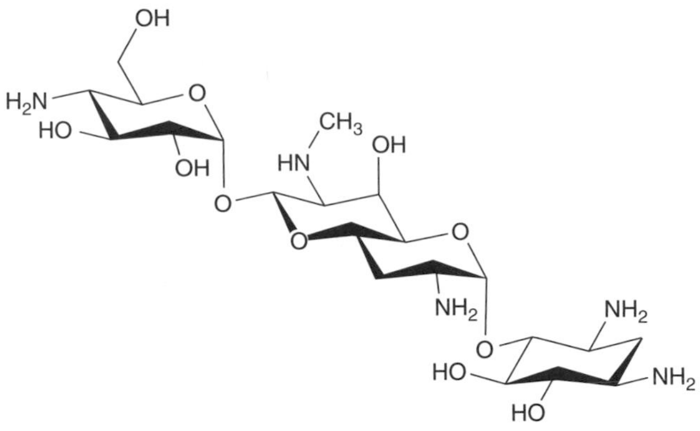

FIGURE 10.38
Apramycin.

group removed, chemically, from kanamycin. Gentamicin C is not inactivated by most of the enzymes modifying the amino sugar linked to the 4-position (and therefore is also active against *P. aeruginosa*), although it is still susceptible to the enzymes modifying the aminocyclitol (AAC(3)) or the amino sugar linked to the 6-position (ANT(2″)).

Gentamicin C is produced together with gentamicin B, which does contain both the 3′- and 4′-hydroxyl groups and is thus susceptible to inactivation by common enzymes. However, a semisynthetic derivative, isepamicin (Figure 10.37), which has on its 1-amino group a substituent similar to that present on amikacin – that is, the 3-amino-2-hydroxypropyl group – is not modified by most commonly occurring enzymes.

## Apramycin and Fortimicin A

Apramycin and fortimicin A have structures very different from the common aminoglycosides. In apramycin, only the 4-position of the aminocyclitol is substituted (Figure 10.38). Apramycin is inert to inactivation by most enzymes, but it is still inactivated by at least one type of AAC(3). In many countries, its application is limited to veterinary use; however, this practice appears to select for R plasmids coding for AAC(3), which, when transferred to human pathogens, can inactivate aminoglycosides used for human therapy.

Fortimicin A (Figure 10.39), introduced into clinical use in Japan, has an aminocyclitol moiety different from 2-deoxystreptamine. It is based on 1,4-diaminocyclitol, which has only one amino sugar substituent at the 6-position. The compound was isolated from *Micromonospora* in 1977. Because its structure is radically different from that of all other aminoglycosides currently in use, it is not inactivated by most of the aminoglycoside-modifying enzymes specified by the resistance plasmids that are commonly found at present among Gram-negative bacteria in hospitals. AAC(3)-I constitutes a rare exception, but the enzyme modifies the 1-amino group rather than the 3-position of the cyclitol, where an amino group does not exist. Fortimicin A is not used in the United States, presumably because there, amikacin is still active against more than 90% of the Gram-negative isolates

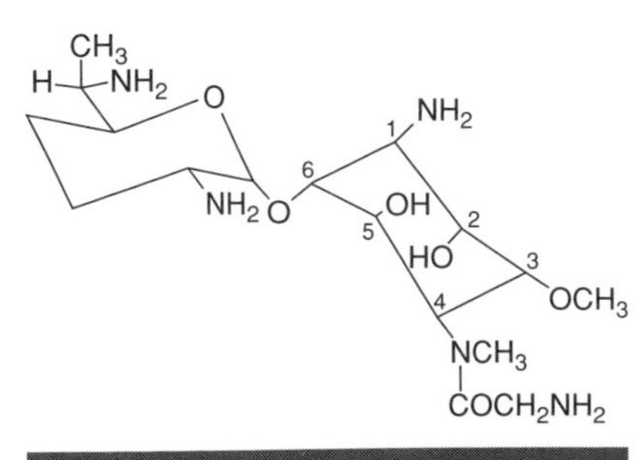


FIGURE 10.39
Fortimicin A.

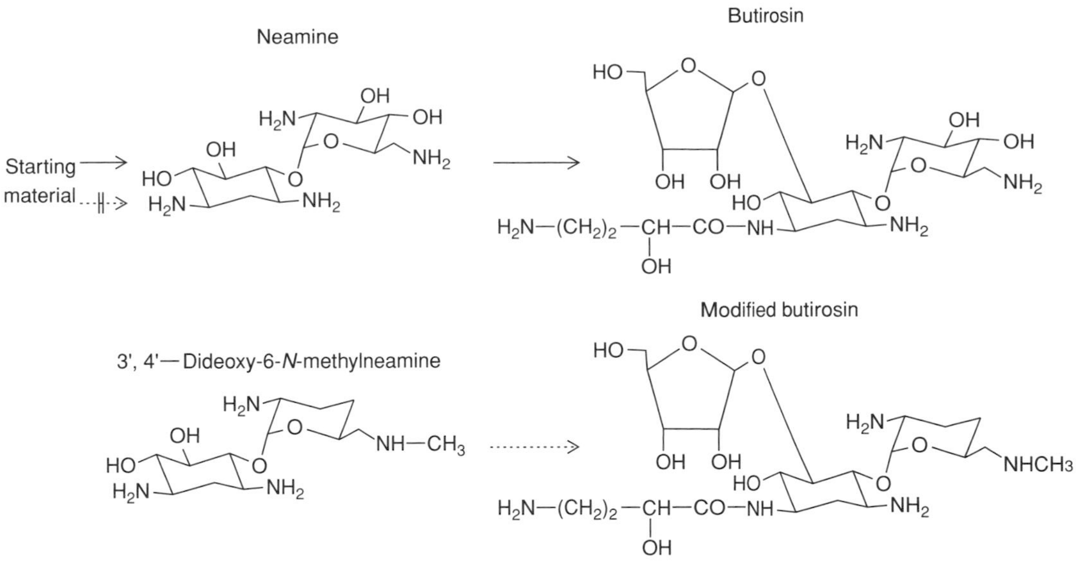

resistant to other aminoglycosides and also because fortimicin A does not have good activity against *P. aeruginosa*. In contrast, it may become more important in Japan, where amikacin has been used heavily in the past and, consequently, amikacin-resistant strains comprise more than 35% of aminoglycoside-resistant isolates (see below).

**FIGURE 10.40**

A new aminoglycoside produced by a mutasynthesis approach. In the wild-type organism (above), an intermediate, neamine, is converted into butirosin. Scientists used a mutant *Bacillus* strain blocked in the biosynthesis of neamine (*broken arrow*) and fed it a synthetic analog of neamine (below). As a result, the bacteria synthesized an analog of butirosin that lacked several potential inactivation sites.

Redrawn based on artwork from the first edition (1995), published by W.H. Freeman.

## USE OF MUTASYNTHESIS FOR DEVELOPING NEW AMINOGLYCOSIDES

Mutasynthesis, which was proposed in the 1960s, is a method for developing new aminoglycosides that takes advantage of the fact that some of the enzymes involved in antibiotic biosynthesis may not have an absolutely strict substrate specificity and may incorporate structural analogs of an intermediate into the final product. If a wild-type organism is used to make the antibiotic, the normal intermediate is present in its cells and competes successfully against the analog supplied. Thus, one must use mutant strains that are unable to synthesize the natural substrate. In one example, chemically synthesized 3′,4′-dideoxy-6-*N*-methylneamine was added to a *Bacillus* strain that was defective in the biosynthesis of neamine and therefore required neamine in the medium in order to synthesize the aminoglycoside butirosin. The resulting compound (Figure 10.40) was missing the 3′-hydroxyl group and had a bulky substituent on the amino nitrogen at the 6′-position. Thus, it was resistant to two modifying enzymes to which butirosin is susceptible, APH(3′) and AAC(6′).

## ORIGIN AND PATTERN OF AMINOGLYCOSIDE RESISTANCE

Table 10.1 lists only the most common aminoglycoside-modifying enzymes. Still, the number and variety of these enzymes are almost bewildering. Most of them are coded for by genes on resistance plasmids, although a few

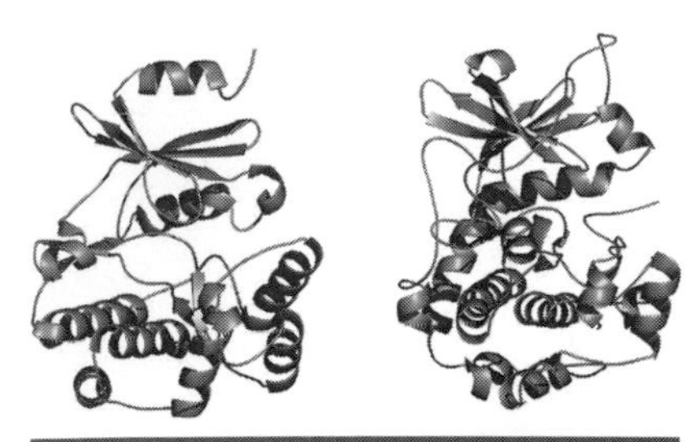

FIGURE 10.41

Comparison of the three-dimensional folding patterns of APH(3′)-IIIa (left) and cAMP-dependent protein kinase (right). The figures were drawn with PyMol using PDB files 1J7L and 1ATP.

bacterial species, such as *Serratia marcescens* and *P. aeruginosa*, are known to contain many strains that possess chromosomal genes coding for some of them.

Even the strong selective pressure imposed by the extensive clinical use of aminoglycosides is unlikely to account for the evolution of so many enzymes so quickly. Thus, Julian Davies suggested that the original sources of these genes were probably the organisms producing the target antibiotics; he reasoned that the antibiotic-producing organisms had to protect themselves against their own products. The cloning and sequencing of genes for aminoglycoside-modifying enzymes from resistance plasmids and from the producing actinomycete strains have, in fact, revealed an extensive homology between the DNA sequences of some genes (e.g., AAC(3) and APH(3′)) from these two sources. The crystallization of several aminoglycoside-inactivating enzymes further showed that APH(3′) and some protein kinases have remarkably similar folding patterns (Figure 10.41), suggesting that the modifying enzyme has derived from protein kinases, although there is no significant sequence similarity. The folding pattern of AAC(3) is also similar to that of protein acetylases, such as histone acetyltransferase.

The cloning of various genes for aminoglycoside-modifying enzymes now enables researchers to classify the enzyme, at the gene level, by using DNA probe hybridization assay (see Chapter 3). Such a study using bacterial strains collected between 1987 and 1991 showed that the most prevalent enzymes found in Gram-negative bacteria were AAC(3)-I, AAC(6′)-I, and APH(3′)-I; see Table 10.1. Because assaying for enzyme activity and determining the specificity of an enzyme are not easily carried out for a large number of strains, little data existed on the occurrence of various enzymes before the DNA probes became available. One source is a Bristol-Myers study, conducted in the 1970s, that analyzed the enzyme patterns for aminoglycoside-resistant isolates as a service to hospitals in the United States. The most prevalent enzymes among Gram-negative bacteria were APH(3′), 58%; AAC(6′), 38%; and AAC(3), 18%. Thus, there was no drastic alteration in the relative abundance of various enzymes up to 1990. However, a more recent study showed an increase in the enzymes producing amikacin resistance, such as AAC(6′)-III and AAC(6′)-IV. Researchers have also noted that more recently isolated bacteria usually produce more than one aminoglycoside-inactivating enzyme, a situation that did not occur frequently in the past.

In a study published in 1985, aminoglycoside-resistant bacterial strains obtained in different countries were compared. In strains from Japan, Taiwan, and South Korea, where kanamycin and its derivatives, especially dibekacin and amikacin, were used extensively, 78% of the resistant strains contained AAC(6′), which inactivates these kanamycin derivatives. In contrast, in strains from sources in the United States, where gentamicin was quite popular, the most commonly found enzyme (42%) was ANT(2″), which inactivates gentamicin very efficiently. Furthermore, AAC(3), which is another efficient inactivator of gentamicin but is often ineffective against kanamycin derivatives, was present in 17% of the U.S. isolates but was virtually absent from the isolates from the Far East. Analyzed in terms of the resistance pattern, 99% of the Far Eastern isolates were resistant to dibekacin, probably the most popular aminoglycoside in that region but not licensed for use in the United States, whereas 92% of the U.S. isolates were resistant to gentamicin. These data show beyond any doubt that the usage pattern of antibiotics contributes significantly to the prevalence of various types of drug resistance genes in the local microbial population.

How prevalent is aminoglycoside resistance? Physicians usually must decide on which antibiotic to use before they get data on the identity and resistance pattern of the causative organisms; thus, they are unlikely to use a drug if the prevalence of resistant strains reaches 10% or more in a common pathogen species. Perhaps because aminoglycoside use is generally limited to the treatment of life-threatening infections due to its toxicity, the prevalence of resistant strains seems rather low. Thus, in studies using large numbers of isolates from the United States and Europe in 1997, only 4% of *E. coli* strains were gentamicin-resistant. However, these figures are somewhat deceptive, as the prevalence depends on the bacterial species as well as the locale. Thus, in the same study, gentamicin-resistant strains comprised 12%, 10%, 13%, and 21% of *Klebsiella*, *Enterobacter*, *Serratia*, and *P. aeruginosa* isolates, respectively, from Europe, figures suggesting that gentamicin is not a very useful drug any more for treating Gram-negative infections in general. Furthermore, it is alarming that the fraction of gentamicin-resistant strains showed a clear increase over the study carried out in 1990. In a 1983 survey involving hundreds of U.S. hospitals, the 22% frequency of gentamicin-resistant *P. aeruginosa* already reflected a large increase from the figure of only 8% of isolates in 1977. In a single large London hospital, in 1976 only 3% of *Acinetobacter* isolates were gentamicin resistant, but the frequency increased to between 20% and 39% in the period of 1978 to 1983.

In the 1980s, some U.S. hospitals adopted a program that emphasized the use of amikacin instead of gentamicin. Interestingly, after a few years of amikacin use, the prevalence of gentamicin resistance decreased significantly, confirming again the relationship between antibiotic use and the prevalence of resistance plasmids. However, hospitals then had to begin using gentamicin again to avoid an increase in amikacin resistance, and the long-range outcome of such a "recycling" strategy is not clear at present.

## AMINOGLYCOSIDES USED FOR NONMEDICAL PURPOSES

### Hygromycin B

Hygromycin B (Figure 10.42), like many compounds discussed earlier, is based on 2-deoxystreptamine. However, unlike many other aminoglycosides, the sugar substituent occurs not on the 4- and 6-positions but on the 5-position of the aminocyclitol. Furthermore, the substituent is a 6-hydroxy-6-amino sugar linked to a neutral sugar (Figure 10.42). Interestingly, this compound inhibits not only prokaryotic ribosomes but also eukaryotic ribosomes. This implies strong toxicity, and the practical use of the antibiotic

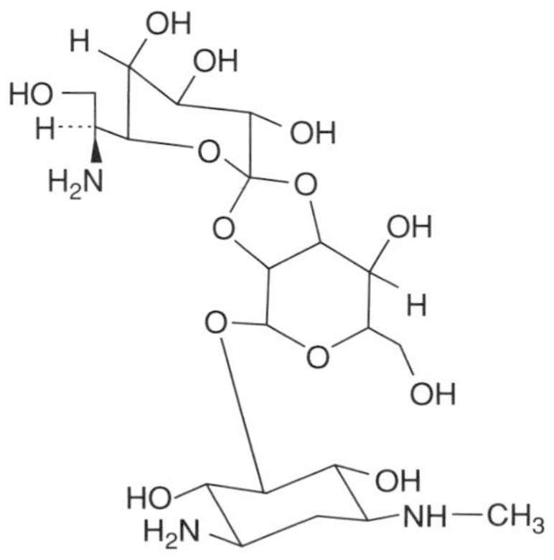

**FIGURE 10.42**
Hygromycin B.

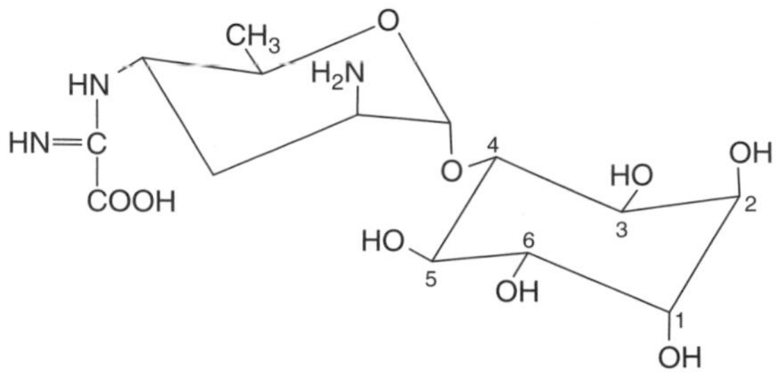

FIGURE 10.43

Kasugamycin.

has been limited to occasional use as an antihelmintic (because worms are eukaryotic) in poultry and swine. In the laboratory, hygromycin has been very useful. If a plasmid contains a hygromycin resistance gene, then the plasmid-containing cell, regardless of whether it is prokaryotic or eukaryotic, can be selected for by the use of this drug. This method has found wide use in shuttle cloning between bacteria and plant cells, for example.

#### Kasugamycin

Kasugamycin (Figure 10.43) does not contain any aminocyclitol, and the cyclitol found is an uncharged D-inositol. The net charge of the antibiotic is only +1, in contrast to all other aminocyclitol antibiotics that are polycationic. Kasugamycin is known to inhibit bacterial protein synthesis, but apparently by a mechanism targeted to translation initiation and different from that used by other aminocyclitol antibiotics. This compound is important in controlling the infection of rice caused by the fungus *Pyricularia oryzae*.

## DEVELOPMENT OF THE β-LACTAMS

β-Lactams are probably the most important class of antibiotics, for several reasons. For one thing, penicillin, the first antibiotic isolated and characterized, is a β-lactam. Since its discovery, an enormous amount of work has been done with the compounds of this class. Second, the inherent toxicity of β-lactams against higher animals is extremely low, yet they are often very effective at killing infecting bacteria. This is because they inhibit peptidoglycan synthesis, a reaction that does not exist in the eukaryotic world. Consequently, they are used widely and represent more than half the monetary value of all the antibiotics used for the treatment of human diseases worldwide. Penicillins and cephalosporins are the classic members of the β-lactam group. In recent years, however, compounds with novel nuclei have been added, such as monobactams and carbapenems.

### PENICILLIN G

The penicillin-producing organism in Alexander Fleming's laboratory was a true fungus, *Penicillium notatum*. During World War II, mycologists at the

Northern Regional Research Laboratory of the U.S. Department of Agriculture found that a strain of *Penicillium chrysogenum* there produced far more penicillin than Fleming's original isolate of *P. notatum*. Further improvements in penicillin-producing strains are described later in this chapter.

**TABLE 10.2 Microorganisms responsible for hospital-acquired infections leading to bacteremia in Boston City Hospital**

| Causative organism | Percent of all bacteremia cases in year 1935 | 1969 |
|---|---|---|
| Streptococci | 62 | 12 |
| Staphylococci | 19 | 19 |
| Enterococci | 0 | 7 |
| *E. coli* | 11 | 11 |
| *Enterobacter/Klebsiella/Serratia* | 0 | 22 |
| *Proteus/Providencia* | 4 | 10 |
| *P. aeruginosa* | 1 | 8 |
| Other Gram-negatives | 3 | 3 |
| Fungi | 0 | 8 |

*Source*: McGowan, Jr., J. E. (1985). Changing etiology of nosocomial bacteremia and fungemia and other hospital-acquired infections at Boston City Hospital. *Reviews of Infectious Disease*, 7, S357–S370.

Because penicillin and other antibiotics have so dramatically changed the nature of the infectious diseases we suffer from, it is not easy nowadays to imagine the impact penicillin had at the time of its introduction. Before the introduction of antibiotics, the major killer diseases were those caused by Gram-positive cocci such as *Streptococcus pneumoniae* (pneumonia), *Streptococcus pyogenes* (scarlet fever, nephritis), and *Staphylococcus aureus* (various purulent infections, sepsis). In the early antibiotic era, some of the most troublesome bacterial infections became those caused by Gram-negative bacteria, which are more resistant to antibiotic treatment, mainly because of their rather impermeable outer membrane and the multidrug efflux pumps of wide substrate specificity. This situation is a direct result of the near-complete success of antibiotic therapy, at least in its early days, in curing most of the "classic" bacterial infectious diseases. Perhaps this trend can best be appreciated by comparing the principal causative organisms of hospital-acquired serious infections (resulting in bacteremia – i.e., multiplication of bacteria in the bloodstream) in the mid-1930s and the late 1960s (Table 10.2). Clearly, cases caused by *Streptococcus* species in the preantibiotic era decreased drastically, and their role was later taken over by Gram-negative rods such as *Enterobacter, Klebsiella, Serratia, Proteus, Providencia,* and *P. aeruginosa*. Penicillin's fame as a "miracle drug" was well deserved because the drug was extremely active against those Gram-positive bacteria that were most important at the time of its introduction. (Gram-negative bacteria currently play a somewhat diminished role in septicemia, presumably because of the availability of antibiotics effective against this group of organisms.)

In the early days of penicillin research, the nature of growth medium was found to influence the substituent at position 6 (Figure 10.44). Thus in the United States, where corn-steep liquor containing phenylacetic acid was used as a carbon source in fermentation, the main product was penicillin G, with its benzyl side chain. In contrast, the product that British workers obtained during their early efforts was penicillin F, with a 2-pentenyl side chain. Various compounds were then tested for their possible incorporation into the penicillin molecule, and penicillin V, with a phenoxymethyl side chain, was produced by the addition of phenoxyacetic acid rather than phenylacetic acid to the medium. Penicillin V better withstands the acidic pH in the stomach, and it can be administered orally.

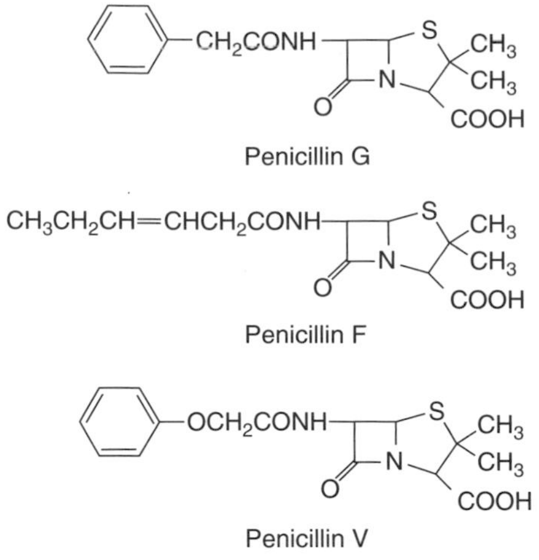

Penicillin G

Penicillin F

Penicillin V

**FIGURE 10.44**

Penicillins found in different culture filtrates.

Statistics from the mid-1970s show that about 20% of the penicillin produced by fermentation was penicillin V and 80% was penicillin G. Only about half the penicillin G was used without further modification, the rest being used as the starting material for production of the semisynthetic penicillins (and cephalosporins) as described below.

## SEMISYNTHETIC PENICILLINS

Although penicillin was indeed an amazingly effective drug, improvement was still desired in two areas. First, after just a few years of its widespread use, some of the *S. aureus* strains had become penicillin-resistant. They produced a penicillinase, a penicillin-hydrolyzing enzyme (Box 10.3), that was coded by a gene on a plasmid. Although the plasmid was not easily transferred from one strain to another, the resistant staphylococci were perfectly suited to cause hospital-acquired infections. Staphylococci are resistant to

**Classification of $\beta$-Lactamases**

Most bacteria that are resistant to penicillins and cephalosporins owe this phenotype to the production of $\beta$-lactamases. They have been classified in detail [e.g., see Bush K., Jacoby G. A., and Medeiros, A. A. (1995) A functional classification scheme for beta-lactamases and its correlation with molecular structure. *Antimicrobial Agents and Chemotherapy*, 39, 1211–1233]. For our purposes, however, the most useful classification is the one based on sequence homology. This system divides commonly occurring $\beta$-lactamases into three homologous groups. Class A includes the staphylococcal penicillinase and most $\beta$-lactamases found in Gram-positive bacteria, as well as the TEM enzyme found commonly in Gram-negative bacteria and coded by a gene on R plasmids. Class C includes many of the chromosomally coded $\beta$-lactamases of Gram-negative bacteria. Both class A and class C enzymes have serine at the active site, but class B enzymes, which include enzymes capable of hydrolyzing carbapenems (discussed later in this chapter), have zinc at the active center. The three-dimensional structure of one of these enzymes is shown later in this chapter (see Figure 10.61).

**BOX 10.3**

drying and thus remain viable in dust particles: In a hospital environment, where all staphylococcal infections are treated with penicillin G, the plasmid-containing strain had a strong selective advantage and spread easily from patient to patient. There was thus an urgent need to produce penicillins that could withstand the staphylococcal penicillinase. Second, penicillin G (like penicillin V) is essentially inactive against Gram-negative bacteria (with the exception of *Neisseria gonorrhoeae*, the causative organism of the sexually transmitted disease gonorrhea). Thus, it was desirable to expand the spectrum of penicillin's targets to include Gram-negative bacteria as well.

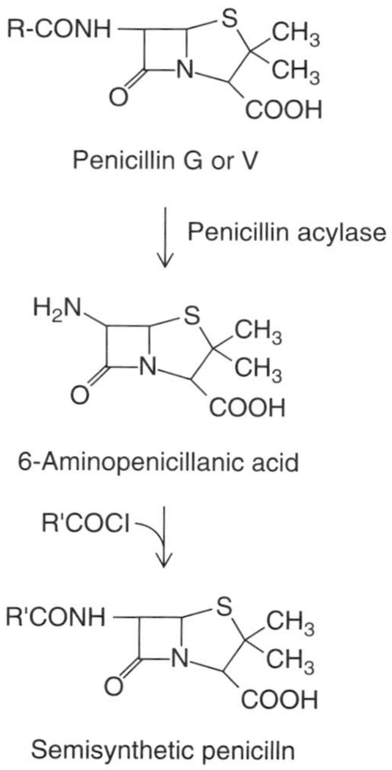

Because penicillin V was much more acid-resistant than penicillin G, scientists hoped that further modification of the acyl substituent at the 6-position would produce compounds with the properties they desired. However, adding potential precursors of acyl substituents to the growth medium produced no other useful antibiotics. Researchers thus turned to the strategy of chemically acylating 6-aminopenicillanic acid, the deacylated derivative of penicillin G, instead (Figure 10.45).

The first issue that arose in implementing this approach was how to produce the starting material, 6-aminopenicillanic acid. In the late 1950s, Beecham scientists found that 6-aminopenicillanic acid was most abundant in fermentation broths that lacked potential precursors of the side chain, and they succeeded in obtaining quantities sufficient for the synthesis of semisynthetic penicillins. Later, in the 1960s and 1970s, a large number of organisms were found to produce penicillin acylase, which cleaves off the 6-acyl side chain of penicillins G and V without touching the rest of the molecule. Furthermore, chemical procedures for splitting the side chain were developed in the early 1970s.

R-CONH S $CH_3$ $CH_3$ N O COOH

Penicillin G or V

Penicillin acylase

$H_2N$ S $CH_3$ $CH_3$ N O COOH

6-Aminopenicillanic acid

R'COCl

R'CONH S $CH_3$ $CH_3$ N O COOH

Semisynthetic penicilln

**FIGURE 10.45**

Production of semisynthetic penicillins. The starting material, usually penicillin G, is cleaved enzymatically by penicillin acylase to yield 6-aminopenicillanic acid. This is then chemically acylated with acyl chlorides (or acid anhydrides) to produce various semisynthetic compounds. When the R′ group contains an amino function, as in the synthesis of ampicillin, this amino group must be blocked first and then deblocked after the acylation reaction.

### Penicillins not Hydrolyzed by the Staphylococcal Penicillinase

By systematically going through various substituents, scientists found that the presence of bulky substituents on the $\alpha$-carbon of the 6-substituents produced compounds that were hydrolyzed more slowly by the staphylococcal enzyme. Methicillin and nafcillin (Figure 10.46) are examples of compounds that were produced by refining these structure–function studies, and they are almost absolutely resistant to enzymatic hydrolysis. In these compounds, the flexible $\alpha$-methylene group of benzylpenicillin is absent, and the $\alpha$-carbon is now a part of a bulky and rigid aromatic ring system. In addition, an $O–CH_3$ or $O–C_2H_5$ group at the ortho position of the ring further contributes to the steric hindrance of the interaction between the penicillinase and these penicillins. Similar stability against the staphylococcal penicillinase enzyme was obtained by the use of 3-phenyl-5-methylisoxazolyl substituents, as in oxacillin, cloxacillin, and the other isoxazolyl penicillins (see Figure 10.46), which have the added advantage of being able to be administered orally.

These drugs remained effective against penicillinase-producing staphylococci for a remarkably long time – from the early 1960s to the late 1980s – but recently, methicillin-resistant staphylococci with altered targets (i.e., altered penicillin-binding proteins) have appeared (see following).

Methicillin

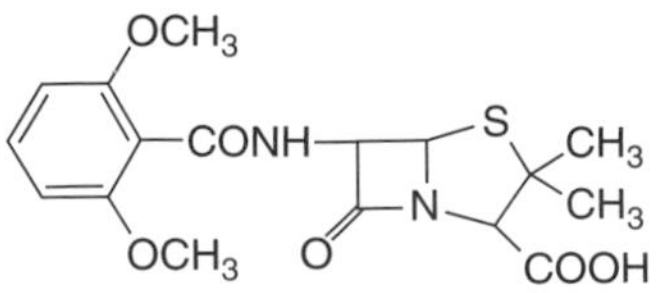

Nafcillin

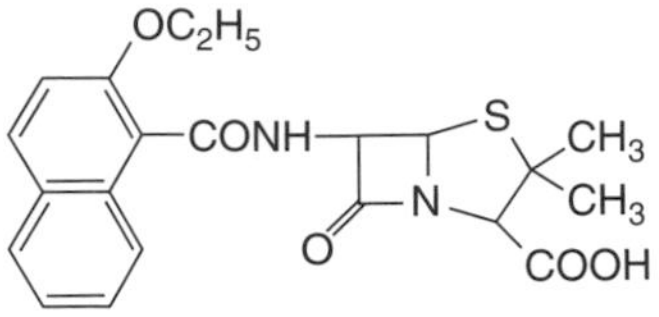

Isoxazolyl penicillins

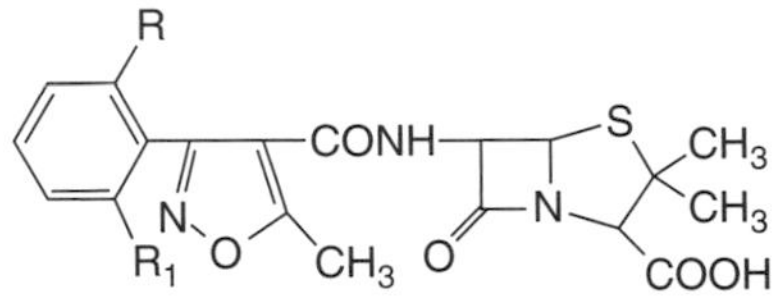

| | R | $R_1$ |
|---|---|---|
| Oxacillin | H | H |
| Cloxacillin | Cl | H |
| Dicloxacillin | Cl | Cl |
| Flucloxacillin | Cl | F |

**FIGURE 10.46**

Penicillins impervious to the action of staphylococcal penicillinase.

### Penicillins with Gram-Negative Activity

Penicillin N, a natural fermentation product that carries an aminoadipyl side chain, has been shown to have lower Gram-positive activity but significantly higher Gram-negative activity than penicillin G. Similarly, introduction of an amino group on the $\alpha$-carbon of the benzyl substituent of penicillin G produced ampicillin (see Figure 10.42), which is about 10 times more active against Gram-negative rods, such as *E. coli*, and retains half the activity of penicillin G against Gram-positive bacteria. Ampicillin is very widely used as a safe, inexpensive, broad-spectrum antibiotic. A close relative of ampicillin is amoxicillin (Figure 10.47), which is absorbed more efficiently than ampicillin upon oral administration.

A major reason why ampicillin is active against Gram-negative bacteria is its rate of diffusion across the outer membrane. As we noted earlier in this chapter, the main difference between Gram-positive and Gram-negative bacteria is that the latter are protected by an additional permeability barrier, the outer membrane. Most antibiotics must cross the outer membrane barrier by diffusing through the narrow, water-filled porin channels. Diffusion through the porin channels of enteric bacteria for zwitterionic compounds

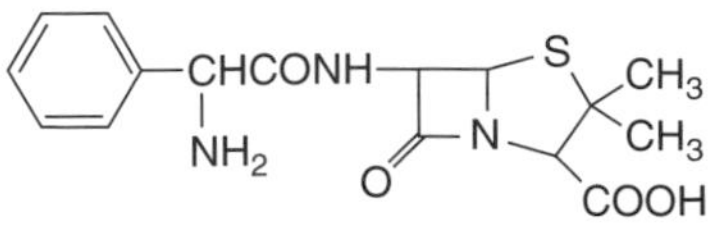

Ampicillin

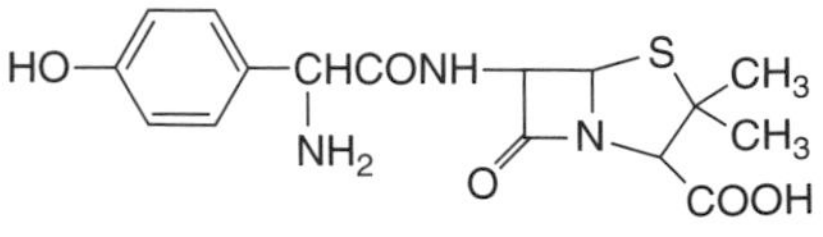

Amoxycillin

**FIGURE 10.47**

Ampicillin and amoxicillin, semisynthetic penicillins active against Gram-negative bacteria.

is faster than for anionic compounds, and the diffusion of hydrophilic compounds is faster than that of hydrophobic ones. Ampicillin is both zwitterionic and more hydrophilic in contrast to benzylpenicillin, which is anionic and quite hydrophobic. Thus, ampicillin can cross the outer membrane barrier much faster. However, this is only a superficial description of a more complex situation.

Measurements of the actual rates of penetration of various β-lactam antibiotics across the *E. coli* outer membrane have shown them to be quite fast: Most compounds achieve half-equilibrium between both sides of the membrane in less than 10 seconds. Thus differences in permeation rate alone cannot explain the differences in the efficacy of various penicillins. Most Gram-negative bacteria, however (*N. gonorrhoeae* is a possible exception), produce β-lactamase, an enzyme that hydrolyzes penicillins and cephalosporins even when they do not contain any resistance plasmids. The β-lactamase is located in the periplasm, the narrow space between the outer and inner (cytoplasmic) membranes. Thus, the small number of β-lactam molecules that succeed in traversing the outer membrane immediately encounter these enzyme molecules and are likely to be hydrolyzed by them. In this synergistic interaction between the outer membrane barrier and the periplasmic, β-lactamase "barrier," ampicillin not only crosses the outer membrane faster, but it is also more resistant to enzymatic hydrolysis than penicillin G.

### Penicillins with Negatively Charged Substituents

Ampicillin is active against *E. coli* and a few other Gram-negative bacteria, but it shows little activity against the many Gram-negative, rod-shaped bacteria that are more and more frequently the causative organisms of hospital-acquired infections. These include *P. aeruginosa, Enterobacter cloacae, S. marcescens*, and some *Proteus* species. In *E. coli*, the level of β-lactamase remains low even after exposure to drugs. These other bacteria, however, produce a very high level of the enzyme whenever they sense the presence of β-lactam in the medium. In other words, the enzyme is inducible in these species. Because of this, only the β-lactams with exceptional stability against β-lactamase can kill these bacteria. It would be even more desirable if those exceptionally stable β-lactams were also poor inducers of the enzyme.

Penicillins with negatively charged groups in the α-position of the 6-substituent fulfill these conditions. Carbenicillin, ticarcillin, and sulbenicillin (Figure 10.48) show at least some activity against such organisms as *P. aeruginosa* and *E. cloacae* (see "Carbenicillin" in Table 10.3).

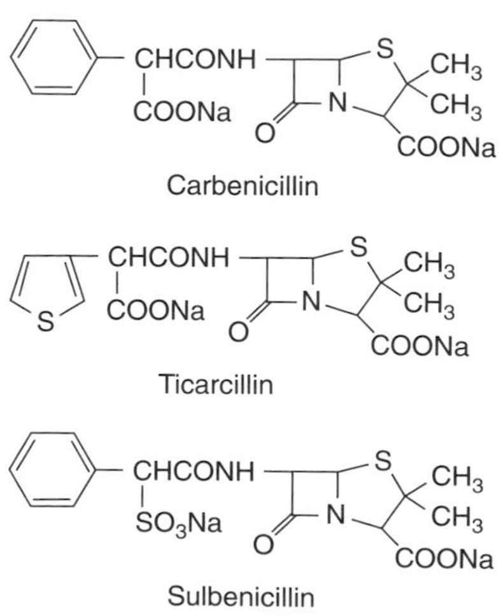

Carbenicillin

Ticarcillin

Sulbenicillin

**FIGURE 10.48**

Carbenicillin, ticarcillin, and sulbenicillin, semisynthetic penicillins with negatively charged substituents.

### Acylampicillins

In a number of semisynthetic penicillins, the amino group of ampicillin is substituted with a carboxyl group connected to a heterocyclic ring (Figure 10.49). These compounds (e.g., azlocillin, mezlocillin, piperacillin, and apalcillin) are similar to carbenicillin in their stability in the presence of the

**TABLE 10.3 Typical MIC (Minimal Inhibitory Concentration) values of some β-lactams (μg/ml)**

| β-Lactam | *S. aureus* (S) | *S. aureus* (R) | *E. coli* (S) | *E. coli* (R plasmid) | *E. cloacae* (inducible) | *E. cloacae* (constitutive) | *P. aeruginosa* |
|---|---|---|---|---|---|---|---|
| Penicillin G | 0.02 | >128 | 64 | >128 | >128 | | >128 |
| Methicillin | 1 | 2 | >128 | >128 | >128 | | >128 |
| Ampicillin | 0.05 | >128 | 2 | >128 | >128 | | >128 |
| Carbenicillin | 1 | 16 | 4 | >128 | 16 | >128 | 64 |
| Azlocillin | 1 | >128 | 16 | >128 | 32 | >128 | 4 |
| Cephalothin | 0.2 | 0.4 | 4 | >128 | >128 | >128 | >128 |
| Cephaloridine | 0.05 | 0.8 | 2 | 16 | >128 | >128 | >128 |
| Cefoxitin | 4 | | 4 | 4 | >128 | >128 | >128 |
| Cefotaxime | 2 | | 0.05 | 0.05 | 0.2 | >128 | 16 |
| Ceftazidime | 8 | | 0.05 | 0.05 | 0.1 | 32 | 2 |
| Cefepime | 1 | 1 | 0.05 | 0.05 | 0.1 | 1 | 2 |
| Aztreonam | >128 | >128 | 0.2 | 0.2 | 0.2 | 16 | 4 |
| Imipenem | 0.01 | 0.02 | 0.1 | 0.2 | 1 | 1 | 2 |

"S" and "R" indicate susceptible and resistant strains, respectively. "R plasmid" denotes strains containing the common R plasmid, producing TEM-type β-lactamase. Strains producing the chromosomally coded β-lactamase in the inducible (as in the wild-type strains) and constitutive manner are shown as "inducible" and "constitutive."

*Sources*: Rolinson G. N. (1986) β-Lactam antibiotics. *Journal of Antimicrobial Chemotherapy*, 17, 5–36; Nikaido H., unpublished data.

chromosomally coded β-lactamases of Gram-negative bacteria, and they are even weaker inducers of these enzymes than is carbenicillin or sulbenicillin. Thus, they have a similar spectrum, but at least some of the acylampicillins are significantly more active against *P. aeruginosa* (see "Azlocillin" in Table 10.3).

## CEPHALOSPORINS

In cephalosporins, the β-lactam ring is fused to a six-member dihydrothiazine ring rather than to the five-member thiazolidine ring found in penicillins (Figure 10.50). The natural product, cephalosporin C, was discovered accidentally in 1955. Edward Abraham's group was studying the products secreted by *Cephalosporium acremonium* (now called *Acremonium chrysogenum*), a very different kind of mold from the classical penicillin producer, *Penicillium*. The major product, penicillin N (Figure 10.50), with its hydrophilic aminoadipyl side chain, was of interest at the time because it had a low but significant activity against Gram-negative bacteria (see page 356). When these researchers tried to purify penicillin N, they found a smaller amount of a second product that turned out to be cephalosporin C. It too had activity against Gram-negative bacteria. It was also more resistant to

**FIGURE 10.49**

Azlocillin, an example of acylampicillins.

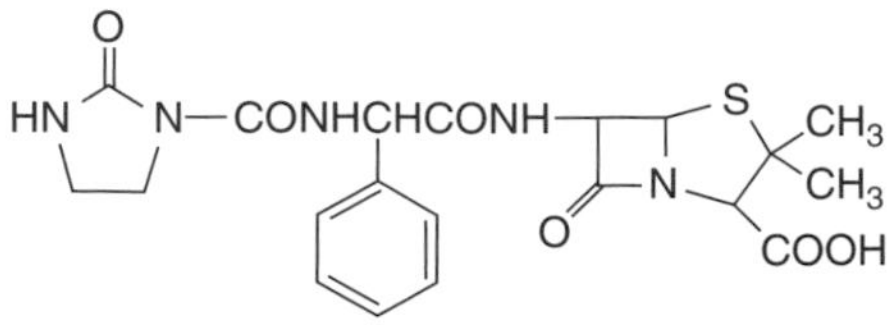

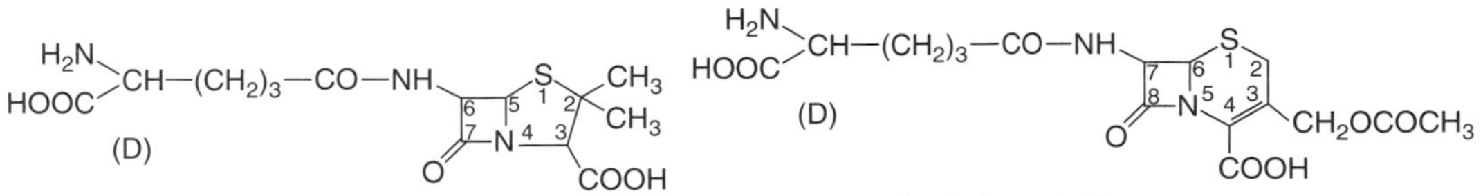

Penicillin N

Cephalosporin C

FIGURE 10.50

Penicillin N and cephalosporin C. In both compounds, the $\alpha$-aminoadipic acid moiety in the side chain has a D-configuration.

hydrolysis by staphylococcal penicillinase, giving Abraham some hope of finding a compound that is active on both Gram-negative bacteria and penicillinase-producing staphylococci. Unfortunately, cephalosporin C itself had only a low antibacterial activity, but its chemical structure, when elucidated, showed that a range of chemical modifications – not only at the 7-position (which corresponds to the 6-position in penicillins) but also at the 3-position – should be possible.

The first step toward synthesizing new cephalosporin compounds was to remove the 7-substituent from cephalosporin C. The resulting 7-aminocephalosporanic acid would then serve as the starting material. Because enzymes that remove the 6-substituent of penicillin G occur in a very large number of organisms, a great deal of effort was devoted to searching for a cephalosporin C acylase. Surprisingly, such an enzyme was not found. Thus, it was Robert B. Morin's 1962 discovery of a chemical method for creating 7-aminodeacetylcephalosporanic acid by the ring expansion from penicillin that opened the way to development of semisynthetic cephalosporins. Subsequently, it became possible to remove the 7-side chain of cephalosporin C by a two-step enzymatic process (Figure 10.51). Some of the penicillins sold as commercial antibiotics are natural fermentation products (penicillins G and V, for example); in contrast, every cephalosporin marketed is a semisynthetic compound.

### "First-Generation" Cephalosporins

The first semisynthetic cephalosporin antibiotics, introduced between 1962 and 1965, included cephalothin, cephaloridine, and cefazolin (Figure 10.52). At that time, there were two major groups of bacteria for which penicillin G had little activity: penicillinase-producing staphylococci and Gram-negative rods. Although methicillin and its relatives were active against the former, and although ampicillin was active against the latter, no penicillin derivative showed activity against both of these groups. The cephalosporins had the advantage of being quite active against both groups, thus fulfilling the promise that Abraham saw in them.

The action of cephalosporins against Gram-negative rods merits some comment. Cephalosporins as a class have some advantage over penicillins in passing through the porin channel, because the cephalosporins are less hydrophobic than the penicillins. Cephaloridine, especially, has a very high rate of diffusion through the outer membrane, thanks to its zwitterionic nature. In order to reach the targets (i.e., penicillin-binding proteins or transpeptidases; see page 333), these compounds have to overcome the second "barrier": periplasmic $\beta$-lactamases. These chromosomally coded (class C,

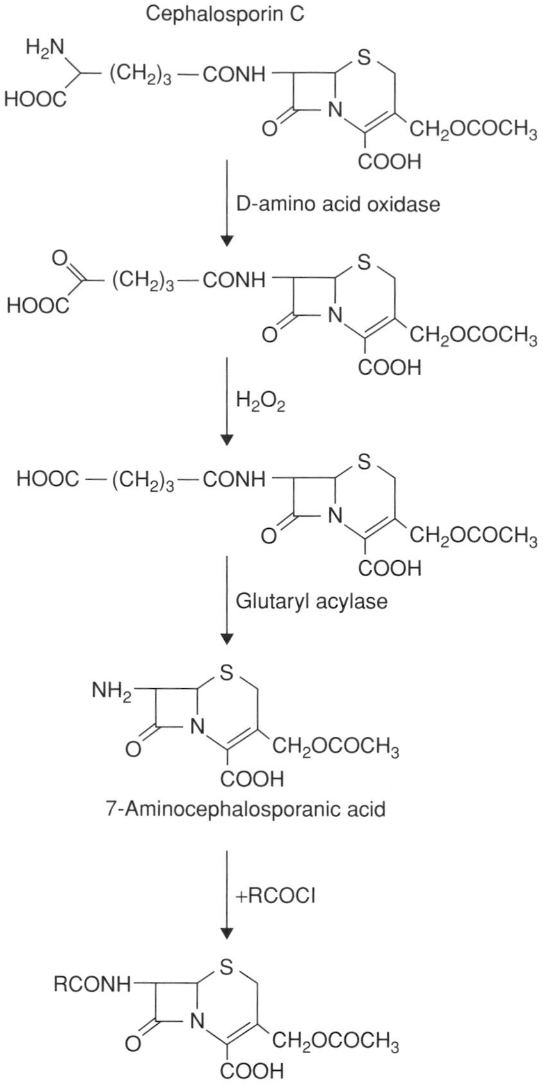

**FIGURE 10.51**

Conversion of cephalosporin C into semi-synthetic cephalosporins. Cephalosporin C is first oxidized by a D-amino acid oxidase, and the resulting keto acid is decarboxylated by hydrogen peroxide, which is generated during the previous reaction. The resulting 7-glutarylcephalosporanic acid is deacylated by a deacylase, and the 7-aminocephalosporanic acid is then modified by the addition of various side chains. In addition to substitutions at the 7-position, the acetoxy substituent at the 3-position can easily be replaced by a nucleophile, opening up additional possibilities for chemical modification of the cephalosporin structure.

Box 10.3) enzymes, which are present in most Gram-negative rods, have been called "cephalosporinases" because they were believed to hydrolyze cephalosporins much better than they do penicillins. Herein lies a paradox: If cephalosporins are hydrolyzed more readily by these ubiquitous enzymes, how could they be more effective than penicillins against Gram-negative bacteria?

The notion that the class C enzymes are cephalosporinases arose because they catalyzed a much more rapid hydrolysis of cephalosporins than penicillins when assays were carried out using very high concentrations (usually 1 to 5 mM) of substrate. Under these conditions, we are comparing maximum velocity ($V_{max}$) values, and indeed $V_{max}$ for cefazolin is more than 10 times higher than that for penicillin G with the *E. coli* enzyme (Table 10.4). However, the efficiency of an enzyme is better compared on the basis of $V_{max}/K_m$ values. On this basis, the enzyme hydrolyzes penicillin almost 80 times more efficiently than it does cefazolin, according to the data given in

**TABLE 10.4 Hydrolysis of β-lactams by *E. coli* enzymes**

| | Chromosomally coded enzyme | | | | Plasmid-coded TEM enzyme | | | |
|---|---|---|---|---|---|---|---|---|
| | | $V_{max}$ | V(5 mM) | V(0.1 μM) | | $V_{max}$ | V(5 mM) | V(0.1 μM) |
| **Antibiotic** | $K_m$ (μM) | (relative rates) | | | $K_m$ (μM) | (relative rates) | | |
| Cefazolin | 1900 | 100 | 100 | 100 | 320 | 100 | 100 | 100 |
| Penicillin G | 1.9 | 7.6 | 10.5 | 7200 | 18 | 880 | 930 | 15,500 |
| Cefoxitin | 0.22 | 0.02 | 0.03 | 120 | 3600 | 0.03 | 0.02 | <0.001 |
| Cefotaxime | 0.16 | 0.007 | 0.01 | 51 | 9500 | 15 | 5.5 | 0.5 |
| Cefepime | 80 | 0.001 | 0.01 | 0.2 | 5000 | 15 | 8 | 0.5 |

V(5 mM) and V(0.1 μM) represent rates of hydrolysis at 5 mM and 0.1 μM substrate concentration, respectively. Values of $V_{max}$, V(5 mM), and V(0.1 μM) were all expressed as relative values by setting the cefazolin rate at 100. The chromosomal enzyme data on cefepime were obtained by using *E. cloacae* enzyme rather than the *E. coli* enzyme.

*Sources*: Nikaido, H., and Normark, S. (1987) Sensitivity of *Escherichia coli* to various beta-lactams is determined by the interplay of outer membrane permeability and degradation by periplasmic beta-lactamases: a quantitative predictive treatment. *Molecular Microbiology*, 1, 29–36; Nikaido, H., et al. (1990). Outer membrane permeability and beta-lactamase stability of dipolar ionic cephalosporins containing methoxyimino substituents. *Antimicrobial Agents and Chemotherapy*, 34, 337–342.

Table 10.4. We can also see the physiological relevance of this conclusion as follows: The targets of β-lactams, penicillin-binding proteins, are irreversibly inactivated at very low drug concentrations, usually in the range of 0.1 to 1 μM. Whether the enzymatic hydrolysis can protect the bacterial cells against attack by β-lactams is therefore decided by the behavior of the β-lactamases at micromolar concentrations of the substrate. The calculated hydrolysis rate of penicillin is indeed 72-fold higher than that of cefazolin at 0.1 μM (see "V(0.1 μM)" in Table 10.4), in spite of the much higher $V_{max}$ for cefazolin. Thus, the Gram-negative activity of cephalosporins owes as much to their lower affinity for the chromosomally coded β-lactamases (which, to avoid misunderstanding, should *not* be called cephalosporinases) as to their higher rate of diffusion across the outer membrane.

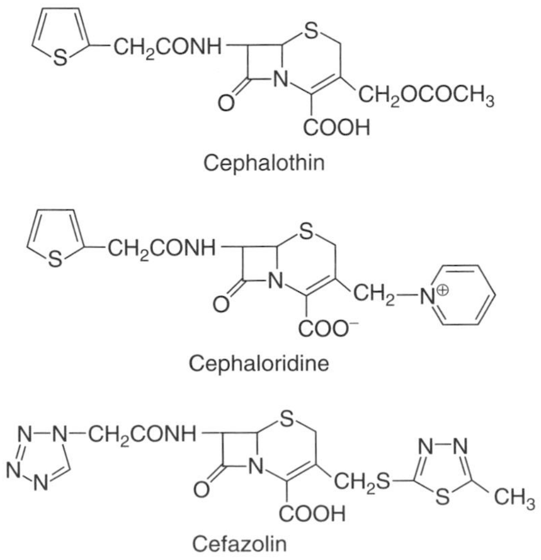

Cephalothin

Cephaloridine

Cefazolin

**FIGURE 10.52**

The "first-generation" cephalosporins.

### Discovery of Cephamycins

Because of their wide spectrum, the first-generation cephalosporins were used extensively, a situation that unfortunately resulted in the selection of Gram-negative bacteria containing R plasmids. The majority of these plasmids, isolated from bacteria all over the world, carried a gene for a class A $\beta$-lactamase called TEM $\beta$-lactamase (these are the initials of the patient from whom it was first isolated). The TEM enzyme has a wide substrate specificity that includes most penicillins and all of the first-generation cephalosporins. Although its affinity toward cephalosporins is not particularly high (see Table 10.4), it can hydrolyze these compounds effectively even at low concentrations because the $V_{max}$ values are high and because the presence of multiple copies of the plasmid, and therefore of the gene, in a cell leads to very high levels of the enzyme.

Compounds that are stable in the presence of TEM $\beta$-lactamase were not found among semisynthetic derivatives of cephalosporin C at that time, so, as happens often in the development of antibiotics, scientists turned to nature for inspiration. Specifically, and significantly, they turned to organisms that are *not* classic producers of $\beta$-lactams. A search through the culture filtrates of *Streptomyces* species, which are eubacteria and thus very far from fungi in phylogenetic terms, yielded cephamycins, with a methoxy group at the 7-$\alpha$-position of the nucleus (Figure 10.53). These compounds were found to be absolutely resistant to the TEM enzyme (see the behavior of the TEM enzyme against cefoxitin, in Table 10.4).

The cephamycin cefoxitin was introduced commercially in the United States in 1978. In Japan, another cephamycin derivative, cefmetazole, was introduced soon after. These compounds, sometimes called second-generation cephalosporins, were hailed as the last word in $\beta$-lactams because they were effective against all the target groups thought to be important at that time, including the penicillinase-producing staphylococci and the R plasmid–containing Gram-negative rods (see Table 10.3).

### "Third-Generation" Cephalosporins

Within a few years after the introduction of cephamycins, physicians were noticing the presence of pathogens such as *E. cloacae*, *S. marcescens*, and *P. aeruginosa*, especially as the causes of hospital-acquired infections. These bacteria are intrinsically resistant to cephamycins (i.e., even without the acquisition of R plasmids or new mutations), but it is doubtful whether the use of cephamycins actually resulted in a strong selection for these organisms in the environment. What probably happened was that once infections caused by most other classes of bacteria could be treated effectively with

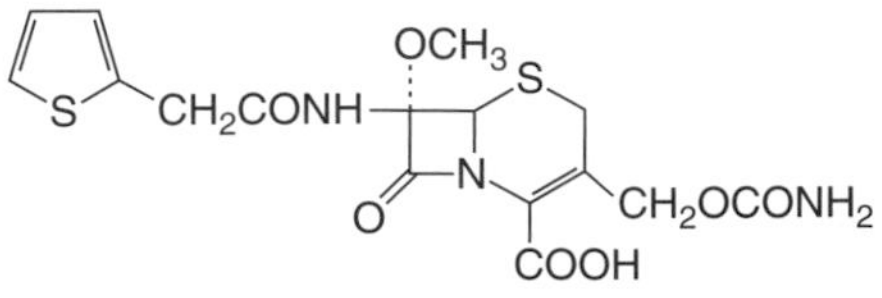

**FIGURE 10.53**

Cefoxitin, an example of a cephamycin.

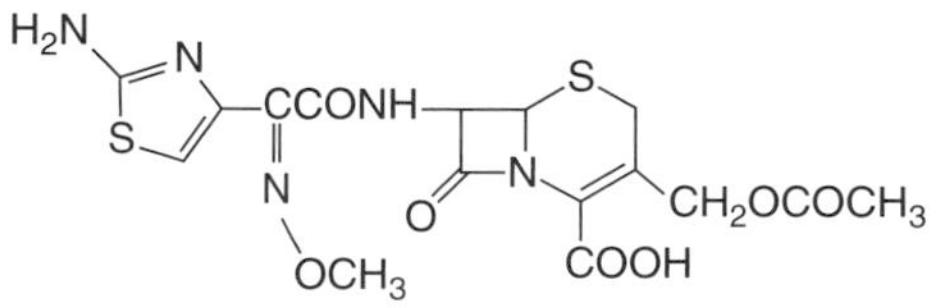

**FIGURE 10.54**

Cefotaxime, an example of a "third-generation" cephalosporin.

cephamycins, physicians became especially aware of the infections caused by cephamycin-resistant bacteria.

In any case, a search for cephalosporins active against these resistant bacteria produced fruitful results, this time from synthetic chemistry. It was discovered in the early 1980s that the addition of a substituted oxime group on the $\alpha$-carbon of the side chain, as well as the use of a hydrophilic aminothiazole group in the side chain, produced cephalosporins with revolutionary efficacy against Gram-negative bacteria, including many of the "intrinsically resistant" bacteria mentioned above. These compounds, often called third-generation cephalosporins, include cefotaxime (Figure 10.54), ceftizoxime, ceftazidime, and ceftriaxone. As illustrated in Table 10.3, cefotaxime and ceftazidime show strong activity against the wild-type strains of *E. cloacae* (strains that have inducible $\beta$-lactamase and are therefore labeled "inducible") and ceftazidime shows excellent activity against *P. aeruginosa*; both of these organisms were totally refractory to cefoxitin therapy.

### "Fourth-Generation" Cephalosporins

The third-generation cephalosporins appeared to have solved practically all the problems remaining in the development of the semisynthetic $\beta$-lactams. These cephalosporins were said to be absolutely impervious both to plasmid-coded TEM $\beta$-lactamase and to chromosomally coded $\beta$-lactamases of Gram-negative bacteria. However, the clinical use of these drugs led to the emergence of highly resistant mutants of *E. cloacae, S. marcescens*, and *P. aeruginosa*; an example is shown in Table 10.3 as "*E. cloacae* (constitutive)." These mutants produced the normally inducible, chromosomally coded $\beta$-lactamase in a constitutive manner and at a very high level. How could a bacterium become resistant to a drug by producing an enzyme that is supposed to be absolutely inactive against that drug? This dilemma led to some creative hypotheses. In retrospect, we can see once again how the practice of using very high, arbitrarily chosen substrate concentrations in laboratory tests of enzyme activity has misled many scientists.

The chromosomally coded $\beta$-lactamases of Gram-negative rods have a very low $V_{max}$ of hydrolysis with the third-generation compounds (see Table 10.4, which shows that the $V_{max}$ value for cefotaxime hydrolysis of the *E. coli* chromosomal enzyme is less than 0.01% of that for cefazolin hydrolysis). However, these enzymes have very strong affinity, or low $K_m$, for the third-generation compounds (Table 10.4 again shows that the $K_m$ value for cefotaxime is more than 10,000-fold lower than that for cefazolin). Thus, with a traditional assay using 5 mM of substrate, for example, one would get the impression that a third-generation compound, cefotaxime, is completely

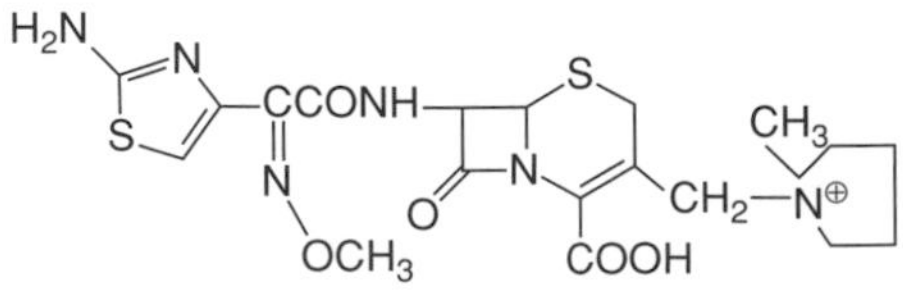

FIGURE 10.55

Cefepime, an example of a "fourth-generation" cephalosporin.

unaffected by the chromosomal enzyme. However, if we rely on $V_{max}/K_m$, or V (0.1 $\mu$M), we see that cefotaxime is hydrolyzed almost as rapidly as cefazolin (see Table 10.4).

We thus understand why the $\beta$-lactamase–constitutive mutants can develop high levels of resistance to third-generation compounds. But why were the inducible parent strains so exquisitely sensitive to the same compounds? In answering this question, we should compare the cephamycins with the third-generation compounds. We now know that cephamycins also are rapidly hydrolyzed at low concentrations by chromosomal enzymes (see Table 10.4). Thus, both cephamycins and third-generation compounds are hydrolyzed rapidly by the chromosomal enzyme, yet the wild-type strains of *E. cloacae* are very resistant to the former and very sensitive to the latter. The major difference in the behavior of these two classes of compounds is that cephamycins are very strong inducers of the chromosomal enzyme, whereas the third-generation compounds have almost no inducer activity. Clearly, cephamycins become hydrolyzed after the induction of these efficient enzymes, but the third-generation compounds remain active because in the absence of induction, the enzyme levels remain extremely low. The constitutive mutants can of course efficiently hydrolyze both cephamycins and the third-generation cephalosporins.

The "fourth-generation" cephalosporins, $\beta$-lactam antibiotics that can better withstand the chromosomal $\beta$-lactamases, have been developed more recently. These compounds (including cefepime, cefpirome, and cefclidin) retain the oxyimino substituents of the third-generation compounds but also contain a 3-substituent with a quaternary nitrogen atom (Figure 10.55).

As Table 10.4 shows, cefepime has a much lower affinity for chromosomally coded enzymes than do the third-generation compounds and is therefore quite stable in these enzymes' presence. It shows good activity against constitutive mutants of *E. cloacae, S. marcescens*, and *P. aeruginosa* (see Table 10.3).

## COMPOUNDS WITH NONTRADITIONAL NUCLEI

As we have seen already, the spectrum of $\beta$-lactam antibiotics has been expanding steadily. At the same time, the introduction of each new "generation" has resulted in the "emergence" of strains that show resistance to these drugs. In fact, there is already a significant threat to the continued efficacy of third- and fourth-generation cephalosporins, because mutants of the TEM enzyme that show drastically increased activity against these compounds have appeared in different parts of the world.

In response to this situation, researchers are developing new compounds on a continuous basis. It is estimated that 50,000 to 100,000 semisynthetic antibiotics have been made and tested already, and an overwhelming majority among them must have been β-lactams. Thus, some workers feel that at this point, only compounds with radically new structures are worth investigating. They may also hope that such compounds will be less likely to lead to a rapid emergence of resistant organisms. Compounds that are not based on the traditional penicillin or cephalosporin nucleus certainly fall into the category of radically new compounds, and in recent years there have been many efforts to develop such compounds.

In some cases, the nucleus has been totally synthetic. On the other hand, many of the newest antibiotics were discovered, as in the past, by the screening of natural products. Of course, the traditional screening methods using growth-inhibition assays are unlikely to turn up anything very new, because most of the compounds that are produced by easily cultivated microorganisms and show strong antibacterial activity have already been discovered. Thus, much more sensitive or narrowly focused assays are now being used – for example, assays that test for the inhibition of cell wall biosynthetic reactions in cell-free systems, the inhibition of β-lactamase, or the inhibition of hypersensitive mutants of bacteria. When a novel compound shows even traces of antimicrobial activity, then semisynthetic approaches can be utilized to produce more active derivatives.

It is also important that in recent years, nontraditional source material has been used for screening. Microorganisms that are only remotely related to the fungi or to *Streptomyces* (the traditional sources of antibiotics) – even unicellular bacteria that show no signs of cellular differentiation – have been examined, and this has resulted in major discoveries. Although the inhabitants of unusual environments have not yet been examined very extensively, the use of acidic soil alone has already led to the discovery of monobactams (see below).

## Carbapenems

Carbapenems, which include thienamycin, are compounds isolated from *Streptomyces* species. They have a carbon atom in place of the sulfur atom of the penicillin nucleus, and they contain a double bond in the five-member ring of the nucleus. (The nucleus of the penicillin is called a *penam*, but when a double bond exists, the nucleus becomes a *penem*). Some of these carbapenems were detected by screening for inhibitors of bacterial cell wall synthesis, others by looking for inhibitors of β-lactamases.

Thienamycin is a potent antibiotic with a broad spectrum. However, thienamycin is chemically unstable, and some ingenious chemical modification was necessary to convert it to a clinically useful compound, imipenem (Figure 10.56). Because imipenem is hydrolyzed by a peptidase present in human tissues, it is administered together with the peptidase inhibitor cilastatin.

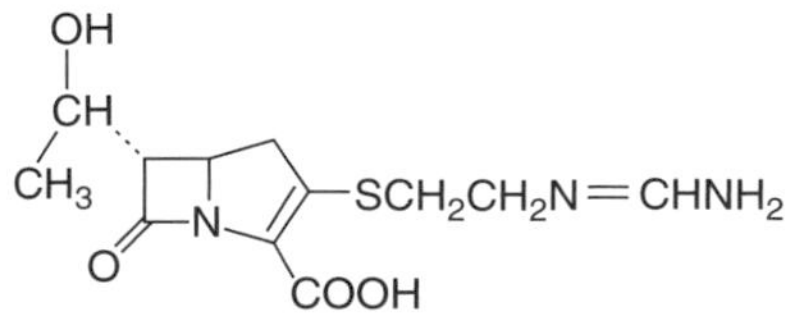

Although the starting material for imipenem synthesis, thienamycin, was a natural product, its mass production by fermentation was hampered by low yield, the instability of the compound, and the production of several related

OH
CH
$CH_3$
N
O
$SCH_2CH_2N{=}CHNH_2$
COOH

**FIGURE 10.56**

Imipenem.

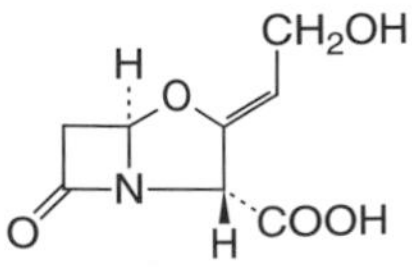

compounds that complicated the isolation process. Chemical synthesis was also difficult because of the presence of three contiguous chiral centers; it has been successfully achieved, however, and imipenem is now produced commercially by a totally synthetic process.

Imipenem is stable in the presence of many $\beta$-lactamases. With others it acts as a suicide inhibitor; these include the chromosomally coded class C enzymes of Gram-negative rods that are responsible for the resistance of *E. cloacae, S. marcescens*, and *P. aeruginosa* mutants to the third-generation cephalosporins. Thus, imipenem has an extremely wide spectrum (see Table 10.3). More recently, however, it was found that the exceptional activity of imipenem against *P. aeruginosa* was largely the result of its rapid penetration through the specific outer membrane channel, which normally functions in the uptake of basic amino acids. Because this channel can be lost by mutation, imipenem resistance can arise very rapidly in *P. aeruginosa* populations, limiting the drug's usefulness against this important pathogen.

**FIGURE 10.57**

Clavulanic acid.

### Clavams

Clavams have no side chain corresponding to the 6-substituent of penicillins. They have oxygen instead of sulfur in the penicillin (penam) nucleus. Clavams were isolated from *Streptomyces* through screening to detect $\beta$-lactamase inhibitors. Because they usually have only very low antibiotic activity, they would not have been detected in the conventional screening for antibiotics. One clavam, called clavulanic acid (Figure 10.57), is an efficient and irreversible inhibitor of class A $\beta$-lactamases, including the staphylococcal enzyme and the TEM enzyme. It is sold in combination with amoxycillin under the trade name Augmentin.

### Nocardicins and Monobactams

Nocardicins and monobactams are characterized by an isolated $\beta$-lactam ring, without the fused thiazolidine or dihydrothiazine ring (Figure 10.58). In nocardicins, the lactam nitrogen is connected to a carbon atom, whereas

Nocardicin A

SQ 28,503 (monobactam)

Aztreonam (monobactam)

**FIGURE 10.58**

Nocardicin and the monobactams SQ 28,503 and aztreonam.

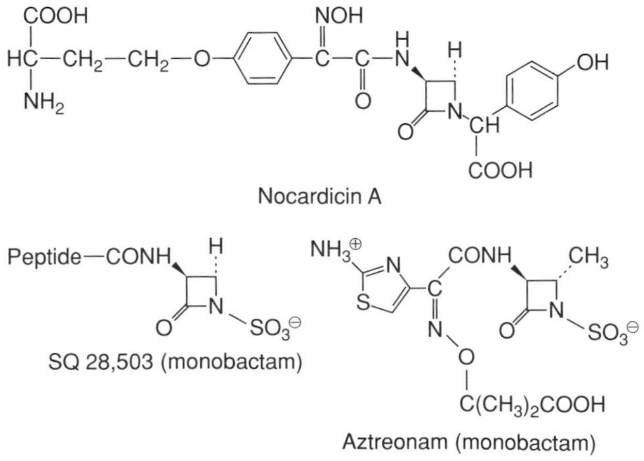

in monobactams it is connected to the sulfur atom of a sulfonate group. Nocardicins are produced by *Nocardia*, and monobactams come from Gram-negative rods that do not go through any developmental cycle. Many of these, such as *Gluconobacter, Acetobacter*, and *Agrobacterium*, belong to the $\alpha$-division of the purple bacteria branch (Chapter 1). Other monobactam producers include members of *Chromobacterium* (in the $\beta$-division of the purple bacteria branch) and even *Flexibacter* species, which belong to the *Bacteroides–Flavobacterium* branch (Flavobacteria group in Figure 1.6).

Nocardicins have low antibacterial activity and so far have not led to any commercially useful product. Monobactams isolated from culture filtrates also have very low antibacterial activity. For example, SQ 28,503 (see Figure 10.58), isolated from a *Flexibacter* strain, showed minimum inhibitory concentration (MIC) values of 50 to 100 $\mu$g/ml against many pathogenic bacteria. Nevertheless, Squibb chemists used the basic monobactam structure as a starting point and after extensive studies of structure–activity relationships, produced a totally synthetic compound, aztreonam, that has good activity against a wide spectrum of Gram-negative bacteria (see Table 10.3).

### Enzymatic Synthesis with Substrate Analogs

Some of the enzymes that catalyze key steps in antibiotic biosynthesis appear to have a rather broad substrate specificity. We can take advantage of this fact to obtain a novel antibiotic, as was described for aminoglycosides. The approach in that case (see "Use of Mutasynthesis for Developing New Aminoglycosides") used intact cells, but *in vitro* methods using purified enzymes are also possible. In the latter case, production of the relevant enzyme in large amounts and in reasonable purity needs to be achieved by gene amplification via cloning.

An example of this approach is the use of isopenicillin N synthase (Figure 10.59) for the production of a totally new kind of $\beta$-lactam. This enzyme, which catalyzes the crucial ring-closure step in the synthesis of penicillins, cephalosporins, and cephamycins, has been cloned from various sources. From the cephalosporin-producing fungus *A. chrysogenum*, for example, it was cloned by using synthetic oligonucleotides based on the amino acid sequence of the purified enzyme. Figure 10.60 shows that by adding to this enzyme an allyl analog of the tripeptide $\alpha$-aminoadipyl-cysteinyl-valine, one can generate a whole new set of $\beta$-lactams, most of which contain nuclei of novel structure (shaded in Figure 10.60).

## RATIONAL DESIGN OF NEW $\beta$-LACTAMS AND $\beta$-LACTAMASE INHIBITORS

The three-dimensional structures of many $\beta$-lactamases and several penicillin-binding proteins have been determined by x-ray crystallography. Remarkably, these proteins, which do not show extensive homology in amino acid sequences, exhibit overall folding patterns that are very much alike (Figure 10.61), and the substrate-binding cavities are lined by similar amino acid

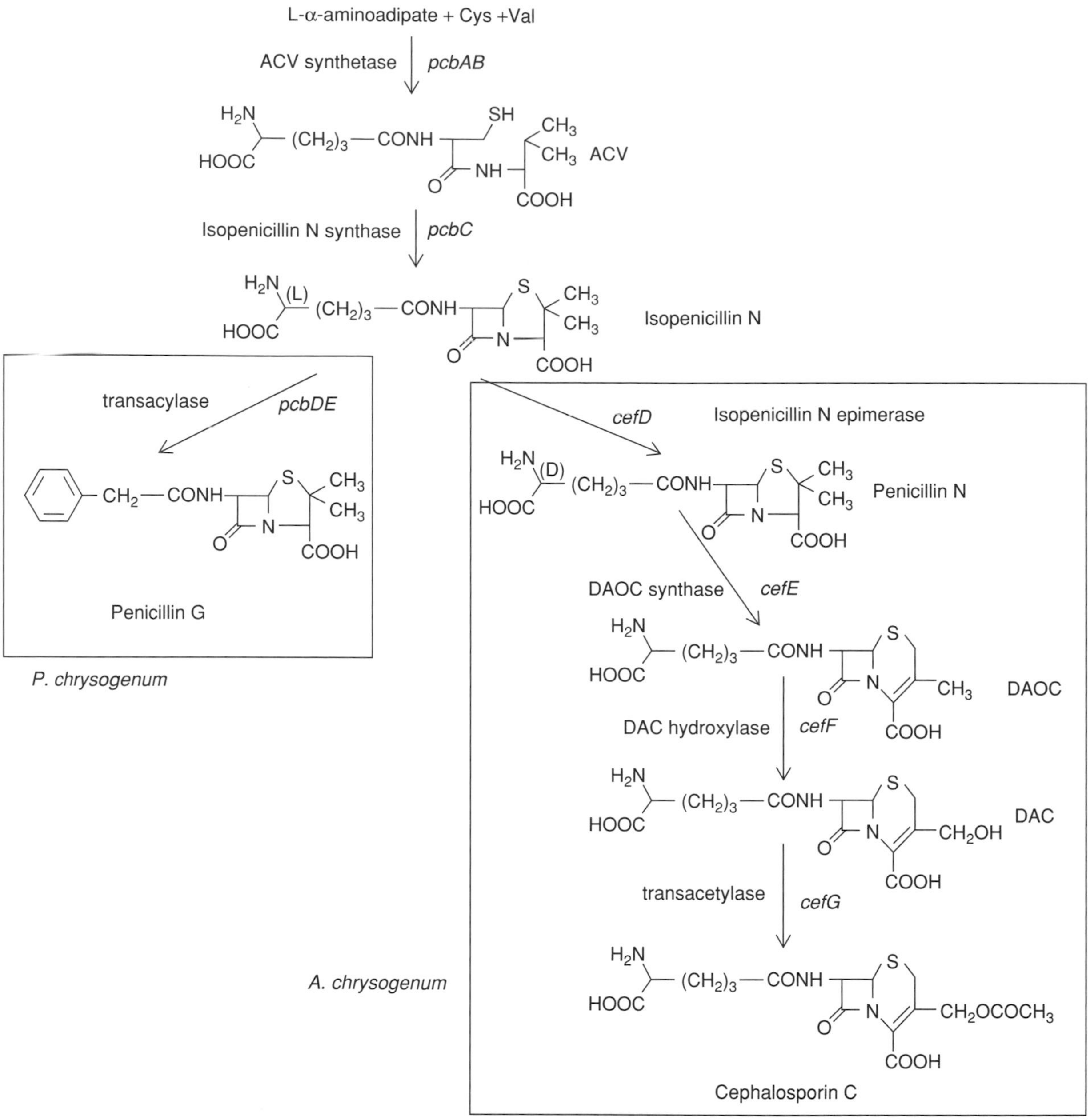

**FIGURE 10.59**

Synthetic pathway of penicillins and cephalosporins. In the portion common to all $\beta$-lactams, L-$\alpha$-aminoadipic acid, L-cysteine, and L-valine are converted into a tripeptide, ACV, by the nonribosomal peptide synthetase coded by *pcbAB*. The tripeptide L-$\alpha$-aminoadipyl-L-cysteinyl-D-valine (LLD-ACV) is then cyclized by the enzyme isopenicillin N synthase, or cyclase, to produce a key intermediate isopenicillin N. From this point, exchange of the side chain will produce penicillin G in *P. chrysogenum*. In *A. chrysogenum*, several enzymes that do not exist in *P. chrysogenum* (shown in the *large rectangle* on the right) will convert isopenicillin N into cephalosporin C, through penicillin N, deacetoxycephalosporin C (DAOC), and deacetylcephalosporin C (DAC) as intermediates. Genes *cefE* and *cefF* are fused into a single gene in *A. chrysogenum* but exist as two independent genes in *Streptomyces clavigerus*.

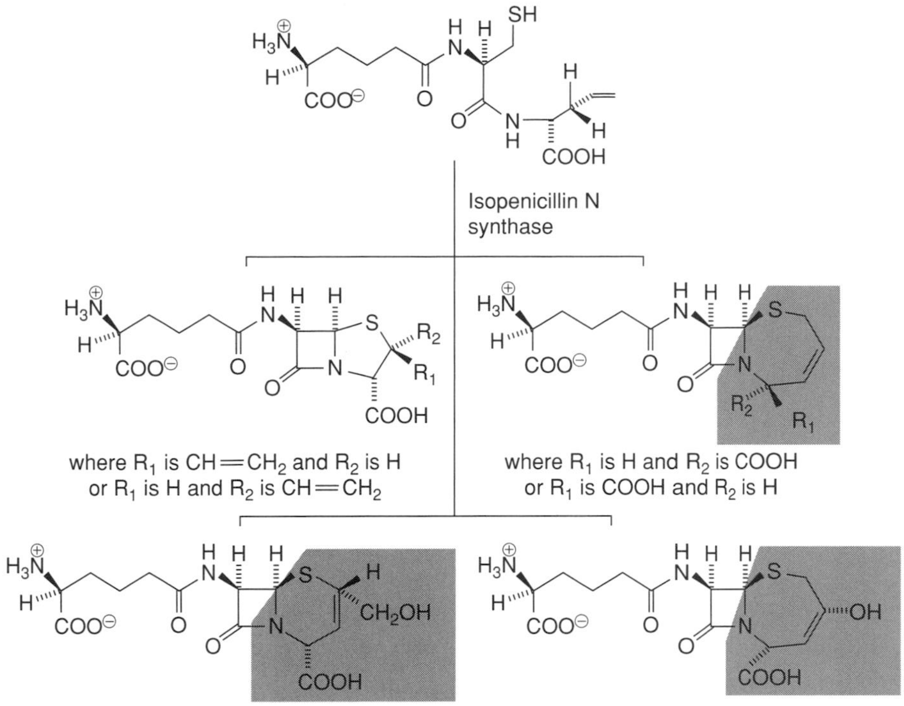

residues. Thus, the right half of the molecule as depicted is composed of a $\beta$-sheet containing five strands, protected by two $\alpha$-helices in front and a few $\alpha$-helices in the back. The left half of the molecule consists of a cluster of $\alpha$-helices. We can already explain from these structures why $\beta$-lactams are hydrolyzed by $\beta$-lactamases, but not by penicillin-binding proteins. Thus, both proteins are acylated at the active site serine (shown as *solid spheres* in Figure 10.61) by $\beta$-lactams. But with the class A $\beta$-lactamase (*left*), the acyl-serine is rapidly hydrolyzed by the activated water presented by the optimally placed glutamic acid residue (shown by a stick model). In contrast, such activated water molecules are not present in penicillin-binding proteins, and they remain permanently inactivated by the $\beta$-lactams. The logical extension of these studies is to produce $\beta$-lactam compounds that more effectively withstand degradation by the $\beta$-lactamases prevalent among the hospital isolates or inhibitors of such $\beta$-lactamases, and these studies are in progress in several laboratories. Knowledge about the structure of a cell wall-synthesizing enzyme (see Figure 10.61) may one day also lead to the design of a drug that will bind to the target transpeptidase yet avoid being captured by $\beta$-lactamases.

Isopenicillin N synthase

where $R_1$ is $CH{=}CH_2$ and $R_2$ is H or $R_1$ is H and $R_2$ is $CH{=}CH_2$

where $R_1$ is H and $R_2$ is COOH or $R_1$ is COOH and $R_2$ is H

**FIGURE 10.60**

Examples of compounds synthesized by feeding an unnatural substrate to isopenicillin N synthase. The analog introduced here, instead of the natural substrate $\alpha$-aminoadipyl-cysteinyl-valine, contained an allyl group in the valine residue. [Based on Floss, H. G. (1987) Hybrid antibiotics–the contribution of the new gene combinations. *Trends in Biotechnology*, 5, 111–115.]

## PRODUCTION OF ANTIBIOTICS

As we mentioned at the beginning of this chapter, antibiotic research is unusual among the fields of microbial biotechnology in that a large part of the scientific effort has always gone into the development of new

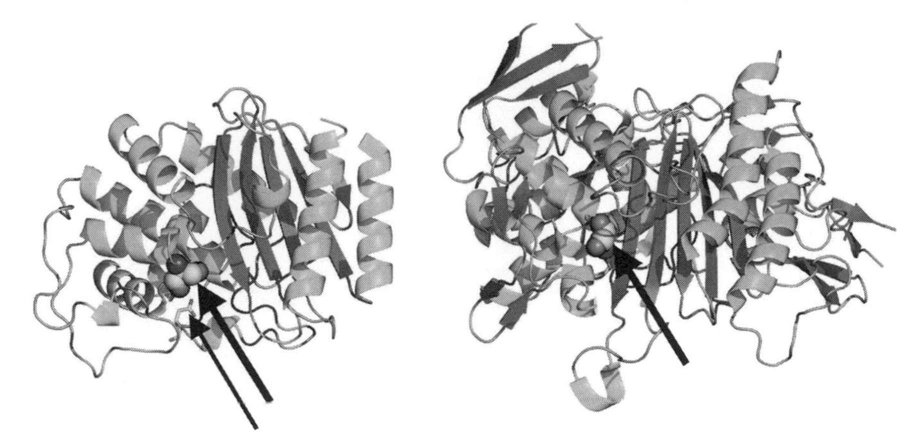

**FIGURE 10.61**

The folding patterns of a class A $\beta$-lactamase (TEM; left) and a penicillin-binding protein (PBP2x from *S. pneumoniae*; right). Note the very similar overall folding patterns. *Thick arrows* point to the active site serine residues (shown in the CPK model), and the *thin arrow* in the left panel shows the glutamic acid residue providing activated water molecules for the hydrolysis of the acyl-enzyme (shown in the stick model). This figure was drawn using PyMol based on PDB files 1BTL and 1QME. See the color plate, this chapter.

compounds. The foregoing examination of aminoglycosides and $\beta$-lactams provided numerous examples. It also showed, however, that most antibiotics are still produced by fermentation or by the chemical modification of fermentation products. It follows, therefore, that research devoted to improving the fermentation processes can have a significant beneficial impact.

## GENETICS OF ANTIBIOTIC PRODUCTION

The efficiency of production of antibiotics is determined largely by the genetic makeup of the producing strains, and much effort has been spent on improving these strains.

### Traditional Method of Strain Improvement

The traditional genetic approach to improving the yield of an antibiotic-producing organism depends entirely on random mutagenesis and screening of high producers. Until recently, the screening had to be done by growing each progeny clone in liquid media and assaying for the antibiotic in the culture filtrate. Because such screening was laborious and slow, only a small number of progeny could be tested in one experiment. Introduction of a large number of mutations – many of them with deleterious effects on the growth of the organism – through heavy mutagenesis was necessary in order to boost the probability that the small number of strains tested would contain interesting mutants. In spite of these disadvantages, practically all the improvements achieved until quite recently in the strains that produce commercially important antibiotics have come about in this manner. A part of the genealogy of penicillin-producing strains is shown in Figure 10.62. Interestingly, the magnitude of the improvement per step was greater during the early stages. Nevertheless, current yields in fermentation are said to be about 40 g/L, almost a magnitude higher than seen in Figure 10.62, as a result of constant improvement in both the producing strain and the fermentation conditions. Not a great deal is known about the biochemical and genetic nature of the improvements that have been made, but many overproducing strains appear to have gene duplications (see below).

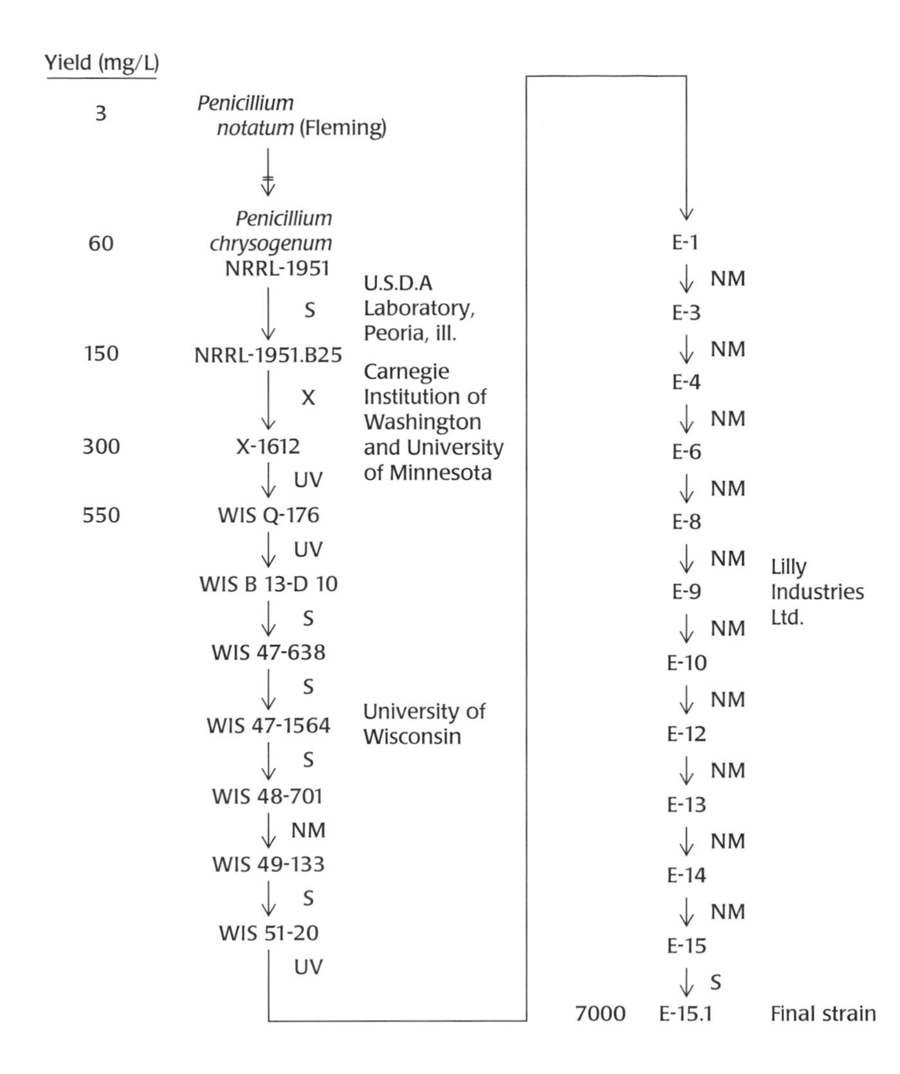


**FIGURE 10.62**

The improvement of penicillin-producing strains. The mutagenesis procedures used for each step: S, spontaneous; X, x-ray; UV, ultraviolet light; and NM, nitrogen mustard. [Modified from Aharonowitz, Y., and Cohen, G. (1981). The microbial production of pharmaceuticals. *Scientific American*, September, 141–152.]

### Methods of Classical Genetics

In the ideal scenario, a scientist would identify desirable mutant alleles of various genes that increase antibiotic production and recombine them into a single organism, just as was done with amino acid–overproducing strains. Unfortunately, this goal has been difficult to attain; our knowledge about the genetics of antibiotic-producing organisms was too limited. (Today, however, we are in a new era. Not only have many genes of antibiotic biosynthesis been identified through cloning, but complete genome sequences are available for a few strains of the prokaryotic *Streptomyces*; see later.)

The most effective use of classical genetics in the past was the backcrossing of overproducing strains with parent strains to improve the vigor of the mutant strains. Because the traditional approach to strain improvement involves many steps of heavy, random mutagenesis, each step introduced many unwanted mutations into the organism, and the overproducing strain that resulted was invariably a weakened strain, one that grew poorly. After backcrossing such a strain with the wild type, one can hope to find progeny cells that have inherited the overproducing traits from the mutant parent and the wild-type hardiness and vigor from the wild-type parent.

At first, this did not seem possible with *Penicillium*, which does not have a true sexual cycle. However, a parasexual cycle resulting in the production of heterokaryons was discovered in *Penicillium* in 1958 and was used to improve the strains.

Some strains of *Streptomyces*, a eubacterium, carry out conjugational transfer of DNA. David Hopwood and coworkers, who have made extensive studies of *Streptomyces* genetics, found that in many cases, genes of antibiotic synthesis are clustered together. Recombinant DNA methods have shown that this is also true of the genes involved in the biosynthesis of commercially important antibiotics (see below).

Scientists also used protoplast fusion, a technique that achieves the efficient intercellular transfer of genetic material in species that, like many antibiotic-producing organisms, have no natural mechanism for conjugation. This method was used often in the backcrossing, described above, to restore "vigor" to the line of the producing strains. In this technique, bacterial or fungal cells are converted to protoplasts by dissolving the cell wall with lytic enzymes. The membranes of two protoplasts are then encouraged to fuse together by the addition of high concentrations (typically 20% to 35%) of polyethyleneglycol or other agents. Then cell walls are regenerated in the progeny in suitable protective media. Usually the chromosomes from the two parental protoplasts undergo recombination, and the redundant material is eventually discarded in the process of successive cell division. Thus, many of the progeny cells recovered at the end of this process have only one equivalent of genetic material, which consists of a mixture of genes from both parents.

Most of the antibiotic biosynthesis genes *of Streptomyces* are located on the chromosome, which is linear, unlike that in most other bacterial species. Furthermore, many antibiotic biosynthesis genes are located close to the ends of the linear chromosome, a section called "arms." Interestingly, in the genome of *Streptomyces*, very large-scale deletions involving several hundred kilobases occur at a high frequency in the arm region, and thus these deletions sometimes involve the antibiotic production genes.

The antibiotic producers often have to protect themselves against the toxic effects of their own products. This need is especially acute for eubacterial *Streptomyces* that produce antibiotics with antieubacterial activity. Many antibiotic-producing species thus have mechanisms that inactivate or pump out the antibiotics or produce resistance by modifying the target. In most cases studied, the genes responsible for resistance are part of the cluster of antibiotic synthesis genes. As we will see later, this arrangement has been useful in the cloning of antibiotic synthesis genes.

### Targeted Mutagenesis

As we saw above, the major problems plaguing the classic strain improvement procedure, which is based on random mutagenesis, were the very low probability of introducing mutations into relevant genes and the high rate of unwanted mutations in other, unrelated genes. These problems can be avoided by mutagenizing only the relevant pieces of DNA.

Because many antibiotic production genes have now been cloned, targeted mutagenesis of the cloned DNA can be performed *in vitro*, followed by transformation of the recipient organism. Although intact cells of *Streptomyces* are difficult to transform, their protoplasts, without the cell wall barrier, are easily transformed.

### Rational Selection

In traditional strain improvement methods, the progeny produced after the mutagenesis have to be *screened* – looked at one by one – to find a line that produces more of the antibiotic. If methods were available for *selecting* an improved producer out of a very large population of progeny – say, 100 million cells – then the strain improvement process would be infinitely more efficient (for the distinction between screening and selection, see Box 3.3). Although nobody has yet been able to devise a general protocol for such a selection, some indirect selection methods have been used in certain cases. For example, some antibiotics, notably penicillins and tetracyclines, are chelators of heavy metal ions. The more of these antibiotics an organism produces, the more resistant it will be to heavy metals in the medium. Thus, selection for mutants resistant to heavy metals was used in the improvement of the penicillin producers.

## GENOMICS AND CLONING

The complete genomes of *Streptomyces coelicolor* (a "model" *Streptomyces* that has been studied in detail) and *Streptomyces avermitilis* (the producer of the antiparasitic drug avermectin) have been sequenced. These sequences give fascinating insights into the biology of *Streptomyces*. (1) The genome is very large, containing 8.7 to 9.0 million bases. This is about twice the size of the *E. coli* genome (4.7 million bases) and approaches the sizes of the genomes of eukaryotic microbes – for example, the *Saccharomyces cerevisiae* genome (12 million bases). (2) Gene expression is regulated in a complex manner. An astounding number (60 to 65) of sigma factors (which recognize different classes of promoters) were identified, and this can be contrasted with *E. coli*, which produces only seven. (3) About 800 secreted proteins were found, in contrast to only one or two proteins secreted by *E. coli* K-12. (4) There are 21 (*S. coelicolor*) and 30 (*S. avermitilis*) gene clusters involved in the production of secondary metabolites, including antibiotics.

Genes involved in the production of a single antibiotic are usually grouped together in a closely linked cluster. As early as 1984, Francisco Malpartida and David Hopwood succeeded in cloning all the genes for the biosynthetic pathway of actinorhodin into a single plasmid. The complete gene clusters coding for the production of numerous commercially important antibiotics have since been cloned.

Cloning of the production genes also clarified the changes brought about by the traditional strain improvement. For example, the wild-type strain of *P. chrysogenum* NRRL-1951 contained a single copy each of genes involved in penicillin biosynthesis, but penicillin-overproducing strains developed after many steps of mutagenesis and selection (Figure 10.62) contained up to 50 copies of these genes, amplified on the chromosome.

The cloning of antibiotic production genes has had a profound impact on the field. First, the cloning and sequencing of these genes gave us new and important insights into the nature of antibiotics. Most antibiotics, and indeed most secondary metabolites, are now known to be produced by the polyketide synthesis pathway, the nonribosomal peptide synthesis pathway, or the isoprenoid pathway. The first two pathways proceed by using large, multifunctional enzymes, to which the intermediates are bound as thioesters to the sulfhydryl of the pantetheine groups of enzymes. As we mentioned at the beginning of this chapter, the structures of the known antibiotics are bewilderingly diverse. However, macrolides (see Figure 10.19), tetracyclines (see Figure 10.21), polyenes (see Figure 10.28), FK-506 (see Figure 10.14B), and anthracyclines (see Figure 10.2), are produced by the polyketide synthesis pathway, and classical peptide antibiotics (bacitracin, polymyxin; see Figure 10.23) as well as $\beta$-lactams, cyclosporin A (see Figure 10.14A), vancomycin (see Figure 10.24), and daptomycin (see Figure 10.29) are produced mainly by the nonribosomal peptide synthesis pathway. Some antibiotics, such as rifamycins (see Figure 10.20) and bleomycin (see Figure 10.4), are produced by collaboration of the two types of pathways. Both of these pathways participate in the production of important nonantibiotic products: for example, the polyketide pathway is only a minor modification of the fatty acid biosynthetic pathway and produces pigments such as melanin, and nonribosomal peptide synthesis is involved in making siderophores such as enterobactin and ferrichrome. These considerations show that most antibiotics are produced by minor changes in the biosynthetic pathways for "normal" metabolites.

### Polyketide Biosynthesis Pathway

In fatty acid synthesis, an acetyl group is added at each round of synthesis to produce a long chain without branches, and the carbonyl group introduced at the condensation step in each round is usually reduced to the level of $-CH_2-$. In the biosynthesis of polyketide antibiotics, the unit added is often larger than an acetyl, yet each condensation step adds two carbon atoms to the elongating main chain, so the remaining part of the unit protrudes from the main chain as a branch. This is seen in the structure of macrolides, for example, in which a propionate unit is added at each step so that every second carbon in the main chain bears a methyl branch (see Figure 10.19). Some of the carbonyl groups are not reduced at all, and others are reduced only to the level of CHOH (Figure 10.19). In spite of these differences, cloning the macrolide biosynthesis genes and sequencing them revealed a remarkable

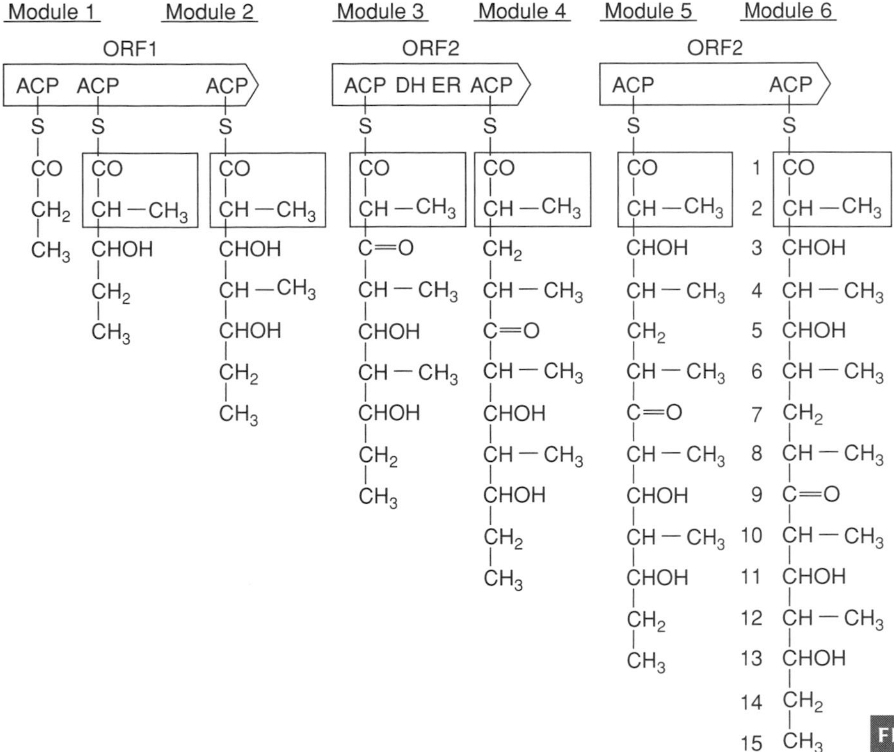


**FIGURE 10.63**

The organization of genes for erythromycin A biosynthesis in *Saccharopolyspora erythrea*. The 30-kb region of DNA is divided into three open reading frames (ORFs), each of which codes for a large, complex enzyme molecule. Each enzyme is divided into two modules. Each module contains an acyl-carrier protein domain (ACP), an acyl transferase domain, and a $\beta$-ketoacyl-ACP synthase domain (for clarity, only the ACP domain is shown). Each module adds a new propionic acid unit (shown in *boxes*) to the growing chain. All the modules except module 3 also contain a $\beta$-ketoacyl-ACP reductase domain. Thus, the keto group of the newly added unit is reduced (usually to CHOH) except in the unit added by module 3 (C9 in the completed open-chain structure, far right, where the numbering scheme for the carbon atoms is shown). Module 4 contains dehydratase (DH) and enoyl-ACP reductase (ER) domains in addition to all four domains mentioned already; these two additional domains convert the original keto group in the propionic acid moiety added in the preceding step (by module 3) into a methylene group (C7 in the completed chain, far right). [Based on the results of Donadio, S., et al. (1991) Modular organization of genes required for complex polyketide biosynthesis. *Science*, 252, 675–679.]

homology between these polyketide synthesis genes and the fatty acid synthetase genes.

Fatty acid synthesis is carried out either by type I systems, which are prevalent in eukaryotes and contain all necessary enzymatic activities and the carrier protein as domains of a large, single protein, or by type II systems, which are usually found in bacteria and are composed of many proteins, each carrying out a single function. Similarly, polyketide synthases are either type I or type II, the latter often catalyzing the synthesis of compounds with polyaromatic ring structures, such as daunomycins (see Figure 10.2).

The production of the erythromycin macrocyclic ring by *Saccharopolyspora erythrea* (*Streptomyces erythreus*, in older literature) is catalyzed by the type I modular system, involving only three giant proteins, each of which consists of two modules catatalyzing the two-carbon extension of the macrolide skeleton. Thus, the entire enzyme complex contains six "modules," each of which in turn contains acyl-carrier protein, acyl transferase, ketoacyl-ACP synthase, and other domains necessary for the biosynthetic process (Figure 10.63). In erythromycin A, only $C_7$ is reduced to the level of $CH_2$ (seen in Figure 10.19), and only module 4 contains the domains needed for this reduction: sequences homologous to the $\beta$-hydroxyacyl-ACP dehydratase and enoyl-ACP reductase. Thus, we can deduce that each of modules 1 through 6 catalyzes, in precise succession, the addition of C3 units and that the fate of the keto group depends on the enzyme content of the particular module. These findings have far-reaching implications. Although a large number of polyketide antibiotics exist, it appears that the specific structure of each compound is determined by the substrate specificity and composition of each enzyme module. This means that the apparent diversity of the structure of polyketide antibiotics is rather deceptive and that the producing organisms can create an impressive variety of antibiotics by introducing minor variations in the biosynthetic apparatus. Indeed, the cloning of the erythromycin biosynthetic gene cluster produced explosive research activity to modify the various catalytic units of various modules of synthase. It also allowed the exchange of modules between gene clusters producing different macrolides, which led to the production of hybrid antibiotics. Studies of this type resulted in the production of several hundred new macrolide structures within only five years.

Much of this research was carried out in *S. coelicolor*, a "model" *Streptomyces*. Recently, the expression of erythromycin biosynthesis cluster was achieved in *E. coli*. By expressing the genes coding for the biosynthesis of precursors that are not made by *E. coli*, it became possible to let the plasmid-containing *E. coli* produce the macrolide portion of erythromycin. Because gene manipulation is so much easier in *E. coli*, this development is expected to speed up, even more, the effort to produce novel macrolides, a feat that is difficult to achieve by chemical synthesis because of the complexity of their structure.

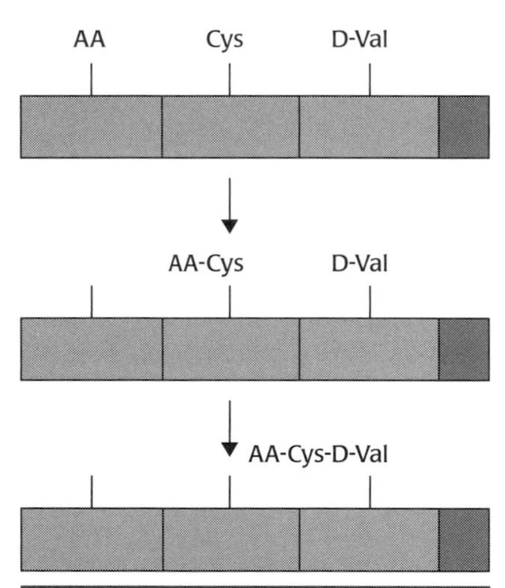


**FIGURE 10.64**

Model of ACV synthetase. ACV synthetase, catalyzing an important step in penicillin and cephalosporin synthesis, is a giant protein with three similar domains, which activate the three amino acids L-$\alpha$-aminoadipic acid (AA), cysteine (Cys), and valine (Val). The valine activation domain additionally contains an epimerization domain (*black rectangle*), and Val is converted into D-valine. The three amino acids are attached to the sulfhydryl groups of 4′-phosphopantetheine. Most likely there will be a sequential extension of the peptide as shown in the figure, culminating in the final release of ACV tripeptide.

## Nonribosomal Peptide Synthesis Pathway

Nonribosomal peptide synthases are also organized into modules, each of which catalyzes the activation and addition of a single amino acid. The modules are usually organized into a single (or at most a few) giant proteins, a situation reminiscent of the type I polyketide synthase (Figure 10.64). The growing peptide is transferred sequentially from one phosphopantetheine SH to another one. This mechanism explains the fact that many antibiotics synthesized by enzymes of this type often contain "unnatural" amino acids (D-amino acids, ornithine, methylamino acids, aminoadipic acid, diaminobutyric acid): all that is needed is an amino acid activation domain with specificity for these compounds (see Figure 10.64).

In principle, modification of the specificity of nonribosomal peptide synthase should produce peptides of novel structure, and successful pilot experiments have been reported. However, this area has not been pursued as vigorously as it has in the modification of polyketide synthases, primarily because peptides can be produced easily by chemical synthesis, unlike polyketides.

The genes coding for the first two enzymes of $\beta$-lactam biosynthesis, ACV synthetase (a single gene called, rather confusingly, *pcbAB*, because earlier data suggested two separate biochemical steps) and isopenicillin N synthase (*pcbC*), were found to exist next to each other in both the bacterial and the eukaryotic producers of $\beta$-lactams. In addition, strong sequence similarity was found among homologous genes of these clusters from diverse organisms. These observations suggest that the capacity to produce $\beta$-lactams was spread from a prokaryote into eukaryotic fungi by a horizontal gene-transfer process fairly recently on an evolutionary time scale (370 million years ago, according to one estimate).

### Other Enzymes of Antibiotic Biosynthesis Pathways

Biosynthesis of most antibiotics involves enzymes other than polyketide synthetase and nonribosomal peptide synthetase. The last decade witnessed many improvements in antibiotic producers involving genes of these other enzymes. For example, 7-aminocephalosporanic acid, the starting material for many semisynthetic cephalosporin drugs, is produced from cephalosporin C first by converting the $\alpha$-aminoadipyl side chain with D-amino acid oxidase into a glutaryl group, and then by removing this group with a glutaryl acylase (see Figure 10.51). An attempt was made to accomplish this conversion in the producing fungal cell by the introduction of relevant exogenous genes.

7-Aminodeacetoxycephalosporanic acid (7-ADOCA), the starting material for another group of semisynthetic cephalosporins, is usually produced by the removal of the side chain, starting from phenylacetyl-7-ADOCA, obtained by chemical ring expansion from penicillin G, a costly and environmentally unfriendly process. However, in a more recent process, scientists introduced into *P. chrysogenum cefD* and *cefE* genes from *Streptomyces* (see Figure 10.59). The resultant strain has a truncated cephalosporin biosynthetic pathway and produces aminoadipyl-7-ADOCA under suitable conditions. The aminoadipyl side chain can then be removed by the standard two-step enzymatic process (see Figure 10.51). This method is reported to be operating at an industrial scale in Europe.

### Increasing the Yield and Decreasing the Production Cost

Gene-cloning studies also revealed possible ways to increase the yield of fermentation. For example, it is possible to enhance antibiotic production by increasing the copy number of strategic genes, such as a gene whose low level of expression creates a "bottleneck" in the pathway. An approach of this kind was successfully used with the cephalosporin C–producing organism *A. chrysogenum*. Cephalosporin is produced by the pathway shown in Figure 10.59. Because this organism produced about one third as much penicillin N as cephalosporin C, investigators hypothesized that a bottleneck was occurring at the conversion of penicillin N to a cephalosporin by "expandase" (deacetoxycephalosporin C synthetase/deacetylcephalosporin synthetase), an oxygenase catalyzing conversion of the penicillin nucleus to the cephalosporin nucleus (see Figure 10.59). Researchers introduced the cloned expandase gene into the organism, where it apparently was integrated into a chromosome, thereby doubling the copy number. The engineered strain produced significantly larger (by 20% to 40%) amounts of cephalosporin C. This was accompanied by a drastic decrease in the secretion of penicillin N by the engineered strain, perhaps a clearer indicator of the experiment's success.

A possible beneficial outcome of cloning research could be that when an antibiotic is synthesized by a relatively simple pathway, the enzymes of that pathway can be overproduced from cloned genes and then placed in reactor columns to make antibiotics, or at least intermediates of antibiotic biosynthesis, *in vitro*. Investigators have done this successfully on an experimental scale, producing penicillins and cephalosporins in immobilized enzyme reactors by using crude extracts of producing organisms. The starting materials for $\beta$-lactam biosynthesis are common, inexpensive amino acids, so the overproduction of critical enzymes such as isopenicillin N synthase and expandase might one day make this process a commercially viable one.

## PHYSIOLOGY OF ANTIBIOTIC PRODUCTION

Antibiotics are small molecules whose synthesis often requires dozens of enzymes. Enzyme activities are of necessity closely regulated in such complex pathways. It is therefore important to understand the physiology of the producing organisms in order to maximize the fermentative production of antibiotics.

### Secondary Metabolism and Its Control

What sets the fermentative production of antibiotics apart from other types of fermentation is that antibiotics are typical *secondary metabolites*. Primary metabolites are produced during the entire growth phase of a culture. Thus, continuous processes are often feasible and preferable for these compounds. In contrast, the production of secondary metabolites is regulated in a complex manner and often occurs only when the cultures are entering, or are already in, a stationary phase. Thus, in most cases, continuous fermentation (Box 10.4) is not an option.

**Continuous Fermentation**

In the usual fermentation process, called *batch fermentation*, a given amount of medium in a tank is sterilized and inoculated with the microorganism. The culture goes through lag and exponential phases of growth and finally reaches the stationary phase at which there is little or no net increase in the density of the organisms. In contrast with this closed system, *continuous fermentation* is an open system: Fresh, sterile medium is constantly added, and the same amount of medium containing microorganisms and products is constantly taken out. Continuous fermentation is advantageous because the culture can always be at a highly productive concentration; in batch fermentation, much time is wasted waiting for the culture to attain productive concentrations. In spite of this advantage, only a few products are currently made by continuous fermentation on an industrial scale, partly because avoiding contamination in such an open system is difficult.

BOX 10.4

Because many antibiotics are produced by spore-forming organisms (*Streptomyces* among prokaryotes and filamentous fungi among eukaryotes), and because both antibiotic production and sporulation occur close to the beginning of the stationary phase, one may suspect that these two processes are regulated by an overlapping mechanism. Molecular genetic studies of antibiotic-producing prokaryotes, such as *Streptomyces*, are finally shedding some light on the complex regulatory mechanisms that operate in these processes. Thus, the antibiotic production and sporulation are both modulated by intercellular signaling molecules. These signals include both peptides and membrane-permeable lactones similar to the acyl homoserine lactones known to operate as quorum-sensing signals in Gram-negative bacteria (Figure 10.65).

However, in some cases there is no tight connection between the spore formation and antibiotic production. This becomes clear in antibiotic production by nonsporulating organisms. For example, a typical Gram-negative, quorum-sensing signal *N*-hexanoyl homoserine lactone (Figure 10.65) induces the production of carbapenem by *Erwinia carotovora* (which is related to *E. coli*) by directly binding to the repressor protein of the

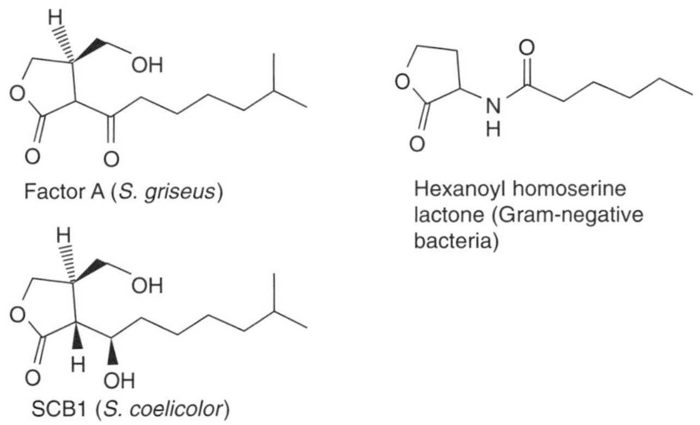

Factor A (*S. griseus*)

Hexanoyl homoserine lactone (Gram-negative bacteria)

SCB1 (*S. coelicolor*)

**FIGURE 10.65**

Lactones that act as intercellular signals in *Streptomyces* (left) and in Gram-negative bacteria (right).

carbapenem production operon. Also, in some species of *Streptomyces*, cytosolic receptors for lactones directly activate the transcription of genes for antibiotic production in a similar manner.

### Catabolite Repression

The requirement for diffusible, quorum-sensing signals at least partially explains the fact that antibiotic production is limited to the stationary phase, in which the cell density becomes higher. One can hypothesize that at low cell densities, rapid growth and therefore primary metabolism are the first priority and that only when the growth is slowed down at high cell densities will the cells devote much of their energy to the production of secondary metabolites such as antibiotics. Indeed, many antibiotic-producing organisms are less productive in the presence of an excess carbon source, such as glucose. This is reminiscent of the catabolite repression phenomenon that is well known in *E. coli*. To overcome catabolite repression, carbon sources must be added to the culture medium in carefully adjusted, small increments.

### Nitrogen and Phosphate Repression

In many cases, the presence of excess nitrogen compounds or phosphate in the fermentation medium decreases antibiotic production severely; the ecological advantage of such regulation is probably similar to that of catabolite repression. Phosphate has been shown to inhibit the transcription of some of the genes of antibiotic synthesis, and this regulation is abolished in mutants with deletion of the PhoR-PhoP two-component regulatory system.

### Feedback Regulation

Some scientists suspect that the antibiotics themselves, as end products, may exert negative-feedback regulation on their synthesis. Supporting data come from experiments in which penicillin added to the culture of penicillin-producing fungi apparently inhibited the synthesis of the antibiotic.

Furthermore, the level of exogenous penicillin required for this inhibition was much higher with overproducers of penicillin, suggesting that resistance to this feedback inhibition was at least a factor in the overproduction in these strains.

### Effects of Precursors

Secondary metabolites have to be synthesized from primary metabolites. Thus, the efficient production of antibiotics requires a steady flow of their precursors. In many cases, the production of these precursors is regulated by known mechanisms.

An interesting example of how the supply of precursors is regulated and how it affects production of the antibiotic is the effect of culture conditions on the production of $\alpha$-aminoadipic acid, a precursor for $\beta$-lactam biosynthesis. In fungi, $\alpha$-aminoadipic acid is an intermediate in the pathway of lysine biosynthesis. Because lysine is the end product of a biosynthetic pathway, a high level of lysine in the medium shuts off that biosynthesis by inhibiting the first enzyme of the pathway (feedback inhibition; see Chapter 9). This results in a shortage of all the intermediates on the pathway, including $\alpha$-aminoadipic acid. Thus, the presence of excess lysine strongly inhibits penicillin production in *P. chrysogenum* fermentations. In striking contrast, the addition of excess lysine *stimulates* the production of cephamycin C in *Streptomyces*. This is because $\alpha$-aminoadipic acid is synthesized by a totally different route in eubacteria, lysine being the precursor (Figure 10.66).

In addition to $\alpha$-aminoadipic acid, the biosynthesis of penicillin or cephalosporin requires the presence of cysteine and valine (see Figure 10.59). The way cysteine is made is different in different species and even in different strains. In *P. chrysogenum*, much of the sulfur atom of cysteine is

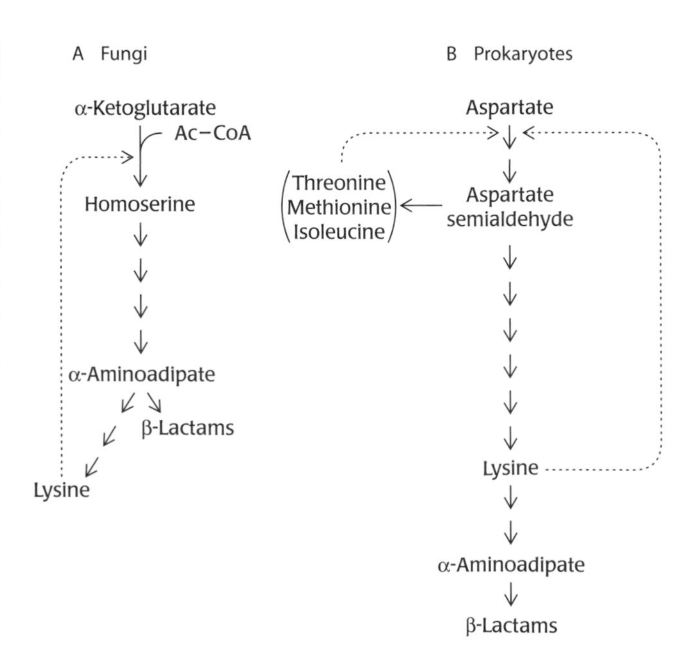


**FIGURE 10.66**

Pathways for the biosynthesis of $\alpha$-aminoadipic acid. **A.** In fungi, $\alpha$-aminoadipic acid is a precursor of lysine. Because lysine regulates the first step of the pathway (*dotted line*) by negative feedback, the addition of lysine inhibits $\beta$-lactam synthesis by decreasing the supply of $\alpha$-aminoadipic acid. **B.** In contrast, bacteria such as *Streptomyces* make $\alpha$-aminoadipic acid from lysine. Although the addition of lysine inhibits its own synthesis (*dotted line*), exogenous lysine is converted efficiently into $\alpha$-aminoadipic acid and stimulates the synthesis of $\beta$-lactam.

Redrawn based on artwork from the first edition (1995), published by W.H. Freeman.

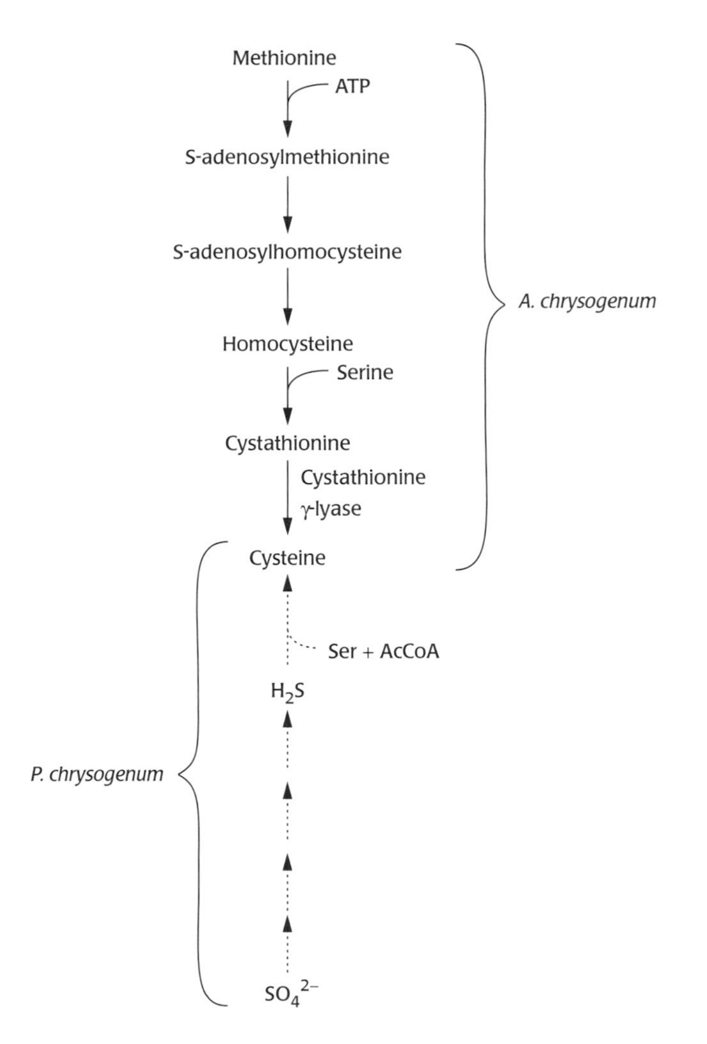


**FIGURE 10.67**

Major pathways for the biosynthesis of cysteine, which in turn is used in $\beta$-lactam synthesis. Available data suggest that in *A. chrysogenum*, much of the cysteine is made by transsulfuration via cystathionine (*solid arrows*). In contrast, in *P. chrysogenum*, the conversion of sulfate into cysteine (*broken arrows*) appears to play a much more important role.

derived from inorganic sulfate in the medium. In contrast, *A. chrysogenum*, which produces cephalosporin, derives much of its cysteine from methionine via a transsulfuration reaction (Figure 10.67). In this case, the addition of methionine to the medium strongly stimulates cephalosporin production, at least partly by increasing the supply of cysteine. Furthermore, when several strains producing higher amounts of cephalosporin C were examined, a proportional relationship emerged between the level of the cystathionine $\gamma$-lyase, an enzyme involved in cysteine production, and the yield of the drug.

### Conditions for Fermentation

Numerous empirical improvements have doubtless been made in the conditions for fermentative production of antibiotics, but most of them constitute proprietary knowledge. Apparently, many fermentation processes are run in two stages. The first stage starts from spores, usually of established strains, because antibiotic production is often an unstable trait that can be lost. The organisms are cultured under submerged conditions, with sufficient aeration and a generous supply of nutrients so that they will attain near-maximal density in a short time. In the second stage, when the culture reaches the stationary phase or stops growing and begins to produce antibiotics, the concentration of the key nutrients, such as carbon source, phosphate, and nitrogen source, must be controlled carefully by continuous-feed processes.

The fermentative production of penicillin is much better described in the literature than that of other antibiotics. Published data indicate that with the currently available strain of *P. chrysogenum*, a large fraction of the carbon in the glucose added finds its way into penicillin G. Particular attention must be paid to supplying just-sufficient amounts of the side-chain precursor phenylacetic acid, which is toxic and therefore must be added slowly by a controlled-feed process.

## PROBLEM OF ANTIBIOTIC RESISTANCE

A unique feature of the antibiotic field is the constant need to develop new agents to keep pace with the constant increase in the frequency of resistant isolates. Because physicians must begin antibiotic therapy before the causative microorganism is identified and its drug susceptibility pattern is determined, if the frequency of resistance for a given drug in any given pathogen species exceeds a certain level (usually around 10%), they will essentially stop using that drug: the drug thus will become "useless."

Antibiotics are manufactured on a very large scale (100,000 tons annually worldwide, according to one estimate) and have been used for human therapy and other purposes since the mid-twentieth century. This practice has had a profound impact on the ecology of bacteria on Earth. More and more strains in any given species have become resistant to antibiotics. Figure 10.68 shows just one example of penicillin resistance among *S. pneumoniae*, the classical causative organism of pneumonia. A similar trend can be found for almost any organism for almost any drug.

As a result of this trend, some strains have become resistant to practically all of the commonly available agents. A notorious case is MRSA, which

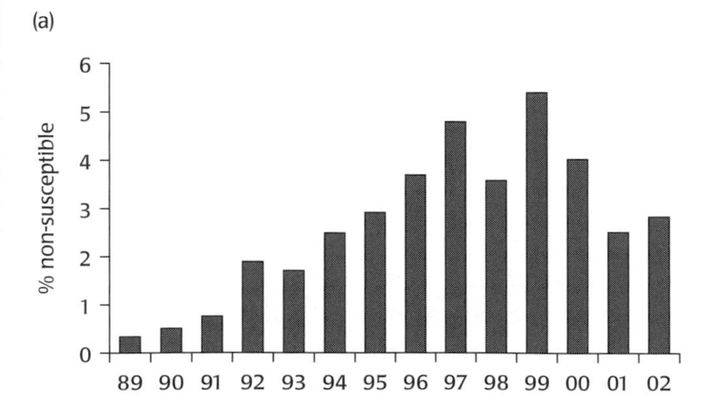


**FIGURE 10.68**

Prevalence of penicillin resistance (or more correctly, nonsusceptibility) in invasive *Streptococcus pneumoniae* isolated in most clinical laboratories in England and Wales. [From Livermore, D. M. (2004) The need for new antibiotics. Clinical Microbiology and Infectious Diseases, 10(suppl 4), 1–9; with permission from the author and the publisher.]

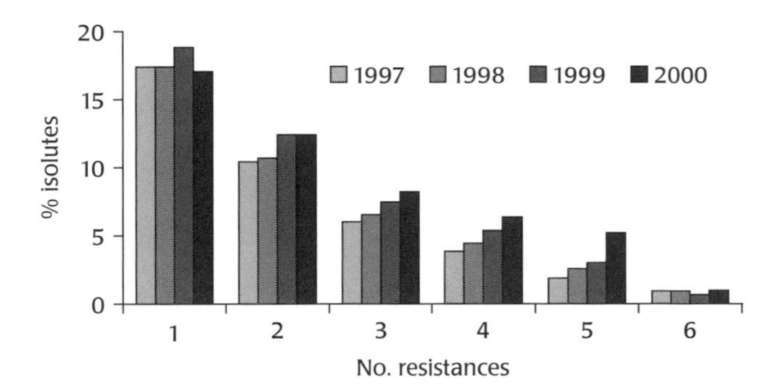


**FIGURE 10.69**

Proportion of *P. aeruginosa* isolates resistant to one to six agents among those that are currently the most effective against this organism (ceftazidime, piperacillin, imipenem, amikacin, gentamicin, and ciprofloxacin). The year in which the strains were isolated is indicated by the shading of the bars. The data are those reported from more than 250 laboratories to the TSN-Database. [From Livermore, D. M. (2004) The need for new antibiotics. *Clinical Microbiology and Infectious Diseases*, 10(suppl 4), 1–9; with permission from the author and the publisher.]

emerged in the early 1990s and is usually resistant not only to methicillin (which was developed to fight against penicillinase-producing *S. aureus* as described in the section on $\beta$-lactams) but also to aminoglycosides, macrolides, tetracycline, chloramphenicol, and lincosamides. Because such strains are also resistant to disinfectants, MRSA has become a major source of hospital-acquired infections. An old antibiotic, vancomycin, was "resurrected" for treatment of infections caused by MRSA. However, transferable resistance to vancomycin is now quite common in another Gram-positive group, *Enterococcus*, and finally found its way to MRSA in 2002, although such strains are still rare as of this writing.

Some specialists feel that an even more serious threat is the emergence of Gram-negative pathogens that are resistant to essentially all of the available antibiotics. The threat of MRSA and the likely outcome that it will eventually acquire vancomycin resistance have been known for some time, and researchers had time to react against this threat. Thus, there are newly developed agents that can be used against vancomycin-resistant MRSA. These include linezolid (a totally synthetic drug active against only Gram-positive bacteria), quinupristin/dalfopristin (natural antibiotics of the streptogramin class), and daptomycin (mentioned in this chapter). However, the emergence of "pan-resistant" Gram-negative species occurred more recently, after most major pharmaceutical companies dropped their development programs for antibacterial agents. Hence there are very few or no agents that could possibly be used against these strains. In terms of numbers, *P. aeruginosa* is a major cause of hospital-acquired infections. This species was "intrinsically" resistant to $\beta$-lactams of earlier generations (see Table 10.3) as well as to agents such as tetracycline and chloramphenicol because of the presence of a rather impermeable outer membrane and of effective multidrug efflux pumps (see below). In addition, many strains carry chromosomally coded aminoglycoside-inactivating enzymes. Thus, the choice was limited only to $\beta$-lactams of more recent generations, aminoglycosides that withstand common mechanisms of inactivation (i.e., amikacin or gentamicin), and fluoroquinolones. Most alarmingly, isolates showing resistance to four or five among six such useful agents are increasing very rapidly (Figure 10.69). Although lower in incidence, *Acinetobacter baumannii*, an environmental organism that causes hospital-acquired infections, is closer to achieving the status of pan-resistance. It has been generally resistant to most common antibiotics for some time, leaving imipenem as the only choice. However, the frequency of imipenem resistance (including intermediate-level resistance) had already increased to about 15% in 2001 in this species.

## BIOCHEMICAL MECHANISMS OF RESISTANCE

Frequently found resistance mechanisms are as follows:

**1.** Enzymatic inactivation of the drug. This is a common resistance mechanism for antibiotics of natural origin, as we have seen with aminoglycosides (enzymatic phosphorylation, acetylation, or adenylation) and $\beta$-lactams (enzymatic hydrolysis by $\beta$-lactamases). Genes coding for such enzymes will make the bacterial cell resistant when they are present as additional genetic components on plasmids. This may be one of the reasons why such genes have become so prevalent on R plasmids.

**2.** Mutational alteration of the target protein. Completely man-made compounds, such as fluoroquinolones, are unlikely to become inactivated by the enzymatic mechanisms just described. However, bacteria can still become resistant through mutations that make the target protein less susceptible to the agent. Fluoroquinolone resistance is mainly the result of mutations in the target proteins, DNA topoisomerases (although high-level resistance also requires increased active efflux, discussed below). Resistance of this type, caused by alteration of chromosomal gene(s), is not easily transferred to other cells because the recipient will still retain the drug-susceptible, unaltered target enzyme and consequently will remain susceptible. Nevertheless, such mutants will become more and more prevalent in the presence of selective pressure. Fluoroquinolone resistance is rapidly increasing in almost all groups of pathogens. Figure 10.70 shows the situation with Gram-negative Enterobacteriaceae.

**3.** Acquisition of genes for less susceptible target proteins from other species. Scientists discovered, by sequencing the genes coding for the targets of penicillin, DD-transpeptidase or penicillin-binding proteins, that penicillin resistance among *S. pneumoniae*, which has become more frequent in recent years (see Figure 10.68), is largely the result of the production of mosaic proteins, parts of which come from other organisms

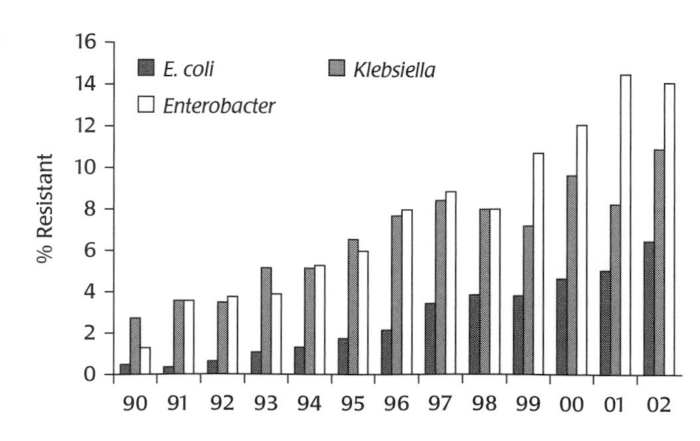


**FIGURE 10.70**

Growing proportions of *E. coli* (black), *Klebsiella* species (gray), and *Enterobacter* species (white) resistant to ciprofloxacin isolated in the year indicated on the abscissa. Based on reports to the U.K. Health Protection Agency for bacteremia cases from most clinical laboratories in England and Wales. [From Livermore, D. M. (2004) The need for new antibiotics. *Clinical Microbiology and Infectious Diseases*, 10(suppl 4), 1–9; with permission of the author and the publisher.]

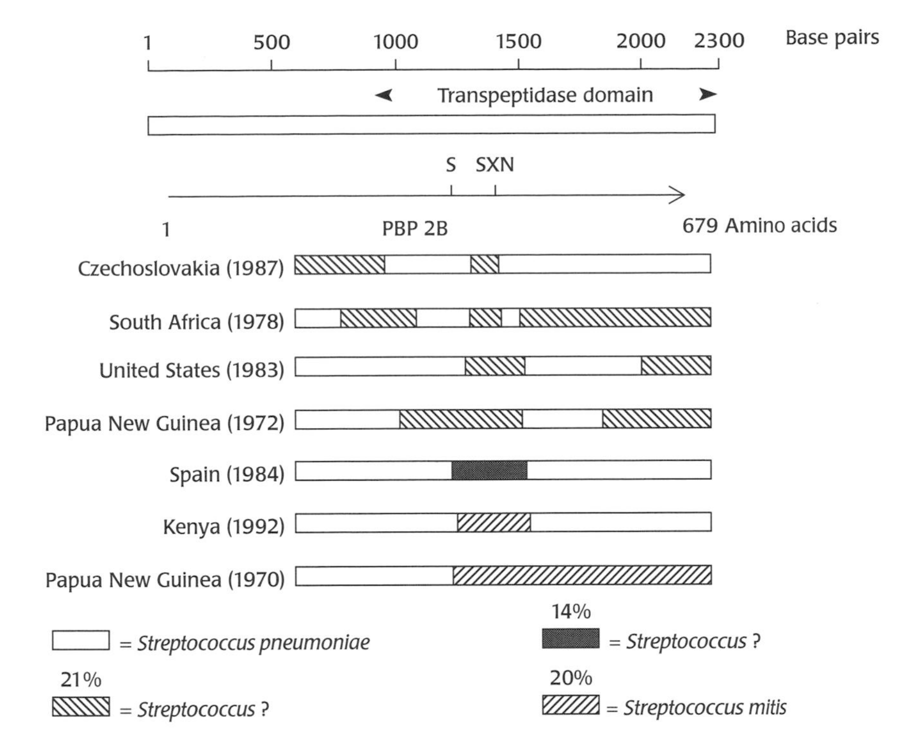


**FIGURE 10.71**

Mosaic structure of the transpeptidase domain of PBP-2B in penicillin-resistant clinical isolates of *S. pneumoniae*. As shown in the *top bar*, PBP-2B, a major target of penicillin in *S. pneumoniae*, is composed of the N-terminal transglycosidase domain and the C-terminal, 486-residue transpeptidase domain (shown with the *horizontal arrow*). "S" shows the position of the active site serine residue (Ser192 in the transpeptidase domain). "SXN" is a conserved motif. In the transpeptidase domains of PBP-2B from resistant isolates, shown in the five *shorter bars*, various regions of this domain are replaced by sequences from organisms other than *S. pneumoniae*, as shown by the *shaded rectangles*. Especially important are the substitutions in residues 232 to 238, occurring soon after the active site serine residue, which lower the affinity of this enzyme to penicillin. [From Spratt, B. (1994). *Science*, 264, 388–393; with permission.]

(Figure 10.71). Most likely this development is related to the fact that *S. pneumoniae* is an organism capable of natural transformation (see Chapter 3) and can readily import DNA of other organisms. Interestingly, penicillin resistance due to the production of mosaic penicillin-binding proteins was also found in another organism capable of natural transformation, *Neisseria meningitidis*.

An extreme case in this scenario is the generation of MRSA. MRSA strains contain a new methicillin-resistant penicillin-binding protein, called PBP-2A or 2′, whose expression is often induced by methicillin and other $\beta$-lactams. The gene for this new penicillin-binding protein is located in a large (30- to 60-kb) segment of DNA, which apparently came from an organism other than *S. aureus* and also contains other genes coding for resistance to macrolides and aminoglycosides. (The presence of a plasmid with genes coding for the conventional $\beta$-lactamase and for tetracycline resistance makes these strains resistant to almost all available agents, except vancomycin.) *S. aureus* is not naturally transformable, yet its genome sequence revealed the presence of a number of other genes apparently imported from distantly related organisms.

**4.** Bypassing of the target. Vancomycin, a fermentation product from streptomycetes, has an unusual mode of action. Instead of inhibiting an enzyme, it binds to a substrate, the lipid-linked form of disaccharide pentapeptide, which is a precursor of cell wall peptidoglycan. Because of this mechanism, some scientists assumed that it would be impossible to generate resistance against vancomycin. However, vancomycin resistance is now prevalent among enterococci. Because enterococci, normal

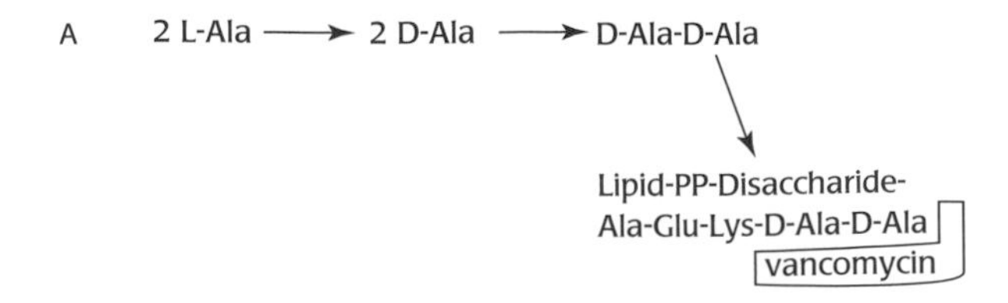


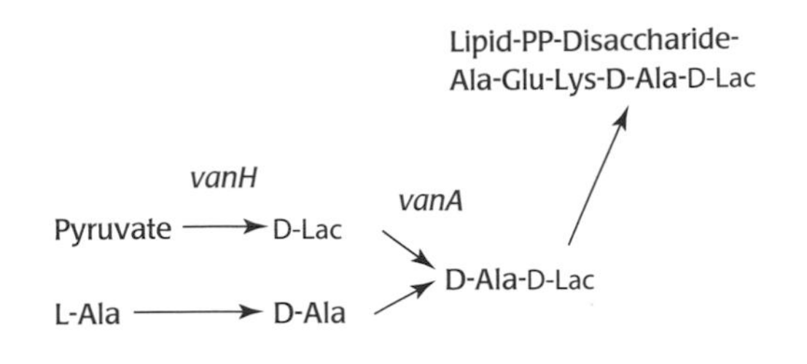


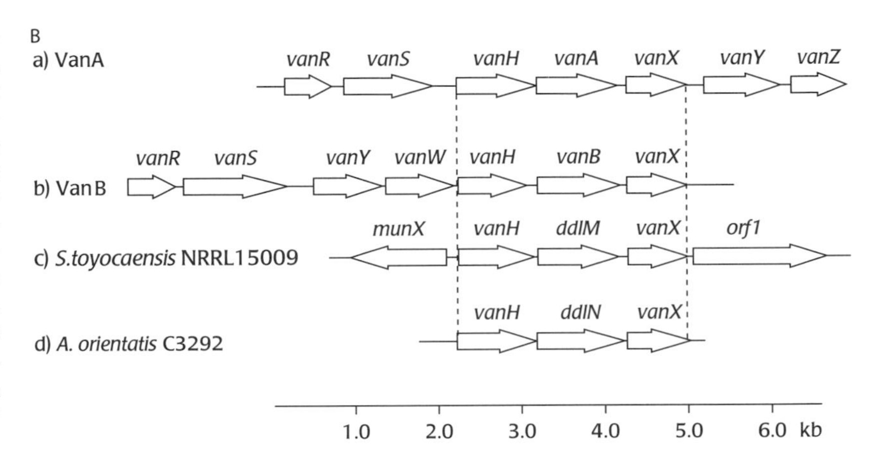


**FIGURE 10.72**

Mechanism of vancomycin resistance. (**A**) The biochemical mechanism. In the wild-type strains (top), the terminal D-Ala-D-Ala portion of the peptidoglycan-biosynthetic intermediate lipid-PP-disaccharide-pentapeptide becomes bound to vancomycin, and the peptidoglycan synthesis becomes inhibited. In the resistant strains expressing *vanA* and *vanH* genes (bottom), this terminal dipeptide is replaced by the D-Ala-D-Lac structure, which does not bind to vancomycin and yet can function as an intermediate for peptidoglycan synthesis. (**B**) Genes involved in vancomycin resistance in pathogens are not only homologous to the vancomycin resistance genes found in drug producer organisms (*Streptomyces toyocaensis* and *Amycolatopsis orientalis*), but are also arranged in a similar manner. [Part **B** is from Marshall, C. G., et al. (1998) Glycopeptide resistance genes in glycopeptide-producing organisms. *Antimicrobial Agents and Chemotherapy*, 42, 2215–2220; with permission from the publisher.]

inhabitants of our intestinal tract, are naturally resistant to $\beta$-lactams, aminoglycosides, macrolides, and tetracycline, these vancomycin-resistant strains of enterococci become prevalent in a hospital environment, colonize the patients, and cause infections that are difficult to treat. Study of the resistance mechanism (Figure 10.72A) showed that the end of the pentapeptide, D-Ala-D-Ala, where vancomycin binds, was replaced by an ester structure, D-Ala-D-lactic acid (Lac) in the resistant strain. This structure still allows the formation of cross-links in peptidoglycan, but it is not bound by vancomycin. Production of this altered structure requires the participation of several imported genes, apparently derived ultimately from the vancomycin-producing organisms (Figure 10.72B).

**5.** Preventing drug access to targets. Drug access can be reduced by an active efflux process, discovered first with tetracycline, or, at least in Gram-negative bacteria, by decreasing the influx across the outer membrane barrier. The latter mechanism is somewhat detrimental to bacterial growth because the influx of nutrients is also reduced. Nevertheless, it is now found in some species of enteric bacteria as a means of "last-resort" resistance to the more recent versions of $\beta$-lactams that withstand inactivation by most common $\beta$-lactamases.

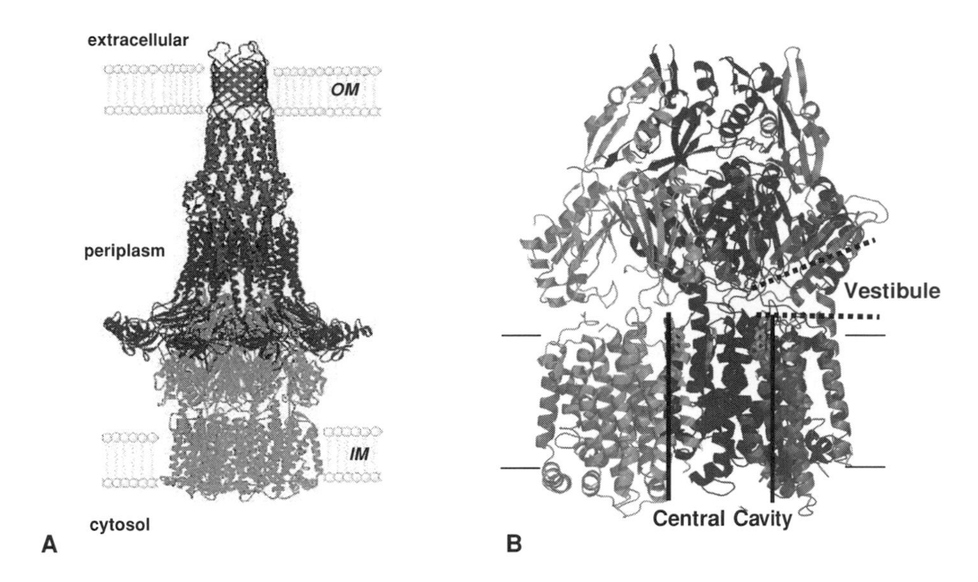


FIGURE 10.73

(**A**) The model of the AcrB-AcrA-TolC tripartite complex that exports drugs directly into the medium. The transmembrane domain of the AcrB pump trimer (green) is embedded in the cytoplasmic membrane, whereas its periplasmic domain is connected to the TolC channel (red) with a number of periplasmic AcrA linker proteins (blue). [From Eswaran, J., et al. (2004) Three's company: component structures bring a closer view of tripartite drug efflux pumps. *Current Opinion in Structural Biology*, 14, 741–747.] (**B**) AcrB trimer with the bound ciprofloxacin molecules. Each protomer is shown in cyan, mauve, and blue, and ciprofloxacin molecules are shown in the green stick models. The large central cavity is connected to the periplasm through vestibules between protomers. The proximal portion of the structure was cut away to reveal the presence of the vestibule, through which drug molecules are thought to be captured from the periplasm. The figure was drawn using PyMol with PDB coordinate 1OYE. Recent elucidation of X-ray structures of asymmetric trimers of AcrB (Murakami S., et al. (2006) Crystal structures of a multidrug transporter reveal a functionally rotating mechanism. *Nature*, 443, 173–179; Seeger M. A., et al. (2006) Structural asymmetry of AcrB trimer suggests a peristaltic pump mechanism. *Science* 313, 1295–1298) suggests how the substrate binds to a binding site within AcrB periplasmic domain and how it is exported into the TolC channel. See the color plate, this chapter.

The efflux process is now known to play an important role in resistance to many drugs. This realization came about as a result of the discovery of multidrug efflux pumps. Multidrug pumps were first identified as plasmid-coded proteins in *S. aureus* that made them resistant to antiseptics like cationic dyes and quaternary ammonium compounds. However, it was found later that such pumps play a much more important role in Gram-negative bacteria because of several reasons: (i) Most Gram-negative bacteria contain chromosomal genes coding for such pumps, and their constitutive expression gives the characteristic "intrinsic" resistance to many antibiotics. For example, *E. coli* is intrinsically resistant to penicillin G, oxacillin, cloxacillin, nafcillin, macrolides, novobiocin, linezolid, and fusidic acid. All this resistance goes away if one constitutively expressed pump, AcrB, is deleted, decreasing the oxacillin MIC, for example, by a factor of 512. (ii) Some of these pumps, especially those belonging to the RND (resistance-nodulation-division) family, show an extremely wide substrate specificity. *E. coli* AcrB can pump out not only most of the common antibiotics but also dyes, detergents, and even solvents. Although AcrB cannot pump out aminoglycosides, there are homologs that carry out this function. (iii) Thus, increased production of these pumps through regulatory mechanisms makes the bacteria more resistant to practically all antimicrobial agents in one single step. (iv) RND family pumps, and sometimes pumps of other families, exist as a multiprotein complex spanning both the cytoplasmic membrane and the outer membrane (Figure 10.73A). This construction allows the extrusion of drug molecules all the way into the external medium (rather than to periplasm, which would be the case for simple pumps located in the cytoplasmic membrane). Because the reentry of the drugs involves penetration across the outer membrane barrier, the efflux functions synergistically with the outer membrane, making the drug efflux an extremely efficient process. (v) Finally, RND pumps often prefer to capture the drugs either in the periplasmic space or from the location that is in equilibrium with the periplasm. This allows the pumps not only to prevent the entry of drugs into the cytoplasm, but also to extrude agents, such as $\beta$-lactams, that have targets in the periplasm and sometimes do not easily diffuse across the cytoplasmic membrane.

We mentioned earlier that fluoroquinolone resistance in *P. aeruginosa* is rapidly increasing. Efflux makes a major contribution in most of such resistant strains. A significant portion of aminoglycoside-resistant strains of Gram-negative bacteria was earlier classified as being the result of "decreased permeability." These strains are now known to owe their resistance to increased active efflux. In view of the important roles multidrug efflux processes play in resistance, many scientists are concerned with the trend of disinfectants being added more and more to household products, such as soaps, because such compounds may select for pump overproduction mutants.

## SOURCES OF THE RESISTANCE GENES

### Producing Organisms

We have described the evidence showing that many of the genes coding for aminoglycoside resistance are derived from streptomycetes that produce these antibiotics. A similar case is the genes coding for vancomycin resistance. Resistance here requires the production of several new enzymes (Figure 10.72B), and it is unlikely that all the genes coding for these enzymes evolved in the few decades after vancomycin was introduced into clinical use. Indeed, the genes in the vancomycin-resistant clinical isolates of enterococci were found to be homologs of those found in the vancomycin-producing streptomycetes, organized in exactly the same manner (Figure 10.72B), an observation that leaves no doubt as to the origin of resistance genes in the clinically relevant species of bacteria.

### Microorganisms in the Environment, Especially Soil

Some resistance genes are found in the chromosomes of environmental bacteria. A classical case is the gene for the class C $\beta$-lactamase in the environmental genera of Enterobacteriaceae, such as *Enterobacter, Serratia,* and *Proteus,* and in the soil organism *P. aeruginosa.* These genes do not show signs that they have been "imported" in the recent past, such as GC content or codon usage that differs from the rest of the genome. Thus, they must have been transferred in the ancient past, as a result of the selective advantage they would confer in an environment full of antibiotic-producing organisms, such as soil.

During the recent age of extensive antibiotic usage, these genes could also have become sources of plasmid-borne resistance. Indeed one of the $\beta$-lactamase genes often found on plasmids apparently came from one of these environmental organisms.

## ACCUMULATION AND TRANSFER OF RESISTANCE GENES

Regardless of the ultimate origin of various genes involved in resistance, it is remarkable that many such genes are usually clustered together on a single R plasmid in drug-resistant clinical isolates of pathogenic bacteria so that resistance to many agents can be transferred to a susceptible bacteria in a single conjugation event. This was already the situation when R plasmids were discovered in Japan in 1950s. Comparison of sequences of many plasmids has now shown how this clustering occurred.

First, many resistance genes become organized in a single operon with the same orientation (of transcription) under a strong promoter in an organization called *integrons*. Integrons contain a gene coding for *integrase*, which catalyzes the insertion of resistance genes at a predetermined site downstream from a strong promoter (Figure 10.74). Once integrated, the resistance gene becomes marked so that it can easily become integrated into another integron, perhaps containing a different set of resistance genes. It is unclear how the modern integrons evolved, but there is a rather similar mechanism that builds the assembly of many genes in *Vibrio cholerae*, which is known to live in very different kinds of environment in its life cycle.

Second, these integrons are usually inserted into transposons, which allow the entire array of the resistance genes to hop between different plasmids and between plasmid and chromosome, increasing their chances of dissemination.

Finally, these genes can become a part of a large transposon of a special class called conjugative transposons. They use a $\lambda$-phage–like mechanism of transposition, which generates a circular DNA (like the replicative form of many phage DNAs), as an intermediate. This intermediate can insert at a predetermined location in the plasmid or chromosomal DNA, but it can also behave like a conjugative plasmid and can transfer its copy into another bacterium by conjugation. Remarkably, in the transposons containing the tetracycline resistance gene, the transfer function can be induced strongly by the presence of tetracycline through a complex regulatory mechanism. Thus, in these transposons, the drug resistance genes are not just passive passengers but are a part of an integrated regulatory mechanism of conjugational transfer; these conjugative transposons are finely tuned weapons for the dissemination of resistance genes. Indeed, such transfer is known to occur easily between distantly related bacteria – for example, between Gram-positive and Gram-negative bacteria.

## WHAT CAN WE DO TO PREVENT THE INCREASE IN DRUG RESISTANCE?

Antibiotic use on a very large scale obviously produces a very strong selective pressure for the survival of any resistant microorganism. The very strong correlation between antibiotic usage and the emergence of resistance can be seen, for example, in the case of penicillin-resistant pneumococci in Europe

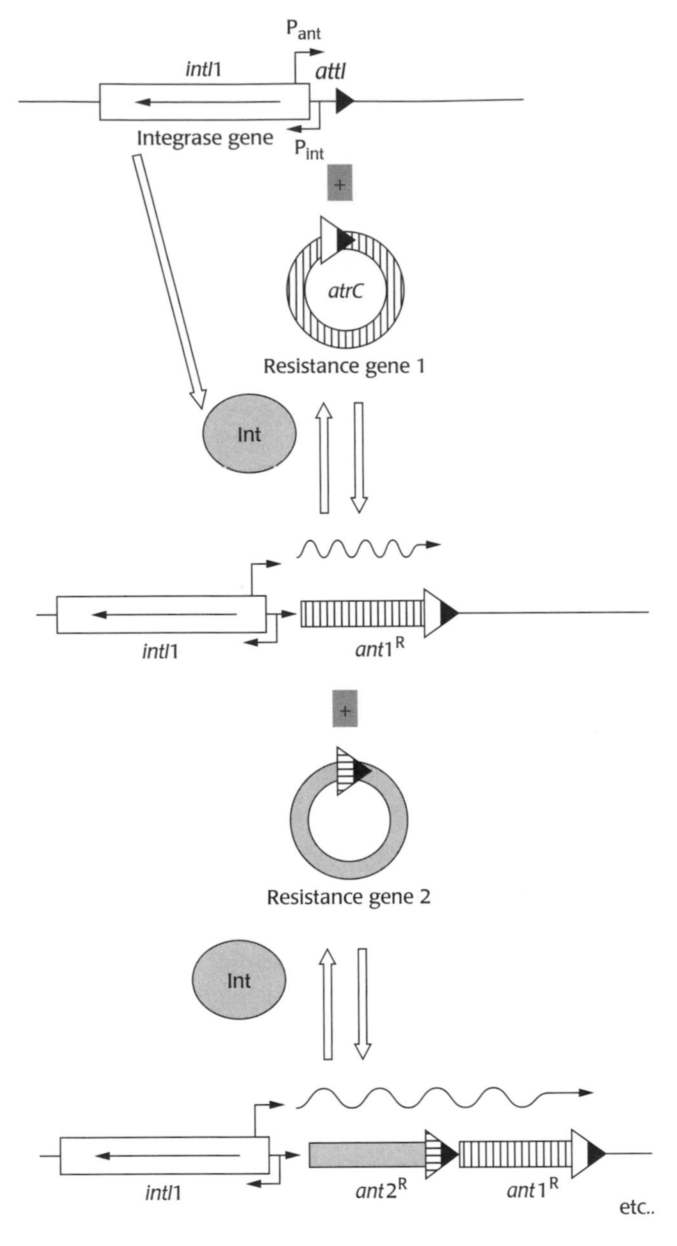


FIGURE 10.74

Presumed mechanism of resistance gene capture by integrons. From [Rowe-Magnus, D. A., and Mazel, D. (1999) Resistance gene capture. *Current Opinion in Microbiology*, 2, 483–488.]

(Figure 10.75). A logical conclusion is that minimizing the pressure would decrease the selection and, hence, would decrease the incidence of resistant strains. This was indeed proven to be the case in human therapy. Many hospitals restricted the use of certain aminoglycosides, as they were important as the last-resort drug for patients suffering from serious infections by hospital-resident Gram-negative bacteria (e.g., *P. aeruginosa*) that are resistant to most other types of antibiotics. The frequency of resistance to the restricted drugs indeed decreased significantly. On a larger scale, the frequency of penicillin-resistant pneumococci in Britain showed an unexpected decrease

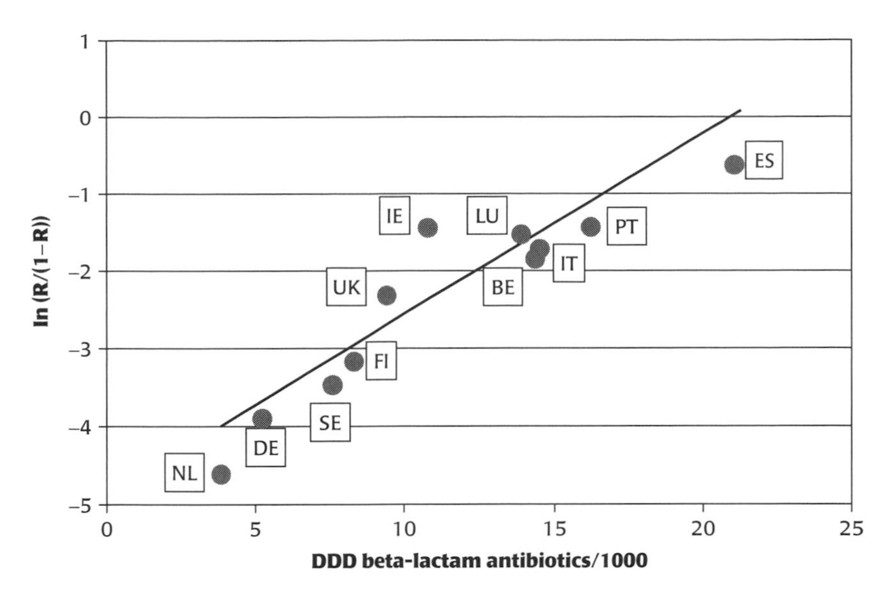


**FIGURE 10.75**

Correlation between the use of penicillin and the occurrence of penicillin-resistant pneumococci in European countries, 1998 to 1999. The ordinate represents the probability of resistance, plotted as ln {R/(1-R)}, where R is the fraction of pneumococcal strains found to be resistant. For example, in Spain, R was 0.34 (34%), giving the ln {R/(1-R)} value of −0.66. The abscissa shows the use of $\beta$-lactams in each country as the defined daily dose (DDD) per 1000 population per day. Abbreviations: BE, Belgium; DE, Germany; FI, Finland; IE, Ireland; IT, Italy; LU, Luxembourg; NL, the Netherlands; PT, Portugal; ES, Spain; Se, Sweden; UK, United Kingdom. [From Bronzwaer, S. L. A. M., et al. (2002). A European study on the relationship between antimicrobial use and antimicrobial resistance. *Emerging Infectious Diseases*, 8, 278–282.]

in the period 1999 to 2001 (see Figure 10.68); this is likely the result of a decrease in prescriptions for penicillin for cases of respiratory infection, subsequent to the publication of several reports on increased drug resistance in pathogenic bacteria.

It is impossible, however, to completely abolish the use of antibiotics for human therapy. It seems prudent, therefore, to limit antibiotic use in other areas. One area is agriculture. Antibiotics are used as "growth promoters" of farm animals. The European Union banned this use of drugs that are also used for human therapy, but this use is still common in the United States. Hard figures for the amounts of antibiotics used for animals are difficult to obtain. One reliable source is DANMAP, which provides the statistics for all antibiotics prescribed in Denmark. The data for 2003 show that 102 tons of antibiotics were used for animals, in contrast to only 44 tons used for humans. Considering that in Denmark antibiotics cannot be used for growth promotion of animals, this seems to suggest that worldwide, the use of antibiotics for farm animals far exceeds the use for human therapy. Indeed the use of antibiotics in nonhuman areas is expanding as these agents are now widely used in aquaculture. (On a smaller scale, antibiotics – 66,000 pounds in 1997 – are also used for spraying orchards in the United States.)

The European banning of "human-use" antibiotics for animal growth promotion led to the development of special antibiotics for veterinary use. However, this approach also has had its problems. Avoparcin, which is closely related to vancomycin in structure, was widely used in Europe for animals and selected for avoparcin-resistant enterococci in animals. Now avoparcin-resistant strains are also vancomycin resistant. This happened before vancomycin became the most important drug for treating MRSA infections in humans. Now the avoparcin-selected–resistant enterococci could act as a huge reservoir of vancomycin resistance genes.

In terms of human therapy, it is obviously necessary to limit the use of antibiotics, especially those with a wider spectrum, as much as possible. So far, this has been difficult because the physician does not know the identity of the causative organism when the treatment has to be started. However, with the improvement of molecular methods that rely on PCR, a more rapid diagnosis may become possible (Chapter 3). We should, in any case, make more of an effort to prevent the emergence of resistant strains. For example, we should discourage single-drug therapy and encourage the simultaneous use of more than one drug, as has been done for many years in the treatment of tuberculosis. The simultaneous administration of an inhibitor of multidrug efflux pumps significantly lowers the MIC of various drugs in Gram-negative bacteria; indeed, this approach has prevented the emergence of resistant organisms, at least in an experimental setting.

## CONTINUED NEED FOR NEW ANTIBIOTICS

Clearly, the need for new antibiotic agents is more acute than ever. In view of the fact that currently available antibiotics target only a few proteins in the microbial cell, there have been serious efforts to find novel targets (and novel drugs that inhibit them) by taking advantage of the genome information on numerous pathogens. Furthermore, as a source for genes coding for antibiotic synthesis, in addition to cultured microorganisms, direct cloning from soil DNA was used to obtain genes coding for novel secondary metabolites in 2000, and presumably such a practice is now widely used.

However, when new compounds are derived from natural products, there is always a possibility that enzymes capable of inactivating these compounds already exist in the producing microorganisms and that the genes coding for these enzymes could quickly become widespread among pathogenic microorganisms through the lateral transfer of plasmids and phages. In contrast, resistance mechanisms based on enzymatic hydrolysis or inactivation have not appeared for the totally synthetic agents, such as fluoroquinolones, sulfonamides, and trimethoprim (although bacteria can develop resistance to these agents by altering the drugs' targets, sometimes quite easily). However, in comparison with the great diversity of natural products, the organic compounds synthesized so far represent a more limited and biased collection of structures.

For this reason, *combinatorial* approaches that would expand the range of compounds may be important, especially because screening for their activity can now be carried out rapidly through the so-called high-throughput processes. Combinatorial chemistry, in which a large number of compounds can be synthesized simultaneously (Figure 10.76), is now an essential tool in the production of synthetic and semisynthetic antimicrobial agents. In another approach, a phage display is utilized. Thus, a potential target protein in a pathogenic microorganism is prepared by gene cloning and overexpression and is immobilized on a matrix. Then peptides with random sequences are generated from synthetic oligonucleotides, and expressed as fusion proteins on the surface of a filamentous bacteriophage vector (see Figure 3.15). Finally, these phages are then added to an affinity matrix holding the target

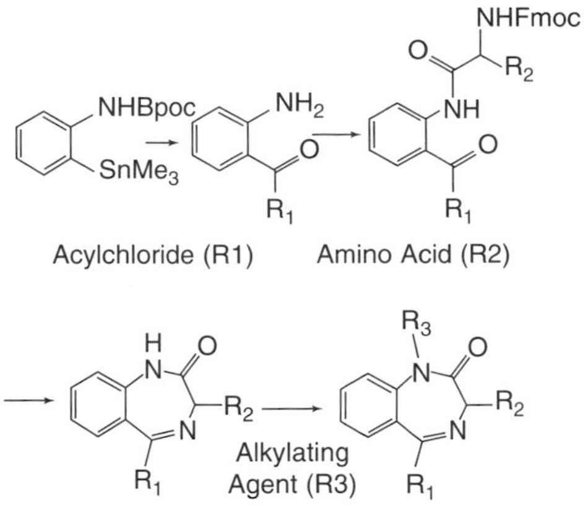

FIGURE 10.76

An example of a combinatorial chemistry process. This version, developed by J. A. Ellman's group, starts with acylation using acylchloride (R1; *first arrow*). The amino group, deprotected after the first reaction, is then acylated using an amino acid ester R2 (*second arrow*). After cyclization to produce 1,4-benzodiazepine (*third arrow*), the NH group is modified by an alkylating agent (R3). If one uses a set of 50 different reagents at each step, one may end up synthesizing 125,000 different compounds after only three steps. Although not shown in the figure, the starting material is anchored on a solid support, so isolation of intermediates becomes trivial. Bpoc and Fmoc are protective groups for $-NH_2$. [Based on Bunin, B. A., Plunkett, M. J., and Ellman, J. A. (1996). Synthesis and evaluation of 1,4-benzodiazepine libraries. *Methods in Enzymology*, 267, 448–465.]

protein. Any phage expressing a peptide that binds tightly to the target protein will become adsorbed to this affinity matrix. The peptide produced by these phages thus will give us a clue as to the shape of the "peptidomimetic" inhibitors to be synthesized. As we have already seen, we can combine genes from different sources to achieve "combinatorial" biosynthesis of antibiotics, although the range of compounds generated will be much more limited. It is encouraging that innovative methods such as these are being developed in response to the challenge of countering drug resistance among pathogenic microorganisms.

## SUMMARY

Antibiotics are secondary metabolites of microorganisms and are usually produced in the stationary phase of growth. The majority of antibiotics come from eukaryotic fungi or from streptomycetes, a prokaryotic group with a complex life cycle reminiscent of that of fungi; a few are produced by unicellular bacteria not known to have developmental cycles. Many of the genes of antibiotic biosynthesis have been cloned and sequenced. In many cases, they occur as a cluster. This arrangement explains the apparent occurrence of the horizontal transfer of these genes, during evolution, between distantly related organisms. The gene sequence also explains (e.g., with the producers of polyketide antibiotics) why a single antibiotic producer often secretes many related antibiotics that are different from one another in minor details. Cloning of production genes is opening up ways to produce novel antibiotics by the combination of defined genes.

Antibiotics have been extremely effective in our battle against bacterial infections and they have drastically changed the ecological balance between humans and various pathogenic microorganisms. Yet every introduction of a new antibiotic into clinical practice has been followed by the emergence of resistant organisms. Very often these organisms have harbored R plasmids that have contained resistance genes specifying the degradation, inactivation, or pumping out of the antibiotic molecules. Sometimes the resistant strains have been opportunistic wild-type organisms that are intrinsically resistant to antibiotics but have not been thought of as significant pathogens. Such wild-type strains resistant to $\beta$-lactams unexpectedly contain $\beta$-lactamase genes on their chromosomes; this situation presumably arose because these species, normally inhabitants of soil, have been in contact with natural $\beta$-lactam compounds in their environment throughout their evolutionary history.

Production of new antibiotics that are effective against resistant organisms has therefore been one of the most important aspects of antibiotic research. Two approaches have played prominent roles in this endeavor. First, discovery of new antibiotics through the screening of natural products led to many new classes of compounds, including cephalosporins, cephamycins, carbapenems, and monobactams among $\beta$-lactam antibiotics, and gentamicin, tobramycin, and fortimicin A among aminoglycosides. Second, chemical modification of natural products led to a number of compounds with increased stability to enzymatic degradation or inactivation mechanisms. Examples include methicillin and the third-generation cephalosporins among $\beta$-lactams, and dibekacin among aminoglycosides. In spite of these successful efforts, the frequency of occurrence of resistance genes among bacterial populations is increasing, and new mechanisms of resistance, such as target modification and the multiple drug efflux pump, are becoming clinically important. New approaches are vital in our fight against the emergence of antibiotic resistance. The combination of cloned genes for the production of new antibiotics and the screening of combinatorial libraries for affinity to defined target structures are examples of these novel methods.

## SELECTED REFERENCES

### General

Davies, J. (1990). What are antibiotics? Archaic functions for modern activities. *Molecular Microbiology*, 4, 1227–1232.

Rehm, H.-J., and Reed, G. (eds.) (1996). Products of secondary metabolism. In *Biotechnology*, 2nd Edition, Volume 7, H. Kleinkauf and H. von Döhren (eds.), Weinheim, Germany: VCH.

Walsh, C. (2003). *Antibiotics: Actions, Origins, Resistance*, Washington, DC: ASM Press.

Bryskier, A. (ed.) (2005). *Antimicrobial Agents: Antibacterials and Antifungals*, Washington, DC: ASM Press.

### Secondary Metabolites Other Than Antibiotics

Serizawa, N., Hosobuchi, M., and Yoshikawa, H. (1997). Biochemical and fermentation technological approaches to production of pravastatin, a HMG-CoA reductase inhibitor. In *Biotechnology of Antibiotics*, W. Strohl, (ed.), pp. 779–805, New York: Marcel Dekker.

Patchett, A. A. (2002). Natural products and design: interrelated approaches in drug discovery. *Journal of Medicinal Chemistry*, 45, 5609–5616.

### Entry of Antimicrobial Agents into Gram-Negative Bacteria

Nikaido, H. (2003). Molecular basis of bacterial outer membrane permeability revisited. *Microbiology and Molecular Biology Reviews*, 67, 593–650.

### Aminoglycosides

Umezawa, H., and Hooper, I. R. (eds.) (1982). *Aminoglycoside Antibiotics*, Berlin: Springer-Verlag.

Davis, B. D. (1987). Mechanism of bactericidal action of aminoglycosides. *Microbiological Reviews*, 51, 341–350.

Wright, G. D. (1999). Aminoglycoside-modifying enzymes. *Current Opinion in Microbiology*, 2, 499–503.

Miller, G. H., Sabatelli, F. J., Hare, R. S., Glupczynski, Y., Mackey, P., Shlaes, D., Shimizu, K., and Shaw, K. J. (1997). The most frequent aminoglycoside resistance mechanisms – changes with time and geographic area: a reflection of aminoglycoside usage patterns? Aminoglycoside Resistance Study Groups. *Clinical Infectious Disease*, 24(suppl 1), S46–S62.

Schmitz, F.-J., Verhoef, J., Fluit, A. C., and the SENTRY Participants Group (1999). Prevalence of aminoglycoside resistance in 20 European university hospitals participating in European SENTRY antimicrobial surveillance programme. *European Journal of Clinical Microbiology and Infectious Disease*, 18, 414–421.

Poole, K. (2005). Aminoglycoside resistance in *Pseudomonas aeruginosa*. *Antimicrobial Agents and Chemotherapy*, 49, 476–487.

Over, U., Gur, D., Unal, S., and Miller, G. H. (2001). The changing nature of aminoglycoside resistance mechanisms and prevalence of newly recognized resistance mechanisms in Turkey. *Clinical Microbiology and Infection*, 7, 470–478.

Gerding, D. N. (2000). Antimicrobial cycling: lessons learned from the aminoglycoside experience. *Infection Control and Hospital Epidemiology*, 21(1 suppl), S12–S17.

Ogle, J. M., Carter, A. P., and Ramakrishnan, V. (2003). Insights into the decoding mechanism from recent ribosome structures. *Trends in Biochemical Science*, 28, 259–266.

Poehlsgaard, J., and Douthwaite, S. (2005). The bacterial ribosome as a target for antibiotics. *Nature Reviews Microbiology* 3, 870–881.

### β-Lactams

Page, M. I. (ed.) (1992). *The Chemistry of β-Lactams*, Glasgow, UK: Blackie Academic and Professional.

Matagne, A., Lamotte-Brasseur, J., Dive, G., Knox, J. R., and Frère, J.-M. (1993). Interactions between active-site-serine β-lactamases and compounds bearing a methoxy side chain on the α-face of the β-lactam: kinetic and molecular modelling studies. *Biochemical Journal*, 293, 607–611.

Martin, J. F. (1998). New aspects of genes and enzymes for β-lactam antibiotic biosynthesis. *Applied Microbiology and Biotechnology*, 50, 1–15.

Elander, R. P. (2003). Industrial production of β-lactam antibiotics. *Applied Microbiology and Biotechnology*, 61, 385–392.

Thykaer, J., and Nielsen, J. (2003). Metabolic engineering of β-lactam production. *Metabolic Engineering*, 5, 56–69.

Nestrovich, E. M., Danelon, C., Winterhalter, M., and Bezrukov, S. M. (2002). Designed to penetrate: time-resolved interaction of single antibiotic molecules with bacterial pores. *Proceedings of the National Academy of Sciences U.S.A.*, 99, 9789–9794.

Tondi, D., Morandi, F., Bonnet, R., Costi, M. P., and Shoichet, B. K. (2005). Structure-based optimization of a non-β-lactam lead results in inhibitors that do not up-regulate β-lactamase expression in cell culture. *Journal of the American Chemical Society*, 127, 4632–4639.

### Other Classes of Antibiotics

Omura, S. (ed.) (2002). *Macrolide Antibiotics: Chemistry, Biology, and Practice*, 2nd Edition, Amsterdam: Academic Press.

Nelson, M., Hillen, W., and Greenwald, R. A. (eds.) (2001). *Tetracyclines in Biology, Chemistry and Medicine*, Basel: Birkhäuser Verlag.

Hooper, D. C., and Rubinstein, E. (eds.) (2003). *Quinolone Antimicrobial Agents*, 3rd Edition, Washington, DC: ASM Press.

**Antibiotic Production**

Embley, T. M., and Stackebrandt, E. (1994). The molecular phylogeny and systematics of the actinomycetes. *Annual Review of Microbiology*, 48, 257–289.

van Lanen, S. G., and Shen, B. (2006). Microbial genomics for the improvement of natural product discovery. *Current Opinion in Microbiology*, 9, 252–260.

Ikeda, H., et al. (2003). Complete genome sequence and comparative analysis of the industrial microorganism *Streptococcus avermitilis*. *Nature Biotechnology*, 21, 526–531.

Chater, K. F., and Horinouchi, S. (2003). Signalling early developmental events in two highly diverged *Streptomyces* species. *Molecular Microbiology*, 48, 9–15.

Bibb, M. J. (2005). Regulation of a secondary metabolism in streptomycetes. *Current Opinion in Microbiology*, 8, 208–215.

Takano, E. (2006). $\gamma$-Butyrolactones: Streptomyces signalling molecules regulating antibiotic production and differentiation. *Current Opinion in Microbiology*, 9, 287–294.

Welch, M., Todd, D. E., Whitehead, N. A., McGowan, S. J., Bycroft, B. W., and Salmond, G. P. C. (2000). N-acyl homoserine lactone binding to the CarR receptor determines quorum-sensing specificity in *Erwinia*. *EMBO Journal*, 19, 631–641.

Martin, J. F. (2004). Phosphate control of the biosynthesis of antibiotics and other secondary metabolites is mediated by the PhoR-PhoP system: an unfinished story. *Journal of Bacteriology*, 186, 5197–5201.

McDaniel, R., Ebert-Khosla, S., Hopwood, D. A., and Khosla, C. (1993). Engineered biosynthesis of novel polyketides. *Science*, 262, 1546–1550.

Baltz, R. H. (2006). Molecular engineering approaches to peptide, polyketide and other antibiotics. *Nature Biotechnology*, 24, 1533–1540.

Pfeiffer, B. A., Admiraal, S. J., Gramajo, H., Cane, D. E., and Khosla, C. (2001). Biosynthesis of complex polyketides in a metabolically engineered strain of *E. coli*. *Science*, 291, 1790–1792.

Grünewald, J., and Marahiel, M. A. (2006). Chemoenzymatic and template-directed synthesis of bioactive macrocyclic peptides. *Microbiology and Molecular Biology Reviews*, 70, 121–146.

Butler, M. J., Bruheim, P., Jovetic, S., Marinelli, F., Postma, P. W., and Bibb, M. J. (2002). Engineering of primary carbon metabolism for improved antibiotic production in *Streptomyces lividans*. *Applied and Environmental Microbiology*, 68, 4731–4739.

**Antibiotic Resistance**

Nikaido, H. (1994). Prevention of drug access to target: resistance mechanisms in bacteria based on permeability barriers and active efflux. *Science*, 264, 382–388.

Spratt, B. G. (1994). Resistance to antibiotics mediated by target alterations. *Science*, 264:388–393.

Courvalin, P., and Trieu-Cuot, P. (2001). Minimizing potential resistance: the molecular view. *Clinical Infectious Diseases*, 33(suppl 3), S138–S146.

Hiramatsu, K., Cui, L., Kuroda, M., and Ito, T. (2001). The emergence and evolution of methicillin-resistant *Staphylococcus aureus*. *Trends in Microbiology*, 9, 486–493.

Rice, L. B. (2006). Antimicrobial resistance in Gram-positive bacteria. *American Journal of Medicine*, 119, S11–S19.

Liebert, C. A., Hall, R. M., and Summers, A. O. (1999). Transposon Tn21, flagship of the floating genome. *Microbiology and Molecular Biology Reviews*, 63, 507–522.

Livermore, D. M. (2005). Minimising antibiotic resistance. *Lancet Infectious Diseases*, 5, 450–459.

Rowe-Magnus, D. A., Guerout, A.-M., Ploncard, P., Dychinco, B., Davies, J., and Mazel, D. (2001). The evolutionary history of chromosomal super-integrons provides an ancestry for multiresistant integrons. *Proceedings of the National Academy of Sciences U.S.A.*, 98, 652–657.

Brazas, M. D., and Hancock, R. E. W. (2005). Using microarray gene signatures to elucidate mechanisms of antibiotic action and resistance. *Drug Discovery Today*, 10, 11245–1292.

**The Quest for New Antibiotics**

Livermore, D. M. (2004). The need for new antibiotics. *Clinical Microbiology and Infection*, 10(suppl 4), 1–9.

Brown, E. D., and Wright, G. D. (2005). New targets and screening approaches in antimicrobial drug discovery. *Chemical Reviews*, 105, 759–774.

Wijkmans, J. C. H. M., and Beckett, R. P. (2002). Combinatorial chemistry in anti-infective research. *Drug Discovery Today*, 7, 126–132.

Wang, G.-Y.-S., Grazian, E., Waters, B., Pan, W., Li, X., McDermott, J., Meurer, G., Saxena, G., Andersen, R. J., and Davies, J. (2000). Novel natural products from soil DNA libraries in a streptomycete host. *Organic Letters*, 2, 2401–2404.

Walsh, C. T. (2004). Polyketide and nonribosomal peptide antibiotics: modularity and versatility. *Science*, 303, 1805–1810.

Clardy, J., Fischbach, M. A., and Walsh, C. T. (2006). New antibiotics from bacterial natural products. *Nature Biotechnology*, 24, 1541–1550.

ELEVEN

# Biocatalysis in Organic Chemistry

In carrying out their metabolic processes, microorganisms interconvert diverse organic compounds. These "biotransformations" are catalyzed with high specificity and efficiency by enzymes. The active site of an enzyme, where substrate binding and catalysis are carried out, is an asymmetric surface whose special geometry frequently guarantees that the enzyme-catalyzed reaction will yield a particular stereoisomer as the sole product. Such stereospecific, or enantioselective, reactions may be difficult or impossible to achieve by purely chemical means. The terms used to describe the stereochemistry of organic compounds are defined in Box 11.1, and the determination of enantioselectivity is described in Box 11.2.

Even when an organic compound *can* be synthesized chemically, the process may require many steps, whereas a single enzyme-catalyzed reaction can often achieve the same end. Also, enzymes can catalyze reactions at ambient temperature, away from extremes of pH, and at atmospheric pressure. Undesired isomerization, racemization, epimerization, and rearrangement reactions that are frequently encountered during chemical processes are generally avoided. The absence of such side reactions is a particular advantage when the desired product is rather labile. Finally, enzymes can accelerate the rates of chemical reactions by factors of $10^8$ to $10^{12}$. For all these reasons, biotransformations by microorganisms, or by enzymes purified from microorganisms, are highly useful in preparative organic chemistry.

Certain disadvantages do limit the use of enzymes in organic chemical processes, but these limitations have frequently proved to be surmountable challenges rather than impenetrable barriers. One problem is that most enzymes exhibit low or no activity in most anhydrous organic solvents and are denatured both at high temperatures and under strongly acidic or basic conditions. Some of these difficulties can be overcome by immobilizing the enzymes on a solid support or carrying out the reaction in a two-phase solvent system. Another limitation is that enzymes are frequently subject to product and substrate inhibition, which may require that the reaction be performed at low substrate and/or product concentrations to achieve optimal reaction rates. Sophisticated systems have been developed that continuously

**Stereochemistry of Organic Compounds: A Glossary of Terms**

**Enantiomer** Compounds that are *mirror images* of each other are called *enantiomers*.

**Optically active compounds** A compound is optically active, rotating the plane of polarization of plane-polarized light, *if it is not superimposable with its mirror image*, that is, with its enantiomer. A simple diagnostic test for superimposability is the presence of a plane or a center of symmetry. The presence of such symmetry indicates lack of optical activity; the absence indicates optical activity.

**Chiral center** The optical activity of many organic molecules of biological interest is caused by an asymmetric carbon atom with four different groups around it. This type of atom is called a *chiral center* or a *stereogenic center*.

**Prochiral center** The carbon atom in $CR_2R'R''$ is called *prochiral*. Although it is not optically active, because it is bound to two identical groups (and thus has a plane of symmetry), it is *potentially* chiral because one of the R groups could be chemically replaced by another group (not = R′ or R″).

**Diastereoisomers** Isomeric molecules with *two or more* chiral centers that are not mirror images of one another are called *diastereoisomers*.

***Meso* isomer** An optical isomer whose lack of optical activity is the result of internal compensation.

**Absolute molecular asymmetry** Absolute molecular asymmetry is denoted according to the *RS convention*. In this convention, the groups around the chiral carbon atom are assigned an order of priority according to three basic rules:

**1.** Assign priority to functional groups in order of decreasing atomic number. For isotopes, the higher mass number has priority. For example,

$$O > N > C > H$$

$$^{3}H > ^{2}H > ^{1}H$$

$$C\text{-}OH > C\text{-}CH_2Cl > C\text{-}CH_2OH > C\text{-}CH_3 > C\text{-}H.$$

Unsaturated centers should be treated as though carbon atoms were attached.

**2.** Orient the center under examination so that either (a) the viewer is farthest away from the lowest priority substituent in a tetrahedral projection or (b) the lowest priority substituent occupies the bottom position in a Fischer projection.

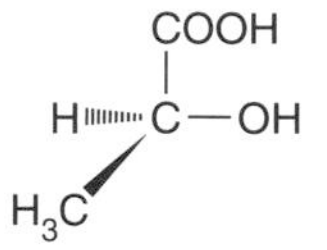

Tetrahedral projection (H behind plane of paper)

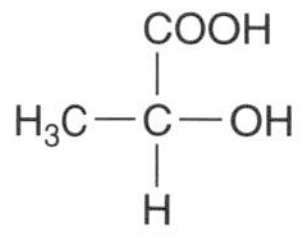

Fischer projection (H at bottom)

**3.** With the center thus oriented, count around the remaining three substituents in order of decreasing priority.
If these *three* substituents thereby describe a *clockwise turn*, the center is designated **R** (for *rectus*, Latin "right[handed]").
If these *three* substituents thereby describe an *anticlockwise turn*, the center is designated **S** (for *sinister*, Latin "left[handed]").

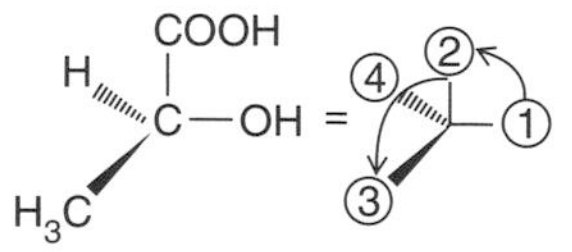

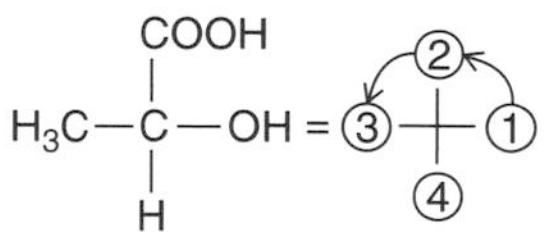

**BOX 11.1**

feed substrate to a reaction mixture at an appropriately low concentration and continuously remove the product. Ingenious approaches have also been developed to regenerate cofactors and replenish cosubstrates when these are costly or must be constantly replaced.

**Enantiomeric Excess and Enantiomeric Ratio**

The enantiomeric purity of a compound is expressed in terms of its *enantiomeric excess*, *ee* value, defined as:

$$\text{For } R > S \quad ee_R = (R - S)/(R + S) \quad \text{and} \quad \%\, ee_R = [(R - S)/(R + S)] \times 100,$$

where *R* and *S* are the concentrations of the (*R*)- and (*S*)-enantiomers, respectively.
For a racemic compound, *ee* = 0, and for an enantiomerically pure compound, *ee* = 1 (or 100% *ee*).

The parameter E, the *enantiomeric ratio*, describes the stereoselectivity or enantioselectivity of an enzyme-catalyzed reaction. E is defined as the product of the ratio of catalytic constants and the (reciprocal) ratio of the Michaelis–Menten constants for the pure enantiomers:

$$\mathrm{E} = \left(k_{\mathrm{cat}}^{\mathrm{R}}/k_{\mathrm{cat}}^{\mathrm{S}}\right) \times \left(K_{\mathrm{M}}^{\mathrm{S}}/K_{\mathrm{M}}^{\mathrm{R}}\right)$$

When starting with the racemic (*R,S*)-substrate, the enantioselectivity can be calculated from the relationship $\mathrm{E} = \ln[(1 - c).\,[1 - ee(S)]]/\ln[(l - c).\,[1 + ee(S)]]$, where *c* is the fraction of (*R,S*)-substrate converted to product and *ee*(S) is ([S] − [R]/[S] + [R]), where [S] and [R] correspond to the concentrations of *S* and *R* isomers remaining after completion of the enzyme reaction. A nonselective reaction has an E value of 0. For an acceptable resolution, the E value must be above 20.

**BOX 11.2**

## MICROBIAL TRANSFORMATION OF STEROIDS AND STEROLS

A retrospective look at the contribution of enzymes to the production of therapeutically valuable steroids, such as cortisone, demonstrates that the power of biocatalysis was already fully appreciated more than 50 years ago.

The oxidation and reduction reactions that microorganisms perform on steroid and sterol substrates provide particularly impressive examples of regioselective and stereospecific biotransformations and also showcase the ability of enzymes to promote reactions at unactivated centers in hydrocarbons. Virtually any position in the carbon skeleton of a steroid nucleus (Figure 11.1) can be hydroxylated stereospecifically by enzymes present in some microorganism. Steroid hydroxylases are named according to the

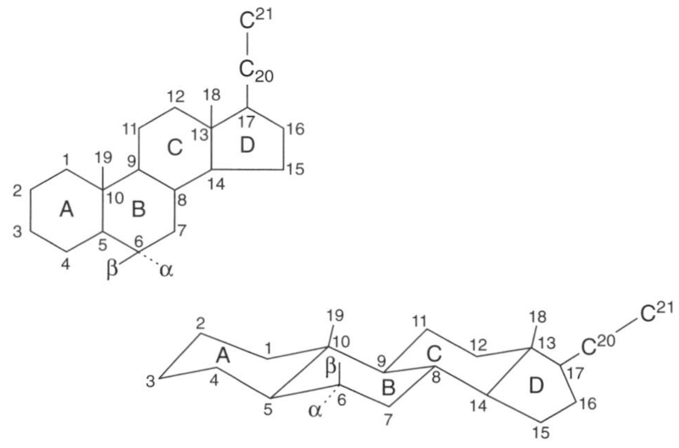

**FIGURE 11.1**

Structure, stereochemistry, and numbering of the nucleus of adrenocorticosteroids. The four rings, A through D, do not lie in a flat plane as conventionally represented by the upper structure but rather have the configuration shown by the lower structure. The biological activity of steroids depends on the orientation of the groups attached to the ring system. As shown at C6, groups that project above the plane of the steroid are designated $\beta$. Their connection to the ring system is shown by *solid-line bonds*. Those that project below the plane are designated $\alpha$, and their connection to the ring is shown by a *dotted-line bond*.

position they attack on the rings or the side chain of the steroid nucleus. There are three primary carbon atoms (C18, C19, and C21). An enzyme that catalyzes hydroxylation at C21, for example, is designated as 21-hydroxylase. There are 18 secondary carbon atoms. At the secondary carbon atoms within the ring system, there are two alternative ways, designated $\alpha$ and $\beta$, to attach the –OH group. The $\alpha$ (equatorial) position lies below the plane of the steroid ring, and the $\beta$ (axial) position lies above the plane (Figure 11.1). Every one of the 18 secondary carbon atoms can be hydroxylated, in either the $\alpha$ or $\beta$ configuration, each by a different known microbial hydroxylase (Tables 11.1 and 11.2). In addition to hydroxylations, certain microbial enzymes can aromatize ring A, reduce double bonds in the rings, and reduce specific ketone substituents. The microbial transformations of steroids and sterols (Figure 11.2) have dramatically lowered the cost of manufacturing steroid hormones.

**TABLE 11.1 Selective microbial hydroxylation of steroids**

| Position of hydroxylation | Stereochemistry of incoming hydroxyl group | Position of hydroxylation | Stereochemistry of incoming hydroxyl group |
|---|---|---|---|
| 1 | $\alpha$ | 10 | $\beta$ |
| 1 | $\beta$ | 11 | $\alpha$ |
| 2 | $\alpha$ | 12 | $\beta$ |
| 2 | $\beta$ | 13 | $\alpha$ |
| 3 | $\alpha$ | 14 | $\alpha$ |
| 3 | $\beta$ | 15 | $\alpha$ |
| 4 | $\alpha$ | 15 | $\beta$ |
| 4 | $\beta$ | 16 | $\alpha$ |
| 5 | $\alpha$ | 16 | $\beta$ |
| 6 | $\alpha$ | 17 | $\alpha$ |
| 6 | $\beta$ | 17 | $\beta$ |
| 7 | $\alpha$ | | |
| 7 | $\beta$ | | |
| 9 | $\alpha$ | | |

*Source*: Davies, H. G., Green, R. H., Kelly, D. R., and Roberts, S. M. (1989). *Biotransformations in Preparative Organic Chemistry. The Use of Isolated Enzymes and Whole Cell Systems in Synthesis*, pp. 175–176, London: Academic Press.

In the early 1930s, Edward C. Kendall of the Mayo Foundation and Tadeus Reichstein of the University of Basel isolated cortisone, a steroid secreted by the adrenal gland. In 1949, Philip S. Hench of the Mayo Foundation found that administration of cortisone led to remission in patients with acute rheumatoid arthritis. Discovery of the anti-inflammatory effects of cortisone had a profound impact on the medical world and earned Kendall, Reichstein, and Hench the Nobel Prize in 1950.

The demand for large amounts of cortisone spurred the development of a chemical synthesis for the hormone. It was an elaborate synthesis, requiring 31 steps, and its final yield was extremely low. A starting batch of 615 kg of deoxycholic acid (purified from beef bile) was converted to 1 kg of cortisone acetate. The market price for the synthetic hormone was $200 per gram.

A major complication in the synthetic route from deoxycholic acid to cortisone is the need to shift the C12$\beta$ hydroxyl in deoxycholic acid to C11. In the chemical synthesis, this required nine steps. In 1952, however, researchers at Upjohn Company discovered that an aerobically grown bread mold, *Rhizopus arrhizus*, could hydroxylate progesterone (another steroid and an early intermediate in cortisone synthesis) at C11$\alpha$, and workers at the Squibb Institute found that another common mold, *Aspergillus niger*, carried out the same reaction. By exploiting microbial hydroxylation at C11, industrial cortisone synthesis was shortened from 31 to 11 steps. Moreover, the microbial hydroxylation of progesterone had economic benefits beyond those

**TABLE 11.2 Examples of steroid hydroxylations by different fungi**

| Hydroxylation position | Substrate | Product | Microorganism |
|---|---|---|---|
| 1$\alpha$ | Androst-4-ene-3,17-dione | 1$\alpha$-Hydroxyandrost-4-ene-3,17-dione | *Penicillium* sp. |
| 1$\beta$ | Androst-4-ene-3,17-dione | 1$\beta$-Hydroxyandrost-4-ene-3,17-dione | *Xylaria* sp. |
| 3$\alpha$ | Androstane-7,17-dione | 3$\alpha$-Hydroxyandrostane-7,17-dione | *Diaporthe celastrina* |
| 3$\beta$ | 17$\beta$-Hydroxyandrostan-11-one | 3$\beta$,17$\beta$-Dihydroxyandrostan-11-one | *Wojnowicia graminis* |
| 11$\alpha$ | Progesterone | 11$\alpha$-Hydroxyprogesterone | *Rhizopus* sp. |
| 11$\beta$ | 11-Deoxycortisone | Hydrocortisone | *Curvularia lunata* |
| 12$\beta$ | 17$\beta$-Hydroxy-estr-4-ene 3-one | 12$\beta$,17$\beta$-Dihydroxy-estr-4-ene-3-one | *Colletotrichum derridis* |

*Source*: Neidleman, S. L. (1991). Industrial chemicals: fermentation and immobilized cells. In *Biotechnology. The Science and the Business*, V. Moses and R. E. Cape (eds.), pp. 306–307, Chur, Switzerland: Harwood Academic Publishers.

resulting from the abbreviation of the chemical synthesis. This biotransformation takes place at 37°C in aqueous solution at atmospheric pressure, conditions that are much less expensive than the high temperature and pressure and nonaqueous solvents required for the equivalent steps in chemical synthesis. The commercial price of cortisone dropped to $6 per gram shortly after these discoveries.

Further reductions in the cost of cortisone came from introducing inexpensive sterols, instead of deoxycholate, as the starting material. Two such sterols, stigmasterol and sitosterol, are generated in large amounts as by-products in the production of soybean oil; a third, diosgenin, comes from the roots of the Mexican barbasco plant. To make steroids from these plant sterols, the side chain beyond C21 must be removed. Although chemical degradation can accomplish this step, it is achieved much more economically by mycobacteria, aerobic Gram-positive eubacteria that can utilize sterols as a carbon and energy source. To prevent mycobacteria from breaking the sterols down totally, mutant strains have been developed that are unable to degrade the sterols beyond the desired stage. The introduction of these process changes brought the price of cortisone in the United States down to 46 cents per gram by 1980, a 400-fold reduction from the original price without even adjusting for inflation!

In addition to being used to treat rheumatoid arthritis, steroids are prescribed for allergies and other inflammatory diseases (especially of the skin), contraception, and hormonal insufficiencies. Various steroids useful for these purposes are produced with the aid of microbes capable of modifying the steroid nucleus in specific ways. Worldwide bulk sales of the four major steroids – cortisone, aldosterone, prednisone, and prednisolone – amount to more than 700,000 kg/year.

## ASYMMETRIC CATALYSIS IN THE PHARMACEUTICAL AND AGROCHEMICAL INDUSTRIES

Biocatalysis represents an important general approach to the synthesis of *chiral synthons* (optically active building blocks) for a wide variety of

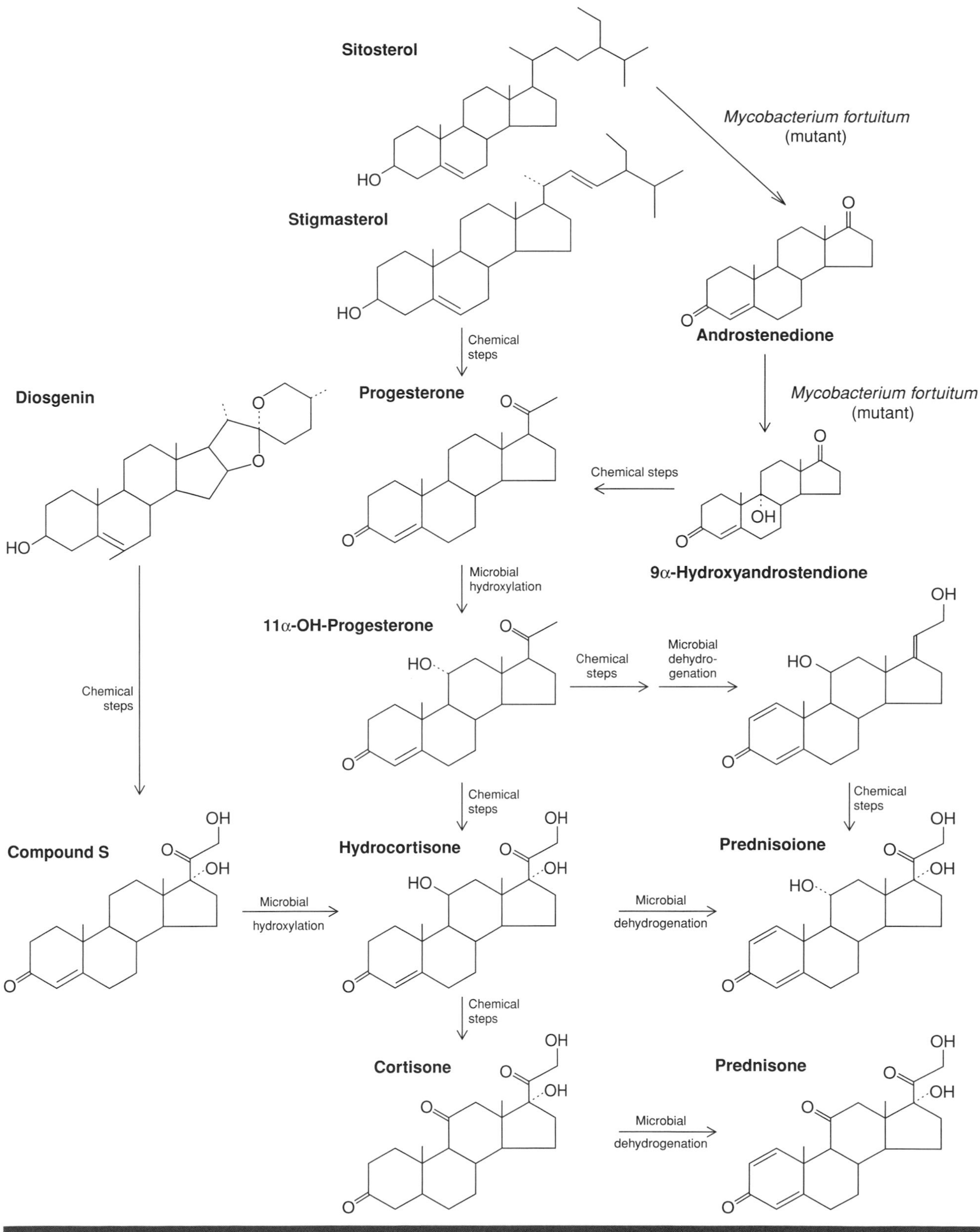


**FIGURE 11.2**

Chemical and microbial transformations in the production of therapeutically useful steroids. [Primrose, S. B. (ed.) (1987). *Modern Biotechnology*, p. 76, Oxford: Blackwell Scientific Publications; Hogg, J. A. (1992) Steroids, the steroid community, and Upjohn in perspective: a profile of innovation. *Steroids*, 57, 593–616.]

**Importance of Chirality in the Action of Synthetic Drugs**

"Perhaps looking-glass milk isn't good to drink?"
— Lewis Carroll

Drugs, herbicides, and pesticides frequently act by interacting with receptors, enzymes, carrier molecules, and the like. All such interactions are highly stereospecific. About 25% of the drugs now in use are chiral. The tacit assumption that only one of the isomers is active and the other inactive is dangerous. One stereoisomer of a drug may interact tightly with a particular receptor, whereas the other stereoisomer may have a different target altogether, as indicated in the following examples.

- The dextrorotatory isomer of the antituberculosis drug ethambutol, 2,2′-(ethylenediimino)-di-1-butanol dihydrochloride, has potent antitubercular activity, whereas the levorotatory isomer causes degeneration of the optic nerve, leading to blindness.
- The dextrorotatory isomer of propoxyphene, $\alpha$-(+)-4-(dimethylamino)-3-methyl-1,2-diphenyl-2-butanol propionate, is an analgesic, whereas the levorotatory isomer is a cough suppressant.

BOX 11.3

pharmaceuticals, herbicides, and pesticides. Optically pure drugs frequently show fewer side effects than do racemates (Box 11.3), and there is considerable regulatory pressure on the pharmaceutical industry to market homochiral drugs. Occasionally, enantiomers have completely different biological activities, and there are instances in which toxicity of a racemic drug has been linked to one of the stereoisomers, the one without the desired pharmacological activity.

The examples that follow illustrate the application of biocatalysis to the generation of chiral synthons for a pharmaceutical agent with several stereogenic centers, and of a chiral synthon utilized in the synthesis of widely used herbicides.

## CHIRAL INTERMEDIATES FOR THE SYNTHESIS OF $\beta_3$-RECEPTOR AGONISTS

$\beta_3$-Adrenergic receptors are found on the cell surface of adipocytes and are key components of signaling pathways that affect lipolysis, thermogenesis, and relaxation of intestinal smooth muscle. Selective $\beta_3$-receptor agonists may prove effective in the treatment of gastrointestinal disorders, type II diabetes, and obesity. We describe below the biocatalytic syntheses of two different chiral intermediates required for the total synthesis of a $\beta_3$-receptor agonist (Figure 11.3).

In the first of the transformations, whole cells of *Sphingomonas paucimobilis* SC 16113 are used to effect the reduction of 4-benzyloxy-3-methanesulfonylamino-2′-bromoacetophenone to the corresponding (*R*)-alcohol in a yield greater than 85% and ee values greater than 98%. In the second biocatalytic reaction, an enantioselective amidase within wet cells of *Mycobacterium neoaurum* effected an enzymatic resolution of racemic $\alpha$-methyl-4-methoxyphenylalanine amide, generating the desired chiral amino acid (Figure 11.3, compound 5).

## BIOCATALYTIC SYNTHESIS OF (*S*)-2-CHLOROPROPIONIC ACID

(*S*)-2-Chloropropionic acid serves as a homochiral intermediate in the large-scale syntheses of aryloxyphenoxypropionic acid derivatives. These widely used herbicides block the conversion of acetyl-CoA to malonyl-CoA by inhibiting the activity of the enzyme acetyl-CoA carboxylase in the stroma of plastids. The resulting inhibition of fatty acid synthesis leads selectively

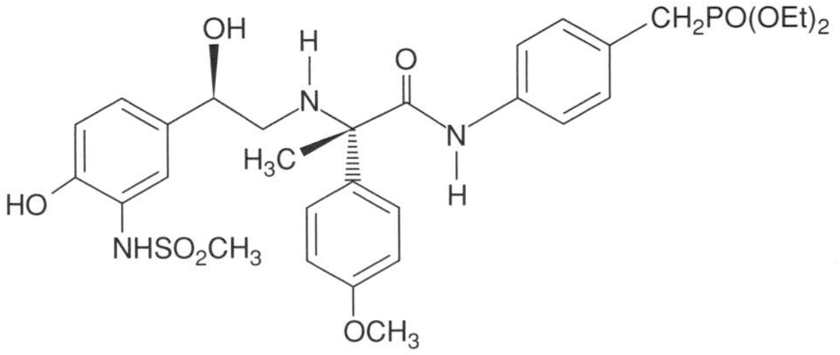

β3-adrenergic receptor agonist
1

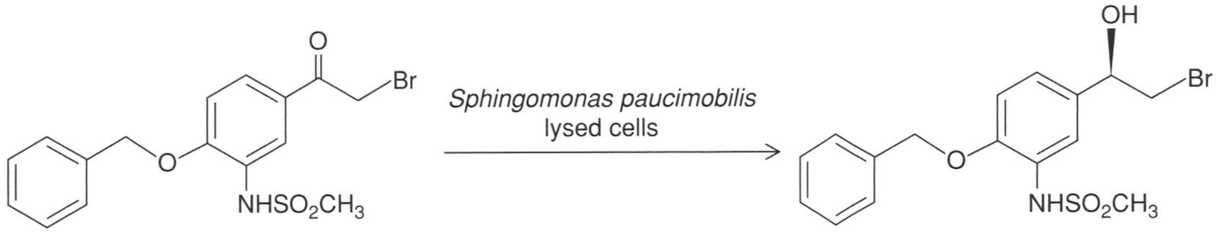


Racemic ketone
2

(*R*)-Alcohol
3

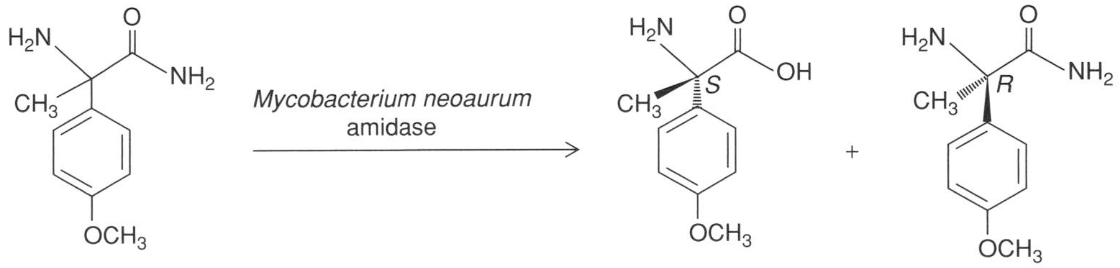


Racemic amide
4

(*S*)-Acid
5

**FIGURE 11.3**

Enzymatic synthesis of chiral intermediates for the production of a $\beta_3$-adrenergic receptor agonist (**1**): enantioselective reduction of 4-benzyloxy-3-methanesulfonylamino-2′-bromoacetophenone (**2**) to the (*R*)-alcohol (**3**); enantioselective hydrolysis of $\alpha$-methyl-4-methoxyphenylalanine amide (**4**) to the (*S*)-acid (**5**).

to the death of plants with an acetyl-CoA carboxylase sensitive to these herbicides.

In the earlier route to (*S*)-2-chloropropionic acid, glucose was fermented to (*R*)-lactic acid. The lactic acid was extracted from the fermentation broth, purified, esterified, and the ester chlorinated with thionyl chloride. The esterification is necessary to protect the acid group during the chlorination reaction.

The current route is based on enantioselective hydrolytic dehalogenation. More than 3800 different kinds of organohalogen compounds are known products of biological processes as well as of abiogenic ones, such as volcanic eruptions. Many microorganisms possess a variety of enzymes that act on such compounds. Among these enzymes are 2-haloacid dehydrogenases. The biotransformation route starts with racemic 2-chloropropionic acid, a low-cost commodity chemical.

A bacterium capable of dehalogenating both (*R*)- and (*S*)-2-chloropropionic acid was isolated from soil adjacent to a factory using this chemical. Nitrosoguanidine mutagenesis was employed to isolate a mutant unaltered in its ability to convert the (*R*)-isomer quantitatively to (*R*)-lactic acid

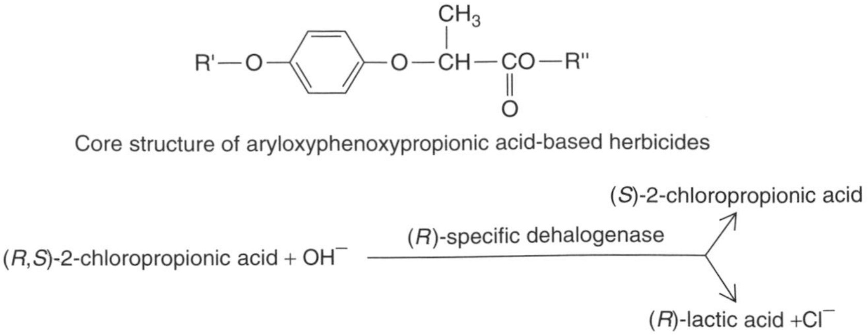

R'—O—$C_6H_4$—O—CH($CH_3$)—CO—R" (C=O)

Core structure of aryloxyphenoxypropionic acid-based herbicides

(*R*,*S*)-2-chloropropionic acid + $OH^-$ —(*R*)-specific dehalogenase→ (*S*)-2-chloropropionic acid + (*R*)-lactic acid + $Cl^-$

**FIGURE 11.4**

The core structure of aryloxyphenoxypropionic acid–based herbicides is shown in the upper part of the figure, and the resolution of (*R*,*S*)-2-chloropropionic acid by hydrolysis of (*R*)-2-chloropropionic acid by an enantioselective dehalogenase is shown in the lower part.

but inactive toward the (*S*)-isomer (Figure 11.4). Further genetic engineering of the mutant enhanced its activity toward the (*R*)-isomer 10-fold.

The reaction can be performed with whole cells. Upon completion of the dehalogenation of the (*R*)-isomer, the reaction mixture is acidified to precipitate the microbial cells, which are then filtered off. The (*S*)-2-chloropropionic acid is extracted with an organic solvent and purified by distillation.

The activities of three different kinds of enzymes are exploited in the above examples of enantioselective biocatalysis, a reductase, an amidase, and a dehalogenase. The striking commonality between these examples is that with these particular substrates, there was no need to use highly purified enzymes. Whole cells were employed in each instance. This observation holds true for many cases of regio- or enantioselective synthesis of diverse organic compounds.

## MICROBIAL DIVERSITY: A VAST RESERVOIR OF DISTINCTIVE ENZYMES

Prokaryotes and fungi colonize virtually every ecological niche. Rich assemblages of microorganisms are found in environments (biotopes) characterized by extremes of temperature, pH, salinity, pressure, chemical composition, light intensity and quality, and so on. The properties of enzymes made by an organism adapted to a particular biotope are compatible with the need to function in the physical and chemical conditions within that biotope. In other words, each of these organisms produces the enzymes it needs to survive in its particular environment and utilize whatever nutrients are available there. Collectively, therefore, microbial enzymes catalyze an enormous variety of chemical reactions under widely varying conditions that transform both naturally occurring and human-made organic compounds. This immense reservoir of enzymes can be explored for biocatalysts with desired specificities either by screening microorganisms available in pure culture or – by culture-independent methods – screening clones derived from environmental DNA samples. In principle, the means are now available to acquire virtually any enzyme in the living world and produce it on a large scale in heterologous host cells. Consequently, the number of microbial and microbially produced enzymes used for the industrial manufacture of chemicals is growing rapidly.

The physical and catalytic properties of an enzyme that catalyzes a particular reaction frequently vary in important ways, depending on the source organism. A great deal of attention has been given to enzymes from organisms that flourish in extreme environments (Table 11.3). Taq DNA polymerase is without a doubt the best known and most valuable example of an enzyme obtained from an extremophile, *Thermus aquaticus*, isolated from a hot spring in Yellowstone National Park in 1976. The activity of Taq polymerase has a half-life of 1.6 hours at 95°C. Because this thermostable enzyme withstands the alternating cycles of heating and cooling that enable the PCR to amplify target DNA, it was of great importance in making PCR extremely rapid and efficient. Some of the DNA polymerases subsequently isolated from other extremophiles turned out to have properties that compared favorably with those of Taq polymerase. For example, *Thermococcus littoralis* DNA polymerase (Vent) has a half-life of seven hours at 95°C. However, both Taq and Vent lack $3' \rightarrow 5'$ exonuclease activity, whereas Pfu DNA polymerase, isolated from the hyperthermophile *Pyrococcus furiosus* has $3' \rightarrow 5'$ exonuclease activity. As expected, polymerases lacking exonuclease activity show higher error rates than those with exonuclease activity. Reported error rates for Taq DNA polymerase range from $1 \times 10^{-4}$ to $1 \times 10^{-5}$/bp, whereas Pfu DNA polymerase has a much lower error rate of about $1.5 \times 10^{-6}$/bp.

**TABLE 11.3 Classification of extremophiles**

| Type | Growth conditions |
|---|---|
| Hyperthermophiles | >80°C |
| Thermophiles | 60–80°C |
| Mesophiles | 20–45°C |
| Psychrophiles | <15°C |
| Halophiles | High salt (e.g., 5 M NaCl) |
| Alkaliphiles | pH >9 |
| Acidophiles | pH <3 |
| Baro- or piezophiles | High pressures (up to 130 Mpa) |

## HIGH-THROUGHPUT SCREENING OF ENVIRONMENTAL DNA FOR NATURAL ENZYME VARIANTS WITH DESIRED CATALYTIC PROPERTIES: AN EXAMPLE

In organic chemistry, the aldol condensation is a classic carbon–carbon bond-forming reaction. *Escherichia coli* 2-deoxyribose-5-phosphate aldolase (DERA) catalyzes the reversible reaction of acetaldehyde with D-glyceraldehyde-3-phosphate to form 2-deoxyribose-5-phosphate with a $K_{eq}$ of $4.2 \times 10^3\ M^{-1}$ for the condensation reaction (Figure 11.5). DERA is the only known aldolase enzyme reported to date to condense two aldehydes. Other aldolases use ketones as their aldol donors and aldehydes as their acceptors. DERA can also catalyze the sequential and stereoselective condensation of three aldehydes to form 2,4-dideoxyhexoses. The appropriate enzymatic lactol product can be readily transformed to a lactone to serve as a precursor with two desired stereogenic centers in the synthesis of the widely used cholesterol-lowering drugs Lipitor and Crestor, the 3-hydroxy-3-methylglutaryl-CoA reductase inhibitors known as statins (Figure 11.5).

The synthesis first reported with *E. coli* DERA was ill suited to large-scale production. The enzyme–substrate ratio required was 1:5, by weight. The enzyme was inhibited at concentrations of the limiting reagent, chloroacetaldehyde, greater than 100 mM, and the process yielded the desired product at only 85 mg/L/hour. The quest for a superior DERA explored a large environmental DNA library. DNA was purified from environmental samples collected from a variety of habitats, randomly fragmented, and the fragments inserted directly into an expression vector so that the expression of

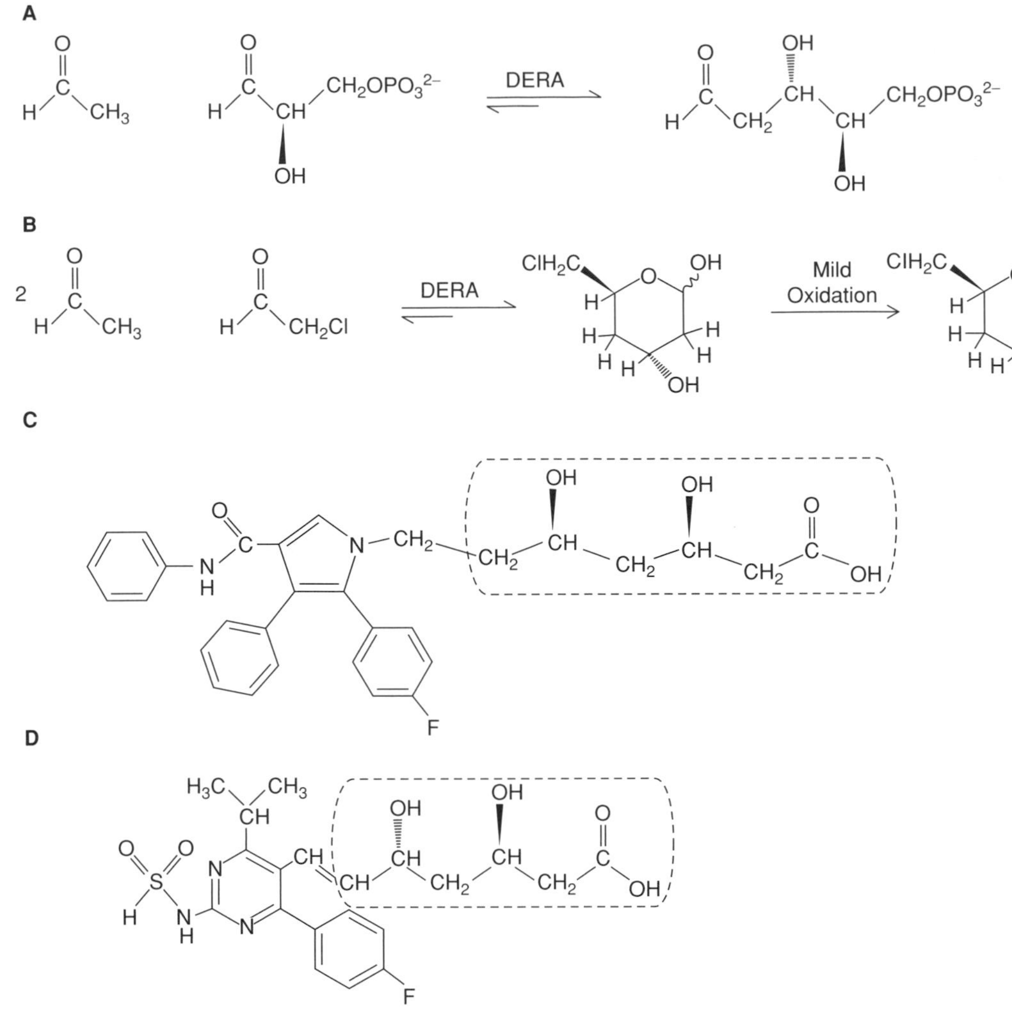

**FIGURE 11.5**

Use of D-2-deoxyribose-5-phosphate aldolase (DERA) in the synthesis of precursors to cholesterol-lowering drugs (statins). (**A**) DERA-catalyzed aldol reaction between the natural donor, acetaldehyde, and the acceptor, D-glyceraldehyde-3-phosphate, to form D-2-deoxyribose-5-phosphate. (**B**) DERA-catalyzed tandem aldol reaction in which two moles of acetaldehyde react consecutively with cloroacetaldehyde to form a lactol product. This product is converted by mild oxidation to (3*R*,5*S*)-6-chloro-2,4,6-trideoxy-*erythro*-hexonolactone with an enantiomeric excess of greater than 99.9% and a dioastereoisomeric excess of 96.6%. The latter compound is the precursor to the portions enclosed by *dotted lines* of the statin drugs Lipitor (**C**) and Crestor (**D**).

genes on the insert was under the control of *cis*-acting vector-based promoters. To generate the library, a standard laboratory strain of *E. coli* was transformed with these constructs. An appropriately diluted culture was then incubated in a microtiter plate format at 37°C in a growth medium in which each clone would replicate to at least $10^4$ clones in 24 hours. The clone array was then subjected to high-throughput expression screening, and DERA-expressing clones were detected using a fluorogenic substrate analog designed to release a fluorescent molecule, 4-methylumbelliferone, upon DERA-catalyzed hydrolysis.

Screening of the environmental DNA library yielded a DERA (from an unknown source organism) with properties much improved over those of the *E. coli* enzyme. The inhibition by chloroacetaldehyde was avoided by a process modification – slow feeding of substrates at a constant ratio of 2:1 acetaldehyde to chloroacetaldehyde. The catalyzed formation of the two carbon–carbon bonds giving rise to the two stereogenic centers led to an enantiomeric excess of greater than 99.9% and a diastereomeric excess of

96.6%. A satisfactory reaction rate was achieved at an enzyme substrate ratio of 1:50, and the process yielded product at a rate of 30.6 g/L/hour, a 360-fold improvement.

## APPROACHES TO OPTIMIZATION OF THE "BEST AVAILABLE" NATURAL ENZYME VARIANTS

Our understanding of protein folding and of structure–function relationships is insufficiently complete to design from-scratch proteins with desired biological and physical properties. However, various methods are available whereby new enzymes can be engineered by using existing enzymes as the starting points for the generation of desired versions. The methods described below have been very successful at generating enzymes with properties optimized for particular applications. It is becoming rare for any process to utilize a native microbial enzyme.

### *IN VITRO* EVOLUTION OF ENZYMES

Very frequently, one would like to optimize the physical and/or catalytic properties of an enzyme for which no tertiary structure information or knowledge of the reaction mechanism is available. A number of approaches address this need, and we will discuss three of these: DNA shuffling, saturation mutagenesis, and error-prone mutagenesis.

#### DNA Shuffling

In many instances, the objective is to improve particular properties of an enzyme for which there is no information about tertiary structure and only limited understanding of the mechanism of action. DNA shuffling has proven to be a very successful approach to evolving enzymes with desired alterations in substrate specificity, pH-activity profile, specific activity, enantioselectivity, thermostability, tolerance to organic solvents, solubility, crystallizability, and so on. In the most commonly used form of DNA shuffling, two or more related genes are fragmented and then reassembled, as shown in Figure 11.6.

First, a pool of homologous DNA sequences is randomly fragmented with pancreatic deoxyribonuclease I. This enzyme introduces single-strand scissions into double-stranded DNA. When the DNA scissions are within a few nucleotides of each other, the DNA duplexes fragment. Such fragments have protruding 5′-ends that can serve as templates for DNA polymerases.

The fragments are denatured, and homologous sequences from the different DNAs are allowed to hybridize. The fragments are then reassembled into a full-length gene by cycling in the presence of Taq DNA polymerase under the conditions of the PCR, *but in the absence of primers.* The crossovers between fragments derived from different DNAs are locked in place by the DNA polymerase extensions. Under these conditions, the average size of the

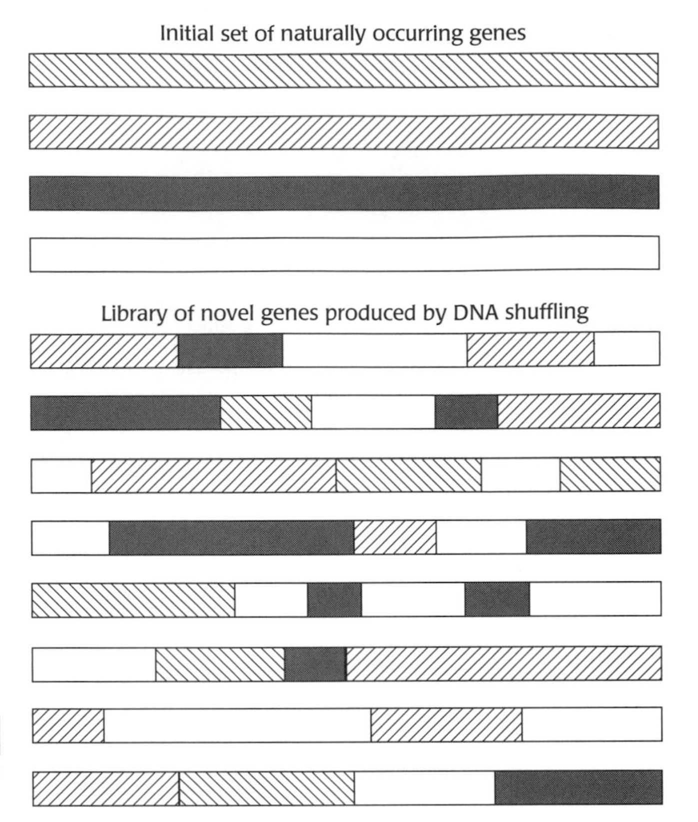


**FIGURE 11.6**

Production of a library of novel genes by DNA shuffling.

PCR products increases gradually through the priming of one product on another. The cycle of fragment reassembly is repeated several times. Flanking primers are then added and additional cycles of PCR are performed to rescue full-length genes. This typically results in the generation of a single product of the correct size. A library of full-length chimeric sequences that have undergone homologous recombination results from this process (Figure 11.6). The library of DNA sequences is inserted into an expression vector and transferred into a host cell for expression. The clones are screened for improvement in the properties of interest. A pool of the sequences from the best clones is then subjected to another round of DNA shuffling. The process can be repeated several times as needed. Important variables that influence the outcome of DNA shuffling include the length of the shuffled DNA, the number and molar ratio of the different sequences in the starting gene pool, the degree of sequence homology, and the number of crossovers.

The directed evolution of a glyphosate (Roundup) tolerance gene provides an outstanding example of successful DNA shuffling. Globally, about 75% of genetically modified crops are engineered for herbicide tolerance. Roundup Ready crops, tolerant to glyphosate occupy the greatest acreage. As described elsewhere in this book, glyphosate inhibits *enol*pyruvyl-shikimate-3-phosphate synthase (EPSPS), an enzyme localized in the chloroplasts that functions in the pathway of aromatic amino acid biosynthesis. In commercial glyphosate-tolerant crops, microbial EPSPS

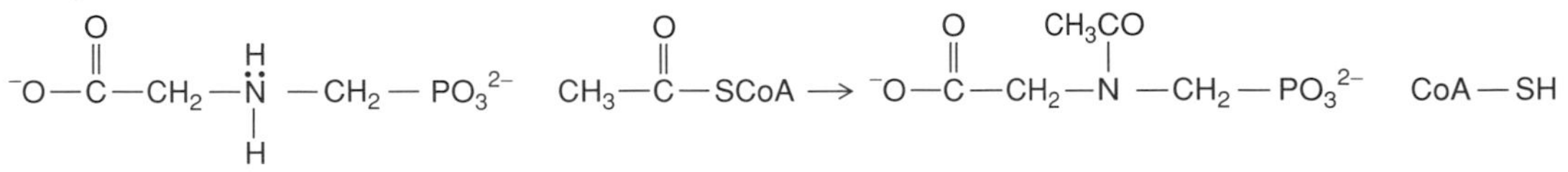

$$^{-}O-\overset{O}{\overset{\|}{C}}-CH_2-\underset{H}{\overset{H}{\ddot{N}}}-CH_2-PO_3^{2-} \quad CH_3-\overset{O}{\overset{\|}{C}}-SCoA \rightarrow {}^{-}O-\overset{O}{\overset{\|}{C}}-CH_2-\overset{CH_3CO}{\overset{|}{N}}-CH_2-PO_3^{2-} \quad CoA-SH$$

**FIGURE 11.7**

*N*-Acetylation of glyphosate catalyzed by a glyphosate acetyltransferase (GAT).

transgenes provide the sole means of conferring tolerance to this herbicide. In such plants, glyphosate accumulates in the meristems, where it may interfere with reproductive development and may lower crop yield.

*N*-Acetylglyphosate is not herbicidal and is a very poor inhibitor of EPSPS. A soluble enzyme with high affinity for glyphosate, and that would rapidly acetylate this secondary amine, would be a strong alternative candidate for conferring tolerance, without product accumulation. No such enzyme had been identified until 2004, when Castle and coworkers reported the successful creation of such an "optimized" enzyme by directed evolution. These researchers screened hundreds of isolates of *Bacillus* species for the ability to acetylate glyphosate. Cultures were grown to stationary phase, permeabilized, and incubated with glyphosate and acetyl-CoA (Figure 11.7). Supernatants were analyzed for *N*-acetylglyphosate by a sensitive mass spectrometry method. Three strains (ST401, B6, and DS3) of the common saprophytic bacterium *Bacillus licheniformis* exhibited *N*-acetylglyphosate transferase (GAT) activity. The responsible enzymes were cloned and expressed in *E. coli* and characterized. These 146-residue soluble enzymes were 94% identical in sequence. The acetylation of glyphosate by these enzymes had a rate constant ($k_{cat}$) of 1.0 to 1.7 $min^{-1}$. Their affinity for glyphosate was low ($K_m$ 1.2 to 1.8 mM at pH 6.8 and 21°C) but high for acetyl-CoA ($K_m$ 1 to 2 $\mu$M), indicating that acetyl-CoA is the native acetyl substrate. Their average $k_{cat}/K_m$ was 0.81 $min^{-1}$ $mM^{-1}$. Expression of these enzymes in transgenic tobacco or *Arabidopsis* did not confer herbicide tolerance. Improvement in GAT catalytic properties was then sought through directed evolution.

BLAST search of databases for sequences related to the three *B. licheniformis* GAT enzymes revealed a number of homologs in other bacteria. The three *B. licheniformis* genes were then subjected to 11 iterations of DNA shuffling. During the shuffling process, at the end of several cycles, amino acid diversity from the predicted sequences of four other hypothetical *N*-acetyltransferase proteins ranging in identity with GAT from 59% to 28% (Table 11.4) was incorporated into the library. The best GAT enzyme obtained by this directed molecular evolution process had a $k_{cat}/K_m$ of 8320 $min^{-1}$ $mM^{-1}$, an improvement in enzyme efficiency of about 10,000-fold over the *B. licheniformis* ST401 GAT. The sequence of the evolved enzyme differed in more than 30 positions from that of the parent enzyme. The evolved enzyme conferred high glyphosate tolerance to *E. coli*, *Arabidopsis*, tobacco, and maize, and caused no adverse symptoms.

## Gene Site Saturation Mutagenesis

The term *gene site saturation mutagenesis* describes a technique whereby each amino acid of a protein is replaced with each of the other 19 naturally

**TABLE 11.4 Known and hypothetical glyphosate N-acetyltransferase proteins from diverse bacteria[a]**

| Organism and protein | % amino acid sequence identity |
|---|---|
| *Bacillus licheniformis* ST401 GAT | 100 |
| *Bacillus subtilis* YITI[b] | 59 |
| *Bacillus cereus* YITI | 49 |
| *Listeria inocua* NAT | 38 |
| *Zymomonas mobilis* NAT | 28 |

[a] The genes encoding the known 94% identical glyphosate acetyltransferase proteins from *B. licheniformis* ST401, B66, and DS3, and those for the more distantly related proteins were used in directed molecular evolution of GAT.

[b] *B. subtilis* YITI, a hypothetical *N*-acetyltransferase predicted from the genomic sequence, was expressed in recombinant *E. coli*. The protein is capable of acetylating glyphosate, but with lower efficiency than GAT.

*Source*: Castle, L. A., et al. (2004). Discovery and directed evolution of a glyphosate tolerance gene. *Science*, 304, 1151–1154.

occurring amino acids. This can be accomplished by using PCR amplification protocols in which the codons for the selected amino acids are randomized. To obtain primers so that all three nucleotides in a randomized codon contain a statistical representation of C, G, A, and T, a mixture of 64 different forward primers and 64 different reverse primers would be used. If desired, the third nucleotide of the codon can be restricted to either G or C. This eliminates stop codons and reduces the degeneracy of the genetic code with an increase in the probability that rarely encoded amino acids, such as tryptophan, will be adequately represented. In this situation, a mixture of 32 forward and reverse degenerate primers is employed in a reaction resulting in one randomized codon. The number of possible variants of a protein with $n$ residues that can theoretically arise from gene site saturation mutagenesis is $20^n$. The size of the library that needs to be screened to ensure a high probability of including each mutant can be calculated statistically.

The example that follows describes a two-step process leading to the creation of a particular highly enantioselective enzyme. In the first step, the most highly enantioselective enzyme is identified in a set of environmental DNA libraries. In the second step, gene site saturation mutagenesis is used to find an improved variant.

For a particular enzyme, the number of variants from organisms available in pure culture is limited. Environmental DNA libraries provide access to hundreds of variants. The account of the identification and isolation of a highly enantioselective nitrilase provides an excellent illustration of the potential value of environmental DNA. In nature, nitrilases function in pathways for the synthesis and degradation of both naturally occurring and xenobiotic nitriles by catalyzing the cleavage of the C–N bond shown below, where the asterisk indicates an asymmetric center:

$$\underset{(R,S)\text{-nitrile}}{\text{R-C}^{*}\text{H(OH)-CN}} + 2\text{H}_2\text{O} \xrightarrow{\text{nitrilase}} \underset{(R,S)\text{-carboxylic acid}}{\text{R-C}^{*}\text{H(OH)-COOH}} + \text{NH}_3.$$

Nitriles are inexpensive, attractive precursors for carboxylic acids. The ability of nitrilases to catalyze the C–N bond cleavage under mild conditions – and more importantly, their stereospecificity – is important to the synthesis of pure enantiomers, particularly those needed as precursors of drugs. However, the usefulness of the available nitrilases was generally limited by problems of inadequate stability and specificity.

This problem was addressed by the screening of 651 biotope-specific environmental DNA (eDNA) libraries. The environments sampled included active volcanoes, deep-sea hydrothermal vents, coral reefs, deserts, and hazardous waste sites. High-quality, high molecular weight DNA was isolated from each of the environmental samples, and fragments ranging from 1 to 10 kb were generated either by shearing or by digestion with restriction enzymes. Phagemid gene libraries were constructed in *E. coli* for the fragments derived from each eDNA sample (Box 11.4). The resulting clones were assayed for the ability to grow on a medium supplemented with a nitrile substrate as a sole source of organic nitrogen. Growth was dependent on the ability of the organism to metabolize the substrate and generate ammonia.

The phagemid DNA from cultures exhibiting growth was isolated and sequenced to confirm the presence of a nitrilase gene and to determine its sequence. One to three unique nitrilases were discovered per $10^6$ to $10^{10}$ clones derived from a single eDNA library. As an interesting aside, although most of the nitrilases came from samples collected in soil or in aquatic or marine sediments, there was no clear biogeographic relationship between the source of the sample and the phylogenetic relationships indicated by a tree derived from the sequences of the enzymes.

Of the nitrilase genes with unique sequences, 137 were subcloned into an expression vector and the enzymatic properties of the protein products were characterized. An objective of the study was to find a nitrilase that would enantioselectively convert 3-hydroxyglutaryl nitrile to (*R*)-4-cyano-3-hydroxybutyric acid (Figure 11.8A). The ethyl ester of (*R*)-4-cyano-3-hydroxybutyric acid is an intermediate in the enantioselective synthesis of the cholesterol-lowering drug Lipitor. Of 137 nitrilases, 110 hydrolyzed 3-hydroxyglutaryl nitrile. The most selective enzyme generated (*R*)-4-cyano-3-hydroxybutyric acid with an enantiomeric excess (*ee*) greater than 95% at 100 mM substrate concentration (see Box 11.2).

For the intended production process, the enzyme-catalyzed hydrolysis would need to be performed at a substrate concentration of 3 M. Unfortunately, when this enzyme was assayed at 0.5, 1, 2, and 3 M substrate, the enantiomeric excesses determined upon completion of the reaction were 92.1%, 90.7%, 89.2%, and 87.6%, respectively. For enzyme-catalyzed reactions, the factors that come into play at very high organic substrate concentrations include enzyme instability with associated lower activity and selectivity, and substrate and/or product inhibition.

A comprehensive library of variant genes of the best wild-type nitrilase described above, each encoding a single-site enzyme mutant, was generated and expressed in *E. coli*. To find the desired stereoselective enzyme, a high-throughput mass spectrometric assay was devised that used a chiral $^{15}N$-labeled substrate, $^{15}N$-(*R*)-3-hydroxyglutaryl nitrile. Hydrolysis of this substrate with an (*R*)-selective nitrilase removes the $^{15}N$, whereas upon hydrolysis with an (*S*)-selective nitrilase, the $^{15}N$ remains in the product (Figure 11.8B).

This assay led to the identification of a mutant nitrilase (Ala190His) significantly improved over the wild-type enzyme. At 2.25 M substrate concentration, the wild-type nitrilase converted the substrate to product in 24 hours with 87.8% *ee*, while the mutant nitrilase (Ala190His) required only 15 hours to yield product with 98.1% *ee*. Thus, the mutant enzyme satisfied the demands of a process for high-volume synthesis of

**Phagemid**

A phagemid is a defective phage whose genome contains a plasmid that can be excised by coinfection of the host with an f1 or M13 helper phage. The helper phage provides the proteins the defective phage is missing and needs to complete its life cycle. Phagemids do not cause lysis of the host cell. In the analysis of the putative nitrilase clones, a lambda insertion type DNA cloning vector, λZAP, was used. The lambda library constructed in λZAP was then converted through *in vivo* excision in *E. coli* into a phagemid library. Details of the construction of λZAP and of the expression of coding sequences in the phagemid clones excised from λZAP are provided in Short, J. M., Fernandez, J. M., Sorge, J. A., and Huse, W. D. (1988). λZAP: a bacteriophage λ expression vector with *in vivo* excision properties. *Nucleic Acids Research*, 16, 7583–7600.

**BOX 11.4**

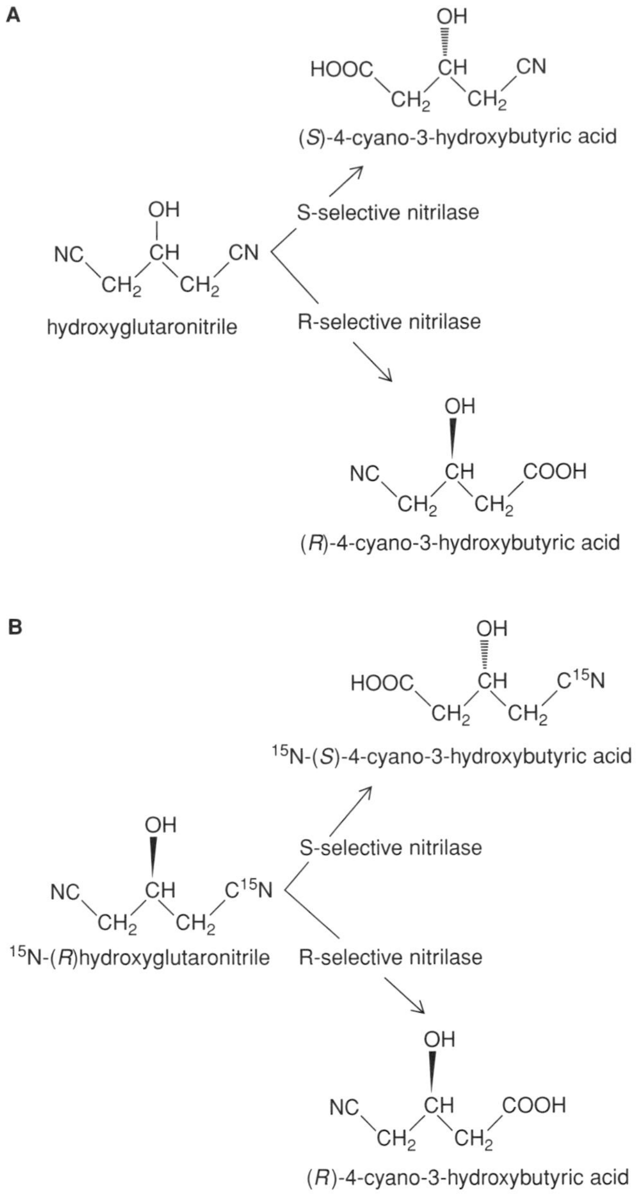

**FIGURE 11.8**

(**A**) Conversion of the symmetric substrate hydroxyglutaronitrile to (*S*)- or (*R*)-4-cyano-3-hydroxybutyric acid by hypothetical nitrilases with either a 100% *S* or 100% *R* enantiomeric specificity. (**B**) Mass spectrometric assay for the *S* versus *R* selectivity of a nitrilase by utilizing a chiral substrate, $^{15}N$-(*R*)-hydroxyglutaronitrile. The $^{15}N$-(*S*)-product has an *m/z* of 130, whereas the (*R*)-product has an *m/z* of 129.

(*R*)-4-cyano-3-hydroxybutyric acid, by acting at an acceptable rate at very high substrate concentrations to generate a product with high enantiomeric excess.

## Error-Prone Mutagenesis

There are protocols for the PCR reaction that introduce errors in DNA amplification that result in few or many amino acid substitutions in the encoded protein. Error-prone PCR (epPCR) has frequently been used as a first stage in the directed evolution of enzymes. A brief look at the challenges of screening of mutants generated by epPCR is provided by the following calculation performed for a *Pseudomonas aeruginosa* lipase. This enzyme consists of

285 amino acids. The size of the mutant library ($N$) can be calculated using the algorithm

$$N = 19^M \times 285!/[(285 - M)! \times M!]$$

as a function of $M$, the number of sites of amino acid substitution per enzyme molecule. For $M = 1$, the library would theoretically contain 5415 unique sequences. However, because of the degeneracy of the genetic code and the problems with the properties of certain sequences, the actual number would be significantly smaller. When $M = 2$, the theoretical number is approximately 15 million, and when $M = 3$, the theoretical number of *P. aeruginosa* variants exceeds 50 billion. This is a vivid example of the demands that these types of methods place on high-throughput screening systems for the sought-after characteristic, whether improved enantioselectivity, altered substrate specificity, or modified stability.

Here is the manner in which this formidable problem was confronted in searching for a *P. aeruginosa* lipase with improved enantioselectivity using the model reaction shown in Figure 11.9. This reaction was chosen because screening for one of the products, *p*-nitrophenol, can be performed very simply by absorbance spectroscopy. The wild-type enzyme exhibited very low enantioselectivity. The conditions for epPCR were chosen to introduce an average of one amino acid substitution per enzyme molecule per cycle of mutant generation; 2000 to 3000 mutants were screened and the process was repeated on the most enantioselective mutant from each cycle. Mutant A (S149G) showed an $E$ value of 2.1 in favor of the $S$-isomer as compared to $E =$ 1.1 for the wild-type enzyme. The results for the next three cycles were as follows: mutant B (S149G, S155L), $E = 4.4$; mutant C (V47G, S155L, S149G), $E = 9.4$; and mutant D (F259L, V47G, S155L, S149G), $E = 11.3$. Interestingly,

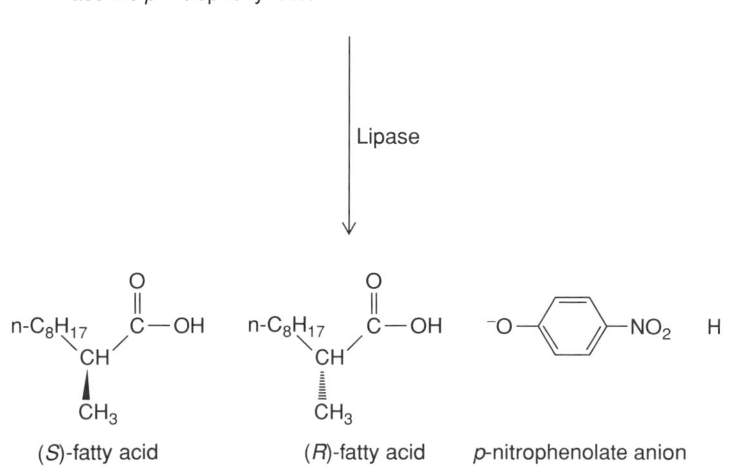


**FIGURE 11.9**

Assay for the enantioselectivity of a lipase by measuring the hydrolytic kinetic resolution of a racemic mixture of *p*-nitrophenyl esters. Enzyme-catalyzed hydrolysis of the racemic substrate yields two enantiomeric fatty acids and a *p*-nitrophenylate anion that absorbs strongly at 405 nm. To measure the enantioselectivity of each lipase variant, the assay was performed by comparing separately and pairwise the rate of *p*-nitrophenylate formation from the (*S*)- and (*R*)-substrates.

most of these mutations are at sites distant from S82, the active site of the enzyme.

Ultimately, further experiments involving epPCR of the wild-type gene at a higher mutation rate of three amino acid substitutions per enzyme molecule, followed by DNA shuffling of mutant genes, led to the isolation of a mutant (D20N, S53P, S155M, L162G, T180I, and T234S) with $E > 51$. In addition to the greatly improved enantioselectivity, the mutant enzyme is two orders of magnitude more active than its wild-type counterpart. A detailed molecular modeling and quantum mechanical study indicated that the greatly enhanced (*S*)-enantioselectivity of the mutant was attributable to the cooperative contribution of two of the mutations, S53P and L162G. Again, these residues are remote from the active site. These mutations indirectly result in the formation of a chiral pocket that accommodates the (*S*)-ester and also provides additional stabilization of the transition state in the enzyme-catalyzed hydrolysis of the ester substrate. The latter effect is responsible for the higher activity of the mutant enzyme. This is a striking example of an enzyme variant generated by directed evolution that could not have been discovered by traditional protein design.

## RATIONAL METHODS OF PROTEIN ENGINEERING

More and more frequently, the specificity of an enzyme, its interaction with cofactors, inhibitors, or other proteins, and its catalytic mechanism are all understood with precision on the basis of sequence, the known three-dimensional structure of the enzyme, and relevant complexes with various ligands, as well as through kinetic and thermodynamic studies. Here, site-directed mutagenesis is frequently the method of choice. Widely described variants of current techniques for site-specific mutagenesis all utilize PCR with primers containing mutant codon(s) of interest.

### SITE-SPECIFIC MUTAGENESIS OF *E. COLI* 2-DEOXYRIBOSE-5-PHOSPHATE ALDOLASE

The studies directed at altering the substrate specificity and catalytic activity of an enzyme discussed earlier in this chapter, *E. coli* DERA, offer an elegant example of the power of site-specific mutagenesis. The crystal structures of DERA and of its complex with 2-deoxyribose-5-phosphate have been determined at the very high resolution of 0.99 Å and 1.05 Å, respectively. These studies have provided a detailed understanding of the tertiary structure of the enzyme, the determinants of its substrate specificity, and its catalytic mechanism.

Wild-type DERA has a very strong preference for phosphorylated substrates and accepts a limited range of substrates. Extending the range of unnatural substrates for DERA is central to expanding the utility of this enzyme in organic synthesis. The 1.05-Å structure of the DERA–substrate

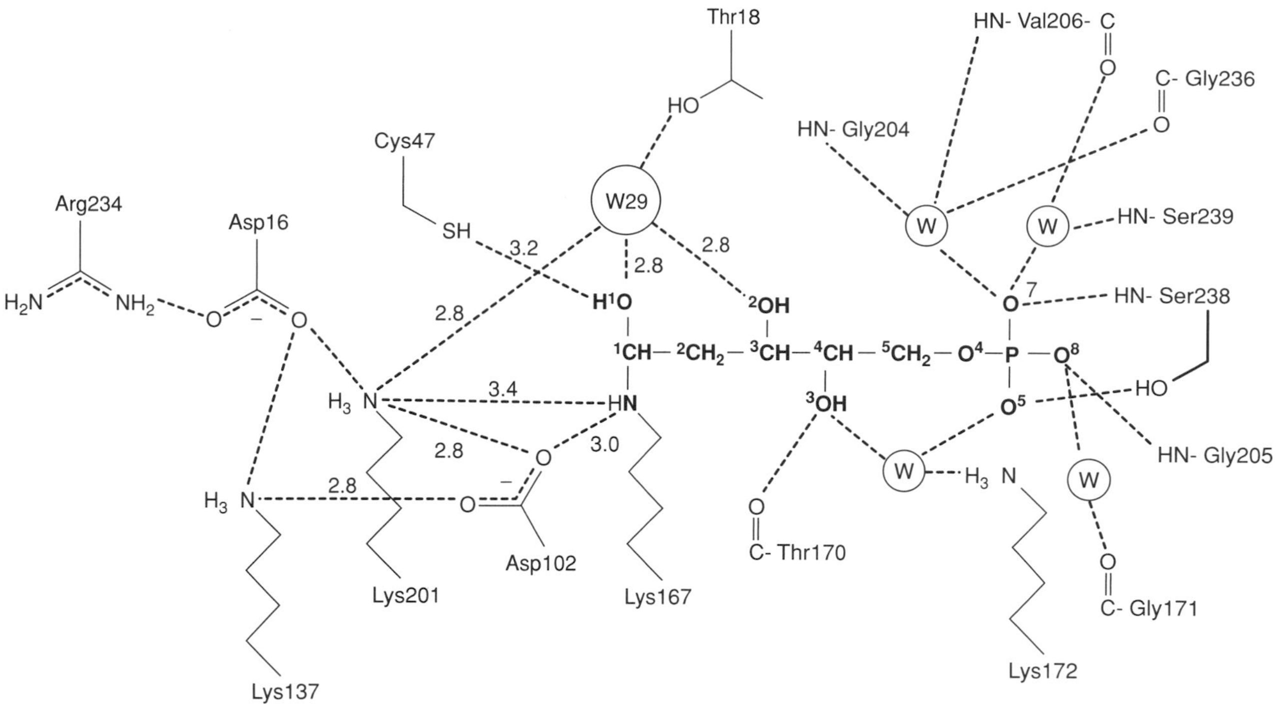

**FIGURE 11.10**

Wild-type D-2-deoxyribose-5-phosphate aldolase interactions with D-2-deoxyribose-5-phosphate, as seen in the covalent carbinolamine intermediate in the enzyme–substrate complex at 1.05 Å resolution. Hydrogen bonds are indicated by *dotted lines* and lengths are given in angstroms. [Heine, A., DeSantis, G., Luz, J. G., Mitchell, M., Wong, C-H., and Wilson, I. A. (2001). Observation of covalent intermediates in an enzyme mechanism at atomic resolution. *Science*, 294, 369–374.]

complex provided indispensable guidance for the site-directed mutagenesis studies. A number of residues form the phosphate-binding pocket, including Gly171, Lys172, Gly204, Gly205, Val206, Arg207, Gly236, Ser238, and Ser239 (Figure 11.10). Notably, only the side chain of Ser238 forms a direct hydrogen-binding contact with the phosphate portion of 2-deoxyribose-5-phosphate. The mutant Ser238Asp proved to be the most valuable of site-specific mutants of five different residues that were examined. It was expected that the introduction of a negative charge in very close proximity to the phosphate group would result in electrostatic repulsion and a marked decrease in the affinity of the enzyme for its natural substrate. This was indeed the case. For the reverse (retro-aldol) reaction catalyzed by the Ser238Asp mutant, the $K_m$ for D-2-deoxyribose decreased by about 30% and the $k_{cat}$ doubled. At the same time, the $k_{cat}$ for the phosphorylated substrate dropped 100-fold, and the $K_m$ increased more than 50-fold. Overall, the Ser238Asp mutant showed a 2.5-fold enhancement in catalytic rate toward 2-deoxyribose. Further experiments showed that these data were predictive of the synthetic capabilities of the mutant DERA. The improvements in the aldol reaction activity paralleled the retro-aldol kinetic data.

An unanticipated bonus was that the Ser238Asp mutant enzyme showed an enhanced tolerance for unnatural substrates. It catalyzed a novel sequential aldol reaction using 3-azidopropinaldehyde as the first acceptor and two moles of acetaldehyde as donors to form an azidopyranose, a key intermediate in the synthesis of Lipitor (see Figure 11.5C). 3-Azidopropinaldehyde is not a substrate for the wild-type enzyme.

## LARGE-SCALE BIOCATALYTIC PROCESSES

The examples given below describe products produced in the largest amount by biocatalytic processes. Two of these examples are drawn from the food industry, one from polymer chemistry, and one from the chemical industry. These biotransformations are notable for their simplicity, efficiency, and minimization of waste by-products.

### PRODUCTION OF HIGH-FRUCTOSE SYRUPS

The dairy, baking, and brewing industries, as well as other major food industries, have long depended on enzymes from animals, plants, and microorganisms for producing cheese, bread, and malt beverages, clarifying fruit and vegetable juices, and tenderizing meats (Table 11.5). The dominant use of partially purified or pure microbial enzymes continues to be in the production of glucose and fructose syrups for the confectionary and soft drink industries. High-fructose syrups are produced in an annual amount in excess of a million tons and are the largest-volume product of a biocatalytic process.

The process for the manufacture of glucose and fructose syrups from starch is inexpensive, and the products compete successfully in price with sucrose. Glucose has only approximately 75% of the sweetness of sucrose, whereas its isomer, fructose, has about twice the sweetness of sucrose (Figure 11.11). Consequently, fructose is the preferred sweetener in low-calorie foods, providing twice the sweetness of sucrose at half the weight. The first step, *liquefaction*, involves the conversion of starch to low–dextrose equivalent (DE) maltodextrins (Figure 11.12; Table 11.6). This step is performed at near-neutral pH for a relatively short period of time at temperatures ranging from 95°C to 107°C. Under the stabilizing influence of very low concentrations of $Ca^{2+}$ ions and of the substrate, starch, *B. licheniformis* $\alpha$-amylase is able to withstand these extreme temperatures and converts starch to low-DE maltodextrins. Conversion of the low-DE maltodextrins to glucose, *saccharification*, is performed with a mixture of two enzymes. Glucoamylase from the fungus *A. niger* rapidly splits $\alpha$-1,4 linkages with stepwise release of glucose molecules from the nonreducing ends of the starch chains. This enzyme also splits $\alpha$-1,6 linkages, but slowly; however, the second enzyme, *Bacillus* species pullulanase, splits them rapidly. Saccharification is performed under mildly acidic conditions and at a lower temperature, to prevent the formation of psicose (which is favored by alkaline pH) and to avoid forming colored products by caramelization of glucose at high temperatures.

The development of techniques for the large-scale *isomerization* of glucose to fructose offered several challenges. The well-known metabolic route for conversion of glucose to fructose proceeds by several steps: glucose is phosphorylated to glucose-6-phosphate; glucose-6-phosphate is isomerized to fructose-6-phosphate, and the latter is dephosphorylated to fructose. Richard O. Marshall and Earl Kooi in 1957 were the first to report the discovery of a microbial enzyme capable of converting glucose directly to fructose. Marshall had observed that the bacterium *Aerobacter cloacae* (now

**TABLE 11.5 Microbial enzymes with industrial-scale applications in the food industry and some of their sources**

| Enzyme | Source | Action | Application |
|---|---|---|---|
| α-Amylase | *B. subtilis*<br>*B. licheniformis*<br>*Aspergillus oryzae* | Endohydrolysis of α-1,4-glucosidic linkages | Starch processing |
| Glucoamylase | *Aspergillus oryzae*<br>*Aspergillus niger*<br>*Rhizopus oryzae* | Removes glucose from nonreducing end of starch, also splits α-1,6 linkages at branch points, but more slowly | Starch processing; brewers' and distillers' mashes |
| Pullulanase | *Klebsiella aerogenes* | Splits α-1,6 linkages in amylopectin | Starch processing |
| Glucose isomerase | *Bacillus coagulans*<br>*Streptomyces albus* | Converts D-glucose to D-fructose. This enzyme is actually a xylose isomerase. | Production of high-fructose syrups |
| β-Glucanase | *B. subtilis*<br>*A. niger*<br>*Penicillium emersonii* | Degrades β-glucan by cleaving β-(1,3)- or -(1,4)-glucosidic linkages | Brewing |
| Invertase | *Saccharomyces cerevisiae* | Splits sucrose to glucose and fructose | Confectionary industry; baking |
| Lactase | *Saccharomyces lactis*<br>*A. oryzae*<br>*A. niger*<br>*R. oryzae* | Splits lactose to glucose and galactose | Dairy industry (treatment of milk and whey) |
| Pectinase | *A. oryzae*<br>*A. niger*<br>*R. oryzae* | Degrades pectin (α-1,4-linked anhydrogalacturonic acid with some of the carboxyl groups esterified as the methyl esters) | Clarification of fruit juices and wines |
| Neutral protease | *B. subtilis*<br>*A. oryzae* | Hydrolyzes peptide bonds in proteins | Flavoring of meat and cheese; baking |
| Rennin (chymosin) | *Mucor miehei* spp.<br>Recombinant enzyme produced in *E. coli* and fungi | Hydrolyzes a specific peptide bond in κ-casein leading to coagulation of milk proteins | Cheese making |
| Lipase | *A. oryzae*<br>*A. niger*<br>*R. oryzae* | Hydrolyzes ester linkages in fats | Dairy industry |

*Enterobacter cloacae*), when grown on xylose, was able to convert glucose to fructose in the presence of arsenate and magnesium chloride. This conversion was catalyzed by xylose isomerase, whose metabolic role is to convert D-xylose to D-xylulose (Figure 11.13; Box 11.5). Numerous disadvantages

**FIGURE 11.11**

Structure of sucrose and the α and β anomers of fructose.

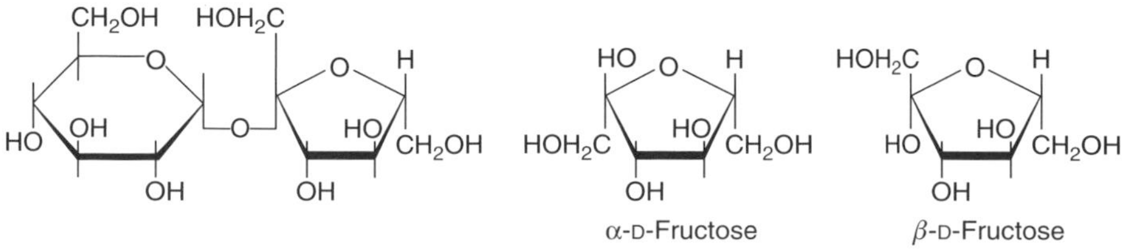

α-D-Fructose β-D-Fructose

**Sucrose: α-D-glucosyl-(1→2)-β-D-fructoside**

**Fructose**

**TABLE 11.6 Properties and industrial applications of hydrolyzed starch products**

| Type of syrup | DE[a] | Composition (%) | Properties | Applications |
|---|---|---|---|---|
| Low-DE maltodextrins | 15–30 | 1–20 D-glucose<br>4–13 maltose<br>6–22 maltotriose<br>50–80 higher oligomers | Low osmolarity | Clinical feed formulations; raw materials for enzymic saccharification; thickeners, fillers, stabilizers, glues, pastes |
| Maltose syrups | 40–45 | 16–20 D-glucose<br>41–44 maltose<br>36–43 higher oligomers | High viscosity, reduced crystallization, moderately sweet | Confectionary, soft drinks; brewing and fermentation; jams, jellies, ice cream, sauces |
| High-maltose syrups | 48–55 | 2–9 D-glucose<br>48–55 maltose<br>15–16 maltotriose | Increased maltose content | Hard confectionary; brewing and fermentation |

[a] Dextrose equivalent (DE). Hydrolysis of glucosidic bonds leads to exposure of reducing aldehyde groups (Figure 11.12). The hydrolysis of starch to a mixture of sugar monomers and oligomers is monitored by measuring their specific chemical-reducing power relative to the same amount of pure glucose (dextrose); the latter is assigned a DE value of 100. A fully hydrolyzed sample of starch has a DE value of 100.
*Source*: Kennedy, J. F., Cabalda, V. M., and White, C. A. (1988). Enzymic starch utilization and genetic engineering. *Trends in Biotechnology*, 6, 184–189.

prevented commercial exploitation of this initial discovery, however, including the high cost of xylose, the inducer of xylose isomerase synthesis in the source organism; the low affinity of xylose isomerase for glucose; fructose yields of only 33%; and long reaction times. Moreover, the use of arsenate was particularly undesirable in a step leading to a food product.

A search for xylose isomerases in other organisms led to the discovery that certain *Streptomyces* species produced xylose isomerases that did not require arsenate. Investigation of streptomycetes also resolved the problem of the high expense of xylose as an inducer for enzyme production. Several *Streptomyces* species produced an extracellular xylanase as well as intracellular xylose isomerase (henceforth referred to as glucose isomerase). Xylanase

**FIGURE 11.12**

Structure of starch, indicating points of cleavage by $\alpha$- and $\beta$-amylase, glucoamylase, and pullulanase.

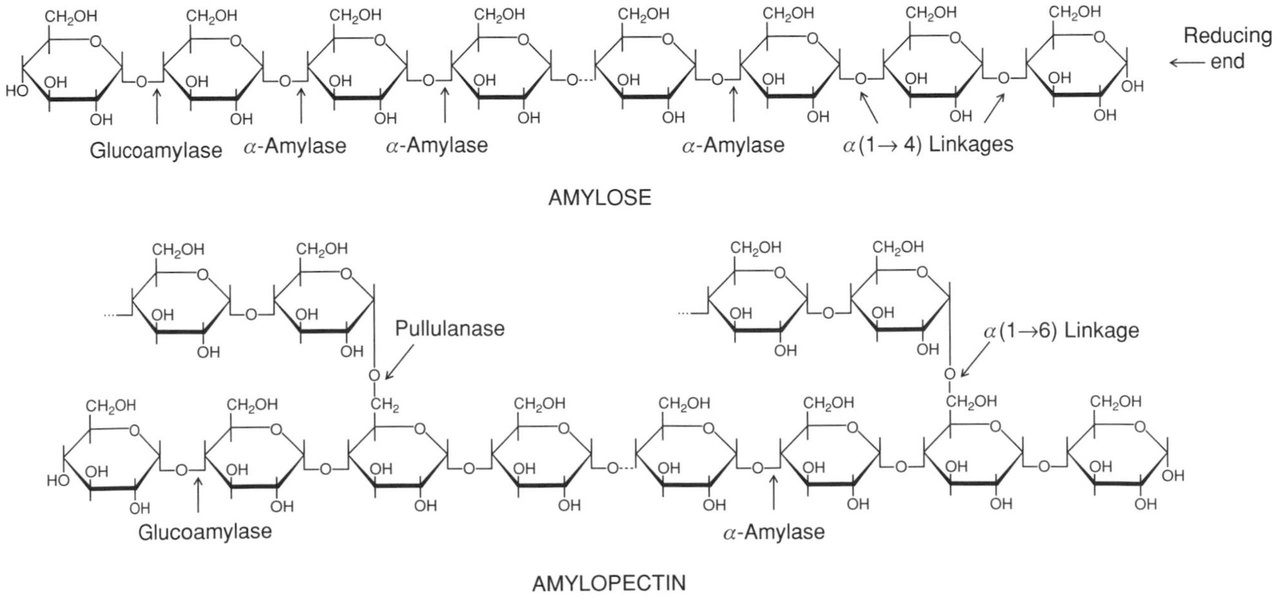

degrades polymeric xylan to xylose, and xylan can be obtained cheaply from straw or wood. By using xylan, the inducer costs for enzyme preparation were reduced dramatically. The yield of fructose from glucose achieved with the *Streptomyces* enzyme was in the 40% to 50% range. This improvement was important because the sweetness of syrups containing 42% fructose and 58% glucose is equivalent to the sweetness of sucrose.

The early commercial processes for enzymic conversion of glucose to fructose in batches used stirred tank reactors. Crude glucose isomerase preparations were added to high-DE sugar in large quantities to compensate for the low substrate affinity of the enzyme. However, the enzyme preparations were expensive, and the subsequent refining step destroyed them. Later improvements in the process sought to conserve the enzymes. One solution was to use continuous-flow reactors containing immobilized bacterial cells. These later gave way to reactors in which pure enzyme was bound either covalently or noncovalently to a solid support. A reactor containing 1 kg of immobilized enzyme allows the commercial production of over 18 metric tons of 42% fructose syrup solids.

The equilibrium for the glucose isomerization reaction lies at a fructose concentration of about 55%. In practice, the conversion is carried out only to about 42% fructose, because attainment of equilibrium is slow. Syrups with higher fructose levels can be obtained by cation-exchange chromatography under appropriate conditions; fructose is more strongly adsorbed than glucose, and a syrup containing 42% fructose can be enriched to 85% fructose by a single pass through a column.

D-Xylose ⇌ D-Xylulose

D-Glucose ⇌ D-Fructose

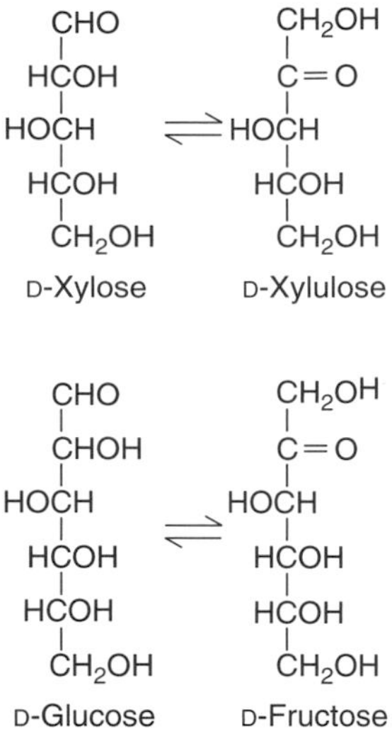

**FIGURE 11.13**

Isomerization reactions catalyzed by xylose isomerase. This enzyme is frequently referred to as glucose isomerase.

**Metabolic Function of Xylose Isomerase**

Many bacteria utilize D-xylose as an energy source. Active transport systems in the cytoplasmic membrane bring the sugar into the cell, where xylose isomerase converts it to D-xylulose. D-Xylulose is phosphorylated by xylulose kinase to form D-xylulose-5-phosphate. This phosphorylated sugar is then metabolized in the pentose phosphate and the glycolytic pathways.

**BOX 11.5**

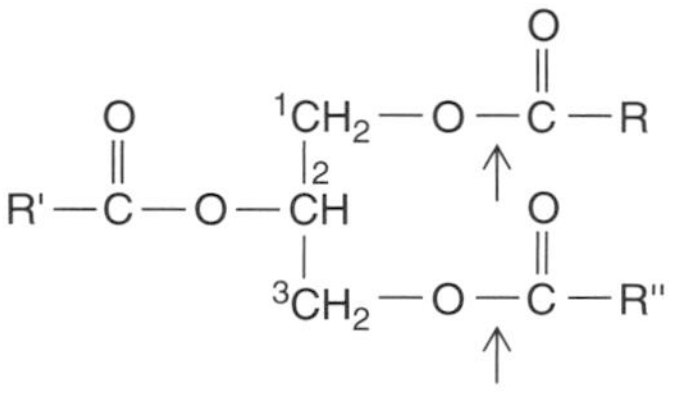

## PROPERTIES AND APPLICATIONS OF LIPASES

Lipases (triacylglycerol acylhydrolases) are present in all organisms to catalyze the synthesis or hydrolysis of fats. Depending on the source, these enzymes vary widely in pH optima and thermostability, positional specificity, and selectivity with regard to the structure or length of the fatty acid chains that they either hydrolyze off or utilize for esterification. Some lipases show high regiospecificity for the 1- and 3-positions of a triglyceride (Figure 11.14), whereas others show no positional preference. Other lipases show an intermediate level of specificity.

Lipases catalyze the following types of reactions:

1. Ester hydrolysis and synthesis

   Ester hydrolysis

$$\text{R-COOR}' + H_2O \rightarrow \text{R-COOH} + \text{R}'\text{-OH}$$

   Ester synthesis

$$\text{R-COOH} + \text{R}'\text{-OH} \rightarrow \text{R-COOR}' + H_2O$$

2. Transesterification

   Transesterification by acidolysis

$$R_a\text{-COOR}' + R_b\text{-COOH} \rightarrow R_b\text{-COOR}' + R_a\text{-COOH}$$

1, 2, 3-Triacyl-sn-glycerol (triglyceride)

**FIGURE 11.14**

Structure and conventional numbering of a triacyl glycerol. *Arrows* indicate bonds whose cleavage is catalyzed by a 1,3-regiospecific lipase.

Redrawn based on artwork from the first edition (1995), published by W.H. Freeman.

Transesterification by alcoholysis

$$\text{R-COOR}'_a + \text{R}'_b\text{-OH} \rightarrow \text{R-COOR}'_b + \text{R}'_a\text{-OH}$$

Ester exchange (interesterification)

$$\text{R}_1\text{-COOR}'_a + \text{R}_2\text{-COOR}'_b \rightarrow \text{R}_1\text{-COOR}'_b + \text{R}_2\text{-COOR}'_a$$

**3.** Aminolysis

$$\text{R-COOR}'_a + \text{R}_b\text{-NH}_2 \rightarrow \text{R-CONH-R}_b + \text{R}'_a\text{-OH}$$

The food industry exploits lipase-catalyzed reactions to manufacture fats of defined composition and to improve the flavor of food. Lipases are also used in many organic syntheses that require the resolution of racemic mixtures.

## TRANSESTERIFICATION OF FATS AND OILS: PRODUCTION OF COCOA BUTTER

The reaction enthalpy of triglyceride hydrolysis is exceptionally small, and the net free energy change in transesterification reactions is zero. Consequently, both hydrolysis and resynthesis of triacylglycerols occur when lipases are incubated with fats and oils. The resulting interchange of fatty acyl groups between triacylglycerol molecules gives rise to transesterified products.

The regiospecificity of lipases makes it possible to produce triacylglycerol mixtures that cannot be obtained by conventional chemical methods (Figure 11.15). Numerous microorganisms, including the bacteria *Pseudomonas fluorescens* and *Chromobacterium vinosum* and the fungi *A. niger* and *Humicola lanuginosa* excrete 1,3-regiospecific lipases into their growth medium to catalyze the degradation of lipids. These enzymes can thus be produced on a large scale by fermentation and have come to be widely used.

Cocoa butter, the edible natural fat of the cacao bean extracted during the preparation of chocolate and cocoa powder, is widely used in the confectionary industry. The unique triglyceride composition of this vegetable fat (Table 11.7) results in a very narrow melting temperature range. An industrial process for the production of high–commercial value cocoa butter substitutes uses a fungal lipase to catalyze the transesterification of readily available inexpensive oils – for example, 1,3-dipalmitoyl-2-oleyl glycerol from palm oil – with stearic acid to produce 1,3-distearoyl-2-oleyl glycerol, the major component of cocoa butter (Table 11.7). More than 10,000 tons of cocoa butter are manufactured annually. The industrial transesterification process has two features of general interest to the field of biocatalysis. First, the reaction takes place in a two-phase system: the reactants and products are present in a water-immiscible liquid organic phase and the hydrated enzyme protein is present in the small volume of an aqueous phase. Second, in this process, a hydrolytic enzyme is used to catalyze the reverse of its natural reaction.

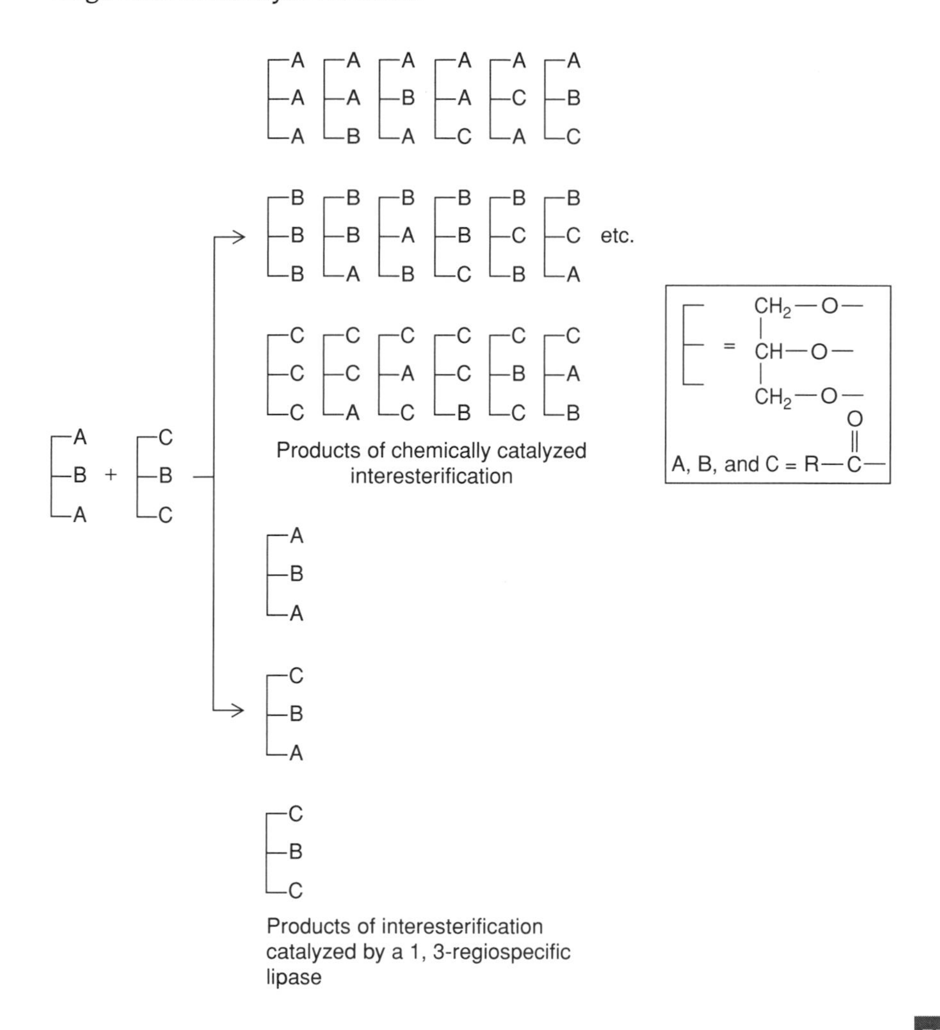


**FIGURE 11.15**

Comparison of the products of chemical and enzymatic catalysis of interesterification derived from a mixture of two triacylglycerols. Because the 1- and 3-positions of triacylglycerols are not equivalent, the compound ABC is different from CBA. This complication, however, was ignored in this figure to simplify the scheme.

Redrawn based on artwork from the first edition (1995), published by W.H. Freeman.

## LIPASE-CATALYZED SYNTHESIS OF POLYESTERS

The formation of linear polyesters with a regular structure from equimolar mixtures of unactivated diacids and diols represents a remarkable example of a lipase-catalyzed synthesis. This example is particularly noteworthy because the reaction is carried out in the absence of water.

The term *polyester* comprises all polymers with ester functional groups in the polymer backbone. Polyesters occupy a special place in the history of polymer science. Pioneering studies of polyesters in the 1930s by Wallace Hume Carothers and his team of organic chemists at DuPont led to fundamental advances in polymer chemistry that culminated in the invention of nylon and the appearance of a wide array of consumer products made from this extraordinary polyamide. A common route to polyesters is the stepwise condensation reaction between unactivated difunctional monomers, diols and diacids. Carothers's research investigated the properties of the products of reactions between aliphatic diols and diacids of various lengths with the aim of producing a synthetic replacement for silk. These polyesterification reactions require temperatures in excess of 200°C.

**TABLE 11.7 Composition of a cocoa butter substitute obtained from the transesterification of palm oil fraction (1 part) and stearic acid (0.5 parts), catalyzed by *Mucor miehei* 1,3-regiospecific lipase**

| Triacylglycerol | Amount in cocoa butter (%) | Amount in transesterification product (%) |
|---|---|---|
| SSS | 1.0 | 3.0 |
| POP | 16.3 | 16.2 |
| POS | 40.8 | 38.5 |
| SOS | 27.4 | 28.5 |
| SLnS | 7.5 | 8.0 |
| SOO | 6.0 | 4.0 |
| Others | 1.0 | 1.0 |

S, stearoyl; P, palmitoyl; O, oleyl; Ln, linoleyl; POP, etc., triacylglycerol with the specified acyl groups in the 1, 2, and 3 positions.

Two unfavorable properties of the polyester products compromise their value in many applications. Their molecular weights do not exceed 5000, and their lack of structural regularity leads to materials of low degrees of crystallinity and weak mechanical properties. These disadvantageous properties are avoided by employing lipases to catalyze the polymerization reaction. A careful study examined lipase-catalyzed polyester synthesis from equimolar amounts of adipic acid [$HOOC–(CH_2)_4–COOH$] and butane-1,4-diol [$HO–(CH_2)_4–OH$]. The butane-1,4-diol functioned both as a building block and the water-free solvent. Polymerization, performed at 60°C, was initiated by addition of *Candida antarctica* lipase immobilized on a macroporous acrylic resin (Novozym 435; 13% enzyme by weight) as the catalyst. Water formed in the reaction was removed by evaporation under a moderate vacuum. The reaction mixture is heterogeneous. The adipic acid is very poorly soluble in the butane-1,4-diol and, of course, the immobilized enzyme is insoluble. Substantial evidence supported the following reaction mechanism under these unusual conditions.

Reaction 1. Adipic acid (A) acylates the active site serine on the lipase to form the acyl-enzyme:

$$A + \text{lipase} \rightarrow \text{lipase–A} + H_2O.$$

Reaction 2. Attack on the acyl enzyme by butane-1,4-diol (B) forms a key synthon, 6-carboxy-11-hydroxy-7-oxaundecanoic acid (AB):

$$\text{lipase-A} + B \rightarrow \text{lipase} + AB.$$

Reaction 3. Enzyme-catalyzed esterification of AB with butane-1,4-diol forms BAB.

Reaction 4. Growth of the polymer takes place by stepwise addition of AB to B(AB) to give $B(AB)_2$, $B(AB)_3$, and so on.

The strongest evidence for the proposed mechanism is that the AB is the only acid-ended moiety among the oligomers produced during the polymerization, and that the different oligomers differ from each other by AB increments. The reaction goes to completion and gives rise to a high molecular weight polymer (up to 15,000) with a regular structure that crystallizes over a narrow temperature range. Because of the special physical properties that result from their highly regular structures, these polyesters are well suited to coating and adhesive applications.

## PRODUCTION OF ACRYLAMIDE

The industrial manufacture of acrylamide represented the first successful example of a biotransformation process for the manufacture of a commodity

**TABLE 11.8 Comparison of the catalytic and enzymatic processes for the conversion of acrylonitrile to acrylamide**

| | Catalytic process | Enzymatic process |
|---|---|---|
| Reaction temperature | 70°C | 0–5°C |
| One-pass reaction yield | 70–80% | 99.99% |
| Product concentration | ~30% | 48–50%[a] |
| Residual acrylonitrile | >30%[b] | Trace |
| By-products | Multiple[c] | None |
| Purification | Removal of $Cu^{2+}$; decoloring | Decoloring |
| Energy consumption in megajoules per kilogram acrylamide | | |
| Steam | 1.6 | 0.3 |
| Electric power | 0.3 | 0.1 |

[a] The product concentration is so high that concentration is not required.
[b] The unreacted acrylonitrile has to be recovered.
[c] Acrylic acid is a substantial by-product. The rate of acrylic acid formation is higher than that of acrylamide formation. Other by-products arising from addition reactions to the double bond in both the substrate and product include nitrilotrispropionamide and ethylene cyanohydrin. Polymerization also occurs at the double bond of both substrate and product.
*Sources*: Yamada, H., Shimizu, S., and Kobayashi, M. (2001). Hydratases involved in nitrile conversion: screening, characterization and application. *The Chemical Record*, 1, 152–161; Organization for Economic Co-operation and Development. (2001). *The Application of Biotechnology to Industrial Sustainability*, pp. 71–75, Paris: OECD.

chemical. More than 200,000 tons of acrylamide per year are used as a flocculant, a soil conditioner, and a component in synthetic fibers, as well as in petroleum recovery. In amount, acrylamide is second only to high-fructose syrups as a commercial product of a biocatalytic process.

In the chemical synthesis of acrylamide, acrylonitrile is hydrated in a reaction catalyzed by copper salts:

$$CH_2{=}CHCN + H_2O \xrightarrow[\text{or nitrile hydratase}]{Cu^{2+}\text{ catalyst}} CH_2{=}CHCONH_2.$$

The production of acrylamide from acrylonitrile by biotransformation employs *Rhodococcus rhodochrous* J1 nitrile hydratase. This nitrile hydratase contains cobalt as a cofactor. The acrylamide production process operates at pH 7.5 to 8.5, at 0°C to 5°C. *R. rhodochrous* J1 cells containing very high levels of nitrile hydratase were immobilized on a cationic acrylamide-based gel within a bioreactor. Acrylonitrile passed through the reactor was converted to acrylamide in a 99.99% yield with virtually no by-products. The immobilized cells could be reused repeatedly.

The chemical and biotransformation processes are compared in Table 11.8. In addition to the many advantages of the enzymatic process over the chemical one that are evident from the information in this table, the enzymatic process uses less energy and generates little toxic waste.

## SUMMARY

Prokaryotes and fungi colonize virtually every ecological niche. Rich assemblages of microorganisms are found in environments (biotopes) characterized by variation in extremes of temperature, pH, salinity, pressure, chemical composition, light intensity and quality, and so on. The properties of enzymes made by an organism adapted to a particular biotope are compatible with the need to function in the physical and chemical conditions within that biotope. Collectively, microbial enzymes catalyze an enormous variety of chemical reactions under widely varying conditions that transform countless naturally occurring and human-made organic compounds. This immense reservoir of enzymes can be explored for biocatalysts with desired specificities either by screening microorganisms available in pure culture or by culture-independent methods – by high-throughput screening of clones derived from environmental DNA samples.

Enzymes are extraordinarily versatile and effective at catalyzing regioselective and stereospecific biotransformations difficult or impossible to achieve by purely chemical means. Undesired isomerization, racemization, epimerization, and rearrangement reactions that are frequently encountered during chemical processes, are generally avoided. Finally, enzymes can accelerate the rates of chemical reactions by factors of $10^8$ to $10^{12}$.

Asymmetric catalysis is particularly important in generating chiral synthons for products of the pharmaceutical and agrochemical industries, as illustrated by the case studies of enzymatic synthesis of two chiral intermediates in the synthesis of a $\beta_3$-adrenergic receptor agonist, and of (*S*)-2-chloropropionic acid, an intermediate in the synthesis of widely used aryloxyphenoxypropionic acid–based herbicides. In all three cases, whole microbial cells perform the biotransformations.

Genes encoding naturally occurring enzymes serve as the starting material for directed *in vitro* evolution leading to enzyme variants optimized for a particular reaction or process. Very frequently, there is a need to optimize the physical and/or catalytic properties of an enzyme for which no tertiary structure information or knowledge of the reaction mechanism is available. A number of approaches address this need, and three of these – DNA-shuffling, saturation mutagenesis, and error-prone mutagenesis – are discussed and illustrated with detailed examples. DNA shuffling, in particular, has proven to be a very successful approach to evolving enzymes with desired alterations in substrate specificity, pH-activity profile, specific activity, enantioselectivity, thermostability, tolerance to organic solvents, solubility, and crystallizability.

Frequently, the specificity of an enzyme, its interaction with cofactors, inhibitors, or other proteins, and its catalytic mechanism are all understood with precision on the basis of sequence, the known three-dimensional structure of the enzyme, and its relevant complexes with various ligands, as well as through kinetic and thermodynamic studies. Here, site-directed mutagenesis is frequently the method of choice, and a case study of this approach is examined in detail.

Organic chemists have employed biocatalysis for more than 60 years. The use of biocatalysts has been in a period of continuing explosive growth in recent years, propelled by two major forces. The first is the influence of green chemistry, with its central emphases on atom economy, toxic waste minimization, and energy conservation. The second is the impact of the twin abilities to create immense libraries of genes encoding enzymes of interest from environmental DNA, to screen these very rapidly, and to subject sequences of interest to optimization by *in vitro* manipulation.

A few detailed case studies sample the enormous range of current applications of biocatalysis in industry. In the food industry, high-fructose syrups are the highest-volume product dependent on biocatalysis. The lipase-mediated transesterification of fats and oils to produce cocoa butter substitute offers another example. Lipase-catalyzed synthesis of polyesters provides a fascinating glimpse of the value of biocatalysis in polymer chemistry. The enzyme-catalyzed production of high-purity acrylamide forms the basis of a spectacularly successful manufacturing process for a commodity chemical used on a very large scale.

## REFERENCES


### General

Faber, K. (2004). *Biotransformation in Organic Chemistry*, Berlin: Springer-Verlag.

Schmidt, E., and Blaser, H.-U. (2003). *Asymmetric Synthesis on Industrial Scale: Challenges, Approaches, and Solutions*, New York: Wiley-VCH.

Mattlack, A. S. (2001). Biocatalysis and biodiversity. In *Introduction to Green Chemistry*, pp. 241–289. New York and Basel: Marcel Dekker.

Organization for Economic Co-operation and Development. (2001). *The Application of Biotechnology to Industrial Sustainability*, Paris: OECD.

Liese, A., Seelbach, K., and Wandrey, C. (2000). *Industrial Biotransformations – A Comprehensive Handbook*, Weinheim: Wiley-VCH.

Straathof, A. J. J., Adlercreutz, P. (eds.) (2000). *Applied Biocatalysis*, 2nd Edition, Amsterdam: Harwood Scientific Publishers.

### Asymmetric Catalysis in the Pharmaceutical and Agrochemical Industries

Straathof, A. J. J., Panke, S., and Schmid, A. (2002). The production of fine chemicals by biotransformations. *Current Opinion in Biotechnology*, 13, 548–556.

Patel, R. N. (2001). Biocatalytic synthesis of intermediates for the synthesis of chiral drug substances. *Current Opinion in Biotechnology*, 12, 587–604.

Taylor, S. C. (1988). D-2 haloalkanoic halidohydrolase. U.S. Patent 4,758,518.

### Microbial Diversity: A Vast Reservoir of Distinctive Enzymes

Atomi, H. (2005). Recent progress towards the application of hyperthermophiles and their enzymes. *Current Opinion in Chemical Biology*, 9, 166–173.

Robertson, D. E., et al. (2004). Exploring nitrilase sequence space for enantioselective catalysis. *Applied and Environmental Microbiology*, 70, 2429–2436.

Van den Burg, B. (2003). Extremophiles as a source of novel enzymes. *Current Opinion in Microbiology*, 6, 213–218.

Demirjian, D. C., Moris-Varas, F., and Cassidy, C. S. (2001). Enzymes from extremophiles. *Current Opinion in Chemical Biology*, 15, 144–151.

Greenberg, W. A., et al. (2004). Development of an efficient, scalable, aldolase-catalyzed process for enantioselective synthesis of statin intermediates. *PNAS*, 101, 5788–5793.

### *In Vitro* Evolution of Enzymes

Bloom, J. D., Meyer, M. M., Meinhold, P., Otey, C. R., MacMillan, D., and Arnold, F. H. (2005). Evolving strategies for enzyme engineering. *Current Opinion in Structural Biology*, 15, 447–452.

Yuan, L., Kurekl, I., English, J., and Keenan, R. (2005). Laboratory-directed protein evolution. *Microbiology and Molecular Biology Reviews*, 69, 373–392.

Otten, L. G., and Quax, W. J. (2005). Directed evolution: selecting today's biocatalysts. *Biomolecular Engineering*, 22, 1–9.

Eijsink, V. G. H., Gåseidnes, S., Borchert, T. V., and van den Burg, B. (2005). Directed evolution of enzyme stability. *Biomolecular Engineering*, 22, 21–30.

Antikainen, N. M., and Martin, S. F. (2005). Altering protein specificity: techniques and applications. *Bioorganic and Medicinal Chemistry*, 13, 2701–2716.

Reetz, M. T. (2004). Controlling the enantioselectivity of enzymes by directed evolution: practical and theoretical ramifications. *PNAS*, 101, 5716–5722.

Stemmer, W., and Holland, B. (2003). Survival of the fittest molecule. *American Scientist*, 91, 526–533.

Zhang, Y-X., Vinci, V. A., Powell, K., Stemmer, W. P. C., and del Cardayré, S. B. (2002). Genome shuffling leads to rapid phenotypic improvement in bacteria. *Nature*, 415, 644–646.

Lutz, S., and Benkovic, S. J. (2000). Homology-independent protein engineering. *Current Opinion in Biotechnology*, 11, 319–324.

Castle, L. A., et al. (2004). Discovery and directed evolution of a glyphosate tolerance gene. *Science*, 304, 1151–1154.

Ness, J. E. (1999). DNA shuffling of subgenomic sequences of subtilisin. *Nature Biotechnology*, 17, 893–896.

DeSantis, G., et al. (2002). An enzyme library approach to biocatalysis: development of nitrilases for enantioselective production of carboxylic acid derivatives. *Journal of the American Chemical Society*, 124, 9024–9025.

De Santis, G., et al. (2003). Creation of a productive, highly enantioselective nitrilase through gene site saturation mutagenesis (GSSM). *Journal of the American Chemical Society*, 125, 11476–11477.

### Rational Methods of Protein Engineering

Heine, A., Luz, J. G., Wong, C-H., and Wilson, I. A. (2004). Analysis of the class I aldolase binding site architecture based on crystal structure of 2-deoxyribose-5-phosphate aldolase at 0.99Å resolution. *Journal of Molecular Biology*, 343, 1019–1034.

DeSantis, G., Liu, J., Clark, D. P., Heine, A., Wilson, I. A., and Wong, C-H. (2003). Structure-based mutagenesis approaches towards expanding the substrate specificity of 2-deoxyribose-5-phosphate aldolase. *Bioorganic and Medicinal Chemistry*, 11, 43–52.

### Large-Scale Biocatalytic Processes

Kirk, O., Borchert, T. V., and Fuglsang, C. C. (2002). Industrial enzyme applications. *Current Opinion in Biotechnology*, 13, 345–351.

Ogawa, J., and Shimizu, S. (2002). Industrial microbial enzymes: their discovery by screening and use in large-scale production of useful chemicals in Japan. *Current Opinion in Biotechnology*, 13, 367–375.

Ghanem, A., and Aboul-Enein, H. Y. (2005). Application of lipases in kinetic resolution of racemates. *Chirality*, 17, 1–15.

Binns, F., Harffey, P., Roberts, S. M., and Taylor, A. (1998). Studies of lipase-catalyzed polyesterification on an unactivated diacid/diol system. *Journal of Polymer Science: Part A: Polymer Chemistry*, 36, 2069–2080.

Yamada, H., Shimizu, S., and Kobayashi, M. (2001). Hydratases involved in nitrile conversion: screening, characterization and application. *The Chemical Record*, 1, 152–161.

TWELVE

# Biomass

> We all have dreamed of producing an abundance of useful food, fuel, and chemical products from the cellulose in urban trash and the residues remaining from forestry, agricultural, and food-processing operations. Such processes potentially could: 1) help solve modern waste-disposal problems; 2) diminish pollution of the environment; 3) help alleviate shortages of food and animal feeds; 4) diminish man's dependence on fossil fuels by providing a convenient and renewable source of energy in the form of ethanol; 5) help improve the management of forests and range lands by providing a market for low-quality hardwoods and the other "green junk" that develops on poorly managed lands; 6) aid in the development of life-support systems for deep space and submarine vehicles; and 7) increase the standard of living – especially of those who develop the technology to do the job! At present, all of these aspirations are frustrated by two major features of natural cellulosic materials, crystallinity and lignification.
>
> – Cowling, E. B., and Kirk, T. K. (1976). Properties of cellulose and lignocellulosic materials as substrates for enzymatic conversion processes. *Biotechnology and Bioengineering Symposium*, 6, 95–123.

Biomass can have broader definitions, but in the context of biotechnology, it is generally taken to mean "all organic matter that grows by the photosynthetic conversion of solar energy." The sun, either directly or indirectly, is the principal source of energy on earth, its power converted to a usable organic form – biomass – by green plants, algae, and photosynthetic bacteria. The biomass produced *annually*, by photosynthesis on land and in the oceans, contains an estimated 4500 exajoules ($4.5 \times 10^{21}$ joules; Box 12.1) of energy, some 10 times the yearly worldwide human energy consumption. Of this primary biomass resource, about seven exajoules per year are used in modern energy conversion processes for the production of electricity, steam, and biofuels.

Oil, coal, and natural gas represent the concentrated energy resource capital of the earth, the accumulated residue of billions of years of photosynthesis. Projections of world rates of petroleum consumption and estimates of recoverable crude oil reserves indicate that the petroleum supply will be exhausted some time in the twenty-first century. Known coal and oil shale reserves, meanwhile, are adequate for centuries into the future, and processes for the conversion of coal to liquid hydrocarbon fuels are well established. However, exploitation of coal and oil shale deposits has a serious negative environmental impact, and massive use of fossil fuels adds to the carbon dioxide in the atmosphere, increasing global warming through the so-called greenhouse effect. The use of biomass is promoted as a partial alternative, a large-scale renewable source of liquid fuels and chemical industry feedstocks. As long as biomass regeneration matched biomass use, there would be no net increase in the atmospheric content of carbon dioxide from this source.

**Energy Units, Abbreviations, and Conversion Factors**

| Unit | Prefixes | Unit | Abbreviations |
|---|---|---|---|
| kilo (k; $10^3$) | tera (T; $10^{12}$) | Terajoule | TJ |
| mega (M; $10^6$) | peta (P; $10^{15}$) | Gigacalorie | Gcal |
| giga (G; $10^9$) | exa (E; $10^{18}$) | Million tons oil equivalent | Mtoe |
| | | Million British thermal units | Mbtu |
| | | Gigawatt-hour | GWh |

**Energy Conversions**

| | TJ | Gcal | Mtoe | Mbtu | GWh |
|---|---|---|---|---|---|
| TJ | 1 | 238.8 | $2.388 \times 10^{-5}$ | 947.8 | 0.2778 |
| Mtoe | $4.1868 \times 10^4$ | $10^7$ | 1 | $3.968 \times 10^7$ | 11,630 |
| Mbtu | $1.0551 \times 10^{-3}$ | 0.252 | $2.52 \times 10^{-8}$ | 1 | $2.931 \times 10^{-4}$ |
| GWh | 3.6 | 860 | $8.6 \times 10^{-5}$ | 3412 | 1 |

*Source:* International Energy Agency, Energy Statistics Manual, Annex 3, Units and Conversion Equivalents, pp. 177–183, http://www.iea.org/Textbase/publications/free_new_Desc.asp?PUBS_ID=1461.

BOX 12.1

How large is the store of world biospheric organic carbon compounds, and where is it found? *Forests*, which cover some 10% of the total land area and account for about 21% of the net carbon fixed in the biosphere each year (Table 12.1), contain about 90% of the biomass carbon of the earth. Tropical forests make the largest contribution to this total. *Cultivated land* occupies a similar portion of total land area (about 9%) but accounts for about 13% of mean annual net carbon fixation. Marine sources, savannah, and grasslands are also large contributors to standing carbon reserves and carbon fixation in the biosphere.

Current uses of biomass – food, fuel, fibers, building materials, and many other products – account for only a small fraction of the earth's annual production. Forests and tree plantations are particularly rich sources of excess biomass, whereas in some parts of the world, agriculture can produce more food than is consumed locally or exported. This surplus capacity allows the cultivation of "energy crops" whose constituents can be converted to alcohol fuels or industrial chemicals (Table 12.2). Energy crops include plants with high sugar (e.g., sugarcane, sugar beet, and sweet sorghum) or starch content (e.g., cassava), those with high cellulose content (e.g., kenaf, elephant grass), and those with a high hydrocarbon content (e.g., jojoba, milkweed, and vegetable oilseeds such as sunflower and rapeseed). Anaerobic bacterial digestion of aquatic plants, such as marine and freshwater algae or the fast-growing water hyacinth, produces high yields of methane, making these plants energy crops as well.

Current uses of biomass also generate large amounts of organic wastes: agricultural wastes, such as wheat straw, corn cobs, oat hulls, and sugarcane bagasse (a fibrous residue left after extraction of juice); residues from logging and timber milling, such as wood chips and sawdust; spoiled produce and food-processing wastes; and urban solid waste such as paper, cardboard,

**TABLE 12.1 Estimates of the distribution of world mean net primary productivity (NPP)**

| System | Area (million $km^2$) | Mean NPP (kg C/$m^2$/year) | Total mean NPP (kg C/year $\times 10^{-6}$) |
|---|---|---|---|
| Marine | 349.3 | 0.15 | 52.4 (37.5%) |
| Inland water | 10.3 | 0.36 | 3.7 (2.6%) |
| Forest/woodlands | 42.2 | 0.68 | 28.7 (20.5%) |
| Dry land | 60.9 | 0.26 | 15.8 (11.3%) |
| Island | 9.9 | 0.54 | 5.3 (3.7%) |
| Mountain | 33.2 | 0.42 | 13.9 (9.9%) |
| Polar | 23.0 | 0.06 | 1.4 (1%) |
| Cultivated | 35.6 | 0.52 | 18.5 (13.2%) |
| Global total | | | 139.7 (100%) |

C, carbon.
*Source*: Conditions and Trends Working Group Report, C. SDM Summary, Millennium Ecosystem Synthesis Report, 2005. Chapter 1, p. 31. http:www.millenniumassessment.org.

and kitchen and garden refuse. The bioconversion of these waste products to fuels or protein does not decrease food production.

The conversion of biomass to fuel alcohol is the central topic of Chapter 13. We set the stage for it here with an examination of lignocellulose, the only major, nearly universal, component of biomass, making up about half of all matter produced through photosynthesis. It consists of three types of polymers: cellulose, hemicellulose, and lignin. Each is intimately associated with the others by physical and chemical linkages, and all are degraded in the natural environment by bacteria and fungi.

## MAJOR COMPONENTS OF PLANT BIOMASS

In the cell walls of the vascular tissues of higher land plants, cellulose fibrils are embedded in an amorphous matrix of lignin and hemicelluloses. These three kinds of polymers bind strongly to each other by noncovalent forces as well as by covalent cross-links, making a composite material that is known as *lignocellulose*. It represents over 90% of the dry weight of a plant cell. The quantity of each of the polymers varies with the species and age of a plant and from one part of the plant to another. Usually, softwoods have a higher content of lignin than do hardwoods (Box 12.2; Table 12.3). Hemicellulose content is highest in the grasses. In trees, on average, lignocellulose consists of 45% cellulose, 30% hemicelluloses, and 25% lignin (Table 12.4). The earth's estimated annual production of lignocellulose ranges from 2 to $5 \times 10^{12}$ metric tons.

### CELLULOSE

Cellulose is the most abundant organic compound on Earth. Every year, plants make more than $10^{11}$ metric tons of cellulose. *In situ*, a cellulose polymer is a linear chain of thousands of glucose molecules linked by $\beta$-(1:4)-glycosidic bonds. The basic repeating unit is *cellobiose* (Figure 12.1). Consecutive glucose units in cellulose are rotated through 180° with respect to their neighbors along the axis of the chain, and the terminal cellobiose can thus appear in one of two stereochemically different forms. The cellulose polymer chain has a flat, ribbonlike structure stabilized by internal hydrogen bonds. Other hydrogen bonds between adjacent chains cause them to interact strongly with one another in parallel arrays of many chains that all

**TABLE 12.2 Current and feasible biomass productivity, energy ratios, and energy yields for various crops**

| Crop | Yield[a] (Dry tons/hectare/year) | Energy ratio[b] | Net energy yield (Gigajoules/hectare/year) |
|---|---|---|---|
| Short rotation | | | |
| Woody crops[c] | 10–12 | 10:1 | 180–200 |
| Tropical plantations[d] | 2–10 | 10:1 | 30–180 |
| Wood (commercial forestry)[e] | 1–4 | 20–30:1 | 30–80 |
| Switchgrass | 10–12 | 12:1 | 180–200 |
| Rapeseed | 4–7 | 4:1 | 50–90 |
| Sugarcane[f] | 15–20 | 18:1 | 400–500 |
| Sugar beet | 10–16 | 10:1 | 30–100 |

[a] The biomass productivity might be increased substantially by a combination of improvements in irrigation, fertilizer application, and plant genetic modification.
[b] The net energy yields represent estimates of output minus energy inputs for agricultural operations.
[c] Examples are willow and hybrid poplar.
[d] For example, eucalyptus.
[e] The energy used to transport biomass over land averages around 0.5 megajoules per ton-kilometer.
[f] Inclusion of the energy expended on transportation and processing of sugarcane to produce ethanol leads to an estimated energy ratio of 7.9:1.
GJ, gigajoule (equivalent to one thousand megajoules or one billion joules).
*Source*: United Nations Development Programme (2001). *World Energy Assessment: Energy and the Challenge of Sustainability*, New York: UNDP.

have the same polarity. The resulting very long, largely crystalline aggregates are called *microfibrils*. The microfibrils (250 Å wide) are combined to form larger fibrils. These are then organized in thin layers (lamellae) to form the framework of the various layers of the plant cell wall. Cellulose fibrils have regions of high order, *crystalline regions*, and regions of less order, *amorphous regions*. The fraction of crystalline cellulose varies with the source or with the way the material is prepared. Cellulose is water insoluble, has a high tensile strength, and is much more resistant to degradation than are other glucose polymers, such as starch.

## HEMICELLULOSES

The components of hemicelluloses are complex polysaccharides that are structurally homologous to cellulose because they have a backbone made up of 1,4-linked $\beta$-D-pyranosyl units. Whereas cellulose is a linear homopolymer with little variation in structure from one species to another, hemicelluloses are highly branched, generally noncrystalline heteropolysaccharides. The sugar residues found in the hemicelluloses include pentoses (D-xylose, L-arabinose), hexoses (D-galactose, L-galactose, D-mannose, L-rhamnose, L-fucose), and uronic acids (D-glucuronic acid). These residues are variously modified by acetylation or methylation. Hemicelluloses show a much lower degree of polymerization (<200 sugar residues) than cellulose. Simplified structures of the three most common types of hemicelluloses are shown in Figure 12.2.

**Softwoods and Hardwoods**

For commercial purposes, woods are divided into hardwoods and softwoods. Hardwoods are the woods of dicotyledons irrespective of whether they are hard or soft. Dicotyledons are a class of flowering plants generally characterized as having two cotyledons (seed leaves). Softwoods are the woods of conifers.

**BOX 12.2**

**TABLE 12.3 Percentage composition of lignocellulose in vascular plants**

| Source | Lignin | Cellulose | Hemicellulose |
|---|---|---|---|
| Grasses | 10–30 | 25–40 | 25–50 |
| Softwoods | 25–35 | 45–50 | 25–35 |
| Hardwoods | 18–25 | 45–55 | 24–40 |

The name *hemicellulose* was introduced by E. Schultze in 1891 to describe the fraction of plant cell wall polysaccharides extractable with dilute alkali. Unfortunately, solubility behavior is a wholly inadequate criterion for assessing the hemicellulose content of various woods. Dilute alkali solubilizes both polysaccharides with a $\beta$-(1,4)-linked xylan backbone and galactoglucomannans, but glucomannans, which are a quantitatively important component of softwood hemicelluloses (Table 12.4), are insoluble under this condition and are thus left behind, still tightly associated with the cellulose fibers (Figure 12.3). *The hemicelluloses are better defined as those polysaccharides noncovalently associated with cellulose.*

## Softwood Hemicelluloses

There are three major softwood hemicelluloses: glucomannan, galactoglucomannan, and arabinoglucuronoxylan. The two mannose-containing polymers differ greatly in galactose content. Their approximate sugar compositions (galactose:glucose:mannose) are 0.1:1:4 and 1:1:3, respectively. Their backbone consists of 1,4-linked $\beta$-D-glucopyranose and $\beta$-D-mannopyranose units. $\alpha$-D-Galactopyranose units are linked to the backbone by 1,6 bonds. The backbone sugars are acetylated at C-2 or C-3 with one acetyl group per three to four units (Figure 12.2).

Arabinoglucuronoxylan has a backbone of 1,4-linked $\beta$-D-xylopyranose units partially substituted at C-2 by 4-*O*-methyl-$\alpha$-D-glucuronic acid residues and at C-3 with $\alpha$-L-arabinofuranose units (Figure 12.2).

## Hardwood Hemicellulose

The major hardwood hemicellulose is glucuronoxylan, with a backbone of 1,4-linked $\beta$-D-xylopyranose units, the majority of which are acetylated at C-2 or C-3. About one 4-*O*-methyl-$\alpha$-D-glucuronic acid, attached to the backbone at C-2, is present for every 10 xylose units (Figure 12.2).

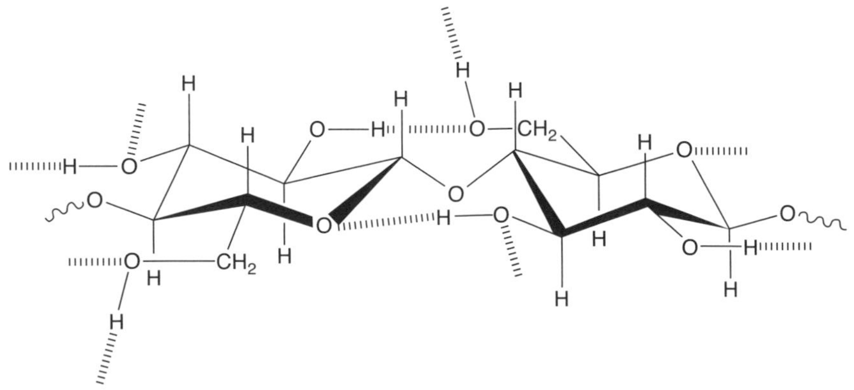

**FIGURE 12.1**

A cellobiose unit, the basic repeating disaccharide of the cellulose chain, is made up of $\beta$-1,4-linked glucose residues in the chair conformation, rotated through 180° with respect to their neighbors in the chain. The conformation shown is stabilized by intramolecular hydrogen bonds (*dashes*) within each chain. Intermolecular hydrogen bonds contribute to the interaction of adjacent chains within a microfibril.

**TABLE 12.4 Composition of the lignocellulose of various wood species[a]**

| Species | Common name | Lignin | Cellulose | Glucomannan[b] | Hemicellulose glucuronoxylan[c] |
|---|---|---|---|---|---|
| Softwoods | | | | | |
| *Pseudotsuga menziesii* | Douglas fir | 29.3 | 38.8 | 17.5 | 5.4 |
| *Tsuga canadensis* | Eastern hemlock | 30.5 | 37.7 | 18.5 | 6.5 |
| *Juniperus communis* | Common juniper | 32.1 | 33.0 | 16.4 | 10.7 |
| *Pinus radiata* | Monterey pine | 27.2 | 37.4 | 20.4 | 8.5 |
| *Picea abies* | Norway spruce | 27.4 | 41.7 | 16.3 | 8.6 |
| *Larix sibirica* | Siberian larch | 26.8 | 41.4 | 14.1 | 6.8 |
| Hardwoods | | | | | |
| *Acer saccharum* | Sugar maple | 25.2 | 40.7 | 3.7 | 23.6 |
| *Fagus sylvatica* | Common beech | 24.8 | 39.4 | 1.3 | 27.8 |
| *Betula verrucosa* | Silver birch | 22.0 | 41.0 | 2.3 | 27.5 |
| *Alnus incana* | Gray alder | 24.8 | 38.3 | 2.8 | 25.8 |
| *Eucalyptus globules* | Blue gum | 21.9 | 51.3 | 1.4 | 19.9 |
| *Ochroma lagopus* | Balsa | 21.5 | 47.7 | 3.0 | 21.7 |

[a] All values are given as percent of the wood dry weight.
[b] Includes galactose and acetyl substituents in softwoods (see Figure 12.2).
[c] Includes arabinose and acetyl substitutents in hardwoods (see Figure12.2).
*Source*: Sjöström, E. (1981). *Wood chemistry: Fundamentals and Applications*, Appendix, New York: Academic Press.

## LIGNIN

Lignin is found in the cell walls of higher plants (gymnosperms and angiosperms), ferns, and club mosses, predominantly in the vascular tissues specialized for liquid transport. It is not found in mosses, lichens, and algae that have no tracheids (long tubelike cells peculiar to xylem). The increased mechanical strength conferred on woody tissues by *lignification* allows huge trees several hundred feet tall to remain upright. Lignification is the process whereby growing lignin molecules fill up the spaces between the preformed cellulose fibrils and hemicellulose chains of the cell wall.

**FIGURE 12.2**

Heteropolysaccharide fragments derived from common types of hemicelluloses. Galactoglucomannans and glucuronoarabinoxylans are the principal hemicelluloses in softwoods. Glucuronoxylan is a major hemicellulose in hardwoods. Abbreviations: Ac, acetyl; Ara*f*, L-arabinofuranose; Glc*p*, D-glucopyranose; Glc*p*A, D-glucuronopyranose; Man*p*, D-mannopyranose; Me, *O*-methyl; Xyl*p*, D-xylopyranose.

→4)-β-D-Man*p*-(1→4)-β-D-Glc*p*-(1→4)-β-D-Man*p*-(1→4)-β-D-Man*p*-(1→4)-β-D-Glc*p*-(1→4)-β-D-Man*p*-(1→
2 | Ac; 3 / Ac

Glucomannan

→4)-β-D-Xyl*p*-(1→4)-β-D-Xyl*p*-(1→4)-β-D-Xly*p*-(1→4)-β-D-Xyl*p*-(1→4)-β-D-Xly*p*-(1→4)-β-D-Xly*p*-(1→4)-β-D-Xly*p*-(1→
2 | Ac; 3 / Ac; 3 2 Ac Ac; 3 2 Ac ↑ 1 α-D-MeGlcA

Glucuronoxylan

→ 4)-β-D-Xyl*p*-(1→4)-β-D-Xyl*p*-(1→4)-β-D-Xyl*p*-(1→4)-β-D-Xyl*p*-(1→
α-D-Glc*p*A-(1→2) α-L-Ara*f*-(1→3)
α-L-Ara*f*-(1→2)-α-L-Ara*f*-(1→3)

Glucuronoarabinoxylan

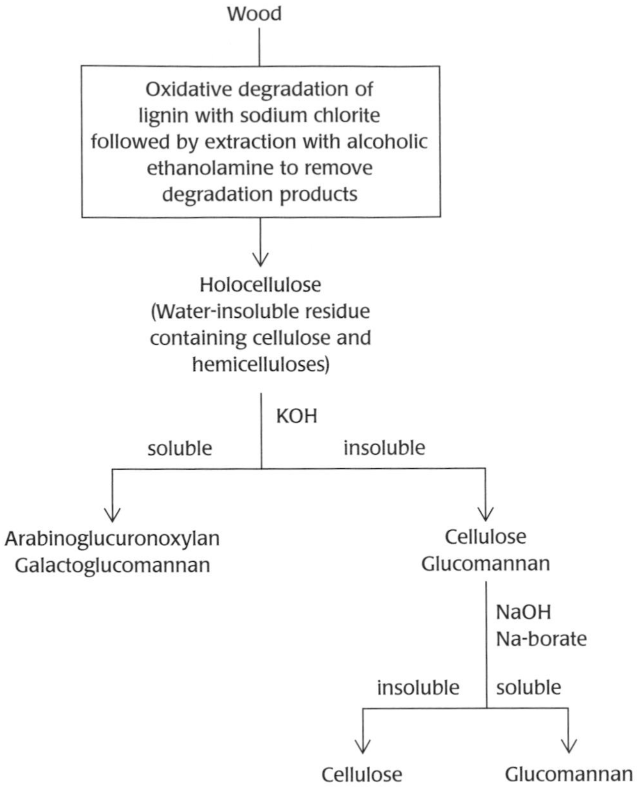

Wood

Oxidative degradation of lignin with sodium chlorite followed by extraction with alcoholic ethanolamine to remove degradation products

Holocellulose (Water-insoluble residue containing cellulose and hemicelluloses)

KOH

soluble

insoluble

Arabinoglucuronoxylan
Galactoglucomannan

Cellulose
Glucomannan

NaOH
Na-borate

insoluble

soluble

Cellulose

Glucomannan

**FIGURE 12.3**

A procedure for the separation of the polysaccharide components of softwood. The dilute alkali extraction method of Schultze (see text) leaves glucomannan hemicelluloses in the insoluble fraction.

Redrawn based on artwork from the first edition (1995), published by W.H. Freeman.

## Building Blocks of Lignin

Lignin, the most abundant aromatic polymer on Earth, is a random copolymer built up of phenylpropane ($C_9$) units. The direct precursors of lignin are three alcohols – the *lignols* – derived from *p*-hydroxycinnamic acid: coniferyl, sinapyl, and *p*-coumaryl alcohols (Figure 12.4). Lignins are described as softwood lignin, hardwood lignin, or grass lignin, on the basis of the relative amounts of these building blocks. A typical softwood (gymnosperm) lignin contains building blocks originating mainly from coniferyl alcohol, some from *p*-coumaryl alcohol, but none from sinapyl alcohol. Hardwood (angiosperm) lignin is composed of equal amounts (46% each) of coniferyl- and sinapylpropane units and a minor amount (8%) of *p*-hydroxyphenylpropane units (derived from *p*-coumaryl alcohol). Grass lignin is composed of coniferyl-, sinapyl-, and *p*-hydroxyphenylpropane

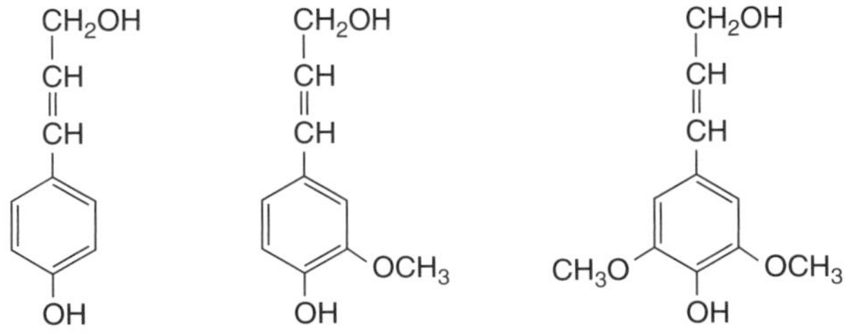

$CH_2OH$ / CH = CH / OH

*p*-Coumaryl alcohol

$CH_2OH$ / CH = CH / $OCH_3$ / OH

Coniferyl alcohol

$CH_2OH$ / CH = CH / $CH_3O$ / $OCH_3$ / OH

Sinapyl alcohol

**FIGURE 12.4**

Three *p*-hydroxycinnamyl alcohols (monolignols) are direct biosynthetic precursors of lignin.

units, with *p*-coumaric acid (5% to 10% of the lignin) mainly esterified with the terminal hydroxyl group of *p*-coumaryl alcohol side chains.

### Biosynthesis of Lignin

The biosynthesis of lignin deserves particular attention because both the formation of this extraordinary polymer and its biodegradation depend on an extensive array of free radical–mediated reactions.

The synthesis of the building blocks of lignin utilizes the *shikimic acid pathway* of aromatic amino acid biosynthesis. In fact, phenylalanine is the precursor in the biosynthesis of the lignols. The biosynthetic pathway is illustrated in Figure 12.5.

The lignin precursors are synthesized in the Golgi apparatus and/or the endoplasmic reticulum of plant cells and then transported to the cell wall by vesicles. The precursors are then polymerized to lignin by a series of dehydrogenative reactions catalyzed by a peroxidase bound to the cell walls. Complex $H_2O_2$-requiring peroxidase-catalyzed reactions convert the coniferyl, sinapyl, and *p*-coumaryl alcohols to phenoxy radicals, each of which can form a number of highly reactive resonance structures. Random coupling of such radicals leads to various quinone methide derivatives, which are converted to dilignols by the addition of water (Figure 12.6) or by intramolecular nucleophilic attack by primary alcohol or quinone groups on the benzyl carbons. Peroxidase-catalyzed dehydrogenation of the dilignols in turn leads to generation of radicals, which form the lignin polymer by radical couplings followed by nucleophilic attack on the benzyl carbons of the oligomeric quinone methides by water, and by aliphatic and phenolic hydroxyl groups of lignols. In parallel, nucleophilic attack by hydroxyl groups of sugar residues of cell wall polysaccharides forms lignin–hemicellulose cross-links (Figure 12.7). There is roughly one bond to carbohydrate for every 40 phenylpropane units. Lignin also forms covalent bonds with the glycoproteins of the cell wall. Laccases (*p*-diphenol:$O_2$ oxidoreductases), encoded by multigene families in plants, may also be important in lignification. Laccases are copper-containing glycoproteins localized in the cell walls in lignifying cells. These enzymes use $O_2$ instead of $H_2O_2$ to oxidize the monolignols.

The principal modes of linkage between monomeric phenylpropane units and their relative abundance in softwood lignin are shown in Figure 12.8, and a simplified qualitative structural model of softwood lignin in Figure 12.9.

## ARCHITECTURE AND COMPOSITION OF THE WOOD CELL WALL

The wood cell wall (Figure 12.10) is made up of several morphologically distinct concentric layers, all of which contain cellulose, hemicelluloses, and lignin in varying proportions. The *middle lamella*, 0.2 to 1.0 $\mu$m in thickness, fills the space between cells and serves to bind them together. In mature wood (*latewood*), the middle lamella is highly lignified. The *primary wall*, on the outside of the cell, is a layer only 0.1 to 0.2 $\mu$m thick made up of cellulose, hemicelluloses, pectin, and proteins completely embedded in lignin. Just

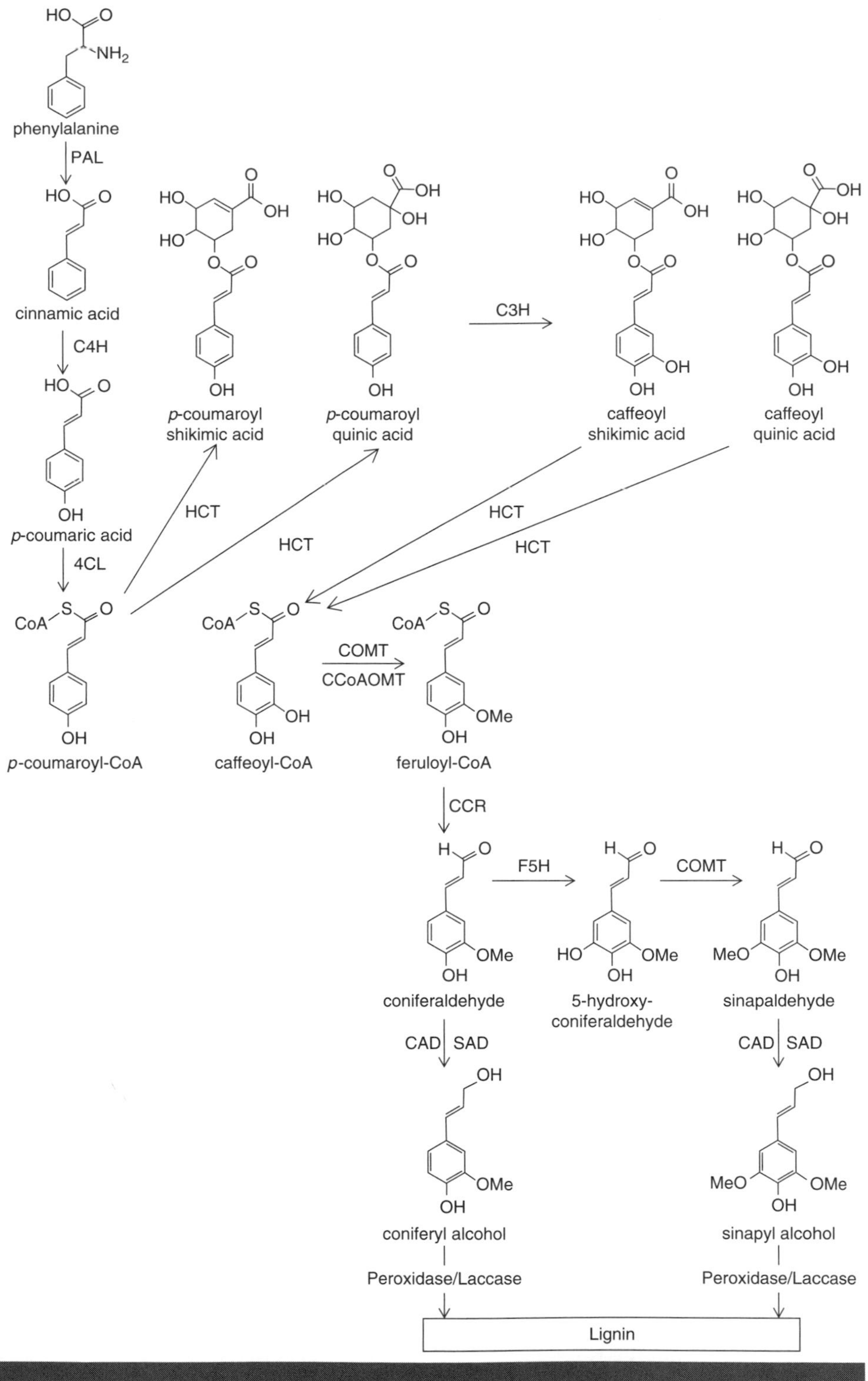

**FIGURE 12.5**

Monolignol biosynthetic pathways most favored in angiosperms. The conversion of *p*-coumaroyl-CoA to *p*-caffeoyl-CoA proceeds through the formation of *p*-coumaroyl esters with shikimic and quinic acid catalyzed by a reversible acyltransferase *p*-hydroxycinnamoyl-CoA:D-quinate *p*-hydroxycinnamoyltransferase (HCT). These esters are hydroxylated by *p*-coumarate 3-hydroxylase (C3H) to caffeoyl shikimic acid and caffeoyl quinic acid, respectively. HCT then catalyzes the conversion of the hydroxylated esters to caffeoyl-CoA. Abbreviations for the other enzymes: CAD, cinnamyl alcohol dehydrogenase; 4CL, 4-coumarate:CoA ligase; C4H, *p*-coumarate 4-hydroxylase; CCoAOMT, caffeol-CoA *O*-methyltransferase; CCR, cinnamoyl-CoA reductase; COMT, caffeic acid *O*-methyltransferase; F5H, ferulate 5-hydroxylase; PAL, phenylalanine ammonia lyase; SAD, sinapyl alcohol dehydrogenase. [Based on Figure 3 in Boerjan, W., Ralph, J., and Baucher, M. (2003). Lignin biosynthesis. *Annual Review of Plant Biology*, 54, 519–546.]

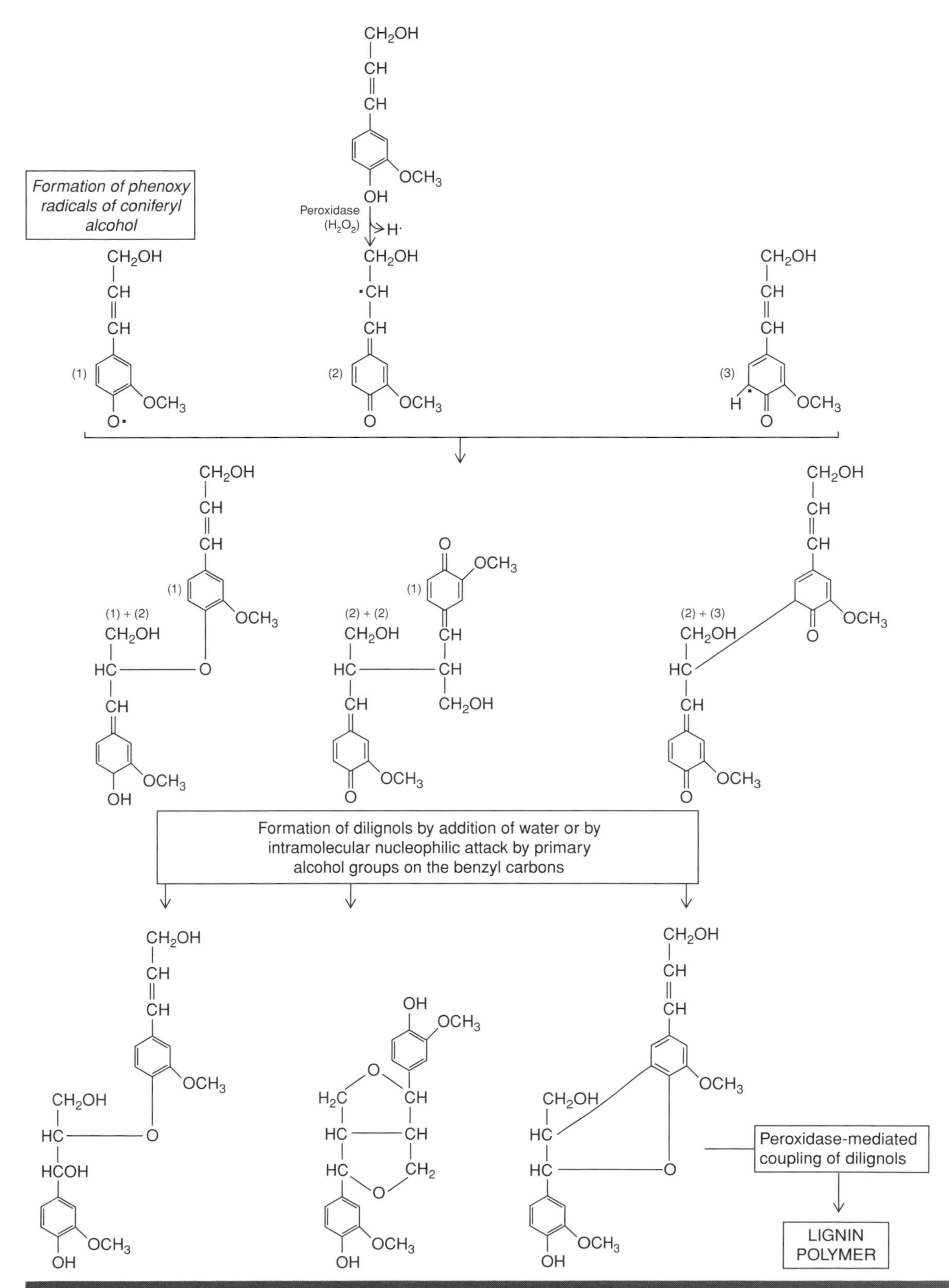


**FIGURE 12.6**

Coupling of free radicals generated from *p*-coniferyl alcohol during lignin biosynthesis to form dilignols.

Redrawn based on artwork from the first edition (1995), published by W.H. Freeman.

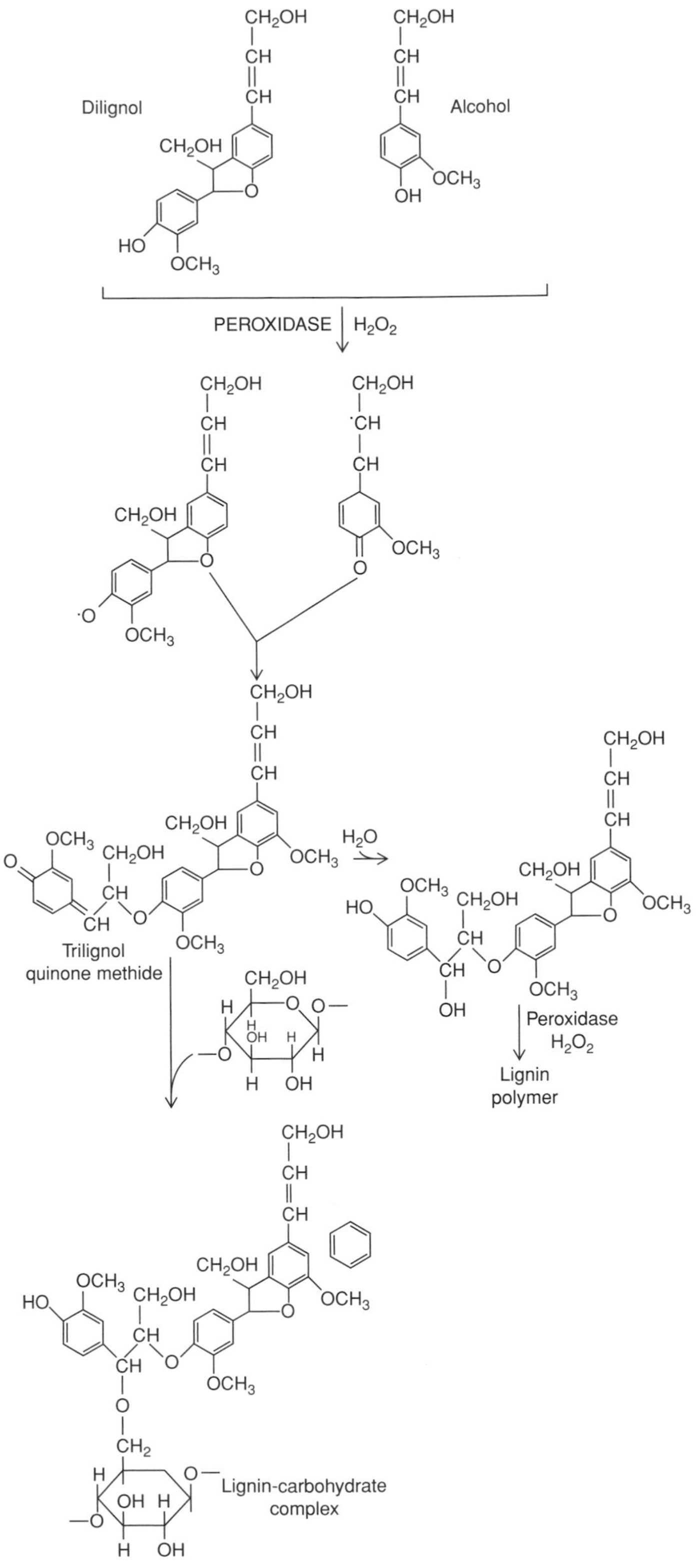

Diliignol Alcohol

PEROXIDASE $H_2O_2$

Trilignol quinone methide

$H_2O$

Peroxidase $H_2O_2$

Lignin polymer

Lignin-carbohydrate complex

**FIGURE 12.7**

Peroxidase-catalyzed reactions of dilignols leading to the formation of the lignin polymer and lignin–carbohydrate cross-links via an oligolignol quinone methide intermediate. Lignol and lignin radicals can undergo *β-O-*4 coupling to form quinone methide intermediates. Quinone methides are structurally related to benzoquinones with one of the carbonyl oxygens replaced by a methylene group. This makes them more reactive. These intermediates are stabilized by the addition of nucleophiles such as alcohols, neutral sugars, or water. The nucleophilic attack at the methylene carbon converts these compounds to phenols.

Redrawn based on artwork from the first edition (1995), published by W.H. Freeman.

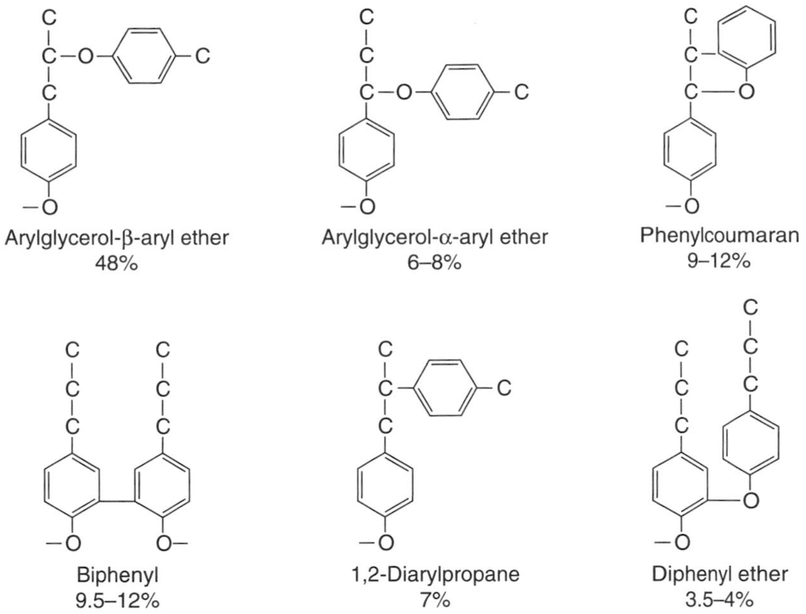

Arylglycerol-β-aryl ether 48%
Arylglycerol-α-aryl ether 6–8%
Phenylcoumaran 9–12%
Biphenyl 9.5–12%
1,2-Diarylpropane 7%
Diphenyl ether 3.5–4%

**FIGURE 12.8**

Major types of bonds between monomeric phenylpropane units in a softwood (spruce) lignin. [Betts, W. B., Dart, R. K., Ball, A. S. and Pedlar, S. L. (1991). Biosynthesis and structure of lignocellulose. In *Biodegradation: Natural and Synthetic Materials*, W. B. Betts (ed.), pp. 140–155, London: Springer-Verlag.]

inside that is the *secondary wall*, consisting of three layers designated $S_1$ (the *outer layer*, 0.2 to 0.3 $\mu$m thick), $S_2$ (the *middle layer*, 1 to 5 $\mu$m thick), and $S_3$ (the *inner layer*, about 0.1 $\mu$m thick). These secondary wall layers, made up of variously oriented cellulose microfibrils embedded in a matrix of lignin and hemicelluloses, represent phases in cellulose synthesis and localization within the original protoplast. *As long as the native structure of the plant cell wall is preserved, the lignin in it hinders degradative enzymes from reaching the cellulose microfibrils.*

## DEGRADATION OF LIGNOCELLULOSE BY FUNGI AND BACTERIA

Because plant tissues are the main repositories of organic matter in the biosphere, it is not surprising that the ability to decompose and obtain carbon and energy from cellulose, hemicellulose, and lignin is widespread among fungi and bacteria.

### DEGRADATION OF WOOD BY FUNGI

The wood-decaying fungi are the primary contributors to the degradation of wood in nature, secreting extracellular enzymes (oxidoreductases) that degrade the polymeric components of the wood cell walls. They are classified as *white rot, brown rot*, or *soft rot* fungi on the basis of morphological aspects of the decay. Of the more than 2000 species of wood-rotting fungi known worldwide, over 90% are white rot fungi. Once a fungus has managed to invade the wood, it expands by growth of its hyphae in the lumina of the parenchyma and vascular cells. The hyphae penetrate from one cell to

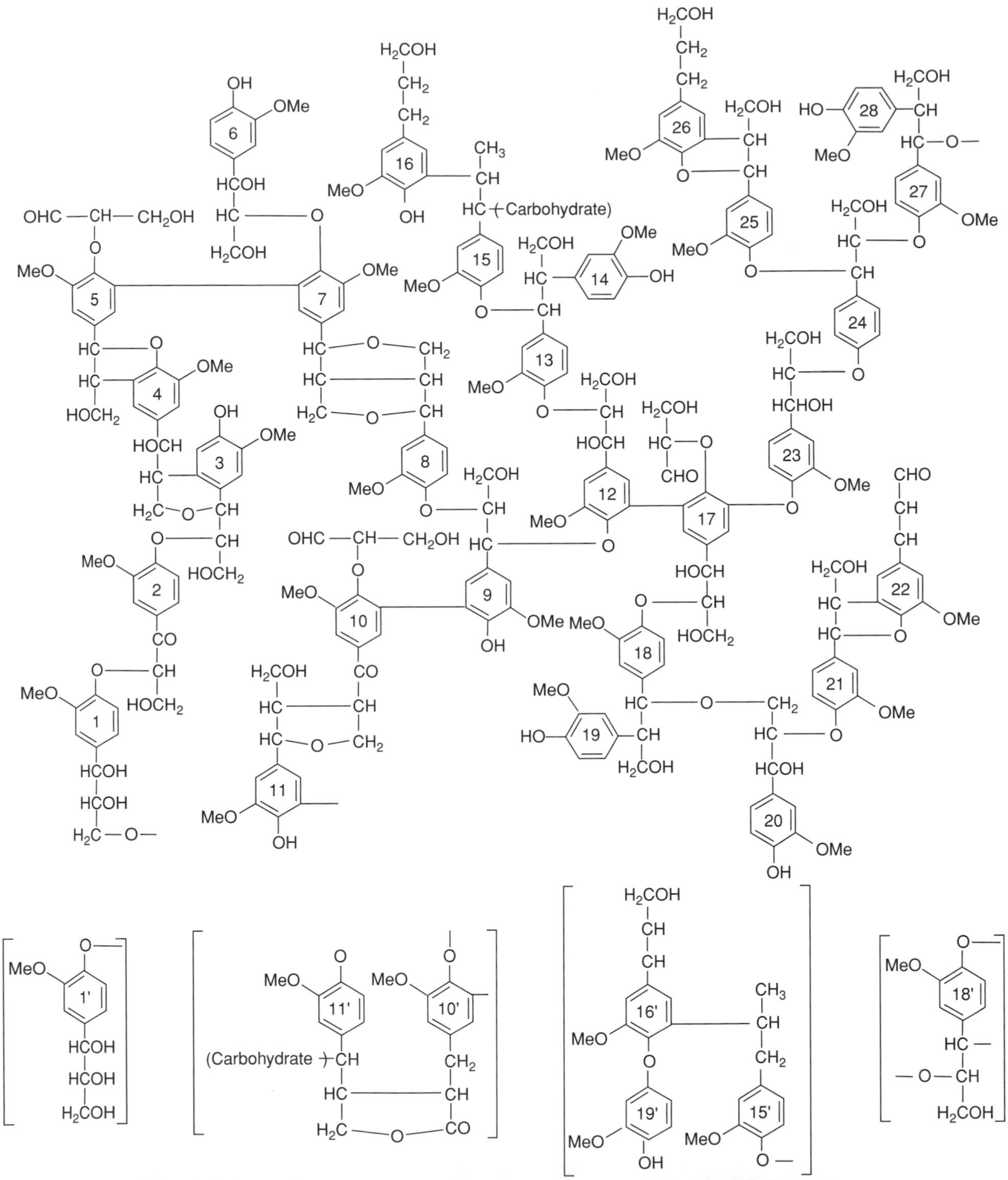

FIGURE 12.9

A simplified qualitative model of softwood lignin. Additional units that may be found within lignin are indicated in *parentheses* in the lower part of the figure. [Sakakibara, A. (1983). Chemical structure of lignin related mainly to degradation products. In *Recent Advances in Lignin Biodegradation Research*, T. Higuchi, H.-M. Chang, and T. K. Kirk (eds.), pp. 12–33, Tokyo: UNI Publisher.]

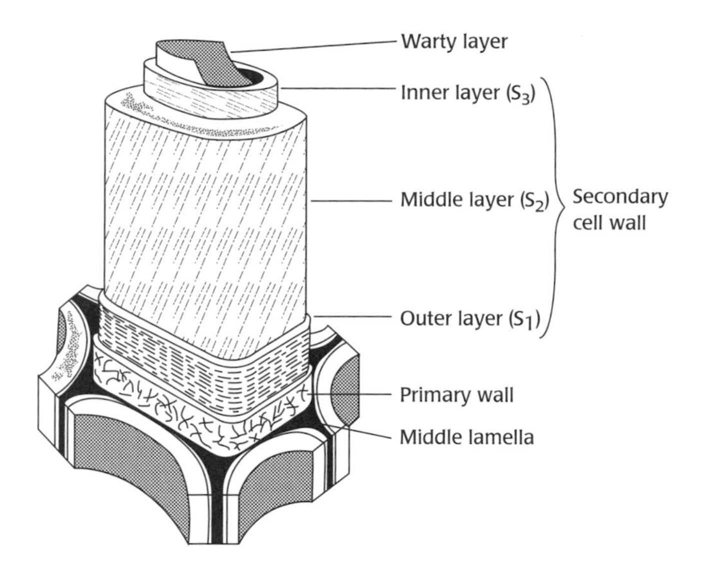


**FIGURE 12.10**

A simplified representation of the cell wall layers of a wood cell.

Redrawn based on artwork from the first edition (1995), published by W.H. Freeman.

the next either through pits or through cell walls. The hyphae of some fungi grow within the middle lamellae or the secondary walls.

White rot and brown rot fungi are filamentous fungi belonging to the subdivision Basidiomycetes. White rots typically cause bleaching of the wood, giving it a fibrous or spongy consistency. Some white rot fungi degrade lignin selectively, giving rise to "sequential decay," whereas others degrade the lignin and polysaccharides at a similar rate, and most of them prefer hardwoods. Brown rot fungi preferentially degrade the polysaccharide components of the wood and cause little degradation of lignin. The decaying wood becomes brown and brittle. Most brown rot fungi attack softwoods.

Soft rots, a group of fungi belonging to the Ascomycetes and Deuteromycetes, are a major cause of decay in wood that is exposed to moisture. Soft rots degrade primarily cellulose and hemicelluloses, and modify lignin only to a slight extent. The relative rates of degradation of these polymers differ among different species of soft rot fungi. Soft rots are found in both softwoods and hardwoods, and their action leads to a softening of the wood.

## ACTION OF BACTERIA ON LIGNOCELLULOSE

Bacteria are poor degraders of lignin in the presence of oxygen, and no bacterial degradation is seen under anaerobic conditions. For example, aspen wood, in which the lignin was selectively biosynthetically labeled with $^{14}C$, was not significantly degraded during six months of incubation in various soils. However, many bacteria produce cellulases as well as enzymes capable of cleaving the various bonds in hemicelluloses. Once these polysaccharides are exposed by fungal delignification, fungi and bacteria together accomplish further degradation.

## DEGRADATION OF LIGNIN

We saw above that lignin is a complex polymer of irregular structure formed by free radical–mediated condensation reactions and lacking in readily hydrolyzable bonds (Figure 12.9). The degradation of this extraordinary polymer likewise exploits free radical chemistry. The mechanism has been most extensively studied in the white rot fungus.

### *PHANEROCHAETE CHRYSOSPORIUM*

Wood-decaying fungi are generally mesophilic, with an optimum temperature for growth between 20°C and 30°C. However, *P. chrysosporium*, which grows in environments such as self-heating piles of wood chips, has an optimum temperature for growth of 40°C and continues to grow at up to 50°C. Lignin degradation by *P. chrysosporium* in culture is triggered by nitrogen limitation. Such nutrient limitation mimics the situation in nature, because wood is low in organic nitrogen compounds. *P. chrysosporium* is an exceptionally efficient lignin decomposer capable of degrading the lignin in wood pulp at a rate approaching 3 g of lignin per gram of fungal cell protein per day.

#### A Look at the Genome of *P. chrysosporium*

*P. chrysosporium* is the most intensively studied white rot fungus. It completely degrades all major components of the plant cell wall. The *P. chrysosporium* genome, the first basidiomycete genome to be sequenced, is approximately 30 Mb organized in 10 chromosomes with 11,777 putative protein-coding genes. The genome sequence reveals extensive genetic diversity within gene families producing oxidative and hydrolytic enzymes most likely involved in the efficient degradation of lignin and other wood components. These gene families encode peroxidases, oxidases, glycosyl hydrolases, and cytochrome P450s. There are 10 isozymes of lignin peroxidase and five of manganese-dependent peroxidase. It has been suggested that the biochemical diversity may allow optimum growth under varying environmental conditions of temperature, pH, and ionic strength, or that similar but not identical enzymes are needed to break down polymeric structures that vary from one plant to another in chemical detail, physical state, or accessibility.

### LIGNIN PEROXIDASE

A breakthrough in the study of lignin degradation was the discovery in 1983 of *P. chrysosporium* lignin peroxidase. This hydrogen peroxide–requiring extracellular enzyme with an acidic pH optimum catalyzes the oxidative breakdown of lignin and of model compounds designed to represent portions of the lignin structure. Lignin peroxidase catalyzes the cleavage of the arylpropane side chains, ether bond cleavage, aromatic ring opening, and

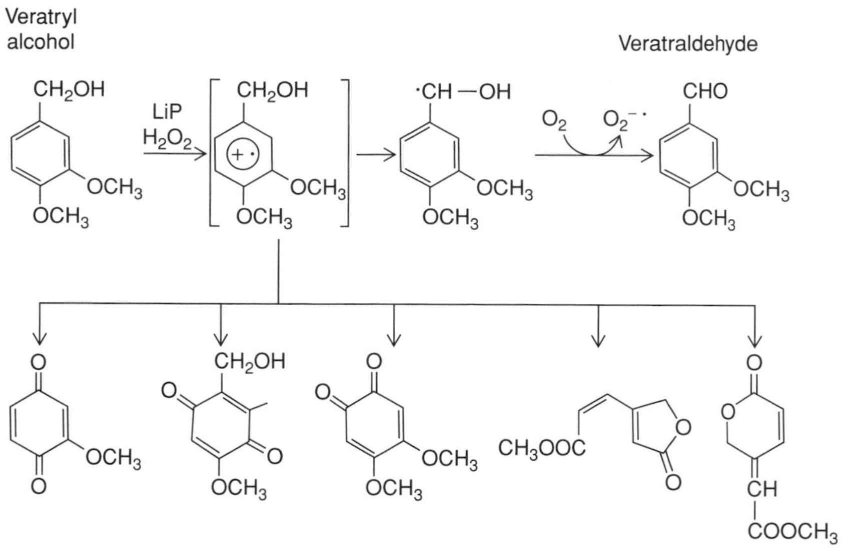

**FIGURE 12.11**

A model for lignin degradation: oxidation of veratryl alcohol by lignin peroxidase (LiP) in the presence of $H_2O_2$. Veratryl alcohol is synthesized by various white rot fungi. It stabilizes LiP and protects it from inactivation by $H_2O_2$. It is also a substrate for LiP that is oxidized primarily to veratraldehyde, as well as to a variety of other products, as illustrated in the figure.

Redrawn based on artwork from the first edition (1995), published by W.H. Freeman.

hydroxylation. All these reactions can be adequately explained by a mechanism that begins with the catalyzed abstraction of an electron from aromatic nuclei in lignin to form unstable cation radicals. Subsequently, various products are formed by nonenzymatic reaction of the radical cations with water, other nucleophiles, and oxygen (Figures 12.11 and 12.12).

The catalytic cycle of lignin peroxidase explains its ability to catalyze one-electron oxidations (Figure 12.13). The native enzyme contains a protoporphyrin prosthetic group with a high-spin ferric iron, Fe(III). Oxidation by hydrogen peroxide removes two electrons and converts the prosthetic group

**FIGURE 12.12**

Pathways and products of degradation of a model compound for a component of lignin substructure, 4-ethoxy-3-methoxyphenylglycerol-$\beta$-vanillin-$\gamma$-benzyl diether, by *P. chrysosporium* lignin peroxidase (LiP). [Higuchi, T. (1990). Lignin biochemistry: biosynthesis and biodegradation. *Wood Science and Technology*, 24, 23–63.]

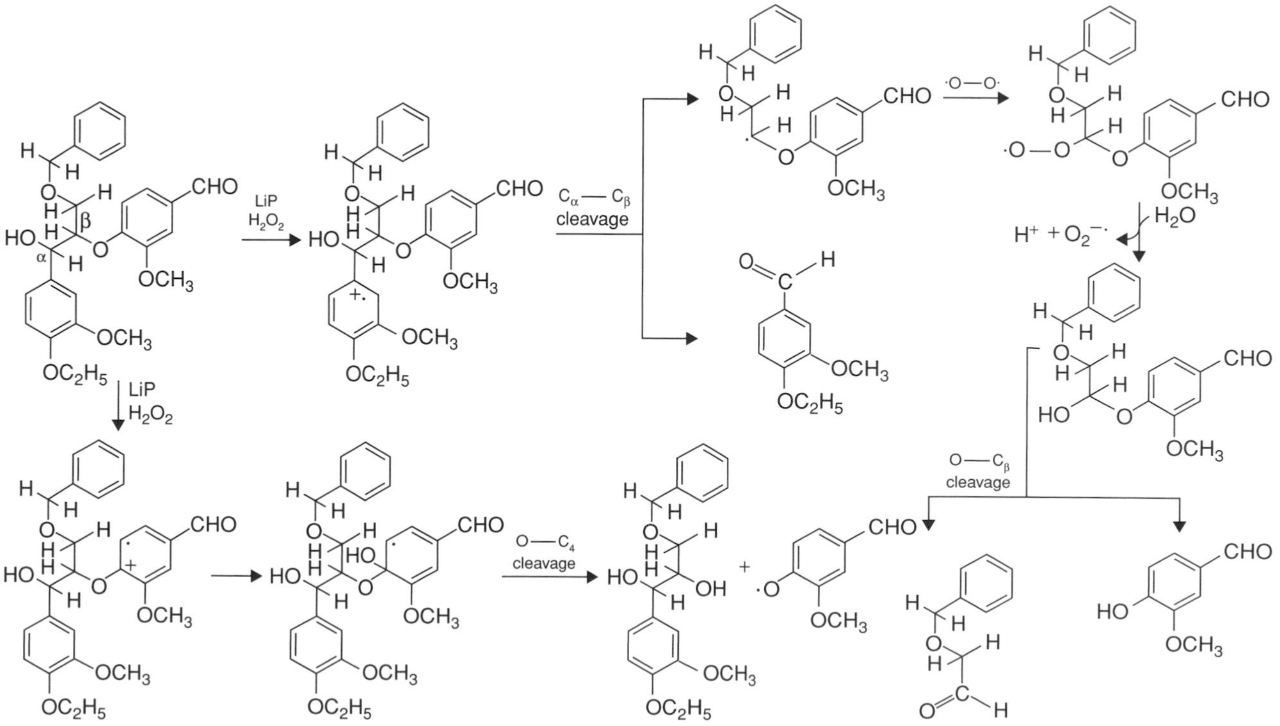

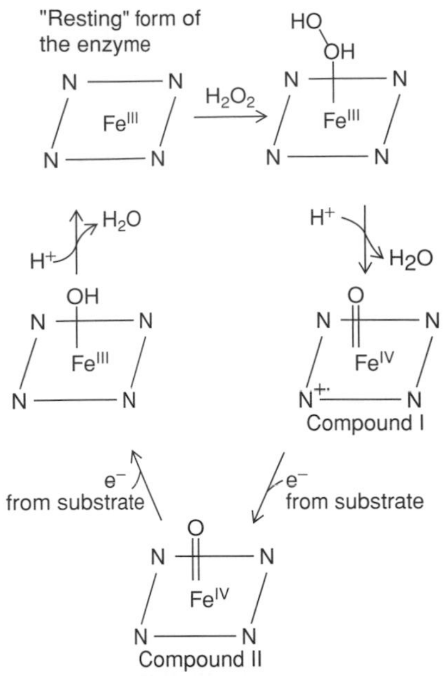

FIGURE 12.13

The catalytic cycle of lignin peroxidase. Compound I is an iron IV-porphyrin $\pi$ cation radical in which one of the electrons has been removed from the iron and one from the porphyrin ligand.

Redrawn based on artwork from the first edition (1995), published by W.H. Freeman.

of the enzyme to an oxo-iron(IV) porphyrin radical cation – an enzyme form designated "compound I." A one-electron reduction by abstraction of an electron from a donor molecule (e.g., an aromatic nucleus) produces an enzyme with an oxo-iron(IV) porphyrin – a form designated "compound II." A second one-electron reduction completes the catalytic cycle by regenerating the native ferric enzyme.

## OTHER ENZYMES WITH ROLES IN LIGNIN BREAKDOWN

As noted above, in addition to lignin peroxidase, ligninolytic cultures of *P. chrysosporium* produce other enzymes that function either in the breakdown of lignin or in modification of the breakdown products.

### Mn(II)-Dependent Peroxidases

Successful degradation of lignin requires attacks on both the nonphenolic and phenolic lignin components. The extracellular Mn(II)-dependent peroxidases oxidize phenolic components of lignin; they cannot oxidize the nonphenolic substrates of lignin peroxidase, such as veratryl alcohol, or nonphenolic model compounds of lignin substructure. The prosthetic group of this family of isoenzymes, a protoheme IX with high-spin ferric iron, is the same as that of lignin peroxidase. However, in addition to $H_2O_2$, these enzymes require Mn(II) and an organic acid such as malonate or oxalate. Mn(II) serves as a reductant for compound II (see Figure 12.13). The resulting Mn(III) is chelated by the organic acid, diffuses into the lignin, and oxidizes phenolic moieties.

### Quinone Reductases

Quinones are among the products of lignin degradation by the peroxidases, and *P. chrysosporium* produces both intracellular and extracellular quinone reductases. The extracellular cellobiose–quinone oxidoreductase is active only in the presence of cellulose. This enzyme uses cellobiose as a hydrogen donor to reduce quinones to hydroquinones. The intracellular quinone reductases use NAD(P)H as a cofactor. The fungus rapidly metabolizes the hydroquinones. Perhaps the role of the quinone reductases is to remove products of lignin degradation that might otherwise rapidly repolymerize.

## METABOLISM OF LIGNIN

The major reactions leading to lignin biodegradation are summarized in Figure 12.14. Degradation is initiated by lignin peroxidase and the phenol oxidases, which catalyze the oxidative degradation of the polymer to oligolignols and thence to mono- and dilignols. Mediators produced by the action of these enzymes may also contribute to the oxidation of lignin.

*Mediators* are reactive low molecular weight species, such as chelated Mn(III) derived from the Mn(II)-dependent peroxidase or the veratryl

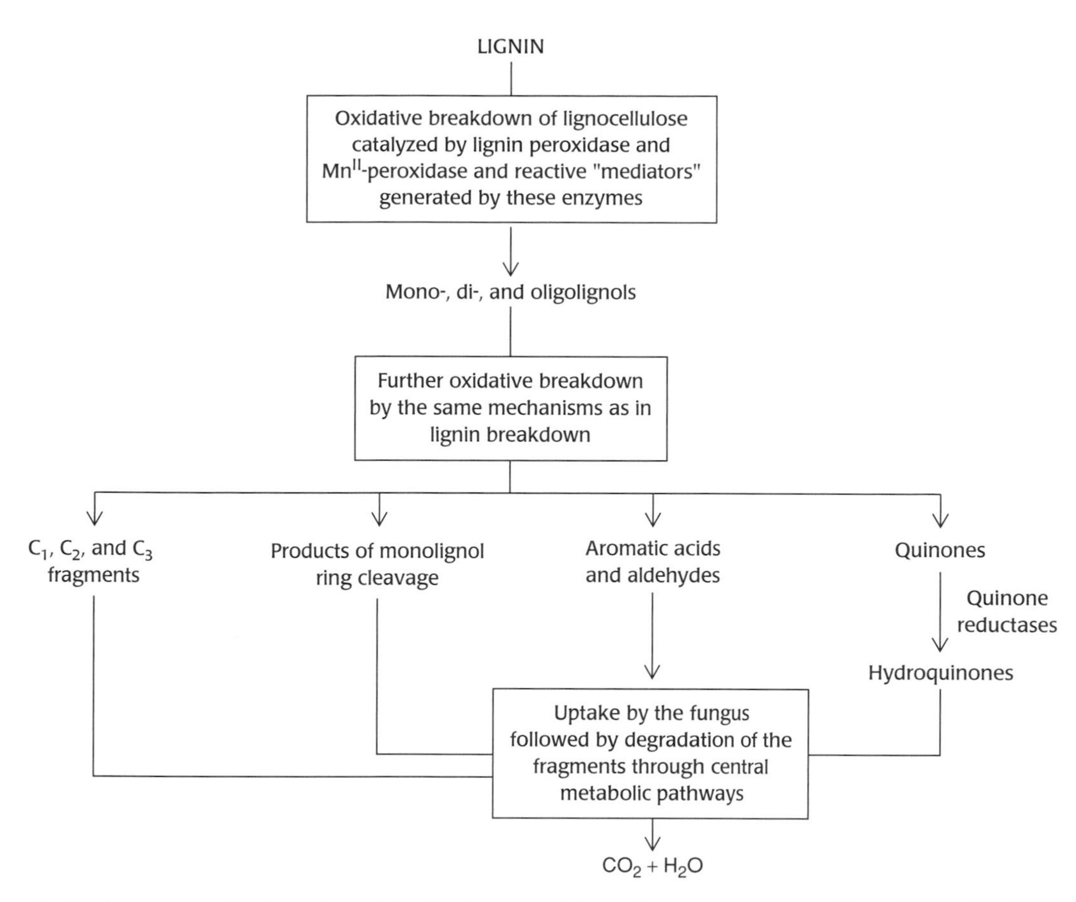


**FIGURE 12.14**

Schematic representation of the major steps in lignin degradation by ligninolytic cultures of the fungus *P. chrysosporium*.

Redrawn based on artwork from the first edition (1995), published by W.H. Freeman.

alcohol cation radical produced by lignin peroxidase. They may act as diffusible one-electron oxidants capable of penetrating and reacting within internal regions of the lignin polymer. Further oxidative cleavage of mono-, di-, and trilignols catalyzed by the peroxidases leads to the production of an array of low molecular weight compounds: $C_1$, $C_2$, and $C_3$ fragments derived from the propane side chains of the phenylpropane building blocks, aromatic acids and aromatic aldehydes, products of aromatic ring cleavage, and quinones. Metabolism of these compounds by combinations of reduction and oxidation reactions leads to their complete degradation to carbon dioxide and water. The $C_2$ and $C_3$ fragments serve as substrates for *glyoxal oxidase*. Their degradation by this extracellular enzyme is accompanied by the generation of $H_2O_2$, needed for the peroxidase-mediated depolymerization of lignin. Aromatic acids, such as vanillic and syringic acid, derived from lignin breakdown are taken up by the mycelium and degraded further by dioxygenases that function in the metabolism of simple aromatic acids. Quinones are rapidly reduced to hydroquinones by cellobiose–quinone oxidoreductase, with concomitant formation of cellobionic acid. Hydroquinones are rapidly taken up by the fungus and degraded.

In the initial steps of lignin breakdown, the various lignols could undergo a reverse reaction, peroxidase-induced polymerization, to give a lignin polymer of “unnatural” structure. Such reactions must take place to some extent. Presumably, the rapid uptake, reduction, and degradation of smaller fragments by the fungal cells ensures that the equilibrium strongly favors degradation.

## DEGRADATION OF CELLULOSE

Hundreds of species of fungi and bacteria are able to degrade cellulose. These organisms include aerobes and anaerobes, mesophiles and thermophiles. They are both widespread and abundant in the natural environment. However, although many microorganisms can grow on cellulose or produce enzymes that can degrade amorphous cellulose, relatively few produce the entire complement of extracellular cellulases able to degrade crystalline cellulose *in vitro*. Among the latter organisms, the most extensively studied sources of cellulolytic enzymes have been the fungi *Trichoderma* and *Phanerochaete* and the bacteria *Cellulomonas* (an aerobe) and *Clostridium thermocellum* (an anaerobe).

### A ROTTING CARTRIDGE BELT

Scientific study of the degradation of cellulose became a practical necessity during World War II, when the clothing and equipment of American military units stationed in the jungles of the South Pacific were found to rot at an alarming rate. All cotton gear – tents, uniforms, webbing, and knapsacks – deteriorated rapidly. Replenishment of this equipment usurped most of the cargo space needed for military supplies. It was easy to see that the damage was being done by fungi, but the particular culprits had to be identified.

In 1944, a Tropical Deterioration Research Laboratory was established in Philadelphia, Pennsylvania, to determine the nature of the deterioration processes, to identify the organisms responsible and their mechanism of action, and to develop methods of control that did not require the use of fungicides. Because cotton consists almost entirely of cellulose, a central focus of this effort was to develop an understanding of cellulose degradation. More than 4000 strains of fungi isolated from soil samples or damaged materials were subjected to study. The screening of these microorganisms was laborious. Thousands of microorganisms were grown on tens of thousands of textile strips, and each strip was tested for its loss in tensile strength. Those fungi found most active by this test were then grown in shake flasks with cellulose (ground filter paper) as a substrate and compared for their ability to produce the extracellular enzymes responsible for the degradation of cotton.

This screen uncovered a powerful cellulose-destroying strain of *Trichoderma viride*, isolated from a rotting cartridge belt found in New Guinea. The culture filtrates from this strain were the most active of any tested in their ability to convert crystalline cellulose to glucose. This discovery focused attention on *Trichoderma* strains. To this day, mutants derived from *Trichoderma reesei* QM6a (formerly *T. viride* QM6a), a strain isolated from a Bougainville Island cotton duck shelter, are among the most efficient known producers of extracellular cellulases.

The studies of cellulolytic activities of particular organisms in the laboratory were not necessarily predictive of the importance of the organism to the destruction of cotton in the field. Samples of deteriorated cotton materials were sent to the United States from 24 military bases in South and Southwest Pacific areas, extending from Hawaii to New Guinea, and from localities in China, Burma, and India as well. Among 4500 strains of fungi isolated from these exposed cotton fabrics, *Trichoderma* strains numbered 385 (8.6%) and were found in samples from most locations. Yet, direct examination of many hundreds of samples of decaying cotton fabrics from all these areas did not show any *fruiting structures* of *T. viride*, although they might have been lost during handling. This observation suggests that, although *T. viride* was not *growing* on them, it was commonly present (probably as spores) on these fabrics. There is no convincing evidence, therefore, that it played any role in the deterioration of these materials in the field.

## *TRICHODERMA* CELLULASES

The cellulolytic system of *T. reesei* is typical of such systems among the filamentous fungi. *T. reesei* produces three types of cellulolytic enzymes that cooperate in the degradation of cellulose: endoglucanases (EG I to EG V), cellobiohydrolases (CBH I and CBH II), and $\beta$-glucosidases (BGL I and BGL II). The endoglucanases hydrolyze internal bonds in disordered regions along cellulose fibers. The ends generated in this manner are then attacked by the cellobiohydrolases, with release of the disaccharide cellobiose. CBH I has a preference for the reducing and CBH II for the nonreducing ends of cellulose chains of microcrystalline cellulose. As the hydrolysis proceeds, the cellobiohydrolases also, apparently, begin to disrupt chain–chain interactions in the crystalline regions of the cellulose fiber. Finally, cellobiose is hydrolyzed to glucose by the $\beta$-glucosidases (Figure 12.15).

## RELATIONSHIP OF STRUCTURE TO FUNCTION IN THE FUNGAL CELLULASES

Four genes, coding for the two endoglucanases and the two cellobiohydrolases, have been cloned from *T. reesei*. The enzymes encoded by these genes are each composed of two distinct domains joined by an extended, flexible "hinge" region (Figure 12.16). Note that the order of the domains in CBH I and EG I is the reverse of that seen in CBH II and EG II. The larger domain, composed of about 500 amino acids, includes the active site. The hinge region, a heavily glycosylated sequence 24– to 44–amino acids long, rich in prolyl and hydroxyamino acid residues, joins the catalytic domain to an approximately 33-residue cellulose-binding domain.

The flexible hinge allows the cellulose-binding domain to attach the enzymes to the cellulose fiber with little restriction on the interaction of the catalytic domain with the cellulose. In this manner, the cellulose-binding domain contributes to maintaining a high concentration of cellulase on the substrate. Studies on the three-dimensional structure of the cellulose-binding domain indicate that it interacts preferentially with the crystalline

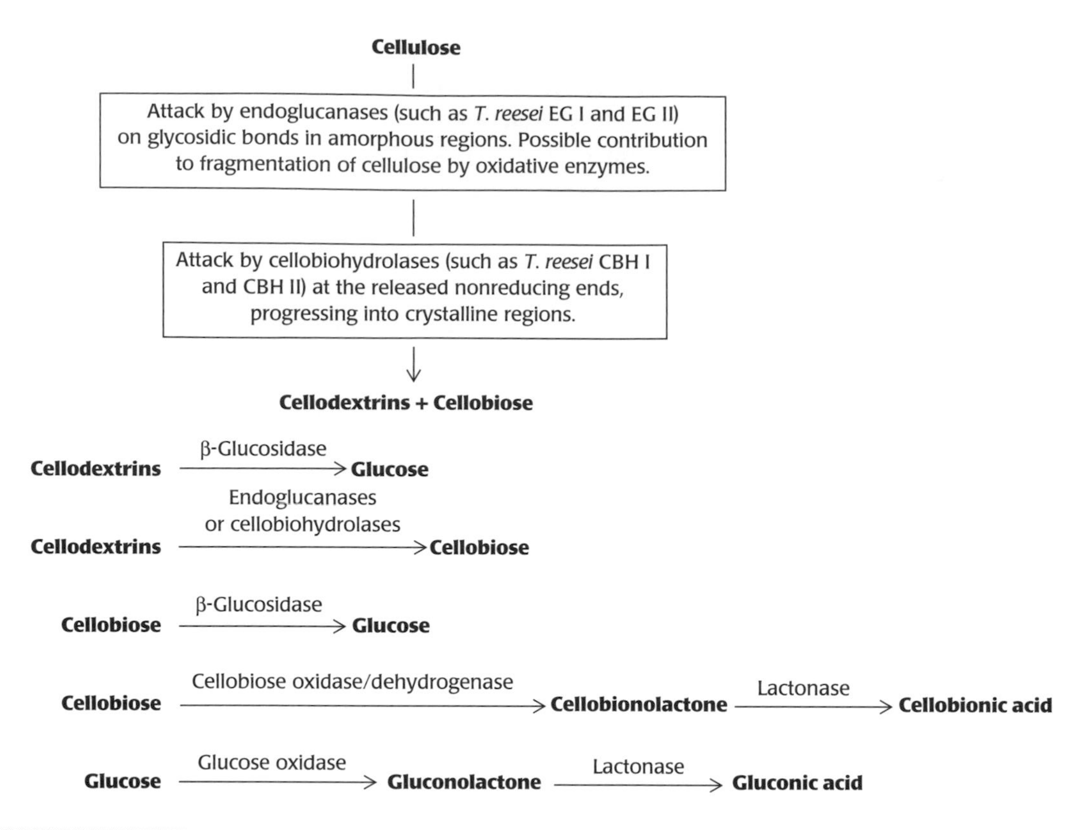


**FIGURE 12.15**

Proposed mechanism of cellulose degradation by fungi. [Coughlan, M. P. (1990). Cellulose degradation by fungi. In *Microbial Enzymes and Biotechnology*, 2nd Edition, W. M. Fogarty and C. T. Kelly (eds.), pp. 1–36, London: Elsevier Applied Science.]

Redrawn based on artwork from the first edition (1995), published by W.H. Freeman.

regions in cellulose. The catalytic domain, on the other hand, has a high affinity for amorphous regions and low affinity for the crystalline regions. In the intact enzyme, the low affinity of the catalytic domain for the crystalline regions is offset by the high affinity of the binding domain, with the result that many more intact enzyme molecules are bound in the crystalline regions, increasing the probability of glycosidic bond cleavage in these regions.

When the small terminal domain is removed from CBH I and CBH II by proteolysis, the affinity and activity of both enzymes toward insoluble cellulose is significantly reduced, whereas their activity toward small soluble substrates is unchanged. This suggests that the cellulose-binding domain contributes both to the binding of the cellulase to the cellulose surface and to the altering of the susceptibility of the substrate to degradation. The interaction of the cellulose-binding domain with crystalline cellulose may lead to local destabilization of interactions between chains, a disruption that in turn may lead to enhanced activity of the intact enzyme toward crystalline cellulose.

The classical approach to the study of enzyme specificity is to examine their action on small synthetic substrates. As illustrated in Figure 12.17, however, such small substrates do not differentiate clearly between the specificities of exo- and endocellulases. Comparative biochemical studies of many cellulolytic enzymes, combined with examination of high-resolution crystal structures of the catalytic domain of *T. reesei* CBH II and a bacterial cellulase, have provided more specific information about the molecular determinants of exo- versus endocellulolytic cleavage. These studies indicate that

Number of Amino Acid Residues

| | 1. Signal sequence | 2. Catalytic domain | 3. Hinge ("linker") | 4. Cellulose-binding domain |
|---|---|---|---|---|
| CBH I | 17 | 425 | 26 | 33 |
| CBH II | 24 | 385 | 44 | 33 |
| EG I | 22 | 363 | 29 | 33 |
| EG II | 21 | 327 | 34 | 33 |

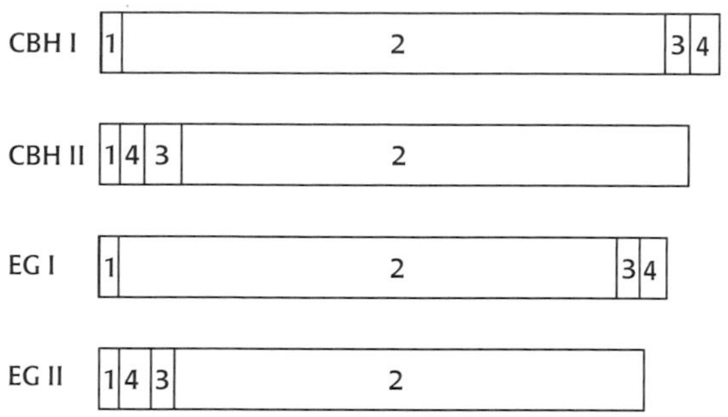

CBH I | 1 | 2 | 3 | 4 |

CBH II | 1 | 4 | 3 | 2 |

EG I | 1 | 2 | 3 | 4 |

EG II | 1 | 4 | 3 | 2 |

**FIGURE 12.16**

The *upper part* of the figure shows a tabulation of the lengths of the various domains of the preprotein forms of *Trichoderma reesei* cellobiohydrolases I and II (CBH I and CBH II) and endoglucanases I and II (EG I and EG II). The *lower part* shows a schematic representation of those lengths and of the domains' relative location (determined by small-angle x-ray scattering). Note that the order of the cellulose-binding domain, hinge ("linker"), and catalytic domain in CBH I and EG I is the opposite of that in CBH II and EG II. [Data from Gilkes, N. R., Henrissat, B., Kilburn, D. G., Miller, R. C., Jr., and Warren, R. A. J. (1991). Domains in microbial $\beta$-1,4-glycanases: sequence conservation, function, and enzyme families. *Microbiological Reviews*, 55, 303–315.]

in the exocellulases, the active site is in an enclosed tunnel through which a cellulose chain threads. Two aspartyl residues, believed to be the catalytic residues, are located in the middle of the tunnel. The architecture of the active site is such that productive substrate binding is predicted to lead to the release of disaccharide units (cellobiose). The active site in the endocellulases lies in an open groove in which there is no restriction on the position of cleavage of the $\beta$-1,4-glycosidic linkage in a cellulose chain.

## BACTERIAL CELLULASES

The ability to degrade crystalline cellulose is widespread among both aerobic and anaerobic bacteria. The extracellular endo- and exoglucanases produced by a variety of bacteria – for example, the soil bacterium *Cellulomonas fimi* – show the same general structural features as those of *T. reesei*. The bacteria either secrete their cellulases as soluble extracellular enzymes or, in anaerobic cellulolytic bacteria, assemble them into large complexes called *cellulosomes* that are attached to the bacterial cell surface. These complexes are equipped to degrade different types of polysaccharides in plant cell walls. In addition to cellulases, cellulosomes include different types of hemicellulases, such as xylanases, mannanases, arabinofuranosides, and pectin lyases.

The best-studied cellulosomes, those of the thermophilic anaerobic bacterium *Clostridium thermocellum*, are each made up of 14 to 18 polypeptides and have a mass of about $2 \times 10^6$. This cellulosome contains several endoglucanases, as well as exoglucanases and xylanases. Cellulose-binding domains are absent from *C. thermocellum* glucanases. The cellulosome also contains polypeptides with no enzymatic activity that function in the organization of the cellulosome or help attach it to the cell surface and to cellulose. The latter

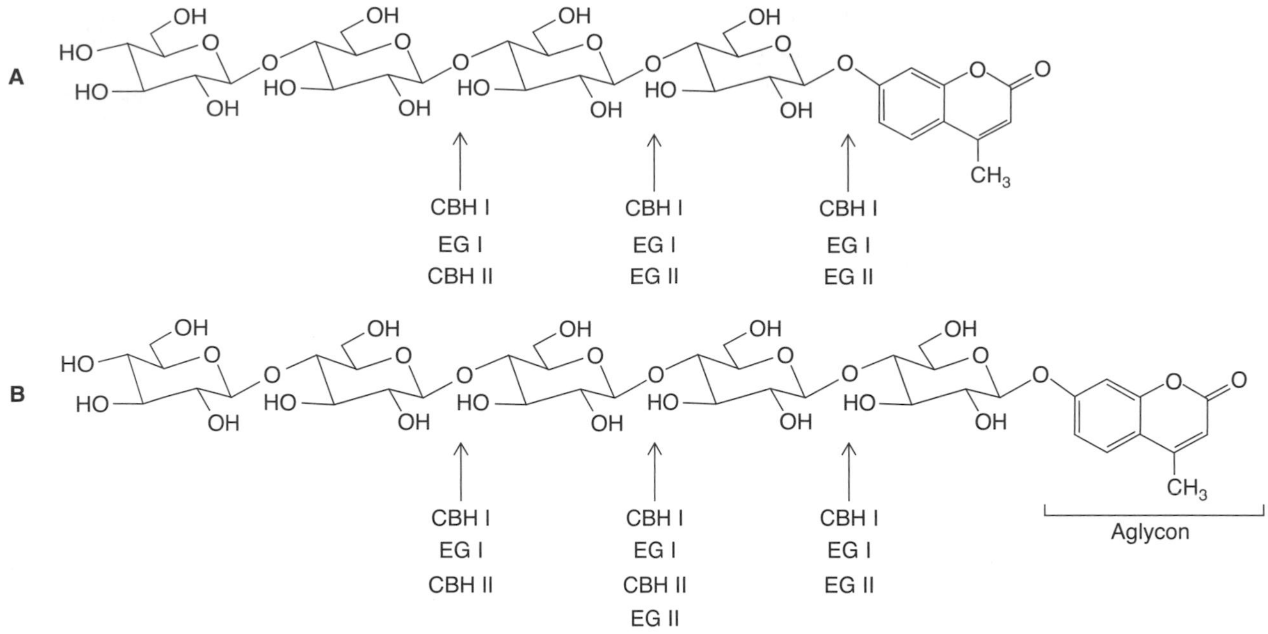

FIGURE 12.17

Cleavage of $\beta$-glycosidic bonds in cellotetraose (**A**) and cellopentaose (**B**) by *T. reesei* endoglucanases (EG I and EG II) and cellobiohydrolases (CBH I and CBH II). For each of the substrates, abbreviations below the *arrows* specify which enzymes cleaved a particular bond. The aglycon, 4′-methylumbelliferyl, was chosen to facilitate detection of substrates and degradation products by detection of the released 4′-methylumbelliferone (4-methyl-7-hydroxycoumarin) by absorption or fluorescence spectroscopy. [Data from Claeyssens, M., and Henrissat, B. (1992). Specificity mapping of cellulolytic enzymes: Classification into families of structurally related proteins confirmed by biochemical analysis. *Protein Science*, 1, 1293–1297.]

ability bestows a competitive advantage in environments in which numerous organisms may compete for the soluble products of cellulose degradation.

## ARCHITECTURE OF THE CELLULOSOME

The cellulosome is a highly active enzyme assembly capable of completely degrading cellulose. It has the unusual property of being more active against crystalline cellulose than against disordered (amorphous) cellulose. During the exponential phase of bacterial growth, the cellulosomes are primarily found attached to the cell surface. As the cells enter their stationary phase, a large percentage of these complexes are released into the medium. A model for a cell-bound cellulosome is shown in Figure 12.18.

*C. thermocellum* cells are bounded by three layers: the cytoplasmic membrane, the peptidoglycan layer, and a surface protein layer (S-layer). A large noncatalytic protein called *scaffoldin* (CipA) is an organizing protein in the cellulosome and the component that interacts with cell surface–anchoring proteins. CipA contains nine reiterated sequences – *cohesin* modules – that bind with very high affinity to an approximately 70-residue *dockerin* domain within each of the different enzymes incorporated into the cellulosome and in other cellulosomal components, which contains two duplicated conserved 23-residue tandemly repeated sequences. Because the cohesins cannot distinguish between the dockerins on the various enzymes, the CipA in any individual cellulosome may have a distinctive combination of enzymes attached to it. In the *C. thermocellum* genome, there are more than 60 dockerin-containing proteins. The scaffoldin includes a cellulose-binding domain (CBD), also referred to as a carbohydrate-binding module (CBM), that mediates the recognition of crystalline cellulose. A single CBM targets

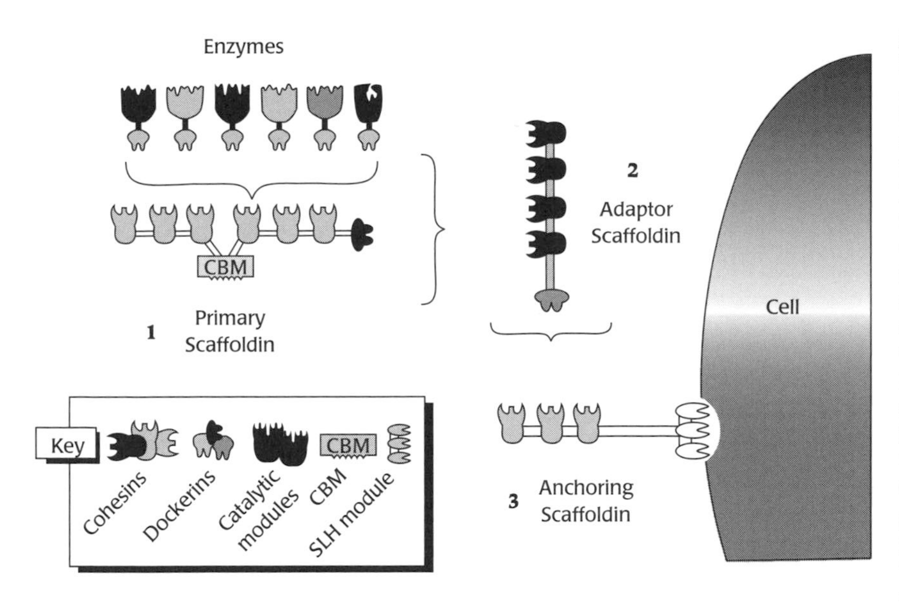


**FIGURE 12.18**

A simplified schematic model proposed for the *Acetivibrio cellulolyticus* cellulosome. *A. cellulolyticus,* a mesophilic cellulolytic bacterium closely related to clostridia, was isolated from sewage sludge. This cellulosome has three scaffoldins. Dockerin-containing enzymes are attached to the primary scaffoldin through interactions with cohesin modules (see text). An adaptor scaffoldin, with four cohesins and a dockerin, bridges between the primary scaffoldin and an anchoring scaffoldin through an interaction with a dockerin on the former with a cohesin on the latter. The S-layer homology (SLH) module on the anchoring scaffoldin attaches the cellulosome to the cell surface. A single cellulose-binding domain (CBM) in the primary scaffoldin targets the cellulosome, and thereby the entire cell, to the cellulose substrate. [Based on Figure 1 in Bayer, B. A., Belaich, J-P., Shoham, Y., and Lamed, R. (2004). The cellulosomes: multienzyme machines for degradation of plant cell wall polysaccharides. *Annual Review of Microbiology*, 58, 521–554.]

the attachment of the entire cellulosome to the cellulosic substrate. The model of the *Acetivibrio cellulolyticus* cellulosome shown in Figure 12.19 displays the key features described above for the *C. thermocellum* cellulosome.

There are three general categories of CBMs: type A, those that bind strongly to the surfaces of insoluble polysaccharides such as crystalline cellulose; type B, those that bind to soluble glycan chains; and type C, those that bind to small saccharides. Whereas scaffoldins all have type A CBMs, the various cellulosomal enzymes have CBMs of all three categories that target the parent enzyme to the appropriate substrate. It seems likely that the spacing of the hydrolytic enzyme sites on the cellulosome is such as to allow near-simultaneous multisite cleavage of bonds in the highly ordered crystalline cellulose substrate. This would account for the cellulosome's preference for crystalline over amorphous cellulose. These molecular characteristics make the cellulosome a highly effective system for hydrolyzing plant cell wall polysaccharides.

## DEGRADATION OF HEMICELLULOSES

Most studies of enzymatic hemicellulose degradation have focused on xylans, to the virtual exclusion of galactoglucomannans and glucomannans. Xylans are the major hemicelluloses in wood from angiosperms (flowering plants), where they account for 15% to 30% of the total dry weight. In the gymnosperms (the ferns, the conifers, and their allies), however, xylans contribute only 7% to 12% to the total dry weight.

Because of the high content of xylans in lignocellulose, economically feasible conversion of biomass to alcohol requires a practicable method for their fermentation. The issue of xylan degradation arises in two other significant contexts as well. Paper manufacture generates effluents from wood

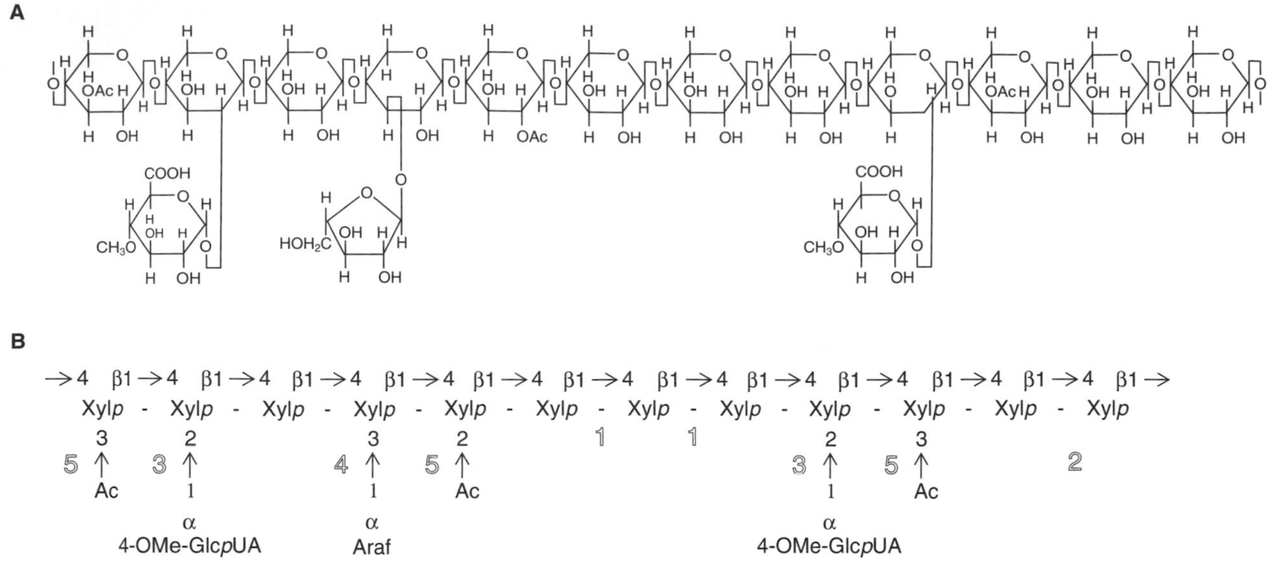

A

B

→4 β1→4 β1→4 β1→4 β1→4 β1→4 β1→4 β1→4 β1→4 β1→4 β1→4 β1→4 β1→

Xylp - Xylp - Xylp - Xylp - Xylp - Xylp - Xylp - Xylp - Xylp - Xylp - Xylp - Xylp

Ac; α 4-OMe-GlcpUA; α Araf; Ac; α 4-OMe-GlcpUA; Ac

**FIGURE 12.19**

Sites of attack on a fragment of a glucuronoarabinoxylan by microbial xylanolytic enzymes. The *upper part* of the figure shows the chemical structure of glucuronoarabinoxylan and the *lower part* a representation of the structure using the conventional abbreviations defined in the legend to Figure 12.2. *Outlined numbers* indicate which type of enzyme cleaves a particular bond: 1, endo-$\beta$-1,4-xylanase; 2, $\beta$-xylosidase; 3, $\alpha$-glucuronidase; 4, $\alpha$-L-arabinofuranosidase; 5, acetylxylan esterase.

pulping and pulp processing that contain large amounts of xylans. These waste streams frequently pollute streams and rivers. Agricultural residues also contain appreciable amounts of xylans.

The structure of xylans is complex, and their complete biodegradation requires the concerted action of several enzymes. Bacteria and fungi isolated from many different habitats contain various xylanases. The domain type of organization described for *T. reesei* cellulases is seen in xylanases as well.

The complete conversion of a glucuronoxylan into its building blocks requires the combined actions of an endo-1,4-$\beta$-xylanase and a $\beta$-xylosidase, as well as an $\alpha$-glucuronidase, an $\alpha$-L-arabinofuranosidase, and an acetylxylan esterase (Figure 12.19). Examples of microorganisms able to both degrade xylan and convert the resulting xylose to ethanol and other products are given in Table 12.5.

## THE PROMISE OF ENZYMATIC LIGNOCELLULOSE BIODEGRADATION

The first step in the utilization of biomass for fuel, chemical feedstocks, or food is to release the carbohydrate and aromatic building blocks locked up in the structure of lignocellulose. This can be achieved by chemical means, but at the expense of producing undesirable side products and chemical waste at the same time.

Fortunately, we can exploit the many bacterial and fungal enzyme systems capable of degrading the polymers of lignocellulose. Large amounts of these enzymes are readily available through cloning and overexpression in bacterial and fungal host cells. Organisms that overproduce particular enzymes or that produce a mix of enzymes appropriate for a particular application are also available. For example, conventional mutation and strain selection have created strains of *T. reesei* that produce up to 40 g/L of extracellular protein, mostly components of the cellulase system, in relative proportions similar to those of the parent wild-type strain, and genetic engineering has been used to produce *T. reesei* strains with modified amounts and proportions of EG I, EG II, CBH I, and CBH II.

**TABLE 12.5 Various bacteria and fungi able to ferment xylan to ethanol**

| Microorganism | Products |
|---|---|
| Bacteria | |
| *Clostridium thermocellum* | Ethanol, acetic acid, lactic acid |
| *Clostridium thermohydrosulfuricum* | Ethanol, acetic acid, lactic acid |
| *Clostridium thermosaccharolyticum* | Ethanol, acetic acid, lactic acid |
| *Thermoanaerobacter ethanolicus* | Ethanol, acetic acid, lactic acid |
| *Thermoanaerobium brockii* | Ethanol, acetic acid, lactic acid |
| *Thermobacteroides acetoethylicus* | Ethanol, acetic acid, lactic acid |
| Fungi | |
| *Neurospora crassa* | Ethanol |
| *Pichia stipitis* | Ethanol |

The availability of known sequences for members of enzyme families, such as ligninase peroxidases, glucanases, and xylanases, from many different microorganisms opens the way to genetic engineering of these enzymes so as to improve their functional and physical properties, leading, for instance, to minimization of product inhibition and an increase in thermostability. As an example, a xylanase identified in a DNA library derived from a sample of fresh bovine manure exhibited a thermal denaturation transition ($T_m$) of 61.4°C. This enzyme was subjected to directed evolution technologies. Gene site saturation mutagenesis (discussed in Chapter 11) followed by gene reassembly, a technology producing a random combination of single beneficial amino acid substitutions designed to allow identification of protein variants with optimal fitness, was used in this case to generate variants with increased thermostability with retention of optimal catalytic activity. One of the evolved xylanase mutants exhibited a $T_m$ of 95.6°C. Such ongoing efforts will make it possible to convert more of the lignocellulose into useful products, and to do so faster and more economically.

## SUMMARY

Plant cell walls contain three types of biopolymers: cellulose, hemicelluloses, and lignin. These polymers are the major storage form both of energy trapped by photosynthesis and of organic matter in the biosphere. Cellulose is a linear polymer of thousands of glucose molecules. The basic repeating unit is cellobiose, glucose-$\beta$-1,4-glucose. Hemicelluloses are highly branched heteropolysaccharides of less than 200 sugar residues. The sugars in hemicellulose include pentoses, hexoses, and uronic acids. The sugars in hemicellulose are variously modified by acetylation or methylation. Lignin is a very high molecular weight irregular polymer made up of oxygenated $C_9$ (phenylpropane) units linked in many different ways. Lignin is formed by peroxidase-catalyzed generation and subsequent reaction of phenoxy radicals of (methoxy-) substituted *p*-hydroxycinnamyl alcohols – *p*-coumaryl, coniferyl, and sinapyl alcohols.

White rot, brown rot, and soft rot fungi are the primary wood degraders in nature. They secrete extracellular enzymes, lignin peroxidases, that mediate lignin depolymerization. Both fungi and bacteria produce extracellular enzymes that degrade cellulose and hemicelluloses. The enzymatic machinery for cellulose and hemicellulose degradation by certain anaerobic bacteria is packaged in intricate complexes, cellulosomes, bound to the cell surfaces of these bacteria.

The genes for lignin peroxidases, cellulases, and various hemicellulose-degrading enzymes have been cloned and the enzymes expressed in fungi and *Escherichia coli*. Moreover, genetically engineered strains producing modified amounts and proportions of the degradative enzymes have been constructed. These advances show promise of efficient utilization of lignocellulose as an abundant source of sugar substrates for fermentation processes such as the production of ethanol.

## REFERENCES AND ONLINE RESOURCES

### General

Siu, R. G. H. (1951). *Microbial Decomposition of Cellulose*, New York: Reinhold Publishing Co.

Reese, E. T. (1976). History of the cellulase program at the U.S. Army Natick Development Center. *Biotechnology and Bioengineering Symposium*, 6, 9–20.

Sjöström, E. (1981). *Wood Chemistry. Fundamentals and Applications*, New York: Academic Press.

Fengel, D., and Wegener, G. (1984). *Wood. Chemistry, Ultrastructure, Reactions*, Berlin: Walter de Gruyter.

Rayner, A. D. M., and Boddy, L. (1988). *Fungal Decomposition of Wood. Its Biology and Ecology*, Chichester, U.K.: J. Wiley and Sons.

Jain, S. M. and Minocha, S. C. (2000). *Molecular Biology of Woody Plants*, Volumes 1 and 2, Dordrecht: Kluwer. [Series: *Forestry Science*, Volumes 64 and 66].

Lynd, L. R., Weimer, P. J., van Zyl, W. H., and Pretorius, I. S. (2002). Microbial cellulose utilization: fundamentals and biotechnology. *Microbiology and Molecular Biology Reviews*, 66, 506–577.

Rose, J. (ed.) (2003). *The Plant Cell Wall*, Oxford: Blackwell Publishing.

### Lignocellulose: Structure, Biosynthesis, Degradation

Wood, W. A., and Kellogg, S. T. (eds.) (1988). *Biomass. Part A. Cellulose and Hemicellulose. Methods in Enzymology*, Volume 160, San Diego: Academic Press.

Wood, W. A., and Kellogg, S. T. (eds.) (1988). *Biomass. Part B. Lignin, Pectin, and Chitin. Methods in Enzymology*, Volume 161, San Diego: Academic Press.

Boerjan, W., Ralph, J., and Baucher, M. (2003). Lignin biosynthesis. *Annual Review of Plant Biology*, 54, 519–546.

Martinez, D., et al. (2004). Genome sequence of the lignocellulose degrading fungus *Phanerochaete chrysosporium* strain RP78. *Nature Biotechnology*, 22, 695–700.

Martinez, Á., et al. (2005). Biogradation of lignocellulosics: microbial, chemical, and enzymatic aspects of the fungal attack of lignin. *International Microbiology*, 8, 195–204.

Cotinho, P. M., and Hendrissat, B. (1999). Carbohydrate-Active Enzymes server at URL http://afmb.cnrs-mrs.fr/CAZY/.

Carrard, G., Koivula, A., Söderlund, H., and Béguin, P. (2000). Cellulose-binding domains promote hydrolysis on different sites on crystalline cellulose. *PNAS*, 97, 10342–10347.

Bourne, Y., and Hendrissat, B. (2001). Glycoside hydrolases and glycosyltransferases: families and functional modules. *Current Opinion in Structural Biology*, 11, 593–600.

Davies, G. J., Gloster, T. H., and Hendrissat, B. (2005). Recent structural insights into the expanding world of carbohydrate-active enzymes. *Current Opinion in Structural Biology*, 15, 637–645.

### Cellulosomes

Carvalho, A. L., et al. (2003). Cellulosome assembly revealed by the crystal structure of the cohesin-dockerin complex. *PNAS*, 100, 13809–13814.

Bayer, B. A., Belaich, J-P., Shoham, Y., and Lamed, R. (2004). The cellulosomes: multienzyme machines for degradation of plant cell wall polysaccharides. *Annual Review of Microbiology*, 58, 521–554.

Doi, R. H., and Kosugi, A. (2004). Cellulosomes: plant-cell-wall-degrading enzyme complexes. *Nature Reviews Microbiology*, 2, 541–551.

Gilbert, H.J. (2007). Cellulosomes: microbial nanomachines that display plasticity in quaternary structure. *Molecular Microbiology* 63, 1568–1576.

### Directed Evolution of a Xylanase

Palackal, N. et al. (2004). An evolutionary route to xylanase process fitness. *Protein Science*, 13, 494–503.

THIRTEEN

# Ethanol

The potential quantity of ethanol that could be produced from cellulose is over an order of magnitude larger than that producible from corn. In contrast to the corn-to-ethanol conversion, the cellulose-to ethanol route involves little or no contribution to the greenhouse effect and has a clearly positive net energy balance (five times better). As a result of such considerations, microorganisms that metabolize cellulose have gained prominence in recent years.

– Demain, A. L., Newcomb, M., and Wu, J. H. D. (2005). Cellulase, clostridia, and ethanol. *Microbiology and Molecular Biology Reviews*, 69, 124–154.

The preceding chapter described the major components of plant biomass – cellulose, hemicelluloses, and lignin – and their natural pathways of biodegradation. Many view the sugars locked up in cellulose and the hemicelluloses as an immense storehouse of renewable feedstocks for the fermentative production of fuel alcohol. This chapter begins with a discussion of the conversion of such sugars to ethanol and ends with an assessment of the future impact of fermentation alcohol as a fuel.

In microbiology, *fermentation* is defined as a metabolic process leading to the generation of ATP and in which degradation products of organic compounds serve as hydrogen donors as well as hydrogen acceptors. Oxygen is not a reactant in fermentation processes. In the words of Louis Pasteur, *"[l]a fermentation est la vie sans l'air"* – fermentation is life without air. The long history of brewing and wine making has produced highly refined technologies for large-scale fermentation and for the recovery of ethanol. In addition to its role as a beverage, ethanol can serve as a fuel and as a starting material for the manufacture of chemicals such as acetic acid, acetaldehyde, butanol, and ethylene (itself a key intermediate in the petrochemical industry).

Whereas the current industries producing fuels and organic chemicals require fossil feedstocks (petroleum and natural gas) as raw materials, the dramatic increases in the price of oil since the 1970s and the evidence of declining reserves have prompted assessment of alternative sources. Ethanol seems a particularly promising alternative. Anhydrous ethanol was already in use as a fuel in internal combustion engines by the late nineteenth century.

In fact, in 1906 the U.S. Congress removed the tax on alcohol to encourage farmers to produce their own engine fuels, in an attempt to assist the mechanization of farms.

With appropriate pretreatment, various forms of biomass can serve as substrates for alcohol fermentation. Since 1975, the Brazilian National Alcohol Program – the most determined effort yet to replace gasoline with alcohol – has used sucrose, obtained directly from sugarcane, as the substrate for fermentation. More than 50 billion liters of ethanol were produced during the program's first decade of operation; by 1989, Brazil produced 12 billion liters of ethanol per year, enabling 4.2 million cars to run on hydrated ethanol (95% ethanol, 5% water) and 5 million on a blend of 78% gasoline and 22% ethanol. In 1996, Brazilian production rose to 13.9 billion liters of ethanol, representing the energy equivalent of 136,000 barrels of petroleum per day. As a transportation fuel, ethanol has a number of advantages over gasoline. In particular, it burns more cleanly and with higher efficiency.

The United States is the world's second largest alcohol producer. In 1980, the government passed the Energy Security Act, which included the Biomass Energy and Alcohol Fuels Act, to provide loans and loan guarantees to alcohol and other biomass energy projects. The objective of this program is to encourage the addition of 10% alcohol to gasoline, creating "gasohol." Gasohol is a lead-free fuel whose combustion produces lower amounts of nitrogen oxides and carbon monoxide than the combustion of regular gasoline. During 1988, about 840 million gallons of ethanol, prepared by fermenting corn and other starch-rich grains, were sold in gasohol blends, primarily in the midwestern United States. These blends accounted for some 7% of total nationwide gasoline sales. The Clean Air Act Amendments, passed in 1990, established the Oxygenated Fuels Program and the Reformulated Gasoline Program to decrease generation of carbon monoxide and ground-level ozone. These programs require oxygen levels of 2.7% by weight for oxygenated fuel and 2% for reformulated gasoline. These requirements are met either by blending gasoline with ethanol or by adding methyl tertiary butyl ether (MBTE). The use of this additive is likely to be eliminated because leakage from MBTE storage tanks has been shown to result in persistent contamination of ground and surface water. A ban on the use of MBTE, a suspected carcinogen, will boost the use of ethanol. Almost all the fuel ethanol in the United States is produced by fermentation from corn using less than 4% of the crop. More than $12.5 \times 10^9$ L were produced in 2003 at a yield of about 0.37 L of ethanol per kilogram of dry corn kernels (at 12% to 15% moisture).

Figure 13.1 presents a flow chart describing the conversion of various feedstocks to alcohol. In the first stage, the polymeric substrates are broken down to monosaccharides through physical, chemical, or enzymatic techniques, as appropriate. In the second stage, microbial (generally yeast) fermentation converts the sugars to alcohol. In the third stage, alcohol is recovered by distillation (as a constant-boiling mixture of 95.6% ethanol and 4.4% water, by volume). Further distillation procedures are needed to obtain anhydrous ethanol.

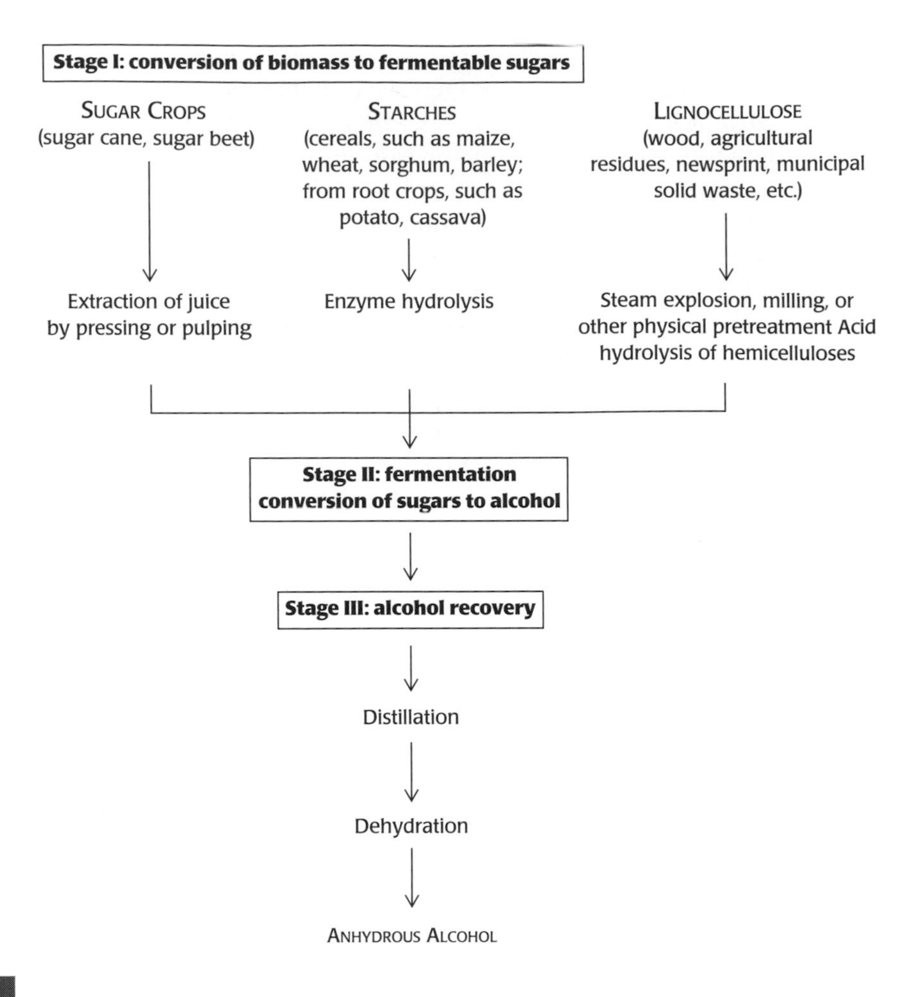


FIGURE 13.1

Stages in the conversion of biomass to ethanol.

Redrawn based on artwork from the first edition (1995), published by W.H. Freeman.

The most important question that must be addressed by any production process for an alternative fuel is the question of energy balance (the energy output–input ratio). Is the energy present in the alternative fuel greater than the nonrenewable energy used in producing the fuel? For ethanol, nonrenewable energy is required to grow corn, harvest and transport it, subject it to dry or wet milling, convert the starch in the corn kernels into ethanol, and recover the ethanol by distillation and dehydration. Widely used 2002 estimates of the energy balance for the corn-to-ethanol conversion are modestly positive, with values ranging from 1.10 to 1.34. The higher value is obtained when credits are assigned to coproducts: stillage (the residue from the fermentation used to produce a high-quality nutritious livestock feed – "dried distillers grains and solubles"), corn oil, corn gluten meal, and corn gluten feed. The $CO_2$ released during the fermentation is captured and sold for carbonating beverages and the manufacture of dry ice.

A number of recent improvements in agricultural practices as well as in the recovery of ethanol are estimated to increase the net energy balance of the corn–ethanol conversion to 1.89. Improved crop yields and corn composition have resulted from a combination of no-till farming, genetically modified corn, slow-release fertilizers, and high fermentable starch varieties. For ethanol production from sugarcane in Brazil, the energy balance is in the range of 3.7 to 7.8. In Brazilian distilleries, all the energy required for operation is supplied by the burning of bagasse, the sugarcane waste left after the juice is extracted.

The cost of the alternative fuel also must be considered. If alcohol is to be competitive with other sources of fuels and chemical feedstocks, all three stages in its production must be simple and inexpensive, irrespective of the starting substrate. In the United States, the government subsidizes ethanol production. Brazil also subsidized ethanol production until 1999. The price of the feedstock is important. Lignocellulose-rich feedstocks are much cheaper than corn or sugarcane. The energy balance for the production of ethanol from lignocellulosic material is strongly positive.

In this chapter, we examine relevant aspects of the biochemistry, microbiology, and technology of the first two stages – the breakdown of polymeric sugars to monosaccharides and their conversion to alcohol. Methodological improvements in those stages have an impact on the third stage, ethanol recovery, as well, but the techniques employed in that stage belong largely to the realm of process engineering.

## STAGE I: FROM FEEDSTOCKS TO FERMENTABLE SUGARS

As indicated in Figure 13.1, in stage I, the carbohydrate-containing raw materials are pretreated in ways that make the sugars they contain readily available to microorganisms. The principal biomass substrates for the production of alcohol by microbial fermentation are sugars, starches, and cellulose. The chemical structures of sugars and starches are described in Chapter 11, and that of cellulose in Chapter 12.

### SUGARS

Sucrose (glucose-α-1,2-fructose; see Figure 11.11) is the most common sweetener used for human consumption. Sugarcane and sugar beets contain up to 20% sucrose by weight, the other major components being water (about 75%), cellulose (5%), and inorganic salts (about 1%). The substrate for fermentation is obtained by extraction of sucrose with water after mechanically crushing the cane or stripping and pulping the beets. Sucrose from sugarcane (the Brazilian approach) is a particularly favorable substrate for the production of alcohol by yeast fermentation. Yeast produces the enzyme invertase in both a cytoplasmic and a secreted form, and this enzyme hydrolyzes sucrose to glucose and fructose, which are then fermented by the yeast cells. The conversion of other substrates to alcohol involves additional pretreatment. *Saccharomyces cerevisiae* and related yeasts can take up and metabolize many other sugars as well, including glucose, fructose, galactose, mannose, maltose (glucose-α-1,4-glucose), and maltotriose.

## STARCHES

Cornstarch, the major feedstock for the production of fuel alcohol In the United States, consists of a water-soluble fraction, amylose (20%), and a water-insoluble higher molecular weight fraction called amylopectin (80%). The structures of these glucose polymers are shown in Figure 11.12. To obtain cornstarch, dry corn is milled, water is added, and the slurry (watery suspension) is sent to a cooker. Heating the slurry solubilizes the starch and makes it vulnerable to enzyme hydrolysis. A thermostable $\alpha$-amylase is added to liquefy the starch, and in the final prelude to fermentation, glucoamylase is added to catalyze saccharification, hydrolysis of the starch polymers to glucose (see page 420).

## CELLULOSE

Cellulose is the most abundant component of lignocellulose (see Chapter 12). Pretreatment of the lignocellulose makes the cellulose more accessible to hydrolytic enzymes and begins to disrupt, at least in part, the highly crystalline structure of the cellulose fibers. Numerous raw material pretreatment processes have been designed to achieve these objectives. Processes developed by the Iotech Corporation Limited of Ottawa, Canada (the Techtrol/Iotech process), and by the U.S. Army Natick Laboratories (the Natick process) are presented as examples.

In the Techtrol/Iotech process, small wood chips are charged with steam in a heated pressure vessel to about 500°F and maintained at that temperature for about 20 seconds, at which point the vessel is rapidly decompressed. Pressure in the vessel reaches 600 psi before release. After the explosive decompression, the cellulose in the wood is susceptible to enzyme hydrolysis.

This "steam explosion" treatment has the further effect of solubilizing the hemicellulose in the wood so it can be washed away with water. Then the cellulose and lignin can be separated from one another in one of two ways. In one method, the lignin is extracted in high yield with methanol or with dilute sodium hydroxide before the cellulose is degraded. In the other method, the cellulose is converted to glucose by hydrolytic enzymes while the lignin remains insoluble and is subsequently removed by filtration. The glucose solution is then fermented to alcohol.

The process developed at the Natick Laboratories consists of extensive physical disruption of the wood followed by enzymatic degradation of cellulose. In this process, the lignocellulose is fragmented by milling and suspended in water. A mixture of *Trichoderma reesei* cellulases (see Chapter 12, pp. 449–451) is isolated from large amounts of culture broth of *T. reesei* grown on cellulosic materials. Addition of the enzymes to the slurry of milled biomass results in a 45% conversion of cellulose to glucose. The volumes are adjusted to obtain a 10% solution of glucose, which is then transferred to a fermentation vessel for conversion to alcohol.

**TABLE 13.1 Some yeasts and bacteria that produce significant quantities of ethanol, and the major carbohydrates utilized as substrates**

| Yeast or Bacteria | Substrates |
|---|---|
| Yeast | |
| *Saccharomyces* spp. | |
| *S. cerevisiae* | Glucose, fructose, galactose, maltose, maltotriose, xylulose |
| *S. carlsbergensis* | Glucose, fructose, galactose, maltose, maltotriose, xylulose |
| *S. rouxii* (osmophilic) | Glucose, fructose, maltose, sucrose |
| *Kluyveromyces* spp. | |
| *K. fragilis* | Glucose, galactose, lactose |
| *K. lactis* | Glucose, galactose, lactose |
| *Candida* spp. | |
| *C. pseudotropicalis* | Glucose, galactose, lactose |
| *C. tropicalis* | Glucose, xylose, xylulose |
| Bacteria | |
| *Zymomonas mobilis* | Glucose, fructose, sucrose |
| *Clostridium* spp. | |
| *C. thermocellum* (thermophilic) | Glucose, cellobiose, cellulose |
| *C. thermohydrosulfuricum* (thermophilic) | Glucose, xylose, sucrose, cellobiose, starch |
| *Thermoanaerobium brockii* (thermophilic) | Glucose, sucrose, maltose, lactose, cellobiose, starch |
| *Thermobacterioides* acetoethylicus (thermophilic) | Glucose, sucrose, cellobiose |

## STAGE II: FROM SUGARS TO ALCOHOL

The second stage utilizes the simple sugars, released from polysaccharides in stage I, as substrates in microbial fermentations to produce alcohol.

### YEASTS

Many yeasts, but few bacteria, carry out near-quantitative conversion of glucose to alcohol. Industrial processes use primarily yeasts in the genus *Saccharomyces*. Although yeasts have many of the attributes of an ideal ethanol producer, they have significant limitations, such as a narrow substrate range and limited tolerance to alcohol. Below, we consider the various facets of yeast fermentations.

#### Substrate Range

The greatest constraint in employing yeasts as agents of fermentation is the limited range of substrates they are able to use (Table 13.1). For instance, yeasts do not ferment most of the oligosaccharides formed during the hydrolysis of starch. These resistant compounds include maltodextrins longer than maltotriose, and isomaltose (an $\alpha$-1,6-linked dimer of glucose). Yeasts thus require the addition of glucoamylases (see Figure 11.12 and Table 11.5) to utilize starch completely. Nor can yeast cells utilize cellulose, hemicellulose, cellobiose, or most pentoses. Their inability to ferment a diversity of cheaper and readily available substrates is the major obstacle confronting attempts to lower the cost of alcohol production. The search for ways to use more substrates is a major focus of research into improving ethanol fermentations.

*S. cerevisiae* ferments a number of common substrates, including the disaccharides sucrose and maltose. The substrates are handled by one of two mechanisms. The disaccharides are either hydrolyzed by extracellular enzymes and the monosaccharides transported into the cell or the disaccharides are also transported into the cell and then hydrolyzed by intracellular enzymes. The uptake and metabolism of various substrates in a mixture occur in an order determined by regulatory mechanisms at the level of gene expression. For example, glucose is the preferred substrate. If it is present, the permeases for the other substrates, such as maltose and maltotriose, are not induced until the glucose disappears. As a result, these substances are fermented sequentially rather than simultaneously. Di- and trisaccharides, once internalized, are hydrolyzed by an $\alpha$-glucosidase. Fermentations must be run long enough to allow induction of the enzyme systems and full use of the various substrates. If so, they will generally go to completion with little fermentable substrate left at the end.

*Saccharomyces* strains are responsible for almost all the current industrial production of alcohol by fermentation. *Saccharomyces* converts glucose by the glycolytic pathway to high yields of ethanol and carbon dioxide (Figure 13.2). Only two ATPs are produced per mole of glucose metabolized, and the yeast cells use them for growth. Ethanol is recovered in 90% to 95% of the theoretical yield.

## Substrate Utilization

An understanding of the arithmetic of ethanol production begins with the equation established by Gay-Lussac in 1810 for the fermentative conversion of glucose to ethanol by yeast:

$$\underset{180\text{ g}}{C_6H_{12}O_6} \rightarrow \underset{92\text{ g}}{2C_2H_5OH} + \underset{88\text{ g}}{2CO_2}.$$

From this, the theoretical yield of alcohol from glucose is calculated to be 51.1% by weight. However, the production of alcohol by yeast is actually a by-product of the yeast's growth, and some of the substrate is utilized to produce more cells: rapidly growing, fermenting microorganisms produce about 10 g dry weight of cells for each mole of ATP they synthesize. As shown in Figure 13.2, each mole of glucose fermented to ethanol produces two moles of ATP, so the theoretical yield of cells is 20 g dry weight. The carbon content of these microbial cells is close to 50%. Because glucose is the sole carbon source in the fermentation medium, and the carbon content of glucose is 40%, 25 g of glucose is needed to provide 10 g of cell carbon. Therefore, allowing for cell growth, the maximum yield of alcohol is expected to be about 86%. How

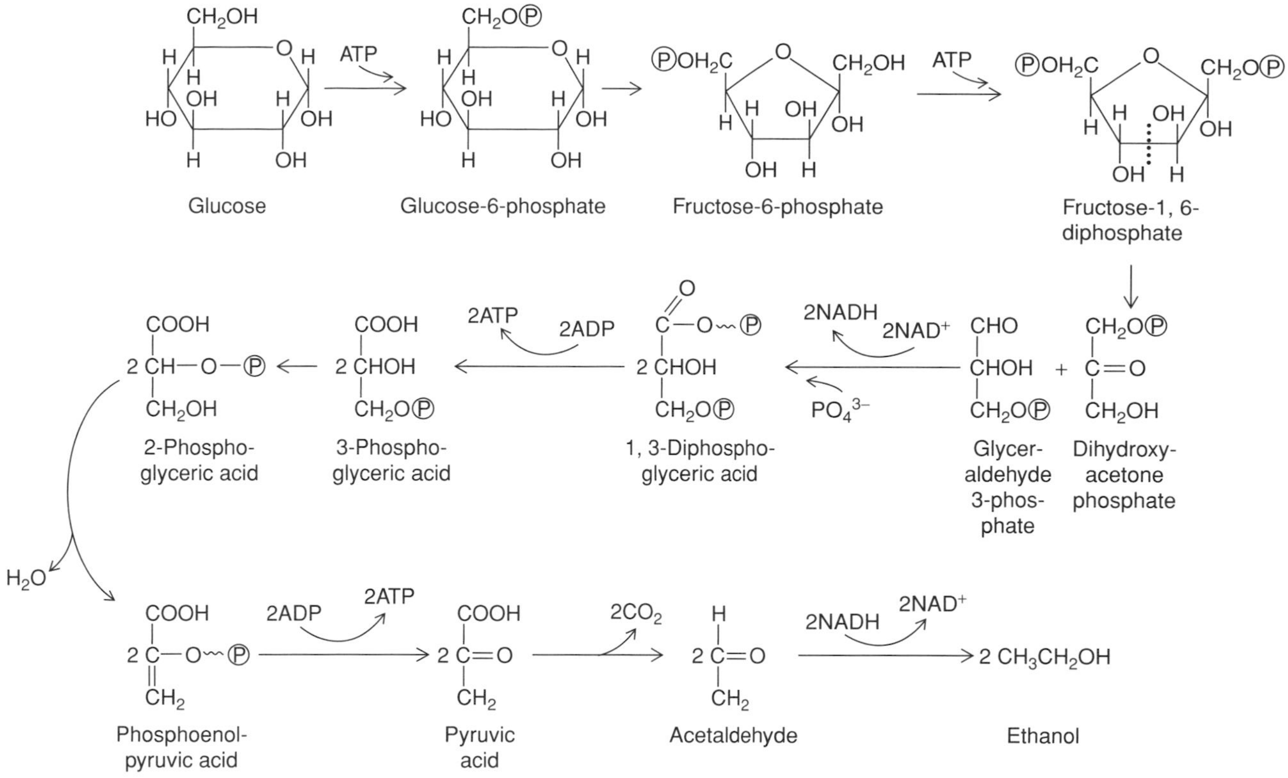

**FIGURE 13.2**

Formation of ethanol and carbon dioxide from glucose by the glycolytic (Embden–Myerhoff) pathway.

then do yeast fermentations result in higher yields of ethanol, 92% to 95% of theoretical?

The reason for the extremely high yield of ethanol is that not all of the ATP is used to produce new cells. Some energy goes for other cellular functions (called "maintenance," for want of a better term) regardless of the rate of cell growth. This proportion is smaller during rapid growth, but when growth slows, as it does during inhibition by ethanol or nutrient limitation, the maintenance-energy requirement does not decrease. In fact, because the presence of ethanol may cause ion leakage across cell membranes, some of the cells' maintenance energy requirements may actually increase at high ethanol concentrations. Thus, as ethanol builds up during batch fermentations, the cells use an increasing share of the ATP for maintenance at the expense of reproduction, and the fraction of glucose carbon in the fermenter represented by cells decreases from 10% to 5% or less of the starting glucose, with a corresponding increase in the yield of ethanol.

Substrate represents the largest component of the cost of ethanol production. Small improvements in the efficiency of conversion can have a significant impact on costs. Increasing yield from 90% to 92% can lower the product cost by 1% or more.

As indicated by the Gay-Lussac equation, carbon dioxide and alcohol are produced from glucose in equimolar amounts. Additional reactions also take place in the fermenter, leading to small amounts of minor by-products such as glycerol, fusel oils, acetic acid, lactic acid, succinic acid, acetaldehyde, furfural, and 2,3-butanediol.

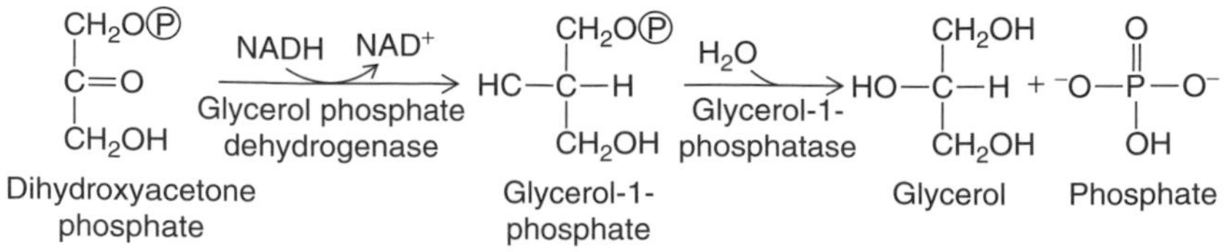

**FIGURE 13.3**

Conversion of dihydroxyacetone phosphate to glycerol.

$$\underset{\text{Dihydroxyacetone phosphate}}{CH_2O(P)-C(=O)-CH_2OH} \xrightarrow[\text{Glycerol phosphate dehydrogenase}]{NADH \rightarrow NAD^+} \underset{\text{Glycerol-1-phosphate}}{CH_2O(P)-HC(H)-CH_2OH} \xrightarrow[\text{Glycerol-1-phosphatase}]{H_2O} \underset{\text{Glycerol}}{CH_2OH-HO-C(H)-CH_2OH} + \underset{\text{Phosphate}}{{}^-O-P(=O)(OH)-O^-}$$

Of these, glycerol accumulates in the largest amount. Industrial fermentations produce up to 5 g of glycerol for every 100 g of ethanol. Glycerol is formed by the reduction of the glycolytic intermediate, dihydroxyacetone phosphate (Figure 13.2) to glycerol-1-phosphate, which is dephosphorylated to glycerol (Figure 13.3). *Saccharomyces* synthesizes glycerol as an osmoregulatory metabolite in response to the high osmotic pressure of the sugar solution in the fermenter. Osmoregulatory metabolites are organic compounds produced and accumulated by many living organisms to regulate their internal osmotic pressure in response to changes in extracellular water activity.

The choice of glycerol as an osmoregulatory metabolite in *Saccharomyces* may in part be influenced by the fact that the pathway leading to glycerol, like the Embden–Meyerhof pathway, directly regenerates NAD. When *Saccharomyces* is growing in media high in sugar content, the highest proportion of glycerol relative to ethanol is produced early in the fermentation, when the osmotic pressure and fermentation rate are at their highest. At high glycolysis rates, the terminal reductive steps in glycolysis (see Figure 13.2) appear to be rate limiting, and the high concentrations of NADH that accumulate under these conditions favor the formation of glycerol. Even though glycerol is produced in significant amounts, and is a valuable chemical, its recovery from the residues left at the end of the process in pure form is not economically feasible. In simultaneous saccharification and fermentation processes (discussed below), in which the steady-state concentration of sugars is relatively low, or in other processes that result in lower cell-growth rates, there should be lower amounts of glycerol formed at the expense of ethanol.

Fusel oils are a mixture of higher alcohols, primarily amyl and butyl alcohols. These compounds are produced from the degradation of amino acids, which in turn come from proteolysis of the proteins in the feedstock. When feedstocks are low in protein – sugarcane juice is an example – less fusel oil is produced. In contrast, fusel oils may represent as much as 0.5% of the crude distillate from starch feedstocks from grains.

Yeasts are little affected by pH in the range of 4 to 6. However, if the pH in the fermenter is allowed to rise above 5, the conditions favor the growth of *Lactobacillus*. These bacteria ferment glucose to produce lactate and acetate as well as ethanol. To avoid contamination, the pH in the fermenter is maintained below 5 by the addition of small amounts of acid.

### Ethanol Tolerance

Because the separation of ethanol from water (during stage III, alcohol recovery) accounts for much of the energy used in the overall production process,

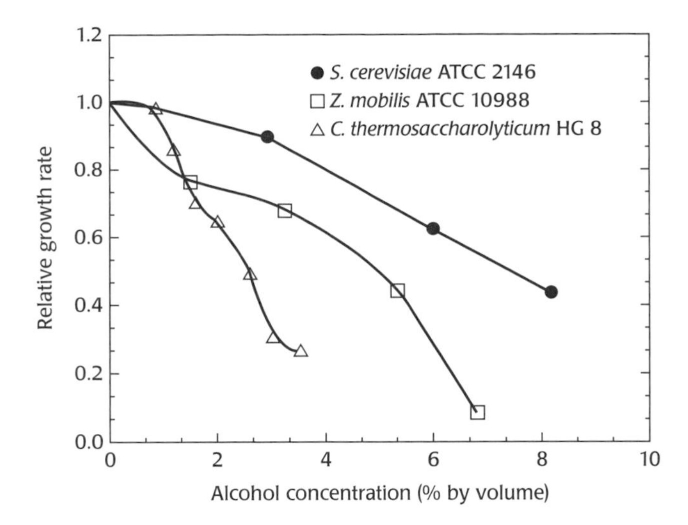


**FIGURE 13.4**

Comparison of the ethanol tolerance of the yeast *Saccharomyces cerevisiae* to that of the bacteria *Zymomonas mobilis* and *Clostridium thermosaccharolyticum*. Growth data are for steady-state continuous cultures to which ethanol has been added at the concentration indicated. [Data from Hogsett, D. A., Ahn, H-J., Bernandez, T. D., South, C. R., and Lynd, L. R. (1992). Direct microbial conversion. Prospects, progress, and obstacles. *Applied Biochemistry and Biotechnology*, 34/35, 527–541.]

the higher the concentration of ethanol in the fermenter, the lower the distillation cost per liter of product. However, ethanol is toxic to yeast cells at concentrations ranging between 8% and 18% by weight, depending on the strain of the yeast and the metabolic state of the culture (Figure 13.4). Yeast fermentation is totally inhibited by ethanol concentrations of about 11% by volume. To understand why, we must consider the properties of cytoplasmic membranes.

### The Structure and Function of Cytoplasmic Membranes

The cytoplasmic membranes of yeast and bacteria consist of a lipid bilayer with embedded protein complexes. Some of these transmembrane proteins serve as channels through which the interior of the cell interacts with the external milieu. Others are electron transport complexes responsible for setting up proton and ion gradients between the interior and exterior of the cytoplasmic membrane. An example is the ubiquitous $F_0F_1$ ATPase, the enzyme complex that utilizes a proton gradient across the cytoplasmic membrane for the synthesis or hydrolysis of ATP.

The transport of nutrients into the cell is dependent on a class of transmembrane proteins or protein complexes called the permeases, which mediate the passage of hydrophilic molecules through the cell membrane. Glucose permease, for example, facilitates the transport of glucose into the cell down a concentration gradient by a “facilitated diffusion” process. It depends on the enzyme hexokinase rapidly phosphorylating any free glucose inside the cells and thus keeping intracellular glucose levels very low. Other permeases catalyze the active transport of nutrients *against* internal nutrient concentrations that are orders of magnitude higher than those in the extracellular medium. Many of these active transport proteins use the pH and ion concentration gradient (“proton motive force”) across the cytoplasmic membrane to provide the energy for active transport. Other permeases use ATP as the energy source. The transport of maltose, amino acids, and ammonium ion into yeast cells depends on active transport by membrane-embedded protein transporters.

The cytoplasmic membrane provides a barrier to the diffusion of protons, other ions, and small polar molecules from the cell to the outside and vice versa. Without such a highly efficient boundary, the internal homeostasis essential to the survival of living cells could not be maintained.

### The Effect of Ethanol on Membrane Structure and Function

Although the cytoplasmic membrane's lipid bilayer is an efficient barrier to hydrophilic molecules, it does allow small amphiphilic molecules such as ethanol to pass through freely without the need for a specific permease. Consequently, the concentration of alcohol within the cell is the same as in the surrounding medium. As the alcohol concentration increases, it disrupts the structure of water, and as a result, the entropic contribution to the stabilization of lipid bilayer membranes is lower in alcohol–water mixtures than in water alone. Moreover, as alcohol partitions into the interior of the membrane, it disturbs lipid–lipid and lipid–protein interactions. Therefore, as alcohol levels increase, the membrane becomes progressively more and more leaky. The ion gradients that give rise to the proton motive force across the membrane gradually collapse, and small molecules leak out of the cell. In some yeast strains, a 50% decrease in the rate of uptake of sugars, ions, and amino acids is seen in the presence of ethanol concentrations as low as 4% by volume.

### The Influence of Membrane Composition on Resistance to Ethanol

Some yeast strains show greater resistance than others to increased concentrations of ethanol. In some cases, the membrane lipid bilayer structure is stabilized by the presence of longer hydrocarbon chains in the lipid molecules. This increases the interaction between neighboring chains. In other cases, leakage is minimized because of a high sterol content in the membrane, since sterols decrease the nonspecific permeability of phospholipid bilayers.

Most yeast cells grown in the presence of ethanol show a small but significant increase in the average chain length of their membrane fatty acids. However, membranes rich in lipids with longer-chain *saturated* fatty acids tend to "freeze" and become rigid at the growth temperature of yeasts. Thus, the longer-chain acids are produced as *unsaturated* fatty acids. For example, the membrane lipids of yeast grown in the presence of 7.5% ethanol contain 34% oleic acid ($\Delta^9$Z-$C_{18:1}$; Figure 13.5), whereas the lipids of cells grown without added ethanol contain only 17%. Much of this increase occurs at the expense of the shorter-chain, saturated fatty acid palmitate ($C_{16:0}$). Oxygen is required for ethanol tolerance. Because yeasts, like higher eukaryotes, make unsaturated fatty acids by using $O_2$ and NADH, yeast cells cannot be grown completely anaerobically in the presence of ethanol. Oxygen is also required for the production of ergosterol, a cytoplasmic membrane sterol that also

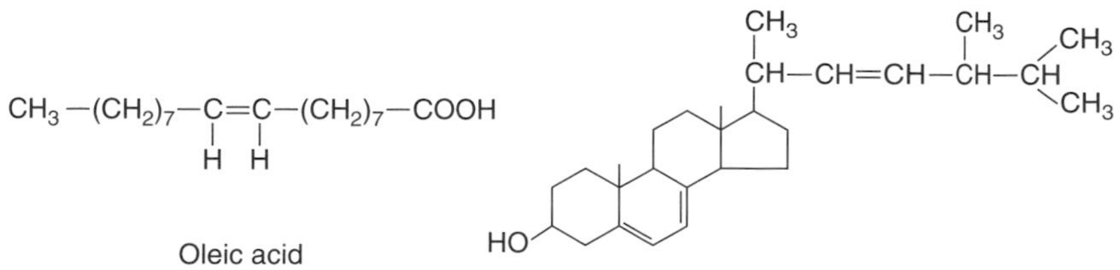

$CH_3-(CH_2)_7-CH=CH-(CH_2)_7-COOH$

Oleic acid
(*cis*-9-octadecenoic acid)

Ergosterol

FIGURE 13.5
Oleic acid and ergosterol.

contributes to membrane stability. Consequently, growth under anaerobic conditions in the presence of ethanol requires addition to the medium of unsaturated, longer-chain fatty acids, as well as of ergosterol (see below).

Microaerophilic growth in the presence of ethanol also leads to an increase in the membrane content of ergosterol (Figure 13.5). Ethanol induces the production of cytochrome P-450, a component of a monooxygenase system responsible for the demethylation of lanosterol to ergosterol in yeasts (and also for the conversion of saturated to unsaturated fatty acids).

It is likely therefore that the differences in the level of tolerance to ethanol among different strains of microorganisms can be attributed in the main to the makeup of their cytoplasmic membranes and to their ability to vary their membrane composition in response to increasing ethanol concentration.

### Temperature

The conversion of glucose to ethanol and $CO_2$ is an exothermic reaction: the complete fermentation of an 18%-by-weight glucose solution would raise the temperature of the medium by more than 20°C. Every 5°C increase in temperature increases the evaporative loss of ethanol by 1.5. Yeast metabolism rates also increase with temperature up to an optimum at 35°C; they then decrease gradually between 35°C and 43°C, and drop abruptly above 43°C. These considerations impose a cooling requirement so that the fermenter operating temperature is maintained below 35°C.

### Flocculence and Cell Recycling

The objectives of a fermentation process are to convert substrates to alcohol as rapidly as possible, to minimize the cost of alcohol recovery, and to decrease the amount of yeast cells produced as a by-product of the process. Exploitation of cell recycling and of the tendency of yeast cells to *flocculate* (clump) contributes to the attainment of these objectives.

Cells can be collected at the end of one batch fermentation to be used as the inoculum for the next. This procedure is termed *cell recycle*. By using large amounts of yeast cells recovered in this manner, cell concentrations in the fermenter can be raised from a few grams dry weight per liter to tens of grams per liter. Because of this increase in the cell concentration, cell recycle may increase the amount of alcohol produced per unit volume even when inhibition by ethanol decreases the specific productivity of individual cells. A constraint is the need to keep the cells viable during the collection process by

**TABLE 13.2 Major components of stillage remaining after alcohol distillation from fermented sugar cane juice**

| Component | Amount remaining (g/L) |
|---|---|
| Organic matter | 40–65 |
| Nitrogen | 0.7–1.0 |
| Phosphorus | 0.1–0.2 |
| Potassium | 4.5–8.0 |

providing adequate nutrition. Because collecting cells by centrifugation or filtration costs more in equipment and attention than any savings that result from such recycling, the key has been finding inexpensive ways of separating yeasts from the fermentation broth. Flocculation provides a partial solution.

Flocculation is a property in yeasts dependent on the gene *flo1*. Cells with this property stick together because they possess a *flo1*-encoded wall protein that binds in a calcium ion–dependent manner to the wall mannans of other cells. As a result, the cells form clumps that sediment rapidly and are easily removed from the fermentation mixture for recycling. Nonflocculating strains, which do not form clumps, are called "powdery." Although synthesis of the Flo1 protein is normally repressed by anaerobic growth, mutants expressing the protein during fermentation are readily found.

Continuous rapid fermentation mixtures require a great deal of agitation to ensure uniform suspension of cells. Moreover, a high rate of $CO_2$ production in the fermentation mixture in itself creates a great deal of agitation. Consequently, even flocculent strains are not necessarily easy to separate without additional equipment, such as centrifuges or cross-membrane filters. Any marginal improvements in productivity resulting from cell recycle and the use of flocculent strains are usually counterbalanced by the extra expense of carrying out the cell separation. This approach does increase the amount of alcohol produced relative to the amount of yeast biomass generated.

### Stillage

Stillage is the residue from the first distillation of fermented substrate (corn mash, sugarcane juice, etc.; see Figure 13.1). With sugarcane, about 12 L of stillage are produced for each liter of ethanol. Such stillage contains 40 to 65 g of organic matter per liter (Table 13.2). Depending on what is done with it, stillage is either a serious water-polluting waste or a source of valuable by-products. Questions of stillage disposal will be taken up later in this chapter. Processes that yield lower ratios of stillage-to-ethanol volumes have lower costs.

## *ZYMOMONAS MOBILIS* – AN ALTERNATIVE ETHANOL PRODUCER

The specifications for an ideal fermentation alcohol producer would include the following important characteristics:

- ability to ferment a broad range of carbohydrate substrates rapidly
- ethanol tolerance and the ability to produce high concentrations of ethanol
- low levels of by-products, such as acids and glycerol
- osmotolerance (ability to withstand the high osmotic pressures encountered at high concentrations of sugar substrates)

- temperature tolerance
- high cell viability for repeated recycling
- appropriate flocculation and sedimentation characteristics to facilitate cell recycling

For all of their shortcomings, described above, *Saccharomyces* strains come closer to meeting these specifications than any other organisms known to produce ethanol. Only two other ethanol producers have attracted serious attention – *Zymomonas mobilis* and certain thermophilic clostridia.

Bacteria of the genus *Zymomonas* attracted the notice of microbiologists in 1912 as contributors to "cider sickness" – spoilage of fermented apple juice. Subsequently, *Zymomonas* was isolated from other fermenting sugar-rich plant juices: agave sap in Mexico, palm saps in various parts of Africa and Asia, and sugarcane juice in Brazil. These fermentations produce alcoholic beverages such as pulque (from agave sap), palm wines, and so on. Zymomonads are anaerobic, Gram-negative rods, 2 to 6 $\mu$m in length and 1 to 1.5 $\mu$m in width, flagellated, but lacking spores or capsules.

*Z. mobilis* takes up glucose and produces ethanol some three to four times more rapidly than yeast, with ethanol yields of up to 97% of the theoretical maximum value. Moreover, unlike yeast, *Zymomonas* requires no oxygen for growth. The organism grows in minimal medium with no organic compound requirements. Many *Z. mobilis* strains grow at 38°C to 40°C.

Zymomonads have a high osmotolerance, and most strains grow in solutions containing 40% by weight glucose, but their salt tolerance is low. No strains are able to grow in 2% NaCl, whereas many yeasts can tolerate even higher salt concentrations. *Z. mobilis* strains are also alcohol tolerant, with fermentation yields of up to 13% alcohol by volume at 30°C. Few bacteria are able to survive such high levels of alcohol.

In spite of these favorable characteristics, *Z. mobilis* has not displaced yeast as a large-scale alcohol producer. Some of the reasons emerge from a detailed consideration of carbohydrate metabolism in *Zymomonas*.

### Carbohydrate Utilization in *Zymomonas*

*Zymomonas* can utilize only three carbohydrates: glucose, fructose, and sucrose. The metabolism of each of these sugars has distinctive features. Therefore, after discussing the overall picture of glucose fermentation in *Zymomonas*, we will consider the distinctive reactions that come into play when fructose or sucrose is the substrate. The pathways of carbohydrate metabolism in *Zymomonas* are charted in Figure 13.6.

### Import of Glucose and Subsequent Fermentation by the Entner–Doudoroff Pathway

Glucose enters the *Zymomonas* cell by means of a stereospecific, low-affinity, high velocity facilitated diffusion transport system. A constitutive

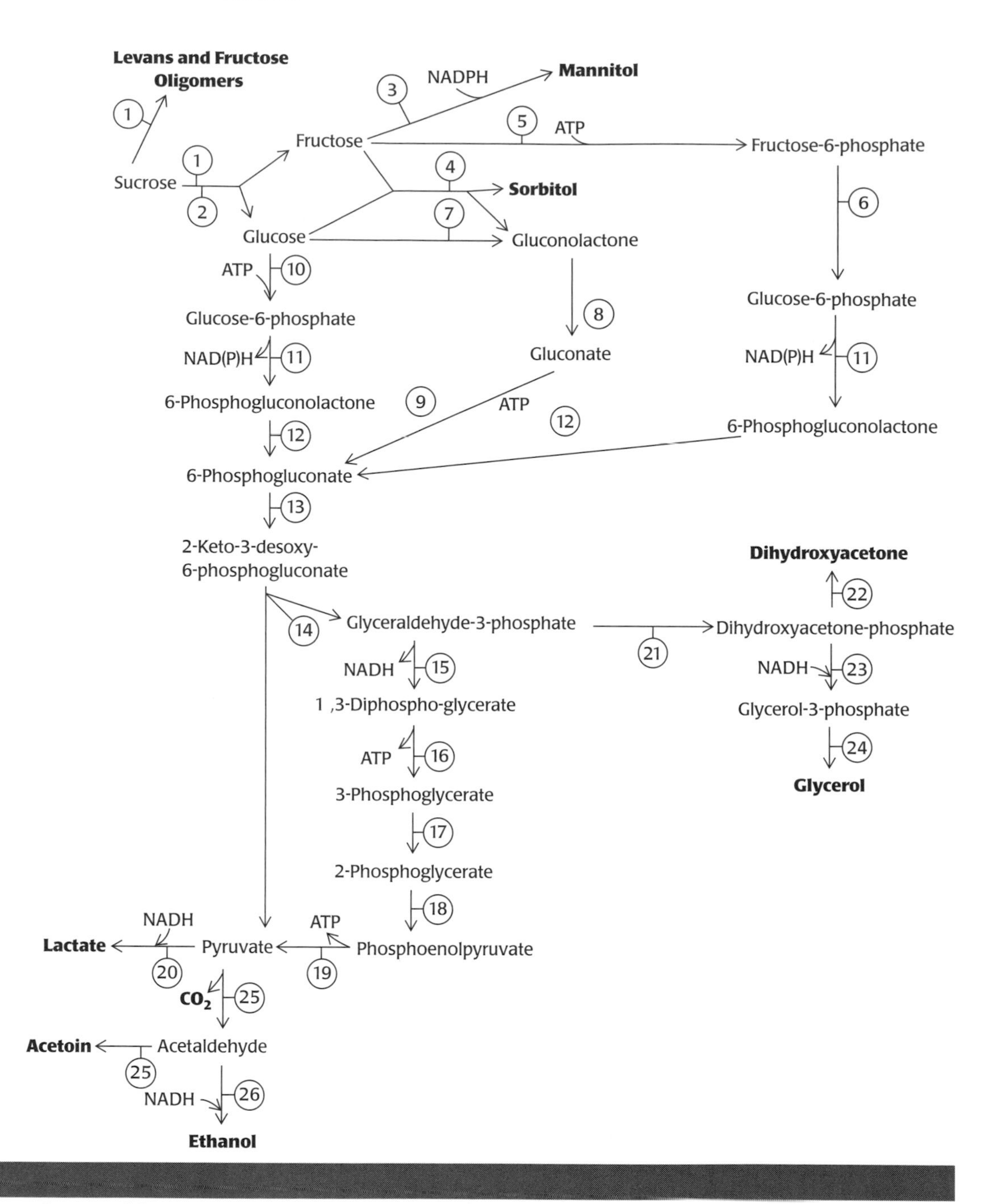


FIGURE 13.6

Metabolism of sucrose, glucose and fructose in *Zymomonas mobilis*. Enzymes (indicated by *circled numbers*) are as follows: 1, levansucrase; 2, invertase; 3, mannitol dehydrogenase; 4, glucose-fructose oxidoreductase; 5, fructokinase; 6, glucose-6-phosphate isomerase; 7, glucose dehydrogenase; 8, gluconolactonase; 9, gluconate kinase; 10, glucokinase; 11, glucose-6-phosphate dehydrogenase; 12, 6-phosphoglucono-lactonase; 13, 6-phosphogluconate dehydratase; 14, keto-deoxy-phosphogluconate aldolase; 15, glyceraldehyde 3-phosphate dehydrogenase; 16, phosphoglycerate kinase; 17, phosphoglycerate mutase; 18, enolase; 19, pyruvate kinase; 20, lactate dehydrogenase; 21, triose-phosphate isomerase; 22, phosphatase; 23, glycerol 3-phosphate dehydrogenase; 24, phosphatase; 25, pyruvate decarboxylase; 26, alcohol dehydrogenase. [Bringer-Meyer, S., and Sahm, H. (1988). Metabolic shifts in *Zymomonas mobilis* in response to growth conditions. *FEMS Microbiological Reviews*, 54, 131–142.]

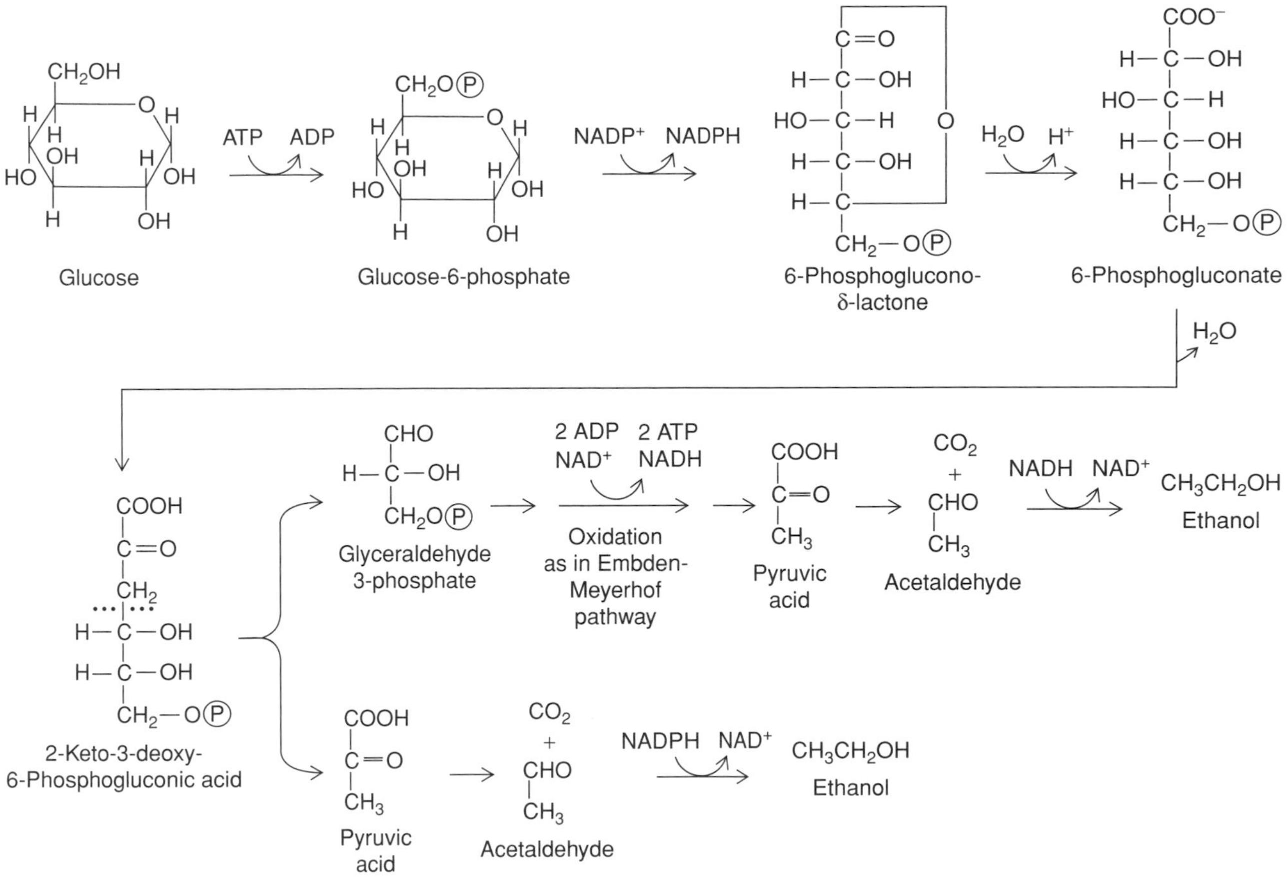

glucokinase then converts the glucose to glucose-6-phosphate, after which *Zymomonas* departs from the glycolytic pathway characteristic of other glucose fermenters, such as yeasts, and instead utilizes the Entner–Doudoroff pathway, shown in Figures 13.6 and 13.7.

In this pathway, glucose-6-phosphate dehydrogenase catalyzes the conversion of glucose-6-phosphate to 6-phosphoglucono-$\gamma$-lactone, with concomitant reduction of NAD to NADH. A lactonase with a very high catalytic activity rapidly hydrolyzes the lactone and 6-phosphogluconate dehydratase then converts the resulting 6-phosphogluconate to 2-keto-3-deoxy-6-phospho-gluconate, the unique intermediate of the Entner–Doudoroff pathway. This compound is cleaved by a specific aldolase to pyruvate and glyceraldehyde-3-phosphate, the latter being converted to a second molecule of pyruvate by a series of reactions that are part of the common glycolytic pathway (Figure 13.2). The pyruvate is then converted to acetaldehyde and carbon dioxide by an unusual pyruvate decarboxylase that, unlike the enzyme in yeast, does not require the cofactor thiamine pyrophosphate for its catalytic activity. Finally, two alcohol dehydrogenases (ADH I and ADH II) reduce the acetaldehyde to ethanol, with concomitant stoichiometric oxidation of NADH. ADH I is a tetrameric enzyme with zinc at the active site,

**FIGURE 13.7**

Formation of ethanol from glucose by *Zymomonas mobilis* using a fermentative version of the Entner–Doudoroff pathway.

Redrawn based on artwork from the first edition (1995), published by W.H. Freeman.

like most of the commonly encountered alcohol dehydrogenases. ADH II is unusual in that it contains iron in its active site. At low alcohol concentrations, the $V_{max}$ of ADH I is about twice as high for acetaldehyde reduction as for alcohol oxidation, whereas ADH II shows a much higher rate of alcohol oxidation. In the absence of ethanol, both contribute equally to the enzymatic catalysis of the reduction of acetaldehyde. At high ethanol concentrations, ADH I is strongly inhibited, but ethanol production continues because the rate of acetaldehyde reduction by ADH II is significantly increased under these conditions.

The Entner–Doudoroff pathway produces two moles of NADH for each mole of glucose consumed: one mole in the reaction catalyzed by glucose-6-phosphate dehydrogenase and one in the reaction catalyzed by glyceraldehyde-3-phosphate dehydrogenase. The two moles of pyruvate produced from each mole of glucose are converted to two moles of acetaldehyde. Reduction of the acetaldehyde to ethanol, catalyzed by the alcohol dehydrogenases, stoichiometrically regenerates NAD.

*A key difference between the glycolytic and Entner–Doudoroff pathways is that the glycolytic pathway results in the net production of two moles of ATP per mole of glucose fermented, and the Entner–Doudoroff pathway produces only one.* In spite of its dependence on such an inefficient pathway for ATP generation, however, *Zymomonas* competes successfully with other microorganisms in the natural environment. It manages so well because of the very high levels of glycolytic and ethanologenic (pyruvate decarboxylase, alcohol dehydrogenase) enzymes it produces, which ensure a rapid flux of substrate through the pathway. Together, these enzymes represent about half the mass of the cytoplasmic proteins in *Z. mobilis*.

## Fructose Metabolism

*Zymomonas* takes up fructose, like glucose, by a constitutive facilitated diffusion transport system. The fructose is then phosphorylated to fructose-6-phosphate by a constitutive kinase, highly specific for fructose and ATP and strongly inhibited by glucose ($K_i$ 0.14 mM; where $K_i$ is the concentration of glucose at which it occupies half of the substrate-binding sites on the enzyme). Glucose phosphate isomerase converts fructose-6-phosphate to glucose-6-phosphate, at which point this pathway merges with that for the metabolism of glucose (Figure 13.7).

*When glucose is the carbon source, the loss of carbon to the formation of by-products is insignificant. When fructose is the carbon source, the results are very different.* Experiments using the same *Zymomonas* strain under similar conditions of batch fermentation show that the yield of ethanol from glucose is 95% of theoretical and from fructose only 90% of theoretical. The cell yield is also lower, showing that the decrease in the yield of ethanol does not reflect greater utilization of fructose for cell growth. A number of side reactions that occur in the presence of fructose are responsible for the difference.

Table 13.3 lists the products of a batch fermentation of fructose, at an original concentration of 15%. The major by-products are dihydroxyacetone, mannitol, and glycerol. *Zymomonas* contains an NADPH-dependent mannitol dehydrogenase with a high $K_m$ (0.17 mM) for fructose. At high fructose concentrations, this enzyme catalyzes the formation of mannitol from fructose at the expense of the NADPH generated when glucose-6-phosphate dehydrogenase (which utilizes NADP in preference to NAD) oxidizes glucose-6-phosphate to 6-phosphogluconolactone. The resulting depletion of NAD(P)H (particularly acute in the early stages of the fermentation) leads to a buildup of acetaldehyde and, indirectly, to accumulation of intermediates at the triose phosphate level. With the buildup of these intermediates, other competing reactions come significantly into play. The equilibrium between glyceraldehyde-3-phosphate and dihydroxyacetone phosphate favors the latter compound, which in its turn is converted to dihydroxyacetone by dephosphorylation or to glycerol phosphate by reduction. Glycerol phosphate is dephosphorylated to glycerol. Because fermentation through the Entner–Doudoroff pathway generates only one ATP per hexose, these side reactions, leading to the formation of dihydroxyacetone or glycerol, each waste the total energy produced in the reaction cycle. Moreover, both dihydroxyacetone and acetaldehyde inhibit cell growth. Such energy losses and inhibitory effects probably account for the lowered cell yield in fructose batch fermentations.

**TABLE 13.3 Products of the fermentation of fructose by *Zymomonas mobilis* strain VTT-E-78082**

| Product | Percent yield (by weight)[a] |
|---|---|
| Ethanol | 45.0 |
| Cells | 0.9 |
| Dihydroxyacetone | 4.0 |
| Mannitol | 2.5 |
| Glycerol | 1.7 |
| Acetic acid | 0.4 |
| Sorbitol | 0.3 |
| Acetoin | 0.3 |
| Acetaldehyde | 0.2 |
| Lactic acid | 0.1 |

[a] Starting fructose concentration was 148 g/L. Final alcohol concentration was 66.7 g/L.

Data from Viikari, L. (1988). Carbohydrate metabolism in *Zymomonas. CRC Critical Reviews of Biotechnology*, 7, 237–261.

## Sucrose Metabolism

*Zymomonas* produces an enzyme, levansucrase ($\beta$-2,6-fructan: D-glucose-1-fructosyltransferase), that hydrolyzes sucrose to glucose and fructose. This enzyme has been detected both in the culture medium and within the cells. In addition to its hydrolytic activity, levansucrase possesses a transfructosylating activity that leads to the formation of high molecular weight sugar polymers (up to $10^7$ daltons) called *levans* when *Zymomonas* is grown on sucrose (Figure 13.6). Substantial amounts of low molecular weight fructooligosaccharides are also formed (Figure 13.8). The main fructofuranosyl linkages in these compounds are $2 \rightarrow 6$ and $2 \rightarrow 1$. The formation of levans and fructose oligomers competes with the fermentation of fructose to ethanol. Fortunately, levan formation is greatly diminished at higher temperatures (37°C), and thus the competition can be minimized.

## Sorbitol Formation

Sorbitol is another by-product that lowers the yield of ethanol when *Zymomonas* ferments sucrose as a carbon source instead of glucose. It is the product of an abundant cytosolic enzyme, *glucose-fructose oxidoreductase*, which converts fructose to sorbitol using glucose as the reductant (Figure 13.6). The enzyme contains tightly bound NADP and does not require

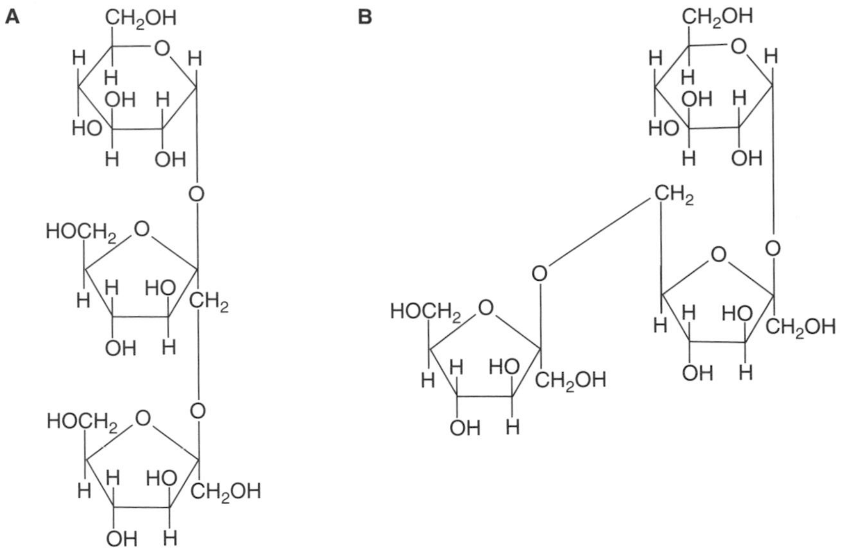

FIGURE 13.8

Structures of oligosaccharides formed by *Z. mobilis* during growth on sucrose. (**A**) $1^F$-$\beta$-Fructosylsucrose-[*O*-$\alpha$-D-glucopyranosyl-(1→2)-*O*-$\beta$-D-fructofuranosyl-(1→2)-$\beta$-D-fructofuranoside]; (**B**) $6^F$-$\beta$-fructosylsucrose-[*O*-$\alpha$-D-glucopyranosyl-(1→2)-*O*-$\beta$-D-fructofuranosyl-(6→2)-$\beta$-D-fructofuranoside].

added cofactors. The second product of the reaction catalyzed by glucose-fructose oxidoreductase, gluconolactone (Figure 13.6), is rapidly hydrolyzed by a gluconolactonase to gluconate and then converted by gluconate kinase into 6-phosphogluconate, an intermediate of the Entner–Doudoroff pathway. When a mixture of glucose and fructose is added to the fermentation medium, or when they appear in the medium as a result of sucrose hydrolysis, *Z. mobilis* converts as much as 11% of the initial carbon sources to sorbitol, which cannot be used as a carbon source and, hence, merely accumulates in the medium.

## Tolerance to Ethanol

As the ethanol concentration increases in the growth medium, most microorganisms begin to experience some impairment of membrane integrity. In *Z. mobilis*, however, unusual features of the cell membrane composition enable the organism to tolerate high levels of ethanol (up to 16% vol/vol) in the medium.

Although the *Z. mobilis* cell membrane contains the usual assortment of phospholipids, with phosphatidylethanolamine as the most abundant, these phospholipids are exceptionally rich (up to 70%) in the monounsaturated fatty acid *cis*-vaccenic acid (18:1; see Figure 13.9). Moreover, the average hydrocarbon chain length in the membrane is greater by about one $-CH_2-$ group than in most other Gram-negative bacteria.

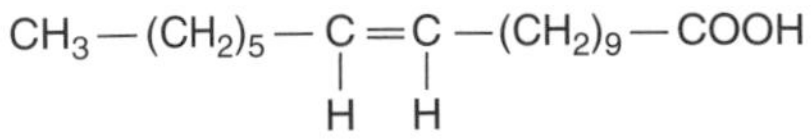

$$CH_3-(CH_2)_5-\underset{H}{\underset{|}{C}}=\underset{H}{\underset{|}{C}}-(CH_2)_9-COOH$$

FIGURE 13.9

Vaccenic acid (*cis*-11-octadecenoic acid).

*Z. mobilis* also profits from the presence of compounds known as hopanoids (Figure 13.10; Box 13.1) in the cell membrane. These pentacyclic triterpenoids, found in various prokaryotes, are functional analogs of sterols (Figure 13.11), but the biosynthesis of sterols involves oxygen-dependent

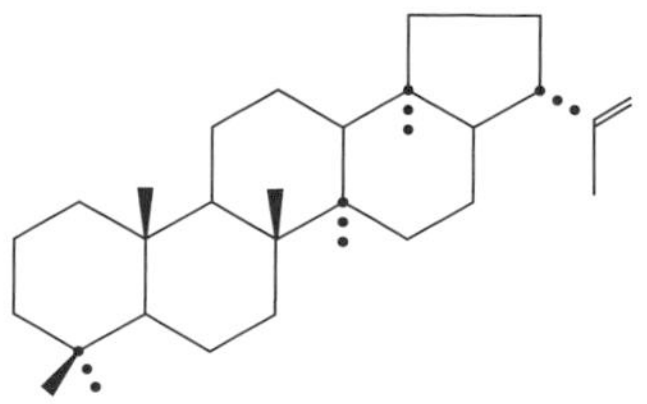

Hopane

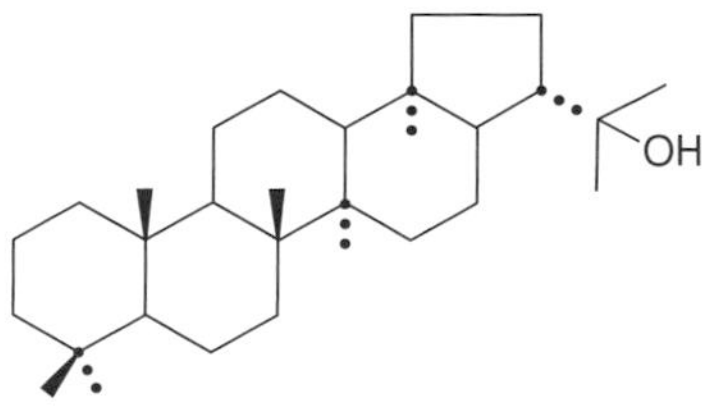

Diplopterol

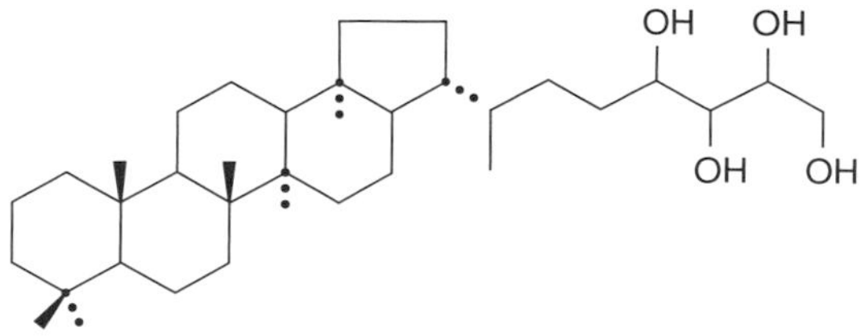

Bacteriohopanetetrol

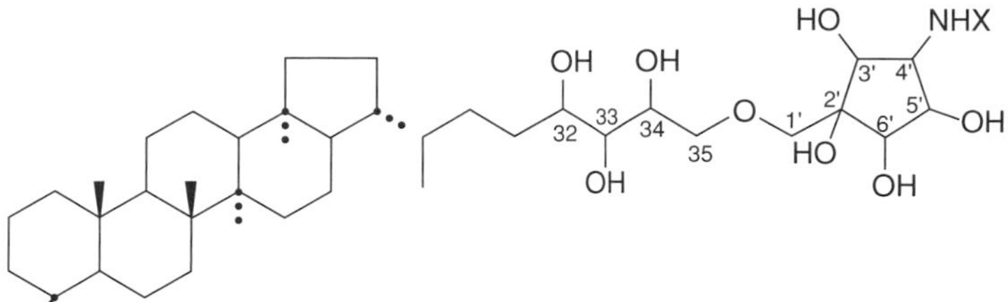

Bacteriohopanetetrol ether

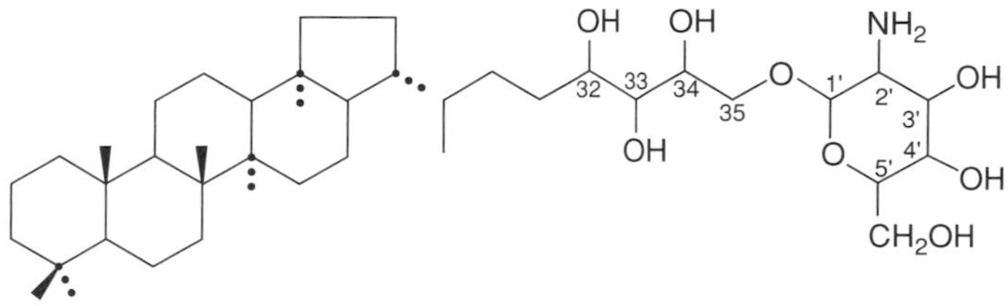

Glucosaminyl–bacteriohopanetetrol

**FIGURE 13.10**

Hopanoids found in *Zymomonas mobilis*.

reactions and that of hopanoids does not. Sterols in the cell membrane contribute to the alcohol tolerance of yeasts.

The fraction of cell lipid represented by hopanoids is strongly influenced by ethanol concentration. When a culture was grown at different constant alcohol concentrations, bacteriohopanetetrol represented 2.5% of the total

**Hopanoids**

The existence of the hopanoid family of bacterial lipids was unsuspected until their geochemical transformation products (*geohopanoids*) were discovered in the 1960s as universal constituents of geological sediments, young and old. The chemical diversity of geohopanoids is considerable, and unrelated sediments usually contain different sets of geohopanoids. Thus, "fingerprinting" of sediments by their geohopanoid patterns is useful in oil exploration. *The total amount of geohopanoids is on the order of $10^{12}$ tons, an amount similar to that estimated for the total mass of organic carbon in all current living organisms.* Surveys of microorganisms for the presence of hopanoids have shown these polyterpenes to be widely distributed among prokaryotes.

**BOX 13.1**

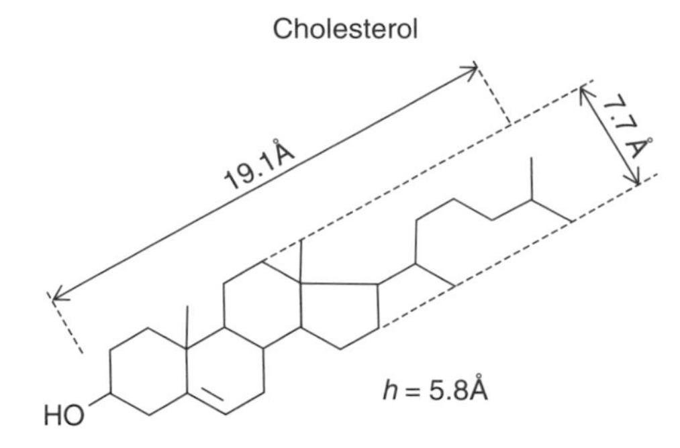


**FIGURE 13.11**

Structures of cholesterol and bacteriohopanetetrol. [Based on Figure 2 in Ourisson, G., and Albrecht, P. (1992). Hopanoids. 2. Biohopanoids: a novel class of bacterial lipids. *Accounts of Chemical Research*, 25, 403–408.]

lipid at 0.5% ethanol, 21% at 6.3% ethanol, and 36.5% at 16% ethanol. The levels of the other hopanoids follow the same trend. The hopanoids substitute for a portion of the phospholipids in the cell membrane; the absolute amount of phospholipids decreases as the amount of hopanoids increases. The hopanoids, just like the sterols, decrease membrane fluidity and, presumably, in this manner counteract the permeabilizing influence of ethanol.

This interpretation is reinforced by the observation that raising the temperature of the medium causes changes in the composition of the *Z. mobilis* cell membrane that parallel those induced by increasing the ethanol concentration. A change in growth temperature from 30°C to 37°C leads to an increase in the relative hopanoid content comparable to the increase seen at high ethanol concentrations. The membrane protein concentration also increases in response to an increase in the alcohol concentration or growth temperature, suggesting that ethanol tolerance in *Z. mobilis* is achieved by coordinated shifts in the relative amounts of cell membrane phospholipids, hopanoids, and proteins, all of which contribute to membrane stability.

### Genome Sequence of *Zymomonas mobilis*

The complete sequence of the DNA of the approximately 2 million–base pair circular chromosome of *Z. mobilis* ZM4 shows 1998 open reading frames and reveals significant metabolic limitations. There are no recognizable genes for 6-phosphofructokinase. In the absence of this enzyme, glycolysis is blocked.

Most of the enzymes for the pentose phosphate pathway are missing as well. The Entner–Doudoroff pathway represents the sole route for glucose fermentation in this organism.

### *Zymomonas mobilis* as an Industrial Ethanol Producer

The low yield of energy in the Entner–Doudoroff pathway of one ATP per mole of glucose limits the growth of *Z. mobilis* and allows ethanol yields as high as 97% with glucose as substrate, as compared with less than 95% for yeast. *Z. mobilis* is so well suited to ethanol production from glucose that strong arguments were advanced by some that it should replace *S. cerevisiae* in starch fermentation. In spite of successful trials on an industrial scale, the well-established use of yeasts has continued.

Wild-type *Z. mobilis* is not well suited for use in ethanol production from biomass feedstocks because it ferments only glucose, fructose, and sucrose. However, the many favorable features of this organism are retained in recombinant strains engineered to ferment xylose and arabinose. Four *Escherichia coli* genes (*xylA, xylB, tktA, talB*), introduced and expressed in *Z. mobilis*, enabled the recombinant strains to ferment xylose. Xylulose isomerase (*xylA*) and xylulose kinase (*xylB*) convert xylose into xylulose-5-phosphate. Xylulose-5-phosphate is next converted by transketolase (*tktA*) and transaldolase (*talB*) to fructose-6-phosphate and glyceraldehyde-3-phosphate. Both these products are intermediates in the Entner–Doudoroff pathway. A strain that ferments arabinose was engineered using the same approach in constructing a *Z. mobilis* strain that expresses five *E. coli* genes (*araA, araB, araD, tktA, talB*). L-Arabinose isomerase (*araA*), L-ribulose kinase (*araB*), and L-ribulose-5-phosphate-4-epimerase (*araD*) convert arabinose to xylulose-5-phosphate, whose conversion into intermediates of the Entner–Doudoroff pathway is described above. Such strains have been shown to function better than wild-type *S. cerevisiae* in the conversion of a poplar wood hydrolysate to ethanol by the simultaneous saccharification and fermentation (SSF) process examined in the next section.

## SIMULTANEOUS SACCHARIFICATION AND FERMENTATION: STAGES I AND II COMBINED

In SSF processes, the hydrolysis of the substrate and the fermentation of monosaccharides to ethanol are carried out in a single vessel. The hydrolytic enzyme(s) are produced in a separate reactor and introduced into the fermenter along with the substrate and yeast. For example, when partially hydrolyzed corn syrup is the substrate, *Aspergillus* species glucoamylase is added to the fermenter with the syrup and yeast cells. Starch hydrolysis produces glucose, which is immediately converted by the yeast to ethanol. With low steady-state concentrations of glucose, much smaller amounts of enzyme are able to achieve adequate rates of substrate hydrolysis, because product inhibition of the glucoamylase reaction is modest at low concentrations of glucose. When cellulose is the substrate, a mixture of *T. reesei* cellulases is used in place of glucoamylase. When lignocellulose is the substrate of an SSF process, pentose (xylose, arabinose) fermentation must be incorporated to maximize ethanol yields by utilizing both cellulose and hemicellulose. Here, engineered strains of *S. cerevisiae*, *E. coli*, or *Z. mobilis* able to ferment both glucose and pentoses come into use. We describe below an SSF process that employs an engineered *S. cerevisiae* strain. SSF-type processes have drastically reduced ethanol production costs.

## GENETICALLY ENGINEERED YEAST CAPABLE OF COFERMENTING GLUCOSE AND XYLOSE

Lignocellulosic biomass – such as hardwood, grasses, agricultural residues such as rice and wheat straw, sugarcane bagasse, corn stover, and corn fiber, yard and wood wastes, and wastes from paper mills – contains D-glucose and D-xylose, in a glucose-to-xylose ratio of 2 to 3:1, as the major fermentable sugars. With lignocellulosic feedstocks, for economic reasons, it is essential that both these sugars be converted to ethanol. No known naturally occurring organisms simultaneously convert both glucose and xylose to ethanol in high yield. A yeast strain genetically engineered to perform this fermentation was successfully constructed in the Laboratory of Renewable Resource Engineering at Purdue University in Indiana.

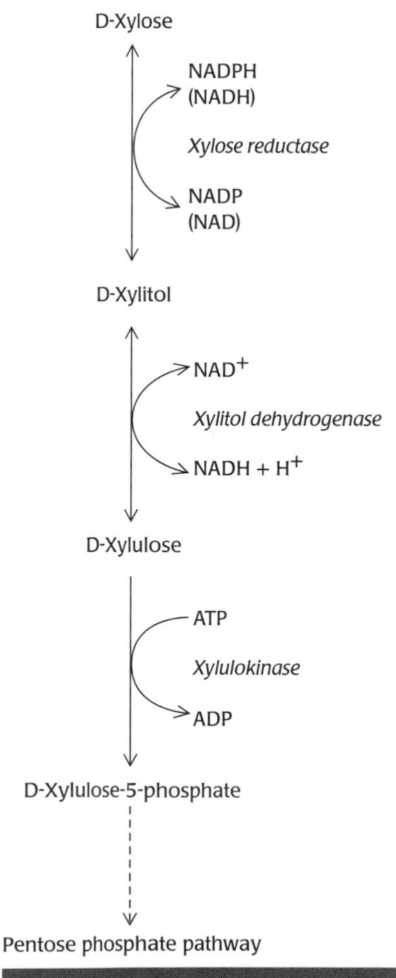


FIGURE 13.12

Xylose metabolic pathway constructed in *Saccharomyces* strain 424A(LNH-ST). For details, see text.

A stable recombinant *Saccharomyces* strain, 424A(LNH-ST), was engineered to ferment xylose by transferring the genes for xylose reductase and xylitol dehydrogenase – for the conversion of xylose to xylulose – from *Pichia stipitis*, the best naturally occurring xylose-fermenting non-*Saccharomyces* yeast, as well as additional copies of the *S. cerevisiae* xylulokinase gene (in addition to the endogenous active xylulokinase gene). The cloned *xyl* genes were stably integrated into its chromosome in high copy number. *Saccharomyces* strain 424A(LNH-ST) was thus able to utilize the pathway shown in Figure 13.12 to ferment xylose to ethanol. Several major problems were successfully addressed in the design of *Saccharomyces* strain 424A(LNH-ST).

### Directing the Metabolic Flux Toward the Formation of Ethanol and Away from Xylitol

Two factors favor the accumulation of xylitol in natural xylose-fermenting yeasts such as *P. stipitis*. First, xylose reductase can use either NADPH or NADH as a cofactor but has a much higher affinity for NADPH. Xylitol dehydrogenase utilizes NAD as its sole cofactor. In yeast, particularly under anaerobic conditions, there is no direct enzymatic interconversion between NAD and NADPH. Second, the equilibrium in the xylitol dehydrogenase–catalyzed interconversion of xylitol and xylulose favors the formation of xylitol. Finally, in wild-type *Saccharomyces* species, the level of xylulokinase activity is low. Consequently, in a recombinant strain with only the xylose reductase and xylitol dehydrogenase genes overexpressed, imbalances in cofactor availability would favor production of xylitol from xylose over production of ethanol. This limitation was overcome in *Saccharomyces* strain 424A(LNH-ST) by overexpressing xylulokinase as well. The rapid conversion of D-xylulose to D-xylulose-5-phosphate also accelerates its subsequent conversion to ethanol through the pentose phosphate and glycolytic pathways. This change in metabolic flux resolves both the cofactor imbalance and the bias toward xylitol production.

### Elimination of Catabolite Repression

In catabolite repression, the expression of genes for various enzymes is inhibited when the cells are growing in a medium containing an energy source. This effect was first observed with glucose as the preferred energy source and hence in such instances has been termed the *glucose effect*. Thus, when both glucose and xylose are present in the medium, the glucose is metabolized first, and the expression of *xyl* genes is induced only upon glucose exhaustion. Moreover, xylose is required for the expression of the *xyl* genes. With a feedstock containing both glucose and xylose, there is a lag before xylose is fermented. Elimination of the glucose effect and of the requirement for xylose induction would speed up the conversion of lignocellulose sugars to ethanol. The regulatory sequences containing the transcriptional and translational signals controlling gene expression that mediate the glucose effect and xylose-dependent induction are at the 5′-end of the *xyl* genes. These sequences were eliminated in *Saccharomyces* strain 424A(LNH-ST) by replacing the 5′ noncoding sequences of the *xyl* genes with 5′-promoter sequences from constitutively expressed *S. cerevisiae* genes encoding glycolytic enzymes (e.g., the pyruvate kinase gene). As a result, the recombinant strain simultaneously coferments glucose and xylose to ethanol.

### *Saccharomyces* Strain 424A(LNH-ST) in Industrial Production of Ethanol

In laboratory experiments, the cellulose and hemicellulose in corn fiber and corn stover were hydrolyzed by autoclaving for one hour at 121°C. Subsequent fermentation with *Saccharomyces* strain 424A(LNH-ST) gave ethanol yields greater than 75% of the theoretical yield of consumed sugars.

In 2004, Iogen Corporation of Ottawa, Ontario, Canada, began industrial production of ethanol using this recombinant yeast strain to make ethanol from straw collected from farms. The feedstock is subjected to steam explosion to disrupt the lignocellulose and solubilize the hemicellulose, then enzymes are added to hydrolyze the polysaccharides. Subsequent fermentation with *Saccharomyces* strain 424A(LNH-ST) leads to the production of about 75 gallons of ethanol per ton of straw. The ethanol is subsequently blended into gasoline at a Canadian petroleum refinery.

## CLOSTRIDIAL FERMENTATIONS: ONE-STEP CONVERSION OF CELLULOSE AND HEMICELLULOSE TO ALCOHOL

The direct conversion of cellulosic biomass to ethanol by anaerobic bacteria is potentially cheaper than a process that combines the actions of fungal cellulases and alcohol-producing yeasts. Three clostridial strains – *Clostridium thermocellum*, *Clostridium thermosaccharolyticum*, and *Clostridium thermohydrosulfuricum* – are regarded as candidate organisms for single-step conversion processes, and a great deal of effort is directed toward that end. These bacteria are all thermophilic, Gram-positive, strict anaerobes.

*C. thermocellum* was first isolated in 1926 from manure and subsequently from many other anaerobic environments rich in organic nutrients. It has an optimum growth temperature of 60°C. The cellulose-degrading complexes of *C. thermocellum*, cellulosomes, which form when the bacteria grow on cellulose, are described in Chapter 12. The cellulase activity levels found in culture supernatants of *C. thermocellum* match those produced by *T. reesei*, long considered the most efficient cellulase producer. Strains of *C. thermocellum* are able to ferment cellulose, xylans, and monomeric sugars (such as glucose, mannose, arabinose, and xylose). The products are ethanol, acetic acid, lactic acid, $CO_2$, and $H_2$.

*C. thermosaccharolyticum* has an optimum growth temperature of 55°C to 60°C, with an upper limit of 67°C. It ferments diverse carbohydrates, including glycogen, starch, and pectin, to produce ethanol, acetate, butyrate, lactate, $CO_2$, and $H_2$. This organism has been studied thoroughly because it causes spoilage of canned foods: contaminated cans blow open under the pressure of the gases produced by the fermentation. Batch cultures of resting cells of *C. thermosaccharolyticum* at pH 7.0 produce up to 1.2 moles of ethanol per mole of glucose.

*C. thermohydrosulfuricum* was first isolated in 1965 from the extraction juices of a beet sugar factory in Austria and subsequently from many other locations. This organism ferments a range of carbohydrates very similar to that of *C. thermosaccharolyticum*, but at an optimum temperature about 10°C higher and an upper limit of cell growth at 74°C to 76°C. Under appropriate conditions, *C. thermohydrosulfuricum* fermentation yields up to 1.5 moles of ethanol per mole of glucose, along with acetate, lactate, $CO_2$, and $H_2$.

Clostridial strains have the advantage over yeasts and *Zymomonas* of being excellent cellulase producers, thus allowing the conversion of cellulose and hemicelluloses to ethanol by a single organism. This eliminates the need for a second organism (one to produce cellulase and one to carry out the fermentation) and the feedstock needed to support it. However, clostridia are unlikely to become commercially important ethanol producers in the near future. The by-products of clostridial fermentation include large amounts of organic acids. The highest reported molar ratio of ethanol to acids is about 2.3. These organisms also produce some $H_2S$ from sulfur-containing amino acids. Furthermore, their ethanol tolerance is lower than that of yeasts and *Zymomonas*.

The clostridial cellulosome is a unique, complicated macromolecular complex of more than a dozen enzymes that work together in the efficient degradation of cellulose and hemicelluloses (see Figure 12.19). In view of this, the clostridia may become more important in the future as sources of enzymes for polysaccharide degradation than as direct ethanol producers.

## PROSPECTS OF FUEL ETHANOL FROM BIOMASS

Energy can be extracted from biomass either by direct combustion or by first converting the biomass to another fuel (ethanol, methanol, or methane) and then combusting it. Most commonly, biomass is simply burned to generate heat; of the 2.8 billion tons of biomass consumed annually worldwide as material for burning, wood represents about 50%, crop residues some 33%, and dung most of the remainder. Currently, the fraction of biomass converted to other fuels, such as ethanol, is very small.

A 2005 study sponsored jointly by the U.S. Departments of Energy and Agriculture examined the feasibility of sustainable use of biomass to produce one third of the current U.S. demand for transportation fuels. Looking just at forest and agricultural land showed a potential annual supply of more than 1.3 billion dry tons of biomass, a supply of feedstocks more than sufficient to achieve the desired objective (Table 13.5). This amount would represent a more than sevenfold increase in the amount of biomass currently processed. Full utilization of the potential of these resources would require a major new biorefinery infrastructure that would not be completely in place until around the mid–twenty-first century. Biorefineries are envisaged to operate in a manner analogous to petroleum refineries. In a petroleum refinery, the crude oil serves as a source of multiple products. Analogously, in the biorefinery concept, the feedstock would be converted to a wide range of products. These would include fuels, electric power and heat energy, animal feed, and a variety of chemicals, such as succinic acid, 1,4-butanediol, and ethyl lactate, that would in turn serve as precursors for the manufacture of plastics, paints, detergents, and so on. The biorefinery would thus be an industrial park with the aggregate capability to utilize biomass – and the sugars obtained from it – in many different ways, depending on the market need for particular end products.

Examination of the assumptions of the 2005 study indicates clearly that success in meeting the biofuel goals rests on highly efficient processes for very large-scale lignocellulose-to-ethanol conversion. The future projected fractional contribution of cornstarch, the currently dominant feedstock in the United States, is estimated to be very modest (Table 13.5). The stakes for research on improving the energy output–input ratio and lowering the costs of the conversion of lignocellulosic biomass to ethanol are very high.

**TABLE 13.4 Contributions of various energy sources to U.S. energy consumption in 2004**

| Energy source | Fractional contribution (%) |
|---|---|
| Petroleum | 39 |
| Natural gas | 24 |
| Coal | 23 |
| Nuclear | 8 |
| Renewable energy | |
| Biomass | 3 |
| Hydroelectric, geothermal, wind, solar | 3 |

Data from Perlach, R. D. et al. (2005). Biomass as Feedstock for a Bioenergy and Bioproducts Industry. The Technical Feasibility of a Billion-Ton Annual Supply. ORNL/TM-2005/66. April 2005. Available electronically at http://www1.eere.energy.gov/biomass/publications.html#ethanol.

## SUMMARY

Cellulose, hemicelluloses, and starches are a vast renewable source of sugars convertible to ethanol by microbial fermentation. Production of ethanol from the polysaccharides of biomass proceeds in three stages: stage I, degradation of polysaccharides to fermentable sugars; stage II, fermentation; and stage III, alcohol recovery. Disruption of the physical structure of lignocellulose makes cellulose and hemicelluloses accessible to enzymatic attack. This is achieved either by steam explosion, by acid hydrolysis, or by ball

**TABLE 13.5 Summary of the potential sustainable forest and agricultural resources for the U.S. bioenergy and bioproducts industry**

| Biomass resource | Million dry tons/year |
|---|---|
| Forest resources | |
| Logging and other residue[a] | 64 |
| Fuel treatments[b] | 60 |
| Urban wood residues | 47 |
| Wood processing residues | 70 |
| Pulping liquor | 74 |
| Fuel wood | 52 |
| Agricultural resources | |
| Perennial grasses and perennial woody crops | 377 |
| Crop residues (e.g., corn stover, small grain straw) | 446 |
| Process residues | 87 |
| Grain-to-ethanol (corn and soybeans) | 87 |
| Total | *1364* |

[a] Logging residues from conventional harvest operations and residues from forest management and land clearing operations.
[b] Removal of excess biomass (fuel treatments) from timberlands and other forestlands.
Data from Perlach, R. D. et al. (2005). Biomass as Feedstock for a Bioenergy and Bioproducts Industry. The Technical Feasibility of a Billion-Ton Annual Supply. ORNL/TM-2005/66. April 2005. Available electronically at http://www1.eere.energy.gov/biomass/publications.html#ethanol.

milling. *Saccharomyces* strains are responsible for almost all of the current industrial production of alcohol by fermentation. These yeasts produce high concentrations of alcohol with low levels of by-products, and have the high cell viability and flocculation characteristics needed for repeated cell recycling. A serious limitation is that the yeasts ferment only a narrow range of carbohydrate substrates. This limitation has been addressed successfully by genetically engineering yeasts that are capable of simultaneous fermentation of glucose and xylose.

*Z. mobilis*, a bacterium isolated from fermenting sugar-rich plant juices, takes up glucose and produces ethanol some fourfold faster than does yeast, with alcohol yields of up to 97% of the theoretical maximum value. However, *Zymomonas* utilizes only three substrates: glucose, fructose, and sucrose. Growth on fructose and sucrose leads to conversion of 10% or more of the substrate to products other than alcohol, such as dihydroxyacetone, mannitol, and glycerol. Genetically engineered strains of *Zymomonas* are capable of fermenting glucose and pentoses (xylose and arabinose). In spite of its very high ethanol tolerance and these added capabilities, *Z. mobilis* has not yet been utilized in the routine, high-volume production of ethanol.

Simultaneous saccharification and fermentation processes combine stages I and II. In SSF processes, enzymatic hydrolysis of the polysaccharide substrate and the fermentation of monosaccharides to alcohol are carried out in a single vessel. With partially hydrolyzed starch, for example, *Aspergillus* species glucoamylase is added to the fermenter along with yeast cells. One-step clostridial conversion of cellulose to alcohol represents another version of an SSF process. Thermophilic clostridia that produce extracellular cellulases are able to convert cellulose all the way to alcohol. Unfortunately, clostridial fermentations also produce large amounts of organic acids and some hydrogen sulfide. A major effort is under way to modify *C. thermocellum* genetically to allow its adoption as a large-scale ethanol producer.

The future of ethyl alcohol as an alternative fuel depends on further improvements in the efficiency and economics of the conversion of lignocellulose to alcohol.

## REFERENCES AND ONLINE RESOURCES

### General

U.S. DOE (2006). *Breaking the Biological Barriers to Cellulosic Ethanol: A Joint Research Agenda*. DOE/SC-0095. Available electronically at http://www.doegenomestolife.org/biofuels/

Hamelinck, C. N., van Hooijdonk, G., and Faaij, A. P. C. (2005). Ethanol from lignocellulosic biomass: techno-economic performance in short-, middle- and long-term. *Biomass and Bioenergy*, 28, 384–410.

Dias de Oliveira, M. E., Vaughan, B. E. and Rykiel, E. J., Jr. (2005). Ethanol as fuel: energy, carbon dioxide balances, and ecological footprint. *Bioscience*, 55, 593–602.

Klass, D. L. (2004). Biomass for renewable energy and fuels. In *Encyclopedia of Energy*, Volume 1, C. J. Cleveland (ed.), pp. 193–212, Amsterdam and Boston: Elsevier Academic Press.

Lynd, R. R., Weimer, P. J., van Zyl, W. H., and Pretorius, I. S. (2002). Microbial cellulose utilization: fundamentals and biotechnology. *Microbiology and Molecular Biology Reviews*, 66, 506–577.

Shapouri, H., Duffield, J. A., and Wang, M. (2002). *The Energy Balance of Corn Ethanol: An Update*, Agricultural Economic Report No. 814, Washington, DC: U.S. Department of Agriculture.

World Energy Assessment: Energy and the Challenge of Sustainability. (2001). New York: United Nations Development Programme (UNDP).

Kheshgi, H. S., Prince, R. C., and Marland, G. (2000). The potential of biomass fuels in the context of global climate change: focus on transportation fuels. *Annual Review of Energy and the Environment*, 25, 199–214.

Wyman, C. E. (1999). Biomass ethanol: technical progress, opportunities, and commercial challenges. *Annual Review of Energy and the Environment*, 24, 189–226.

Lynd, L. R. (1996). Overview and evaluation of fuel ethanol from cellulosic biomass: technology, economics, the environment, and policy. *Annual Review of Energy and the Environment*, 21, 404–465.

**Relationship of Coproducts to the Ethanol Production Process**

RFA (Renewable Fuels Association) (2004). How ethanol is made. http://ethanolrfa.org/prod_process.html.

**Microbial Tolerance to Alcohols**

You, K. M., Rosenfield, C-M., and Knipple, K. C. (2003). Ethanol tolerance in the yeast *Saccharomyces cerevisiae* is dependent on the cellular oleic acid content. *Applied and Environmental Microbiology*, 68, 1499–1503.

Ingram, L. O. (1986). Microbial tolerance to alcohols: role of the cell membrane. *Trends in Biotechnology*, 4, 40–44.

Curtain, C. C. (1986). Understanding and avoiding alcohol inhibition. *Trends in Biotechnology*, 4, 110.

**Yeasts**

Sedlak, M., and Ho, N. W. C. (2004). Production of ethanol from cellulosic biomass hydrolysates using genetically engineered *Saccharomyces* yeast capable of cofermenting glucose and xylose. *Applied Biochemistry and Biotechnology*, 113–116, 403–416.

Ho, N. W. C., Chen, Z., Brainard, A. P., and Sedlak, M. (2000). Genetically engineered *Saccharomyces* yeasts for conversion of cellulosic biomass to environmentally friendly transportation fuel ethanol. In *ACS Symposium Series 767*, P. T. Anastas, G. H. Heine, and T. C. Williamson (eds.), pp. 143–159, Washington, DC: American Chemical Society.

Ho, N. W. C., Chen, Z., and Brainard, A. P. (1998). Genetically engineered *Saccharomyces* yeast capable of effective cofermentation of glucose and xylose. *Applied and Environmental Microbiology*, 64, 1852–1859.

Shigechi, H., et al. (2004). Direct production of ethanol from raw corn starch via fermentation by use of a novel surface-engineered yeast strain codisplaying glucoamylase and $\alpha$-amylase. *Applied and Environmental Microbiology*, 70, 5037–5040.

Dien, B. S., Cotta, M. A., and Jeffries, T. W. (2003). Bacteria engineered for fuel ethanol production: current status. *Applied Microbiology and Biotechnology*, 63, 258–266.

### *Zymomonas*

Seo, J-S., et al. (2005). The genome sequence of the ethanologenic bacterium *Zymomonas mobilis* ZM4. *Nature Biotechnology*, 23, 63–68.

Jeffries, T. W. (2005). Ethanol fermentation on the move. *Nature Biotechnology*, 23, 40–41.

Montenecourt, B. S. (1985). *Zymomonas*, a unique genus of bacteria. In *Biology of Industrial Microorganisms*, A. L. Demain and N. A. Solomon (eds.), pp. 261–289, Menlo Park, CA: Benjamin/Cummings Publishing Co., Inc.

Viikari, L. (1988). Carbohydrate metabolism in *Zymomonas*. *CRC Critical Reviews of Biotechnology*, 7, 237–261.

Swings, J., and De Ley, J. (1977). The biology of *Zymomonas*. *Bacteriological Reviews*, 41, 1–4.

### Hopanoids

Ourisson, G., and Albrecht, P. (1992). Hopanoids. 1. Geohopanoids: the most abundant natural products on Earth? *Accounts of Chemical Research*, 25, 398–402.

Ourisson, G., and Albrecht, P. (1992). Hopanoids. 2. Biohopanoids: a novel class of bacterial lipids. *Accounts of Chemical Research*, 25, 403–408.

### Clostridia

Demain, A. L., Newcomb, M., and Wu, J. H. D. (2005). Cellulase, clostridia, and ethanol. *Microbiology and Molecular Biology Reviews*, 69, 124–154.

Zhang, Y-H. P., and Lynd, L. R. (2005). Cellulose utilization by *Clostridium thermocellum*: bioenergetics and hydrolysis product assimilation. *PNAS*, 102, 7321–7325.

Lynd, L. R., van Zyl, W. H., McBride, J. E., and Laser, M. (2005). Consolidated bioprocessing of cellulosic biomass: an update. *Current Opinion in Biotechnology*, 16, 577–583.

### Prospects for Fuel Ethanol from Biomass

Perlach, R. D. et al. (2005). Biomass as Feedstock for a Bioenergy and Bioproducts Industry. The Technical Feasibility of a Billion-Ton Annual Supply. ORNL/TM-2005/66. April 2005. Available electronically at http://www1.eere.energy.gov/biomass/publications.html#ethanol.

FOURTEEN

# Environmental Applications

Since the middle of the last century, while the population doubled, water use has tripled. At the same time, we have dirtied that water with human, industrial, and agricultural wastes.

– Ward, D. R. (2002). *Water Wars*, p. 3, New York: Riverhead Books.

Microorganisms play a fundamental role in the global recycling of matter by releasing carbon, nitrogen, phosphorus, and sulfur from an immense variety of complex organic compounds for reuse by living organisms and in generating energy. The ability of the life support systems of the planet to function depends on the activities of these organisms. The spectacular metabolic versatility of prokaryotes and fungi is displayed in the major areas of environmental microbiology examined in this chapter. We examine sewage and wastewater treatment and the degradation of xenobiotics and petrochemicals, and end with a consideration of biomining. Microorganisms, primarily chemolithotrophic prokaryotes, are effective in the bioleaching of low-grade ores and in the removal of toxic heavy metals from the environment.

## DEGRADATIVE CAPABILITIES OF MICROORGANISMS AND ORIGINS OF ORGANIC COMPOUNDS

Microorganisms excel at using organic substances, natural or synthetic, as sources of nutrients and energy. These include certain synthetic compounds – detergents, solvents (trichloroethane, toluene, xylenes), and transformer fluids (polychlorobiphenyls) – that seem very different from any natural compounds such an organism would be likely to encounter. The explanation for this remarkable range of degradative abilities is that prior to the advent of humans, microorganisms had already coexisted for billions of years with an immense variety of organic compounds. The vast diversity of potential substrates for growth led to the evolution of enzymes capable of transforming many unrelated natural organic compounds by many different catalytic mechanisms. The resulting giant library of microbial enzymes serves as raw material for further evolution whenever a new synthetic organic

**Petroleum Constituents**

The major constituents of crude oils (petroleum) are paraffins (30% to 50%), cycloparaffins (20% to 65%), and aromatics (6% to 14%). The actual amounts depend on the origin of the crude oil.

Paraffins include compounds such as *n*-heptane, *n*-hexane, 2-methylpentane, and 2,3-dimethylhexane. Cycloparaffins include cyclohexane, methylcyclopentane, methylcyclohexane, and trimethylcyclopentane. Toluene and benzene are major aromatic components. Petroleum contains many other organic compounds.

**BOX 14.1**

chemical becomes available: Mutations in existing enzymes generate catalysts capable of utilizing new substrates; the possessor of such an enzyme capable of degrading a substrate that other organisms cannot gains a growth advantage or a new ecological niche.

How many natural organic chemicals are there, and how are these compounds generated? Scientists have already identified many thousands of them, but the list is far from complete. In addition to the organic constituents typical of all living organisms, the natural world contains an abundance of chemicals produced either by metabolic pathways unique to specific groups of organisms or by biogeochemical processes. For example, coal, petroleum, and natural gas, mixtures of highly reduced carbon compounds, are produced by the combined effects of high pressure and heat on buried remains of plants and phytoplankton.

Petroleum originates primarily from phytoplankton residues that accumulated in the depressions of shallow seas. It consists of a complex array of gaseous, liquid, and solid *n*-alkanes; branched paraffins, cyclic paraffins, and substituted cycloparaffins; aromatic compounds; sulfur compounds including benzo(*b*)thiophene and dibenzothiophene; and many other organic compounds (Box 14.1). These compounds re-enter the biosphere as a result both of upward seepage through porous rocks and sediments and of various kinds of geological upheavals. About a million metric tons of petrochemicals annually are transported into the oceans and the surface soil of the continents in these ways, there to be fed upon by microorganisms. Microorganisms even grow in certain underground petroleum reservoirs, when a steady seepage of groundwater brings them sufficient supplies of mineral nutrients and oxygen. In fact, it is estimated that some 10% of global oil deposits have been destroyed in this manner.

Coal, derived chiefly from terrestrial plant matter, is another common reservoir of highly reduced carbon compounds. It is made up of aromatic rings fused into different small polycyclic clusters that are linked by aliphatic structures. The aromatic rings carry phenolic, hydroxyl, quinone, and methyl substituents (Figure 14.1). Coal, too, contains a variety of sulfur compounds. Many microorganisms can utilize the carbon compounds in coal as substrates and can metabolize the sulfur compounds also.

Synthetic organohalogen compounds such as polychlorobiphenyls (PCBs), dichlorodiphenyltrichloroethane (DDT), and trichloroethane are widespread and long-lived contaminants. In recent years, researchers have discovered that a wide variety of organohalogen compounds also occur naturally in many parts of the world. Soil samples collected from far-flung sites that are unpolluted, or at least only minimally contaminated by long-range atmospheric transport of industrially produced compounds, contain organohalogens in significant amounts. Indeed, the ratio of carbon-bound halogen to total organic carbon ranges from 0.2 to 2.8 mg of chloride per gram of carbon in soil worldwide, and the global contribution of human-made organohalogen compounds can constitute only a small fraction of this amount. Various marine organisms and many plants produce halogenated organic compounds. So do fungal haloperoxidases, through action on soil compounds derived from lignin degradation. Thus, microorganisms have long faced the challenge of breaking down many natural organohalogen compounds. Not surprisingly, the discovery of the abundance of natural organohalogens has been accompanied by the discovery of microorganisms able to degrade all but the most highly halogenated synthetic organic compounds.

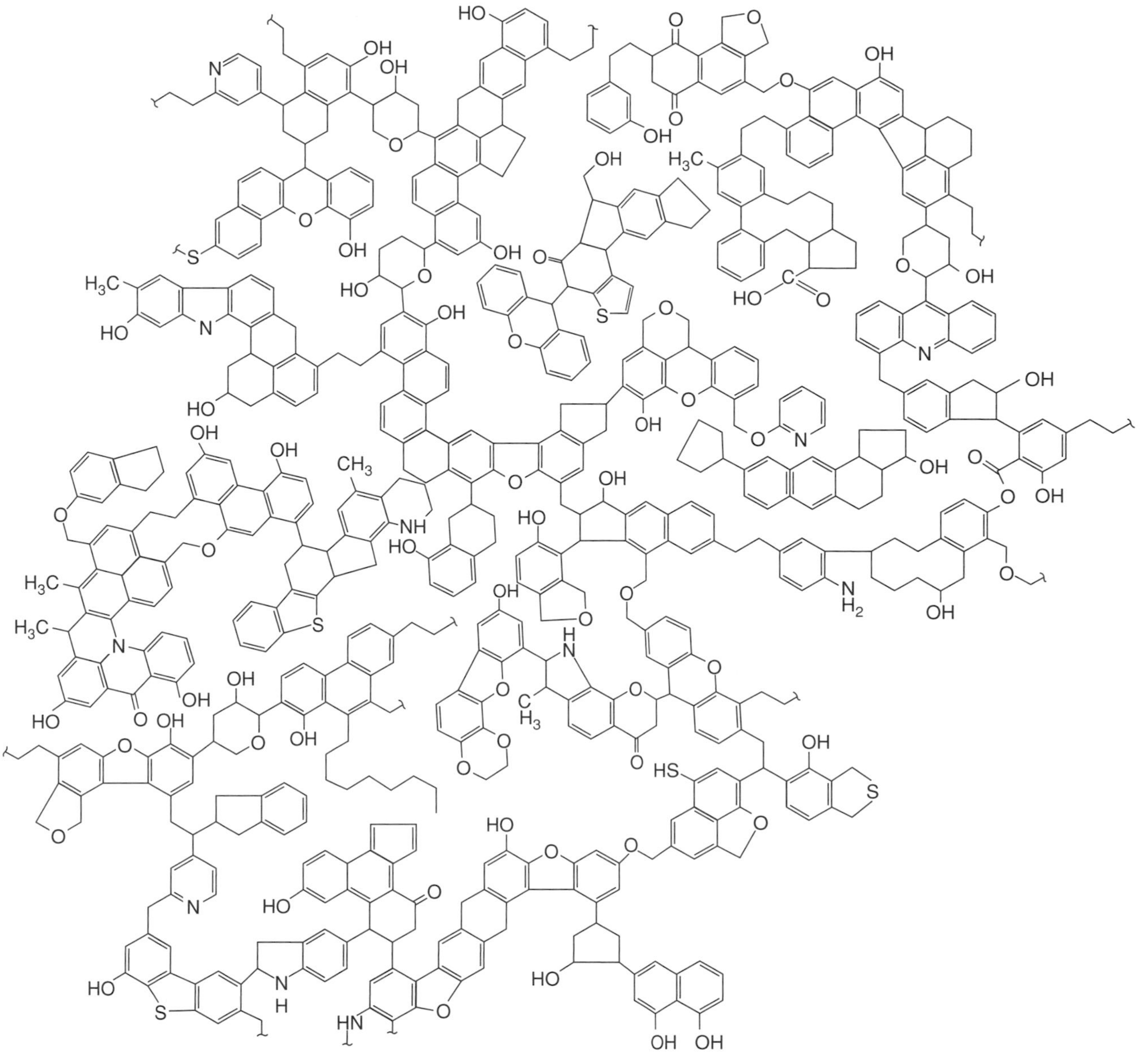

**FIGURE 14.1**

Model of the structure of bituminous coal. [From Crouch, G. R. (1990). Biotechnology and coal: a European perspective. In *Bioprocessing and Biotreatment of Coal,* D. L. Wise (ed.), pp. 29–55. New York: Marcel Dekker.]

## DESCRIBING THE BEHAVIOR OF ORGANIC COMPOUNDS IN THE ENVIRONMENT

Organic compounds are often classified as biodegradable, persistent, or recalcitrant. A *biodegradable* organic compound is one that undergoes a biological transformation. A *persistent* organic compound does not undergo biodegradation in certain environments. Finally, a *recalcitrant* compound resists biodegradation in a wide variety of environments. Note that biodegradation is not always desirable. In some instances, degradation may convert an innocuous compound into a toxic one, or alter an already toxic compound to generate a product toxic to even more organisms.

By itself, the term *biodegradable* does not imply any particular extent of degradation. The transformation may involve one or several reactions, and the effect may be slight or significant. *Primary biodegradation* usually refers to alteration by a single reaction, whereas *partial biodegradation* means a more extensive chemical change. In common parlance, however, when people say that a compound is biodegradable, they mean that it can be mineralized. *Mineralization* is complete degradation to the end products of $CO_2$, water, and other inorganic compounds.

## WASTEWATER TREATMENT

Arguably the most important, if least glamorous, practical application of microbiology is in the treatment of sewage and wastewater. The amounts of waste materials discharged into bodies of water around the world are rising steadily. Population increase, the dependence of agriculture on massive amounts of fertilizers and pesticides, expansion of the food industry, and growth of other industrial processes all contribute to the volume of sewage and wastewater and their content of undesirable substances. The volume of water affected by human activities is huge. Water use worldwide appropriates about 10% of the average runoff in all continental river basins. Of this usage, irrigation consumes some 70%, industry about 20%, and livestock, domestic, and other uses the rest.

The challenge of wastewater treatment is to remove (a) compounds with a high biochemical oxygen demand, (b) pathogenic organisms and viruses, and (c) a multitude of human-made chemicals. Wherever wastewater treatment is nonexistent or inadequate, the consequences are grave for the environment and for human health. Where untreated human excreta are discharged into rivers, estuaries, or coastal waters, the enrichment with organic compounds leads to explosive growth of both the native bacterial community and some of the microorganisms in the sewage. The resulting oxygen depletion and infections lead to die-off of aquatic organisms. The excreta are also the source of multiple human pathogens that cause common water-borne diseases. These include bacterial diseases such as typhoid, cholera, and gastroenteritis; viral ones such as infectious hepatitis; and protozoal infections such as cryptosporidiosis, giardiasis, and amoebic dysentery. Wherever reused water contributes to drinking water supplies, critical public health issues arise. Partially treated effluents, rich in nitrate and phosphate, cause eutrophication resulting in blooms of cyanobacteria and algae. In stratified bodies of water, the subsequent death and decay of these organisms lead to drastic decreases in the oxygen levels in the hypolimnion,

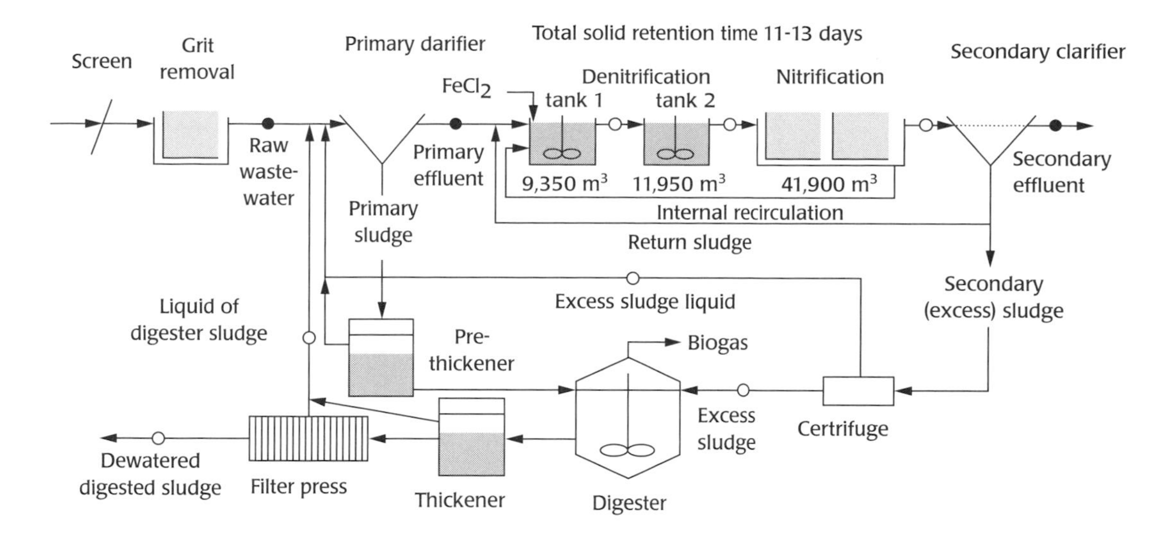


FIGURE 14.2

Flow scheme of the municipal sewage treatment plant in Wiesbaden, Germany. [From Andersen, H., Siegrist, H., Halling-Sørensen, B., and Ternes, T. A. (2003). Fate of estrogens in a municipal sewage treatment plant. *Environmental Science and Technology*, 37, 4021–4026.]

leading to fish mortality. The term *hypolimnion* refers to the lower, noncirculating water in a thermally stratified lake.

How great is the challenge posed by the need for adequate treatment of sewage? To give one anecdotal measure, the quantity of wastewater as sewage treated in a developed country, the United Kingdom, reported for 1984 for a population at the time of around 56 million, totaled 6.5 billion metric tons per year. The shocking fact is that only about 15% of the wastewater produced worldwide is currently treated.

## OPERATION OF A MODERN SEWAGE TREATMENT PLANT

This discussion examines the outcome of the various stages in the treatment of wastewater and follows the flow diagram of a sewage treatment plant shown in Figure 14.2. The wastewater that enters the plant is first subjected to a *mechanical treatment*. This treatment utilizes, consecutively, a screen, an aerated grit-removal tank, and a primary clarifier. Large aggregates are removed by the sieve and by the grit-removal tank. Suspended solids settle to the bottom of the primary clarifier, or, in the case of fats and oils, float to the top. The mechanical treatment removes up to 30% of the waste material.

The primary sludge collected in the primary clarifier is prethickened and pumped into the mesophilic digester, whereas the primary effluent is pumped into tank 1. This initiates the *biological treatment* – the activated sludge process. The wastewater treatment plant discussed here is designed for biological nitrogen removal. Tanks 1 and 2 are not aerated and rapidly become anoxic. The following sequence of transformations takes place in these tanks. Organic matter in wastewater has the approximate composition $C_{18}H_{19}O_9N$. With dissolved oxygen concentration at greater than 2% saturation, the overall process of microbial oxidation of organic matter is represented by the equation

$$C_{18}H_{19}O_9N + 17.5O_2 + H^+ \rightarrow 18CO_2 + 8H_2O + NH_4^+.$$

This equation also describes the overall carbon substrate oxidation process in the aerated tanks. When the dissolved oxygen concentration is depleted to less than 2% of saturation but adequate carbon substrates remain to act as electron donors, microorganisms use nitrate as a terminal electron acceptor to oxidize the organic matter producing carbon dioxide, water, and ammonium ion, thereby reducing most of the nitrate to $N_2$ through a reaction sequence called *denitrification*. If one assumes that all the energy produced by nitrate-dependent respiration is used for growth and that the organism assimilates ammonium, the overall equation that describes this process is

$$0.52C_{18}H_{19}O_9N + 3.28NO_3^- + 0.48NH_4^+ + 2.80H^+$$
$$\rightarrow C_5H_7NO_2 \text{ (new biomass)} + 1.64N_2 + 4.36CO_2 + 3.8H_2O.$$

The maximum yield of new microbial biomass from aerobic heterotrophic metabolism is 0.5 kg biomass per kilogram of organic matter. With nitrate as the terminal electron acceptor, the corresponding yield is 0.4 kg. This difference is a consequence of the higher amount of energy made available by aerobic respiration as compared with that resulting from respiration with nitrate as the terminal energy acceptor.

For phosphate removal, ferrous chloride, $Fe(II)Cl_2$, is added to the first denitrification tank to allow efficient mixing with the contents before oxidation to Fe(III) and subsequent precipitation as ferric phosphate in the aerated tanks. Following transfer to the aerated tank, the residual organic matter is oxidized with oxygen as the terminal electron acceptor to $CO_2$ and $H_2O$. At the same time, ammonia is oxidized to nitrate, which is then recycled to the anoxic tanks (see "Internal recirculation" in Figure 14.2).

In the secondary clarifier, the activated sludge, which consists mainly of microbial biomass, settles out and the treated wastewater (secondary effluent) is discharged – in the case of this treatment plant, into the Rhine River – after its suspended solids have been reduced in a rotary sleeve (not shown in the diagram). Part of the activated sludge is returned to the inlet of the first denitrification tank (see "Return sludge" in Figure 14.2), and the remainder ("Secondary (excess) sludge") is dewatered by centrifugation to a concentration of about 5% total solids and pumped into an anaerobic mesophilic digester operated at 33°C with a 20-day retention time for the contents.

Sewage treatment plants host a complex community of organisms: heterotrophic bacteria, lithotrophic bacteria, fungi, algae, and grazing fauna. The grazing fauna includes protozoa, rotifers, aquatic larvae of various species, and crustaceans. The grazing fauna feed preferentially on free-floating bacterial cells and keep the numbers of potential pathogens, such as *Escherichia coli*, low in the treated wastewater. In this manner, the community is enriched for aggregate-forming bacteria such *Zoogloea* species that promote settling of the activated sludge. In addition to *Zoogloea*, dominant bacteria in the sludge are the filamentous *Leucothrix* and *Thiothrix*, along with less prominent representatives of numerous other bacterial genera.

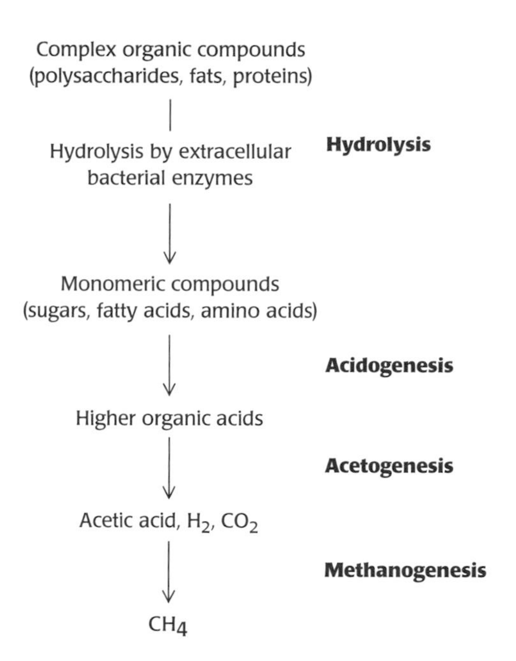


**FIGURE 14.3**

The major metabolic stages in the decomposition of organic wastes into methane and carbon dioxide in an anaerobic digester.

Sludge recycling is a critical part of the wastewater treatment process. The efficiency of the system is controlled by variation in the balance between sludge recycling and removal of excess sludge. Sludge recycling ensures high settling efficiency because those organisms that settle readily are selectively retained in the treatment plant. It also greatly increases the amount of active microbial biomass in tanks 1 and 2 and the aerated tanks, with a resulting increase in the rate of oxidation of the incoming substrate. Finally, as a result of sludge recycling, the organisms in the sludge remain in the treatment plant for weeks, whereas the retention time for water is less than 12 hours.

The excess sewage sludge is subjected to anaerobic treatment in the digester. During a retention time of three to five weeks at 30°C to 35°C, the microbial biomass and other biodegradable components of the sludge are converted to methane and $CO_2$. This mixture, called *biogas*, contains 65% to 70% methane, 25% to 30% $CO_2$, and traces of nitrogen and $H_2S$. The biogas may be used to maintain the temperature of the digester or to generate electricity. The residual digested solids are dewatered to 20% dry solids in a filter press, and the pressed cake disposed in a landfill.

The distinctive biochemical processes that take place in the digester reflect the absence of oxygen and nitrate. These degradative and fermentative reactions can be divided into two stages, acid forming and methane forming (Figure 14.3), and proceed by the cooperative activities of different kinds of mostly strictly anaerobic prokaryotes: the acid-forming bacteria and the methanogenic archaea.

In the acid-forming stage, complex organic polymers, including carbohydrates, fats, and proteins, are hydrolyzed by extracellular hydrolytic enzymes (polysaccharidases, lipases, proteases) and converted to volatile (short-chain) fatty acids, alcohols, and ketones. The fatty acids are fermented to acetate, carbon dioxide, and hydrogen (Table 14.1). Interestingly, some of

**TABLE 14.1 Examples of reactions occurring in an anaerobic digestion tank and their free energies under standard and "typical" conditions**

| Reaction | Reactants | Products | Free energy[a] (kcal/reaction) $\Delta G^{o\prime}$ | $\Delta G'$ |
|---|---|---|---|---|
| Conversion of glucose to $CH_4$ and $CO_2$ | Glucose + $3H_2O$ | $3CH_4 + 3HCO_3^- + 3H^+$ | −96.5 | −95.3 |
| Conversion of glucose to acetate and $H_2$ | Glucose + $4H_2O$ | $2CH_3COO^- + 2HCO_3^- + 4H^+ + 4H_2$ | −49.3 | −76.1 |
| Methanogenesis from acetate | $CH_3COO^- + H_2O$ | $CH_4 + HCO_3^-$ | −7.4 | −5.9 |
| Methanogenesis from $H_2$ and $CO_2$ | $4H_2 + HCO_3^- + H^+$ | $CH_4 + 3H_2O$ | −32.4 | −7.6 |
| Acetogenesis from $H_2$ and $CO_2$ | $4H_2 + 2HCO_3^- + H^+$ | $CH_3COO^- + 2H_2O$ | −25.0 | −1.7 |
| Amino acid oxidation | Leucine + $3H_2O$ | Isovalerate + $HCO_3^- + NH_4^+ + 2H_2$ | +1.0 | −14.2 |
| Butyrate oxidation to acetate | Butyrate + $2H_2O$ | $2CH_3COO^- + H^+ + 2H_2$ | +11.5 | −4.2 |
| Propionate oxidation to acetate | Propionate + $3H_2O$ | $CH_3COO^- + HCO_3^- + H^+ + 3H_2$ | +18.2 | −1.3 |
| Benzoate oxidation to acetate | Benzoate + $7H_2O$ | $3CH_3COO^- + HCO_3^- + 3H^+ + 3H_2$ | +21.4 | −3.8 |
| Reductive dechlorination | $H_2 + CH_3Cl$ | $CH_4 + H^+ + Cl^-$ | −39.1 | −29.0 |

[a] For the calculation of $\Delta G^{o\prime}$, the standard conditions are solutes, 1 M; gases, 1 atm; 25°C; pH 7. For the calculation of $\Delta G'$, the "typical" conditions for an anaerobic digestion tank were estimated to be 37°C and pH 7, and the concentrations of products and reactants were as follows: glucose, leucine, benzoate, and methyl chloride ($CH_3Cl$), 10 $\mu$M; acetate, butyrate, propionate, and isovalerate, 1 mM; $HCO_3^-$ and $Cl^-$, 20 mM; $CH_4$, 0.6 atm; and $H_2$, $10^{-4}$ atm.
*Source*: Zinder, S. H. (1984). Microbiology of anaerobic conversion of organic wastes to methane: recent developments. *ASM News*, 50, 294–298.

the reactions listed in Table 14.1 – for example, the oxidation of butyrate by $H_2O$ – have a large positive $\Delta G^{o\prime}$, and would not have been predicted to occur. These unusual fermentation reactions occur in the anaerobic sludge because the partial pressure of $H_2$, one of the products in these fermentation reactions, is kept exceptionally low, at $10^{-4}$ to $10^{-5}$ atm, by the consumption of $H_2$ by $CO_2$-reducing methanogens and hydrogen-consuming acetogenic bacteria (see below).

In the methane-forming stage, strict anaerobes of the genera *Methanobacterium*, *Methanobacillus*, *Methanococcus*, and *Methanosarcina* convert the acetate, hydrogen, and carbon dioxide produced by the fermenters to methane (Box 14.2). Under the conditions prevailing in an anaerobic digester, methanogenic bacteria derive energy by converting hydrogen plus carbon dioxide and acetate to methane by the following reactions:

$$4H_2 + HCO_3^- + H^+ \rightarrow CH_4 + 3H_2O \qquad \Delta G^{o\prime} = -32.4 \text{ kcal/reaction and } \Delta G' = -7.6 \text{ kcal/reaction}$$

$$CH_3COO^- + H_2O \rightarrow CH_4 + HCO_3^- \qquad \Delta G^{o\prime} = -7.4 \text{ kcal/reaction and } \Delta G' = -5.9 \text{ kcal/reaction.}$$

The $\Delta G'$ refers to the reactant and product concentrations in an anaerobic digestion tank at 37°C at pH 7 (see Table 14.1). Note that the first of these reactions is so strongly favored that $\Delta G'$ remains negative, even at $10^{-4}$ atm of $H_2$. The cleavage of acetate to $CH_4$ plus $CO_2$ is referred as the acetoclastic reaction.

**Methanogens**

Methane ($CH_4$) is produced in many natural anaerobic habitats in which organic matter is degraded by microorganisms and carbon dioxide is the only available electron acceptor – in other words, in the absence of oxygen, sulfate, or nitrate. Alessandro Volta (1745–1827), the Italian physicist famous for his invention of the electric battery, was the first to discover, in 1776, that "combustible air" ($CH_4$) was generated in sediments rich in organic matter at the bottom of marshes, lakes, and rivers. All methanogens, microorganisms responsible for $CH_4$ production, are archaebacteria that derive their energy by reduction of several $C_1$ compounds (such as $CO_2$, acetate, or methanol) by $H_2$ with the formation of $CH_4$.

Methanogens are found in abundance in diverse natural anaerobic habitats, such as the rumen and intestinal tract of animals, freshwater and marine sediments, and waterlogged soils, and near volcanic hot springs and deep-sea hydrothermal vents. Many are thermophiles with growth temperature optima between 60°C and 95°C.

Between 500 million and 800 million tons of biologically generated $CH_4$ are released into the atmosphere per year. This is equivalent to about 2% of the carbon dioxide fixed annually by photosynthesis. A cow produces about 200 liters of $CH_4$ per day, which it releases by belching.

BOX 14.2

An important practical feature of anaerobic digestion is that most of the free energy present in the substrate is conserved in the methane that is produced. Using the methane for energy generation offsets the cost of the sewage treatment process.

## The Anammox Process: A Major Advance in Nitrogen Removal from Wastewater

Anammox stands for "anaerobic ammonium oxidation." In anammox bacteria, this is a two-step process. In the first step, under oxygen-limiting conditions, partial nitrification takes place

$$NH_4^+ + 1.5O_2 \rightarrow NO_2^- + H_2O + 2H^+$$

In the second step, anammox bacteria conserve the energy produced in the anoxic oxidation of ammonium with nitrite for autotrophic growth

$$NH_4^+ + NO_2^- \rightarrow N_2 + 2H_2O$$

The actual metabolic reaction sequence for the second step was deduced by identifying enzymes encoded by genes in the near complete genome of an anammox bacterium, *Kuenenia stuttgartiensis*. Hydrazine is an intermediate in this sequence (where $e^-$ represents an electron)

$NO_2^- + 1e^- + 2H^+ \rightarrow NO + H_2O$ (catalyzed by nitrite-nitric oxide oxidoreductase)

$NO + NH_4^+ + 3e^- + 2H^+ \rightarrow N_2H_4 + H_2O$ (catalyzed by hydrazine hydrolase)

$N_2H_4 \rightarrow N_2 + 4e^- + 4H^+$ (catalyzed by hydrazine dehydrogenase)

Hydrazine is a highly reactive chemical and is toxic to bacteria. It is not known to be a product of any other biological reaction. The enzymes, hydrazine hydrolase and hydrazine dehydrogenase, are unique to the anammox bacteria. In these bacteria, the chemical reactions involving hydrazine are confined within a unique intracellular vesicle called the anammoxosome. The impermeable membrane of this vesicle is composed mainly of specific lipids, *ladderanes*. These lipids are also unique to the anammox bacteria.

Anammox bacteria are ubiquitous, and are found in fresh water, in marine sediments, in the world's open oceans, and in wastewater treatment plants. They belong to the phylum Planctomycetes and phylogenetic analysis shows that they are most closely related to the intracellular parasites within the phylum Chlamydiae. They preferentially utilize carbon dioxide as the sole carbon source. Their importance in the global nitrogen cycle is immense. They are estimated to contribute up to 50% to the removal of fixed nitrogen from the oceans.

The anammox bacteria are effective in wastewater treatment. Paques, a Dutch company, in 2002 developed an efficient anammox plant with a capacity of 500 kg N/day to treat the rejection water of sludge digestion in a municipal wastewater treatment plant in the Netherlands. Compared to the removal of fixed nitrogen by the conventional nitrification/denitrification cycles, such as those described above for a modern sewage treatment plant, the anammox plant did not require aeration, or an organic carbon-containing energy source. Rather than producing $CO_2$, the anammox bacteria consume it. Relative to a conventional nitrification/denitrification process, the anammox process produced 88% less $CO_2$. As a consequence, power consumption was reduced by 60%, very little sludge was produced, and the overall operating cost of was lower by 90%. Anammox wastewater plants have many important potential applications beyond the treatment of reject water from sewage treatment plants. Fertilizer factories, and petroleum refineries exemplify other sources of waste streams with high levels of ammonia.

## Discharge Limits for Wastewater Treatment Plants

Discharge standards have been set for discharges of treated wastewater to fresh waters or to coastal waters. Table 14.2 shows the standards set by the European Communities. The biochemical oxygen demand ($BOD_5$) is a measure of the concentration of biodegradable organic matter present in the wastewater. It is the amount of oxygen (in milligrams $l^{-1}$) taken up by microorganisms in degrading the organic matter in a known amount of sample in five days at 20°C (Figure 14.4). The chemical oxygen demand (COD) measures the content of biodegradable and nonbiodegradable compounds in the wastewater. The COD test is performed by refluxing the sample for two hours with a known amount of potassium dichromate in concentrated sulfuric acid with silver sulfate as catalyst and mercuric sulfate to remove by precipitation any chlorides present that could interfere with the reaction.

Although nearly all the organic compounds in wastewater are oxidized under these conditions, complete oxidation of some aromatic compounds – such as benzene, toluene, or pyridine – is not achieved. The amount of oxidizable organic matter in the sample is proportional to the amount of potassium dichromate consumed in the reaction.

The $BOD_5$ values for urban domestic waste range from 200 to 600. Effluents from agriculture and the food industry have much higher values. For example, the $BOD_5$ of whey ranges from 40,000 to 50,000. As expected, COD values are higher than the corresponding BOD values. For example, the normal COD–BOD ratio for domestic sewage is 2:1. A high ratio (e.g., >4:1) suggests the presence of compounds toxic to bacteria in the sewage, such as heavy metals, whose presence leads to lower $BOD_5$ values. Such a situation would require investigation.

**TABLE 14.2 Discharge limits for urban wastewater treatment plants[a]**

| Parameter | Requirements for treatedwater discharge | Minimum percentage reduction |
|---|---|---|
| $BOD_5$ | 25 mg $O_2$ $l^{-1}$ | 70–90 |
| COD | 125 mg $O_2$ $l^{-1}$ | 75 |
| Suspended solids | 35 mg $l^{-1}$ | 90 |
| Total phosphorus | 2 mg $l^{-1}$[b] | 80 |
| Total nitrogen | 15 mg N $l^{-1}$[b] | 70–80 |

[a] Discharge limits set for wastewater treatment plants by the European Communities Council Directive concerning Urban Wastewater Treatment (91/271/EEC). Official Journal of the European Communities (1991) **L135**, 21.5.1991, 40–52.

[b] Applies to greater than 100,000 PE. The amount of organic matter produced per capita each day, expressed in kilograms $BOD_5$ per capita per day, is known as population equivalent (PE), given by PE = mean flow (l) × mean $BOD_5$ (mg $l^{-1}$)/$10^6$.

## Degradation of Synthetic Organic Compounds during Sewage Treatment: The Case of Alkylbenzene Sulfonates

Many human-made organic compounds are degraded during sewage treatment. The extent of degradation depends critically on the rate of biodegradation of the compound in question. If the rate is slow, the duration of residence of the compound in the sewage treatment plant may be too short for complete degradation. The fate of alkylbenzene sulfonate detergents serves as an illustration of this point.

A highly branched alkylbenzene sulfonate (BAS) was first introduced as a surfactant for synthetic household detergents in the 1940s and has been used worldwide since the 1950s. Usage of BAS was approximately 1.2 billion pounds in 1994. Because of the environmental concerns that we will describe below, a linear alkylbenzene sulfonate (LAS) was developed in the 1960s (Figure 14.5). Worldwide usage of LAS is about 3.2 billion pounds per year.

Alkylbenzene sulfonates enter lakes, rivers, and oceans as components of household sewage (at concentrations of 1 to 20 ppm in the sewage). Aqueous solutions of alkylbenzene sulfonates will foam at these concentrations when agitated, causing problems in sewage treatment plants and at their outfalls (Box 14.3). Moreover, at levels of 1 to 5 ppm in clean water, surfactants such as alkylbenzene sulfonates are toxic to some fish. These problems were solved by the substitution of LAS for BAS.

In spite of the fact that BAS and LAS are compounds foreign to the natural world, they are biodegradable. The rate-limiting step in the microbial decomposition of these detergents is the cleavage of the alkyl chain from the benzenesulfonate head group. Thereafter, biodegradation of the resulting

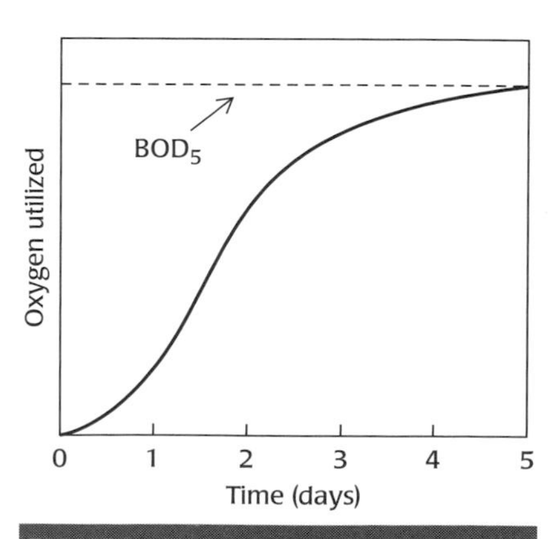


**FIGURE 14.4**

Determination of $BOD_5$.

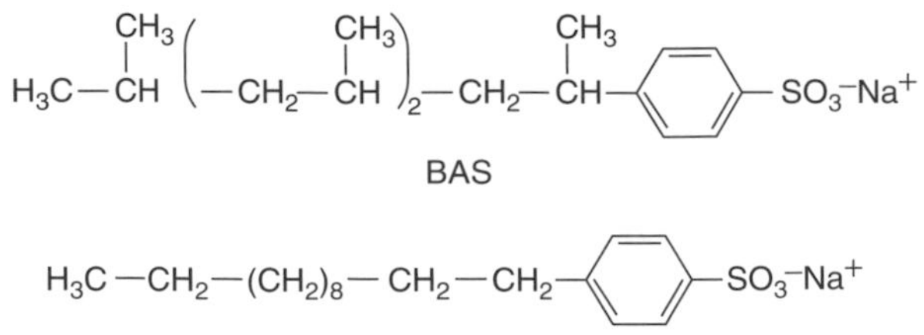

$$H_3C{-}CH(CH_3)(-CH_2{-}CH(CH_3))_2{-}CH_2{-}CH(CH_3){-}C_6H_4{-}SO_3^-Na^+$$

BAS

$$H_3C-CH_2-(CH_2)_8-CH_2-CH_2-C_6H_4-SO_3^-Na^+$$

LAS

**FIGURE 14.5**

Structures of the sodium salts of a branched (BAS) and a linear (LAS) alkylbenzene sulfonate.

fatty acid proceeds by the pathways used for the $\beta$-oxidation of naturally occurring fatty acids to $CO_2$ and water. The benzene sulfonate is degraded to $CO_2$, water, and sulfate. Because of the branching of its alkyl group, the initial step in the degradation of BAS is more than one order of magnitude slower than that of LAS. Consequently, a secondary aerobic treatment time that suffices to decrease the BOD of sewage by over 90% is grossly insufficient for the satisfactory degradation of BAS, but it does allow for the virtually complete degradation of LAS.

The case of the alkylbenzene sulfonates highlights two very important points. The first is that the complex microbial populations present in the aerobic and anaerobic sewage treatment tanks can degrade both naturally occurring compounds and synthetic ones. In fact, the degradation of a number of other industrial synthetic waste products (such as phenols and chlorobenzenes) has been shown to be particularly efficient in the presence of large amounts of sewage sludge. The second point is that it is important to conduct studies of biodegradability under real-world conditions. It is not sufficient to establish that a compound is biodegradable. It must also be shown that the compound is degraded sufficiently rapidly in the treatment facility to ensure its removal from the environment. In many products and processes, it may be possible to substitute a compound that biodegrades slowly with one that is more readily decomposed. This point, which now seems self-evident, was first appreciated when abundant foam was recognized as an obvious signature of contamination by detergent.

My first awareness of this subject came when, in the mid-1950s, I was released from the armed services and returned to a woodland near my home to observe the spring bird migration. A small river wound its way through the woodland, and to my surprise I observed suds developing wherever there was a small waterfall in the river. The suds floated downstream in swan-like masses, sometimes forming a blanket which covered many square metres of the stream. Several miles upstream I found the source of the suds: a small sewage treatment plant with an outfall in the river. As I later learned, the suds were a consequence of the use of branched-chain alkylbenzene surfactants in the manufacture of synthetic detergents. It was all too evident that the detergents were passing through the treatment plant without being broken down.

— R. T. Wright (1987). Microbial degradation of organic compounds in soil and water, in *Essays in Agricultural and Food Microbiology*, J. R. Norris and G. L. Pettipher (eds.), pp. 75–103, London: John Wiley & Sons Ltd.

**BOX 14.3**

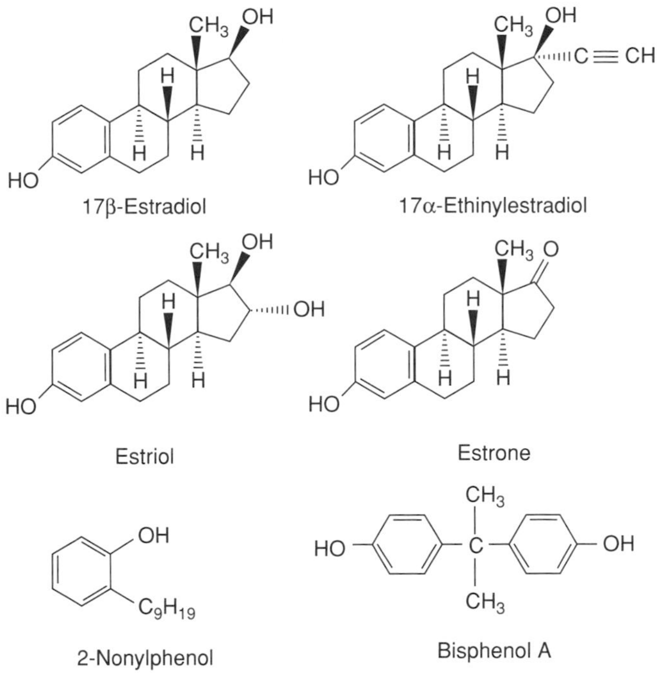

17β-Estradiol

17α-Ethinylestradiol

Estriol

Estrone

2-Nonylphenol

Bisphenol A

FIGURE 14.6

Structures of natural and synthetic compounds with estrogenic activity.

**Potential Contaminants of Drinking Water**

A recent compendium of industrially produced chemicals noted that more than 30,000 chemicals are on the market in quantities greater than one ton. Of these, more than 5000 chemicals are produced in quantities greater than 100 tons. Food additives number some 8700. More than 3300 compounds are used as drugs or in human medicine. All these compounds are present in the environment to a varying extent and may enter drinking water.

Of the many contaminants detected in drinking water, particular attention has focused in recent years on "environmental hormones." These are chemicals found in the environment that interfere with the hormonal systems of humans and animals. Compounds with estrogen activity have attracted special interest. These include the natural female sex hormone estradiol, its metabolic transformation products, estrone and estriol, and the synthetic analog ethinylestradiol, the active ingredient commonly used in oral contraceptives (Figure 14.6). These natural and synthetic estrogens are excreted by humans and partially degraded in wastewater treatment plants, but the remainder is released in the effluent. Certain chemicals used in very large amounts in industrial detergents, such as nonylphenol, or in plastics, such as bisphenol A, have estrogenic activity. The relative estrogenic activities of estradiol, estrone, estriol, and ethinylestradiol are 1:0.474:0.003:1, respectively. The average concentration of estradiol in the effluent discharge from a wastewater plant, such as that illustrated in Figure 14.2, is 2.0 ng/L. Nonylphenol has an estrogenic activity about $2.5 \times 10^{-5}$ that of estradiol. However, it is discharged at concentrations 1000-fold higher than those of estradiol and consequently contributes hormonal activity equivalent to 0.025 ng estradiol per liter. This example illustrates the complexity of assessing the potential biological outcome of simultaneous exposure to different chemical pollutants that may be present in drinking water. For most widely used organic compounds, the potential range of biological effects has yet to be explored.

## MICROBIOLOGICAL DEGRADATION OF XENOBIOTICS

As discussed at the beginning of this chapter, natural biological and geochemical processes produce enormous quantities of organic compounds with a great diversity of structures. Nearly every one of these compounds can be utilized by some microorganism as a source of energy and/or of cell building blocks. However, many of the tens of thousands of organic compounds produced artificially by chemical synthesis for industrial or agricultural purposes have no obvious counterparts in the natural world. Such synthetic novel compounds are called *xenobiotics* (*xenos* means "foreign" in Greek), and many of these compounds are stable in the environment both under aerobic and anaerobic conditions.

The ever-growing list of xenobiotics released into the environment on a large scale includes numerous halogenated aliphatic and aromatic compounds, nitroaromatics, phthalate esters, and polycyclic aromatic hydrocarbons. These compounds enter the environment through many different paths. Some, as components of fertilizers, pesticides, and herbicides, are distributed by direct application. Others, such as the polycyclic aromatic hydrocarbons, dibenzo-*p*-dioxins, and dibenzofurans, are released by combustion processes. And of course many kinds of xenobiotics are found in the waste effluents produced by the manufacture and consumption of all the commonly used synthetic products.

Various xenobiotics are found in particular environments in concentrations ranging from parts per thousand (ppt) to parts per billion (ppb). The local concentration depends on the amount of the compound released, the rate at which it is released, the extent of dilution in the environment, the mobility of the compound in a particular environment (e.g., in soil), and its rate of degradation, both biological and nonbiological. Many toxic xenobiotics present in the environment in parts per billion, levels at which toxicity cannot be demonstrated, are nevertheless strictly regulated. Such regulation is necessary when a compound becomes progressively more concentrated in each link of a food chain – a process called *biomagnification*.

The first study to measure biomagnification was carried out in Clear Lake in northern California. In 1949, Clear Lake had been treated with the persistent pesticide dichlorodiphenyldichloroethane (DDD, a close relative of DDT) at 0.01 to 0.02 ppm of water to control the gnat *Chaoborus astictopus*. By 1954, western grebes, ducklike birds, began dying around the lake. The body fat levels of DDD in the grebes were found to be 1600 ppm, some 100,000 times higher than the DDD concentration in lake water. The DDD was accumulating in progressively higher concentrations first in the plankton in the water, then in the fish that ate the plankton, and finally in the grebes that

**TABLE 14.3 Environmental protection agency (EPA) priority pollutant list[a,b]**

| | | |
|---|---|---|
| PURGEABLE (VOLATILIZABLE) ORGANIC COMPOUNDS | | |
| Acrolein | 1,1,2-Trichloroethane | Bromoform |
| Acrylonitrile | 1,1,2,2-Tetrachloroethane | Dichlorobromomethane |
| **Benzene**[c] | Chloroethane | Trichlorofluoromethane |
| **Toluene** | 2-Chloroethyl vinyl ether | Dichlorodifluoromethane |
| Ethylbenzene | Chloroform | Chlorodibromomethane |
| Carbon tetrachloride | 1,2-Dichloropropane | **Tetrachloroethylene** |
| Chlorobenzene | 1,3-Dichloropropene | **Trichloroethylene** |
| 1,2-Dichloroethane | **Methylene chloride** | Vinyl chloride |
| **1,1,1-Trichloroethane** | Methyl chloride | 1,2-*trans*-Dichloroethylene |
| 1,1-Dichloroethane | Methyl bromide | bis(Chloromethyl) ether |
| 1,1-Dichloroethylene | *1,2-Dibromoethane* | |
| COMPOUNDS EXTRACTABLE INTO ORGANIC SOLVENT UNDER ALKALINE OR NEUTRAL CONDITIONS | | |
| 1,2-Dichlorobenzene | Di-*n*-octyl phthalate | Benzo(k)fluoranthene |
| 1,3-Dichlorobenzene | Dimethyl phthalate | Benzo(a)pyrene |
| 1,4-Dichlorobenzene | Diethyl phthalate | Indeno(1,2,3-c,d)pyrene |
| Hexachloroethane | **Di-*n*-butyl phthalate** | Dibenzo(a,h)anthracene |
| Hexachlorobutadiene | Acenaphthylene | Benzo(g,h,i)perylene |
| Hexachlorobenzene | Acenaphthene | 4-Chlorophenyl phenyl ether |
| 1,2,4-Trichlorobenzene | Benzyl butyl phthalate | 3,3′-Dichlorobenzidine |
| bis(2-Chloroethoxy)methane | Fluorene | Benzidine |
| **Naphthalene** | Fluoranthene | bis(2-Chloroethyl) ether |
| 2-Chloronaphthalene | Chrysene | 1,2-Diphenylhydrazine |
| Isophorone | Pyrene | Hexachlorocyclopentadiene |
| Nitrobenzene | **Phenanthrene** | *N*-Nitrosodiphenylamine |
| 2,4-Dinitrotoluene | **Anthracene** | *N*-Nitrosodimethylamine |
| 2,6-Dinitrotoluene | Benzo(a)anthracene | *N*-Nitrosodi-*n*-propylamine |
| 4-Bromophenyl phenyl ether | Benzo(b)fluoranthrene | bis(2-Chloroisopropyl) ether |
| **bis(2-Ethylhexyl)phthalate** | *3,3′-Dichlorobenzidine* | *bis(2-Chloro-1-methylethyl) ether* |
| COMPOUNDS EXTRACTABLE INTO ORGANIC SOLVENT UNDER ACID CONDITIONS | | |
| **Phenol** | 4,6-Dinitro-*o*-cresol | 2,4-Dichlorophenol |
| 2-Nitrophenol | Pentachlorophenol | **2,4,6-Trichlorophenol** |
| 4-Nitrophenol | 4-Chloro-*m*-cresol | 2,4-Dimethylphenol |
| 2,4-Dinitrophenol | 2-Chlorophenol | *2,3,4,6-Tetrachlorophenol* |
| PESTICIDES, POLYCHLOROBIPHENYL (PCBS) AND RELATED COMPOUNDS | | |
| $\alpha$-Endosulfan | 4,4′-DDE | Toxaphene |
| $\beta$-Endosulfan | 4,4′-DDD | Aroclor 1016[d] |
| Endosulfan sulfate | 4,4′-DDT | Aroclor 1221 |
| $\alpha$-BHC | Endrin | Aroclor 1232 |
| $\beta$-BHC | Endrin aldehyde | Aroclor 1242 |
| $\gamma$-BHC | Heptachlor | Aroclor 1258 |
| Aldrin | Heptachlor epoxide | Aroclor 1254 |
| Dieldrin | Chlordane | Aroclor 1260 |
| *$\alpha,\beta,\gamma,\delta$-Lindane* | – | 2,3,7,8-Tetrachlorodi |
| *Camphechlor* | | benzo-*p*-dioxin (TCDD) |

(*continued*)

**TABLE 14.3 (*continued*)**

| | | |
|---|---|---|
| METALS | | |
| Antimony | Copper | Selenium |
| Arsenic | Lead | Silver |
| Beryllium | Mercury | Thallium |
| Cadmium | Nickel | Zinc |
| Chromium | | |
| MISCELLANEOUS | | |
| Cyanides | Asbestos (friable) | |

[a] The EPA list of priority pollutants was developed as a consequence of a court consent decree on June 7, 1978, in the settlement of a suit brought against the EPA by several plaintiffs (Natural Defense Council, Inc.; Environmental Defense Fund, Inc.; Businessmen for the Public Interest, Inc.; National Audubon Society, Inc.; and Citizens for a Better Environment) for failing to implement portions of the Federal Water Pollution Control Act (P.L. 92–500). Pollutants shown in *italics* were added to the list after 1978. The list is current as of March 2006.
[b] The priority pollutants are divided into groups on the basis of properties that are relevant to the analysis of these compounds in industrial wastewaters.
[c] Compounds shown in **boldface** were found in 10% or more of over 2600 samples of wastewater from 32 different industrial categories analyzed in August 1978.
[d] Aroclor designations are explained in Box 14.4.

ate the fish. Many other persistent fat-soluble organic compounds become increasingly concentrated as they travel up the food chain. Prominent examples are the phthalate esters and the PCBs.

## PRIORITY POLLUTANTS AND THEIR HEALTH EFFECTS

The U.S. Environmental Protection Agency's list of priority pollutants (Table 14.3) includes widely used industrial solvents, building blocks of plastics, PCBs, pesticides, and certain potent carcinogens. Some of these compounds are or have been produced in massive amounts. For example, billions of pounds of *o*-phthalic and terephthalic acids have been used in the plastics and textile industries. Phthalic acid esters are the most important class of plasticizers for cellulose and vinyl plastics and are also used in insect repellents, munitions, and cosmetics and as pesticide carriers. It is estimated that more than 50 million pounds of phthalate esters enter the environment yearly in the United States by leaching out of solid plastic wastes and as a result of direct application (e.g., as pesticide carriers). Phthalate contamination goes hand in hand with the pervasive use of plastics. In the early 1970s, phthalate esters were discovered in blood that had been collected for transfusion and stored in plastic bottles.

Commercial PCBs are mixtures prepared by partial chlorination of biphenyl (Box 14.4). They have a wide range of physical properties, chemical stability, and miscibility with organic solvents, all as the result of the degree of biphenyl chlorination in a given mixture. Thus, PCB formulations have served a wide range of purposes, including use as hydraulic fluids, plasticizers, adhesives, lubricants, flame retardants, and dielectric fluids in capacitors and transformers. Widespread PCB pollution was first detected in 1966 during analysis of environmental samples for DDT, and their manufacture

**Polychlorobiphenyls**

PCBs are a class of 209 distinct synthetic chemical compounds, in which one to 10 chlorine atoms are attached to biphenyl.

3 2 2' 3'
4 4'
5 6 6' 5'
BIPHENYL

The empirical formula for PCBs is $C_{12}H_{10-n}Cl_n$, where n = 1–10. Closely related compounds such as these are called *congeners*.

PCB *isomers* are compounds with the same number of chlorine atoms – for example, 2′,3,4-trichlorobiphenyl and 2′,4,4′-trichlorobiphenyl.

PCBs were manufactured and sold as complex mixtures differing in their average chlorination level. The manufacturers attached numbers to the trade name of their product that conveyed information about the weight percent chlorine in the mixture. For example, Aroclor 1242 (Monsanto, USA; see Table 14.3) indicates 12 carbon atoms and 42% chlorine by weight. Of the 209 possible congeners, only about half were actually produced in the synthesis of PCBs because of steric hindrance.

BOX 14.4

stopped by 1977. Between 1929 and 1977, however, about 1.2 billion pounds of PCBs were produced in the United States, and it is estimated that several hundred million pounds have been released into the environment. PCBs biodegrade very slowly and will persist in the environment for decades.

Thousands of toxic waste sites in the United States release compounds on the priority pollutant list, as well as many others, into the environment. Substantial contamination of land, surface water, groundwater, and air in virtually every part of the country has been massively documented. Where the population has been exposed to high levels of such contaminants, there is convincing evidence of adverse health effects. Certain pollutants have achieved particular notoriety because of acute ill effects of accidental high-level exposures.

There are strong reasons for minimizing the future release of chemicals into the environment and eliminating those already present. It is well documented, for example, that chemicals not known to be harmful to humans can be highly toxic to other organisms. And, as for DDT, many toxic chemicals enter food chains at low levels and through biomagnification reach concentrations sufficiently high to cause health problems for humans and other living organisms. Furthermore, health effects may surface a long time after the exposure, when the cause-and-effect relationship will be difficult to prove.

## THE MICROBIOLOGICAL BASIS OF BIODEGRADATION

Natural microbial communities are complex assemblages in which the various microorganisms (as in Table 14.4) are highly interdependent. This interdependence is evident in the high frequencies of commensalism and

**TABLE 14.4 Population densities of soil microorganisms in samples obtained from common soils in a temperate region**

| Organism | Cells per gram |
|---|---|
| Bacteria[a] | $10^6$–$10^9$ |
| Actinomycetes | $10^5$–$10^8$ (based on spore number) |
| Filamentous hyphae | $10^1$–$10^2$ meters (hyphal length) |
| Yeasts | $10^3$ |
| Algae and cyanobacteria | $10^2$–$10^4$ |
| Protozoans | $10^4$–$10^6$ |

[a] Bacteria other than actinomycetes or cyanobacteria.
*Source*: Based on data in Table 11 from Yanagita, T. (1990). *Natural Microbial Communities. Ecological and Physiological Features*, Tokyo: Japan Scientific Societies Press.

mutualism that are observed. *Commensalism* is an interactive association between two populations of different species that live together in which one population benefits from the association while the other is not affected. *Mutualism* is a symbiosis, an interaction in which two organisms of different species live in close physical association to their mutual benefit.

Commensalism takes different forms. Many microorganisms isolated from soils require amino acids or vitamins for growth, whereas many other species produce those compounds. For example, some 19% of such isolates require thiamine, whereas approximately 36% of the isolates secrete it. The corresponding numbers for vitamin $B_{12}$ are about 7% and about 20%, respectively. Thus, *cross feeding* is a general feature of natural microbial communities. The interaction between organisms through the production and consumption of oxygen is of particular importance; the ability of anaerobes to survive in surface layers of soil depends on the efficient consumption of oxygen by aerobes. Organic acids produced by fungal decomposition of cellulose are utilized as nutrients by bacteria. Ethanol produced from sugars in fruit by yeasts is oxidized to acetic acid by *Acetobacter* species. Methane, formed by methane-producing bacteria (*methanogens*) by anaerobic transformation of various organic compounds, is oxidized aerobically by methane-oxidizing bacteria (*methylotrophs*).

Mutualistic interactions are likewise diverse. Where nitrogen fixers and cellulose decomposers coexist, for example, each organism utilizes compounds produced by the other. The anaerobic *Desulfovibrio* uses $SO_4^{2-}$ as a terminal electron acceptor in its energy-producing respiratory pathway and converts it to $H_2S$; purple sulfur bacteria, which use sunlight for photosynthetic production of ATP, utilize the $H_2S$ as an electron donor and oxidize it to $SO_4^{2-}$. *Lactobacillus arabinosus* and *Streptococcus faecalis* depend on each other to satisfy nutritional requirements; *L. arabinosus* makes folic acid required by *S. faecalis*, whereas *S. faecalis* makes phenylalanine required by *L. arabinosus*.

As a general rule, organic compounds are more effectively degraded in environments containing many microorganisms than in a pure culture of a single organism. This is the result of several factors. The range of degradative capabilities represented in a complex community of many bacteria and fungi is far greater than the capabilities of any single organism alone. Second, the product of partial biodegradation of a xenobiotic by one organism may serve as a substrate for another organism. The concerted action of several different organisms may lead to complete mineralization of the xenobiotic. A microbial community is also likely to be more resistant to a toxic product of biodegradation, because one of its members may be able to detoxify it.

A microbial community is dynamic: its composition responds to environmental conditions, adapting with time to exploit available nutrients in the most effective manner. Thus, when a new biotransformable organic compound is presented at a constant level to such a community, a period of adaptation ensues, after which the rate of biotransformation of the organic compound is generally seen to be much increased. This reflects a selective enrichment within the community for organisms resistant to the new organic compound and able to utilize it or transform it.

The fate of organic compounds introduced into the soil is determined by a combination of physical, chemical, and biological factors. A particular molecule may be removed by volatilization or leaching, or be strongly adsorbed and remain near the site of entry for a long time. The molecule may be degraded photochemically, or it may undergo abiotic oxidation or hydrolysis. Finally, the molecule may undergo biodegradation through the action of bacteria and fungi. In some instances, the products of nonbiological and biological degradation are identical. In other instances, nonbiological degradation is very slow relative to biotransformation and gives rise to different products.

Laboratory studies of the fate of single organic compounds do not provide clear, accurate forecasts of the persistence of xenobiotics in the environment because, in the real world, microbial communities are likely to be exposed to mixtures of organic compounds and heavy metals. Many strains are unable to grow in the presence of heavy metals; the general order of resistance, from most to least resistant, is fungi, actinomycetes, Gram-negative bacteria, Gram-positive bacteria. Thus, the bacteria that are able to degrade certain persistent organic chemicals, such as polycyclic aromatic hydrocarbons or chlorinated organic compounds, are likely to disappear from the environment when heavy metals are also present. As a consequence, those organic chemicals will persist much longer in such an environment than simple laboratory experiments might suggest.

Two other key factors influencing the degradation of xenobiotics are the phenomena of gratuitous biodegradation and cometabolism. *Gratuitous biodegradation* describes the situation in which an enzyme is able to transform a compound other than its natural substrate. The prerequisites are that the unnatural substrate is able to bind to the active site of the enzyme and to do so in such a manner that the enzyme can exert its catalytic activity. And as we have seen, bacteria and fungi are so diverse in their metabolic capabilities that they produce enzymes able to act on a wide range of organic molecules. *Cometabolism*, on the other hand, is the ability of an organism to transform a *nongrowth substrate* as long as a growth substrate or other transformable compound is also present. This requirement distinguishes cometabolism from gratuitous metabolism. A nongrowth substrate is one that cannot serve as the sole source of carbon and energy for a pure culture of a bacterium and hence cannot support cell division.

## TYPES OF BIOREMEDIATION

*Bioremediation* is defined as a spontaneous or managed process in which biological (especially microbiological) catalysis acts on pollutants and thereby remedies or eliminates environmental contamination. Bioremediation is currently being used to decrease the organic chemical waste content of soils, groundwater, effluent from food processing and chemical plants, and oily sludge from petroleum refineries. Bioremediation techniques fall into four categories: *in situ* treatment, composting, landfarming, and aboveground reactors.

### *In situ* Bioremediation

In Latin, *in situ* means "in the original place." Thus, *in situ* bioremediation relies on the indigenous microbial fauna of subsurface soils and groundwater. It rests on the premise that the microorganisms already present in a contaminated site have adapted to the organic chemical wastes there and are able to degrade some or all of the components of these wastes.

The degradation by these adapted organisms will proceed until some nutrient or electron acceptor reaches a limiting concentration. Oxygen level is most often the limiting factor, but nitrate and phosphate limitation frequently plays a role too. The stimulation of natural biotransformations by adding such nutrients to the environment is called *enhanced* in situ *bioremediation.*

The cleanup of the *Exxon Valdez* oil spill in Alaska provided a large-scale field test of the effectiveness of enhanced *in situ* bioremediation. In March 1989, the supertanker *Exxon Valdez* ran aground on Bligh Reef in Prince William Sound. The resulting spill of about 11 million gallons of crude oil severely affected 350 miles of shoreline in the sound. In this case, fertilizers were used to accelerate the removal of oil from the beaches, supplying extra nutrients that would otherwise have diminished to limiting concentrations. A single application of inorganic fertilizer was shown to speed the disappearance of oil by a factor of two to three over the rate on untreated shoreline. Moreover, the accelerated rate was maintained for several weeks, even after nutrient concentrations returned to background level. Samples of oil taken at the end of that time from surfaces of treated beaches showed changes in composition consistent with extensive biodegradation. Enhanced *in situ* bioremediation offers several potential advantages in the elimination of hazardous wastes: it is cheaper than incineration, and workers are not exposed to the risks associated with excavation and removal of contaminated soils. It is well suited to treating large areas contaminated with low levels of wastes.

### *In situ* Remediation of Oil Spills

Seepage from natural oil fields accounts for continuous release of oil into the oceans, but most of the oil comes from human activities. Tanker accidents lead to massive releases of oil and, as described above, cause major environmental damage. Such accidental releases, coupled with deliberate discharge from tankers and processing sites, amount to about 1.3 million tons of petroleum annually. Over time indigenous bacteria slowly degrade the oil in the affected areas. The degradation can be accelerated several-fold by provision of nitrogen and phosphorus compounds readily available in appropriate fertilizers. The identity of the microorganisms that play key roles in the *in situ* biodegradation was not known. Recently, *Alcanivorax borkumensis*, a ubiquitous marine $\gamma$-proteobacterium that preferentially utilizes alkanes as substrates, was shown to play a major role in the bioremediation of oil-contaminated marine environments. *A. borkumensis* utilizes a broad spectrum of petroleum constituents, including *n*- and branched alkanes, *n*-alkylcycloalkanes, and *n*-alkylbenzenes, and also degrades pristane. Pristane, a branched alkane, 2,6,10,14-tetramethylpentadecane, whose structure is shown below,

$$(CH_3)_2CH(CH_2)_3CH(CH_3)(CH_2)_3CH(CH_3)(CH_2)_3CH(CH_3)_2$$

produced by some marine zooplankton, is an important natural substrate for *A. borkumensis*.

*A. borkumensis* is present in low numbers in unpolluted environments, but becomes abundant in petroleum-contaminated seawater, particularly when nitrogen and phosphorus nutrients are supplied. In open-ocean and costal waters, this organism may represent up to 90% of the oil-degrading community. The complete *A. borkumensis* genome reveals features that make this marine organism such a key player in oil degradation. Many marine bacteria degrade hydrocarbons. It is that *A. borkumensis* lives exclusively on alkanes that sets it apart. This organism does not compete for commonly used substrates, such as sugars and amino acids, or the light aromatic fractions of petroleum, utilized by many other marine bacteria. *A. borkumensis* is equipped with many and diverse alkane hydroxylases that efficiently degrade branched alkanes. The genome also specifies nutrient uptake systems, particularly those for organic and inorganic nitrogen compounds, capabilities for biofilm formation at the oil-water interface, production of biosurfactants that most likely increase the bioavailability of oil constituents, and the capacity for niche-specific stress responses. Understanding the factors that underlie the success of *A. borkumensis* is of broad value in pointing to ways in which to improve bioremediation efforts.

## Composting

Compost, a mixture of soil, partially decayed plants, and sometimes manure and commercial fertilizer, is very rich in microorganisms. Composting has long been used by farmers and gardeners to make soils more fertile and improve crop yields, and shows promise in the treatment of high concentrations of resistant chemical wastes, as illustrated by a recent application to the degradation of explosives.

In the past, waste streams from the explosives industry were often discharged to settling basins or lagoons. The waste streams contained 2,4,6-trinitrotoluene (TNT), hexahydro-1,3,5-trinitro-1,3,5-triazine (RDX), octahydro-1,3,5,7-tetranitro-1,3,5,7-tetraazocine (HMX), and *N*-methyl-1,2,4,6-tetranitroaniline (Tetryl) (Figure 14.7). At ambient temperature, these compounds are solids, sparingly soluble in water, and as a result, they have

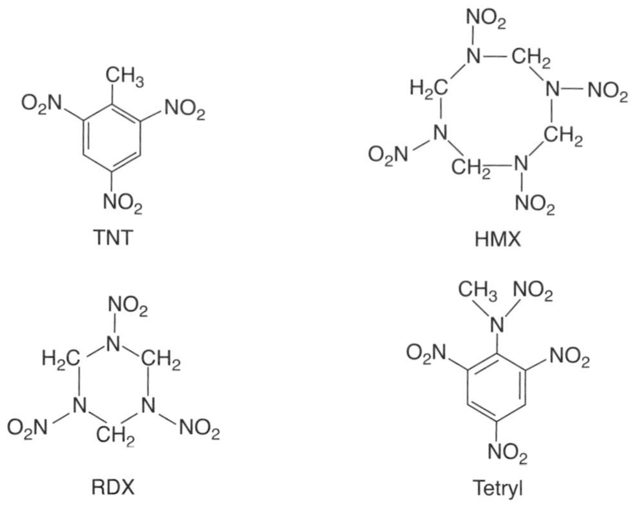

FIGURE 14.7
Structures of explosives.

largely remained in the soil at the discharge sites. The current most common strategy for decontaminating soils that contain explosives is incineration, a process that costs at least $300 per ton of soil to be treated. For the estimated 5,200,000 tons of soil requiring decontamination, this treatment would cost in excess of $1.5 billion. Thus, there is considerable incentive to develop cheaper methods, and composting is emerging as a potential low-cost alternative. Studies in the 1970s showed that anaerobic bacteria are able to degrade RDX and HMX, whereas TNT can be biodegraded under both aerobic and anaerobic conditions. For example, *Desulfovibrio* species can use RDX or HMX as a sole nitrogen source, whereas *Klebsiella pneumoniae* degrades RDX to formaldehyde, $CO_2$, and $H_2O$. Biodegradation or biotransformation of over 90% of these explosives was achieved within 80 days in a compost pile maintained at 55°C. After 150 days, a starting concentration of 18,000 mg of explosives per kilogram of soil was reduced to 74 mg/kg.

### Landfarming

Landfarming is used to dispose of oily sludge from petroleum refinery operations. In this process, oily sludge from refinery wastes is mixed with soil and subjected to enhanced *in situ* bioremediation. The sludge may be pretreated or not. Biological pretreatment of refinery effluents will partially mineralize the organic waste components; the residual solid waste (sludge) then has a high content of aromatic hydrocarbon compounds and is low in aliphatic hydrocarbon compounds (5% to 10% by weight). In contrast, untreated settled solids (like the solids from tank bottoms) contain high amounts of aliphatic hydrocarbons (30% to 50%) and inorganic solids (silt).

The terrain of a landfarm must be flat to minimize runoff, the soil should be light and loamy for adequate aeration, and a clay layer should underlie the porous surface soil to reduce the possibility of groundwater contamination through seepage. The landfarm is graded to a very gentle slope to prevent standing water from collecting after rain, and is surrounded by a moat to contain runoff. The geographic location of the landfarm is also chosen with precipitation and temperature in mind. The air-filled pores in the light landfarm soil ensure rapid access of oxygen to the organisms there, but excessive rain would waterlog the soil and eliminate the air-filled pores. About 20% water saturation of the soil is sufficient for maximal oil degradation. The optimal temperature range for biodegradation is 20°C to 30°C, whereas most activity ceases below 5°C.

Inorganic fertilizer is applied to the site to provide fixed nitrogen, and phosphate and pulverized limestone ($CaCO_3$) are added to raise the pH of the soil–waste mixture to about 7.8. For untreated sludge, maximal oil biodegradation rates in soil are achieved at a hydrocarbon load of 5% to 10% by weight, that is, about 100 to 200 metric tons of hydrocarbon per hectare (an area of about 12,000 square yards). Under favorable conditions, a 5% application can be repeated at intervals of approximately four months. In such landfarms, approximately 50% to 70% of the applied organic waste is degraded before the next batch of sludge is applied.

A disadvantage of landfarming is that the process is slow and incomplete. Moreover, the heavy metal constituents of the sludge gradually accumulate in the landfarm soil. Consequently, a plot of land used intensively as a landfarm cannot later be used for growing crops or grazing livestock.

### Aboveground Bioreactors

Aboveground bioreactors are based on the same technology as fermenters. They are used for the treatment of either excavated soil or groundwater containing high levels of contaminants (e.g., chemical landfill leachates). Contaminated soil is mixed with water and the slurry is introduced into the reactor. Granulated charcoal, plastic spheres, glass beads, or diatomaceous earth provide a large surface area for microbial growth in such bioreactors. The large surface area of the microbial biofilm that forms on such supports leads to a rapid rate of biodegradation. The microbial inoculum may come from an indigenous population at the contaminated site, from activated sludge from a sewage treatment plant, or from a pure culture of an appropriate organism. Because the reactors are enclosed, the use of genetically engineered organisms as inocula is also feasible. Bioreactors can be used in series to accomplish different kinds of degradation. For example, the first reactor can be operated in an anaerobic mode and its effluent transferred to a second reactor operated in an aerobic mode. Some biotransformations, such as dehalogenations of certain compounds, proceed optimally under anaerobic conditions, whereas mineralization requires aerobic conditions.

## CHALLENGES OF EVALUATING *IN SITU* BIODEGRADATION

As we have seen, the fate of an organic compound introduced into the environment depends on the properties of both the compound and the site. The compound may be tightly adsorbed to the soil at the site, it may be loosely bound and leach away from the application site into deeper layers of the soil and into groundwater, it may be photooxidized or decomposed by other abiotic processes, or it may be taken up by plants or transformed by microorganisms. Any or all of these events may contribute to the real or apparent time-dependent disappearance of the compound from the site.

Attempts to decontaminate the IBM Dayton hazardous waste site in New Jersey illustrate the strength with which organic chemicals can adhere to soil particles and the potential difficulty of extracting even highly soluble contaminants by running water through the soil. Water flows preferentially through high-permeability zones in the soil and equilibrates slowly with any water present in the low-permeability zones. The groundwater at the IBM Dayton site had been contaminated with about 400 gallons of 1,1,1-trichloroethane and tetrachloroethylene. The maximum concentrations recorded were 9600 ppb of 1,1,1-trichloroethane and 6130 ppb of tetrachloroethylene. Between 1978 and 1984, pumping water through the site at an average on-site extraction rate of 300 gallons/minute, lowered the concentrations of these compounds in the water to below 100 ppb. The pumping was suspended in 1984, and by 1988 the tetrachloroethylene concentrations in the groundwater had risen to 12,560 ppb. In effect, the chlorohydrocarbons adsorbed to the soil, or sequestered in fine pores, represented a practically inexhaustible slow-release reservoir of pollutant.

Laboratory simulations of on-site conditions are rarely authentic enough to provide trustworthy insights into a field situation. It would take an extraordinary effort at simulating real-life environmental conditions to answer central questions such as: What fraction of a pollutant is strongly adsorbed to soil or sediment? How much of the pollutant is destroyed by abiotic processes and how much is biodegraded? What are the rates of these processes at different pollutant concentrations? What is the response of the microbial community to the introduction of the pollutant? What is the habitat of the organisms that contribute most decisively to the degradation of the pollutant? Is the transformation or degradation favored by oxygen-rich or oxygen-poor conditions? What is the impact of natural variation in other conditions, such as temperature or pH? There are very few studies that can answer even a few of these questions about the *in situ* fate of an important pollutant. A well-designed quantitative study of pentachlorophenol degradation in an experimental channel fed by Mississippi River water, described below, represents an attempt at a comprehensive analysis.

## Degradation of Pentachlorophenol in an Artificial Freshwater Stream

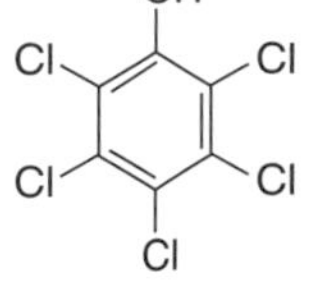

Pentachlorophenol (PCP; Figure 14.8) is a compound generally toxic to living organisms. First introduced during the 1930s as a wood preservative, it has proved effective as a general-purpose killer of algae, bacteria, fungi, weeds, mollusks, and insects in a variety of agricultural and industrial settings (though commercial wood treatment remains the major application). Worldwide production of PCP is about 50 million kilograms per year. PCP is commonly present in streams and groundwater in concentrations of

**FIGURE 14.8**

Pentachlorophenol.

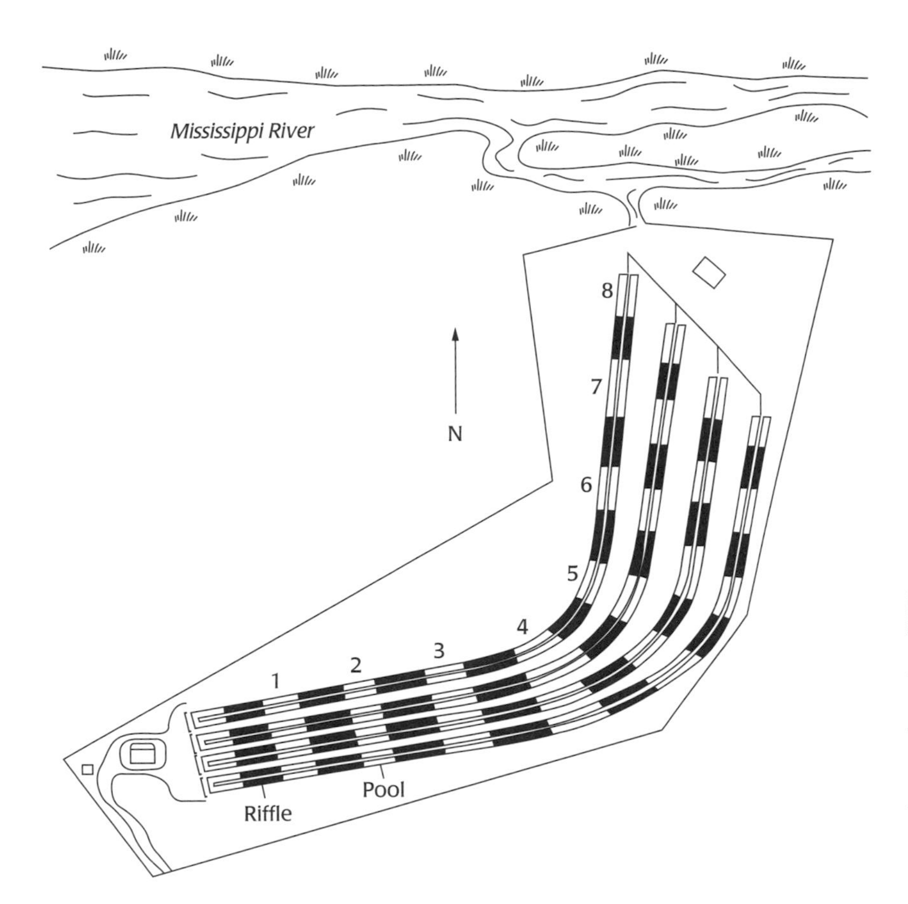


FIGURE 14.9

Location and configuration of outdoor artificial streams at the Monticello Ecological Research Station of the U.S. Environmental Protection Agency at Duluth, Minnesota. [Adapted from Arthur, J. W., Zischke, J. A., and Erickson, G. L. (1982). Effect of elevated temperature on macroinvertebrate communities in outdoor experimental channels. *Water Research*, 16, 1465–1474.]

micrograms per liter. The purpose of the study was to predict the fate of PCP in these natural aquatic systems. This was a field study carried out during the summers of 1982 and 1983 at the Monticello Ecological Research Station of the U.S. Environmental Protection Agency at Duluth, Minnesota. The station has outdoor experimental channels, each 488 meters in length, fed on a year-round basis by water pumped from the Mississippi River. The stretch of each channel that is utilized for experiments consists of eight ponds with mud at the bottom alternating with eight coarse gravel riffles (Figure 14.9). A riffle is a rocky sandbar lying just below the surface of a waterway. Of the various water plants (macrophytes) colonizing the ponds, the predominant species included *Potamogeton crispus*, a rooted pondweed, and *Lemna minor*, a floating (nonrooted) aquatic plant.

These channels are well suited for the study of factors that contribute to the degradation of xenobiotics. They are fed from a natural source of water and contain a diversity of microbial habitats: the water column, the microaerophilic sediment surface of the mud at the bottom of the ponds, the anoxic deeper layers of sediment, the surfaces of water plants, and the rock surfaces in the riffles. The channels were treated continuously for a period of 88 days with PCP, introduced as a concentrated solution of its sodium salt. The effects discussed below were observed in a channel treated with 144 $\mu$g of PCP per liter of water. Rates of photodegradation of PCP by sunlight were determined by suspending glass vials containing the appropriate PCP solution at known depths in the channel pool. Analyses of PCP degradation rates by microorganisms in different habitats were performed using the following samples:

- *Rock surfaces.* Rocks collected individually from the riffles were placed in beakers in a known volume of water containing a known concentration of PCP from the same riffle.
- *Sediment cores.* Cores, 3 cm in diameter, were removed from the pool bottom. The degradation of PCP was then measured under different conditions: aerobic (bubbling air above the sediment surface), microaerophilic (cores left open to air, but otherwise undisturbed), and anaerobic (bubbling high purity nitrogen through the cores).
- *Macrophyte surfaces.* The top portions of *Potamogeton* plants were collected from a pool and *Lemna* were scooped up from the surface. These plants were then carefully submerged in beakers of PCP-containing water from the same site.
- *Microorganisms floating free in the water column and those attached to particles.* Samples of water from various locations were divided into two equal portions. One portion was passed through a 1-$\mu$m filter to remove suspended particles but leave free-floating bacteria in the filtrate; the other portion was unfiltered.

The following observations were made:

- Microbial degradation of PCP in the treated channel became significant about three weeks after the PCP was first introduced, as indicated by (a) a sharp decline in PCP concentration down the length of the dosed channel; (b) rapid degradation of PCP by microorganisms in samples removed after week 4 from a PCP-dosed, but not a control, channel; (c) the appearance in the dosed channel of bacteria capable of mineralizing uniformly-labeled [$^{14}$C]PCP with release of $^{14}CO_2$; (d) a large decline in the PCP concentrations in the sediment between weeks 3 and 5.
- Laboratory studies showed that the timing of the appearance of PCP-degrading activity conformed to the time that would have been necessary for the selective enrichment of an initially low population of PCP-degrading microorganisms in the channel. Many of the pure cultures of bacteria isolated from dosed channels were able to use PCP as a sole source of carbon and energy for growth. One such organism grew on PCP at concentrations as high as 100 mg/L, releasing all of the organically bound chlorine as $Cl^-$.
- After the microbial community had fully adapted, water passing through the channel was cleansed of 50% to 60% of its PCP. Microbes, especially those attached to rock and plant surfaces, were responsible for most of the observed degradation.
- The rate of PCP disappearance in the water column above the sediment cores was more rapid under aerobic than under anaerobic conditions.

- The rate of PCP degradation was virtually temperature independent during the summer, when water temperatures ranged from 19°C to 30°C. However, degradation gradually slowed at lower temperatures and ceased at 4°C.
- Photodegradation of PCP was rapid at the water's surface but decreased rapidly with depth, owing to the attenuation of light by suspended particles and dissolved materials in the channel water. Depending on sunlight, photodegradation accounted for a 5% to 28% decline from the initial PCP concentration during the water's passage down the channel. Adsorption, sedimentation, or volatilization of PCP and its uptake by living organisms accounted for less than a 5% decrease in unacclimated water (i.e., in water immediately after the initial addition of PCP).

The study concluded that PCP is degraded in the aquatic environment and that microorganisms attached to surfaces are responsible for most of this degradation. The biodegradation of PCP requires an adaptation period on the order of weeks, but once the microbial populations have adapted, the degradation process is quite rapid, with a PCP half-life of less than 12 hours. PCP mineralizing activity is greatly reduced at low temperatures, and stream temperatures in northern climates may be too low for biodegradation to occur during much of the year. Other investigations had shown that contamination with PCP is ubiquitous, and low background levels of PCP were found in Mississippi River sediments and macrophytes and also in the sediments of the control channels. It is thus possible that the channels were "primed" with PCP before the start of dosing and that some enrichment for PCP degraders preceded the start of the study.

## GENETIC AND METABOLIC ASPECTS OF BIODEGRADATION

Much of the understanding of microbial genetics and metabolic regulation has come from extensive studies of the enteric bacterium *E. coli*. This organism utilizes a wide range of substrates for growth. In general, the transcription of an operon encoding a particular catabolic pathway in *E. coli* is induced only when a relevant substrate is present. This is a common control mechanism in microorganisms that occupy ecological niches in which the type and availability of substrates vary in space and time.

Soil is an example of an important and extensive habitat that varies from one spot to the next and from one time to another in the nature of the organic compounds present. The territory that a particular microorganism can colonize may be limited to a patch of a few square feet and to a depth of an inch or less. An organism capable of using many different organic compounds as a sole source of carbon and energy for growth is likely to flourish in more patches than one fastidious about its diet. Gram-negative, rod-shaped, polarly flagellated, nonsporulating bacteria – traditionally grouped as "pseudomonads" – exemplify such versatile organisms common in soil. Taxonomically the pseudomonads form a very large and heterogeneous group (see Chapter 1 and Box 14.5). One of their hallmarks is the ability to grow on any one of a large number of organic compounds, including aromatic

> With the advent of 16S RNA–based phylogeny, members of the traditional group of pseudomonads were found to be scattered among the $\alpha$-, $\beta$-, and $\gamma$-proteobacteria. At present, the genus *Pseudomonas* is restricted to species phylogenetically related to its type species *Pseudomonas aeruginosa*, a $\gamma$-proteobacterium. All the other pseudomonads belonging to the $\alpha$- or $\beta$-proteobacteria have been allocated to new genera, such as *Brevundimonas*, *Burkholderia*, *Comamonas*, *Ralstonia*, and *Sphingomonas*.
>
> **BOX 14.5**

hydrocarbons. Consequently, such bacteria are important in the purification of wastewater and the cleanup of oil spills.

Some of the versatility and adaptability of soil microorganisms stems from their possession of *catabolic plasmids*, plasmids that specify a degradative pathway. Bacteria carrying catabolic plasmids have been isolated principally from soil, and the majority of these strains had been classified as pseudomonads, although catabolic plasmids have also been found in organisms belonging to many other genera isolated from a wide variety of environments. The majority of catabolic plasmids are *self-transmissible*, and many have a broad host range. That is, transmissible catabolic plasmids represent a pool of metabolic potential available to many strains in a microbial community through interspecies transfer of genetic information.

The proliferation of catabolic plasmids is analogous to that of R plasmids, which carry genes conferring antibiotic resistance. R plasmids spread through bacterial populations under antibiotic selection pressure. Similarly, the ability to utilize a novel source of nutrient or to eliminate a potentially toxic compound will promote the spread of a catabolic plasmid. For example, unrestricted use of a pesticide will frequently result in the development of microbial populations capable of degrading it. The ability to degrade the pesticide spreads among the different bacterial strains in the natural population by interspecies transfer of a catabolic plasmid that carries the genes for the degradative enzymes.

Catabolic plasmids have been found that encode enzymes for degrading naturally occurring compounds such as camphor, octane, naphthalene, salicylate, and toluene. Other plasmids allow the degradation of various synthetic compounds, including certain widely used herbicides and insecticides (Figure 14.10; Table 14.5). The degradative capabilities of microorganisms

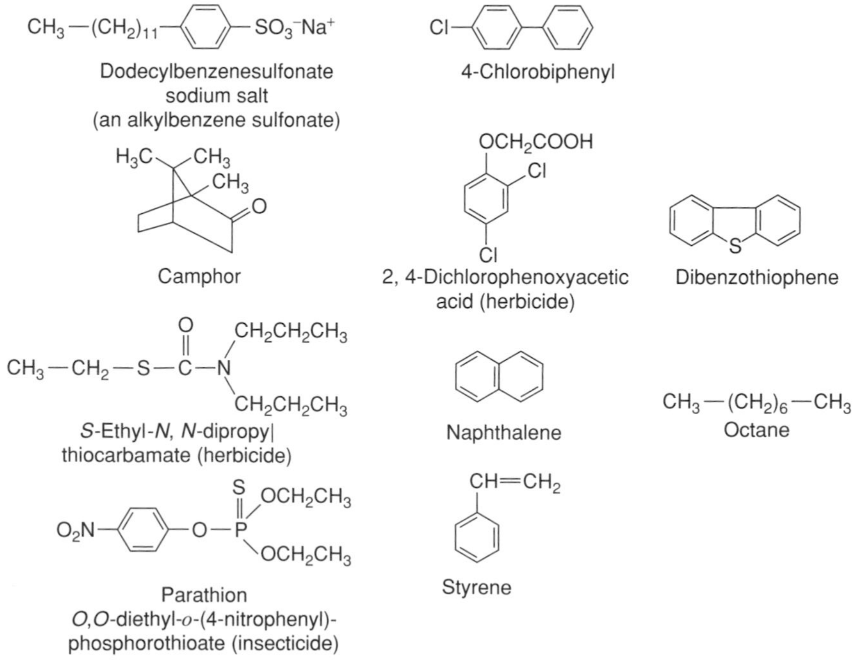

**FIGURE 14.10**

Examples of herbicides and insecticides subject to degradation by enzymes encoded on catabolic plasmids.

$CH_3-(CH_2)_{11}-C_6H_4-SO_3^-Na^+$
Dodecylbenzenesulfonate sodium salt (an alkylbenzene sulfonate)

4-Chlorobiphenyl

Camphor

2, 4-Dichlorophenoxyacetic acid (herbicide)

Dibenzothiophene

*S*-Ethyl-*N*, *N*-dipropyl thiocarbamate (herbicide)

Naphthalene

$CH_3-(CH_2)_6-CH_3$
Octane

Parathion
*O*,*O*-diethyl-*o*-(4-nitrophenyl)-phosphorothioate (insecticide)

Styrene

TABLE 14.5 Examples of naturally occurring transmissible catabolic plasmids

| Primary substrate[a] | Plasmid | Size (kb) | Bacterial host strain |
|---|---|---|---|
| Alkylbenzene sulfonate | ASL | 91.5 | *Pseudomonas testosteroni* |
| Benzoate | pCB1 | 17.4 | *Alcaligenes xylosoxidans* subsp. *denitrificans* PN-1 |
| Biphenyl | pBS241 | 195 | *P. putida* BS893 |
| Camphor | PpG1(CAM) | ~500 | *Pseudomonas* sp. |
| 4-Chlorobiphenyl | pSS50 | 53.2 | *Alcaligenes* spp. |
| 2,4-Dichlorophenoxyacetate | pJP1 | 87 | *Acrossocheilus paradoxus* Jmp116 |
| Dibenzothiophene | NL1 | ~180 | *Sphingomonas aromaticivorans* F199 |
| S-Ethyl-*N*,*N*-dipropyl-thiocarbamate | –[b] | 75.7 | *Rhodococcus* sp. TE1 |
| Naphthalene | Nah7 | 83 | *P. putida* PpG7 |
| Octane | OCT | ~500 | *Pseudomonas oleovorans* |
| Parathion ATCC27551 | pPDL243 | | *Flavobacterium* |
| Styrene | pEG | 37 | *Pseudomonas fluorescens* PAW340 |
| Toluene | pWW0 (TOL) | 117 | *P. putida* mt-2 |

[a] The structures of the substrates are shown in Figure 14.9.
[b] This plasmid has not been named. .
*Source*: Sayler, G. S., Hooper, S. W., Layton, A. C., and King, J. M. H. (1990). Catabolic plasmids of environmental and ecological significance. *Microbial Ecology*, 19, 1–20.

carrying catabolic plasmids result from a cooperative interaction between the genes carried on the plasmid and those on the chromosome of the host cell. *Such interactions are particularly important when a single compound is to be used as a sole source of carbon and energy.* Many plasmids encode only part of the catabolic pathway for a given compound. The products of the transformations mediated by the plasmid-encoded enzymes must be those that can then be utilized by chromosomally encoded enzymes functioning in the central energy-producing metabolic pathways of the cell.

It is not surprising that many catabolic plasmids carry pathways for the degradation of aromatic compounds. The benzene ring is second only to glucose as a building block in nature. Whereas glucosyl residues are the monomer units of cellulose, the most abundant organic compound in nature, benzene rings form part of the precursors of lignin, the second most abundant constituent of biomass (Chapter 12).

## Aerobic Biodegradation of Benzene and Other Aromatic Hydrocarbons

The first step in an oxidative microbial attack on benzene is hydroxylation. Bacteria employ a dioxygenase to catalyze the simultaneous incorporation of two atoms of oxygen from an oxygen molecule into the ring. The product, *cis*-1,2-dihydroxy-1,2-dihydrobenzene, is then converted to catechol (1,2-dihydroxybenzene; see Figure 14.11) in a reaction catalyzed by the enzyme *cis*-benzene glycol dehydrogenase. These initial two steps in benzene biodegradation – dioxygenase-mediated hydroxylation followed by dehydrogenation – are common to the pathways of bacterial degradation of numerous other aromatic hydrocarbons (Figure 14.12).

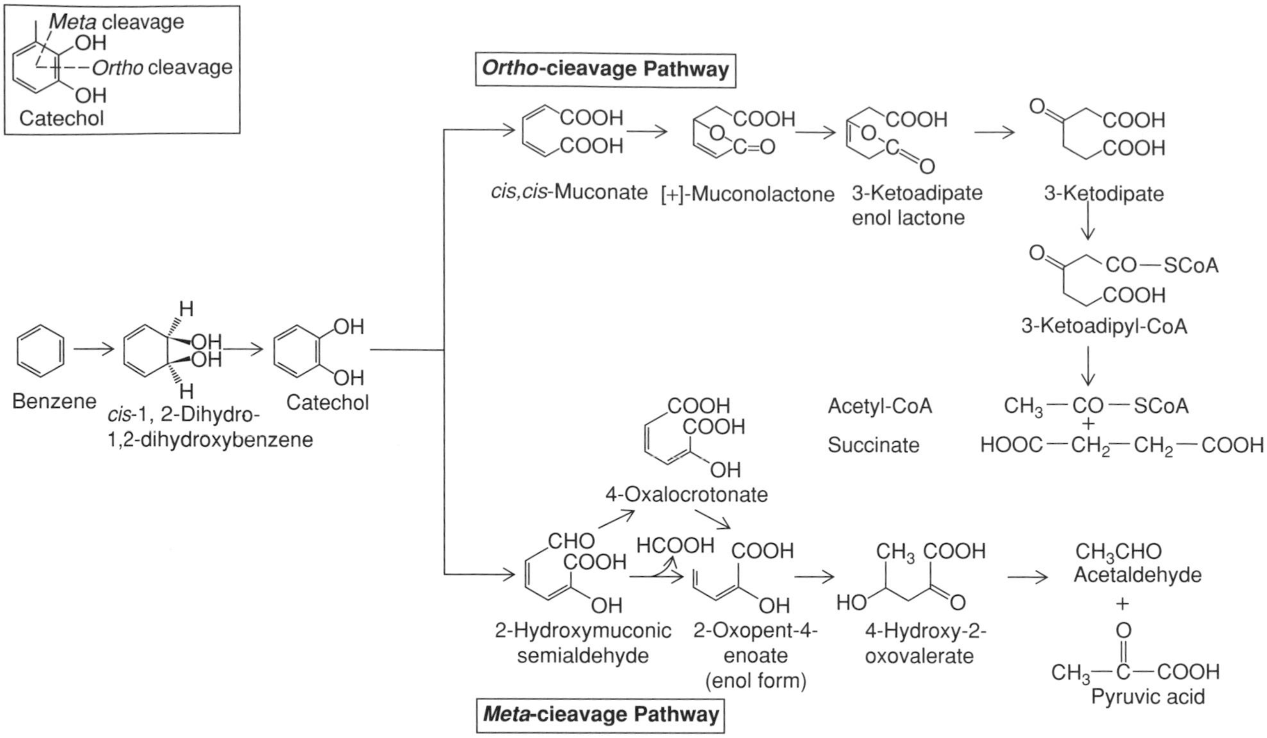

**FIGURE 14.11**

Pathways of benzene degradation by *Pseudomonas* species.

Redrawn based on artwork from the first edition (1995), published by W.H. Freeman.

The subsequent metabolism of catechol follows one of two divergent pathways (Figure 14.11). At the branch point, catechol is either oxidized in a reaction catalyzed by catechol-1,2-dioxygenase, the so-called *ortho* (or intradiol) cleavage, to *cis,cis*-muconate, or by catechol-2,3-dioxygenase, the so-called *meta* (or extradiol) cleavage, to 2-hydroxymuconic semialdehyde. The final products of both pathways are molecules that can enter the tricarboxylic acid cycle.

In addition to benzene itself, the *ortho* and *meta* pathways can catabolize a number of benzene derivatives. As illustrated below in the discussion of the TOL catabolic plasmid, different benzene derivatives induce either one pathway or the other, because the enzymes in the *ortho* and *meta* pathways differ in their ability to utilize particular catechol derivatives as substrates. The possession of two pathways increases the range of benzene derivatives that an organism can utilize as substrates.

### TOL (pWW0) Catabolic Plasmid

Examination of *Pseudomonas putida* mt-2 and its associated transmissible catabolic plasmid, TOL (pWW0; 117 kb), gives a glimpse of the complexity of the pathways whereby bacteria utilize aromatic hydrocarbons and illustrates the interplay between chromosomal and plasmid genes. In *P. putida* mt-2, chromosomal genes encode the *ortho* pathway and the TOL plasmid encodes the *meta* pathway (Figure 14.11). Benzoate, a product of toluene degradation, induces the expression of the genes of the *meta* pathway, whereas

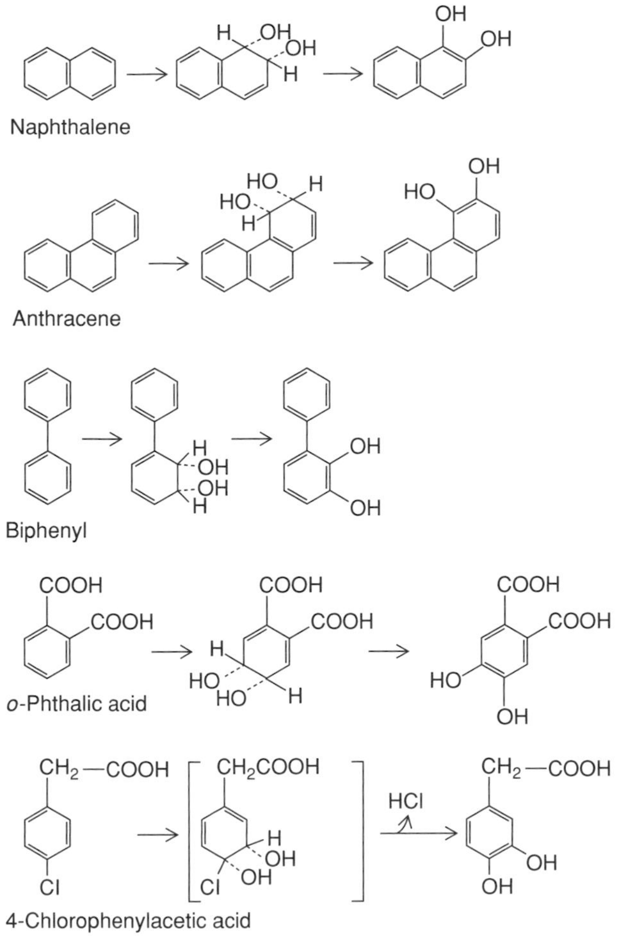

Naphthalene

Anthracene

Biphenyl

*o*-Phthalic acid

4-Chlorophenylacetic acid

**FIGURE 14.12**

Initial steps in the metabolism of various aromatic hydrocarbons.

Redrawn based on artwork from the first edition (1995), published by W.H. Freeman.

catechol – like benzoate, a product of toluene degradation (Figure 14.13) – induces the *ortho* pathway.

The TOL plasmid has been shown to confer on the host the capacity to degrade not only toluene but also *m*- and *p*-xylene and other benzene derivatives. The *xyl* genes of TOL pWW0 are organized into two operons referred to as the *upper* and *lower* (*meta*) pathways (Figure 14.14). The genes encoding catabolic enzymes have been named the *xyl* genes. The upper pathway, *xylUWCMABN*, encodes proteins that function in the uptake and the degradation of toluene and xylenes to benzoate and toluates (methylbenzoates), respectively. The lower pathway, *xylXYZLTEGFJQKIH*, encodes the degradation of benzoate and toluates to acetaldehyde and pyruvate (Figure 14.13 and Table 14.6).

The lower pathway branches at 2-hydroxymuconic semialdehyde, and the branches rejoin at the common product: 2-oxo-4-pentenoate (Figure 14.13). In analogy to the role of the alternate *ortho* and *meta* pathways for the degradation of catechol, this branching broadens the range of substrates that can be utilized by *P. putida* mt-2. For example, *m*-toluate is degraded by the *xylF* branch, whereas benzoate and *p*-toluate are degraded by the *xylGHI* branch. The specificity of the enzymes of these two branches of the

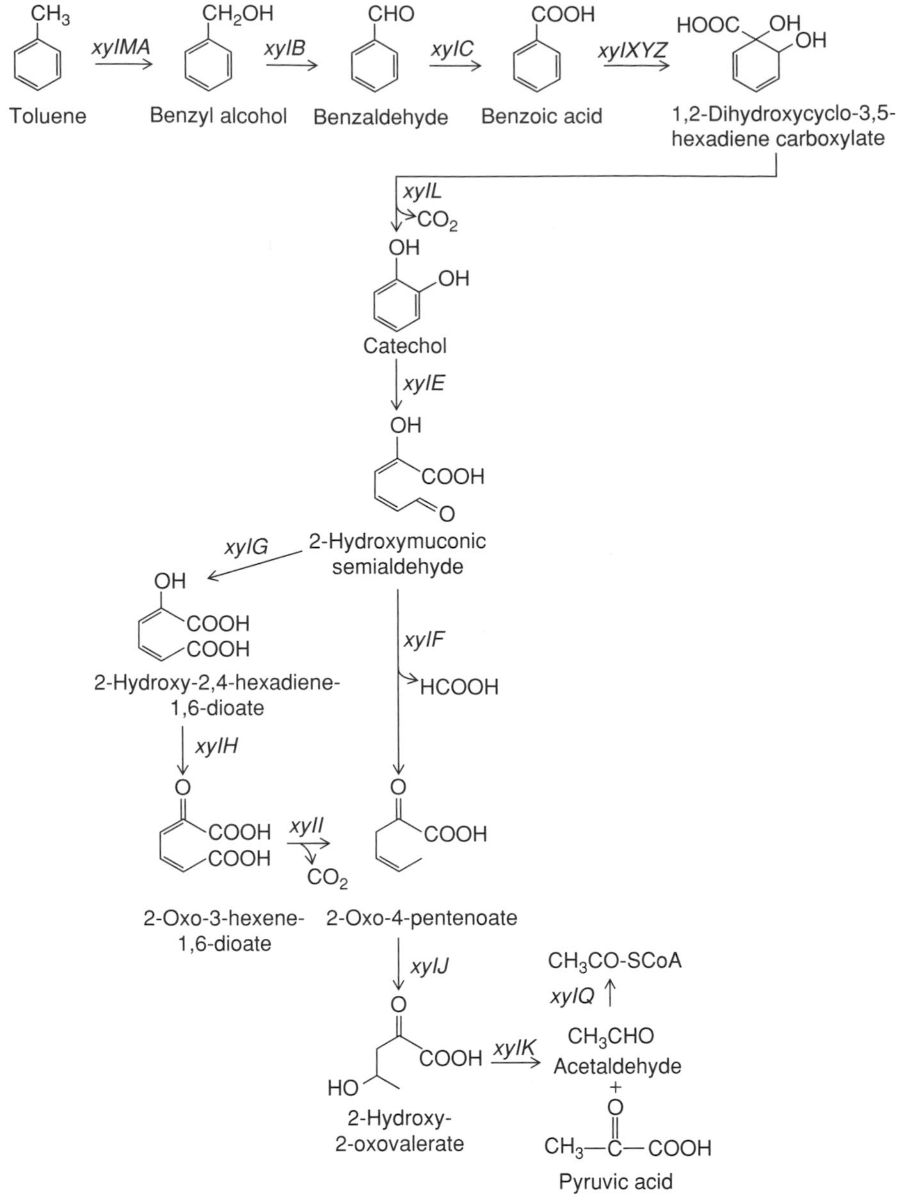

FIGURE 14.13

The pathway of the degradation of benzoate and toluates to acetaldehyde and pyruvate. The *xyl* gene(s) encoding the enzyme(s) that catalyze it are shown for each transformation.

Redrawn based on artwork from the first edition (1995), published by W.H. Freeman.

lower pathway for a particular substrate determines the branch by which the substrate will be catabolized.

## Regulation of TOL Plasmid WW0 *xyl* Gene Expression

DNA array technology allows visualization of the transcriptional response of both plasmid and host genes upon exposure of *P. putida* mt-2 to a substrate such as toluene or xylene. Such analyses reveal that these compounds are sensed both as growth substrates to metabolize and as environmental stressors. Consequently, the organismal response includes both host cell and plasmid regulatory proteins in the transcription of the appropriate *xyl* genes

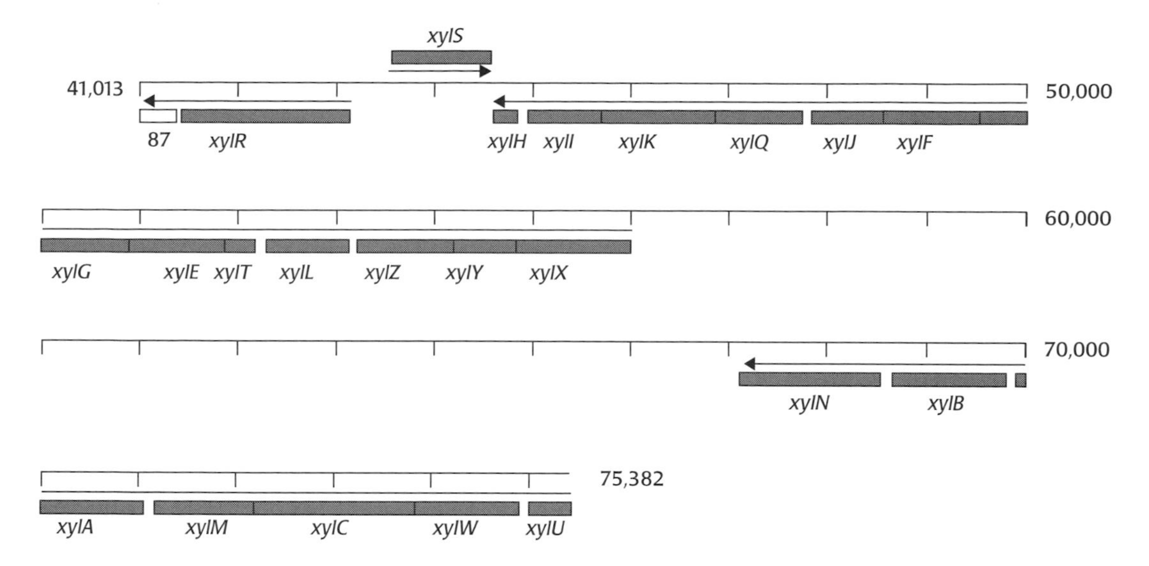


**FIGURE 14.14**

Arrangement and location of toluene/xylene degradation genes in the map of plasmid pWW0 with coordinates from 0 to 116,580 kb. [See Greated, A., Lambertsen, L., Williams, P. A., and Thomas, C. M. (2002). Complete sequence of the IncP-9 TOL plasmid from *Pseudomonas putida*. *Environmental Microbiology*, 4, 856–871.]

and of the arsenal of stress response genes of the host. The regulatory proteins include the proteins encoded by two unlinked regulatory genes, *xylR* and *xylS*, carried by the TOL pWW0 plasmid, and host sigma factors (whose alternative designations are given in parentheses) $\sigma^{54}$ (RpoN), $\sigma^{H}$ (RpoH or $\sigma^{32}$), and $\sigma^{S}$ (RpoS or $\sigma^{38}$).

Sigma factors govern promoter selection by RNA polymerase (RNAP). The RNAP core enzyme is unable to recognize promoters. The RNAP holoenzyme formed upon binding of a particular sigma factor binds to and initiates transcription at the promoters selected by that sigma factor. $\sigma^{54}$ controls operons that must remain silent unless absolutely needed. In addition to operons involved in the degradation of aromatic compounds and xenobiotics, these include those involved in nitrogen assimilation and hydrogenase synthesis. $\sigma^{H}$ responds to heat shock and other stresses manifested by the presence of denatured proteins in the cytoplasm. $\sigma^{S}$ responds to stresses such starvation, high osmolarity, low or high pH, and low or high temperature. To the bacterium, the presence of toluene or xylene signals the potential presence of a whole array of toxic compounds present, for example, in crude coal tar. This insight provides the key to the understanding of the involvement of the particular sigma factors described above in the control of *xyl* gene expression.

A summary of the information on the regulation of the transcription of the *xyl* genes is provided in Figure 14.15. *XylR* is expressed constitutively at a high level and down-regulates its own expression by binding at the promoter $P_r$. When a substrate, such as toluene, enters the cell, it binds to the XylR protein to form a XylR/toluene complex. This complex binds to the $\sigma^{54}$-dependent promoter Pu of the *xylUWCMABN* operon and activates its transcription by enhancing the binding of RNAP holoenzyme. The lower pathway is activated in an analogous manner. Benzoate, the product of the degradation of toluene by the upper pathway, binds to the XylS protein, produced constitutively at a low level from a $\sigma^{54}$-dependent promoter. The XylS/benzoate complex binds to the promoter ($P_m$) of the lower pathway

**TABLE 14.6 Xyl proteins encoded by genes on the TOL plasmid pWW0**

| Gene | Protein function |
|---|---|
| Upper pathway operon (*xylUWCMABN*) | Uptake and conversion of toluene and xylenes to benzoate and toluates |
| *xylN* | Outer membrane protein involved in the transport of *m*-xylene and its analogs across the outer membrane |
| *xylB* | Benzyl alcohol dehydrogenase subunit |
| *xylA* | Xylene oxygenase |
| *xylM* | Xylene monooxygenase hydroxylase component |
| *xylC* | Benzaldehyde dehydrogenase |
| *xylW* | Benzyl alcohol dehydrogenase subunit |
| *xylU* | Function not known |
| Lower (*meta*) operon (*xylXYZLTEGFJQKIH*) | Degradation of benzoate and toluates to acetaldehyde and pyruvate |
| *xylX,Y,Z* | Toluate 1,2-dioxygenase subunits |
| *xylL* | 1,2-Dihydroxycyclohexa-3,4-diene carboxylate dehydrogenase |
| *xylT* | Chloroplast-type ferredoxin. 4-Methylcatechol inactivates XylE by oxidizing the active site iron to ferric. XylT reactivates XylE by reduction of the iron atom to ferrous. |
| *xylE* | Catechol 2,3-dioxygenase |
| *xylG* | 2-Hydroxymuconic semialdehyde dehydrogenase |
| *xylF* | 2-Hydroxymuconic semialdehyde hydrolase |
| *xylJ* | 2-Oxo-4-pentenoate hydratase |
| *xylQ* | Acetaldehyde dehydrogenase |
| *xylK* | 4-Hydroxy-2-oxovalerate aldolase |
| *xylI* | 4-Oxalocrotonate decarboxylase |
| *xylH* | 4-Oxalocrotonate tautomerase |
| | Proteins involved in controlling the transcription of the upper and lower pathway genes |
| *xylR* | Regulatory protein |
| *xylS* | Regulatory protein |

and activates the transcription of its genes. The transcription of these genes is also dependent on the host RNAP holoenzyme containing either $\sigma^H$ or $\sigma^S$. If benzoate, rather than toluene, is introduced as the starting substrate, only the lower (*meta*) pathway genes are induced. This regulation appears appropriate because the upper pathway genes specify enzymes involved in the production of benzoate (Figure 14.13; Table 14.6).

There are additional subtleties to the regulation. The XylR–toluene complex binds to the promoter ($P_s$) of *xylS* as well as to the $P_u$ promoter, thus activating transcription of *xylS* and so achieving simultaneous activation of both the upper- and lower-pathway genes. When the XylS protein is present in an elevated amount, it activates the expression of the lower pathway genes, even in the absence of an inducer such as benzoate.

### Role of the Multiple Catabolic Pathways for Benzene Derivatives in *P. putida* mt-2

It is instructive to summarize the key features enabling the TOL (pWW0)–carrying *P. putida* mt-2 to grow on many different benzene derivatives. The

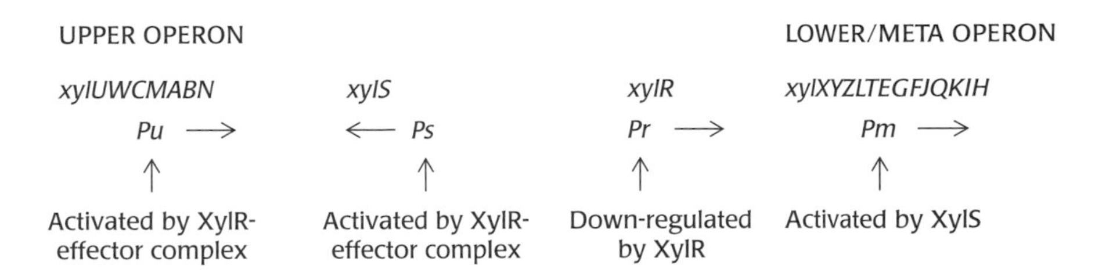


**FIGURE 14.15**

Regulation of the *xyl* genes of the *Pseudomonas putida* mt-2 TOL plasmid pWW0 in the context of the degradation of *m*-xylene (Figure 14.16) as an example. The upper operon encodes the enzymes that convert *m*-xylene to *m*-toluate. The lower (or *meta*) operon encodes the enzymes that degrade *m*-toluate to products that enter the Krebs cycle. The upper operon is transcribed from the promoter *Pu* upon the activation of *Pu* by XylR bound to *m*-xylene (or either of its first two degradation products). The lower operon is transcribed from the *Pm* promoter, which is activated by the XylS–*m*-toluate complex. XylS and XylR are transcribed from divergent, overlapping promoters *Ps* and *Pr*, respectively. The *Ps* promoter is activated by XylR, whereas XylR binds to and down-regulates *Pr*, its own promoter.

oxygenase (XylA) and dehydrogenases (XylB and XylC) that catalyze the conversion of various benzene derivatives to benzoic or toluic acids have a broad substrate specificity and act even on highly substituted compounds, such as 1,2,4-trimethylbenzene. At the level of catechol derivatives, where broad substrate specificity either is not achievable or is inappropriate, different enzymes catalyze the degradation of particular intermediates derived from benzenes with different substitution patterns. The various catechol derivatives are degraded either by the *ortho* pathway specified by chromosomal genes or by the *meta* pathway specified by pWW0 *xyl* genes (Figure 14.14). The choice of pathway is determined by the ability of the enzymes of the pathway to act on a particular catechol derivative. Typically, catechols with alkyl substituents at the 3- or 4-position go through the *meta* cleavage pathway. Similarly, within the *meta* pathway, alternate branches handle different 2-hydroxymuconic semialdehyde derivatives. For example, *m*-toluate is degraded by the *xylF* branch, whereas benzoate and *p*-toluate are degraded by the *xylGHI* branch (Figure 14.16). *The versatile utilization of different benzene derivatives thus depends on the presence of multiple degradative pathways.* Wherever possible, a single enzyme with broad specificity is utilized.

### Atrazine Catabolic Plasmid pADP-1 from *Pseudomonas* Species Strain ADP

Atrazine (6-chloro-*N*-ethyl-*N'*-(1-methylethyl)-1,3,5-triazine-2,4-diamine), a triazine herbicide registered for the control of broadleaf weeds and some grassy weeds, is estimated to be the most heavily used herbicide in the United States. More than 76 million pounds are applied annually. Atrazine is the most commonly detected pesticide in streams, rivers, reservoirs, and groundwater. It is water soluble, with a half-life in freshwater reported to exceed 100 days. Atrazine is a documented endocrine disruptor. In particular, low levels of atrazine have been shown to impair amphibian gonadal development.

We noted above that unrestricted use of a pesticide will frequently result in the development of microbial populations capable of degrading it and that the ability to degrade the pesticide spreads among the different bacterial strains in the natural population by interspecies transfer of a catabolic plasmid that carries the genes for the degradative enzymes. The atrazine catabolic plasmid pADP-1 from *Pseudomonas* species strain ADP offers an excellent example. The six enzymes specified by genes *atzABCDEF* encoded by this plasmid catalyze the mineralization of atrazine (Figure 14.17).

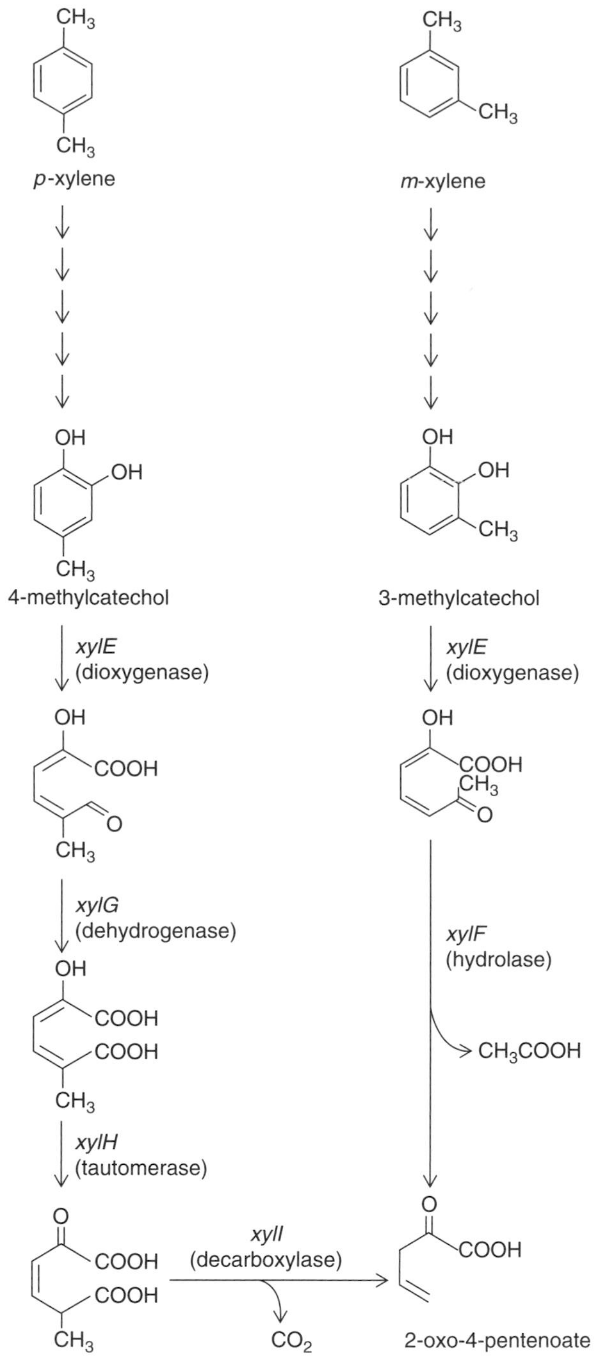

FIGURE 14.16

Portion of the pathways of degradation of *p*-xylene and *m*-xylene to 2-oxo-4-pentenoate. Note that 4-methylcatechol is oxidized to an aldehyde whereas 3-methylcatechol is converted to a ketone. Consequently, different enzymes are needed for the transformation of these compounds to 2-oxo-4-pentenoate.

The AtzA, AtzB, and AtzC, encoded by the closely linked genes *atzABC*, metabolize atrazine to cyanuric acid. Nearly identical *atz* genes are present in *Alcaligenes*, *Agrobacterium*, *Clavibacter*, *Pseudomonas*, *Ralstonia*, and *Rhizobium* strains, strongly suggesting that they spread by horizontal gene transfer.

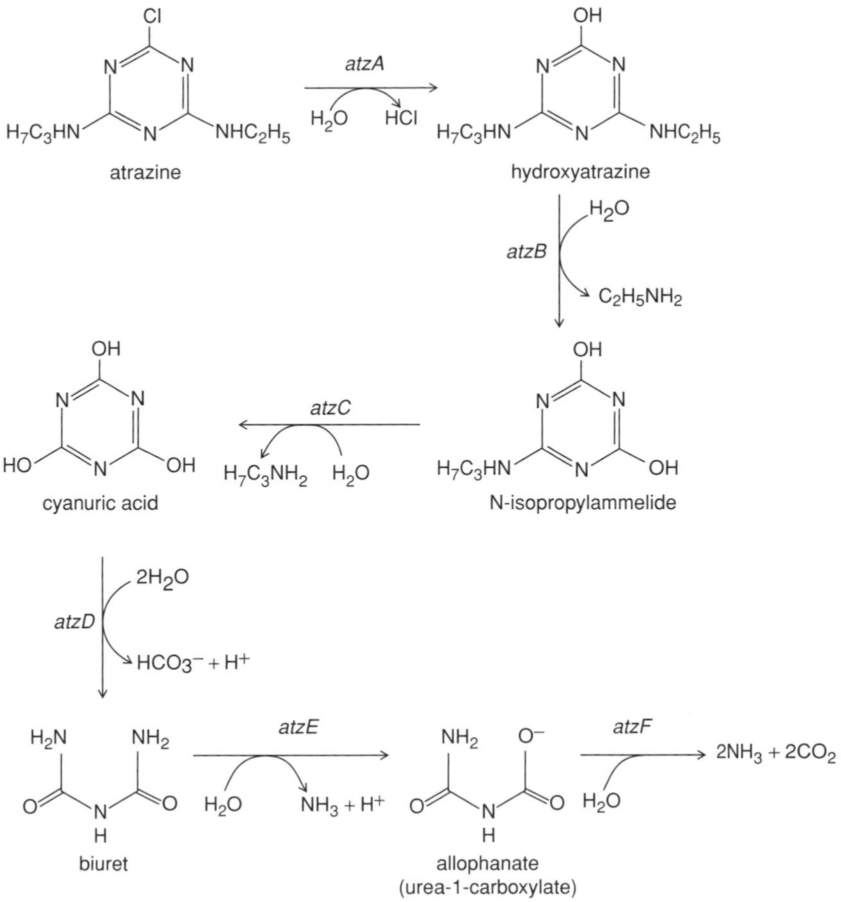

**FIGURE 14.17**

Pathway of atrazine mineralization encoded by genes on the atrazine catabolic plasmid pADP-1. [From Martinez, B., Tomkins, J., Wackett, L. P., Wing, R., and Sadowsky, M. J. (2001). Complete nucleotide sequence and organization of the atrazine catabolic plasmid pADP-1 from *Pseudomonas* sp. strain ADP. *Journal of Bacteriology*, 183, 5684–5697.]

## BIODEGRADATION OF ORGANIC COMPOUNDS IN ANAEROBIC ENVIRONMENTS

Even though the earth's atmosphere contains 20% oxygen by volume, both natural and human-made anaerobic environments abound on our planet. The former include sediments, both freshwater and marine, waterlogged soils, groundwater, and gastrointestinal contents of animals, whereas among the latter are landfills, feedlot wastes, sludge digesters, and bioreactors.

Compounds widely dispersed in the environment that are degraded under anaerobic conditions include petroleum hydrocarbons, nitroaromatic compounds, chlorinated aliphatic and aromatic compounds, pesticides and herbicides, and surfactants. In fact, some xenobiotic compounds, such as tetrachloroethylene, PCBs, and nitro-substituted aromatics, can be efficiently transformed or mineralized only by anaerobic bacteria.

### Anaerobic Biodegradation of Hydrocarbons

Research into the anaerobic degradation of hydrocarbons in the marine environment is strongly motivated by the need to clean up massive oil spills from wrecked supertankers, from which crude oil spreads widely over the

**TABLE 14.7 Gasoline components in contaminated groundwater at two fuel transfer stations, one in Sparks, Nevada, and the other in San Diego, California**

| Contaminant | Nevada mean $\mu g/L^{-1}$ | California mean $\mu g/L^{-1}$ |
|---|---|---|
| Methyl tertiary butyl ether (MTBE) | 330 | 9570 |
| Benzene | 100 | 5770 |
| Ethylbenzene | 7.4 | 140 |
| *m*-, *p*-Xylene | 15 | 570 |
| *o*-Xylene | 5.9 | 290 |
| Toluene | 4.4 | 650 |
| Total petroleum hydrocarbons | 1060 | 16,850 |

*Source*: Stocking, A. J., Deeb, R. A., Flores, A. E., Stringfellow, W., Talley, J., Brownell, R., and Cavanaigh M. C. (2000). Bioremediation of MTBE: a review from a practical perspective. *Biodegradation*, 11, 187–201.

coast and coastal marshes. Aliphatic and aromatic hydrocarbons make up more than 75% of crude oil. Terrestrial sources include gasoline from leaking underground storage tanks and petroleum fuels spilled in pipeline accidents, as well as effluents from metal, paint, varnish, and textile manufacture, wood processing, and the production of organic chemicals. Monoaromatic (BTEX) hydrocarbons – benzene, toluene, ethylbenzene, and xylene – are highly volatile substances commonly found in gasoline (Table 14.7).

### Anaerobic Degradation of Toluene

Of the BTEX components, the anaerobic degradation of toluene is the best understood. Biodegradation of toluene has been demonstrated with nitrate, Mn(IV), Fe(III), sulfate, and $CO_2$ as terminal electron acceptors.

*Geobacter* species are believed to be the most common of known Fe(III)-reducing species in anoxic mesophilic environments and often are the dominant organisms in the Fe(III)-reducing zone of environments contaminated with hydrocarbons. *Geobacter*, a rod-shaped, flagellated, $\delta$-proteobacterium, was first isolated in 1987 from the Potomac River near Washington, D.C. This strain, *Geobacter metallireducens* GS-15, was the first organism in pure culture shown to perform anaerobic degradation of toluene. It completely oxidized toluene to $CO_2$ with the reduction of Fe(III). Several organisms couple anaerobic toluene degradation to nitrate respiration. All these organisms (e.g., members of *Azoarcus* and *Thauera* spp.) are facultative anaerobes and are members of the $\beta$-proteobacteria. Such organisms are commonly isolated from anaerobic sludge or creek sediments. For example, *Azoarcus tolulyticus* was isolated from a gasoline-contaminated aquifer in Michigan.

### Metabolism of Toluene by *Azoarcus* and *Thauera* Species

The metabolic pathway for the mineralization of toluene by *Azoarcus* and *Thauera* species is shown in Figure 14.18. Benzene and the *o*- and *m*-xylenes

toluene + fumarate → (Benzyl succinate synthase) → benzyl succinate

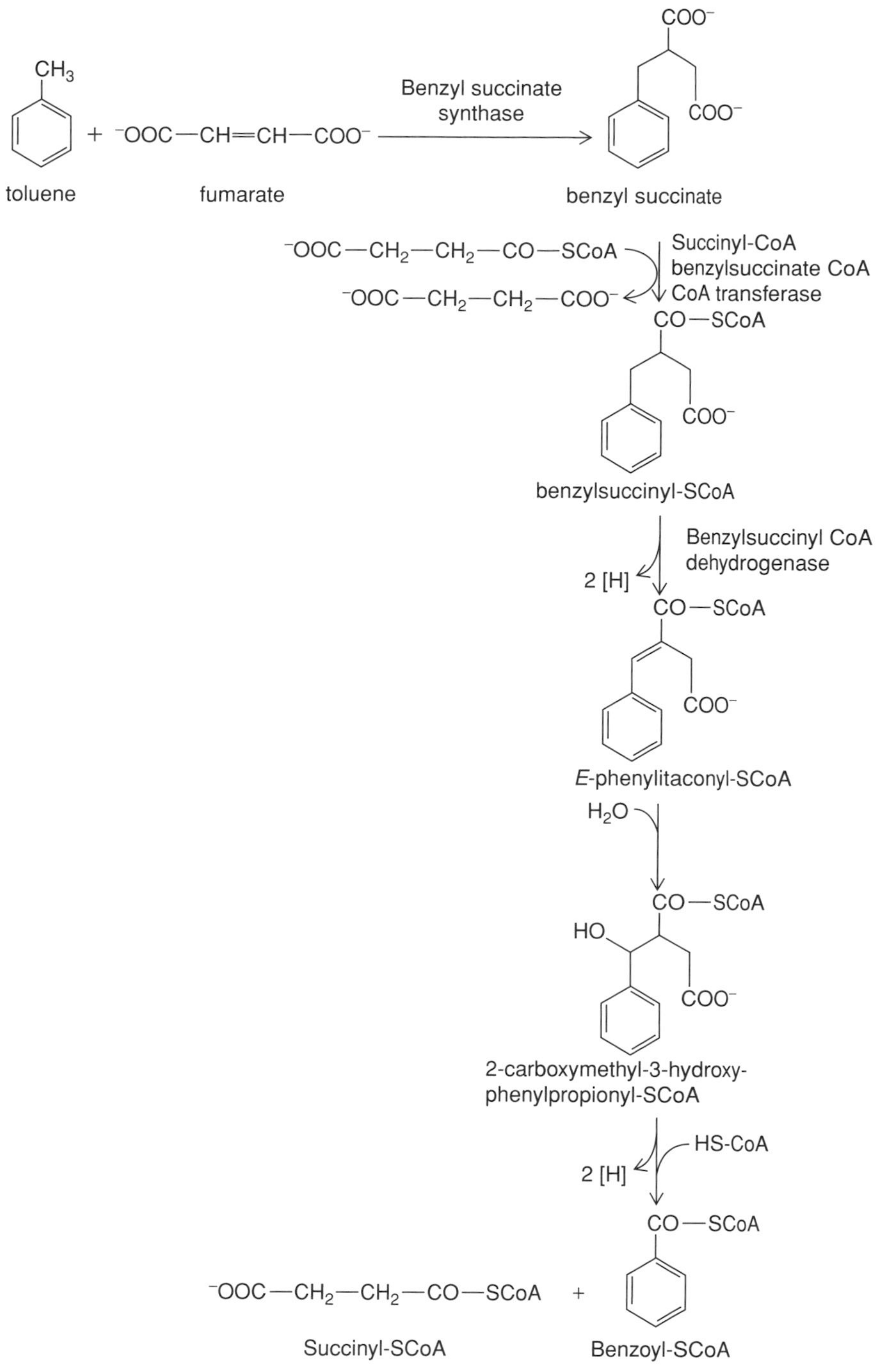

benzyl succinate → (Succinyl-CoA benzylsuccinate CoA CoA transferase; $^{-}OOC—CH_2—CH_2—CO—SCoA$ → $^{-}OOC—CH_2—CH_2—COO^{-}$) → benzylsuccinyl-SCoA

benzylsuccinyl-SCoA → (Benzylsuccinyl CoA dehydrogenase; 2 [H]) → *E*-phenylitaconyl-SCoA

*E*-phenylitaconyl-SCoA → ($H_2O$) → 2-carboxymethyl-3-hydroxy-phenylpropionyl-SCoA

2-carboxymethyl-3-hydroxy-phenylpropionyl-SCoA → (HS-CoA; 2 [H]) → $^{-}OOC—CH_2—CH_2—CO—SCoA$ + Benzoyl-SCoA

Succinyl-SCoA Benzoyl-SCoA

**FIGURE 14.18**

Pathway of anaerobic degradation of toluene by *Azoarcus* and *Thauera* species.

are degraded in a similar manner. The initial reaction in the degradation of toluene is the addition of fumarate onto the toluene methyl group to form benzylsuccinate, catalyzed by a glycyl radical–containing enzyme, benzylsuccinate synthase (Figure 14.18). This activation step is highly conserved among anaerobic organisms capable of toluene degradation, including *G. metallireducens*, which employs dissimilatory Fe(III) reduction, and

*Desulfobacula toluolica*, a dissimilatory sulfate reducer. Both of the latter organisms are $\delta$-proteobacteria.

## BIODEGRADATION OF CHLORINATED ORGANIC COMPOUNDS IN GROUNDWATER UNDER ANAEROBIC CONDITIONS

In many parts of the world, aquifers (water-bearing strata of earth, gravel, or rock) containing essential groundwater supplies are contaminated with toxic chemicals that leach from terrestrial dump sites or enter the ground from other sources because of improper handling or storage. Contamination of groundwater also results from massive treatment of land with fertilizers and pesticides.

Concern about the presence of undesirable organic contaminants (nonhalogenated and halogenated aromatic hydrocarbons, haloalkanes, etc.) in groundwater has led to extensive studies of their biodegradation. One important finding is that some of these compounds are degraded under both aerobic and anaerobic conditions, and some only under aerobic conditions. Still others may be recalcitrant under aerobic conditions but readily degradable under anaerobic conditions. The fate of a particular compound is largely decided by its intrinsic chemical properties and by the metabolic capabilities of microorganisms that have access to it.

A list of the possible reactions that chlorinated aliphatic hydrocarbons can undergo illustrates the influence of chemical properties. Increased chlorination increases the electrophilicity and oxidation state of an aliphatic hydrocarbon, making it more susceptible to dehydrohalogenation and reduction (reactions exemplified by 1 and 2) and less susceptible to substitution and oxidation (reactions exemplified by 3 and 4).

1. Dehydrohalogenation: $CH_3CCl_3 \rightarrow CH_2 = CCl_2 + HCl$
2. Reduction: $CCl_4 + H^+ + 2e^- \rightarrow CHCl_3 + Cl^-$
3. Substitution: $CH_3CH_2CH_2Cl + H_2O \rightarrow CH_3CH_2CH_2OH + HCl$
4. Oxidation: $CH_3CHCl_2 + H_2O \rightarrow CH_3CCl_2OH + 2H^+ + 2e^-$

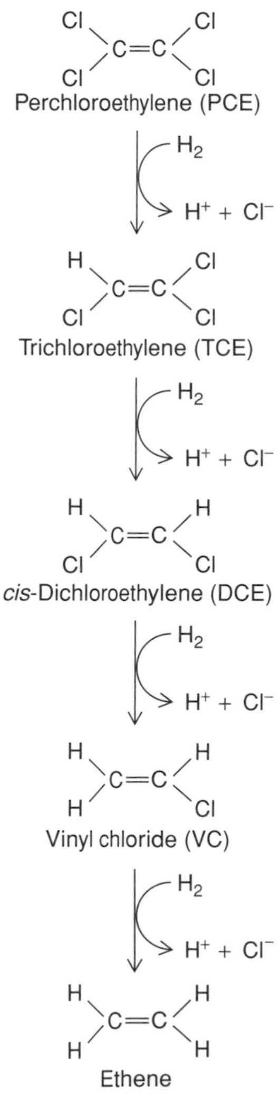

The fate of the dry-cleaning solvent perchloroethylene (tetrachloroethylene; PCE), a common groundwater contaminant, serves as an example. Because of the highly oxidized nature of this compound, it is very stable in the environment under aerobic conditions. No known organism is capable of aerobic degradation of PCE. However, under strictly anaerobic conditions, a pure culture of *Dehalococcoides ethenogenes* is able to dechlorinate PCE completely to ethene, a nontoxic product.

To dechlorinate PCE, *D. ethenogenes* utilizes halorespiration, a mechanism whereby PCE is used as an electron acceptor and hydrogen as an electron donor. The energy obtained from the exergonic dehalogenation reactions is used for bacterial growth. The reaction sequence shown in Figure 14.19 is catalyzed by a set of reductive dehalogenases that contain cobalamin and Fe–S clusters.

$Cl_2C{=}CCl_2$ Perchloroethylene (PCE) — $H_2$ → $H^+ + Cl^-$ → $HClC{=}CCl_2$ Trichloroethylene (TCE) — $H_2$ → $H^+ + Cl^-$ → $HClC{=}CClH$ *cis*-Dichloroethylene (DCE) — $H_2$ → $H^+ + Cl^-$ → $H_2C{=}CHCl$ Vinyl chloride (VC) — $H_2$ → $H^+ + Cl^-$ → $H_2C{=}CH_2$ Ethene

**FIGURE 14.19**

Dechlorination of perchloroethylene by *Dehalococcoides ethenogenes*.

**TABLE 14.8 Examples of metal sulfide minerals**

| Iron sulfides | | Mixed iron sulfides | |
|---|---|---|---|
| Pyrite ("fool's gold") | $FeS_2$ | Chalcopyrite | $CuFeS_2$ |
| Marcasite | $FeS_2$ | Bornite | $Cu_5FeS_4$ |
| Pyrrhotite | $Fe_7S_8$-FeS | Stannite | $Cu_2FeSnS_4$ |
| | | Pentlandite | $(Fe,Ni)_9S_8$ |
| | | Arsenopyrite | FeAsS |
| | | Tetrahydrite | $(Cu,Fe)_{12}Sb_4S_{13}$ |
| | | Cubanite | $CuFe_2S_3$ |
| **Other metal sulfides** | | | |
| Chalcocite | $Cu_2S$ | Millerite | NiS |
| Covellite | CuS | Realgar | AsS |
| Enargite | $Cu_3AsS_4$ | Cinnabar | HgS |
| Cobaltite | CoAsS | Stibnite | $Sb_2S_3$ |
| Galena | PbS | Molybdenite | $MoS_2$ |
| Sphalerite | ZnS | Argentite | $Ag_2S$ |

In summary, because highly chlorinated compounds are highly oxidized, on thermodynamic grounds, they can serve as excellent electron acceptors for bacteria where alternative electron acceptors are absent.

## MICROORGANISMS IN MINERAL RECOVERY

Many minerals of commercial interest are metal sulfides (Table 14.8), and most metal sulfide deposits are of volcanic or magmatic origin. Others are formed biogenically, through the reaction of metal ions with hydrogen sulfide generated by microorganisms. In all cases, the formation of the metal sulfide proceeds by the reaction

$$M^{2+} + S^{2-} \rightarrow MS.$$

The equilibrium of this reaction lies very far to the right because of the extreme insolubility of metal sulfides. For example, the solubility product of CuS is $4 \times 10^{-38}$, that is, $[Cu^{2+}][S^{2-}] = 4 \times 10^{-38}$ (Table 14.9).

Bacterial ore leaching, oxidative solubilization of metals, allows their recovery from low-grade ores without polluting the atmosphere. As sources of high-grade ore become depleted, mining companies must develop techniques for exploiting lower-grade ores profitably and for minimizing the generation of tailings. Metals recovered by leaching include primarily copper, but also cobalt, nickel, zinc, and uranium. The world's copper production was 14.6 million metric tons in 2004. Chile is the world's leading copper producer, with production exceeding 5.4 million metric tons in the same year. Its copper reserves have been estimated at 150 million tons. However, some 47 million tons of this total are contained in low-grade ores, for which metal recovery by concentration and smelting is not commercially profitable. The relative contribution of bacterial leaching to copper recovery is increasing and accounts for some 25% of the approximately 1,160,000 tons

**TABLE 14.9 Solubility products for some metal sulfides**

| | | | |
|---|---|---|---|
| $Ag_2S$ | $1 \times 10^{-51}$ | ZnS | $4.5 \times 10^{-24}$ |
| $Cu_2S$ | $2.5 \times 10^{-50}$ | $CoS_2$ | $7 \times 10^{-23}$ |
| CuS | $4 \times 10^{-38}$ | NiS | $3 \times 10^{-21}$ |
| CdS | $1.4 \times 10^{-28}$ | FeS | $1 \times 10^{-19}$ |

of copper produced in the United States in 2004. Low-grade uranium ores present a similar economic challenge, but with the worldwide availability of high-grade uranium deposits, there is little incentive to use biomining.

The pretreatment process for gold recovery has a different basis. Highly aerated, stirred tanks containing finely ground ore are used in the pretreatment of gold-containing arsenopyrite ores. Bacteria oxidize and dissolve the arsenopyrite, but the gold is unaffected. This pretreatment facilitates access of cyanide to and dissolution of the gold in the insoluble residue.

## DIVERSITY OF BACTERIA THAT CAUSE BIOLEACHING

The prokaryotes that play a primary role in biomining have several properties in common. They are chemolithoautotrophs capable of using ferrous iron and/or reduced sulfur compounds as electron donors in energy-generating pathways, and they fix $CO_2$. They are acidophilic and capable of growth at pH 1.5 to 2.0.

Organisms belonging to at least 11 different prokaryotic divisions cause dissolution of metal sulfides. The longest recognized among these are the extremely acidophilic, mesophilic sulfur and/or Fe(II)-oxidizing Gram-negative $\gamma$-proteobacteria, *Acidithiobacillus ferrooxidans* (formerly *Thiobacillus ferrooxidans*), *Acidithiobacillus thiooxidans* (formerly *Thiobacillus thiooxidans*), and the moderately thermophilic *Acidithiobacillus caldus. At. ferrooxidans* was first isolated from coal mine drainage in 1947, and subsequently found to be almost invariably associated with natural and artificial leaching sites.

*At. ferrooxidans* is a small Gram-negative straight rod, approximately 1.0 $\mu$m long an 0.5 $\mu$m in diameter. It grows best in acidic solutions, at pH 1.5 to 2.5, with an optimum temperature range of 10°C to 30°C and an upper limit of 37°C. *At. ferrooxidans* derives energy from the oxidation of $Fe^{2+}$ to $Fe^{3+}$, and of reduced forms of sulfur to $H_2SO_4$, using oxygen as a terminal electron acceptor. It uses $CO_2$ as a carbon source.

*Acidithiobacillus* strains that also play a role in leaching include *At. thiooxidans, Acidithiobacillus acidophilus*, and *Acidithiobacillus organoparus*. These acidophilic bacteria oxidize sulfur compounds but not $Fe^{2+}$. *At. thiooxidans* and *At. ferrooxidans* cooperate in the leaching of sulfide minerals.

Other acidophilic chemolithotrophic microorganisms believed to be important to the leaching process are *Leptospirillum ferrooxidans, Leptospirillum ferriphilum*, and species belonging to the genus *Sulfolobus*. Acidophilic heterotrophic bacteria growing in association with these autotrophs also contribute to the leaching process. *L. ferrooxidans* is somewhat more acidophilic than *At. ferrooxidans* and grows at pH 1.2 on pyrite at temperatures up to 40°C. These bacteria are highly motile, curved rods capable of forming spirals of joined cells. First isolated in 1972 from a copper deposit in Armenia, they derive energy by the oxidation of $Fe^{2+}$ to $Fe^{3+}$ but are unable to oxidize sulfur compounds. Archaea of *Sulfolobus* species grow autotrophically at pH 1 to 3 and at 50°C to 90°C, deriving energy from the oxidation of sulfur compounds and ferrous iron.

The atoms in crystalline solids are held together in a regular lattice by covalent bonds. Each covalent bond linking two atoms consists of a pair of electrons, called *valence electrons*, one from each atom. Unlike free atoms, in which electrons have discrete levels, the *valence band* of a solid consists of a large number of separate energy levels, one for each valence electron. Together these energy states constitute the valence band.

**BOX 14.6**

## HOW BACTERIA LEACH METALS FROM ORES

For a long time, it was believed that biological metal sulfide oxidation proceeded by enzyme-catalyzed oxidation of the sulfur moiety of the heavy metal sulfide. However, this mechanism does not exist. The pathway leading to the release of the metal cation depends on the reactivity of metal sulfides with protons, that is, acid solubility. Metal sulfides with valence bands (Box 14.6) that are derived only from orbitals of the metal atoms are not susceptible to attack by protons and are acid insoluble. These include metal sulfides such as pyrite ($FeS_2$), molybdenite ($MoS_2$), and tungstenite ($WS_2$). These are exclusively oxidized by the so-called thiosulfate pathway. In this pathway, Fe(III) ions attack the metal sulfide, extract electrons, and are thereby reduced to Fe(II), whereas the metal sulfide crystal releases metal cations ($M^{2+}$) and partially oxidized water-soluble sulfur compounds, notably thiosulfate ($S_2O_3{}^{2-}$). The release of thiosulfate takes place upon completion of six consecutive one-electron oxidation steps. At acidic pH, Fe(II)-oxidizing bacteria, such as *At. ferrooxidans* or *L. ferrooxidans*, convert the Fe(II) to Fe(III), by using it to reduce molecular oxygen through a complex redox chain with the generation of energy. The further oxidation of the thiosulfate via tetrathionate ($S_4O_6{}^{2-}$) and other polythionates ($S_xO_6{}^{2-}$; also known as polysulfane disulfonates) to sulfate, by Fe(III) and oxygen, mediated by *At. ferrooxidans* and *At. thiooxidans*, leads to the production of sulfuric acid.

The overall equations used historically to describe these transformations,

$$\underset{\text{pyrite}}{FeS_2} + 3.5O_2 + H_2O \rightarrow FeSO_4 + H_2SO_4 \quad (14.1)$$

$$FeSO_4 + 0.5O_2 + H_2SO_4 \rightarrow \underset{\text{ferric sulfate}}{Fe_2(SO_4)_3} + H_2O, \quad (14.2)$$

do not reveal the true complexity of the process.

Metal sulfides with valence bands derived from both the metal and the sulfide orbitals, are soluble in acid to varying degrees. These include sphalerite (ZnS), galena (PbS), arsenopyrite (FeAsS), chalcopyrite ($CuFeS_2$), and hauerite ($MnS_2$). Solubilization of these ores involves the "polysulfide pathway." In these metal sulfides, the chemical bonds between the metal and sulfur can be broken by attack by protons with the release of hydrogen sulfide ($H_2S$). In the presence of Fe(III) ions, the sulfur moiety of the metal sulfide is oxidized by a one-electron abstraction concomitant with the proton attack. This is believed to lead to the formation of a hydrogen sulfide cation ($H_2S^+$). The cation spontaneously dimerizes to free disulfide ($H_2S_2$) that is further oxidized through higher polysulfides and polysulfide radicals to elemental sulfur. The various sulfur compounds are oxidized to sulfate by sulfur-oxidizing bacteria such as *At. ferrooxidans* and *At. thiooxidans*.

The overall reactions that describe the oxidation of the metal sulfide crystal, again without revealing the mechanistic complexities of the multiple underlying abiotic and biological transformations, are

$$CuS + 0.5O_2 + 2H^+ \rightarrow Cu^{2+} + S^\circ + H_2O \quad (14.3)$$

$$S^\circ + 1.5O_2 + H_2O \rightarrow H_2SO_4. \quad (14.4)$$

Bacteria solubilize metals from ores by either a "noncontact mechanism" or a "contact" mechanism. In the noncontact mechanism, planktonic bacteria oxidize Fe(II) ions present in solution to Fe(III). The latter ions attack the metal sulfide, extract electrons, and are thereby reduced to Fe(II), whereas the metal sulfide crystal releases metal cations ($M^{2+}$) by the thiosulfate pathway described above.

In the contact mechanism, *At. ferrooxidans* attaches to the mineral particles primarily through electrostatic interactions mediated by bacterial exopolysaccharides that carry Fe(III) ions, each complexed by two uronic acid residues. The residual positive charge in this complex leads to the attachment of the bacteria to the negatively charged surface of the pyrite crystal. The subsequent reactions of the Fe(III) ions with the metal sulfide follow the course described above for the noncontact mechanism.

Ferric sulfate formed by the oxidative processes (see equation 14.2, above) is a strong oxidizing agent, able to dissolve several economically important copper sulfide minerals (Table 14.8) by the reactions

$$\underset{\text{chalcopyrite}}{CuFeS_2} + 2Fe_2(SO_4)_3 \rightarrow CuSO_4 + 5FeSO_4 + 2S^\circ \quad (14.5)$$

$$\underset{\text{chalcocite}}{Cu_2S} + 2Fe_2(SO_4)_3 \rightarrow 2CuSO_4 + 4FeSO_4 + S^\circ \quad (14.6)$$

$$\underset{\text{bornite}}{Cu_5FeS_4} + 6Fe_2(SO_4)_3 \rightarrow 5CuSO_4 + 13FeSO_4 + 4S^\circ. \quad (14.7)$$

Leaching by $Fe_2(SO_4)_3$, shown in equations (14.5) to (14.7), is independent of the presence of oxygen or microbial action. However, such leaching does depend upon the microbe's ability to supply the necessary $Fe_2(SO_4)_3$ by oxidizing $Fe^{2+}$ to $Fe^{3+}$ according to reaction (14.2) above. Iron is the fourth most abundant element in the earth's crust, and iron compounds are present in virtually all types of rock.

In reactions (14.5) to (14.7), the rate of metal extraction from sulfide minerals depends on the ferric iron ($Fe^{3+}$) concentration. At a pH of less than 3.5, the rate of oxidation of ferrous iron becomes independent of pH and is given by the equation

$$-d[Fe^{2+}]/dt = k[Fe^{2+}]\,pO_2, \text{ where } k = 1.0 \times 10^{-7} \text{ atm}^{-1}\text{min}^{-1} \text{ at } 25^\circ\text{C}.$$

Consequently, at the acid pH required for dump leaching (see below), in the absence of catalysis, the oxidation of $Fe^{2+}$ is very slow and leaching would likewise be very slow. *At. ferrooxidans* increases the rate of $Fe^{2+}$ oxidation by $10^6$-fold.

*At. ferrooxidans* also derives energy from the oxidation of elemental sulfur ($S^o$), generated by reactions such as (14.3) and (14.5) to (14.7), to sulfuric acid:

$$2S^\circ + 3O_2 + 2H_2O \rightarrow 2H_2SO_4. \quad (14.8)$$

The sulfuric acid maintains the low pH optimal for the acidophilic *At. ferrooxidans* and suppresses the loss of ferric sulfate by hydrolysis:

$$Fe_2(SO_4)_3 + 2H_2O \rightarrow 2Fe(OH)SO_4 + H_2SO_4. \quad (14.9)$$

The sulfuric acid also leaches various copper oxide minerals, as exemplified by the following reactions:

$$\underset{\text{azurite}}{Cu_3(OH)_2(CO_3)_2} + 3H_2SO_4 \rightarrow 3CuSO_4 + 2CO_2 + 4H_2O. \quad (14.10)$$

$$\underset{\text{chrysocolla}}{CuSiO_3 2H_2O} + H_2SO_4 \rightarrow CuSO_4 + SiO_2 + 3H_2O. \quad (14.11)$$

## RECOVERY OF COPPER BY DUMP LEACHING

Copper ore is typically obtained by open-cut mining. Material containing in excess of 0.5% copper is subjected to smelting, whereas the copper in lower-grade ore is recovered by heap or dump leaching, a process in which the broken rock is piled 100 or more feet high on a relatively impermeable surface and watered. The same water is repeatedly circulated and recirculated through the piles of rock. With time, the pyrite oxidizes, causing the solution to become strongly acidic and rich in ferric sulfate. Continued recirculation then causes the other metal sulfides in the ore to be solubilized, by the processes described above, and the effluent becomes progressively enriched in metals such as copper. Finally, the metal-rich effluent is pumped into a basin called a "launder," and iron scraps are added to precipitate the copper. The precipitation results from the reaction

$$Cu^{2+} + Fe^\circ \rightarrow Cu^\circ + Fe^{2+}. \quad (14.12)$$

The $Fe^{2+}$-rich solution remaining after the copper precipitates is transferred to shallow oxidation ponds, where *At. ferrooxidans* rapidly oxidizes the $Fe^{2+}$ to $Fe^{3+}$ and forms some additional sulfuric acid through the oxidation of sulfur compounds. Much of the $Fe^{3+}$ formed in these oxidation ponds precipitates as ferric hydroxide, $Fe(OH)_3$. The supernatant acidic ferric sulfate solution is then pumped back to the top of the dump.

Acidithiobacilli living in dumps are mostly confined to the top one-meter layer in densities up to $10^8$ bacteria per gram of ore. Once leaching is well under way in a dump, little change is seen in its microbial population. A dump can be viewed as a continuous-flow reactor in which solubilization of metals is performed by the bacteria attached to the ore particles.

An interesting experiment was performed at the Vlaikov Vrah mine in Bulgaria. Researchers there had generated a mutant strain of *At. ferrooxidans*, and laboratory experiments showed it to have a higher leaching activity on the ores from the mine than the wild-type did. This mutant was successfully established in a 100,000-ton section of the leaching dump, but the leaching rate did not increase. Apparently, the main rate limitations on the leaching process were imposed by the availability of oxygen and the fluxes of reactants and products within the ore pile rather than by the rate of oxidation of sulfide minerals by *At. ferrooxidans*.

### URANIUM LEACHING

Uranium ore occurs not as a sulfide but as the oxide $UO_2$, and is frequently associated with pyritic minerals. The uranium is leached from the ore by the mechanism, which, as we saw above, depends on microbial generation of ferric sulfate from pyrite. To initiate the leaching, the tunnels of underground mines are flooded with dilute sulfuric acid solution, allowing the reaction

$$UO_2 + Fe_2(SO_4)_3 + 2H_2SO_4 \rightarrow UO_2(SO_4)_3^{4-} + 2FeSO_4 + 4H^+. \quad (13.1)$$

The ferric ion oxidizes the insoluble $UO_2$, in which the uranium is tetravalent, to the acid-soluble $UO_2(SO_4)_3^{4-}$, in which the uranium is hexavalent. The uranyl salt is then isolated by ion exchange chromatography. The leaching process leads to uranium recoveries ranging from 30% to 90%.

## MICROORGANISMS IN THE REMOVAL OF HEAVY METALS FROM AQUEOUS EFFLUENT

Microorganisms immobilize metal ions by both active and passive processes. For example, bacteria that use sulfate as a terminal electron acceptor actively produce and excrete an ion, sulfide, that forms an insoluble complex with metal ions present in solution causing the ions to precipitate. In contrast, *biosorption* (strong binding of metal ions to bacterial cells and to polymeric substances secreted by the cells) is a passive process seen with both living and dead cells.

On the basis of equal volumes of wet material, biosorbents prepared from bacterial biomass are similar to synthetic ion-exchange resins in their capacity for loading metal ions. In the main, the binding properties of such biosorbents derive from the negatively charged functional groups (carboxylate, phosphate) on the cell walls and exopolymers of the microorganisms used. In addition to ion exchange mechanisms, the biosorbents bind metals by complexation to uncharged sites and by a poorly understood phenomenon in which metal precipitation is nucleated by bound metal ions. Biosorbents effectively remove low concentrations of heavy metal cations (such as $Cu^{2+}$, $Zn^{2+}$, $Cd^{2+}$, $Ni^{2+}$, $Pb^{2+}$) in the presence of high concentrations of alkaline earth metals ($Ca^{2+}$ and $Mg^{2+}$). Nevertheless, the biosorbents have not yet come into common use. Instead, a combination of active and passive microbial metal immobilization mechanisms has been used to remove heavy metal ions from aqueous industrial effluents and from contaminated surface waters.

## PRECIPITATION OF METAL SULFIDES

Lakes, artificially constructed lagoons, and wetlands artificially enriched with appropriate organic nutrients have been shown to act as "biofilters" for the heavy metals present in acid mine drainage. The processes in these environments parallel those naturally occurring in marine sediments.

Chloride is the most abundant anion in seawater, and sulfate is the second. When we compare the availability of the two important electron acceptors, oxygen and sulfate, in seawater, we see that at air saturation, the concentration of sulfate (0.028 M) is some hundredfold higher than that of oxygen. Furthermore, sulfate penetrates over a hundred times deeper into marine sediments than does oxygen. In these anaerobic, sulfate-rich environments, sulfate-reducing bacteria (*Desulfovibrio*, *Desulfotomaculum*, and others) carry out the last stages in the mineralization of organic detritus. To generate energy, these organisms utilize organic acids (higher fatty acids, lactate, acetate, propionate, butyrate, formate), ethanol, benzoate, and $H_2$, all derived from anaerobic degradation of biomass, as hydrogen donors whereas sulfate serves as the terminal electron acceptor and is reduced to $H_2S$. Some 10% of the sulfide produced in the reducing environment of the sediments reacts with metal ions to form insoluble metal sulfides.

Iron is by far the most abundant metal in seawater. Mineral grains coated with iron oxide serve as the source of ferrous ions, which are generated by the reaction

$$2FeOOH + H_2S \rightarrow S^\circ + 2Fe^{2+} + 4OH^-.$$

The free ferrous ions react with $H_2S$ to form amorphous ferrous sulfide:

$$Fe^{2+} + H_2S \rightarrow FeS + 2H^+,$$

and the amorphous ferrous sulfide transforms slowly into a crystalline material called mackinawite ($FeS_{0.9}$), which reacts with elemental sulfur to form pyrite:

$$FeS + S^\circ \rightarrow FeS_2.$$

In natural environments, metal sulfides other than those of iron occur in very low concentrations. However, in the hot brines of the Red Sea and the hydrothermal areas of the East Pacific Rise, sulfide-containing water seeps out of the sea bottom rich in heavy metals and sulfides of zinc and copper are deposited in massive amounts.

In freshwater lakes, the very low level of sulfate ($\sim$0.0001 M) limits the capacity for sulfate-dependent metabolism. However, acidic wastewaters draining from mines and accumulations of tailings contain high concentrations of both heavy metals and sulfate. Enrichment with organic matter enables such wastewaters to support the growth of sulfate-reducing bacteria, which, with their sulfide-producing metabolism, cause heavy metals to precipitate. Moreover, the bicarbonate that is the end product of organic substrate oxidation raises the pH of the water. Applied geochemistry was used in the 1970s to remove metals from mine wastes that contaminated a lake in Manitoba, Canada. Mine and smelter wastes containing Cd, Cu, Fe, Zn, Hg, and sulfate were released into a lake that also received sewage from a nearby town. The abundance of organic matter in the sewage led to profuse cyanobacterial and algal growth, and this increasing biomass bound the heavy metals. As cyanobacteria and algae died and sedimented, heavy metals were transported with them into the anoxic sediments of the lake, where the biomass became a plentiful substrate for sulfate-reducing bacteria. The $H_2S$ produced by the sulfate-reducing bacteria formed the insoluble sulfides CdS, CuS, FeS, HgS, and ZnS, which remained in the bottom sediments of the lake. A growing belief, supported by experience, is that constructed wetlands may likewise provide a relatively simple and inexpensive solution for removing pollutants from mining, urban, agricultural, and industrial effluents.

Microorganisms also remove mercury from wastewater but by a different mechanism – volatilization as dimethylmercury. By methylating the metal, many microorganisms protect themselves from the toxic effects of mercury, but the resulting alkylmercury compounds are very toxic to higher organisms. In 1953, fishermen and their families on the coast of Minamata Bay in Japan were stricken with a disease that produced progressive muscular weakness, loss of vision, impairment of other brain functions, eventual paralysis, and, in some victims, coma and death. The same disease afflicted household cats and the Minamata seabirds. In all cases, the disease was linked to the consumption of substantial amounts of fish taken from the bay, fish discovered to contain high concentrations of methylmercury. The mercury itself was traced to a factory effluent. Anaerobic bacteria in the sediment converted the mercury to its methyl and dimethyl derivatives (using methyl coenzyme $B_{12}$ as the methyl donor)

$$Hg^{2+} \rightarrow CH_3Hg^+ \rightarrow (CH_3)_2Hg$$

that were then concentrated through the food chain to the high levels found in the fish.

## A BIOLOGICAL TREATMENT PLANT FOR DETOXIFICATION OF PRECIOUS METAL–PROCESSING WASTEWATER

Since about 1890, gold and silver have been recovered from ores by leaching the ores with a cyanide solution. Alkaline, aerated solutions of cyanide readily dissolve gold with formation of $Au(CN)^{2-}$:

$$4Au^\circ + 8CN^- + O_2 + 2H_2O \rightarrow 4Au(CN)_2^- + 4OH^-.$$

Metals such as Cd, Co, Cu, and Fe are also leached from the ore as soluble cyanide complexes. The dissolved gold ($Au^o$) is removed from solution by either (1) precipitation with zinc ("cementation") or (2) adsorption on activated carbon followed by stripping and subsequent recovery by electrolysis. When zinc is used, soluble $Zn(CN)_4{}^{2-}$ ion is formed in the gold precipitation reaction. The used cyanide solution is recycled to leach the next batch of ore after gold has been removed, but some of it must be replaced by fresh cyanide solution to prevent buildup of impurities. The discarded solution must be treated to remove cyanide and heavy metals.

The Homestake Mine in Lead, South Dakota, one of the oldest and largest underground gold mines in the Western Hemisphere, operated continuously from 1876 to 2001. At this mine, all the tailings from cyanide leaching were collected in a single impoundment (the Grizzly Gulch Tailing Impoundment). In its final decades, operation of the mine required discharge of 4 million gallons per day of these wastewaters, a combination of the cyanide-containing decant water from this impoundment and water pumped out of the mine from 8000 feet below the surface. The concentrations of cyanide, thiocyanate, and copper in these waters are given in Table 14.9. Before discharge, the water needed to be detoxified and purified to the extent that it did not interfere with the continued operation of a trout fishery in Wildwood Creek, the receiving stream, in which the effluent from the mine wastewater treatment plant could at times make up over 50% of the total flow. Since 1984, the necessary purity was achieved by a process in which microorganisms removed 95% to 98% of the cyanide, thiocyanate, and heavy metals from the wastewater.

In the purification plant, the wastewater flowed through a train of five rotating disks. When the biomass buildup on a disk became excessive, the disk's direction of rotation was temporarily reversed, causing some of the film to slough off. The water was continually aerated to maintain dissolved oxygen at a minimum of 3 to 4 mg/L. The process consisted of two stages. In the first, *Pseudomonas* species, spontaneously attached as a layer to the first two disks, utilized free and metal-complexed cyanide and thiocyanate as sole sources of energy and carbon with production of ammonia and bicarbonate:

$$2CN^- + 4H_2O + O_2 \rightarrow 2NH_3 + 2HCO_3^-$$
$$SCN^- + 2H_2O + 2.5O_2 \rightarrow NH_3 + HCO_3^- + SO_4^-.$$

The resulting biomass removed metal from the solution through biosorption. In the second stage, nitrifying bacteria, attached as films to the next three disks, oxidized ammonia to nitrate, with nitrite as an intermediate. The nitrifying bacteria are strict autotrophs that use $CO_2$ as a sole carbon source:

$$\textit{Nitrosomonas:}\quad NH_4^+ + 1.5O_2 \rightarrow NO_2^- + 2H^+ + H_2O$$
$$\textit{Nitrobacter:}\quad NO_2^- + 0.5O_2 \rightarrow NO_3^-.$$

Residual cyanide, if present at significant levels, was also removed in the second stage.

The treated effluent contained 15 to 50 mg/L of sloughed biomass from the disks. This was removed by settling and by passage through three sand-filter beds. Impressively, the process required no nutrient additions of any sort. As indicated in Table 13.10, the clarified effluent had very low levels of cyanides and heavy metals, and it was nontoxic to fish. For a treatment plant

**TABLE 13.10 Some characteristics of the Homestake Mine wastewaters before and after detoxification[a]**

| | Concentration (mg/L) | |
|---|---|---|
| **Parameter** | **Influent blend[b]** | **Effluent** |
| Thiocyanate | 62 | <0.5 |
| Total cyanide | 4.1 | 0.06 |
| Metal-complexed cyanide | 2.3 | <0.02 |
| Copper | 0.56 | 0.07 |
| Ammonia N | 5.60 | <0.050 |
| Total suspended solids | – | 6 |

[a] Mixture of adjusted proportions of decant water from Grizzly Gulch tailings impoundment and mine water.
[b] The temperature range for the influent blend was 10°C to 25°C, and the pH was 7.5 to 8.5.
*Source*: Whitlock, J. L. (1990). Biological detoxification of precious metal processing wastewaters. *Geomicrobiology Journal*, 8, 241–249.

that handled 4 million gallons of wastewater a day, this was a remarkable achievement.

The design of the process evolved from the observation that a biological degradation of cyanide was taking place at the air–surface water interface in the tailings impoundment but not below the aerobic zone. A *Pseudomonas* species was isolated from the tailings impoundment and cultured in wastewater containing increasing amounts of cyanide. The cyanide-resistant mutant strain selected in this manner was the source of the inoculum for the first stage of the process. Sludge from a sewage treatment plant was the source of the nitrifying bacteria for the second stage. The high levels of cyanide present in the first stage kept the nitrifying bacteria from competing on the disks colonized by the *Pseudomonas* species. The very low cyanide concentration and high ammonia levels of the second stage prevented the development of substantial *Pseudomonas* populations on the last three disks. The process operated continuously from its inception until termination of mining operations, with continuing improvement in performance.

## SUMMARY

Microorganisms are exposed to an immense variety of organic compounds in the natural environment, those derived from living organisms and those generated by geochemical processes. Virtually every one of these myriad naturally occurring compounds is utilized as a source of energy and/or carbon by some microorganism. The spectacular metabolic versatility of bacteria and fungi is exploited in three areas of environmental microbiology: sewage and wastewater treatment, degradation of xenobiotics, and mineral recovery.

In sewage and wastewater treatment, near-complete mineralization of natural products through aerobic and anaerobic degradation by many different microorganisms leads to the reclamation of pure water. Many human-made chemicals (phenol, chlorobenzenes) present in wastewater are also efficiently mineralized by this treatment. Others, however, are not completely degraded. For example, branched alkylbenzene sulfonate detergents are only slowly degraded during sewage treatment. These detergents are toxic to aquatic organisms, even at very low concentrations. This example shows that a compound may be biodegradable yet may not be degraded sufficiently rapidly in a treatment facility or in the environment to prevent undesirable effects.

These considerations apply particularly strongly to massive pollution from oil spills, plumes of toxic chemicals emanating from waste dumps, and widespread contamination by a great many xenobiotics that are released into the environment deliberately, as pesticides (DDT), wood preservatives (pentachlorophenol), or electric transformer oil (PCBs). Some of these compounds persist for very long periods of time in the environment. Their removal by *in situ* bioremediation, composting, landfarming, or aboveground reactors, generally exploits degradation by members of indigenous microbial communities. Rigorous assessment of the physical and biological factors that contribute to the disappearance of xenobiotics in the environment is challenging. The complexities are illustrated by a detailed discussion of the degradation of pentachlorophenol in an artificial freshwater stream.

Catabolic plasmids confer on bacteria the ability to degrade a variety of xenobiotic compounds. Two examples are discussed in detail: the TOL (pWW0) catabolic plasmid, which encodes the degradative pathways for toluene and xylenes, and catabolic plasmid pADP-1, which encodes the pathway for the complete degradation of atrazine, the world's most widely used herbicide.

Microbial degradation under anaerobic conditions is important in the mineralization of hydrocarbon compounds such as benzene, toluene, and the xylenes and of a variety of chlorinated and nitro-aromatic organic compounds.

Indigenous microorganisms (notably *Acidithiobacillus* spp.) are important in the dissolution of metals such as copper and uranium from low-grade ores. These acidophilic bacteria derive energy from the oxidation of ferrous to ferric iron and of sulfide (or sulfur) to sulfate, producing $Fe_2(SO_4)_3$. $Fe_2(SO_4)_3$, a strong oxidizing agent, reacts with several economically important copper sulfide minerals releasing soluble $CuSO_4$. Leaching by $Fe_2(SO_4)_3$ is independent of the presence of oxygen or microbial action. In "contact" leaching, acidithiobacilli become attached to the mineral particles and then enzymes associated with the cell membrane catalyze an oxidative attack on the crystal lattice of the metal sulfide.

The chemistry in the removal of heavy metals from aqueous effluents is the reverse of that operating in the dissolution of ores. Lakes, constructed lagoons, and wetlands artificially enriched with appropriate organic nutrients remove heavy metals from organic compound–rich effluent streams by precipitation as the metal sulfides. The abundance of nitrate and phosphate in the waste stream leads to profuse surface growth of photosynthesizers, cyanobacteria and algae. This increasing biomass binds heavy metals (biosorption). As the cyanobacteria and algae self-shade, die, and sediment, heavy metals are transported with them into the anoxic bottom sediments, where the biomass becomes a substrate for bacteria that use sulfate as a terminal electron acceptor. The $H_2S$ produced by the sulfate-reducing bacteria forms highly insoluble metal sulfides (CdS, CuS, FeS, HgS, and ZnS), which remain in the bottom sediments.

Leaching of ores with a cyanide solution is a longstanding practice. For many years, a treatment plant at the Homestake Mine in Lead, South Dakota, purified 4 million gallons of cyanide-containing wastewater a day by completely converting cyanide to nitrate. In the first stage of this process, *Pseudomonas* species converted cyanide and thiocyanate to ammonia and bicarbonate, and in the second stage, the nitrifying bacteria *Nitrosomonas* and *Nitrobacter* cooperated in converting ammonia to nitrate.

## REFERENCES

### Wastewater and Sewage Treatment

Gray, N. F. (2004). *Biology of Wastewater Treatment*, 2nd Edition, London: Imperial College Press.

Mogens, H., Harremoës, P., Jansen, J. la C., and Arvin, E. (2002). *Wastewater Treatment. Biological and Chemical Processes*, 3rd Edition, Berlin: Springer-Verlag.

Stadelmann, F. X., Killing, D., and Herter, U. (2002). Sewage sludge: fertilizer or waste? *EAWAG News*, 53, 9–11.

Giger, W. (2002). Dealing with risk factors. *EAWAG News*, 53, 3–5.

McArdell, C. S., Alder, A. G., Golet, G. M., Molna, E., Nipales. N. S., and Giger, W. (2002). Antibiotics – the flipside of the coin. *EAWAG News*, 53, 21–23.

Siegrist, H., et al. (2003). Micropollutants – new challenge in wastewater disposal? *EAWAG News*, 57, 7–10.

Andersen, H., Siegrist, H., Halling-Sørensen, B., and Ternes, T. A. (2003). Fate of estrogens in a municipal sewage treatment plant. *Environmental Science and Technology*, 37, 4022–4026.

### Anammox Bacteria

On-line anammox resource http://www.anammox.com/research.html

Strous, M., et al. (2006). Deciphering the evolution and metabolism of an anammox bacterium from a community genome. *Nature* 440, 790–794.

Kuypers, M. M. M., et al. (2005) Massive nitrogen loss from the Benguela upwelling system through anaerobic ammonium oxidation. *Proceedings of the National Academy of Sciences* 102, 6478–6483.

### Bioremediation

Talley, J. W. (ed.) (2006). *Bioremediation of Recalcitrant Compounds*, Boca Raton, FL: CRC Press.

Atlas, R. M., and Philp, J. (eds.) (2005). *Bioremediation. Applied Solutions for Real-World Environmental Cleanup*, Washington, D.C.: ASM Press.

### *In situ* Remediation of Oil Spills

Kasai, Y., Kishira, H., Sasaki, T., Syutsubo, K., Watanabe, K., and Harayama, S. (2002). Predominant growth of *Alcanivorax* strains in oil-contaminated and nutrient supplemented sea water. *Environmental Microbiology*, 4, 141–147.

Hara, A., Syutsubo, K., and Harayama, S. (2003). *Alcanivorax* which prevails in oil-contaminated seawater exhibits broad substrate specificity for alkane degradation. *Environmental Microbiology*, 5, 746–753.

De Lorenzo, V. (2006). Blueprint of an oil-eating bacterium. *Nature Biotechnology*, 24, 952–953.

Schneiker, S., et al. (2006). Genome sequence of the ubiquitous hydrocarbon-degrading marine bacterium *Alcanivorax borkumensis. Nature Biotechnology*, 24, 997–1004.

### Pentachlorophenol Degradation in an Aquatic Environment

Pignatello, J. J., Martinson, M. E., Steiert, J. G., Carlson, R. E., and Crawford, R. L. (1983). Biodegradation and photolysis of pentachlorophenol in artificial freshwater streams. *Applied and Environmental Microbiology*, 46, 1024–1031.

Pignatello, J. J., Johnson, L. K., Martinson, M. E., Carlson, R. E., and Crawford, R. L. (1985). Response of the microflora in outdoor experimental streams to

pentachlorophenol: compartmental contributions. *Applied and Environmental Microbiology*, 50, 127–132.
Pignatello, J. J., Johnson, L. K., Martinson, M. E., Carlson, R. E., and Crawford, R. L. (1986). Response of the microflora in outdoor experimental streams to pentachlorophenol: environmental factors. *Canadian Journal of Microbiology*, 32, 38–46.

### Catabolic Plasmids

Greated, A., Lambertsen, L., Williams, P. A., and Thomas, C. M. (2002). Complete sequence of the IncP-9 TOL plasmid pWW0 from *Pseudomonas putida*. *Environmental Microbiology*, 4, 856–871.
Velázquez, F., Parro, V., and de Lorenzo, V. (2005). Inferring the genetic network of *m* xylene metabolism through expression profiling of the *xyl* genes of *Pseudomonas putida* mt-2. *Molecular Microbiology*, 57, 1557–1569.
Velázquez, F., de Lorenzo, V., and Valls, M. (2006). The *m*-xylene biodegradation capacity of *Pseudomonas putida* mt-2 is submitted to adaptation to abiotic stresses: evidence from expression profiling of *xyl* genes. *Environmental Microbiology*, 8, 591–602.
Martinez, B., Tomkins, J., Wackett, L. P., Wing, R., and Sadowsky, M. J. (2001). Complete nucleotide sequence and organization of the atrazine catabolic plasmid pADP-1 from *Pseudomonas* sp. strain ADP. *Journal of Bacteriology*, 183, 5684–5697.

### Biodegradation of Organic Compounds in Anaerobic Environments

Smith, M. R. (1990). The degradation of aromatic hydrocarbons by bacteria. *Biodegradation*, 1, 191–206.
Spormann, A. M. and Widdel, F. (2000). Metabolism of alkylbenzenes, alkanes, and other hydrocarbons in anaerobic bacteria. *Biodegradation*, 11, 85–105.
Lovley, D. R. (2000). Anaerobic benzene degradation. *Biodegradation*, 11, 107–116.
Chakraborty, R., and Coates, J. D. (2004). Anaerobic degradation of monoaromatic hydrocarbons. *Applied Microbiology and Biotechnology*, 64, 437–446.
Zhang, C., and Bennett, G. N. (2005). Degradation of xenobiotics by anaerobic bacteria. *Applied Microbiology and Biotechnology*, 67, 600–618.
Maymó-Gatell, X., Chien, Y., Gossett, J. M., and Zinder, S. H. (1997). Isolation of a bacterium that reductively dechlorinates tetrachloroethene to ethene. *Science*, 276, 1568–1571.
Hölscher, T., Krajmalnik-Brown, R., Ritalahti, K. M., von Wintzingerode, F., Görisch, H., Löffler, F. E., and Adrian, L. (2004). Multiple nonidentical reductive-dehalogenase homologous genes are common in *Dehalococcoides*. *Applied and Environmental Microbiology*, 70, 5290–5297.

### Microorganisms in Mineral Recovery

Rawlings, D. W. (2002). Heavy metal mining using microbes. *Annual Review of Microbiology*, 56, 65–91.
Rawlings, D. W. (2005). Characteristics and adaptability of iron- and sulfur-oxidizing microorganisms used for the recovery of metals from minerals and their concentrates. *Microbial Cell Factories*, 4:13 (15 pp.), on-line journal published by BioMed Central.
Rohwerder, T., Gehrke, T., Kinzler, K., and Sand, W. (2003). Bioleaching review part A. Progress in bioleaching: fundamentals and mechanisms of bacterial metal sulfide oxidation. *Applied Microbiology and Biotechnology*, 63, 239–248.
Olson, G. J., Brierley, J. A., and Brierley, C. L. (2003). Bioleaching review part B. Progress in bioleaching: application of microbial processes by the minerals industries. *Applied Microbiology and Biotechnology*, 63, 249–257.

Schippers, A., and Sand, W. (1999). Bacterial leaching of metal sulfides proceeds by two indirect mechanisms via thiosulfate or via polysulfides and sulfur. *Applied and Environmental Microbiology*, 65, 319–321.

Touvinen, O. H., and Bhatti, T. M. (2001). Microbiological leaching of uranium ores. In *Mineral Biotechnology: Microbial Aspects of Mineral Beneficiation, Metal Extraction, and Environmental Control*, S. K. Kawatra and K. A. Natarajan (eds.), pp. 101–119, Littleton, CO: Society for Mining, Metallurgy, and Exploration.

**Biological Detoxification of Wastewater from Precious Metal Processing**

Whitlock, J. L. (1990). Biological detoxification of precious metal processing wastewater. *Geomicrobiology Journal*, 8, 241–249.

# Index